PPARs as Key Mediators of Metabolic and Inflammatory Regulation

PPARs as Key Mediators of Metabolic and Inflammatory Regulation

Editors

Manuel Vázquez-Carrera
Walter Wahli

MDPI • Basel • Beijing • Wuhan • Barcelona • Belgrade • Manchester • Tokyo • Cluj • Tianjin

Editors

Manuel Vázquez-Carrera
Pharmacology, Toxicology
and Medicinal Chemistry
University of Barcelona
Barcelona
Spain

Walter Wahli
Center for Integrative Genomics
University of Lausanne
Lausanne
Switzerland

Editorial Office
MDPI
St. Alban-Anlage 66
4052 Basel, Switzerland

This is a reprint of articles from the Special Issue published online in the open access journal *International Journal of Molecular Sciences* (ISSN 1422-0067) (available at: https://www.mdpi.com/journal/ijms/special_issues/PPARs_Metabolic_Inflammatory).

For citation purposes, cite each article independently as indicated on the article page online and as indicated below:

LastName, A.A.; LastName, B.B.; LastName, C.C. Article Title. *Journal Name* **Year**, *Volume Number*, Page Range.

ISBN 978-3-0365-4192-1 (Hbk)
ISBN 978-3-0365-4191-4 (PDF)

Contents

About the Editors

Manuel Vázquez-Carrera

Manuel Vázquez-Carrera obtained his Pharmacy degree at the University of Barcelona in 1990. In 1991, he received a predoctoral grant of the Generalitat de Catalunya (for the period 1991-1994) for young researchers. In December 1994, he finished his Ph.D. and in 1995 he obtained a grant of the European Science Foundation for a postdoctoral stage at the Institut de Biologie Animale, Université de Lausanne, Switzerland, under the supervision of Dr. Walter Wahli. Currently, Manuel Vázquez-Carrera is a Professor in Pharmacology at the Faculty of Pharmacy and Science Food of the University of Barcelona.

Manuel Vázquez-Carrera has been the chief investigator for several grants and he currently has a total of around 150 international publications. The topic of his research group is the study of the effects of Peroxisome Proliferator-Activated Receptors (PPARs) in metabolic diseases, including cardiac metabolism, insulin resistance/type 2 diabetes and metabolic syndrome. This research group is one of the 32 research groups included in the Spanish Biomedical Research Centre in Diabetes and Associated Metabolic Disorders (CIBERDEM), which is a Spanish public research consortium to promote research into diabetes and associated metabolic disorders.

Walter Wahli

Walter Wahli is a Professor emeritus at the Center for Integrative Genomics, University of Lausanne, Switzerland, and Visiting Professor of Metabolic Disease at the Lee Kong Chian School of Medicine, Nanyang Technological University, Singapore. Until recently, he was also the President of the Council of the Nestle Foundation for the Study of Problems of Nutrition in the World.

He is internationally recognized as a leader in the field of molecular endocrinology and metabolism and discovered the nuclear hormone receptors peroxisome proliferator-activated receptors β and γ (PPARβ and PPARγ). He has received several awards for the elucidation of key functions of these receptors, which are activated by fatty acids and fatty acid derivatives. Synthetic PPAR agonists are drugs used mainly for lowering triglycerides and blood sugar. Prior to his present appointments, Walter Wahli was a visiting associate at the National Cancer Institute, National Institutes of Health, Bethesda, USA. He then became a Professor and Director of the Institute of Animal Biology at the University of Lausanne. He was the Vice-Rector for Research and Continuing Education of the university and founded the Center of Integrative Genomics, which he directed for several years. He was also a member of the Swiss National Science Foundation Research Council and presided over the Biology and Medicine Division.

Preface to "PPARs as Key Mediators of Metabolic and Inflammatory Regulation"

Mounting evidence suggests a bidirectional relationship between metabolism and inflammation. Molecular crosstalk between these processes occurs at different levels with the participation of nuclear receptors, including peroxisome proliferator-activated receptors (PPARs). There are three PPAR isotypes, α, β/δ, and γ, which modulate metabolic and inflammatory pathways, making them key for the control of cellular, organ, and systemic processes. PPAR activity is governed by fatty acids and fatty acid derivatives, and by drugs used in clinics (glitazones and fibrates). The study of PPAR action, also modulated by post-translational modifications, has enabled extraordinary advances in the understanding of the multifaceted roles of these receptors in metabolism, energy homeostasis, and inflammation both in health and disease. This Special Issue of *IJMS* includes a broad range of basic and translational articles, both original research reports and reviews, focused on the latest developments in the regulation of metabolic and/or inflammatory processes by PPARs in all organs and the microbiomes of different vertebrate species.

Manuel Vázquez-Carrera and Walter Wahli
Editors

International Journal of
Molecular Sciences

MDPI

Editorial

PPARs as Key Mediators in the Regulation of Metabolism and Inflammation

Manuel Vázquez-Carrera [1,2,*] and Walter Wahli [3,4,5,*]

[1] Department of Pharmacology, Toxicology and Therapeutic Chemistry, Faculty of Pharmacy and Food Sciences, Institute of Biomedicine (IBUB), University of Barcelona, 08007 Barcelona, Spain
[2] Spanish Biomedical Research Center in Diabetes and Associated Metabolic Diseases (CIBERDEM), Instituto de Salud Carlos III, 28029 Madrid, Spain
[3] Lee Kong Chian School of Medicine, Nanyang Technological University, Singapore 308232, Singapore
[4] Center for Integrative Genomics, University of Lausanne, CH-1015 Lausanne, Switzerland
[5] Toxalim, INRAE UMR 1331, ENVT, INP-Purpan, University of Toulouse, Paul Sabatier University, F-31027 Toulouse, France
* Correspondence: mvazquezcarrera@ub.edu (M.V.-C.); walter.wahli@unil.ch (W.W.)

Citation: Vázquez-Carrera, W.; Wahli, W. PPARs as Key Mediators in the Regulation of Metabolism and Inflammation. *Int. J. Mol. Sci.* **2022**, *23*, 5025. https://doi.org/10.3390/ijms23095025

Received: 25 April 2022
Accepted: 29 April 2022
Published: 30 April 2022

Nuclear receptors (NRs) form a large family of ligand-dependent transcription factors that control the expression of a multitude of genes involved in diverse, vital biological processes. Three of these receptors, the peroxisome proliferator-activated receptors (PPARs), were discovered in the 1990s [1,2] and play key roles in the regulation of cellular differentiation, embryonic development, cellular and whole-body metabolism, inflammation, and tumorigenesis in higher organisms. PPARs are activated not only by fatty acids and their derivatives, some of which also signal through membrane receptors, but also by many plant- and marine-derived natural ligands [3]. Furthermore, drugs that target PPARs, such as fibrates and thiazolidinediones, have been developed to treat metabolic diseases. The 'classic' molecular mode of action of PPARs in the control of physiological and metabolic processes is via their direct binding, as PPAR:retinoid X receptor (RXR) heterodimers, to peroxisome proliferator response elements (PPREs) in the regulatory regions of target genes (Figure 1). The activity of PPARs can also be modulated by posttranslational modifications and their transcriptional regulatory capacity may present a circadian pattern, depending on their expression and the availability of ligands [3].

Inflammation plays a key role in the progression of metabolic diseases, which has led to numerous studies on the functions of PPARs in the regulation of immune cells and the resolution of inflammation. In fact, all three PPAR isotypes (PPARα, PPARβ/δ, and PPARγ) demonstrate anti-inflammatory capacities. Multiple direct and indirect mechanisms promote the anti-inflammatory effects of PPARs [4]. A much-used mechanism is transrepression, relying on protein–protein interactions, involving different immune cell types (B cells, T cells, macrophages, and dendritic cells). Briefly, the most frequently observed transrepression mechanism is the repression of NF-κB activity through protein–protein interactions between PPARs and the p65 subunit of NF-κB (Figure 1). Other mechanisms are the upregulation by PPAR of IκB, the tethering of the PPARs to activator protein 1 (AP-1), nuclear factor of activated T cells (NFAT), and signal transducers and activators of transcription (STATs), as well as the stabilization of corepressor complexes by ligand-activated PPARs on the promoter of inflammatory genes, which results in their downregulation [4].

At present, there is strong evidence for bidirectional relationships between metabolic and inflammatory processes, with crosstalk between them occurring at different levels, involving all three PPAR isotypes (α, β/δ, and γ) that modulate both metabolic and inflammatory pathways. The strong interest for the multifaceted roles of these receptors in health and disease has led to extraordinary advances in the understanding of metabolism, energy homeostasis, and inflammation. The aim of this Special Issue on "PPARs as Key

Mediators of Metabolic and Inflammatory Regulation" was to assemble a broad range of basic and translational original and review articles on the latest discoveries in the regulation of metabolic and/or inflammatory processes by PPARs, in the whole organism of different vertebrate species. Below, we provide a general summary of the topics and data presented in the Special Issue. The scheme in Figure 1 indicates the reference numbers of the articles in the Special Issue, which are grouped according to the topic of their content.

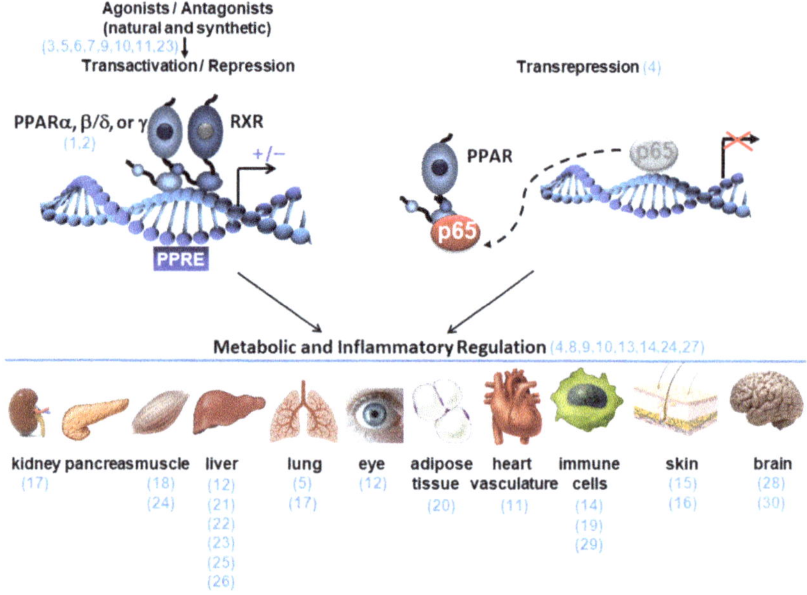

Figure 1. Modes of action of PPARs in transcription transactivation and transrepression (only one of the several mechanisms of transrepression is shown). A non-exhaustive list of the target organs subjected to PPAR-mediated metabolic and inflammatory regulation is shown. The reference numbers (in blue and in parentheses) refer to the articles quoted in this Special Issue (see References); in the figure, they are listed and grouped according to the topic addressed in the corresponding articles. PPRE: peroxisome proliferator hormone response element. p65 subunit of NF-κB. RXR: retinoid X receptor.

The development of synthetic PPAR ligands and new PPAR binding assays that may help to establish alternative therapeutics with reduced side effects for metabolic and inflammatory diseases has been a promising recurrent activity over the last 25 years. The structural basis of ligand binding to the ligand-binding pocket of PPARs has been explored in much depth to help clarify the molecular mode of action of natural and synthetic ligands, not least through the perspective of drug development. Docking studies by Perez Diaz et al. [5], using computational chemistry methods, revealed that an agonist and antagonist can bind simultaneously to the large ligand-binding pocket of PPARβ/δ without affecting the specificity of one another for the binding domain. Agonist binding followed by simultaneous antagonist binding switches the PPARβ/δ mode of action from induction to repression, as shown by studying the effects of LPS-induced inflammation in the pulmonary artery. In a different approach, Yoshikawa et al. [6] developed coumarin-based PPARγ fluorescence probes for competitive binding assays. Compounds incorporating 7-diethylamino (7-Et2N) coumarin are not difficult to synthesize and can be used in PPARγ binding assays. Such compounds can also be applied in live-cell imaging. Of note, the reported coumarin conjugation strategy could be used to synthesize probes for other nuclear receptors. Honda et al. [7] investigated the PPAR α/βδ/γ selectivity of the clinically approved bezafibrate, fenofibric acid, and pemafibrate, using a cellular transactivation assay, a coactivator recruitment assay, and a thermostability assay. Furthermore, cocrystal structures of the PPARβδ/γ-ligand-binding domains (LBD) and the three fibrates are presented. The results of this study underscore both the differences in the PPAR dual/pan agonistic characteristics of the investigated fibrates and their potential for NAFLD therapy. They also show a way for improved fine-tuning of PPAR isotype selectivity.

The PPARα isotype was first characterized as a member of the receptor family mediating the peroxisome proliferation effect of clofibrate in the rodent liver, hence, its name [1]. It was then found to stimulate peroxisomal and mitochondrial fatty acid β-oxidation pathways (reviewed in [8]). The expression pattern of PPARα in different species and tissues and its functions are discussed by Thari-Joutey et al. [9], who also underscored the important modulation of PPARα activity and function by micronutrients and the possible dietary relevance of these effects. Along a similar line, Hassan et al. [10] summarized the potential of polyunsaturated fatty acids, vitamins, dietary amino acids, and phytochemicals to activate or repress PPARs, describing different mechanisms by which these natural molecules modulate PPARs and how they contribute to prevent metabolic disorders. Special attention is given to transition dairy cows, with insights on how the activity of PPARs could be modulated by nutrigenomic interventions to improve energy homeostasis in dairy animals.

The PPARα isotype is implicated in several metabolic regulations in different organs. In the era of precision medicine, there is a need for highly selective agonists to address treatment gaps, such as the correction of atherogenic dyslipidemia that remains an unmet clinical demand [11]. Using the specific PPARα modulator (SPPARMα) pemafibrate (K-877), which selectively and potently activates PPARα, Lee et al. [12] found that, in addition to stimulating liver function that results in elevated serum levels of fibroblast growth factor 21 (FGF21), a neuroprotective hormone in the eye, PPARα also protects against retinal impairment induced by unilateral common carotid artery occlusion. These observations indicate a possibility of using pemafibrate therapy to improve retinal dysfunction in cardiovascular diseases. Since the first observations in 1996 linking PPARα to the control of inflammation [13], the anti-inflammatory role of this receptor has been very well documented. In their review, Grabacka et al. [14] discussed the interplay of PPARα with different pathways in inflammation, transcription, pattern-recognition receptor signaling, and the endocannabinoid system.

There are several inflammatory skin conditions, such as dermatitis and psoriasis, with the latter being an autoimmune disease. In their study of human psoriatic skin, Sobolev et al. [15] found that PPARγ was downregulated in psoriatic lesions and its expression could be normalized by laser treatment. In this skin condition, PPARγ downregulates the expression of genes promoting the development of psoriatic lesions. The PPARβ/δ isotype also regulates inflammatory pathways, keratinocyte proliferation and differentiation, as well as the oxidative stress response. Its involvement in psoriasis and atopic dermatitis, which is less known, is the focus of the review by Blunder et al. [16], which also debates the relevance of targeting PPARβ/δ to alleviate skin inflammation. One of the major cytokines that drives inflammation is the multifunctional transforming growth factor (TGF)β, which also promotes fibrosis. PPARγ dampens these two processes, highlighting crosstalk between TGFβ and PPARγ, which has been implicated in pulmonary arterial hypertension and kidney failure, in which similar, but also different, mechanisms are involved [17]. The therapeutic potential of the PPARγ agonist pioglitazone against these two conditions is discussed.

PPARs have critical roles in the main functions and homeostasis of metabolic organs, such as the muscle, adipose tissue, and liver. Crossland et al. [18] provided an overview, focusing on the effects of PPARβ/δ agonists on the ability of skeletal muscle to contract to generate force and in the regulation of the necessary metabolic support, highlighting observations from in vivo/ex vivo animal models and human volunteers. They also focus on the potential role of PPARγ in reducing muscle inflammation and the metabolic disorders caused by sepsis. Interestingly, synthetic ligands of PPARβ/δ can enhance performance in athletes and are included as S4.5 Metabolic Modulators in the World Anti-Doping Agency's (WADA) Prohibited List. Sibille et al. [19] investigated whether a specific signature in blood T cells could identify the ingestion of the prohibited PPARβ/δ agonist GW0742. PPARβ/δ activation by GW0742 has been shown to stimulate fatty acid oxidation (FAO) in mouse and human T cells, with increased Treg polarization of human primary T cells. Interestingly, PPARβ/δ activation increases FAO in mouse blood T cells too, but this effect is obscured by training, indicating that this signature cannot be used to control doping.

Adipose tissue is a key organ for maintaining healthy energy homeostasis and its dysfunction is often associated with pro-inflammatory, hyperlipidemic, and insulin-resistant environments that promote type 2 diabetes and metabolic syndrome. Sun et al. [20] summarized the roles of the three PPAR isotypes in the metabolic processes and differentiation of white, beige, and brown adipocytes and how they contribute to maintaining metabolic homeostasis in fat.

Non-alcoholic fatty liver disease (NAFLD) is a major health issue all around the world and is often associated with type 2 diabetes and obesity. Initial steatosis, identified by lipid accumulation in hepatocytes, can progress to non-alcoholic steatohepatitis (NASH), which is characterized by inflammation and various levels of fibrosis and is associated with an increased risk of cirrhosis and hepatocellular carcinoma. PPARs regulate many processes that are impaired in NAFLD, such as lipid and glucose metabolism, as well as inflammation. Therefore, PPARs have emerged as attractive clinical targets for NAFLD [21].

In their review, Monroy-Ramirez et al. [22] provided updated information on the critical roles of PPARs in the mechanisms involved in the genesis of several liver diseases and how these receptors could be engaged in therapeutic scenarios. Using a mouse model of metabolic syndrome with an altered expression of PPARα and γ, Cano-Martinez et al. [23] studied the mechanisms underlying the beneficial effects of the polyphenols resveratrol (RSV) and quercetin (QRC) on inflammation in damaged livers. They found a downregulation in the expression of the purinergic receptor P2Y2, neutrophil elastase (NE), and toll-like receptor 4 (TLR4). The repression of these pathways decreased apoptosis and hepatic fibrosis. In the cluster of conditions belonging to metabolic syndrome, type 2 diabetes is a condition with an unmet need for additional treatment options to better control the disease in many patients. PPARβ/δ shows promise, with most of its antidiabetic effects mediated through the activation of AMP-activated protein kinase (AMPK). The review from Aguilar-Recarte et al. [24] outlines the most recent findings on the PPARβ/δ-AMPK antidiabetic pathway, consisting of the upregulation of glucose uptake, fatty acid oxidation, and autophagy, as well as muscle remodeling and the inhibition of endoplasmic reticulum stress and inflammation. Understanding the mechanisms underlying the activation of the PPARβ/δ-AMPK pathway may result in the development of new therapies to prevent and treat the disease and insulin resistance. Along a similar line of thought, Lange et al. [25] provided a general overview of current PPAR-targeting treatments of NAFLD and NASH in patients with type 2 diabetes. This important knowledge on treatments was gained from both clinical trials and observational studies. Lange et al. also considered treatment outcomes on obesity, dyslipidemia, and cardiovascular disease that are often associated with NAFLD/NASH and discussed agonists currently in clinical trials. Finally, sexually dimorphic effects of PPAR-targeting interventions are addressed. There is indeed mounting evidence for sexual dimorphism in NAFLD, with men being more affected than women. Shiffrin et al. [26] studied sexual dimorphism in different mouse models, including those with PPARγ deletion. They found a clear sexual dimorphism in lipodystrophic fat-specific *Pparg*-null mice. The mutant females developed macro- and microvesicular hepatosteatosis, which was lost in gonadectomized mutant mice. In all the tested models, hepatosteatosis strongly impacted sex-biased gene expression in the liver.

Currently, increasing attention is being given to the PPAR epigenetic landscape, as presented by Porcuna et al. [27]. This landscape comprises epigenetic effectors, PPAR regulators, and PPAR-regulated factors, including epigenetic enzymes, DNA methyltransferases, histone modifiers, and non-coding RNAs. The focus of the review of Porcuna et al.

is on PPARα- and PPARγ-related epigenetic regulation in obesity, diabetes, immune disorders, and cancers. The possible therapeutic use of PPAR-controlled epigenetic modulation is also discussed.

Gold [28] summarized the roles of PPARγ in depression, the second largest cause of disability worldwide, according to the World Health Organization. Genetic predisposition alongside recurrent social and other stressors reduce neuronal resilience that can result in the development of depression. In this condition, extreme endoplasmic reticulum stress responses, glutamate toxicity, parainflammation, brain-derived neurotrophic factor (BDNF) function, and the down-regulation of central and peripheral insulin signaling are enhanced. As detailed in the review, the PPARγ system can modulate and dampen all these pathological mechanisms. It is proposed that PPARγ agonists may have significant antidepressant effects, which remain to be explored further.

PPARs are involved in mycobacterial and viral infections. Grabacka et al. [14] summarized PPARα-specific immunomodulatory functions during infections by parasites, bacteria, and viruses, as well as the modulation of processes associated with innate immunity. Tanigawa et al. [29] discussed the advancement in understanding PPARs in host–mycobacteria crosstalk via their impact on the host-dependent mechanisms of lipid metabolism, anti-inflammatory processes, and autophagy during infection. PPARγ is activated in macrophages infected with Mycobacterium leprae or Mycobacterium tuberculosis and regulates some genes involved in the uptake and accumulation of lipids and in cellular metabolism. Mycobacteria use the triacylglycerol (TAG) and cholesterol derived from the host as nutrients and support for evading the host immune system. Layrolle et al. [30] provided an update of the little-known role of PPARγ in viral infections of the brain parenchyma. Viruses can overcome the defensive pathways of host cells to replicate and spread. In these processes, PPARγ becomes a critical target. There is strong evidence for its involvement in brain or neural cells infected by human immunodeficiency virus 1 (HIV-1), the Zika virus, and cytomegalovirus. In fact, PPARγ is a double-edged sword with respect to the triad of neurogenesis, viral replication, and inflammation. In an infected adult brain, PPARγ is beneficial against inflammation, oxidative stress, and viral replication. In this context, PPARγ agonists are considered to be candidate drugs in the treatment of HIV-1-induced brain inflammation to improve neurocognitive outcomes. On the contrary, PPARγ activation is deleterious in neurogenesis in a developing brain, as observed in human cytomegalovirus infections and possibly in Zika viral infections as well. Notably, the activation of PPARγ during an infection of developing brains by human cytomegalovirus promotes viral replication.

In conclusion, the amazing pleiotropy of the three PPAR isotypes, with such diverse effects on different processes and organs, is highlighted once more herein. Future PPAR research, which is needed more than ever, will undoubtedly uncover many other roles of these ligand-activated transcription factors in all vital metabolic and physiological pathways. Further exploration of the vast ensemble of natural ligands, many probably still to be identified, is also needed, which will uncover more about how PPARs contribute to the adaptation of organisms to their environment, in terms of nutrition, toxic substances, infectious agents, and strong temperature fluctuations, just to mention a few of the environmental agents and factors. The use of potent specific synthetic ligands interacting with one or more PPAR isotypes simultaneously offers a very broad avenue for advances in precision medicine.

Author Contributions: Equal contribution from M.V.-C. and W.W. All authors have read and agreed to the published version of the manuscript.

Funding: There was no external funding for the preparation of this Editorial.

Conflicts of Interest: The authors declare no conflict of interest.

References

1. Issemann, I.; Green, S. Activation of a member of the steroid hormone receptor superfamily by peroxisome proliferators. *Nature* **1990**, *347*, 645–650. [CrossRef] [PubMed]
2. Dreyer, C.; Krey, G.; Keller, H.; Givel, F.; Helftenbein, G.; Wahli, W. Control of the peroxisomal beta-oxidation pathway by a novel family of nuclear hormone receptors. *Cell* **1992**, *68*, 879–887. [CrossRef]
3. Tan, C.K.; Zhuang, Y.; Wahli, W. Synthetic and natural Peroxisome Proliferator-Activated Receptor (PPAR) agonists as candidates for the therapy of the metabolic syndrome. *Expert Opin. Ther. Targets* **2017**, *21*, 333–348. [CrossRef] [PubMed]
4. Wahli, W.; Michalik, L. PPARs at the crossroads of lipid signaling and inflammation. *Trends Endocrinol. Metab.* **2012**, *23*, 351–363. [CrossRef]
5. Perez Diaz, N.; Lione, L.A.; Hutter, V.; Mackenzie, L.S. Co-Incubation with PPARβ/δ Agonists and Antagonists Modeled Using Computational Chemistry: Effect on LPS Induced Inflammatory Markers in Pulmonary Artery. *Int. J. Mol. Sci.* **2021**, *22*, 3158. [CrossRef]
6. Yoshikawa, C.; Ishida, H.; Ohashi, N.; Itoh, T. Synthesis of a Coumarin-Based PPARγ Fluorescence Probe for Competitive Binding Assay. *Int. J. Mol. Sci.* **2021**, *22*, 4034. [CrossRef]
7. Honda, A.; Kamata, S.; Akahane, M.; Machida, Y.; Uchii, K.; Shiiyama, Y.; Habu, Y.; Miyawaki, S.; Kaneko, C.; Oyama, T.; et al. Functional and Structural Insights into Human PPARα/δ/γ Subtype Selectivity of Bezafibrate, Fenofibric Acid, and Pemafibrate. *Int. J. Mol. Sci.* **2022**, *23*, 4726. [CrossRef]
8. Feige, J.N.; Gelman, L.; Michalik, L.; Desvergne, B.; Wahli, W. From molecular action to physiological outputs: Peroxisome proliferator-activated receptors are nuclear receptors at the crossroads of key cellular functions. *Prog. Lipid Res.* **2006**, *45*, 120–159. [CrossRef]
9. Tahri-Joutey, M.; Andreoletti, P.; Surapureddi, S.; Nasser, B.; Cherkaoui-Malki, M.; Latruffe, N. Mechanisms Mediating the Regulation of Peroxisomal Fatty Acid Beta-Oxidation by PPARα. *Int. J. Mol. Sci.* **2021**, *22*, 8969. [CrossRef]
10. Hassan, F.U.; Nadeem, A.; Li, Z.; Javed, M.; Liu, Q.; Azhar, J.; Rehman, M.S.; Cui, K.; Rehman, S.U. Role of Peroxisome Proliferator-Activated Receptors (PPARs) in Energy Homeostasis of Dairy Animals: Exploiting Their Modulation. Nutrigenomic Interventions. *Int. J. Mol. Sci.* **2021**, *22*, 12463. [CrossRef]
11. Fruchart, J.C.; Santos, R.D.; Aguilar-Salinas, C.; Aikawa, M.; Al Rasadi, K.; Amarenco, P.; Barter, P.J.; Ceska, R.; Corsini, A.; Després, J.P.; et al. The selective peroxisome proliferator-activated receptor alpha modulator (SPPARMα) paradigm: Conceptual framework and therapeutic potential: A consensus statement from the International Atherosclerosis Society (IAS) and the Residual Risk Reduction Initiative (R3i) Foundation. *Cardiovasc. Diabetol.* **2019**, *18*, 71.
12. Lee, D.; Tomita, Y.; Jeong, H.; Miwa, Y.; Tsubota, K.; Negishi, K.; Kurihara, T. Pemafibrate Prevents Retinal Dysfunction in a Mouse Model of Unilateral Common Carotid Artery Occlusion. *Int. J. Mol. Sci.* **2021**, *22*, 9408. [CrossRef]
13. Devchand, P.R.; Keller, H.; Peters, J.M.; Vazquez, M.; Gonzalez, F.J.; Wahli, W. The PPARalpha-leukotriene B4 pathway to inflammation control. *Nature* **1996**, *384*, 39–43. [CrossRef]
14. Grabacka, M.; Pierzchalska, M.; Płonka, P.M.; Pierzchalski, P. The Role of PPAR Alpha in the Modulation of Innate Immunity. *Int. J. Mol. Sci.* **2021**, *22*, 10545. [CrossRef]
15. Sobolev, V.; Nesterova, A.; Soboleva, A.; Mezentsev, A.; Dvoriankova, E.; Piruzyan, A.; Denisova, E.; Melnichenko, O.; Korsunskaya, I. Analysis of PPARγ Signaling Activity in Psoriasis. *Int. J. Mol. Sci.* **2021**, *22*, 8603. [CrossRef]
16. Blunder, S.; Pavel, P.; Minzaghi, D.; Dubrac, S. PPARdelta in Affected Atopic Dermatitis and Psoriasis: A Possible Role in Metabolic Reprograming. *Int. J. Mol. Sci.* **2021**, *22*, 7354. [CrossRef]
17. Kökény, G.; Calvier, L.; Hansmann, G. PPARγ and TGFβ-Major Regulators of Metabolism, Inflammation, and Fibrosis in the Lungs and Kidneys. *Int. J. Mol. Sci.* **2021**, *22*, 10431. [CrossRef]
18. Crossland, H.; Constantin-Teodosiu, D.; Greenhaff, P.L. The Regulatory Roles of PPARs in Skeletal Muscle Fuel Metabolism and Inflammation: Impact of PPAR Agonism on Muscle in Chronic Disease, Contraction and Sepsis. *Int. J. Mol. Sci.* **2021**, *22*, 9775. [CrossRef]
19. Sibille, B.; Mothe-Satney, I.; Le Menn, G.; Lepouse, D.; Le Garf, S.; Baudoin, E.; Murdaca, J.; Moratal, C.; Lamghari, N.; Chinetti, G.; et al. Gene Doping with Peroxisome-Proliferator-Activated Receptor Beta/Delta Agonists Alters Immunity but Exercise Training Mitigates the Detection of Effects in Blood Samples. *Int. J. Mol. Sci.* **2021**, *22*, 11497. [CrossRef]
20. Sun, C.; Mao, S.; Chen, S.; Zhang, W.; Liu, C. PPARs-Orchestrated Metabolic Homeostasis in the Adipose Tissue. *Int. J. Mol. Sci.* **2021**, *22*, 8974. [CrossRef]
21. Fougerat, A.; Montagner, A.; Loiseau, N.; Guillou, H.; Wahli, W. Peroxisome Proliferator-Activated Receptors and Their Novel Ligands as Candidates for the Treatment of Non-Alcoholic Fatty Liver Disease. *Cells* **2020**, *9*, 1638. [CrossRef]
22. Monroy-Ramirez, H.C.; Galicia-Moreno, M.; Sandoval-Rodriguez, A.; Meza-Rios, A.; Santos, A.; Armendariz-Borunda, J. PPARs as Metabolic Sensors and Therapeutic Targets in Liver Diseases. *Int. J. Mol. Sci.* **2021**, *22*, 8298. [CrossRef]
23. Cano-Martínez, A.; Bautista-Pérez, R.; Castrejón-Téllez, V.; Carreón-Torres, E.; Pérez-Torres, I.; Díaz-Díaz, E.; Flores-Estrada, J.; Guarner-Lans, V.; Rubio-Ruíz, M.E. Resveratrol and Quercetin as Regulators of Inflammatory and Purinergic Receptors to Attenuate Liver Damage Associated to Metabolic Syndrome. *Int. J. Mol. Sci.* **2021**, *22*, 8939. [CrossRef]
24. Aguilar-Recarte, D.; Palomer, X.; Wahli, W.; Vázquez-Carrera, M. The PPARβ/δ-AMPK Connection in the Treatment of Insulin Resistance. *Int. J. Mol. Sci.* **2021**, *22*, 8555. [CrossRef]

25. Lange, N.F.; Graf, V.; Caussy, C.; Dufour, J.F. PPAR-targeted therapies in the treatment of non-alcoholic fatty liver disease in diabetic patients. *Int. J. Mol. Sci.* **2022**, *23*, 4305. [CrossRef]
26. Schiffrin, M.; Winkler, C.; Quignodon, L.; Naldi, A.; Trötzmüller, M.; Köfeler, H.; Henry, H.; Parini, P.; Desvergne, B.; Gilardi, F. Sex Dimorphism of Nonalcoholic Fatty Liver Disease (NAFLD) in Pparg-Null Mice. *Int. J. Mol. Sci.* **2021**, *22*, 9969. [CrossRef]
27. Porcuna, J.; Mínguez-Martínez, J.; Ricote, M. The PPARα and PPARγ Epigenetic Landscape in Cancer and Immune and Metabolic Disorders. *Int. J. Mol. Sci.* **2021**, *22*, 10573. [CrossRef]
28. Gold, P.W. The PPARg System in Major Depression: Pathophysiologic and Therapeutic Implications. *Int. J. Mol. Sci.* **2021**, *22*, 9248. [CrossRef] [PubMed]
29. Tanigawa, K.; Luo, Y.; Kawashima, A.; Kiriya, M.; Nakamura, Y.; Karasawa, K.; Suzuki, K. Essential Roles of PPARs in Lipid Metabolism during Mycobacterial Infection. *Int. J. Mol. Sci.* **2021**, *22*, 7597. [CrossRef] [PubMed]
30. Layrolle, P.; Payoux, P.; Chavanas, S. PPAR Gamma and Viral Infections of the Brain. *Int. J. Mol. Sci.* **2021**, *22*, 8876. [CrossRef] [PubMed]

International Journal of
Molecular Sciences

Article

Co-Incubation with PPARβ/δ Agonists and Antagonists Modeled Using Computational Chemistry: Effect on LPS Induced Inflammatory Markers in Pulmonary Artery

Noelia Perez Diaz [1], Lisa A. Lione [1], Victoria Hutter [1] and Louise S. Mackenzie [1,2,*]

1 School of Life and Medical Sciences, University of Hertfordshire, Hatfield AL10 9AB, UK; n.perez-diaz@herts.ac.uk (N.P.D.); l.lione@herts.ac.uk (L.A.L.); v.hutter@herts.ac.uk (V.H.)
2 School of Pharmacy and Biomolecular Sciences, University of Brighton, Brighton BN2 4GJ, UK
* Correspondence: l.mackenzie2@brighton.ac.uk

Abstract: Peroxisome proliferator activated receptor beta/delta (PPARβ/δ) is a nuclear receptor ubiquitously expressed in cells, whose signaling controls inflammation. There are large discrepancies in understanding the complex role of PPARβ/δ in disease, having both anti- and pro-effects on inflammation. After ligand activation, PPARβ/δ regulates genes by two different mechanisms; induction and transrepression, the effects of which are difficult to differentiate directly. We studied the PPARβ/δ-regulation of lipopolysaccharide (LPS) induced inflammation (indicated by release of nitrite and IL-6) of rat pulmonary artery, using different combinations of agonists (GW0742 or L−165402) and antagonists (GSK3787 or GSK0660). LPS induced release of NO and IL-6 is not significantly reduced by incubation with PPARβ/δ ligands (either agonist or antagonist), however, co-incubation with an agonist and antagonist significantly reduces LPS-induced nitrite production and *Nos2* mRNA expression. In contrast, incubation with LPS and PPARβ/δ agonists leads to a significant increase in *Pdk−4* and *Angptl−4* mRNA expression, which is significantly decreased in the presence of PPARβ/δ antagonists. Docking using computational chemistry methods indicates that PPARβ/δ agonists form polar bonds with His287, His413 and Tyr437, while antagonists are more promiscuous about which amino acids they bind to, although they are very prone to bind Thr252 and Asn307. Dual binding in the PPARβ/δ binding pocket indicates the ligands retain similar binding energies, which suggests that co-incubation with both agonist and antagonist does not prevent the specific binding of each other to the large PPARβ/δ binding pocket. To our knowledge, this is the first time that the possibility of binding two ligands simultaneously into the PPARβ/δ binding pocket has been explored. Agonist binding followed by antagonist simultaneously switches the PPARβ/δ mode of action from induction to transrepression, which is linked with an increase in *Nos2* mRNA expression and nitrite production.

Keywords: nuclear receptor; gene transcription; inflammation; molecular docking; PPARβ/δ; inflammation; lung; pulmonary artery; GW0742; GSK3787; docking; lipopolysaccharide (LPS)

Citation: Perez Diaz, N.; Lione, L.A.; Hutter, V.; Mackenzie, L.S. Co-Incubation with PPARβ/δ Agonists and Antagonists Modeled Using Computational Chemistry: Effect on LPS Induced Inflammatory Markers in Pulmonary Artery. *Int. J. Mol. Sci.* **2021**, *22*, 3158. https://doi.org/10.3390/ijms22063158

Academic Editor: Manuel Vázquez-Carrera

Received: 23 February 2021
Accepted: 16 March 2021
Published: 19 March 2021

1. Introduction

PPARβ/δ are ligand dependent transcription factors that belong to the nuclear receptor family [1]. They are ubiquitously expressed in all cells tested [2] and control key biological functions such as inflammation, metabolism, cell proliferation and migration [3–5]. Consequently, agonists and antagonists for PPARβ/δ have been studied as potential therapies for a wide range of diseases and conditions. However, they have failed to lead to a marketed drug which may be linked to a fundamental lack of understanding of the complexity by which PPARβ/δ controls cell function through gene induction and transrepression.

In vivo studies (mice, rats and rhesus monkeys) indicate that PPARβ/δ agonists induce several favorable pharmacological effects: reduced weight gain, increased metabolism in the skeletal muscle and cardiovascular function, suppression of atherogenic inflammation as well as improvement of the blood lipid profile, all of which are common abnormalities in patients with metabolic syndrome [3,6,7]. These encouraging results led to the first clinical trials on humans. Glaxo Smith Kline (GSK) developed the agonist GW501516 (Endurobol), a promising compound that completed proof-of-concept clinical trials successfully for dyslipidaemia [8] and hypocholesteraemia [9]. Further studies revealed a potential link with tumor development [10,11], and any further clinical trial with GW501516 was suspended.

Nevertheless, the interest in PPARβ/δ continued, and in the last few years several compounds targeting PPARβ/δ were developed and entered clinical trials. The Phase II clinical trial on the PPARβ/δ agonist MBX−8025 for treatment of non-alcoholic steatohepatitis and primary sclerosing cholangitis was terminated early when patients developed early signs of liver damage [12]. This leads to questions on how ligands are binding to the receptor to induce different cellular outcomes.

PPARβ/δ can be activated by numerous endogenous ligands such as eicosanoids, fatty acid, metabolites derived from arachidonic acid and linoleic acid [13–15] as well as exogenous synthetic ligands like GW0742, L−165041, MBX−8025 and GW501516; whereas PPARβ/δ can also be inhibited by two synthetic antagonists GSK3787 (irreversible) and GSK0660 (competitive). There is a great deal of complexity in the manner by which agonists and antagonists control PPARβ/δ signaling, and the resulting changes in gene expression controls the functional outcome of the cell.

The PPARβ/δ endogenous and exogenous ligands control cellular function through changes in very small concentration range. Added to this, in any cell or tissue, the activity of PPARβ/δ may also depend on its promoter activity and relative expression, as well as presence and activity of co-repressor and co-activator proteins. It has been shown that GW0742 is capable of behaving as an agonist activating the transcription pathway at lower concentrations (nM) and antagonist inhibiting this effect at higher concentrations (μM) [16]. In the same line, a study in a model of systemic inflammation in mice showed that higher doses of GW0742 (0.3 mg/kg) triggered a pro-inflammatory response, whereas a lower concentration (0.03 mg/kg) showed an anti-inflammatory trend, although without a significant difference [17]. It was suggested that the large variation in results may be due to the binding of more than one ligand in the large PPARβ/δ ligand binding domain, which requires further investigation.

After ligand activation, PPARβ/δ regulates genes by two different mechanisms, induction and transrepression. In the induction mode, PPARβ/δ forms a complex with the retinoid X receptor (RXR) and together, as a heterodimer, binds the promoter of the target genes (PPRE). In the absence of ligand, co-repressor proteins and histone deacetylases (HDACs) are bound to the heterodimer which tightens the chromatin and prevent it from binding to the PPRE sites [18]. The presence of ligand induces a conformational change of PPARβ/δ which promotes the binding of co-activators, releases the co-repressor proteins, induces histone acetylation and methylation and finally allows the transcription of the target genes [19,20].

In the transrepression mode PPARβ/δ regulates gene expression in a PPRE-independent manner through the regulation (mostly suppression) of other transcription factors, including nuclear factor-κB (NF-κB) [21], activator protein 1 (AP−1) [22] and B cell lymphoma 6 (Bcl6) [23]. There are great discrepancies in the literature about the effect of the ligand-activation of PPARβ/δ in the cell, and both pro- and anti- effects in inflammation [24,25], cell proliferation [26,27] and migration [28,29] have been reported.

In order to isolate the transrepression mode of PPARβ/δ, a novel approach was taken in this project. Tissues were co-incubated with both agonist; this will alter the conformational shape of PPARβ/δ and bind to co-activators and place the PPARβ/δ predominantly into gene induction mode. In theory, addition of an antagonist will then prevent PPRE binding and induction of genes, revealing the effects of PPARβ/δ transrepression.

Studies have generally focussed on the effects of agonists and antagonists separately, and often results in conflicting theories of how PPARβ/δ controls gene expression. Here we show for the first time that co-incubation of agonists and antagonists to PPARβ/δ leads to a significant decrease in LPS-induced inflammation in rat pulmonary artery compared to single applications of each drug type. The mechanism of action may be explained by the binding studies indicated by docking studies of the agonists and antagonists with PPARβ/δ, which confirms that agonist and antagonist co-binding can occur.

2. Results

2.1. PPARβ/δ Expression and Basal NO Production Over Time in Pulmonary Artery

Expression of PPARβ/δ was confirmed by ELISA and calculated to 0.04 pg/mL/μg protein. Rat lung pulmonary arteries were incubated with LPS for different periods of time (8 h, 20 h and 24 h) in order to ascertain the minimum time required to significantly increase NO production. LPS induced a significant increase in NO at 24 h in all tissues tested (Figure A1), subsequently the 24 h incubation time period was used.

2.2. PPARβ/δ Regulation of LPS-Induced Inflammation

Pulmonary arteries exposed to LPS for 24 h significantly increased production of NO and IL-6, markers of innate inflammation (Figure 1). Incubation of either PPARβ/δ agonists (GW0742 or L−165041), or antagonist (GSK3787 or GSK0660) had no effect on nitrite production (a measure of NO release) (Figure 1A,B) or IL-6 release (Figure 1C,D). In contrast, co-incubation with a mixture of both GW0742 (agonist) plus GSK3787 (irreversible antagonist) (Figure 1A,C) or L−165041 (agonist) plus GSK0660 (competitive antagonist) (Figure 1B,D) led to a significant decrease in NO and IL-6 production.

Production of nitrite is a measure of NO release from cells and tissues; the iNOS specific inhibitor 1400W was used to indicate the proportion of NO originating from iNOS as opposed to eNOS or iNOS. In all experiments, nitrite production following LPS incubation was significantly decreased by incubation with 1400W (Figure 1A,B).

2.3. Marker Genes for PPARβ/δ Induction and Transrepression in Pulmonary Artery

In pulmonary artery incubated in LPS, GW0742 significantly increases the transcription of *Pdk−4* mRNA by 7-fold (Figure 2A) and *Angptl−4* mRNA by 3-fold (Figure 2B), which is inhibited by the irreversible antagonist GSK3787. The expression of *Nos2* mRNA, the gene that encodes for iNOS, is significantly increased by LPS, and its expression is significantly inhibited by co-incubation with GW0742 and GSK3787 (Figure 2C).

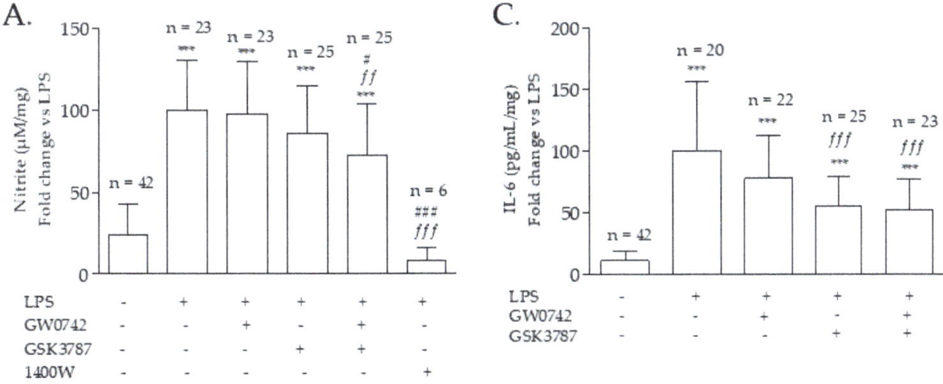

Figure 1. *Cont.*

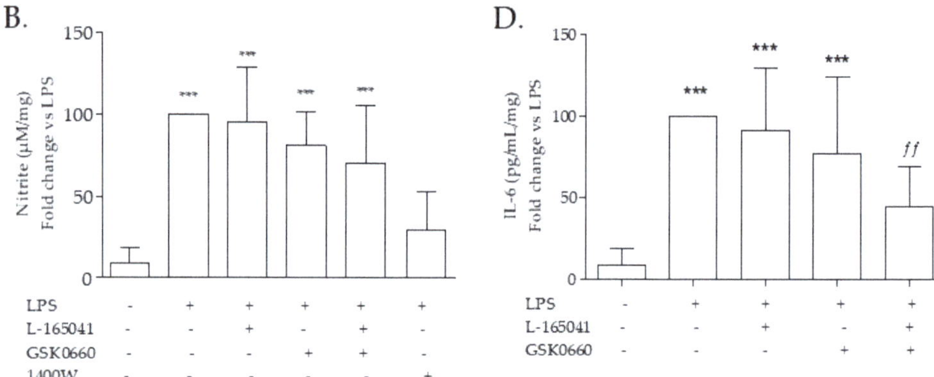

Figure 1. Nitrite (a marker for NO release) and IL−6 production by pulmonary artery. Rat pulmonary artery rings were treated with two combinations of PPARβ/δ agonist-antagonist and iNOS inhibitor 1400W: (**A–C**) 100 nM GW0742—1 µM GSK3787—100 µM 1400W (**B–D**) 1 µM L−165041—1 µM GSK0660—100 µM 1400W. NO and IL−6 production was measured after 24 h and normalized with LPS: (**A**) the average of NO production in lipopolysaccharide (LPS) incubated tissues was used for the normalization of the data and the number of samples per treatment is written at the top of the bar; (**B**) each experiment was normalized with its own LPS treatment ($n = 9$); (**C**) the average of IL−6 production with LPS treatment was used for the normalization of the data and the number of samples per treatment is written at the top of the bar; (**D**) each experiment was normalized with its own LPS treatment ($n = 9$). Data passed the D'Agostino–Pearson normality test; significant difference between treatments was analyzed by one-way ANOVA followed by Bonferroni post-hoc test and the data are presented as mean ± SD. *** = $p < 0.001$ compared with vehicle; *ff* = $p < 0.01$, *fff* = $p < 0.001$ compared with LPS; # = $p < 0.05$, ### = $p < 0.001$ compared with LPS + GW0742.

2.4. Computational Chemistry: PPARβ/δ Docking Analysis

2.4.1. Docking of One PPARβ/δ Ligand

The PPARβ/δ-LBD crystal structure 3TKM has an X-ray resolution of 1.95 Å and was co-crystallized with GW0742, the same agonist that was used during the development of this project, therefore this structure was chosen for our docking experiments.

The two PPARβ/δ agonists GW0742 and L−165041 as well as the two antagonists GSK3787 and GSK0660 were docked into the crystal structure of the LBD of PPARβ/δ. The best eight hits were analyzed by Pymol to identify the residues that form polar interactions with each of the different poses of the ligands (Table 1).

2.4.2. Docking of GW0742

The most stable orientation of GW0742 within the PPARβ/δ binding pocket predicted by Autodock Vina (green) was compared to the real GW0742 present in the crystal structure (pink) (Figure 3A). The more detailed image (Figure 3B) clearly shows that the residues that form polar interactions with GW0742 are His247, His413 and Tyr437 (Pymol), whereas the 2D image created by Ligplot+ showing how the head of GW0742 forms the polar bindings and the tail is surrounded by the hydrophobic amino acids (Figure 3C).

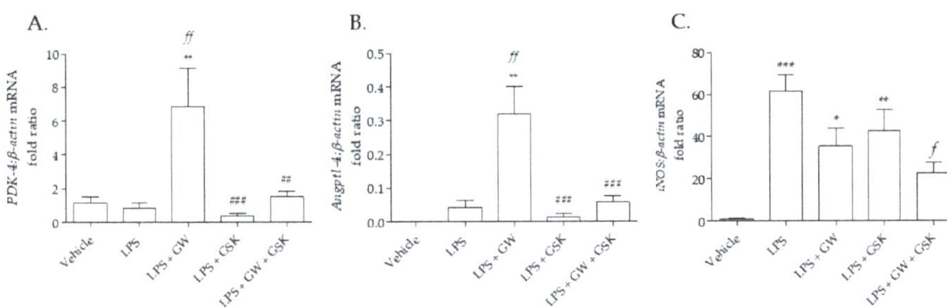

Figure 2. (**A**) *PDK−4*, (**B**) *AngPtl4* and (**C**) *iNOS* mRNA expression in pulmonary arteries following incubation with LPS and PPARβ/δ ligands. The expression of different mRNA was measured following 24 h incubation with treatments: vehicle (0.01% DMSO); 1 μg/mL LPS; 1 μg/mL LPS + 100 nM GW0742; 1 μg/mL LPS + 1 μM GSK3787; and 1 μg/mL LPS + 100 nM GW0742 + 1 μM GSK3787 (n = 4–5). Relative quantitation was calculated with the comparative CtΔΔ method and normalized against β-actin as an endogenous control. The data are presented as mean ± standard deviation; the data was not normally distributed (D'Agostino–Pearson normality test). Significant difference by the Krustal–Wallis test with Dunns post hoc test is indicated by * = $p < 0.05$, ** = $p < 0.01$ and *** = $p < 0.001$ compared with Vehicle; f = $p < 0.05$, ff = $p < 0.01$ compared to LPS; ## = $p < 0.01$ and ### = $p < 0.001$ compared to LPS + GW0742.

Table 1. Best eight docking hits of four ligands into PPARβ/δ (PBD: 3TKM).

Best Fit	**Agonists**				**Antagonists**			
	GW0742		**L-165041**		**GSK3787**		**GSK0660**	
	Affinity (Kcal/mol)	**Aa with Polar Interactions**	**Affinity (Kcal/mol)**	**Aa with Polar Interactions**	**Affinity (Kcal/mol)**	**Aa with Polar Interactions**	**Affinity (Kcal/mol)**	**Aa with Polar Interactions**
1	−11.1	His287 His413 Tyr437	−8.7	His287 His413 Tyr437	−9.1	Thr252 Asn307	−8.6	Arg248 Thr252 Ala306 Asn307
2	−10.8	Thr253 His287 His413 Tyr437	−8	Met192 Thr252 Thr256 Ile290 Ala306	−8.9	Thr252	−8.3	Arg248 Thr252 Ala306
3	9.9	Thr253 His413	−7.6	Thr252 Arg258 Glu259	−8.6	Thr252	−8.1	Thr256 Asn307
4	−9.6	Thr256	−7.5	Thr252 Thr253 Ala306	−8.5	Thr256	−8.1	Asn307
5	−9.3	No bonds	−7.5	Trp228 Thr252 Thr256 Ile290	−8.4	No bonds	−7.9	Thr256 Ala306 Asn307
6	−8.8	Thr253	−7.1	Thr252 Thr256 Ala306	−8.3	Thr252 Thr253	−7.6	Asn307
7	−8.7	Thr252	−6.9	Tyr284 Arg361	−8.3	Thr252	−7.5	Ala306 Asn307
8	−8.4	Arg258	−6.8	Glu255 Asn307	−8.2	Met192	−7.4	Thr256 Asn307

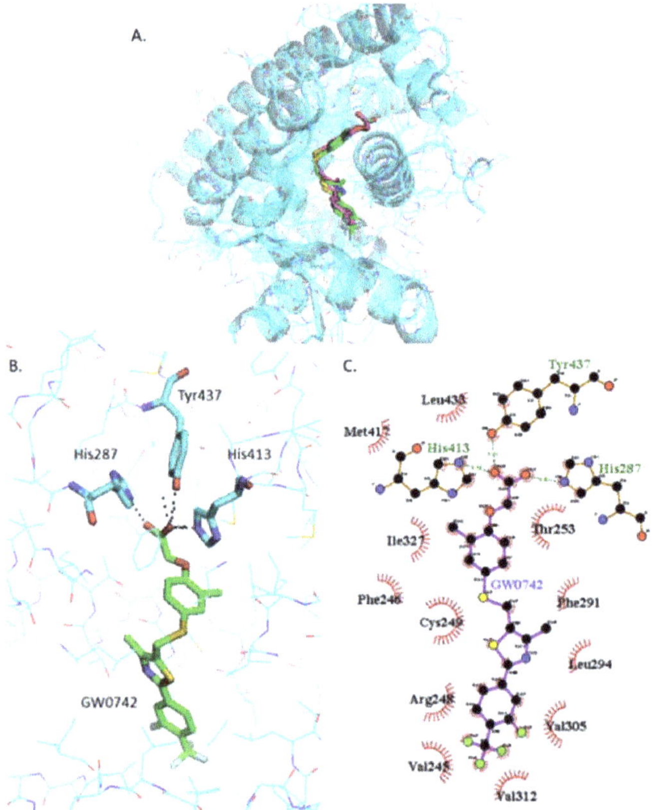

Figure 3. Analysis of GW0742 docked into PPARβ/δ (PBD:3TKM). (**A**) Representation of the most stable GW0742 docking conformation (green) compared to the GW0742 of the crystal structure (pink). (**B**) 3D detail of the amino acids forming polar bindings with GW0742 calculated by Pymol. Colour coding of atoms: red O, blue N, mustard S, white F, pink C of GW0742 from the crystal structure, green C of GW0742 docked into the crystal structure, cyan C from PPARβ/δ. (**C**) Schematic 2D representation of the interaction between PPARβ/δ LBD and GW0742 calculated using Ligplot+. The green dashed lines indicate polar interactions and the red spoked arcs indicate hydrophobic interactions. Colour coding of atoms: red O, blue N, yellow S, green F, black C.

2.4.3. Docking of GSK3787

GSK3787 binds in a slightly different place than GW0742, although there is some overlapping of the binding sites (Figure 4A). Also, the amino acids involved in the polar interaction of GSK3787 predicted by Pymol, Thr252 and Asn307, are different to those of the agonists (Figure 4B) as well as the residues that interact with the hydrophobic tail of GSK3787 (Figure 4C).

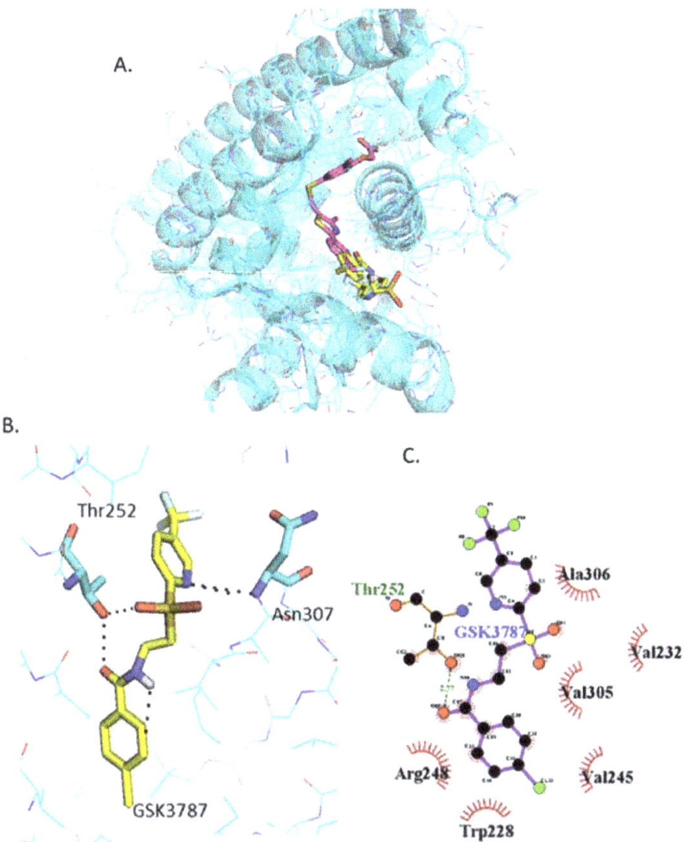

Figure 4. Analysis of GSK3787 docked into PPARβ/δ (PBD:3TKM). (**A**) Representation of the most stable GSK3787 docking conformation (yellow) compared to the GW0742 of the crystal structure (pink). (**B**) 3D detail of the amino acids forming polar bindings with GSK3787 calculated by Pymol. Color coding of atoms: red O, blue N, mustard S, white F, pink C of GW0742 from the crystal structure, yellow C of GSK3787 docked into the crystal structure, cyan C from PPARβ/δ. (**C**) Schematic 2D representation of the interaction between PPARβ/δ LBD and GSK3787 calculated using Ligplot+. The green dashed lines indicate polar interactions and the red spoked arcs indicate hydrophobic interactions. Colour coding of atoms: red O, blue N, yellow S, green F, black C.

2.4.4. Docking of L−165042

The most stable L−165041 orientation predicted by Autodock Vina binds in the same physical place as GW0742 (Figure 5A) and the same three amino acids (His287, His413, Tyr437) form polar interactions with the head of L−165041 (Figure 5B). The tail of L−165041 also forms hydrophobic interactions with a number of residues in common with GW0742, such as Val245, Arg248, Cys249, Thr253, Phe291, Leu294, Val305, Val312, Met417, Leu433 (Figure 5C).

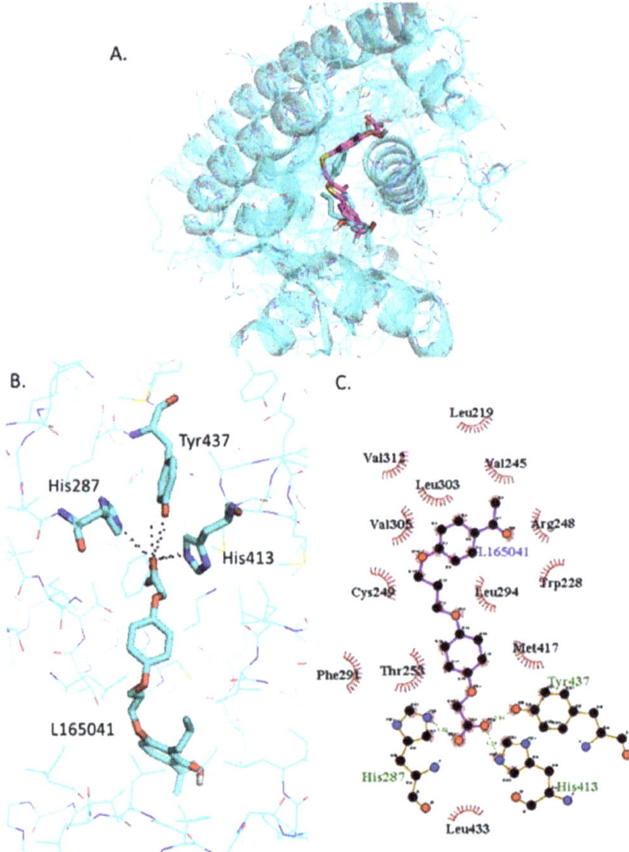

Figure 5. Analysis of L−165042 docked into PPARβ/δ (PBD:3TKM). (**A**) Representation of the most stable L−165041 docking conformation (cyan sticks) compared to the GW0742 of the crystal structure (pink). (**B**) 3D detail of the amino acids forming polar bindings with L−165041 calculated by Pymol. Color coding of atoms: red O, Blue N, mustard S, white F, pink C of GW0742 from the crystal structure, cyan sticks C of L−165041 docked into the crystal structure, cyan lines C from PPARβ/δ. (**C**) Schematic 2D representation of the interaction between PPARβ/δ LBD and -L−165041 calculated using Ligplot+. The green dashed lines indicate polar interactions and the red spoked arcs indicate hydrophobic interactions. Colour coding of atoms: red O, blue N, yellow S, green F, black C.

2.4.5. Docking of GSK0660

GSK0660 binds very close but not in the same binding site as GW0742 (Figure 6A). The amino acids involved in the polar bindings with GSK0660, Thr252, Asn307, Arg248 and Ala306, are again different to those for the agonists, although two of them are common with GSK3787 (Figure 6B). Ligplot+ predicts slightly different polar binding profile (Figure 6C), probably because these two software's use different algorithms for binding prediction, although still show hydrophobic interactions common with GSK3787, such as Trp228, Val305 and Ala306.

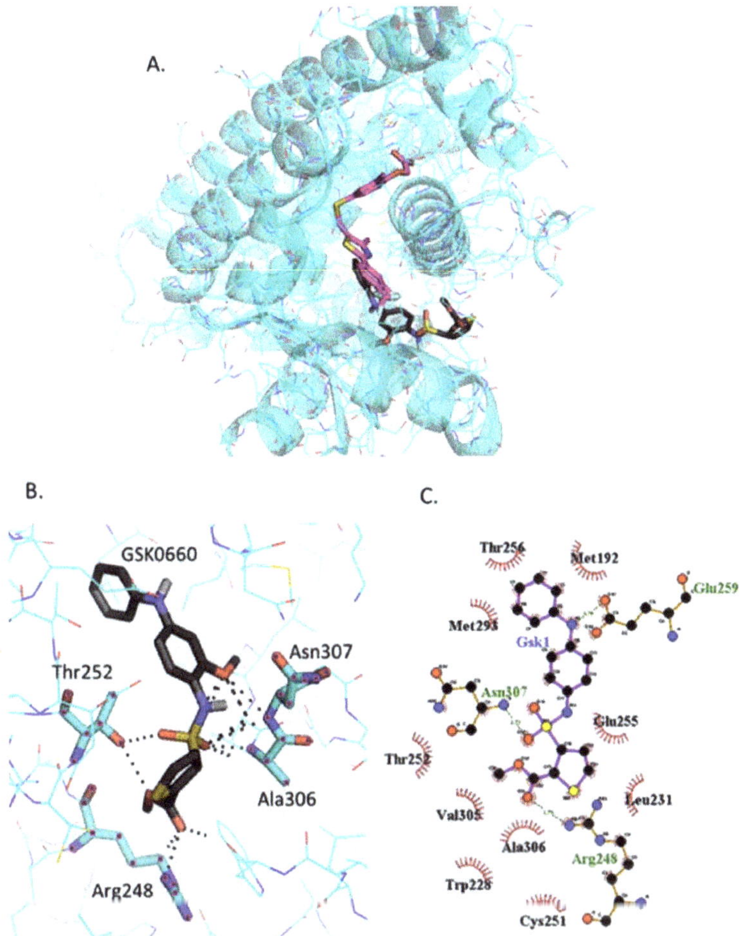

Figure 6. Analysis of GSK0660 docked into PPARβ/δ (PBD:3TKM). (**A**) Representation of the most stable GSK0660 docking conformation (grey) compared to the GW0742 of the crystal structure (pink). (**B**) 3D detail of the amino acids forming polar bindings with GSK0660 calculated by Pymol. Color coding of atoms: red O, Blue N, mustard S, white F, pink C of GW0742 from the crystal structure, grey C of GSK0660 docked into the crystal structure, cyan C from PPARβ/δ. (**C**) Schematic 2D representation of the interaction between PPARβ/δ LBD and GSK0660 calculated using Ligplot+. The green dashed lines indicate polar interactions and the red spoked arcs indicate hydrophobic interactions. Colour coding of atoms: red O, blue N, yellow S, green F, black C.

2.4.6. Docking of Two PPARβ/δ Ligands Simultaneously

In order to investigate the docking of two ligands simultaneously, the first ligand was bound in the most stable orientation first. The best hit from previous docking was assigned Ligand 1, and then a second molecule was docked, assigned Ligand 2. The aim was to mimic the conditions of the experiments performed in this study and predict what could have happened at the molecular level. When the tissues were treated with only one ligand there is only one option for two ligands to bind, but when two different ligands are present at the same time either of them can bind first into the binding pocket. All these ligand-binding possibilities were considered and summarized in Table 2. A further analysis on Pymol was completed for each option.

Table 2. Docking prediction of binding affinities and amino acids forming polar interactions with the PPARβ/δ ligands bound into the LBD.

Ligand 1			Ligand 2		
Molecule	**Affinity (Kcal/mol)**	**Amino Acid with Polar Interactions**	**Molecule**	**Affinity (Kcal/mol)**	**Amino Acid with Polar Interactions**
GW0742	−11.1	His287 His413 Tyr437	GW0742	−8.5	Arg258
			GSK3787	−7.7	Trp228 Thr252
GSK3787	−9.1	Thr252 Asn307	GW0742	−8.1	Trp228 Lys229
			GSK3787	−7.4	Thr253 His413
L−165041	−8.7	His287 His413 Tyr437	L-165041	−8.3	Met192 Cys251 Thr252 Thr256 Ile290 Ala306
			GSK0660	−6.5	Arg198 Asn339
GSK0660	−8.6	Arg248 Thr252 Ala306 Asn307	L-165041	−8.1	Thr253 His413
			GSK0660	−8.9	Thr252 Thr253

2.4.7. Analysis of GW0742 and GSK3787 Docked into GW0742-Bound PPARβ/δ

Once GW0742 is bound in the most stable orientation within the PPARβ/δ-LBD, GW0742 and GSK3787 can still bind at favorable energies (−8.5 kcal/mol and −7.7 kcal/mol respectively), although at very different binding sites to the most stable one and forming polar interactions with different residues, as shown in Figure 7.

2.4.8. Analysis of GW0742 and GSK3787 Docked into GSK3787-Bound PPARβ/δ

GW0742 and GSK3787 can also bind into the binding pocket after GSK3787 at favorable energies (−8.1 kcal/mol and −7.4 kcal/mol respectively). The binding site is also different to the most stable ones (Figure 8), but interestingly the binding site is also different to the previous one, when GW0742 is bound first into the binding pocket instead of GSK3787 (Table 2).

2.4.9. Analysis of L−165041 and GSK0660 Docked into L−165041-Bound PPARβ/δ

When L−165041 is bound to the ligand binding pocket first, a second molecule of L−165041 or GSK0660 can bind with favorable energies (−8.3 kcal/mol and −6.5 kcal/mol respectably) and again, forming polar interactions with different residues. The most interesting finding is that GSK0660, although still in the PPARβ/δ-LBD, has the potential to bind outside the binding pocket (Table 2).

2.4.10. Analysis of L−165041 and GSK0660 Docked into GSK0660-Bound PPARβ/δ

L−165041 and GSK0660 can also bind within the binding pocket at favorable energies (−8.1 kcal/mol and −8.9 kcal/mol respectively) after GSK0660 is bound in the most stable orientation (Table 2), but again the binding site is different to previously when L−165041 was bound first.

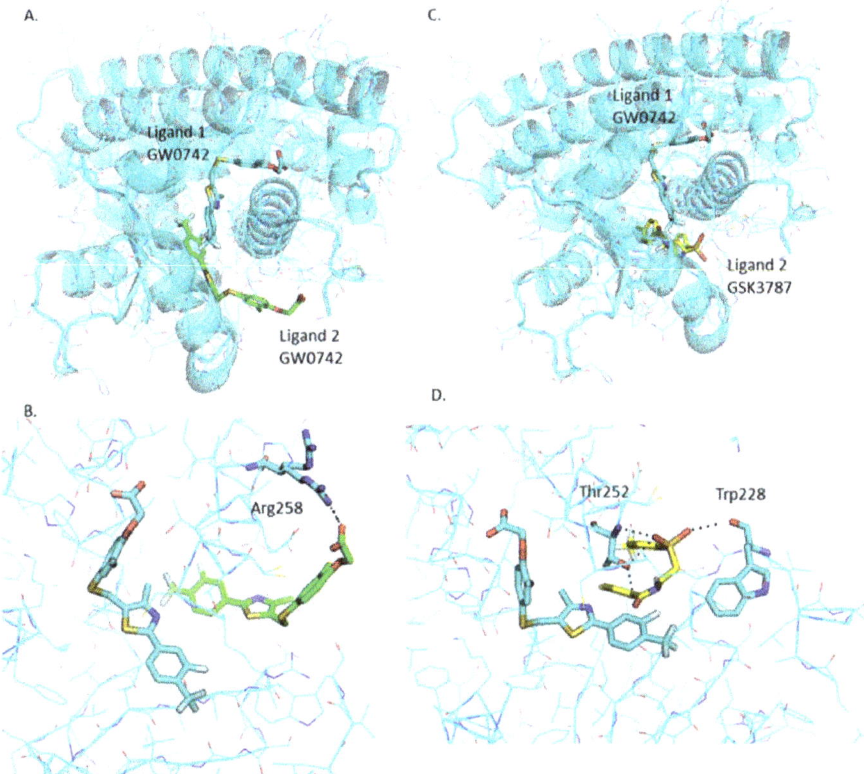

Figure 7. Analysis of GW0742 and GSK3787 docked into PPARβ/δ + GW0742. A second molecule of GW0742 (**A**,**B**) or GSK3787 (**C**,**D**) was docked into the LBD of PPARβ/δ containing one molecule of GW0742. (**A**) Representation of how two GW0742 molecules bind into the PPARβ/δ binding pocket at same time. (**B**) Detail of the amino acids interacting with the second molecule of GW0742. (**C**) Representation of how one molecule of GW0742 first and then one molecule of GSK3787 bind into the PPARβ/δ binding pocket at same time. (**D**) Detail of the amino acids interacting with the second molecule of GSK3787. Colour coding of atoms: red O, blue N, mustard S, white F, cyan C, PPARβ/δ and GW0742 that binds first within the binding pocket, green C of GW0742 that binds second into the binding pocket, yellow C of GSK3787 that binds second into the binding pocket.

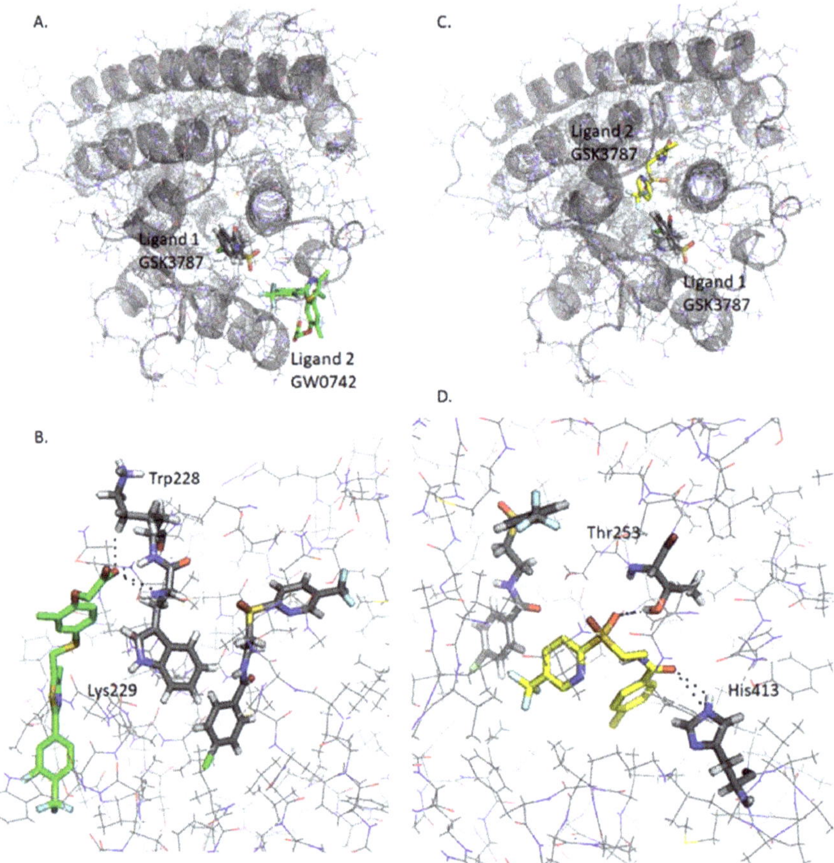

Figure 8. Analysis of GW0742 and GSK3787 docked into PPARβ/δ + GSK3787. A second molecule of GW0742 (**A**,**B**) or GSK3787 (**C**,**D**) was docked into the LBD of PPARβ/δ containing one molecule of GSK3787. (**A**) Representation of how one molecule of GSK3787 first and then one molecule of GW0742 bind into the PPARβ/δ binding pocket at same time. (**B**) Detail of the amino acids interacting with the second molecule of GW0742. (**C**) Representation of how two molecules of GSK3787 bind into the PPARβ/δ binding pocket at same time. (**D**) Detail of the amino acids interacting with the second molecule of GSK3787. Colour coding of atoms: red O, blue N, mustard S, white F, grey C PPARβ/δ and GSK3787 that binds first within the binding pocket, green C of GW0742 that binds second into the binding pocket, yellow C of GSK3787 that binds second into the binding pocket.

3. Discussion

The study confirms previous studies that PPARβ/δ is expressed in pulmonary arteries [30]. Here we show for the first time that dual incubation of PPARβ/δ ligands (agonist plus antagonist) decreases LPS-induced NO and IL-6 release from pulmonary arteries compared to single incubation with either agonist or antagonist alone. This result was repeatable with two combinations of agonists and antagonists. Docking studies using computational chemistry methods indicates that multiple binding of both types of ligand is possible and retain similar binding profiles as when binding alone.

Other in vivo studies in whole mouse have indicated that GW0742 treatment attenuates inflammation, which is significantly reduced by GSK0660 [31–33]. These in vivo studies differ from our ex vivo studies in isolated tissues, as they are not affected by the immune system infiltration of the alveolar cavity. In our ex-vivo pulmonary artery LPS model, PPARβ/δ was shown to regulate NO and IL-6 release only under certain conditions, which required further investigation.

Marker genes were selected to represent the induction mode of action of PPARβ/δ. Khozoie [34] and Adhikary [35] performed a genome wide analysis of genes regulated by PPARβ/δ in mouse keratinocytes and human myofibroblasts respectively. Khozoie [34] cross-linked the two lists of genes regulated by PPARβ/δ and created a new list of 103 genes regulated by PPARβ/δ in both human and mouse models. There is a high possibility that these genes are regulated by PPARβ/δ in rats as well, therefore this list was used to select the PPARβ/δ induction marker genes *Angptl−4* and *Pdk−4*. Here in this study, qRT-PCR of pulmonary artery showed that the activation of PPARβ/δ with GW0742 increases the transcription of *Pdk−4* and *Angptl−4* mRNA indicating that agonist activation of PPARβ/δ triggers induction, which was inhibited by co-incubation with GSK3787.

It has been shown in several studies that *Nos2* mRNA is regulated by PPARβ/δ [33,36,37], although it has not been described whether this regulation is via induction or transrepression. This study indicates that the significant increase in LPS induced *Nos2* mRNA expression is not altered by the presence of either GW0742 or GSK3787 which indicates that direct induction of genes is not the predominant mode of control PPARβ/δ has on *Nos2* mRNA expression. However, *Nos2* mRNA expression is significantly reduced when co-incubated with both agonist and antagonist. With the activation of the receptor followed by inactivation by antagonist, any reduction in expression and activity of iNOS must be due to transrepression of PPARβ/δ on other nuclear receptors.

It has been suggested that the large ligand binding pocket of PPARβ/δ can accommodate more than one ligand, resulting in unusual PPAR:ligand stoichiometries that could trigger inactive receptor conformations [16], a possibility that we further investigated using in silico methods. Firstly, the two PPARβ/δ agonists (GW0742 and L−165042) and two antagonists (GSK3787 and GSK0660) were docked into the PPARβ/δ binding site. It was found that the agonists and antagonists have a different binding profile within the binding pocket: the same three amino acids His287, His413 and Tyr437 form polar interactions with the two agonists tested, but they do not bind the antagonists. Whereas the amino acids Thr252 and Asn307 are more prone to bind the antagonists GSK3787 and GSK0660.

This finding agrees with previous results were GW0742 was docked to another PPARβ/δ crystal structure (PDB: 3GZ9) using another docking software (Glide), and the same three amino acids bound to GW0742 [38]. Furthermore, several studies co-crystallized PPARβ/δ with different agonists both synthetic, such as iloprost [39], the fibrate GW2433 [40], or GW501516 [41] and natural PPARβ/δ agonists such as with eicosapentaenoic acid (EPA) [40], and in all cases the agonists showed polar bindings with the same three amino acids His287, His413 and Tyr437.

It is worth mentioning another study where the authors selected 5 compounds that potentially bound PPARβ/δ and performed a luciferase transactivation assay to biologically test if these compounds activate PPARβ/δ. They further analyzed two of them by docking and molecular dynamics (MD) simulation, one compound that activated PPARβ/δ (Compound **1**) and another one that did not activate PPARβ/δ (Compound **2**). The docking and MD simulation results for the Compound **1** showed an interaction with His287, His413 and Tyr437, and the results for Compound **2** showed an interaction with Thr252 [42]. This suggests the possibility that the different binding profile between agonists and antagonists can provoke a different 3D conformational change that might explain why PPARβ/δ binds to co-repressors instead of co-activators and vice versa.

Our findings clearly indicate that a ligand that shows a high binding affinity and is predicted to form polar bonds with His287, His413 and Tyr437 will most likely behave as agonist. On the other hand, if one ligand shows high binding affinity but it is predicted to bind other residues such as Thr252 and Asn307 it is more likely that it will behave as antagonist.

Co-incubation of pulmonary artery with two types of ligands led to unexpected results, raising the question on whether the agonist or antagonist retained the potential to bind PPARβ/δ in the expected manner. Incubation with only one agonist GW0742 allows two possibilities: the binding of one or two molecules into the ligand-binding domain (LBD). If a second molecule of GW0742 binds to PPARβ/δ, this molecule is predicted to bind not too far from the most stable binding site and with the same binding affinity and residue interaction (Arg258) than the 8th best position predicted for the first molecule. Similarly, when the irreversible antagonist GSK3787 is present, a second molecule of GSK3787 is predicted to bind also not far away from its most stable binding site and with favorable binding affinity. In addition, it is predicted that GSK3787 will still form polar bonds with Thr252, an amino acid that is predicted to bind GSK3787 in five out of eight most stable poses predicted by docking.

When investigating dual occupancy of the PPARβ/δ LBD with both agonist (GW0742) and antagonist (GSK3787) all the options mentioned above still apply but two more options are available: GW0742 binds first and GSK3787 after or GSK3787 binds first and GW0742 after. When GW0742 binds first, GSK3787 can still bind very close to its most stable binding site with a very favorable binding affinity, and what is more, still binds the residue Thr252. If GSK3787 binds first, GW0742 is predicted to bind very far away from the most stable binding site, at the entrance of the binding pocket, and as a consequence it will have a very different binding profile forming polar bonds with Trp228 and Lys229, two residues that did not show any interaction with ligands before. This suggests that if the PPARβ/δ receptor is inhibited by GSK3787, the agonist cannot reverse this inhibition. Similar analysis can be done with the other pair of agonist and antagonist L−165041 and GSK0660.

Taking into account the docking scores and molecular poses of the ligands, all possibilities described above have very favorable energies for it to happen. That opens a whole new scenario of possibilities that could dramatically change the 3D conformation of PPARβ/δ in ways that have not been thought of before, resulting in the binding of different co-regulators, which ultimately could change the PPARβ/δ response from induction to transrepression or vice versa.

To our knowledge, this is the first time that the possibility of binding two ligands simultaneously into the PPARβ/δ binding pocket has been explored. The results suggest that this possibility is very likely to happen with very favorable affinity energies, and it is worth considering when designing and interpreting experiments where PPARβ/δ is ligand-activated and high concentrations of ligands are used. In regards to the conditions set in our study, docking indicates that the co-incubation with both agonist and antagonist does not prevent the specific binding of each other to the large PPARβ/δ binding pocket.

4. Conclusions

In summary, this is a multidisciplinary approach of the study of PPARβ/δ that provides novel information about its functioning at molecular level. In the model used here, the simultaneous co-incubation of pulmonary with both agonist and antagonist potentially opens a window to understand the alternative transrepression PPARβ/δ mode of action as compared to the induction mode of action. Here we show for the first time that there is a characteristic PPARβ/δ-ligand binding profile for agonists and antagonists even in combination; PPARβ/δ agonists form polar bonds with His287, His413 and Tyr437, while antagonists are more promiscuous about which amino acids they bind to, although they are very prone to bind Thr252 and Asn307. In dual predictive docking studies, all the options studied seem feasible with favorable binding energies, suggesting the need for caution when designing and interpreting the results of experiments using PPARβ/δ ligands.

5. Materials and Methods

5.1. Reagents

The PPARβ/δ ligands GW0742, GSK3787, L−165042 and GSK0660 as well as 1400W, LPS O55:B5, sulfanilamide and naphthylethylenediamine dihydrochloride were purchased from Sigma (Gillingham, Dorset, UK). Sodium nitrate, orthophosphoric acid and DMSO were purchased from Fisher Scientific (Loughborough, UK). Primers from Applied Biosystem (Foster City, CA, USA): β-actin (Rn00667869_m1), Pdk−4 (Rn00585577_m1), Angptl−4 (Rn015228817_m1), Nos2 (Rn00561646_m1).

5.2. Animals

Male Wistar rats (350–450 g) were sourced from Charles River (Harlow, UK) and housed in pairs in standard cages (Tecniplast 2000P) with sawdust (Datesand grade 7 substrate) and shredded paper wool bedding with water and food (5LF2 10% protein LabDiet) in the Biological Services Unit at the University of Hertfordshire. The housing environment was maintained at a constant temperature of 22 ± 2 °C, under a 12 h light/dark cycle (lights on: 07:00 to 19:00 h). All testing was conducted under the light phase of the animals' light/dark cycle, and care was taken to randomize treatment sequences to control for possible order effects.

All experiments involving protected animals were carried out in accordance with the University of Hertfordshire animal welfare ethical guidelines and the Animals (Scientific Procedures) Act 1986 and European directive 2010/63/EU. Rats were euthanized according to schedule 1 procedure by CO_2 asphyxiation followed by cervical dislocation. Lungs were removed and immediately placed in physiological saline solution (PSS) buffer (118 mM NaCl, 4.7 mM KCl, 2.5 mM $CaCl_2$, 1.17 mM $MgSO_4$, 1 mM KH_2PO_4, 5.5 mM, glucose, 25 mM $NaHCO_3$ and 0.03 mM Na_2EDTA). Following dissection, tissues were incubated in 1% *v*/*v* penicillin/streptomycin in DMEM (no serum) under 5% *v*/*v* CO_2 at 37 °C in the required treatment (detailed in Table 3) for up to 24 h. Whole arteries can be incubated in serum free media for 4 days without significant loss of contraction and differentiation status or death [43]. After incubation, the culture medium was removed and stored at −20 °C until Greiss assay or IL−6 ELISA analysis. The tissues were stored at −80 °C until needed for mRNA extraction for qRT-PCR.

Table 3. Treatments of tissues with two different combination types of agonists and antagonists: 1 μg/mL LPS; 100 nM GW0742; 10 μM L−165041; 10 μM GSK3782; 10 μM GSK0880; 10 μM 1400 W.

	Combination 1	**Combination 2**
Vehicle	0.01% DMSO	0.01% DMSO
LPS	LPS 1 μg/mL	LPS 1 μg/mL
LPS + agonist	LPS + GW0742	LPS + L-165041
LPS + antagonist	LPS + GSK3787	LPS + GSK0660
LPS + agonist + antagonist	LPS + GW0742 + GSK3787	LPS + L-165041 + GSK0660
LPS + 1400 W	LPS + 1400 W	LPS + 1400 W

5.3. Quantification of PPARβ/δ Expression in Lung Tissues by Enzyme-Linked Immunosorbent Assay (ELISA)

The expression of PPARβ/δ in pulmonary artery from naïve rats was measured using rat PPARβ/δ ELISA kit (Abbkine; Wuhan, China) according to the manufacturer's instructions. Briefly, the dissected tissues were homogenized with liquid N_2 using a mortar and a pestle, and the proteins were extracted in ice-cold phosphate buffered saline (PBS) with proteinase inhibitor cocktail in a ratio 9 mL PBS (137 mM NaCl, 2.7 mM KCl, 10 mM Na_2HPO_4, 1.8 mM KH_2PO_4, pH 7.4) per g tissue. The samples were then sonicated 3 × 30 s and centrifuged for 5 min at 5000 g. The supernatant was collected, and the protein concentration was quantified by BCA assay. Then, 50 μL of standards and samples were added in duplicate to the plate containing pre-coated anti-PPARβ/δ and incubated 45 min at 37 °C. The wells were washed and incubated with 50 μL of the horseradish peroxidase (HRP)-conjugated detection antibody for another 30 min at 37 °C. The wells were washed again and incubated in the dark with the chromogen solution for 15 min at 37 °C. Finally, the reaction was stopped by adding 50 μL of Stop solution and the plate was immediately measured in a microplate reader. The concentrations of PPARβ/δ were determined by comparison of OD_{450} to a standard curve (0–8 ng/mL).

5.4. *Quantification of Nitric Oxide Released by Lung Tissues by the Griess Assay*

An aliquot of the culture medium (50 μL) thawed on ice was mixed with an equal volume of Griess reagent (mixture of equal volumes Griess reagent 1 and Griess reagent 2 containing sulfanilamide 1% *w/v* + orthophosphoric acid 5% *v/v* and naphthylethylenediamine dihydrochloride 0.5% w/v respectively). The concentration was determined by comparison of the OD_{540} to a standard curve of solutions of sodium nitrite (0–1 mM).

5.5. *Quantification of IL−6 Released by Pulmonary Artery by ELISA*

The release of IL−6 by pulmonary arteries to the culture medium was measured using Rat IL−6 DuoSet ELISA kit (R&D Systems, Minneapolis, MN, USA) according to the manufacturer's instructions. Briefly, all samples and standards were conducted in duplicate to a microtiter plate containing the capture antibody. After two hours of incubation the wells were washed and incubated with the detection antibody for another two hours. The wells were washed again and incubated in the dark with Streptavidin-HRP for 20 min. The wells were washed once more and incubated in the dark with substrate solution. After 20 min the stop solution was added and immediately measured in a microplate reader. The readings at 540 nm were subtracted from the readings at 450 nm and the concentrations of samples were determined by comparison with the standard curve (0–8 ng/mL).

5.6. *Quantitative Real Time-Polymerase Chain Reaction (qRT-PCR)*

Total RNA was extracted from tissues using RNeasy Fibrous Tissue Mini Kit (Qiagen, Hilden, Germany). The tissues were first pulverized in a pestle with liquid N_2 and the RNA was then extracted following the manufacturer's instruction and stored at −80 °C until use. The quality and concentration of the RNA was measured using Nanodrop (SimpliNano, GE Healthcare Life Science; Chalfont St Giles, Buckinghamshire, UK) at a wavelength of 260 nm. Further to that, the degradation of RNA was checked in a 1% *w/v* agarose gel, and the DNA contamination of the RNA was checked by PCR. To do that, genomic DNA was also extracted from the tissues using PureLink Genomic DNA mini kit from Invitrogen following the manufacturer's instructions and used as a positive control of the PCR, and two primers for the housekeeping gene *β-actin* were designed (forward CTGGTCGTACCACTGGCATT, reverse AATGCCTGGGTACATGGTGG). The termociclator Mastercycler Nexus gradient (Eppendorf, Hamburg, Germany) was set with the following PCR protocol: 95 °C for 10 min, 40 cycles of 95 °C for 30 s, 56 °C for 30 s and 72 °C for 30 s, 72 °C for 10 min and hold at 4 °C.

cDNA was obtained by reverse transcription (RT) using iScript cDNA synthesis kit (Bio-Rad) following the manufacturer's instructions. The RT was performed using a thermociclator Stratagene Mx3005P (Agilent Technologies, Santa Clara, CA, USA) with the following steps: 5 min at 25 °C, 20 min at 46 °C, 5 min at 95 °C, and hold at 4 °C.

qRT-PCR was performed to analyze mRNA expression using a Taqman System. Briefly, 10 μL of reaction mix containing the primers and cDNA was incubated in a 96 well-plate following the cycle conditions: 95 °C for 10 min, 40 cycles of 95 °C for 15 s and 60 °C for 1 min.

5.7. *Docking*

The ability of drugs to bind into protein active sites was investigated using AutoDock/Vina with Pymol and Ligplot+ as a graphical user interface. For the docking simulations, the PPARβ/δ crystal structure 3TKM was selected for having one of the highest resolutions (1.95 Å). The PDB file was downloaded from the Protein Data Bank. Water molecules, ligands and other hetero atoms were removed from the protein structure, and the addition of hydrogen atoms to the protein was performed using AutoDock Tools version 1.5.6. The grid was set manually to cover the active site. The file was saved as a pdbqt file.

The ligand molecule structures were drawn in ChemSketch, the energy was minimized and saved in PDB format, and converted into a pdbqt file with AutoDock Tools version 1.5.6 (ADT/AutoDockTools—AutoDock (scripps.edu) (accessed on 23 February 2021)).

Molecular docking was performed with the software AutoDock Vina and all parameters set as default. Results with minor calculated free energy variations were analyzed using Pymol version 1.7.4 and LigPlot+ version1.4.5 softwares.

For the docking of two molecules, the 3TKM PDB file without hetero atoms was combined with the best docking result of each ligand in one single PDB file, one PDB file per ligand. These files were opened in Autodock Tools, H_2 were added, the grid was set manually and saved in a new pdbqt file. This file was used for the docking with the second molecule.

5.8. Statistical Analysis

Statistical comparisons were performed on GraphPad Prism 5.0 software using one-way ANOVA with Bonferroni's post hoc analysis for NO and IL−6 detection assays, and for qRT-PCR. The values are expressed as observed mean ± SEM. Data was normalized previously to the statistical analysis. In short, data from NO and IL−6 detection assays was normalized against the group treatment LPS and expressed as % change. The relative quantification of genes analyzed by qRT-PCR was calculated with the comparative CtΔΔ method, β-actin was used as endogenous control and data was normalized against the control group as a fold change.

Values of $p < 0.05$ were considered statistically significant. When the level of probability (p) is less than 0.05 (*), less than 0.01 (**) or less than 0.001 (***), the effect of the difference was regarded as significance.

Author Contributions: Conceptualization, N.P.D. and L.S.M.; methodology, N.P.D., V.H. and L.A.L.; validation and formal analysis, N.P.D.; investigation, N.P.D. and L.S.M.; resources, L.A.L. and V.H.; data curation, N.P.D.; writing—original draft preparation, N.P.D. and L.S.M.; writing—review and editing, L.S.M. and N.P.D.; visualization, L.S.M.; supervision, L.S.M., L.A.L. and V.H.; project administration, L.A.L.; funding acquisition, N.P.D., L.S.M. and L.A.L. All authors have read and agreed to the published version of the manuscript.

Funding: This research received no external funding.

Institutional Review Board Statement: All experiments involving protected animals were approved on 19 October 2016 by the University of Hertfordshire animal welfare ethical review body and conducted in accordance with the guidelines established by the Animals (Scientific Procedures) Act, 1986 and European directive 2010/63/EU (Establishment license PEL7003708).

Informed Consent Statement: Not applicable.

Data Availability Statement: The data presented in this study are openly available in the repository: LM University of Brighton.

Conflicts of Interest: The authors declare no conflict of interest.

Appendix A

The NO produced by pulmonary artery was measured after 8 h, 20 h and 24 h of incubation (Figure A1).

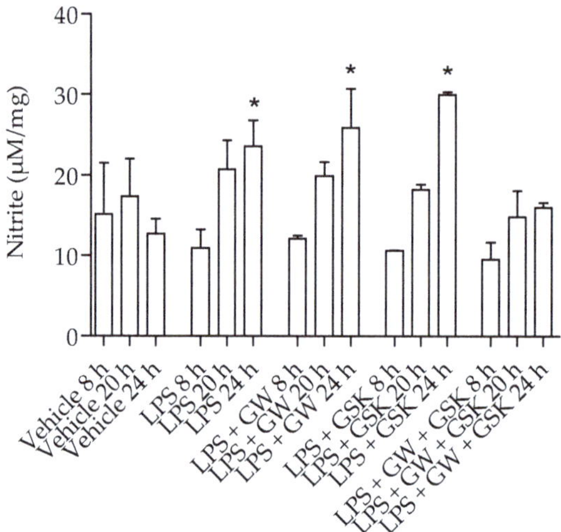

Figure A1. Nitrite (a marker for NO release) production by pulmonary artery over time. Rat pulmonary artery rings were treated with different combinations of 1 μg/mL LPS (to induce innate inflammation), 100 nM GW0742 (PPARβ/δ agonist), 1 μM GSK3787(PPARβ/δ antagonist), and the NO production was measured at 8 h, 20 h and 24 h of incubation (n = 3). The data are presented as mean ± standard deviation; the data was not normally distributed (D'Agostino–Pearson normality test). Significant difference by the Krustal-Wallis test with Dunns post hoc test is indicated by * = $p < 0.05$ compared with vehicle of same time point.

References

1. Issemann, I.; Green, S. Activation of a member of the steroid hormone receptor superfamily by peroxisome proliferators. *Nature* **1990**, *347*, 645–650.
2. Mandard, S.; Patsouris, D. Nuclear control of the inflammatory response in mammals by peroxisome proliferator-activated receptors. *PPAR Res.* **2013**, *2013*, 613864. [CrossRef] [PubMed]
3. Oliver, W.R., Jr.; Shenk, J.L.; Snaith, M.R.; Russell, C.S.; Plunket, K.D.; Bodkin, N.L.; Lewis, M.C.; Winegar, D.A.; Sznaidman, M.L.; Lambert, M.H.; et al. A selective peroxisome proliferator-activated receptor delta agonist promotes reverse cholesterol transport. *Proc. Natl. Acad. Sci. USA* **2001**, *98*, 5306–5311. [CrossRef] [PubMed]
4. Kim, H.J.; Kim, M.Y.; Hwang, J.S.; Kim, H.J.; Lee, J.H.; Chang, K.C.; Kim, J.H.; Han, C.W.; Kim, J.H.; Seo, H.G. PPARdelta inhibits IL-1beta-stimulated proliferation and migration of vascular smooth muscle cells via up-regulation of IL-1Ra. *Cell. Mol. Life Sci. CMLS* **2010**, *67*, 2119–2130. [CrossRef] [PubMed]
5. Quintela, A.M.; Jimenez, R.; Piqueras, L.; Gomez-Guzman, M.; Haro, J.; Zarzuelo, M.J.; Cogolludo, A.; Sanz, M.J.; Toral, M.; Romero, M.; et al. PPARbeta activation restores the high glucose-induced impairment of insulin signalling in endothelial cells. *BJP* **2014**, *171*, 3089–3102. [CrossRef]
6. Tanaka, T.; Yamamoto, J.; Iwasaki, S.; Asaba, H.; Hamura, H.; Ikeda, Y.; Watanabe, M.; Magoori, K.; Ioka, R.X.; Tachibana, K.; et al. Activation of peroxisome proliferator-activated receptor δ induces fatty acid β-oxidation in skeletal muscle and attenuates metabolic syndrome. *Proc. Natl. Acad. Sci. USA* **2003**, *100*, 15924–15929. [CrossRef] [PubMed]
7. Wang, Y.-X.; Lee, C.-H.; Tiep, S.; Yu, R.T.; Ham, J.; Kang, H.; Evans, R.M. Peroxisome-proliferator-activated receptor δ activates fat metabolism to prevent obesity. *Cell* **2003**, *113*, 159–170. [CrossRef]
8. Ooi, E.M.; Watts, G.F.; Sprecher, D.L.; Chan, D.C.; Barrett, P.H. Mechanism of action of a peroxisome proliferator-activated receptor (PPAR)-delta agonist on lipoprotein metabolism in dyslipidemic subjects with central obesity. *J. Clin. Endocrinol. Metab.* **2011**, *96*, E1568–E1576. [CrossRef] [PubMed]
9. Olson, E.J.; Pearce, G.L.; Jones, N.P.; Sprecher, D.L. Lipid effects of peroxisome proliferator-activated receptor-delta agonist GW501516 in subjects with low high-density lipoprotein cholesterol: Characteristics of metabolic syndrome. *ATVB* **2012**, *32*, 2289–2294. [CrossRef]
10. Geiger, L.N.; Dunsford, W.S.; Lewis, D.J.; Brennan, C.; Liu, K.C.; Newsholme, S.J. Rat carcinogenicity study with GW501516, a PPAR delta agonist. *Toxicol. Sci. Off. J. Soc. Toxicol.* **2009**, *108*, 895.

11. Pollock, C.B.; Yin, Y.; Yuan, H.; Zeng, X.; King, S.; Li, X.; Kopelovich, L.; Albanese, C.; Glazer, R.I. PPARdelta activation acts cooperatively with 3-phosphoinositide-dependent protein kinase-1 to enhance mammary tumorigenesis. *PLoS ONE* **2011**, *6*, e16215. [CrossRef]
12. Jones, D.; Boudes, P.F.; Swain, M.G.; Bowlus, C.L.; Galambos, M.R.; Bacon, B.R.; Doerffel, Y.; Gitlin, N.; Gordon, S.C.; Odin, J.A.; et al. Seladelpar (MBX-8025), a selective PPAR-δ agonist, in patients with primary biliary cholangitis with an inadequate response to ursodeoxycholic acid: A double-blind, randomised, placebo-controlled, phase 2, proof-of-concept study. *Lancet Gastroenterol. Hepatol.* **2017**, *2*, 716–726. [CrossRef]
13. Yu, K.; Bayona, W.; Kallen, C.B.; Harding, H.P.; Ravera, C.P.; McMahon, G.; Brown, M.; Lazar, M.A. Differential activation of peroxisome proliferator-activated receptors by eicosanoids. *J. Biol. Chem.* **1995**, *270*, 23975–23983. [CrossRef]
14. Forman, B.M.; Chen, J.; Evans, R.M. Hypolipidemic drugs, polyunsaturated fatty acids, and eicosanoids are ligands for peroxisome proliferator-activated receptors alpha and delta. *Proc. Natl. Acad. Sci. USA* **1997**, *94*, 4312–4317. [CrossRef] [PubMed]
15. Moraes, L.A.; Piqueras, L.; Bishop-Bailey, D. Peroxisome proliferator-activated receptors and inflammation. *Pharmacol. Ther.* **2006**, *110*, 371–385. [CrossRef] [PubMed]
16. Nandhikonda, P.; Yasgar, A.; Baranowski, A.M.; Sidhu, P.S.; McCallum, M.M.; Pawlak, A.J.; Teske, K.; Feleke, B.; Yuan, N.Y.; Kevin, C.; et al. Peroxisome proliferation-activated receptor delta agonist GW0742 interacts weakly with multiple nuclear receptors, including the vitamin D receptor. *Biochemistry* **2013**, *52*, 4193–4203. [CrossRef] [PubMed]
17. Yin, L.; Busch, D.; Qiao, Z.; van Griensven, M.; Teuben, M.; Hildebrand, F.; Pape, H.C.; Pfeifer, R. Dose-dependent effects of peroxisome proliferator-activated receptors beta/delta agonist on systemic inflammation after haemorrhagic shock. *Cytokine* **2018**, *103*, 127–132. [CrossRef] [PubMed]
18. Neels, J.G.; Grimaldi, P.A. Physiological functions of peroxisome proliferator-activated receptor beta. *Physiol. Rev.* **2014**, *94*, 795–858. [CrossRef] [PubMed]
19. Cronet, P.; Petersen, J.F.W.; Folmer, R.; Blomberg, N.; Sjöblom, K.; Karlsson, U.; Lindstedt, E.-L.; Bamberg, K. Structure of the PPARα and -γ ligand binding domain in complex with AZ 242; ligand selectivity and agonist activation in the PPAR family. *Structure* **2001**, *9*, 699–706. [CrossRef]
20. Helsen, C.; Claessens, F. Looking at nuclear receptors from a new angle. *Mol. Cell. Endocrinol.* **2014**, *382*, 97–106. [CrossRef] [PubMed]
21. Rodriguez-Calvo, R.; Serrano, L.; Coll, T.; Moullan, N.; Sanchez, R.M.; Merlos, M.; Palomer, X.; Laguna, J.C.; Michalik, L.; Wahli, W.; et al. Activation of peroxisome proliferator-activated receptor beta/delta inhibits lipopolysaccharide-induced cytokine production in adipocytes by lowering nuclear factor-kappaB activity via extracellular signal-related kinase 1/2. *Diabetes* **2008**, *57*, 2149–2157. [CrossRef] [PubMed]
22. Schnegg, C.I.; Kooshki, M.; Hsu, F.-C.; Sui, G.; Robbins, M.E. PPARδ prevents radiation-induced proinflammatory responses in microglia via transrepression of NF-κB and inhibition of the PKCα/MEK1/2/ERK1/2/AP-1 pathway. *Free Radic. Biol. Med.* **2012**, *52*, 1734–1743. [CrossRef]
23. Fan, Y.; Wang, Y.; Tang, Z.; Zhang, H.; Qin, X.; Zhu, Y.; Guan, Y.; Wang, X.; Staels, B.; Chien, S.; et al. Suppression of pro-inflammatory adhesion molecules by PPAR-delta in human vascular endothelial cells. *ATVB* **2008**, *28*, 315–321. [CrossRef]
24. Di Paola, R.; Crisafulli, C.; Mazzon, E.; Esposito, E.; Paterniti, I.; Galuppo, M.; Genovese, T.; Thiemermann, C.; Cuzzocrea, S. GW0742, a high-affinity PPAR-beta/delta agonist, inhibits acute lung injury in mice. *Shock* **2010**, *33*, 426–435. [CrossRef] [PubMed]
25. Wang, D.; Fu, L.; Ning, W.; Guo, L.; Sun, X.; Dey, S.K.; Chaturvedi, R.; Wilson, K.T.; DuBois, R.N. Peroxisome proliferator-activated receptor delta promotes colonic inflammation and tumor growth. *Proc. Natl. Acad. Sci. USA* **2014**, *111*, 7084–7089. [CrossRef]
26. Zhang, J.; Fu, M.; Zhu, X.; Xiao, Y.; Mou, Y.; Zheng, H.; Akinbami, M.A.; Wang, Q.; Chen, Y.E. Peroxisome proliferator activated receptor delta is up-regulated during vascular lesion formation and promotes post-confluent cell proliferation in vascular smooth muscle cells. *JBC* **2002**, *277*, 11505–11512. [CrossRef] [PubMed]
27. Wang, D.; Wang, H.; Guo, Y.; Ning, W.; Katkuri, S.; Wahli, W.; Desvergne, B.; Dey, S.K.; DuBois, R.N. Crosstalk between peroxisome proliferator-activated receptor delta and VEGF stimulates cancer progression. *Proc. Natl. Acad. Sci. USA* **2006**, *103*, 19069–19074. [CrossRef]
28. Lim, H.J.; Lee, S.; Park, J.H.; Lee, K.S.; Choi, H.E.; Chung, K.S.; Lee, H.H.; Park, H.Y. PPAR delta agonist L-165041 inhibits rat vascular smooth muscle cell proliferation and migration via inhibition of cell cycle. *Atherosclerosis* **2009**, *202*, 446–454. [CrossRef]
29. Ham, S.A.; Yoo, T.; Hwang, J.S.; Kang, E.S.; Lee, W.J.; Paek, K.S.; Park, C.; Kim, J.H.; Do, J.T.; Lim, D.S.; et al. Ligand-activated PPARdelta modulates the migration and invasion of melanoma cells by regulating Snail expression. *Am. J. Cancer Res.* **2014**, *4*, 674–682.
30. Bishop-Bailey, D.; Hla, T. Endothelial cell apoptosis induced by the peroxisome proliferator-activated receptor (PPAR) ligand 15-deoxy-Delta12, 14-prostaglandin J2. *JBC* **1999**, *274*, 17042–17048. [CrossRef]
31. Haskova, Z.; Hoang, B.; Luo, G.; Morgan, L.A.; Billin, A.N.; Barone, F.C.; Shearer, B.G.; Barton, M.E.; Kilgore, K.S. Modulation of LPS-induced pulmonary neutrophil infiltration and cytokine production by the selective PPARbeta/delta ligand GW0742. *Inflamm. Res. Off. J. Eur. Histamine Res. Soc.* **2008**, *57*, 314–321.
32. Bao, X.C.; Fang, Y.Q.; You, P.; Zhang, S.; Ma, J. Protective role of peroxisome proliferator-activated receptor beta/delta in acute lung injury induced by prolonged hyperbaric hyperoxia in rats. *Respir. Physiol. Neurobiol.* **2014**, *199*, 9–18. [CrossRef] [PubMed]
33. Kapoor, A.; Shintani, Y.; Collino, M.; Osuchowski, M.F.; Busch, D.; Patel, N.S.; Sepodes, B.; Castiglia, S.; Fantozzi, R.; Bishop-Bailey, D.; et al. Protective role of peroxisome proliferator-activated receptor-beta/delta in septic shock. *Am. J. Respir. Crit. Care Med.* **2010**, *182*, 1506–1515. [CrossRef]

34. Khozoie, C.; Borland, M.G.; Zhu, B.; Baek, S.; John, S.; Hager, G.L.; Shah, Y.M.; Gonzalez, F.J.; Peters, J.M. Analysis of the peroxisome proliferator-activated receptor-beta/delta (PPARbeta/delta) cistrome reveals novel co-regulatory role of ATF4. *BMC Genom.* **2012**, *13*, 665. [CrossRef] [PubMed]
35. Adhikary, T.; Kaddatz, K.; Finkernagel, F.; Schonbauer, A.; Meissner, W.; Scharfe, M.; Jarek, M.; Blocker, H.; Muller-Brusselbach, S.; Muller, R. Genomewide analyses define different modes of transcriptional regulation by peroxisome proliferator-activated receptor-beta/delta (PPARbeta/delta). *PLoS ONE* **2011**, *6*, e16344. [CrossRef] [PubMed]
36. Di Paola, R.; Briguglio, F.; Paterniti, I.; Mazzon, E.; Oteri, G.; Militi, D.; Cordasco, G.; Cuzzocrea, S. Emerging role of PPAR-beta/delta in inflammatory process associated to experimental periodontitis. *Mediat. Inflamm.* **2011**, *2011*, 787159. [CrossRef] [PubMed]
37. Bojic, L.A.; Burke, A.C.; Chhoker, S.S.; Telford, D.E.; Sutherland, B.G.; Edwards, J.Y.; Sawyez, C.G.; Tirona, R.G.; Yin, H.; Pickering, J.G.; et al. Peroxisome proliferator-activated receptor delta agonist GW1516 attenuates diet-induced aortic inflammation, insulin resistance, and atherosclerosis in low-density lipoprotein receptor knockout mice. *Arterioscler. Thromb. Vasc. Biol.* **2014**, *34*, 52–60. [CrossRef]
38. Perez-Diaz, N.; Zloh, M.; Patel, P.; Mackenzie, L.S. In silico modelling of prostacyclin and other lipid mediators to nuclear receptors reveal novel thyroid hormone receptor antagonist properties. *Prostaglandins Other Lipid Mediat.* **2016**, *122*, 18–27. [CrossRef]
39. Jin, L.; Lin, S.; Rong, H.; Zheng, S.; Jin, S.; Wang, R.; Li, Y. Structural basis for iloprost as a dual peroxisome proliferator-activated receptor alpha/delta agonist. *JBC* **2011**, *286*, 31473–31479. [CrossRef]
40. Xu, H.E.; Lambert, M.H.; Montana, V.G.; Parks, D.J.; Blanchard, S.G.; Brown, P.J.; Sternbach, D.D.; Lehmann, J.M.; Wisely, G.B.; Willson, T.M.; et al. Molecular recognition of fatty acids by peroxisome proliferator-activated receptors. *Mol. Cell* **1999**, *3*, 397–403. [CrossRef]
41. Wu, C.C.; Baiga, T.J.; Downes, M.; La Clair, J.J.; Atkins, A.R.; Richard, S.B.; Fan, W.; Stockley-Noel, T.A.; Bowman, M.E.; Noel, J.P.; et al. Structural basis for specific ligation of the peroxisome proliferator-activated receptor delta. *Proc. Natl. Acad. Sci. USA* **2017**, *114*, E2563–E2570. [CrossRef] [PubMed]
42. Maltarollo, V.G.; Togashi, M.; Nascimento, A.S.; Honorio, K.M. Structure-based virtual screening and discovery of new PPARdelta/gamma dual agonist and PPARdelta and gamma agonists. *PLoS ONE* **2015**, *10*, e0118790. [CrossRef] [PubMed]
43. Mackenzie, L.S.; Lymn, J.S.; Hughes, A.D. Linking phospholipase C isoforms with differentiation function in human vascular smooth muscle cells. *Biochim. Biophys. Acta (BBA) Mol. Cell Res.* **2013**, *1833*, 3006–3012. [CrossRef] [PubMed]

International Journal of
Molecular Sciences

MDPI

Article

Synthesis of a Coumarin-Based PPARγ Fluorescence Probe for Competitive Binding Assay

Chisato Yoshikawa, Hiroaki Ishida, Nami Ohashi and Toshimasa Itoh *

Laboratory of Drug Design and Medicinal Chemistry, Showa Pharmaceutical University, 3-3165 Higashi-Tamagawagakuen, Machida, Tokyo 194-8543, Japan; d1702@g.shoyaku.ac.jp (C.Y.); ishida@ac.shoyaku.ac.jp (H.I.); ohashi@ac.shoyaku.ac.jp (N.O.)
* Correspondence: titoh@ac.shoyaku.ac.jp

Abstract: Peroxisome proliferator-activated receptor γ (PPARγ) is a molecular target of metabolic syndrome and inflammatory disease. PPARγ is an important nuclear receptor and numerous PPARγ ligands were developed to date; thus, efficient assay methods are important. Here, we investigated the incorporation of 7-diethylamino coumarin into the PPARγ agonist rosiglitazone and used the compound in a binding assay for PPARγ. PPARγ-ligand-incorporated 7-methoxycoumarin, **1**, showed weak fluorescence intensity in a previous report. We synthesized PPARγ-ligand-incorporating coumarin, **2**, in this report, and it enhanced the fluorescence intensity. The PPARγ ligand **2** maintained the rosiglitazone activity. The obtained partial agonist **6** appeared to act through a novel mechanism. The fluorescence intensity of **2** and **6** increased by binding to the ligand binding domain (LBD) of PPARγ and the affinity of reported PPARγ ligands were evaluated using the probe.

Keywords: PPARγ ligand; coumarin; fluorescent ligand; screening; crystal structure

Citation: Yoshikawa, C.; Ishida, H.; Ohashi, N.; Itoh, T. Synthesis of a Coumarin-Based PPARγ Fluorescence Probe for Competitive Binding Assay. *Int. J. Mol. Sci.* **2021**, 22, 4034. https://doi.org/10.3390/ijms22084034

Academic Editor: Manuel Vázquez-Carrera

Received: 27 March 2021
Accepted: 11 April 2021
Published: 14 April 2021

1. Introduction

Peroxisome proliferator-activated receptors (PPARs) belong to the nuclear receptor superfamily and are categorized into three subtypes—PPARα, β/δ, and γ [1–3]. PPARγ is an important target molecule for the inflammatory disease and metabolic syndrome. PPARγ agonists were developed as antidiabetic drugs, such as rosiglitazone and pioglitazone [4] but cause adverse effects, such as heart failure, edema, and increased risk of myocardial infarction [2]. PPARγ ligands were thus developed using various strategies and include partial PPARγ agonists [5], selective PPARγ modulators, PPAR α/γ dual agonists, oxidized fatty acid agonists, antagonists, and covalent ligands [6–12] (Figure 1).

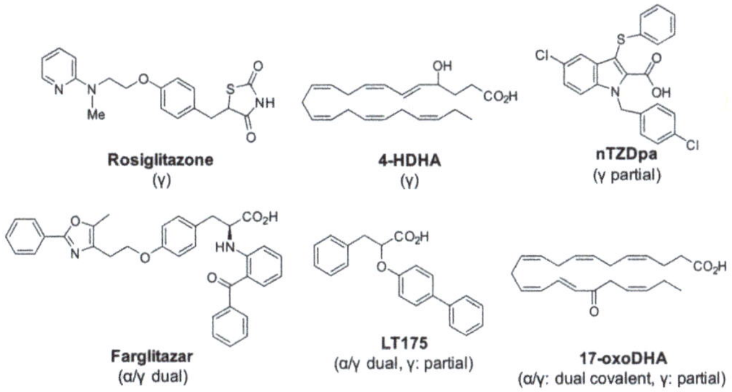

Figure 1. Structures of PPARγ agonists rosiglitazone and 4-HDHA, partial agonist nTZDpa, PPAR α/γ dual agonist farglitazar, dual and partial agonist LT175, and dual covalent and partial agonist 17-oxoDHA.

PPARγ was identified as a novel target of the nonsteroidal anti-inflammatory drugs (NSAIDs) action by direct binding. Like PPAR agonists, NSAIDs are believed to be fatty acid analogues and can suppress the expression of proinflammatory genes via PPARγ activation [13]. A PPARγ ligand inhibited monocyte elaboration of inflammatory cytokines and chemokine expression, and prevented microglial activation [14,15]. A fatty acid-based agonist, 4-HDHA, alleviates the symptoms of DSS-induced colitis [16]. Thus, PPARγ is a possible target molecule for anti-inflammatory as well as metabolic syndrome. The ligand binding cavity of PPARγ is versatile [17,18], allowing the design of a large number of unique compounds.

Nuclear receptor ligands, including PPARγ ligands, are often evaluated by investigating their gene transcription activity [19]. This assay is useful in identifying agonists with strong efficacy, but this is likely to overlook antagonists and partial agonists because of its weak efficacy. Therefore, the binding assay is important for exploring novel ligands. Conventional PPAR binding assays often use competition with a radioligand [20] and such assays provide superior sensitivity, but they are costly, are possible health hazards, and require laborious experimental procedures and special facilities. Fluorescent probes overcome these drawbacks [21]. A fluorescent probe for a PPAR α/γ fluorescent polarization (FP) binding assay, based on the large molecule fluorescein was developed [22,23]. In contrast, coumarins are small, and have a solvatochromic fluorescence property that increases fluorescence intensity in hydrophobic environments and decreases it in hydrophilic environments [24]. Compounds containing coumarin were developed to detect ligand binding to target proteins [25].

The coumarin skeleton is used in various fields [26–28] and many synthetic strategies were reported [29–35], including C–C bond formation reactions that require advanced knowledge and techniques for probe synthesis. In contrast, we recently reported strategies for constructing a coumarin skeleton on a target protein (TCC probe) using small molecules (coumarin precursors) [36,37]. The advantages of using a coumarin precursor in organic synthesis include incorporation of the coumarin skeleton into the ligand in the final step of synthesis in one step, and this incorporation is facile if the precursor of the ligand has a nucleophilic functional group.

We previously demonstrated the usefulness of precursors by synthesizing compound **1**, whose structure incorporates 7-methoxy (7-MeO) coumarin into the rosiglitazone (Scheme 1) [36]. However, the fluorescence property of compound **1** is poor.

Here, we synthesized a rosiglitazone-based fluorescence probe using a coumarin precursor. PPARγ binding assay shows that the probe binds to rosiglitazone. We utilized the properties of solvatochromic fluorophores instead of FP to facilitate the evaluation of the ligands.

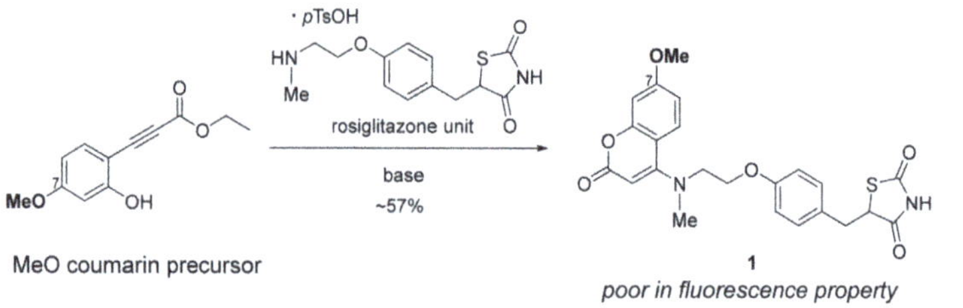

Scheme 1. Synthesis of 7-methoxy coumarin-incorporated rosiglitazone.

2. Results

2.1. Design and Synthesis of a Coumarin-Based PPARγ Ligand

We previously synthesized the rosiglitazone-based model compound **1**, in which the terminal pyridine was replaced by the coumarin skeleton, using a coumarin precursor (Scheme 1). Coumarin in compound **1** was poor in fluorescence property. We, thus, designed compound **2**, which contained a 7-diethylamino (7-Et_2N) coumarin unit

(Figure 2). The coumarin-substituted electron-donating groups at position 7 showed strong fluorescence [38] and thus we expected the designed probe **2** to be sufficiently fluorescent to be useful for PPARγ binding assays. The precursor used was the TBS-protected form **3**, because the Et_2N moiety destabilized the precursor. Since the Et_2N and TBS groups were electron-donating, the reactivity (electro-deficiency) of the alkynone moiety was weak [39]. We, thus, used precursor **3** (Scheme 2) in which CH_2CF_3 provided an electron-withdrawing group at the ester group of the 7-Et_2N coumarin precursor (Scheme 2). When we used Et_3N as a base and solvent, no desired product **2** was isolated and unexpectedly, we obtained compound **5**, which was conjugated to an Et_2N group (Scheme 2; Table 1, entry 1). We, therefore, changed the synthesis conditions from those used to synthesize 7-MeO coumarin incorporating rosiglitazone **1**. The use of DMF and Et_3N as a solvent, gave the desired **2** in 14% yield and **5** was increased to 51% yield. Interestingly, ethylation at the thiazolidinedione (TZD) occurred to give **6** (14%) (entry 2). Yamaguchi et al. reported the conjugate addition of an ynone-containing azulene with a tertiary amine [40]. We considered that products **5** and **6** resulted from a similar mechanism (Schemes S1 and S2) and thus we reduced the amount of Et_3N. When 3.0 equiv. Et_3N in DMF was used, **2** was afforded in moderate yield (57%) (entry 3).

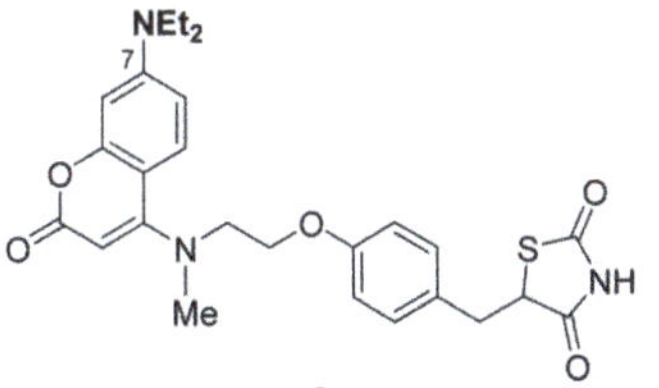

Figure 2. Structure of PPARγ ligand incorporated 7-diethylamino (7-Et_2N) coumarin.

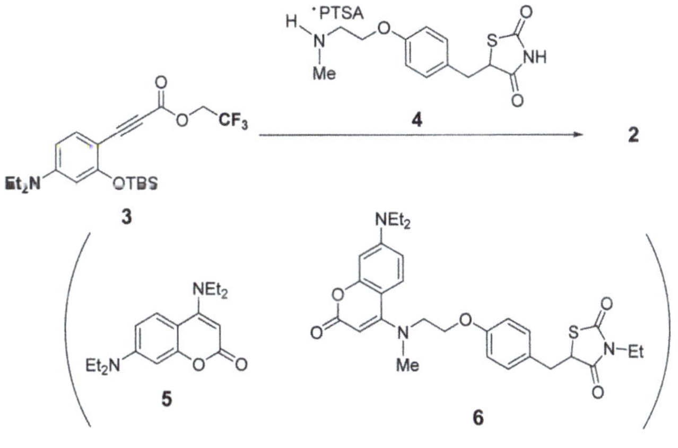

Scheme 2. Synthesis of PPARγ ligand.

Table 1. Investigation of PPARγ ligand synthesis conditions.

Entry	Solvent	Base	Temperature (°C)	Yield (%)		
				2	5	6
1	Et_3N	-	100	-	40	-
2	DMF: Et_3N = 1:1	-	100	14	51	14
3	DMF	Et_3N (3.0 equiv.)	60	57	11	-

2.2. Transcriptional Activity

The transcriptional activities of compound **1** [36], **2**, and **6** were compared for PPARγ activity, using the dual luciferase assay in Cos-7 cells (Figure 3, Table 2). Compounds **1** and **2** had activities comparable to rosiglitazone, showing that the incorporation of coumarin did not reduce ligand agonistic activity, regardless of if it was 7-MeO coumarin or 7-Et_2N coumarin. This result suggests that incorporation of a coumarin unit into a ligand maintains biological activity because of its small structure. In contrast, compound **6** showed partial agonistic activity.

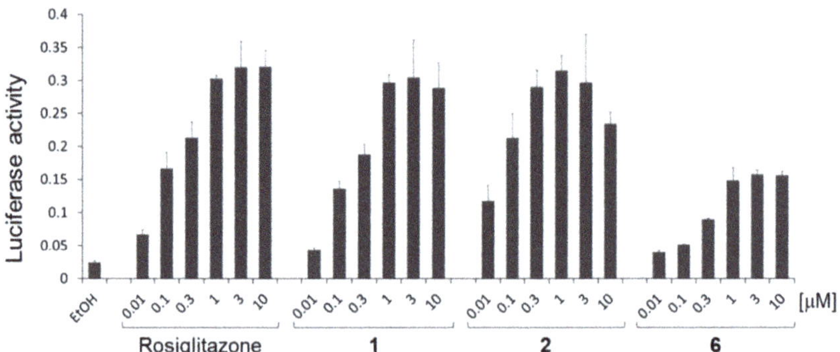

Figure 3. Transcriptional activity of synthetic compounds **1**, **2**, and **6** evaluated in Cos-7 cells using a dual luciferase assay with a GAL4-PPARγ chimera expression plasmid (pSG5-GALhPPARγ), a reporter plasmid (MH100×4-TK-Luc), and an internal control plasmid containing sea pansy luciferase expression constructs (pRL-CMV). The data represent the mean ± SD of three independent experiments.

Table 2. The EC_{50} values of compounds **1**, **2**, and **6**.

Compound	Rosiglitazone	1	2	6
EC_{50} (μM)	0.17	0.18	0.034	0.35

2.3. Fluorescence Spectra

We evaluated the fluorescence properties of PPARγ ligands **2** and **6** (Figure 4) by measuring the fluorescence spectra and quantum yields (Q.Y.) in CH_2Cl_2, THF, MeOH, or H_2O. The Q.Y. of the 7-Et_2N coumarins (**2**: Φ = 0.143–0.541, **6**: Φ = 0.0865–0.764) were much better than that of 7-MeO coumarin (**1**: Φ = 0.0359–0.00461) (Table 3), and exhibited a high fluorescence intensity in organic solvents and a weaker fluorescence in H_2O. Additionally, **2** and **6** showed similar fluorescence spectral shifts, with the emission maximum shifting to a longer wavelength in polar solvent (Figure 4). The difference between Abs_{max} and Em_{max} ($\Delta = Em_{max} - Abs_{max}$) showed that the Δ value of **2** and **6** increased with increasing solvent polarity (Table 3) in the order THF < CH_2Cl_2 < MeOH < H_2O. The dielectric constant of each solvent was THF: 7.4, CH_2Cl_2: 8.9, MeOH: 32.6, and H_2O: 78.5 [41,42]. This correlation clarified that Et_2N coumarins **2** and **6** exhibited a Stokes shift and thus we expected the fluorescence spectra of **2** and **6** to shift upon binding to the hydrophobic binding site of the protein.

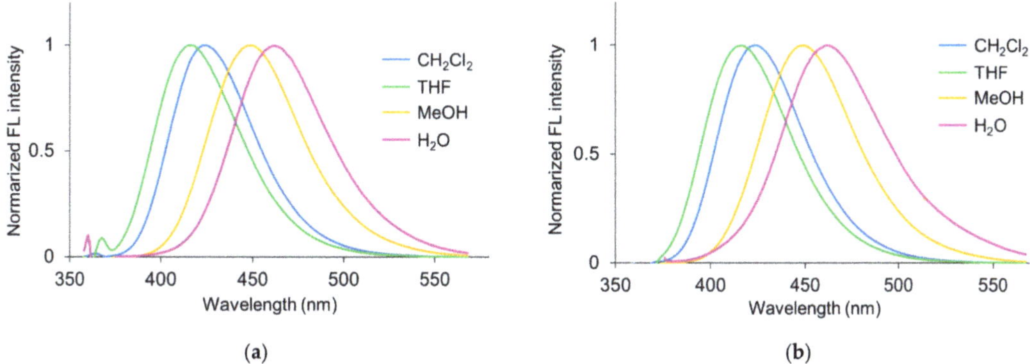

(a) (b)

Figure 4. (**a**) Fluorescence spectrum of **2**, in CH_2Cl_2, THF, MeOH, or H_2O (2 μM). (**b**) Fluorescence spectrum of **6**, in CH_2Cl_2, THF, MeOH, or H_2O (2 μM).

Table 3. Fluorescence properties for PPARγ ligands **1**, **2**, and **6**.

Comp.	In CH_2Cl_2		In THF		In MeOH		In H_2O	
	Abs_{max} [1] Em_{max} Δ [3] (nm)	Q.Y. [2]	Abs_{max} [1] Em_{max} Δ [3] (nm)	Q.Y. [2]	Abs_{max} [1] Em_{ma} Δ [3] (nm)	Q.Y. [2]	Abs_{max} [1] Em_{max} Δ [3] (nm)	Q.Y. [2]
1	316 469 153	0.00359	314 467 153	0.00461	314 392 78	0.0039	312 451 139	0.00369
2	359 423 64	0.541	354 416 62	0.191	359 449 90	0.532	364 462 98	0.143
6	359 423 64	0.764	353 416 63	0.172	358 449 91	0.432	369 461 92	0.0865

[1] See the Supplementary Materials (Figure S1) for absorption spectrum of **2** and **6**. [2] The quantum yields were determined using quinine sulfate in 0.1 M H_2SO_4 (Φ = 0.577) [43]. [3] Δ = Abs_{max} − Em_{max}.

2.4. X-ray Crystal Structure

We attempted to crystallize human PPARγ-LBD complexed with **2** or **6** to identify the binding mode of PPARγ-LBD. Crystals were grown in the presence of each ligand but **6** provided only the apo structure of PPARγ. The complex with **2** provided the co-crystal structure; the crystallographic analysis data are summarized in Table S1. The overall crystal structure of the **2**/PPARγ-LBD complex was similar to that of the rosiglitazone/PPARγ-LBD complex (Figure 5). The TZD moiety formed hydrogen bonds with His323, Tyr473, Ser289, and Gln286, which was identical to that of rosiglitazone in PPARγ-LBD (Figure 6a,b). Furthermore, the coumarin moiety was positioned similar to that of the pyridine moiety in the X-ray crystal structure of the rosiglitazone/PPARγ complex (2PRG.pdb), and thus, we concluded that the transcriptional activity of probe **2** resembled that of rosiglitazone (Figure 6c,d). The N-H group on TZD of byproduct **6** was ethylated (N-Et), and therefore, it could not form a hydrogen bond with Tyr473 on helix12. Indeed, this ethyl group caused steric repulsion with helix12, and this hydrogen bond was important for PPARγ activation, explaining why probe **6** showed partial agonist activity. Importantly, although rosiglitazone is the most commonly used PPARγ ligand, no rosiglitazone-based partial agonist is reported to date. Comparison with rosiglitazone might contribute to the development of PPARγ-targeted drugs.

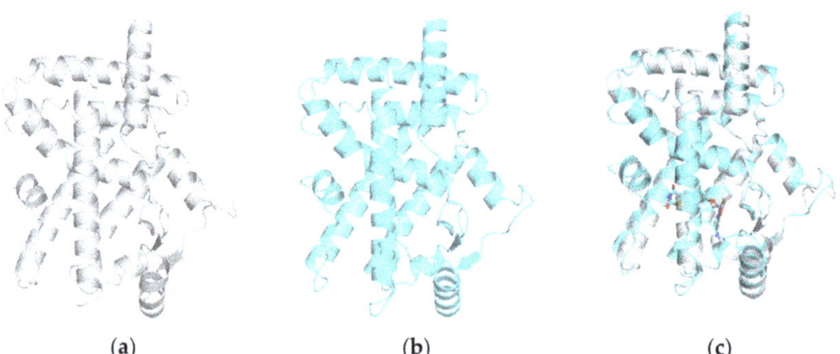

Figure 5. (**a**) The overall crystal structure of the **2**/PPARγ-LBD complex (PDB code: 7EFQ). (**b**) The overall crystal structure of the rosiglitazone/PPARγ-LBD complex (PDB code: 2PRG). (**c**) Comparison of the overall crystal structure of the **2**/PPARγ-LBD complex with the rosiglitazone/PPARγ-LBD complex.

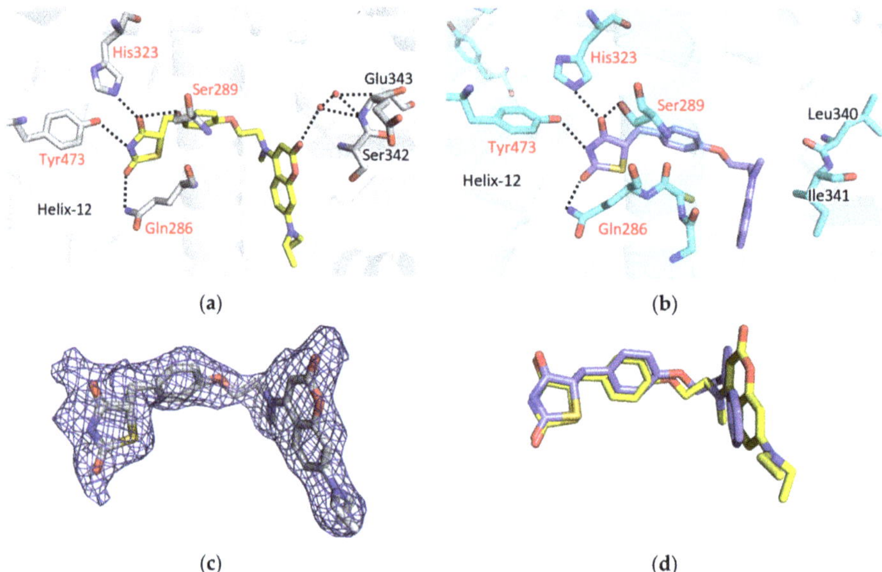

Figure 6. (**a**) Hydrogen bonds between **2** and hPPARγ-LBD (PDB code: 7EFQ). (**b**) Hydrogen bonds between rosiglitazone and hPPARγ-LBD (PDB code: 2PRG). (**c**) Omit map of **2** bound to hPPARγ-LBD. (**d**) Comparison of **2** with rosiglitazone bound to hPPARγ-LBD.

2.5. Application of PPARγ Binding Assay

We examined whether **2** and **6** were useful for PPARγ binding assays. We attempted to determine the K_d value of **2** or **6** with hPPARγ-LBD. The fluorescence spectra were measured by adding hPPARγ-LBD (0.05 to 8.0 μM) in Tris–HCl buffer to **2** (1 μM) (Figure 7a). The fluorescence maxima shifted to shorter wavelength and the fluorescence intensity increased upon increasing the concentration of PPARγ-LBD. We calculated K_d using the fluorescence intensity at 410 nm (**2**: Kd = 1558 ± 93.61 nM, Figure 7b), (**6**: Kd = 4082 ± 712.2 nM, Figure S2). The K_d value showed that the PPARγ binding activity of **6** was weaker than that of **2**, and thus **6** could be useful for screening the lower affinity ligands.

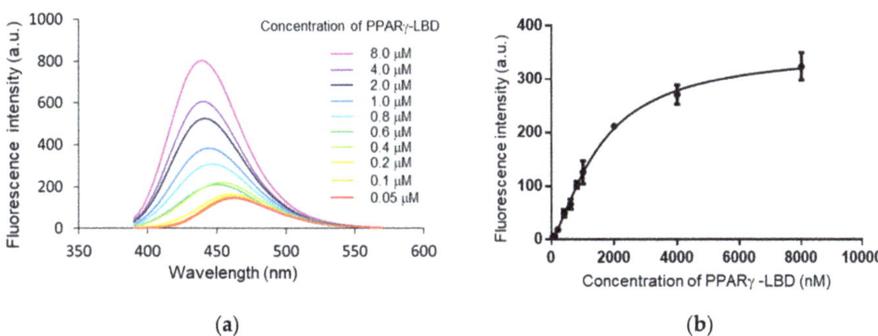

Figure 7. (**a**) Fluorescence spectra of **2** (1 μM) upon addition to hPPARγ-LBD (0.05–8.0 μM) in Tris–HCl buffer with Ex = 367 nm. (**b**) Fluorescence intensity of **2** at 410 nm depending on the concentration of hPPARγ-LBD. Data are mean ± SD (n = 3).

We performed PPARγ competitive binding assays using a fixed concentration of **2** (0.72 μM) and hPPARγ-LBD (0.6 μM). First, we carried out a binding assay using rosiglitazone. The addition of rosiglitazone to **2** and hPPARγ-LBD decreased the fluorescence intensity of **2** and it shifted to a longer wavelength, clearly showing that rosiglitazone replaced **2** bound to hPPARγ-LBD. We determined the value of K_i using the fluorescence intensity at 410 nm (K_i = 1157 ± 1.08 nM, Figure 8, Table 4) and that of farglitazar using the same procedure (K_i = 132.3 ± 1.13 nM, Figure S3, Table 4). Next, we attempted to evaluate the K_i value of pioglitazone, whose affinity was lower than that of rosiglitazone, but were unsuccessful because it required a concentration of pioglitazone above its solubility limit in the buffer. Therefore, we determined the K_i value of pioglitazone using probe **6** (1.44 μM), whose affinity was lower than that of **2**. The binding assay succeeded and we obtained the Ki value (K_i = 5495 ± 3.14 nM, Figure S4, Table 4). The Ki value of the PPARγ partial agonist LT175 was also determined using **6** (K_i = 2334 ± 1.46 nM, Figure S5, Table 4) and thus the order of the calculated K_i values was farglitazar < rosiglitazone < pioglitazone, consistent with the previously reported IC_{50} values for farglitazar, rosiglitazone, and pioglitazone [23] and the reported EC_{50} values [11,44] farglitazar < rosiglitazone < LT175 < pioglitazone. The determined Ki values therefore showed an identical relationship with the reported EC_{50} values, suggesting that **2** and **6**, PPARγ ligands that incorporate 7-Et_2N coumarin, were useful probes for competitive binding assays of PPARγ ligands

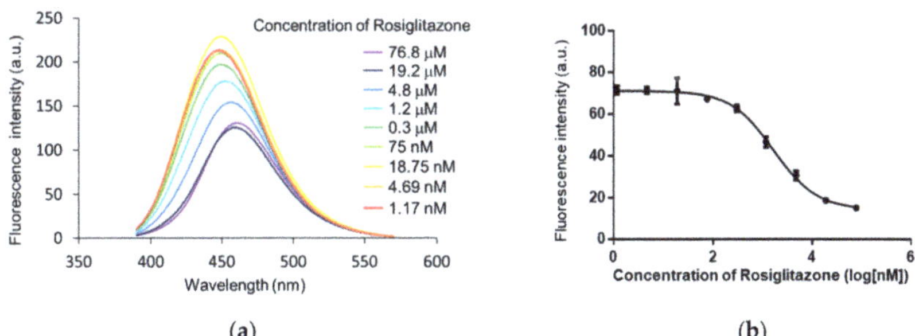

Figure 8. Fluorescence binding assay using the fluorescence intensity of **2** at Ex = 367 nm in the presence of 0.6 μM hPPARγ-LBD in Tris–HCl buffer. (**a**) Fluorescence spectrum of **2** (0.72 μM) and hPPARγ-LBD (0.6 μM) upon addition to rosiglitazone (1.17 nM–76.8 μM) in Tris–HCl buffer at Ex = 367 nm. (**b**) Fluorescence intensity of **2** at 410 nm in the presence of 0.6 μM hPPARγ-LBD at various concentrations of rosiglitazone. Data are mean ± SD (n = 3).

Table 4. Comparison of the K_i value determined using **2** or **6** with reported IC_{50} or EC_{50} values.

Concentration (nM) [ref]	Rosiglitazone	Farglitazar	Pioglitazone	LT175
The calculated K_i value using **2** or **6** [1]	1157 ± 1.08	132.3 ± 1.13	5495 ± 3.14	2334 ± 1.46
The reported IC_{50} value [23]	660 ± 252	90 ± 34	4500 ± 465	-
The reported EC_{50} value [44]	18 ± 4	0.20 ± 0.05	280 ± 42	-
The reported EC_{50} value [11]	40 ± 20	-	-	480 ± 80

[1] Data are mean ± SD (n = 3).

3. Discussion

Here, we reported the facile synthesis of PPARγ ligands incorporating coumarin, using our coumarin precursor instead of the conventional synthetic approach. We previously showed that coumarin was formed by conjugated addition from nucleophiles such as thiols, amines, alcohols, or phenols [39], and thus this strategy might be applicable to other ligands.

We also showed that **6** showed partial agonist activity caused by pushing helix12. Fewer side effects were likely caused by PPARγ partial agonists than by full agonists, such as rosiglitazone, and several partial agonists were synthesized [6] that functioned either as an "active antagonist" or a "passive antagonist". An active antagonist regulated helix12 through direct interaction, such as steric repulsion whereas a passive antagonist interacted indirectly. Most PPARγ partial agonists act via a passive mechanism, through indirect [5] or weak [45,46] interactions or through multiple conformations [47,48]. However, probe **6** was believed to be an "active partial agonist" and pushed helix12. When we tried to co-crystallize PPARγ with **6**, we obtained only apo form crystals (data not shown). Helix12 was believed to be in an active position in the crystal packing but helix12 could not adopt its active position due to the steric repulsion of the ethyl group of **6**, resulting in the apo form that was unfavorable for crystallization. Although **6** resembled rosiglitazone, no mechanisms are reported for PPARγ partial agonism and thus probe **6** was a novel PPARγ ligand.

Competitive binding assays for nuclear receptors were either radiometric or fluorometric assays [49]. A scintillation proximity assay (SPA) was reported for PPAR [50] and an FP assay was reported for ligands containing fluorescein [22,23]. The incorporation of the 7-Et_2N coumarin did not affect the activity of rosiglitazone, as determined by a gene transcriptional assay. We suggest that the compounds incorporating 7-Et_2N coumarins, which are easy to synthesize, could be applied to PPARγ binding assays and did not require fluorescence anisotropy apparatus and techniques. 7-Et_2N coumarin could be used for live cells imaging [51] and thus **2** and **6** were potent probes for cell imaging.

Recently, a coumarin-containing nuclear receptor RXR agonist was reported. The authors demonstrated its utility in a competitive binding assay for several RXR ligands [52]. Coumarins, thus, appear suitable as nuclear receptor ligands. Furthermore, it is not limited to ligands that target nuclear receptors, coumarins were reported to have an established structure for introducing fluorescence into tool compounds for the biochemical studies [53]. This was because studies on the effects of the positions of substituents and the properties of functional groups (electron-withdrawing or electron-donating) on the fluorescent properties of coumarin were widely studied for decades [54]. Moreover, some natural productss with a coumarin skeleton were reported to show PPARγ activity [55]. From this point of view, our coumarin conjugated strategy could be used to synthesize other nuclear receptor probes, photochemical probes, and bioactive compounds. Our coumarin conjugation strategy could be used to synthesize other nuclear receptor probes.

4. Materials and Methods

4.1. General Information for Synthesis

All non-aqueous reactions were performed in an oven-dried or a flame-dried glassware, under nitrogen atmosphere. Unless otherwise mentioned, all reagents were purchased from commercial suppliers and used without further purification. Organic solvents were dried by standard methods. All reactions were monitored by a thin-layer chromatography. Thin-layer chromatography was performed on silica gel 70 F_{254} TLC plates pre-coated with 0.25 mm thickness (FUJIFILM Wako Pure Chemical Corporation, Osaka, Japan). Visualization was done by UV light (254 nm or 365 nm), phosphomolybdic acid (PMA) stain, or Hanessian's stain. Purification on silica gel column chromatography was performed on silica gel 60N (40−50 μm, 63−210 μm, Kanto Chemical Co. Inc., Tokyo, Japan). ^{1}H-NMR spectra were recorded on a Bruker AV300M (300 MHz) or Bruker AV600 (600 MHz) spectrometer in appropriate deuterated solvents. ^{13}C-NMR spectra were recorded at 75 MHz or 150 MHz. Chemical shifts were reported in parts per million (ppm) on the d scale from TMS peak. NMR descriptions: s, singlet; d, doublet; t, triplet; q, quartet; m, multiplet; and br, broad. Coupling constants, *J*, are reported in Hertz (Hz). High-resolution mass spectra (ESI) were obtained from a JEOL AccuTOF LC-plus JMS-T100LP spectrometer (JEOL Ltd., Tokyo, Japan).

Compounds **3** and **4** were synthesized in the same methods as in the reference [36,39]. The structures of compounds were confirmed by ^{1}H-NMR.

4.1.1. 5-(4-(2-((7-(diethylamino)-2-oxo-2H-chromen-4-yl)(methyl)amino)ethoxy)benzyl)thiazolidine-2,4-dione (**2**)

To a solution of **4** (19.8 mg) in DMF (0.9 mL) Et_3N (18 μL, 0.131 mmol, 3.0 equiv.) and **3** (18.8 mg, 0.0438 mmol) in DMF (0.6 mL) was added. The mixture was stirred at 60 °C for 16 h, and then concentrated under reduced pressure. The crude solution was purified by open column chromatography (silica gel: 7 g, 5−100% AcOEt/hexane) to give **2** (12.4 mg, 0.0250 mmol, 57%, from compound **3**) and **5** (1.4 mg, 0.000485 mmol, 11%). ^{1}H NMR (300 MHz, $CDCl_3$) δ [ppm]: 7.63 (d, *J* = 9.0 Hz, 1H), 7.15 (d, *J* = 8.6 Hz, 2H), 6.81 (d, *J* = 8.6 Hz, 2H), 6.56−6.50 (overlapped, 2H), 5.36 (s, 1H), 4.49 (dd, *J* = 8.5, 3.9 Hz, 1H), 4.22 (m, 2H), 3.79 (m, 2H), 3.44−3.34 (overlapped, 5H), 3.18 (dd, *J* = 14.2, 8.6 Hz, 1H), 3.03 (s, 3H), 1.21 (t, *J* = 7.1 Hz, 6H). ^{13}C NMR (75 MHz, $CDCl_3$) δ 173.9, 170.1, 164.1, 161.6, 157.8, 156.7, 150.1, 130.7 (2 carbons), 127.9, 126.3, 114.9 (2 carbons), 107.7, 104.5, 98.2, 90.8, 65.3, 53.9, 53.4, 44.6 (2 carbons), 39.9, 37.7, 12.5 (2 carbons). ESI-HRMS: m/z calcd for $C_{26}H_{30}N_3O_5S$ $[M + H]^+$: 496.19062; found: 496.18896. IR (NaCl): 1749, 1698, 1658, 1245 cm^{-1}.

4.1.2. 4,7-bis(diethylamino)-2H-chromen-2-one (**5**)

^{1}H NMR (300 MHz, $CDCl_3$) δ [ppm]: 7.43 (d, *J* = 9.1 Hz, 1H), 6.53 (d, *J* = 9.1, 2.7 Hz, 1H), 6.48 (d, *J* = 2.6 Hz, 1H), 5.38 (s, 1H), 3.39 (quin, *J* = 7.0 Hz, 8H), 1.23 (t, *J* = 7.1 Hz, 6H), 1.20 (t, *J* = 7.1 Hz, 6H). ^{13}C NMR (75 MHz, $CDCl_3$) δ [ppm]: 164.0, 160.1, 156.7, 149.9, 126.1, 107.5, 105.1, 98.2, 90.7, 45.5 (2 carbons), 44.6 (2 carbons), 12.5 (2 carbons), 12.3 (2 carbons). ESI-HRMS: m/z calcd for $C_{17}H_{24}N_2O_2$ $[M + H]^+$: 289.19160; found: 289.19162. IR (NaCl): 1698 cm^{-1}.

4.1.3. 5-(4-(2-((7-(diethylamino)-2-oxo-2H-chromen-4-yl)(methyl)amino)ethoxy)benzyl)-3-ethylthiazolidine-2,4-dione (**6**)

^{1}H NMR (300 MHz, $CDCl_3$) δ [ppm]: 7.61 (d, *J* = 9.0 Hz, 1H), 7.17−7.12 (overlapped, 2H), 6.86−6.81 (overlapped, 2H), 6.55−6.49 (overlapped, 2H), 5.43 (s, 1H), 4.41 (dd, *J* = 9.0, 3.9 Hz, 1H), 4.21 (t, *J* = 5.6 Hz, 2H), 3.76 (t, *J* = 5.5 Hz, 2H), 3.61 (m, 2H), 3.46−3.35 (overlapped, 5H), 3.09 (dd, *J* = 14.1, 8.9 Hz, 1H), 3.04 (s, 3H), 1.20 (t, *J* = 7.1 Hz, 6H), 1.11 (t, *J* = 7.2 Hz, 3H). ^{13}C NMR (75 MHz, $CDCl_3$) δ [ppm]: 173.8, 170.9, 163.7, 161.5, 157.8, 156.7, 150.1, 130.6 (2 carbons), 128.3, 126.3, 114.7 (2 carbons), 107.6, 104.5, 98.2, 91.0, 65.3, 53.7, 51.6, 44.6 (2 carbons), 39.9, 37.8, 36.9, 12.8, 12.5 (2 carbons). ESI-HRMS: m/z calcd for $C_{28}H_{34}N_3O_5S$ $[M + H]^+$: 524.22192; found: 524.22030. IR (NaCl): 1745, 1682, 1613 cm^{-1}.

4.2. General Information for Biological Experiments

All biological reagents were used without further purification, unless otherwise noted. Sodium dodecyl sulfate-polyacrylamide gel electrophoresis (SDS-PAGE) was carried out with an AE-6530 electrophoresis apparatus. UV-visible absorption spectra were recorded on a V-630BIO spectrophotometer (JASCO, Tokyo, Japan). Fluorescent spectra were measured using a F-7100 fluorescence spectrophotometer (HITACHI, Tokyo, Japan).

4.3. Transactivation Assay

Transactivation in COS-7 cells was measured using a dual luciferase assay according to a previously reported procedure [46]. EC_{50}s were calculated by GraphPad Prism 6 (GraphPad Software, San Diego, USA) (<A>LogEC = $LogEC_{50}$Control <~A>LogEC = $LogEC_{50}$Control + log(EC_{50}Ratio) Y = Bottom + (Top-Bottom)/(1 + 10^((LogEC-X) × HillSlope))).

4.4. Protein Expression and Purifications

PPARγ expression and purification were carried out as previously described [46]. The human PPARγ-LBD (residues 204−477) was expressed using a modified pET30a vector with an N-terminal 6×His tag cleavable by TEV protease. *E. coli* Rosetta (DE3) was freshly transformed with the plasmid and grown 1 L of 2 × TY medium with 34 mg/mL kanamycin and 50 mg/mL chloramphenicol at 37 °C, to an OD at 600 nm of 0.6−1.0. Protein synthesis was then induced with 0.5 mM isopropyl-b-Dthiogalactopyranoside (IPTG), and the cultures were further incubated at 20 °C for 16 h. Cells were harvested and resuspended in 50 mL of lysis buffer (20 mM Tris–HCl pH 8.0, 100 mM NaCl, 1 mM TCEP, 0.5 mM EDTA, and 13 Protease inhibitor cocktail). Cells were lysed by sonication, and the soluble fraction was isolated by centrifugation (18,000× *g* for 30 min). The supernatant was applied to cOmplete His-Tag Purification Resin (Roche, Basel, Switzerland), and the resin was thoroughly washed in wash buffer (20 mM Tris–HCl pH 8.0, 100 mM NaCl, 1 mM TCEP, and 5 mM imidazole). The human PPARγ-LBD was eluted with the elution buffer (20 mM Tris–HCl pH 8.0, 100 mM NaCl, 1 mM TCEP, and 250 mM imidazole). TEV protease was added to the eluate, and the mixture was dialyzed overnight at 4 °C with 500 mL of buffer (20 mM Tris–HCl pH 8.0, 1 mM TCEP, 0.5 mM EDTA). The cleaved protein was passed through complete His-Tag Purification Resin (Roche). The flow-through was loaded onto a Resource Q (6 mL) column (GE Healthcare, Chicago, USA) equilibrated with buffer (20 mM Tris–HCl pH 8.0, 1 mM TCEP, 0.5 mM EDTA). The column was eluted with an NaCl gradient from 0 to 1 M in the starting buffer. The eluted fractions were concentrated and loaded onto a Superdex 75 Increase 10/300 GL gel filtration (24 mL) column (GE Healthcare) equilibrated with buffer (20 mM Tris–HCl pH 8.0, 1 mM TCEP, and 0.5 mM EDTA). Purified human PPARγ-LBD was concentrated in buffer (20 mM Tris–HCl pH 8.0, 1 mM TCEP, and 0.5 mM EDTA) to 6.0 mg/mL, which was estimated by UV absorbance at 280 nm.

4.5. X-Ray Crystallography

Crystals were obtained through co-crystallization in ligand **2**. For the PPARγ, co-crystallization was performed by vapor diffusion using a hanging drop made by mixing 1 μL of the PPARγ-LBD solution (6 mg/mL, in 20 mM Tris–HCl pH 8.0, 1 mM TCEP, 0.5 mM EDTA) with 0.5 equivalent Ligand, (**2**) with 1 μL of reservoir solution (0.8 M sodium citrate and 0.1 M Tris–HCl pH 7.27) and the drops were equilibrated against 300 μL of reservoir solution at room temperature. The mixture was stored at room temperature, and crystals appeared after about 2 weeks. Crystals were flash-cooled in liquid nitrogen, after a fast soaking in a cryoprotectant buffer (LV Cryo Oil (MiTeGen, NY, USA)). Diffraction data sets were collected at 100 K in a stream of nitrogen gas at beamline BL-5A, at the high energy accelerator research organization (KEK, Tsukuba, Japan). Reflections were recorded with an oscillation range per image of 1.0°. Diffraction data were indexed, integrated, and scaled using the program iMOSFLM (MRC-LMB, Cambridge, UK) [56,57]. The ternary complex structures were solved by molecular replacement with the software Phaser [58] in the CCP4 program (Research Complex at Harwell, Oxford, UK) [59] using rat VDR-LBD coordinates (PDB code: 2VV3) [7], and the finalized sets of atomic coordinates were obtained after iterative rounds of model modification with the program Coot (MRC-LMB, Cambridge, UK) [60] and refinement with refmac5 (University of York, York, UK) [61–65].

4.6. UV-Visible Absorption and Fluorescence Spectroscopic Analyses

Stock solutions of model coumarin compounds PPAR ligands **1**, **2**, and **6** were prepared in DMSO, and stored in the dark at −20 °C. The stock solutions were diluted (2 μM) with solvents (CH_2Cl_2, THF, MeOH, and H_2O) and then the UV-visible absorption and fluorescence signals were measured by a spectrometer. The fluorescence quantum yields of coumarin analogues were calculated using quinine sulfate (Φ = 0.577 in 0.1 M H_2SO_4) as a reference standard [43].

4.7. K_d Determination of **2** or **6**

PPARγ-LBD (6 mg/mL) was diluted to the concentration of 0.05, 0.1, 0.2, 0.4, 0.6, 0.8, 1.0, 2.0, 4.0, and 8.0 μM (0.25, 0.5, 1.0, 1.5, 2.0, 3.0, 4.0, 5.0, 10.0, and 15.0 μM, in case of **6**) with the assay buffer (20 mM Tris–HCl pH 7.0, 1 mM TCEP, and 0.5 mM EDTA). Then, each concentration levels of PPARγ-LBD solutions were added **2** or **6** (final concentration of 1 μM), and the fluorescence spectra were measured using the mixture of PPARγ-LBD and **2**, or **6** (200 μL), using a quartz cuvette (5 mm). The assay buffer was measured as the background for fluorescence spectrum. The excitation wavelength of fluorescence spectra was set at 367 nm, and emission was detected from 350 nm to 570 nm. The specific equilibrium binding constant (K_d) was derived using GraphPad Prism6 (Y = B_{max} × X^h/(K_d^h + X^h) (X: concentration of PPARγ-LBD [nM], Y: fluorescence intensity at 410 nm, h: Hill slope).

4.8. Binding Assay of Rosiglitazone or Farglitazar with hPPARγ- LBD Using **2**

PPARγ-LBD (6 mg/mL) was diluted to the concentration of 0.6 μM with the assay buffer, and PPARγ-LBD solution was added **2** (0.72 μM final concentration). Then, four-fold serial dilutions of Rosiglitazone or Farglitazar (Rosiglitazone: final concentration of 1.17 nM to 76.8 μM, 9 concentration levels, Farglitazar: final concentration of 0.59 nM to 9.6 μM, 8 concentration levels) was added to the mixture, and the fluorescence spectra were measured using the mixture of PPARγ-LBD, **2**, and Rosiglitazone or Farglitazar (200 μL), using a quartz cuvette (5 mm). The assay buffer was measured as the background for the fluorescence spectrum. The excitation wavelength of fluorescence spectra was set at 367 nm, and emission was detected from 350 nm to 570 nm. The inhibition constant (K_i) value was derived from K_d of **2**, using GraphPad Prism6 (logEC_{50} = log(10^logK_i × (1 + Radioligand [nM]/HotK_d [nM])) Y = Bottom + (Top-Bottom)/(1 + 10^(X-LogEC_{50}))) (X: concentration of Rosiglitazone or Farglitazar [nM], Y: fluorescence intensity at 410 nm, Radioligand [nM]: concentration of **2**, HotK_d [nM]: the K_d value of **2**).

4.9. Binding Assay of Pioglitazone or LT175 with hPPARγ- LBD Using **6**

PPARγ-LBD (6 mg/mL) was diluted to the concentration of 0.6 μM with the assay buffer, and PPARγ-LBD solution was added **6** (1.44 μM final concentration). Then, four-fold serial dilutions Pioglitazone or LT175 (final concentration of 4.69 nM to 76.8 μM, 8 concentration levels) was added to the mixture. Next, the fluorescence spectra were measured using the mixture of PPARγ-LBD, **6**, and Pioglitazone or LT175 (200 μL) using a quartz cuvette (5 mm). The assay buffer was measured as the background for fluorescence spectrum. The excitation wavelength of fluorescence spectra was set at 367 nm, and emission was detected from 350 nm to 570 nm. The inhibition constant (K_i) value was derived from K_d of **6** using GraphPad Prism6 (logEC_{50} = log(10^logK_i × (1 + Radioligand [nM]/HotK_d [nM])) Y = Bottom + (Top-Bottom)/(1 + 10^(X-LogEC_{50}))) (X: concentration of Pioglitazone or LT175, Y: fluorescence intensity at 410 nm, Radioligand [nM]: concentration of **6**, HotK_d [nM]: the K_d value of **6**).

5. Conclusions

To efficiently evaluate the ligands for PPARγ, a target molecule for metabolic syndrome and inflammatory diseases, we synthesized compound **2** using our method. In the process, we also obtained partial agonist **6**, which appeared to act via a novel mechanism. By utilizing **2** and **6**, we showed that a fluorescence spectrophotometer could be used to evaluate PPARγ binding affinity. These results might contribute to the understanding of metabolic syndrome and inflammation, as well as drug development.

Supplementary Materials: The following are available online at https://www.mdpi.com/article/10.3390/ijms22084034/s1. Scheme S1: The proposed mechanism of reaction for synthesis of **5**. Scheme S2: The proposed mechanism of reaction for synthesis of **6**. Figure S1: Absorption spectrum of **2** and **6**. Figure S2: Kd determination of 6. Figure S3: Binding assay of Farglitazar for hPPARγ—LBD using **2**. Figure S4: Binding assay of Pioglitazone for hPPAR—LBD using **6**. Figure S5: Binding assay of LT175 for hPPARγ—LBD using **6**. Spectra of compounds **2**, **5**, and **6**(^{1}H NMR, ^{13}C NMR). Table S1: Data collection and refinement statistics of the crystal structures.

Author Contributions: Conceptualization, C.Y., H.I., and T.I.; methodology, C.Y., H.I., N.O., and T.I.; software, C.Y., H.I., N.O., and T.I.; formal analysis, C.Y. and N.O; investigation, C.Y.; writing—original draft preparation, C.Y.; writing—review and editing, T.I.; visualization, C.Y. and T.I.; supervision, T.I.; project administration, T.I. All authors have read and agreed to the published version of the manuscript.

Funding: This research received no external funding.

Institutional Review Board Statement: Not applicable.

Informed Consent Statement: Not applicable.

Data Availability Statement: Data are contained within the article or Supplementary Materials.

Acknowledgments: We are grateful to Showa Pharmaceutical University for financial support. Synchrotron radiation experiments were performed at the Photon Factory (Proposal No. 2020G618), and we are grateful for the assistance provided by the beamline scientists at the Photon Factory.

Conflicts of Interest: The authors declare that they have no known competing financial interests or personal relationships that could have appeared to influence the work reported in this paper.

References

1. Ban, S.; Oyama, T.; Kasuga, J.; Ohgane, K.; Nishio, Y.; Morikawa, K.; Hashimoto, Y.; Miyachi, H. Bidirectional fluorescence properties of pyrene-based peroxisome proliferator-activated receptor (PPAR) α/δ dual agonist. *Bioorg. Med. Chem.* **2012**, *20*, 3460–3464. [CrossRef] [PubMed]
2. Lee, W.S.; Kim, J. Peroxisome Proliferator-Activated Receptors and the Heart: Lessons from the Past and Future Directions. *PPAR Res.* **2015**, *2015*, 1–15. [CrossRef] [PubMed]
3. Wahli, W.; Michalik, L. PPARs at the crossroads of lipid signaling and inflammation. *Trends Endocrinol. Metab.* **2012**, *23*, 351–363. [CrossRef]
4. Murphy, G.J.; Holder, J.C. PPAR-γ agonists: Therapeutic role in diabetes, inflammation and cancer. *Trends Pharm. Sci.* **2000**, *21*, 469–474. [CrossRef]
5. Berger, J.P.; Petro, A.E.; Macnaul, K.L.; Kelly, L.J.; Zhang, B.B.; Richards, K.; Elbrecht, A.; Johnson, B.A.; Zhou, G.; Doebber, T.W.; et al. Distinct Properties and Advantages of a Novel Peroxisome Proliferator-Activated Protein γ Selective Modulator. *Mol. Endocrinol.* **2003**, *17*, 662–676. [CrossRef]
6. Capelli, D.; Cerchia, C.; Montanari, R.; Loiodice, F.; Tortorella, P.; Laghezza, A.; Cervoni, L.; Pochetti, G.; Lavecchia, A. Structural basis for PPAR partial or full activation revealed by a novel ligand binding mode. *Sci. Rep.* **2016**, *6*, 34792. [CrossRef]
7. Itoh, T.; Fairall, L.; Amin, K.; Inaba, Y.; Szanto, A.; Balint, B.L.; Nagy, L.; Yamamoto, K.; Schwabe, J.W.R. Structural basis for the activation of PPARγ by oxidized fatty acids. *Nat. Struct. Mol. Biol.* **2008**, *15*, 924–931. [CrossRef]
8. Soares, A.F.; Nosjean, O.; Cozzone, D.; D'Orazio, D.; Becchi, M.; Guichardant, M.; Ferry, G.; Boutin, J.A.; Lagarde, M.; Géloën, A. Covalent binding of 15-deoxy-delta12,14-prostaglandin J2 to PPARγ. *Biochem. Biophys. Res. Commun.* **2005**, *337*, 521–525. [CrossRef]
9. Hashimoto, Y.; Miyachi, H. Nuclear receptor antagonists designed based on the helix-folding inhibition hypothesis. *Bioorg. Med. Chem.* **2005**, *13*, 5080–5093. [CrossRef] [PubMed]

10. Henke, B.R.; Blanchard, S.G.; Brackeen, M.F.; Brown, K.K.; Cobb, J.E.; Collins, J.L.; Harrington, W.W., Jr.; Hashim, M.A.; Hull-Ryde, E.A.; Kaldor, I.; et al. N-(2-Benzoylphenyl)-L-tyrosine PPARγ Agonists. 1. Discovery of a Novel Series of Potent Antihyperglycemic and Antihyperlipidemic Agents. *J. Med. Chem.* **1998**, *41*, 5020–5036. [CrossRef]
11. Montanari, R.; Saccoccia, F.; Scotti, E.; Crestani, M.; Godio, C.; Gilardi, F.; Loiodice, F.; Fracchiolla, G.; Laghezza, A.; Tortorella, P.; et al. Crystal Structure of the Peroxisome Proliferator-Activated Receptor γ (PPARγ) Ligand Binding Domain Complexed with a Novel Partial Agonist: A New Region of the Hydrophobic Pocket Could Be Exploited for Drug Design. *J. Med. Chem.* **2008**, *51*, 7768–7776. [CrossRef]
12. Gilardi, F.; Giudici, M.; Mitro, N.; Maschi, O.; Guerrini, U.; Rando, G.; Maggi, A.; Cermenati, G.; Laghezza, A.; Loiodice, F.; et al. LT175 is a novel PPARα/γ ligand with potent insulin sensitizing effects and reduced adipogenic properties. *J. Biol. Chem.* **2014**, *289*, 6908–6920. [CrossRef] [PubMed]
13. Landreth, G.E.; Heneka, M.T. Anti-inflammatory actions of peroxisome proliferator-activated receptor gamma agonists in Alzheimer's disease. *Neurobiol. Aging* **2001**, *22*, 937–944. [CrossRef]
14. Jiang, C.; Ting, A.T.; Seed, B. PPAR-γ agonists inhibit production of monocyte inflammatory cytokines. *Nature* **1998**, *391*, 82–86. [CrossRef]
15. Kapadia, R.; Yi, J.H.; Vemuganti, R. Mechanisms of anti-inflammatory and neuroprotective actions of PPAR-γ agonists. *Front. Biosci.* **2008**, *13*, 1813–1826. [CrossRef] [PubMed]
16. Yamamoto, K.; Ninomiya, Y.; Iseki, M.; Nakachi, Y.; Kanesaki-Yatsuka, Y.; Yamanoue, Y.; Itoh, T.; Nishii, Y.; Petrovsky, N.; Okazaki, Y. 4-Hydroxydocosahexaenoic acid, a potent peroxisome proliferator-activated receptor γ agonist alleviates the symptoms of DSS-induced colitis. *Biochem. Biophys. Res. Commun.* **2008**, *367*, 566–572. [CrossRef]
17. Nolte, R.; Wisely, G.B.; Westin, S.; Cobb, J.E.; Lambert, M.H.; Kurokawa, R.; Rosenfeld, M.G.; Willson, T.M.; Glass, C.K.; Milburn, M.V. Ligand binding and co-activator assembly of the peroxisome proliferator-activated receptor-γ. *Nature* **1998**, *395*, 137–143. [CrossRef] [PubMed]
18. Uppenberg, J.; Svensson, C.; Jaki, M.; Bertilsson, G.; Jendeberg, L.; Berkenstam, A. Crystal structure of the ligand binding domain of the human nuclear receptor PPARγ. *J. Biol. Chem.* **1998**, *273*, 31108–31112. [CrossRef]
19. Raucy, J.L.; Lasker, J.M. Current In Vitro High Throughput Screening Approaches to Assess Nuclear Receptor Activation. *Curr. Drug Metab.* **2010**, *11*, 806–814. [CrossRef]
20. Lehmann, J.M.; Moore, L.B.; Smith-Oliver, T.A.; Wilkinson, W.O.; Willson, T.M.; Kliewer, S.A. An antidiabetic thiazolidinedione is a high affinity ligand for peroxisome proliferator activated receptor γ (PPARγ). *J. Biol. Chem.* **1995**, *270*, 12953–12956. [CrossRef]
21. Leopoldo, M.; Lacivita, E.; Berardi, F.; Perrone, R. Developments in fluorescent probes for receptor research. *Drug Discov. Today* **2009**, *14*, 706–712. [CrossRef]
22. DeGrazia, M.J.; Thompson, J.; Heuvelb, J.P.V.; Peterson, B.R. Synthesis of a High-Affinity Fluorescent PPARγ Ligand for High-Throughput Fluorescence Polarization Assays. *Bioorg. Med. Chem.* **2003**, *11*, 4325–4332. [CrossRef]
23. Seethala, R.; Golla, R.; Ma, Z.; Zhang, H.; O'Malley, K.; Lippy, J.; Cheng, L.; Mookhtiar, K.; Farrelly, D.; Zhang, L.; et al. A rapid, homogeneous, fluorescence polarization binding assayfor peroxisome proliferator-activated receptors alpha and gamma using a fluorescein-tagged dual PPARα/γ activator. *Anal. Biochem.* **2007**, *363*, 263–274. [CrossRef]
24. Loving, G.S.; Sainlos, M.; Imperiali, B. Monitoring protein interactions and dynamics with solvatochromic fluorophores. *Trends Biotechnol.* **2010**, *28*, 73–83. [CrossRef]
25. Nomura, W.; Ohashi, N.; Okuda, Y.; Narumi, T.; Ikura, T.; Ito, N.; Tamamura, H. Fluorescence-Quenching Screening of Protein Kinase C Ligands with an Environmentally Sensitive Fluorophore. *Bioconjugate Chem.* **2011**, *22*, 923–930. [CrossRef] [PubMed]
26. Bhatia, R.; Pathania, S.; Singh, V.; Rawal, R.K. Metal-catalyzed synthetic strategies toward coumarin derivatives. *Chem. Heterocycl. Compd.* **2018**, *54*, 280–291. [CrossRef]
27. Sekino, E.; Kumamoto, T.; Tanaka, T.; Ikeda, T.; Ishikawa, T. Concise Synthesis of Anti-HIV-1 Active (+)-Inophyllum B and(+)-Calanolide A by Application of (-)-Quinine-Catalyzed Intramolecular Oxo-Michael Addition. *J. Org. Chem.* **2004**, *69*, 2760–2767. [CrossRef]
28. Donnelly, A.C.; Mays, J.R.; Burlison, J.A.; Nelson, J.T.; Vielhauer, G.; Holzbeierlein, J.; Blagg, B.S.J. The Design, Synthesis, and Evaluation of Coumarin Ring Derivatives of the Novobiocin Scaffold that Exhibit Antiproliferative Activity. *J. Org. Chem.* **2008**, *73*, 8901–8920. [CrossRef]
29. Perkin, W.H. XXIII.—On the hydride of aceto-salicyl. *J. Chem. Soc.* **1868**, *21*, 181–186. [CrossRef]
30. Pechmann, H.; Duisberg, C. Ueber die Verbindungen der Phenole mit Acetessigäther. *Ber. Dtsch. Chem. Ges.* **1883**, *16*, 2119–2128. [CrossRef]
31. Pechmann, H. Neue Bildungsweise der Cumarine. Synthese des Daphnetins. I. *Ber. Dtsch. Chem. Ges.* **1884**, *17*, 929–936. [CrossRef]
32. Aydıner, B.; Seferoğlu, Z. Proton Sensitive Functional Organic Fluorescent Dyes Based on Coumarin-imidazo [1,2-*a*]pyrimidine; Syntheses, Photophysical Properties, and Investigation of Protonation Ability. *Eur. J. Org. Chem.* **2018**, *2018*, 5921–5934. [CrossRef]
33. Kaye, P.T.; Musa, M.A.; Nocanda, X.W. Efficient and Chemoselective Access to 3-(Chloromethyl)coumarins via Direct Cyclisation of Unprotected Baylis–Hillman Adducts. *Synthesis* **2003**, *2003*, 531–534. [CrossRef]
34. Mizoroki, T.; Mori, K.; Ozaki, A. Arylation of Olefin with Aryl Iodide Catalyzed by Palladium. *Bull. Chem. Soc. Jpn.* **1971**, *44*, 581. [CrossRef]

35. Heck, R.F.; Nolley, J.P., Jr. Palladium-catalyzed vinylic hydrogen substitution reactions with aryl, benzyl, and styryl halides. *J. Org. Chem.* **1972**, *37*, 2320–2322. [CrossRef]
36. Yoshikawa, C.; Ishida, H.; Itoh, T. Incorporation of a coumarin unit by nucleophilic addition into a PPARγ ligand. *Tetrahedron Lett.* **2020**, *61*, 151842. [CrossRef]
37. Kojima, H.; Fujita, Y.; Takeuchi, R.; Ohashi, N.; Itoh, T. Cyclization Reaction-Based Turn-on Probe for Covalent Labeling of Target Proteins. *Cell Chem. Biol.* **2020**, *27*, 334–349. [CrossRef]
38. Droumaguet, C.L.; Wang, C.; Wang, Q. Fluorogenic click reaction. *Chem. Soc. Rev.* **2010**, *39*, 1233–1239. [CrossRef]
39. Yoshikawa, C.; Ishida, H.; Ohashi, N.; Kojima, H.; Itoh, T. Construction of 7-diethylaminocoumarins promoted by an electron-withdrawing group. *Chem. Pharm. Bull.*. Accepted. [CrossRef]
40. Yamaguchi, J.; Sugiyama, S. Conjugate addition of an ynone containing azulene with a tertiary amine. *Tetrahedron Lett.* **2016**, *57*, 4514–4518. [CrossRef]
41. Raikar, U.S.; Renuka, C.G.; Nadaf, Y.F.; Mulimani, B.G.; Karguppikar, A.M.; Soudagar, M.K. Solvent effects on the absorption and fluorescence spectra of coumarins 6 and 7 molecules: Determination of ground and excited state dipole moment. *Spectrochim. Acta A* **2006**, *65*, 673–677. [CrossRef]
42. Mayer, U.; Gutmann, V.; Gerger, W. The Acceptor Number–A Quantitative Empirical Parameter for the Electrophilic Properties of Solvents. *Mon. Chem.* **1975**, *106*, 1235–1257. [CrossRef]
43. Shiraishi, T.; Kagechika, H.; Hirano, T. 6-Arylcoumarins: Versatile scaffolds for fluorescent sensors. *New J. Chem.* **2015**, *39*, 8389–8396. [CrossRef]
44. Xu, H.E.; Lambert, M.H.; Montana, V.G.; Plunket, K.D.; Moore, L.B.; Collins, J.L.; Oplinger, J.A.; Kliewer, S.A.; Gampe, R.T., Jr.; McKee, D.D.; et al. Structural determinants of ligand binding selectivity between the peroxisome proliferatoractivated receptors. *Proc. Natl. Acad. Sci. USA* **2001**, *98*, 13919–13924. [CrossRef]
45. Bruning, J.B.; Chalmers, M.J.; Prasad, S.; Busby, S.A.; Kamenecka, T.M.; He, Y.; Nettles, K.W.; Griffin, P.R. Partial Agonists Activate PPARγ Using a Helix 12 Independent Mechanism. *Structure* **2007**, *15*, 1258–1271. [CrossRef]
46. Egawa, D.; Itoh, T.; Akiyama, Y.; Saito, T.; Yamamoto, K. 17-OxoDHA Is a PPARα/γ Dual Covalent Modifier and Agonist. *ACS Chem. Biol.* **2016**, *11*, 2447–2455. [CrossRef]
47. Kato, A.; Anami, Y.; Egawa, D.; Itoh, T.; Yamamoto, K. Helix12-stabilization antagonist of vitamin D receptor. *Bioconjug. Chem.* **2016**, *27*, 1750–1761. [CrossRef] [PubMed]
48. Pochetti, G.; Godio, C.; Mitro, N.; Caruso, D.; Galmozzi, A.; Scurati, S.; Loiodice, F.; Fracchiolla, G.; Tortorella, P.; Laghezza, A.; et al. Insights into the mechanism of partial agonism: Crystal structures of the peroxisome proliferator-activated receptor γ ligand-binding domain in the complex with two enantiomeric ligands. *J. Biol. Chem.* **2007**, *282*, 17314–17324. [CrossRef]
49. Inglese, J.; Johnson, R.L.; Simeonov, A.; Xia, M.; Zheng, W.; Austin, C.P.; Auld, D.S. High-throughput screening assays for the identification of chemical probes. *Nat. Chem. Biol.* **2007**, *3*, 466–479. [CrossRef] [PubMed]
50. Sun, S.; Almaden, J.; Carlson, T.J.; Barker, J.; Gehring, M.R. Assay development and data analysis of receptor-ligand binding based on scintillation proximity assay. *Metab. Eng.* **2005**, *7*, 38–44. [CrossRef]
51. Long, L.; Li, X.; Zhang, D.; Meng, S.; Zhang, J.; Sun, X.; Zhang, C.; Zhou, L.; Wang, L. Amino-coumarin based fluorescence ratiometric sensors for acidic pH and their application for living cells imaging. *RSC Adv.* **2013**, *3*, 12204–12209. [CrossRef]
52. Yamada, S.; Kawasaki, M.; Fujihara, M.; Watanabe, M.; Takamura, Y.; Takioku, M.; Nishioka, H.; Takeuchi, Y.; Makishima, M.; Motoyama, T.; et al. Competitive Binding Assay with an Umbelliferone-Based Fluorescent Rexinoid for Retinoid X Receptor Ligand Screening. *J. Med. Chem.* **2019**, *62*, 8809–8818. [CrossRef]
53. Breidenbach, J.; Bartz, U.; Gütschow, M. Coumarin as a structural component of substrates and probes for serine and cysteine proteases. *Biochim. Biophys. Acta Protein Proteonomics* **2020**, *1868*, 140445. [CrossRef]
54. Takadate, A.; Masuda, T.; Murata, C.; Tanaka, T.; Irikura, M.; Goya, S. Fluorescence Characteristics of Methoxycoumarins as Novel Fluorophores. *Anal. Sci.* **1995**, *11*, 97–101. [CrossRef]
55. Li, H.; Yao, Y.; Li, L. Coumarins as potential antidiabetic agents. *J. Pharm. Pharmacol.* **2017**, *69*, 1253–1264. [CrossRef]
56. Leslie, A.G.W.; Powell, H.R. Processing Diffraction Data with Mosflm. In *Evolving Methods for Macromolecular Crystallography*; Read, R.J., Sussman, J.L., Eds.; NATO Science Series II, Physics and Chemistry; Springer: Dordrecht, The Netherlands, 2007; Volume 245, pp. 41–51.
57. Battye, T.G.G.; Kontogiannis, L.; Johnson, O.; Powell, H.R.; Leslie, A.G.W. iMOSFLM: A New Graphical Interface for Diffraction-Image Processing with MOSFLM. *Acta Cryst. Sect. D Biol. Cryst.* **2011**, *67*, 271–281. [CrossRef]
58. McCoy, A.J.; Grosse-Kunstleve, R.W.; Adams, P.D.; Winn, M.D.; Storoni, L.C.; Read, R.J. Phaser Crystallographic Software. *J. Appl. Cryst.* **2007**, *40*, 658–674. [CrossRef]
59. Collaborative Computational Project, Number 4. The CCP4 Suite: Programs for Protein Crystallography. *Acta Cryst. Sect. D Biol. Cryst.* **1994**, *50*, 760–763. [CrossRef] [PubMed]
60. Emsley, P.; Lohkamp, B.; Scott, W.G.; Cowtan, K. Features and Development of Coot. *Acta Cryst. Sect. D Biol. Cryst.* **2010**, *66*, 486–501. [CrossRef]
61. Murshudov, G.N.; Vagin, A.A.; Dodson, E.J. Refinement of Macromolecular Structures by the Maximum-Likelihood Method. *Acta Cryst. Sect. D Biol. Cryst.* **1997**, *53*, 240–255. [CrossRef] [PubMed]
62. Pannu, N.S.; Murshudov, G.N.; Dodson, E.J.; Read, R.J. Incorporation of Prior Phase Information Strengthens Maximum-Likelihood Structure Refinement. *Acta Cryst. Sect. D Biol. Cryst.* **1998**, *54*, 1285–1294. [CrossRef] [PubMed]

63. Winn, M.D.; Isupov, M.N.; Murshudov, G.N. Use of TLS Parameters to Model Anisotropic Displacements in Macromolecular Refinement. *Acta Cryst. Sect. D Biol. Cryst.* **2001**, *57*, 122–133. [CrossRef] [PubMed]
64. Steiner, R.; Lebedev, A.A.; Murshudov, G.N. Fisher's Information in Maximum-Likelihood Macromolecular Crystallographic Refinement. *Acta Cryst. Sect. D Biol. Cryst.* **2003**, *59*, 2114–2124. [CrossRef] [PubMed]
65. Murshudov, G.N.; Skubák, P.; Lebedev, A.A.; Pannu, N.S.; Steiner, R.A.; Nicholls, R.A.; Winn, M.D.; Long, F.; Vagin, A.A. REFMAC 5 for the Refinement of Macromolecular Crystal Structures. *Acta Cryst. Sect. D Biol. Cryst.* **2011**, *67*, 355–367. [CrossRef] [PubMed]

International Journal of
Molecular Sciences

MDPI

Article

Functional and Structural Insights into Human PPARα/δ/γ Subtype Selectivity of Bezafibrate, Fenofibric Acid, and Pemafibrate

Akihiro Honda [1], Shotaro Kamata [1], Makoto Akahane [1], Yui Machida [1], Kie Uchii [1], Yui Shiiyama [1], Yuki Habu [1], Saeka Miyawaki [1], Chihiro Kaneko [1], Takuji Oyama [2] and Isao Ishii [1,*]

1. Department of Health Chemistry, Showa Pharmaceutical University, Machida 194-8543, Tokyo, Japan; d2103@g.shoyaku.ac.jp (A.H.); kamata@ac.shoyaku.ac.jp (S.K.); kajiwara@ac.shoyaku.ac.jp (M.A.); b17112@ug.shoyaku.ac.jp (Y.M.); a17024@ug.shoyaku.ac.jp (K.U.); b16049@ug.shoyaku.ac.jp (Y.S.); b17096@ug.shoyaku.ac.jp (Y.H.); b18107@ug.shoyaku.ac.jp (S.M.); b18034@ug.shoyaku.ac.jp (C.K.)
2. Faculty of Life and Environmental Sciences, University of Yamanashi, Kofu 400-8510, Yamanashi, Japan; takujio@yamanashi.ac.jp

* Correspondence: i-ishii@ac.shoyaku.ac.jp

Abstract: Among the agonists against three peroxisome proliferator-activated receptor (PPAR) subtypes, those against PPARα (fibrates) and PPARγ (glitazones) are currently used to treat dyslipidemia and type 2 diabetes, respectively, whereas PPARδ agonists are expected to be the next-generation metabolic disease drug. In addition, some dual/pan PPAR agonists are currently being investigated via clinical trials as one of the first curative drugs against nonalcoholic fatty liver disease (NAFLD). Because PPARα/δ/γ share considerable amino acid identity and three-dimensional structures, especially in ligand-binding domains (LBDs), clinically approved fibrates, such as bezafibrate, fenofibric acid, and pemafibrate, could also act on PPARδ/γ when used as anti-NAFLD drugs. Therefore, this study examined their PPARα/δ/γ selectivity using three independent assays—a dual luciferase-based GAL4 transactivation assay for COS-7 cells, time-resolved fluorescence resonance energy transfer-based coactivator recruitment assay, and circular dichroism spectroscopy-based thermostability assay. Although the efficacy and efficiency highly varied between agonists, assay types, and PPAR subtypes, the three fibrates, except fenofibric acid that did not affect PPARδ-mediated transactivation and coactivator recruitment, activated all PPAR subtypes in those assays. Furthermore, we aimed to obtain cocrystal structures of PPARδ/γ-LBD and the three fibrates via X-ray diffraction and versatile crystallization methods, which we recently used to obtain 31 structures of PPARα-LBD cocrystallized with 17 ligands, including the fibrates. We herein reveal five novel high-resolution structures of PPARδ/γ–bezafibrate, PPARγ–fenofibric acid, and PPARδ/γ–pemafibrate, thereby providing the molecular basis for their application beyond dyslipidemia treatment.

Keywords: bezafibrate; fenofibric acid; pemafibrate; peroxisome proliferator-activated receptor; dual/pan agonist; X-ray crystallography

Citation: Honda, A.; Kamata, S.; Akahane, M.; Machida, Y.; Uchii, K.; Shiiyama, Y.; Habu, Y.; Miyawaki, S.; Kaneko, C.; Oyama, T.; et al. Functional and Structural Insights into Human PPARα/δ/γ Subtype Selectivity of Bezafibrate, Fenofibric Acid, and Pemafibrate. *Int. J. Mol. Sci.* **2022**, *23*, 4726. https://doi.org/10.3390/ijms23094726

Academic Editors: Walter Wahli and Manuel Vázquez-Carrera

Received: 6 April 2022
Accepted: 22 April 2022
Published: 25 April 2022

1. Introduction

Peroxisome proliferator-activated receptors (PPARs) belong to the nuclear receptor (NR) superfamily and ligand-activated transcription factors that sense intracellular free fatty acids [1]. Three subtypes (PPARα, PPARβ/δ, and PPARγ) with considerable amino acid identity (54–71% in humans) have been identified in mammals. PPARα regulates lipid metabolism mainly in the liver and skeletal muscle and glucose homeostasis via direct transcriptional control of genes involved in peroxisomal/mitochondrial β-oxidation, fatty acid uptake, and triglyceride (TG) catabolism [2]. PPARδ is ubiquitously expressed and controls energy metabolism and cell survival [3]. PPARγ is most highly expressed in white/brown adipose tissues, where it acts as a master regulator of adipogenesis and a potent modulator of whole-body lipid metabolism and insulin sensitivity [4]. Three PPARs, similar to other NRs, comprise amino-terminal domains containing the activation function (AF)-1, DNA-binding domain, hinge region, and ligand-binding domain (LBD) containing the AF-2 region, and carboxyl-terminal domains [1]. Their ligand-binding pockets (LBPs) are relatively large, with a total volume of 1300–1400 Å^3, compared to those found in other NRs that have 600–1100 Å^3 LBPs [5]. Fatty acids and their derivatives from diet, de novo lipogenesis, and TG lipolysis are considered natural PPAR ligands [6,7]. Upon ligand binding, PPARs manifest conformational changes that facilitate corepressor molecule dissociation to enable the spatiotemporally orchestrated recruitment (association) of coactivators to the ligand-bound receptors [8]. Coactivators contain one or more highly conserved LXXLL α-helix motif called an NR box for direct interaction with AF-2 regions in PPARs [8].

Synthetic PPARα agonists "fibrates" have been widely used to treat hypertriglyceridemia; they decrease blood TG levels and increase high-density lipoprotein-cholesterol levels [9]. Bezafibrate, fenofibrate, ciprofibrate, clofibrate, and gemfibrozil were developed about half a century ago and have been clinically used in many countries [9]. Synthetic PPARγ agonists "thiazolidinediones (glitazones)," such as rosiglitazone and pioglitazone, are antidiabetic drugs with potent insulin-sensitizing effects that confer long-term glycemic control, although some of them might induce serious adverse effects, including edema, bone fracture, and heart failure [10]. PPARδ agonists are not yet clinically available but are expected to treat metabolic or cardiovascular diseases [11]. The development of a PPARδ-selective agonist seladelpar (MBX-8025) for non-alcoholic steatohepatitis (NASH) and primary sclerosing cholangitis (PSC) has been once discontinued at phase 2b clinical trial [12] and seladelpar is now only in phase 3 trial for primary biliary cholangitis (ClinicalTrials.gov number: NCT03301506) [13], although the Food and Drug Administration (FDA) lifted clinical holds on seladelpar for investigational new drug applications in NASH and PSC on 23 July 2020 [14]. Furthermore, PPAR dual/pan agonists are expected to treat metabolic diseases including non-alcoholic fatty liver disease (NAFLD) and NASH [15]. Saroglitazar (α/γ dual) and lanifibranor (α/δ/γ pan) are currently in clinical trials [16,17]. However, the development of most PPARα/γ dual agonists (e.g., muraglitazar, tesaglitazar, and aleglitazar) has been abandoned due to serious safety concerns [11], and that of a PPARα/δ dual agonist elafibranor for NASH has been discontinued due to it having no significant benefits [18]. Thus, the management of PPARα/δ/γ selectivity is indispensable.

We have recently revealed 34 novel high-resolution X-ray cocrystal structures of PPARα-LBD and 17 PPARα ligands, including bezafibrate, fenofibric acid, and pemafibrate, using sophisticated cocrystallization techniques [6,19] and further obtained PPARγ-LBD–saroglitazar structures [20]. The aim of this study was to use X-ray crystallography to examine the PPARα/δ/γ selectivity of clinically used fibrates in three independent assays and to provide its structural basis because PPARs have relatively large LBPs to accept 1–4 small molecule ligands [6]. Thus, this study demonstrated the PPAR dual/pan agonistic activities of all of those fibrates and revealed the novel five high-resolution structures of PPARδ/γ-LBD–bezafibrate, PPARγ-LBD–fenofibric acid, and PPARδ/γ-LBD–pemafibrate.

2. Results

2.1. Fibrates Induce Transactivation of Gene Expression via PPARα/δ/γ-LBD

The GAL4–hPPARα/δ/γ-LBD transactivation system [21] was employed to elucidate the impacts of fibrates on PPARα/δ/γ-LBD-mediated gene transcription activation (transactivation) in COS-7 cells. In this system, the *Firefly* luciferase reporter gene was activated only by the GAL4–human PPARα/δ/γ-LBD chimera, and the potentially confounding effects of endogenous receptors were eliminated. Transfection efficiency was normalized by *Renilla* luciferase activity. First, we confirmed that a potent PPARα-selective agonist GW7647 induced PPARα-LBD-mediated transactivation in a concentration-dependent manner with maximal effect (×5.73 of the basal activity) at 0.1 μM and an EC_{50} value of 8.18 nM (Figure 1A); a potent PPARδ-selective agonist GW501516 induced PPARδ-LBD-mediated transactivation with maximal effect (×18.3) at 0.02 μM and an EC_{50} of 1.59 nM (Figure 1B); and a potent PPARγ-selective agonist GW1929 induced PPARγ-LBD-mediated transactivation with maximal effect (×2.26) at 1 μM and an EC_{50} of 18.3 nM (Figure 1C).

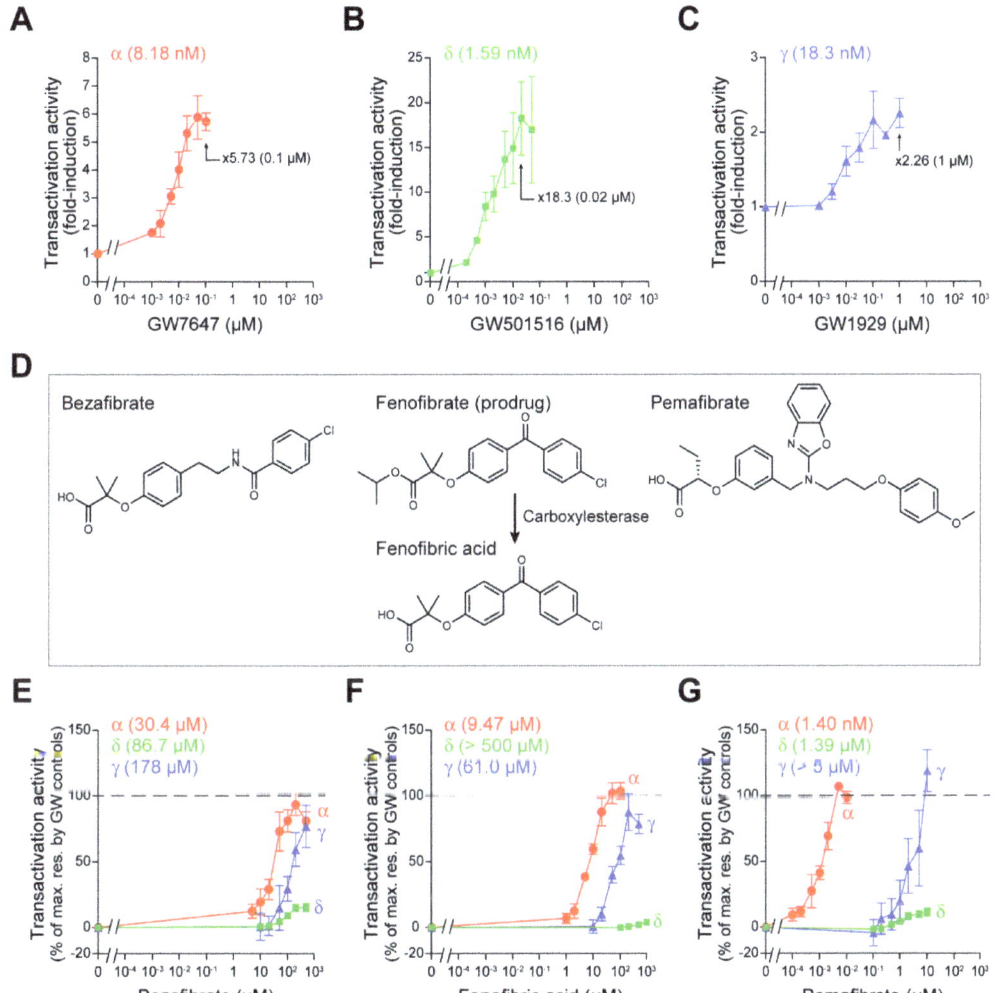

Figure 1. PPARα/δ/γ-LBD-mediated transactivation assay in COS-7 cells. (**A–C**) Control experiments. Human PPARα/δ/γ-LBD-mediated *Firefly* luciferase transactivation was induced by selective PPAR agonists: GW7647 for PPARα (**A**), GW501516 for PPARδ (**B**), and GW1929 for PPARγ (**C**) in a concentration-dependent manner. The maximal responses were observed at 0.1 μM, 0.02 μM, and 1 μM, respectively, which are used as the 100% responses in (**E–G**). (**D**) Chemical structures of fibrates used in this study. Fenofibrate is a prodrug that is metabolized by tissue and plasma carboxylesterases [22] to its active form, fenofibric acid. (**E–G**) PPARα/δ/γ-LBD-mediated transactivation by bezafibrate (**E**), fenofibric acid (**F**), and pemafibrate (**G**). Data are means ± standard error (SE) of three independent experiments with duplicate samples, and calculated EC_{50} values are shown.

We then compared the PPARα/δ/γ-LBD-mediated transactivation by the three fibrates—bezafibrate, fenofibric acid (an active metabolite of its prodrug fenofibrate [22]), and pemafibrate (Figure 1D)—considering the maximal effects by the GW compounds as 100%. Bezafibrate activated all PPAR subtypes, with 93.6% efficacy and 30.4 μM EC_{50} for PPARα, 15.2% efficacy and 86.7 μM EC_{50} for PPARδ, and 77.1% efficacy and 178 μM EC_{50} for PPARγ (Figure 1E). However, fenofibric acid activated PPARα (104% efficacy and 9.47 μM EC_{50}) and PPARγ (87.7% efficacy and 61.0 μM EC_{50}) but not PPARδ (Figure 1F). The concentrations at which pemafibrate (known as a selective PPAR modulator α "SPPARMα") activated PPARα were three orders lower (107% efficacy and 1.40 nM EC_{50}) than those at which it activated PPARδ (11.3% efficacy and 1.39 μM EC_{50}) or PPARγ (119% efficacy and EC_{50} > 5 μM) (Figure 1G).

2.2. Fibrates Induce PPARγ Coactivator 1α (PGC1α) or Steroid Receptor Coactivator 1 (SRC1) Recruitment via PPARα/δ/γ-LBD

In the nucleus, PPARs remain largely in repressed states due to the presence of corepressors, such as nuclear receptor corepressor 1 (NCoR1) and NCoR2 (SMRT), bound to the *cis*-elements (PPREs: PPAR responsive elements) located in the promoter region of their multiple target genes, irrespective of their ligand binding status [8]. Ligand binding initiates a complicated transcription process, which includes the dissociation of the corepressor protein complexes and the association of coactivator protein complexes for linking to the basal transcription machinery [8]. The ligand-induced AF-2 helix 12 formation, which recruits coactivators such as PGC1α and SRC1, is a hallmark of PPAR activation. PGC1α has a preference for PPARα/γ and is highly expressed in brown adipose tissues and cardiac and skeletal muscles, whereas SRC1 has a preference for all PPAR subtypes and is highly expressed in brown and white adipose tissues and the brain [8]. Thus, time-resolved fluorescence resonance energy transfer (TR-FRET)-based detection of the physical association between PPARα/δ/γ-LBD and coactivators becomes a highly sensitive cell-free assay for evaluating PPAR ligand activities. First, we confirmed that GW7647 activates the recruitment of both coactivators in a concentration-dependent manner, with maximal effect (×8.87 of the basal activity) at 1 μM and 44.1 nM EC_{50} for PGC1α and ×6.12 at 1 μM and 81.7 nM EC_{50} for SRC1 (Figure 2A). GW501516 induced a maximal effect of ×13.4 at 1 μM with 8.48 nM EC_{50} for PGC1α and ×3.10 at 1 μM with 5.82 nM EC_{50} for SRC1 (Figure 2B), whereas GW1929 induced a maximal effect of ×5.80 at 1 μM with 32.2 nM EC_{50} for PGC1α and ×9.26 at 1 μM with 75.6 nM EC_{50} for SRC1 (Figure 2C). Next, we again compared the PGC1α/SRC1 recruitment activity of the three fibrates considering the maximal effects by the GW compounds as 100%. Bezafibrate and pemafibrate recruited the PGC1α peptide to all PPARα/δ/γ-LBD; however, fenofibric acid did not recruit PGC1α to PPARδ-LBD (Figure 2D–F). The situation was the same for SRC1 recruitment in that the efficacy and efficiency varied greatly between agonists, PPAR subtypes, and coactivator species (Figure 2G–I).

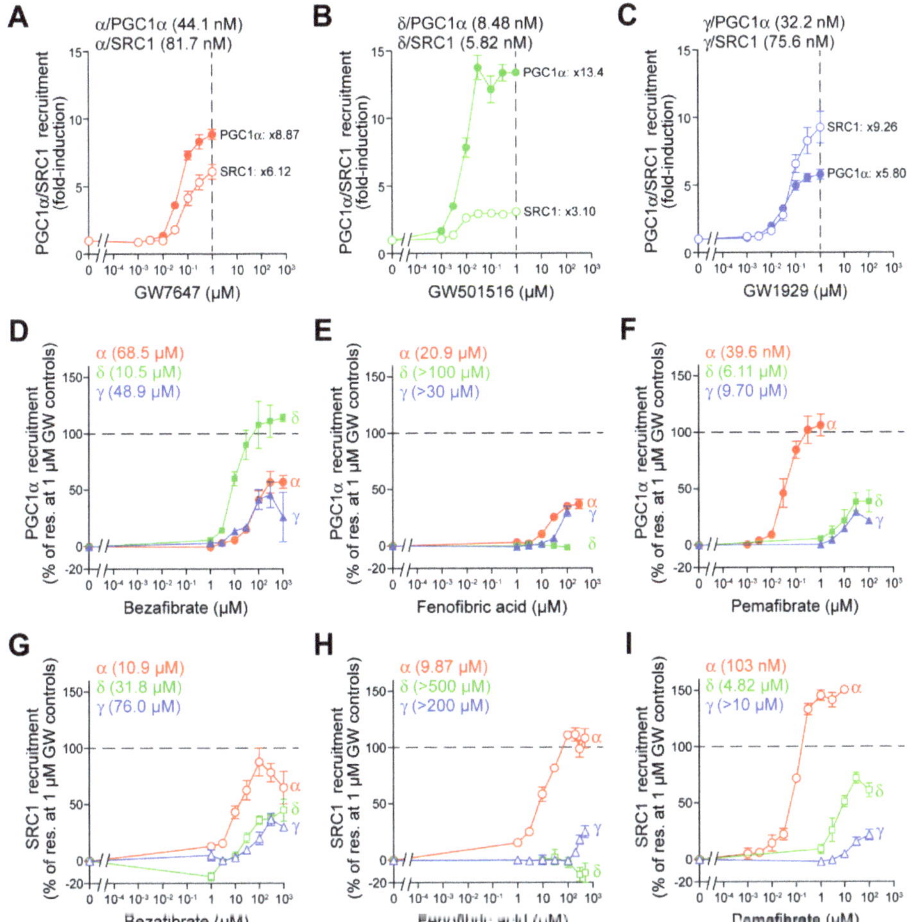

Figure 2. TR-FRET-based PPARα/δ/γ-LBD coactivator recruitment assay. (**A**–**C**) Control experiments. Human PPARα/δ/γ-LBD-mediated recruitment of coactivator peptides, PGC1α (filled symbols) and SRC1 (open symbols), was induced by selective PPAR agonists, GW7647 for PPARα (**A**), GW501516 for PPARδ (**B**), and GW1929 for PPARγ (**C**) in a concentration-dependent manner. Their maximal responses at 1 μM are used as the 100% responses in (**D**–**I**). (**D**–**I**) PPARα/δ/γ-LBD-mediated PGC1α (**D**–**F**) and SRC1 (**G**–**I**) coactivator peptide recruitment was induced by bezafibrate (**D**,**G**), fenofibric acid (**E**,H), and pemafibrate (**F**,**I**) in a concentration-dependent manner. Data are means ± SE of 3–4 independent experiments with duplicate samples, and calculated EC_{50} values are shown.

2.3. Fibrates Induce the Thermostability of PPARα/δ/γ-LBD

PPARs show increased thermostability upon ligand binding, which can be detected using circular dichroism (CD) spectroscopy [6]. Ligand-induced alterations in T_m values at 222 nm were investigated as types of reflection of α-helical stable structures of PPARs because PPAR ligand binding induces stabilization of the LBP [23,24]. The basal (solvent [0.1% DMSO] only) T_m values were 49.54 °C (±0.12 °C) (n = 4), 51.76 °C (±0.17 °C) (n = 6), and 48.95 °C (±0.16 °C) (n = 4) for PPARα-LBD, PPARδ-LBD, and PPARγ-LBD, respectively, and interestingly, all the three fibrates increased T_m values of all PPAR subtypes, although the efficacy and efficiency varied significantly (Figure 3A–C).

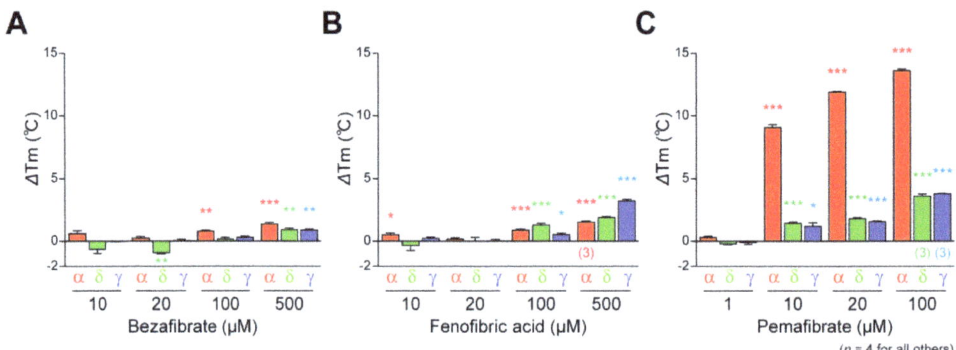

Figure 3. Circular dichroism-based PPARα/δ/γ-LBD thermostability assay. Bezafibrate—(**A**), fenofibric acid—(**B**), and pemafibrate-dependent (**C**) increases in T_m values (ΔT_m) at 222 nm were measured as the reflection of the thermostability of α-helical structures in PPARα/δ/γ-LBD. Data are means ± SE of four independent experiments ($n = 3$ for 500 μM fenofibric acid on PPARα-LBD and 100 μM pemafibrate on PPARδ/γ-LBD). Differences versus basal (0.1% DMSO) levels are significant in * $p < 0.05$, ** $p < 0.01$, and *** $p < 0.001$ in Student's t-test.

2.4. Structures of the PPARα/δ/γ-LBD–Bezafibrate Complexes

We have recently revealed the structures of 34 PPARα-LBD complexed with 17 ligands, including the three fibrates [6] and the structure of PPARγ-LBD–saroglitazar (a PPARα/γ dual agonist in clinical trials for NAFLD treatment) [20]. To gain structural insight into the PPARα/δ/γ selectivity of the fibrates, we aimed to obtain the structures of PPARδ/γ–fibrate complexes by X-ray crystallography and compare them with PPARα complex structures that we obtained (Figure 4A–D; reprinted from Kamata et al. [6]). We first screened various cocrystallization buffer conditions based on previous literature and our experience with PPARα/γ [19] (see Materials and Methods). A PPARδ-LBD–bezafibrate structure was obtained without using any coactivators (Figure 4E–H), whereas PPARγ-LBD–bezafibrate cocrystals were obtained using the SRC1 peptide (Figure 4I–L; SRC1 is indicated by an arrow in Figure 4I). The complex structures of bezafibrate bound to PPARδ/γ-LBD were solved in a monoclinic space group $P2_1$ at 2.09 Å resolution (Figure 4E; deposited in the Protein Data Bank (PDB) with ID: 7WGL) and an orthorhombic space group $P2_12_12_1$ at 2.36 Å resolution (PDB ID: 7WGO), respectively (Supplementary Table S1). The electron density map for bezafibrate in all PPARα/δ/γ-LBD indicated the presence of a single molecule in the protein monomer (Figure 4B,F,J).

The overall structures were identical to the previously reported active conformations that form the AF-2 helix 12, which provided root mean square (RMS) distances of 0.57 Å (219 common Cα positions in PPARα/δ) and 0.61 Å (218 common Cα positions in PPARα/γ) (Figure 4A,E,I). We have previously defined five LBPs (Arm I–III/X and Center) in PPARα-LBD based on our and others' observations of complexes with a total of 38 ligands [19], and four similar pockets (Arm I–III/X) in PPARδ/γ-LBD [20]. The chlorobenzyl moiety of bezafibrate was located in the Arm II region of PPARδ-LBD (Figure 4F) similar to that observed in PPARα-LBD (Figure 4B) [6]; however, it was rotated at an angle of 136.8° to reside in the Arm III region of PPARγ-LBD (Figure 4J). The carboxylic groups of the fibrates may stabilize the AF-2 helix 12 formation through hydrogen bonds (red dotted lines) and electrostatic interactions (blue dotted lines) with the four consensus amino acids, PPARα, Ser280/Tyr314/His440/Tyr464 (Figure 4C) [6]; PPARδ, Thr253/His287/His413/Tyr437 (Figure 4G); and PPARγ, Ser289/His323/His449/Tyr473 (Figure 4K), although a very close proximity (2.3 Å or 1.9 Å < 2.4 Å) was observed in PPARα Y464 (Figure 4C) or PPARγ Y473 (Figure 4K). Hydrogen bonds were also observed between bezafibrate and a water molecule (3.6 Å; Figure 4H) and between PPARγ S289 and the carbonyl group of bezafibrate (3.5 Å; Figure 4L). A single halogen bond was observed between PPARγ R288 and the chlorine atom of bezafibrate (3.3 Å; Figure 4L). Those interactions in PPARγ might stabilize the rotated chlorobenzyl moiety of bezafibrate.

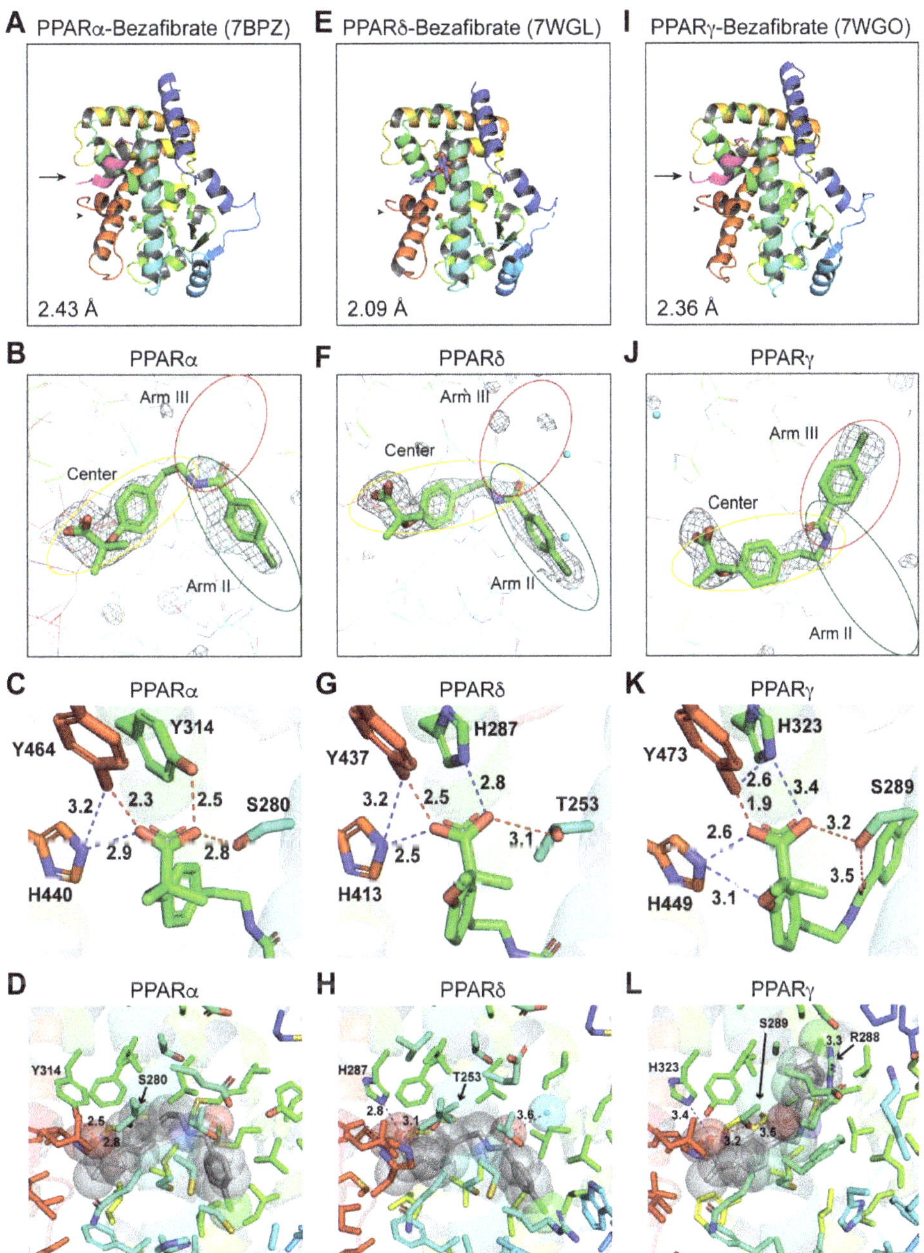

Figure 4. PPARα/δ/γ-LBD–bezafibrate cocrystal structures. Cocrystals of bezafibrate and PPARα-LBD (**A–D**; reprinted from Kamata et al. [6]), PPARδ-LBD (**E–H**), or PPARγ-LBD (**I–L**) were analyzed using X-ray diffraction. (**A**,**E**,**I**) Overall structures of the complexes. The SRC1 peptide (α-helix in magenta) and the AF-2 helix 12 (α-helix in red) are indicated by arrows and arrowheads, respectively.

PDB identities and resolutions are labeled. (**B**,**F**,**J**) Magnified views of bezafibrate located in the Center/Arm II regions of PPARα/δ-LBD (**B**,**F**) or the Center/Arm III regions of PPARγ-LBD (**J**). The electron density is shown in the mesh via F_o-F_c omit maps contoured at +3.0σ. Water molecules are presented as cyan spheres. (**C**,**G**,**K**) Hydrogen bonds and electrostatic interactions between bezafibrate and the four consensus amino acid residues (that recognize the carboxyl moiety of bezafibrate) are indicated by red and blue dotted lines, respectively, with their distances (Å). (**D**,**H**,**L**) Hydrogen bonds and electrostatic interactions between bezafibrate (in van der Waals spheres) and all surrounding amino acid residues located within a distance of 5 Å.

When amino acid residues in the PPARα/δ/γ-LBD are colored by their hydrophobicity (red) and hydrophilicity (white) using a Color h script program [25], all three fibrates (illustrated by their van der Waals spheres) were mainly surrounded by hydrophobic residues of amino acids in PPARα/δ/γ-LBD (Supplementary Figure S1).

2.5. Structures of the PPARα/γ-LBD–Fenofibric Acid Complexes

Cocrystals of PPARδ-LBD–fenofibric acid were not obtained probably because of its low binding affinity (Figures 1F and 2E,H), although increased thermostability was observed at high concentrations (Figure 3B). However, cocrystals with PPARγ-LBD were obtained in the presence of the SRC1 peptide. The complex structure was resolved in the orthorhombic space group $P2_12_12_1$ at 2.53 Å resolution (PDB ID: 7WGP) (Supplementary Table S1). The electron density map for fenofibric acid bound to PPARα-LBD indicated the presence of two molecules in the protein monomer (Figure 5A–D; reprinted from Kamata et al. [19]); however, that of fenofibric acid bound to PPARγ-LBD indicated the existence of three molecules in the protein monomer (Figure 5E–H). Its overall structure was basically identical to the previously reported active conformations that form the AF-2 helix 12 (arrowheads in Figure 5A,E). The two structures were similar, with an RMS distance of 0.57 Å (215 common Cα positions).

In PPARα-LBD, two molecules were located at the Center/Arm I and Arm II/X (Figure 5A,B). In PPARγ-LBD, the first molecule was located beside the Center region, the second molecule at Arm II/III, and the third molecule at Arm II/X (Figure 5F). Only the first molecule was stabilized by the four consensus amino acids (Ser289/His323/His449/Tyr473) via hydrogen bonds (red dotted lines) and electrostatic interactions (blue dotted lines) in PPARγ-LBD (Figure 5G). Hydrogen bonds were also observed between PPARα K257 and the carbonyl group of the first molecule (water-mediated) and between PPARα T279 and the carbonyl group of the second molecule (Figure 5D). No hydrogen bond or electrostatic interaction was observed between PPARγ and the first molecule, but hydrogen bonds were observed between PPARγ R288 and the oxygen atom of the second molecule and between PPARγ S342 and the carboxylic acid of the third molecule. Furthermore, two halogen bonds were observed between PPARγ E259/R280 and the chlorine atom of the third molecule (Figure 5H). Such interactions combined with hydrophobic interactions (Supplementary Figure S1D,E) may stabilize the location of the second and third fenofibric acid molecules.

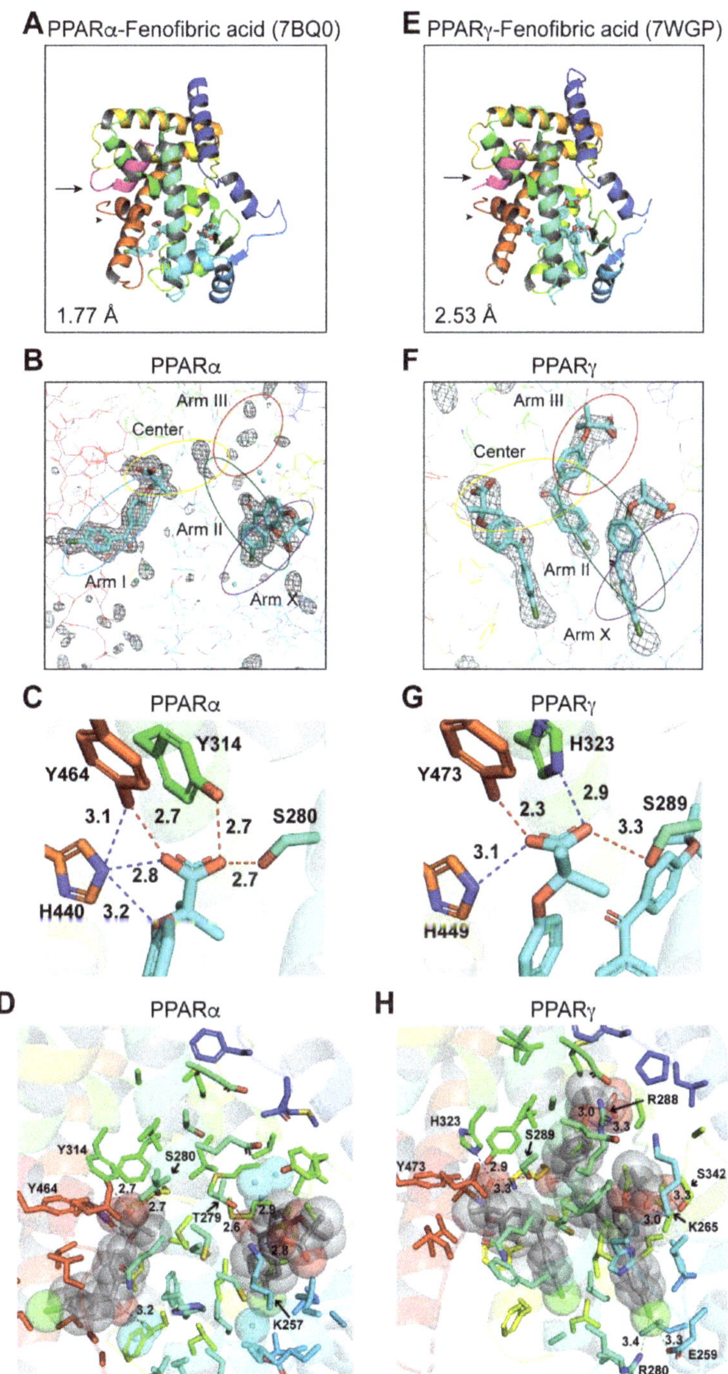

Figure 5. PPARα/γ-LBD–fenofibric acid cocrystal structures. Cocrystals of fenofibric acid and PPARα-LBD (**A–D**; reprinted from Kamata et al. [6]) or PPARγ-LBD (**E–H**) were analyzed using X-ray diffraction. (**A**,**E**) Overall structures of the complexes. The SRC1 peptide (α-helix in magenta) and the

AF-2 helix 12 (α-helix in red) are indicated by arrows and arrowheads, respectively. PDB identities and resolutions are labeled. (**B**,**F**) Magnified views of fenofibric acids located in Arm I/Center regions (single molecule) and Arm II/Arm X regions (another) of PPARα-LBD (**B**) or the Center region (1st molecule), Arm II/Arm III regions (2nd), and Arm II/Arm X (3rd) of PPARγ-LBD (**F**). The electron density is shown in the mesh via F_o-F_c omit maps contoured at +3.0σ. Water molecules are presented as cyan spheres. (**C**,**G**)Hydrogen bonds and electrostatic interactions between fenofibric acid and the four consensus amino acid residues (that recognize the carboxyl moiety of fenofibric acid) are indicated by red and blue dotted lines, with their distances (Å). (**D**,**H**) Hydrogen bonds and electrostatic interactions between fenofibric acid (in van der Waals spheres) and all surrounding amino acid residues located within a distance of 5 Å.

2.6. Structures of the PPARα/δ/γ-LBD–Pemafibrate Complexes

The PPARδ-LBD–pemafibrate cocrystals, similar to the PPARα-LBD–pemafibrate cocrystals (Figure 6A–D; reprinted from Kamata et al. [19]), were obtained without using coactivators (Figure 6E–H), whereas the PPARγ-LBD–pemafibrate cocrystals were obtained only using SRC1 (Figure 6I–L), suggesting that its physical interaction with PPARγ-LBD is the weakest (Figure 2F,I). Their structures were solved in the orthorhombic space group $P22_12_1$ at 1.81 Å resolution (PDB ID: 7WGN) and the orthorhombic space group $P2_12_12_1$ at 2.43 Å resolution (PDB ID: 7WGQ), respectively (Supplementary Table S1). The electron density of pemafibrate in PPARδ/γ-LBD indicated a single molecule in the protein monomer (Figure 6E,I) similar to that observed in PPARα-LBD (Figure 6A). Their overall structures were identical to the active conformations that form the AF-2 helix 12 (arrowheads in Figure 6A,E,I), with RMS distances of 0.75 Å (234 common Cα positions in PPARα/δ) and 0.71 Å (226 common Cα positions in PPARα/γ).

Pemafibrate was located at the almost identical Y-shaped structures comprising the Center and Arm II/III regions in all PPARs (Figure 6B,F,J) although its phenoxyalkyl group seemed to be pushed toward helix 5 in PPARδ-LBD (Figure 6E) and its 2-aminobenzoxazole group seemed to be pushed toward helix 3 in PPARγ-LBD (Figure 6I). Several hydrogen bonds and electrostatic interactions with the four consensus amino acids were observed (Figure 6C,G,K) with a very close proximity (2.0 Å) to PPARγ Y473 (Figure 6K). Hydrogen bonds between PPARα T279 and water molecules were also observed in PPARα (Figure 6D), whereas a water-mediated hydrogen bond or electrostatic interaction with 2-aminobenzoxazole group of pemafibrate was observed in PPARδ-LBD (Figure 6H). Furthermore, no such bonds or interactions were observed in PPARγ-LBD (Figure 6L). Pemafibrate was further stabilized by hydrophobic interactions in all PPARs (Supplementary Figure S1F–H).

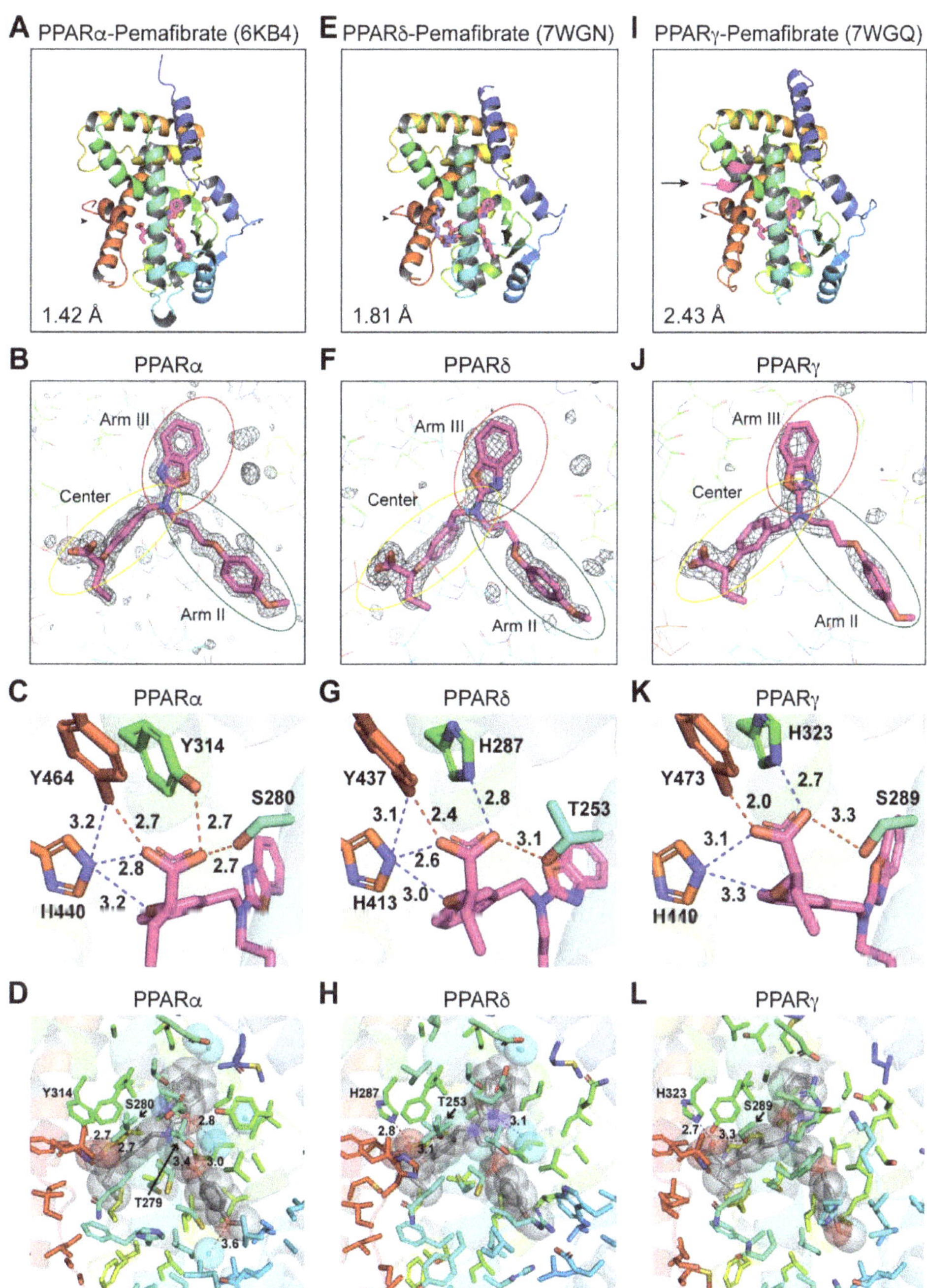

Figure 6. PPARα/δ/γ-LBD–pemafibrate cocrystal structures. Cocrystals of pemafibrate and PPARα-LBD (**A–D**; reprinted from Kamata et al. [6]), PPARδ-LBD (**E–H**), or PPARγ-LBD (**I–L**) were analyzed using X-ray diffraction. (**A,E,I**) Overall structures of the complexes. The SRC1 peptide (α-helix in magenta) and the AF-2 helix 12 (α-helix in red) are indicated by arrows and arrowheads, respectively.

PDB identities and resolutions are labeled. (**B**,**F**,**J**) Magnified views of pemafibrate located in the Center/Arm II/Arm III regions of PPARα/δ/γ-LBD. The electron density is shown in the mesh via F_o-F_c omit maps contoured at +3.0σ. Water molecules are presented as cyan spheres. (**C**,**G**,**K**) Hydrogen bonds and electrostatic interactions between pemafibrate and the four consensus amino acid residues (that recognize the carboxyl moiety of pemafibrate) are indicated by red and blue dotted lines, respectively, with their distances (Å). (**D**,**H**,**L**) Hydrogen bonds and electrostatic interactions between pemafibrate (in van der Waals spheres) and all surrounding amino acid residues located within a distance of 5 Å.

2.7. Various Binding Modes to PPARα/δ/γ-LBD Pockets

We have previously defined five LBPs (Center and Arm I–III/X) in PPARα-LBD [19] (Figure 7A) and four LBPs (Center and Arm I–III) in PPARδ/γ-LBD [20]. Bezafibrate was located at the Center/Arm II regions of PPARα-LBD (Figure 7B), where endogenous fatty acids and many other ligands bind [19]. However, one fenofibric acid molecule was located at the Arm I and another was located at the Arm X of PPARα-LBD (Figure 7B); only the Arm I pocket is known to be occupied by relatively low affinity fibrates, such as ciprofibrate, clofibric acid, and gemfibrozil [19]. Pemafibrate was located at the Y-shape structures comprising the Center and Arm II/III (Figure 7B), similar to GW7647, the potent PPARα-selective agonist [19].

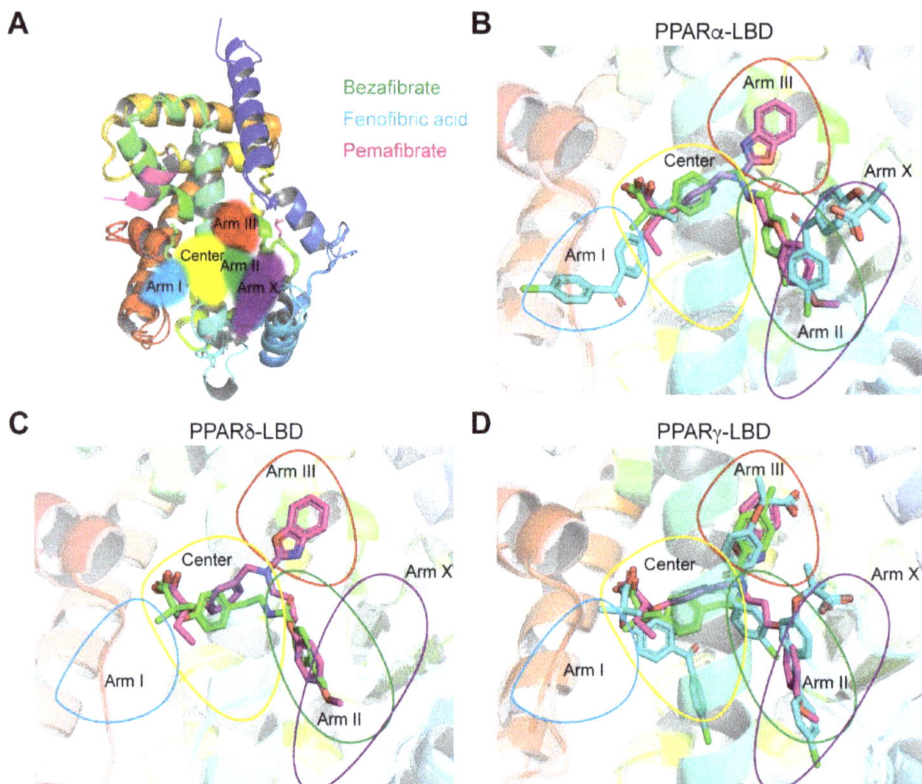

Figure 7. Ligand-binding pocket (LBP) regional localization of the three fibrates in PPARα/δ/γ-LBD. (**A**) LBP comprising the Center and Arms I–III and X of PPARα/δ/γ-LBD [6,17]. (**B**–**D**) Superimposed images of bezafibrate (green), fenofibric acid (blue), and pemafibrate (red) in PPARα/δ/γ-LBD.

In PPARδ-LBD, bezafibrate was located at the Center/Arm II regions (Figure 7C), where the potent PPARδ-selective agonists such as GW501516 and GW0742 bind [26,27], and only pemafibrate was located at the Arm III region (Figure 7C) as in PPARα-LBD (Figure 7B). In PPARγ-LBD, a single bezafibrate was located at the Center/Arm III regions; three fenofibric acids were arranged in a parallel fashion within the Center and Arm II/III/X; and a single pemafibrate was in the Y-shape structures comprising the Center and Arm II/III (Figure 7D). Notably, bezafibrate and fenofibric acid (the second molecule) were located in the Arm III regions only in PPARγ-LBD (Figure 7D).

3. Discussion

Most fibrates were developed in the 1960s–1980s before their molecular target, PPARα, was identified. Therefore, neither the information regarding PPAR agonistic activity (including subtype selectivity) nor the results of molecular docking studies on the three-dimensional structures of PPARs were employed for designing and developing the fibrates. Classical fibrates, such as bezafibrate and fenofibrate, are known to have relatively low PPAR activity and selectivity [28]. Bezafibrate was launched in 1978 by Boehringer Mannheim in Germany (current F. Hoffmann-La Roche AG [Roche], Switzerland) and is currently approved in many countries but not the United States. It is considered to be a "balanced" pan agonist that activates all PPARα/δ/γ at comparable doses and improves dyslipidemia via its actions against PPARα; insulin sensitivity via its actions against PPARγ, and overweightness by enhancing fatty acid oxidation, energy consumption, and adaptive thermogenesis via its actions against PPARδ [29]. Bezafibrate activated all PPARα/δ/γ in the cell-based transactivation assay (Figure 1E), cell-free (both PGC1α and SRC1) coactivator recruitment assay (Figure 2D,G), and cell-free thermostability assay (Figure 3A) by binding to the similar binding pockets (Center/Arm II) of PPARα/δ (Figure 4B,F) or the other binding pocket (Center/Arm III) of PPARγ (Figure 4J). Interestingly, depending on the assays, bezafibrate induced responses with altered affinity against PPARα/δ/γ: the orders of EC_{50} values were α (30.4 μM) < δ (86.7 μM) < γ (178 μM) in transactivation (Figure 1E; equivalent to the values [α (25 μM); δ (95 μM); γ (>100 μM)] in previous transactivation experiments [30]), δ < γ < α in PGC1α recruitment (Figure 2D), α < δ < γ in SRC1 recruitment (Figure 2G), and α < δ = γ in thermostability assay (Figure 3A). In addition, bezafibrate exhibited altered efficacy: the orders of the maximal responses were α > γ > δ in transactivation (Figure 1E), δ > α > γ in PGC1α recruitment (Figure 2D), α > δ > γ in SRC1 recruitment (Figure 2G), and α > δ = γ in the thermostability assay (Figure 3A). In the phase 3 study of "Bezafibrate in Combination with Ursodeoxycholic Acid in Primary Biliary Cirrhosis (BEZURSO; NCT01654731)" completed in December 2016, its dual actions on PPARα/δ were considered important for the improvement of biochemical measures, such as liver stiffness [31]. However, the action of bezafibrate on PPARγ could also be indispensable for the effect. Another phase 2 study investigating the effect of bezafibrate on bipolar depression (NCT02481245) depended on the concept that mitochondrial dysregulation is attributable to bipolar depression, which could be ameliorated by PPAR pan agonists, such as bezafibrate.

Fenofibrate was introduced into clinical practice in 1974 and launched in France in 1975 [32]. More than 200 clinical trials on fenofibrate have been completed and are ongoing, and it has been approved in the United States and many other countries. Unlike bezafibrate, fenofibric acid has been recognized as a PPARα-selective agonist in clinical settings, although it was shown to activate PPARγ in several in vitro experiments [30,33]. Fenofibric acid induced PPARα/γ-LBD-mediated transactivation (Figure 1F), PGC1α/SRC1 recruitment (Figure 2E,H), increases in thermostability (Figure 3B), and bound to PPARα/γ-LBD (Figure 5). Interestingly, fenofibric acid did not induce transactivation (Figure 1F) and PGC1α/SRC1 recruitment (Figure 2E,H) but increased thermostability (Figure 3B) of PPARδ-LBD, and its cocrystals with PPARδ-LBD were not obtained. As observed in the thermostability assay (Figure 3B), fenofibric acid might bind to the non-LBP regions of PPARδ-LBD and stabilize the PPARδ-LBD but fail to functionally activate it. Unlike bezafibrate, fenofibric acid displayed a consistent order ($\alpha > \gamma >> \delta$ = no response) of affinity and efficacy in transactivation (Figure 1F) and PGC1α/SRC1 recruitment (Figure 2E,H), except for thermostability (Figure 3B). To our knowledge, no publication has reported EC_{50} values of fenofibric acid or fenofibrate in PPARδ(-LBD)-mediated reactions; therefore, fenofibric acid is considered a genuine PPARα/γ dual agonist. Pemafibrate (K-877) was approved in 2018 in Japan as a highly selective PPARα agonist that is safe for simultaneous use with statins even in patients with mild adverse effects [34]. Pemafibrate activated PPAR$\alpha/\delta/\gamma$-LBD-mediated transactivation (Figure 1G), PGC1α/SRC1 recruitment (Figure 2F,I), increases in thermostability (Figure 3C), and bound to PPAR$\alpha/\delta/\gamma$-LBD with the similar Y-shaped forms (Figure 6), although it acted at much lower concentrations on PPARα-LBD than on PPARδ/γ-LBD.

This study demonstrated that bezafibrate and pemafibrate could bind to and activate all PPAR subtypes, whereas fenofibric acid could bind to and activate only PPARα/γ. Whether this may happen in clinical settings is an issue of importance and needs to be investigated. Therapeutic doses in Japan are 800 mg, 106.6–160 mg, and 0.2 mg per day for bezafibrate, fenofibrate, and pemafibrate, respectively. The maximal plasma concentration (C_{max}) after the administration of a single dose (200 mg) of bezafibrate was 20.4 μM (7.39 μg/mL) [35], which is roughly equivalent to its EC_{50} values in transactivation (Figure 1E) and PGC1α/SRC1 recruitment (Figure 2D,G); thus, bezafibrate can act as a PPAR pan activator in its clinical doses. According to the package insert of TRICOR (Takeda, Tokyo, Japan), the C_{max} after the administration of a single one-day dose (160 mg) of fenofibrate was 37.0 μM (11.8 μg/mL). Fenofibric acid (30 μM) activated PPARα and slightly activated PPARγ but did not activate PPARδ (Figures 1F and 2E,H). Therefore, fenofibric acid is considered a relatively weak PPARα agonist. According to the package insert of PARMODIA (Kowa, Nagoya, Japan), the C_{max} after the administration of a single one-day maximal dose (400 μg in Japan) of pemafibrate was 7.28 nM (3.57 ng/mL) after seven-day repeats. Pemafibrate (7 nM) induced transactivation only in PPARα (Figure 1G); and the coactivator recruitment assay hardly detected signals produced by <10 nM ligands owing to too many non-responding PPAR-LBDs within a total of 100 or 200 nM proteins. Therefore, pemafibrate is considered a selective agonist of PPARα at its clinical doses.

The global prevalence of NAFLD is estimated to be 25% of the global population, which is consistent with the substantial increases in the number of patients with diabetes and metabolic syndrome. PPAR dual/pan agonists are expected to be some of the most promising therapeutic drugs for NAFLD [36]. Although saroglitazar (α/γ dual agonist; Zydus Discovery, Dubai, UAE) [37] and lanifibranor ($\alpha/\delta/\gamma$ pan agonist; Iventiva Pharma, Daix, France) [17] remain under clinical trials investigating their use against NAFLD/NASH, the development of most of the dual PPARα/γ agonists (mostly against type 2 diabetes)— muraglitazar (Bristol-Myers Squibb/Merck), tesaglitazar (AstraZeneca), aleglitazar (Roche), MK0767 (Kyorin/Banyu/Merck), naveglitazar (Ligand Pharmaceuticals, San Diego, CA, USA), ONO-5219 (Ono Pharma, Osaka, Japan), and DSP-8658 (Sumitomo Dainippon Pharma, Osaka, Japan)—has been discontinued due to PPARγ-related side effects (e.g., heart failure) or no effects [11,38]. Therefore, repositioning of bezafibrate [39] and fenofibrate, which have proven safety, as anti-NAFLD/NASH drugs might be a favorable option. Although three clinical trials investigating the use of fenofibrate against NAFLD/NASH (NCT02781584, NCT00262964, NCT02891408) and a single trial investigating pemafibrate (NCT03350165) have failed, there have been no clinical trials on the use of bezafibrate against NAFLD/NASH. However, bezafibrate has been shown to exert beneficial effects on NASH in tamoxifen- or its analog toremifene-treated breast cancer patients [40,41]. Pemafibrate improved the FibroScan–aspartate aminotransferase (FAST) scores (a novel index of NASH conditions) [42] and the similar scores [43] in NAFLD/NASH patients in some retrospective studies. The differential (in terms of efficiency and efficacy) recruitment of coactivators (PGC1α and SRC1) by bezafibrate (Figure 2D,G) may induce the expression of the specific sets of genes different from those induced by fenofibric acid or pemafibrate [44] in organ-specific manners [45].

In conclusion, this study has highlighted the PPAR dual/pan agonistic aspects of the three approved fibrates—bezafibrate, fenofibric acid, and pemafibrate—using functional and structural analyses. PPAR dual/pan agonists could be a radical remedy for NAFLD/NASH, and our findings contribute toward the fine-tuning of PPAR subtype selectivity.

4. Materials and Methods

4.1. PPAR Activation Assay 1: Transactivation Assay

COS-7 cells (No. RCB0539; Riken BRC Cell Bank, Ibaraki, Japan) were maintained in Dulbecco's modified Eagle's medium (DMEM) supplemented with 10% fetal bovine serum (FBS) and antibiotics at 37 °C in a 5% CO_2/95% air incubator. To evaluate PPARα/δ/γ-mediated transcriptional activation, pSG5-GAL–human PPARα/δ/γ chimera expression plasmids, a MH100(UAS)×4-tk-Luc reporter plasmid, and a pRL-CMV *Renilla* luciferase control plasmid under the control of a cytomegalovirus promoter were cotransfected into COS-7 cells. The pSG5-GAL–hPPARα/δ/γ plasmids express fusion proteins comprising the yeast transcription factor GAL4 DNA-binding domain and each of the human PPARα/δ/γ-LBDs [21,24]. The MH100(UAS)×4-tk-Luc plasmid contains four copies of MH100 GAL4 binding site and the *Firefly* luciferase gene [46]. The cells were seeded on 96-well tissue culture plates at a density of 1.0 × 10^4 per well in 90 μL of DMEM supplemented with 1% FBS. After 24 h, 10 μL mixture containing 20 ng of pSG5-GAL-hPPARα/δ/γ, 80 ng of MH100(UAS)×4-tk-Luc, 30 ng of pRL-CMV, and 0.6 μL ViaFect transfection reagent (Promega, Madison, WI, USA) in Opti-MEM I reduced serum media (Thermo Fisher Scientific, Waltham, MA, USA) were added to each well. After 38 h, cells were treated with PPAR ligands (dissolved in 25 μL of DMEM with no FBS) and cultured for 10 h. Both *Firefly* and *Renilla* luciferase activities were measured using the Dual-Glo Luciferase Assay System (Promega). The transactivation activities were expressed as percentages of the maximal *Firefly* luciferase responses induced by potent/specific PPARα/δ/γ agonists: GW7647 (0.1 μM) for PPARα, GW501516 (0.02 μM) for PPARδ, and GW1929 (1 μM) for PPARγ—after normalization with *Renilla* luciferase responses. GW7647, GW501516, and bezafibrate were purchased from Cayman Chemical (Ann Arbor, MI, USA). GW1929, pemafibrate, and fenofibric acid were purchased from Sigma-Aldrich, ChemScene (Monmouth Junction, NJ, USA), and FujiFilm-Wako (Osaka, Japan), respectively.

4.2. Recombinant PPARα/δ/γ-LBD Expression and Purification

Human PPARα-LBD (amino acids 200–468), PPARδ-LBD (amino acids 170–441), and PPARγ-LBD (amino acids 203–477 in isoform 1) were expressed as amino-terminal His-tagged proteins by the pET28a vector (Merck KGaA (Novagen), Darmstadt, Germany) in Rosetta (DE3) pLysS competent cells (Novagen) and purified using three-step chromatography as described in our PPARα-LBD preparation [6,19]. Transformed cells were cultured at 30 °C in an LB medium (with 15 μg/mL kanamycin and 34 μg/mL chloramphenicol), and 50 mL of overnight culture was seeded in 1 L of a TB medium (with 15 μg/mL kanamycin), which was cultured at 30 °C for 1.5 h and then at 15 °C for 2 h. Protein overexpression was induced by adding 0.5 mM isopropyl β-D-thiogalactopyranoside, which were later cultured at 15 °C for 48 h. The cells were harvested and resuspended in 40 mL of buffer A (20 mM Tris-HCl [pH 8.0], 150 mM NaCl, 1 mM Tris 2-carboxyethylphosphine [TCEP]-HCl, and 10% glycerol) for PPARα/γ or buffer A' (20 mM Tris-HCl [pH 8.0], 500 mM ammonium acetate, 1 mM TCEP-HCl, and 10% glycerol) for PPARδ; both buffers contained a complete EDTA-free protease inhibitor (Sigma-Aldrich). The cells were then lysed by sonication five times, for 2 min each time, using a UD-201 sonicator (Tomy, Tokyo, Japan) at an output of eight; they were clarified by centrifugation at 12,000× *g* at 4 °C for 20 min (these conditions were used throughout the study unless otherwise noted); then, polyethyleneimine, at a final concentration of 0.15% (*v*/*v*), was added to the supernatant to remove nucleic acids. After centrifugation, 35 mL of the supernatant was mixed with 20 g of ammonium sulfate at 4 °C for 30 min using gentle rotation. After centrifugation, the pellet was resuspended in

30 mL of buffer B (for PPARα/γ) or B′ (for PPARδ); these buffers were based on buffer A or A′ supplemented with 10 mM imidazole, respectively. The suspension was loaded on a cobalt-based immobilized metal affinity column (TALON Metal Affinity Resin, Takara Bio (Clontech), Shiga, Japan) equilibrated with buffer B (or B′) and eluted with a linear gradient of 10–100 mM imidazole. The PPARα/δ/γ-LBD-containing elutes were incubated with 33 U/mL thrombin protease (Nacalai Tesque, Kyoto, Japan) to cleave the His-tag and simultaneously dialyzed against buffer A (or A′) overnight at 4 °C using a Slide-A-Lyzer G2 dialysis cassette (20 kDa cutoff, Thermo Fisher Scientific). The sample was later dialyzed against buffer C, which was buffer A without 150 mM NaCl, at 4 °C for 3 h. The sample was then loaded onto a HiTrap Q anion-exchange column (GE Healthcare, Boston, MA, USA) equilibrated with buffer C, and eluted with a linear gradient of 0–150 mM NaCl (or 0–0.5 M ammonium acetate for PPARδ). The elutes were loaded onto a HiLoad 16/600 Superdex 75 pg gel-filtration column (GE Healthcare), which had been equilibrated with buffer A (or A′) and further eluted with buffer A (or A′).

4.3. PPAR Activation Assay 2: PGC1α/SRC1 Coactivator Recruitment Assay

The activation status of each PPARα/δ/γ subtype can also be determined using a TR-FRET assay, which is used to detect physical interactions between His-tagged hPPARα/δ/γ-LBD proteins and a biotin-labeled PGC1α coactivator peptide (biotin-EAEEPSLLKKLLLAPANTQ [amino acids 137–155] synthesized by GenScript) or SRC1 peptide (biotin-CPSSHSSLTERHK ILHRLLQEGSPS [amino acids 676–700] from GenScript) using the LANCE Ultra TR-FRET assay (PerkinElmer) [6]. A 9.5 μL aliquot of PPARα/δ/γ-LBD (200 nM in buffer D for PPARα/γ-LBD or 400 nM in buffer E for PPARδ-LBD), 0.5 μL of a 100× ligand solution (in DMSO), and 5 μL of biotin-PGC1α (4 μM) or biotin-SRC1 peptide (8 μM) were mixed in one well of a Corning 384-well low-volume, white-round-bottom, polystyrene non-binding surface microplate (buffer D comprised 10 mM HEPES-NaOH [pH7.4], 150 mM NaCl, 0.005% Tween 20, 0.1% fatty acid-free bovine serum albumin [BSA]; buffer E comprised 50 mM HEPES-NaOH [pH7.4], 50 mM KCl, 1 mM EDTA, 0.5 mM dithiothreitol, and 0.1% fatty acid-free BSA). Then, 5 μL of 4 nM Eu-W1024-labeled anti-6×His antibody/80 nM ULight-Streptavidin (PerkinElmer) were added to each well, and the microplate was incubated in the dark for 2 h at room temperature. FRET signals were detected at one excitation filter (340/12) and at two emission filters (615/12 and 665/12) using a Varioskan Flash double monochromator microplate reader (Thermo Fisher Scientific). The parameters for the measurements at 615 nm and 665 nm were an integration time of 200 μs and a delay time of 100 μs. The 665 nm emissions were due to ULight-FRET, and the 615 nm emissions were due to Eu-W1024. The 665/615 ratio was calculated and normalized to the negative control reaction using 1% DMSO. The nonlinear fitting and calculation of EC_{50} were performed using Prism 5 software (GraphPad, San Diego, CA, USA).

4.4. PPAR Activation Assay 3: Thermostability Assay Using CD Spectroscopy

PPARα/δ/γ-LBD proteins (10 μM) were incubated with different concentrations of ligands in buffer A. CD spectra were monitored within 200–260 nm at increasing temperatures from 30 °C to 70 °C (2 °C/min) using a J-1500 spectropolarimeter equipped with a PTC-510 thermal controller (JASCO, Tokyo, Japan). The spectra of all PPARs displayed local minima at 208 nm and 222 nm (data not shown), a typical feature of α-helical proteins [47]. The thermal stability of PPARs was investigated by continuously monitoring the ellipticity changes at 222 nm during thermal denaturation [23,24], and a single-site sigmoidal dose-response curve fitting program (Prism 5) was used to obtain the melting temperature (T_m) that corresponds to the midpoint of the denaturation process. The ligand-induced increases in T_m values are defined as ΔT_m.

4.5. Cocrystallization of PPARδ/γ-LBD with the Three Fibrates

Cocrystallization of PPARδ-LBD was performed in hanging-drop mixtures of 0.5 μL of PPARδ-LBD (10 mg/mL in buffer A′), 0.1 μL of 10 mM ligand, 0.3 μL of buffer A′,

0.1 μL of 5% *n*-octyl-β-D-glucoside, and 1 μL reservoir solution (50 mM Bis-Tris propane [pH 8.5], 14% PEG8000, 0.2 M KCl, 6% propanediol, 1 mM EDTA, 1 mM $CaCl_2$ for PPARδ-LBD/pemafibrate; 50 mM Bis-Tris propane [pH 8.5], 14% PEG8000, 0.1 M KSCN, 6% propanediol, 1 mM EDTA, 1 mM $CaCl_2$ for PPARδ-LBD/bezafibrate) at 4 or 20 °C for several weeks. In addition, PPARγ-LBD cocrystallization was performed in hanging-drop mixtures of 0.5 μL PPARγ-LBD (20 mg/mL in buffer A), 0.5 μL ligand (2 mM in buffer A), and 1 μL reservoir solution (0.1 M HEPES-NaOH [pH 7.5], 1.1 M trisodium citrate dihydrate for PPARγ-LBD/pemafibrate/SRC1; 0.1 M Tris [pH 8.0], and 1.1 M trisodium citrate dihydrate for PPARγ-LBD/bezafibrate/SRC1 or PPARγ-LBD/fenofibric acid/SRC1) at 20 °C for several weeks. The SRC1 peptide used for cocrystallization was LTERHKILHRLLQEG (amino acids [683–697] from GenScript). The obtained crystals were briefly soaked in a cryoprotection buffer (reservoir solution + 20% glycerol for PPARδ-LBD crystals and 30% glycerol for PPARγ-LBD crystals); afterward, these were flash cooled in a stream of liquid nitrogen until X-ray crystallography was conducted.

4.6. X-ray Diffraction: Data Collection and Model Refinement

Datasets were collected by BL-5A or BL-17A beamline at the Photon Factory (Ibaraki, Japan) using a synchrotron radiation of 1.0 Å. Diffraction data were collected at 0.1° oscillation per frame, and a total of 1800 frames (180°) were recorded for 1.0 Å X-ray crystallography. Data processing and scaling were carried out using XDS X-ray detector software and AIMLESS, respectively [6]. Resolution cutoff values ($R_{merge} < 0.5$, $R_{pim} < 0.3$, and completeness > 0.9) were set by the highest resolution shell. All structures were determined using molecular replacement in PHASER [48] with PDB ID: 2ZNQ for PPARδ-LBD/bezafibrate, 3GZ9 for PPARδ-LBD/pemafibrate, and 1WM0 for all PPARγ-LBD as the search model. Refinement was performed using iterative cycles of model adjustment in two programs: COOT and PHENIX [6]. The structures were constructed using PyMOL programs (http://www.pymol.org; accessed on 21 April 2020). All collection data and refinement statistics are summarized in Supplementary Table S1.

Supplementary Materials: The following supporting information can be downloaded at: https://www.mdpi.com/article/10.3390/ijms23094726/s1.

Author Contributions: Conceptualization, I.I.; methodology, validation and formal analysis, A.H., S.K. and I.I.; investigation, A.H., S.K., M.A., Y.M., K.U., Y.S., Y.H., S.M. and C.K.; resources, data curation, writing—original draft preparation/review and editing, A.H., S.K., T.O. and I.I.; visualization, A.H., S.K. and I.I.; supervision and project administration, I.I.; funding acquisition, S.K. and I.I. All authors have read and agreed to the published version of the manuscript.

Funding: This research was funded in part by the Platform Project for Supporting Drug Discovery and Life Science Research (Basis for Supporting Innovative Drug Discovery and Life Science Research [BINDS]) from AMED (grant number: JP19am0101071; support number: 1407 [S.K. and I.I.]), Grants-in-Aid for Scientific Research from MEXT (19K16359 to S.K.), and Grants-in-Aid for Young Scientists of Showa Pharmaceutical University (to S.K.). This work was performed under the approval of the Photon Factory Program Advisory Committee (proposal number: 2018G658). A.H. is supported by a scholarship from the Tokyo Biochemical Research Foundation (2021–2022) and Nagai Memorial Research Scholarship from the Pharmaceutical Society of Japan (2022–2024).

Institutional Review Board Statement: Not applicable.

Informed Consent Statement: Not applicable.

Data Availability Statement: Five novel PPARδ/γ-ligand structures reported in this study were deposited in PDB: 7WGL (PPARδ-LBD/bezafibrate), 7WGN (PPARδ-LBD/pemafibrate), 7WGO (PPARγ-LBD/bezafibrate/SRC1), 7WGP (PPARγ-LBD/fenofibric acid/SRC1), and 7WGQ (PPARγ-LBD/pemafibrate/SRC1).

Acknowledgments: We thank Toshimasa Itoh (Showa Pharmaceutical University) for technical advice and helpful discussion.

Conflicts of Interest: The authors declare no conflict of interest.

References

1. Weikum, E.R.; Liu, X.; Ortlund, E.A. The nuclear receptor superfamily: A structural perspective. *Protein Sci.* **2018**, *27*, 1876–1892. [CrossRef] [PubMed]
2. Lefebvre, P. Sorting out the roles of PPAR in energy metabolism and vascular homeostasis. *J. Clin. Investig.* **2006**, *116*, 571–580. [CrossRef] [PubMed]
3. Attianese, G.M.P.G.; Desvergne, B. Integrative and systemic approaches for evaluating PPAR/(PPARD) function. *Nucl. Recept. Signal.* **2015**, *13*, e001. [CrossRef] [PubMed]
4. Ahmadian, M.; Suh, J.M.; Hah, N.; Liddle, C.; Atkins, A.R.; Downes, M.; Evans, R.M. PPARγ signaling and metabolism: The good, the bad and the future. *Nat. Med.* **2013**, *19*, 557–566. [CrossRef]
5. Zoete, V.; Grosdidier, A.; Michielin, O. Peroxisome proliferator-activated receptor structures: Ligand specificity, molecular switch and interactions with regulators. *Biochim. Biophys. Acta* **2007**, *1771*, 915–925. [CrossRef]
6. Kamata, S.; Oyama, T.; Saito, K.; Honda, A.; Yamamoto, Y.; Suda, K.; Ishikawa, R.; Itoh, T.; Watanabe, Y.; Shibata, T.; et al. PPARα ligand-binding domain structures with endogenous fatty acids and fibrates. *iScience* **2020**, *23*, 101727. [CrossRef]
7. Montaigne, D.; Butruille, L.; Staels, B. PPAR control of metabolism and cardiovascular functions. *Nat. Rev. Cardiol.* **2021**, *18*, 809–823. [CrossRef]
8. Yu, S.; Reddy, J. Transcription coactivators for peroxisome proliferator-activated receptors. *Biochim. Biophys. Acta* **2007**, *1771*, 936–951. [CrossRef]
9. Shipman, K.E.; Strange, R.C.; Ramachandran, S. Use of fibrates in the metabolic syndrome: A review. *World J. Diabetes* **2016**, *7*, 74. [CrossRef]
10. Lebovitz, H.E. Thiazolidinediones: The forgotten diabetes medications. *Curr. Diabetes Rep.* **2019**, *19*, 151. [CrossRef]
11. Cheng, H.S.; Tan, W.R.; Low, Z.S.; Marvalim, C.; Lee, J.Y.H.; Tan, S. Exploration and development of PPAR modulators in health and disease: An update of clinical evidence. *Int. J. Mol. Sci.* **2019**, *20*, 5055. [CrossRef] [PubMed]
12. CymaBay Therapeutics Halts Clinical Development of Seladelpar. CymaBay Therapeutics Press Release (25 November 2019). Available online: https://ir.cymabay.com/press-releases/detail/476/cymabay-therapeutics-halts-clinical-development-of-seladelpar (accessed on 21 April 2022).
13. Jones, D.; Boudes, P.F.; Swain, M.G.; Bowlus, C.L.; Galambos, M.R.; Bacon, B.R.; Doerffel, Y.; Gitlin, N.; Gordon, S.C.; Odin, J.A.; et al. Seladelpar (MBX-8025), a selective PPAR-δ agonist, in patients with primary biliary cholangitis with an inadequate response to ursodeoxycholic acid: A double-blind, randomised, placebo-controlled, phase 2, proof-of-concept study. *Lancet Gastroenterol. Hepatol.* **2017**, *2*, 716–726. [CrossRef]
14. FDA Lifts All Clinical Holds on Seladelpar. CymaBay Therapeutics Press Release (23 July 2020). Available online: https://ir.cymabay.com/press-releases/detail/485/fda-lifts-all-clinical-holds-on-seladelpar (accessed on 21 April 2022).
15. Francque, S.; Szabo, G.; Abdelmalek, M.F.; Byrne, C.D.; Cusi, K.; Dufour, J.-F.; Roden, M.; Sacks, F.; Tacke, F. Nonalcoholic steatohepatitis: The role of peroxisome proliferator-activated receptors. *Nat. Rev. Gastroenterol. Hepatol.* **2020**, *18*, 24–39. [CrossRef] [PubMed]
16. Goyal, O.; Nohria, S.; Goyal, P.; Kaur, J.; Sharma, S.; Sood, A.; Chhina, R.S. Saroglitazar in patients with non-alcoholic fatty liver disease and diabetic dyslipidemia: A prospective, observational, real world study. *Sci. Rep.* **2020**, *10*, 21117. [CrossRef]
17. Francque, S.M.; Bedossa, P.; Ratziu, V.; Anstee, Q.M.; Bugianesi, E.; Sanyal, A.J.; Loomba, R.; Harrison, S.A.; Balabanska, R.; Mateva, L.; et al. A randomized, controlled trial of the pan-PPAR agonist lanifibranor in NASH. *N. Engl. J. Med.* **2021**, *385*, 1547–1558. [CrossRef] [PubMed]
18. GENFIT: Announces Results from Interim Analysis of RESOLVE-IT Phase 3 Trial of Elafibranor in Adults with NASH and Fibrosis. GENFIT Press Release (11 May 2020). Available online: https://ir.genfit.com/news-releases/news-release-details/genfit-announces-results-interim-analysis-resolve-it-phase-3/ (accessed on 21 April 2022).
19. Kamata, S.; Oyama, T.; Ishii, I. Preparation of co-crystals of human PPARα-LBD and ligand for high-resolution X-ray crystallography. *STAR Protoc.* **2021**, *2*, 100364. [CrossRef] [PubMed]
20. Honda, A.; Kamata, S.; Satta, C.; Machida, Y.; Uchii, K.; Terasawa, K.; Nemoto, A.; Oyama, T.; Ishii, I. Structural basis for anti-non-alcoholic fatty liver disease and diabetic dyslipidemia drug saroglitazar as a PPAR α/γ dual agonist. *Biol. Pharm. Bull.* **2021**, *44*, 1210–1219. [CrossRef]
21. Lehmann, J.M.; Moore, L.B.; Smith-Oliver, T.A.; Wilkison, W.O.; Willson, T.M.; Kliewer, S.A. An antidiabetic thiazolidinedione is a high affinity ligand for peroxisome proliferator-activated receptor gamma (PPAR gamma). *J. Biol. Chem.* **1995**, *270*, 12953–12956. [CrossRef]
22. Adkins, J.C.; Faulds, D. Micronised fenofibrate: A review of its pharmacodynamic properties and clinical efficacy in the management of dyslipidaemia. *Drugs* **1997**, *54*, 615–633. [CrossRef]
23. Yu, C.; Chen, L.; Luo, H.; Chen, J.; Cheng, F.; Gui, C.; Zhang, R.; Shen, J.; Chen, K.; Jiang, H.; et al. Binding analyses between human PPARgamma-LBD and ligands. *Eur. J. Biochem.* **2004**, *271*, 386–397. [CrossRef]
24. Itoh, T.; Fairall, L.; Amin, K.; Inaba, Y.; Szanto, A.; Balint, B.L.; Nagy, L.; Yamamoto, K.; Schwabe, J.W. Structural basis for the activation of PPARgamma by oxidized fatty acids. *Nat. Struct. Mol. Biol.* **2008**, *15*, 924–931. [CrossRef] [PubMed]
25. Eisenberg, D.; Schwarz, E.; Komaromy, M.; Wall, R. Analysis of membrane and surface protein sequences with the hydrophobic moment plot. *J. Mol. Biol.* **1984**, *179*, 125–142. [CrossRef]

26. Wu, C.C.; Baiga, T.J.; Downes, M.; La Clair, J.J.; Atkins, A.R.; Richard, S.B.; Fan, W.; Stockley-Noel, T.A.; Bowman, M.E.; Noel, J.P.; et al. Structural basis for specific ligation of the peroxisome proliferator-activated receptor δ. *Proc. Natl. Acad. Sci. USA* **2017**, *114*, E2563–E2570. [CrossRef] [PubMed]
27. Batista, F.A.; Trivella, D.B.; Bernardes, A.; Gratieri, J.; Oliveira, P.S.; Figueira, A.C.; Webb, P.; Polikarpov, I. Structural insights into human peroxisome proliferator activated receptor delta (PPAR-delta) selective ligand binding. *PLoS ONE* **2012**, *7*, e33643. [CrossRef]
28. Willson, T.M.; Brown, P.J.; Sternbach, D.D.; Henke, B.R. The PPARs: From orphan receptors to drug discovery. *J. Med. Chem.* **2000**, *43*, 527–550. [CrossRef]
29. Tenenbaum, A.; Motro, M.; Fisman, E.Z.; Adler, Y.; Shemesh, J.; Tanne, D.; Leor, J.; Boyko, V.; Schwammenthal, E.; Behar, S. Effect of bezafibrate on incidence of type 2 diabetes mellitus in obese patients. *Eur. Heart J.* **2005**, *26*, 2032–2038. [CrossRef]
30. Kuwabara, K.; Murakami, K.; Todo, M.; Aoki, T.; Asaki, T.; Murai, M.; Yano, J. A novel selective peroxisome proliferator-activated receptor alpha agonist, 2-methyl-c-5-[4-[5-methyl-2-(4-methylphenyl)-4-oxazolyl]butyl]-1,3-dioxane-r-2-carboxylic acid (NS-220), potently decreases plasma triglyceride and glucose levels and modifies lipoprotein profiles in KK-Ay mice. *J. Pharmacol. Exp. Ther.* **2004**, *309*, 970–977.
31. Corpechot, C.; Chazouillères, O.; Rousseau, A.; Le Gruyer, A.; Habersetzer, F.; Mathurin, P.; Goria, O.; Potier, P.; Minello, A.; Silvain, C.; et al. A placebo-controlled trial of bezafibrate in primary biliary cholangitis. *N. Engl. J. Med.* **2018**, *378*, 2171–2181. [CrossRef]
32. Lalloyer, F.; Staels, B. Fibrates, glitazones, and peroxisome proliferator–activated receptors. *Arterioscler. Thromb. Vasc. Biol.* **2010**, *30*, 894–899. [CrossRef]
33. Dietz, M.; Mohr, P.; Kuhn, B.; Maerki, H.P.; Hartman, P.; Ruf, A.; Benz, J.; Grether, U.; Wright, M.B. Comparative molecular profiling of the PPARα/γ activator aleglitazar: PPAR selectivity, activity and interaction with cofactors. *ChemMedChem* **2012**, *7*, 1101–1111. [CrossRef]
34. Yamashita, S.; Masuda, D.; Matsuzawa, Y. Clinical applications of a novel selective PPARα modulator, pemafibrate, in dyslipidemia and metabolic diseases. *J. Atheroscler. Thromb.* **2019**, *26*, 389–402. [CrossRef] [PubMed]
35. Wei, Z.; Bing-ren, X.; Ying, Z.; Liyan, Y.; Teng, W.; Cai-yun, W. HPLC method for the determination of bezafibrate in human plasma and application to a pharmacokinetic study of bezafibrate dispersible tablet. *J. Chromatogr. Sci.* **2008**, *46*, 844–847. [PubMed]
36. Friedman, S.L.; Neuschwander-Tetri, B.A.; Rinella, M.; Sanyal, A.J. Mechanisms of NAFLD development and therapeutic strategies. *Nat. Med.* **2018**, *24*, 908–922. [CrossRef] [PubMed]
37. Gawrieh, S.; Noureddin, M.; Loo, N.; Mohseni, R.; Awasty, V.; Cusi, K.; Kowdley, K.V.; Lai, M.; Schiff, E.; Parmar, D.; et al. Saroglitazar, a PPAR-α/γ agonist, for treatment of NAFLD: A randomized controlled double-blind phase 2 trial. *Hepatology* **2021**, *74*, 1809–1824. [CrossRef]
38. Kalliora, C.; Drosatos, K. The glitazars paradox: Cardiotoxicity of the metabolically beneficial dual PPARα and PPARγ activation. *J. Cardiovasc. Pharmacol.* **2020**, *76*, 514–526. [CrossRef]
39. Tenenbaum, A.; Motro, M.; Fisman, E.Z. Dual and pan-peroxisome proliferator-activated receptors (PPAR) co-agonism: The bezafibrate lessons. *Cardiovasc. Diabetol.* **2005**, *4*, 14. [CrossRef]
40. Ogawa, Y.; Murata, Y.; Saibara, T.; Nishioka, A.; Kariya, S.; Yoshida, S. Follow-up CT findings of tamoxifen-induced non-alcoholic steatohepatitis (NASH) of breast cancer patients treated with bezafibrate. *Oncol. Rep.* **2003**, *10*, 1473–1478. [CrossRef]
41. Hamada, N.; Ogawa, Y.; Saibara, T.; Murata, Y.; Kariya, S.; Nishioka, A.; Terashima, M.; Inomata, T.; Yoshida, S. Toremifene-induced fatty liver and NASH in breast cancer patients with breast-conservation treatment. *Int. J. Oncol.* **2000**, *17*, 1119–1123. [CrossRef]
42. Hatanaka, T.; Kosone, T.; Saito, N.; Takakusagi, S.; Tojima, H.; Naganuma, A.; Takagi, H.; Uraoka, T.; Kakizaki, S. Effect of 48-week pemafibrate on non-alcoholic fatty liver disease with hypertriglyceridemia, as evaluated by the FibroScan-aspartate aminotransferase score. *JGH Open* **2021**, *5*, 1183–1189. [CrossRef]
43. Ikeda, S.; Sugihara, T.; Kihara, T.; Matsuki, Y.; Nagahara, T.; Takata, T.; Kitao, S.; Okura, T.; Yamamoto, K.; Isomoto, H. Pemafibrate ameliorates liver dysfunction and fatty liver in patients with non-alcoholic fatty liver disease with hypertriglyceridemia: A retrospective study with the outcome after a mid-term follow-up. *Diagnostics* **2021**, *11*, 2316. [CrossRef]
44. Guo, L.; Fang, H.; Collins, J.; Fan, X.-h.; Dial, S.; Wong, A.; Mehta, K.; Blann, E.; Shi, L.; Tong, W.; et al. Differential gene expression in mouse primary hepatocytes exposed to the peroxisome proliferator-activated receptor alpha agonists. *BMC Bioinform.* **2006**, *7* (Suppl. 2), S18. [CrossRef] [PubMed]
45. Feige, J.N.; Auwerx, J. Transcriptional coregulators in the control of energy homeostasis. *Trends Cell Biol.* **2007**, *17*, 292–301. [CrossRef]
46. Kang, T.; Martins, T.; Sadowski, I. Wild type GAL4 binds cooperatively to the GAL1-10 UASG in vitro. *J. Biol. Chem.* **1993**, *268*, 9629–9635. [CrossRef]
47. Kelly, S.M.; Jess, T.J.; Price, N.C. How to study proteins by circular dichroism. *Biochim. Biophys. Acta* **2005**, *1751*, 119–139. [CrossRef] [PubMed]
48. McCoy, A.J.; Grosse-Kunstleve, R.W.; Adams, P.D.; Winn, M.D.; Storoni, L.C.; Read, R.J. Phaser crystallographic software. *J. Appl. Crystallogr.* **2007**, *40*, 658–674. [CrossRef] [PubMed]

International Journal of
Molecular Sciences

MDPI

Review

Mechanisms Mediating the Regulation of Peroxisomal Fatty Acid Beta-Oxidation by PPARα

Mounia Tahri-Joutey [1,2], Pierre Andreoletti [1], Sailesh Surapureddi [3], Boubker Nasser [2], Mustapha Cherkaoui-Malki [1] and Norbert Latruffe [1,*]

[1] Bio-PeroxIL Laboratory, University of Bourgogne Franche-Comté, 21000 Dijon, France; mouniajoutey@gmail.com (M.T.-J.); pierre.andreoletti@u-bourgogne.fr (P.A.); malki@u-bourgogne.fr (M.C.-M.)

[2] Laboratory of Biochemistry, Neurosciences, Natural Resources and Environment, Faculty of Sciences & Techniques, University Hassan I, BP 577, 26000 Settat, Morocco; boubker.nasser@uhp.ac.ma

[3] Office of Pollution Prevention and Toxics, United States Environmental Protection Agency, Washington, DC 20460, USA; surapureddi.sailesh@epa.gov

* Correspondence: Norbert.Latruffe@u-bourgogne.fr

Abstract: In mammalian cells, two cellular organelles, mitochondria and peroxisomes, share the ability to degrade fatty acid chains. Although each organelle harbors its own fatty acid β-oxidation pathway, a distinct mitochondrial system feeds the oxidative phosphorylation pathway for ATP synthesis. At the same time, the peroxisomal β-oxidation pathway participates in cellular thermogenesis. A scientific milestone in 1965 helped discover the hepatomegaly effect in rat liver by clofibrate, subsequently identified as a peroxisome proliferator in rodents and an activator of the peroxisomal fatty acid β-oxidation pathway. These peroxisome proliferators were later identified as activating ligands of Peroxisome Proliferator-Activated Receptor α (PPARα), cloned in 1990. The ligand-activated heterodimer PPARα/RXRα recognizes a DNA sequence, called PPRE (Peroxisome Proliferator Response Element), corresponding to two half-consensus hexanucleotide motifs, AGGTCA, separated by one nucleotide. Accordingly, the assembled complex containing PPRE/PPARα/RXRα/ligands/Coregulators controls the expression of the genes involved in liver peroxisomal fatty acid β-oxidation. This review mobilizes a considerable number of findings that discuss miscellaneous axes, covering the detailed expression pattern of PPARα in species and tissues, the lessons from several PPARα KO mouse models and the modulation of PPARα function by dietary micronutrients.

Keywords: PPARα; peroxisome; β-oxidation; PPRE; ligand; coregulator; micronutrients; PPARα knockout

Citation: Tahri-Joutey, M.; Andreoletti, P.; Surapureddi, S.; Nasser, B.; Cherkaoui-Malki, M.; Latruffe, N. Mechanisms Mediating the Regulation of Peroxisomal Fatty Acid Beta-Oxidation by PPARα. *Int. J. Mol. Sci.* **2021**, *22*, 8969. https://doi.org/10.3390/ijms22168969

Academic Editors: Manuel Vázquez-Carrera and Walter Wahli

Received: 9 July 2021
Accepted: 15 August 2021
Published: 20 August 2021

1. Introduction

As reported in the review by Latruffe and Vamecq [1], peroxisomes are ubiquitous, single membrane-bound organelles. They belong to the fundamental class of intracellular compartments named microbodies. According to the evolutionists, microbodies and eukaryotic cells appeared on Earth around 1.5 billion years ago. Based on their related cell origin, these organelles are defined as glycosomes, glyoxysomes, hydrogenosomes or peroxisomes. Peroxisomes are found in higher vertebrates; glycosomes exist only in trypanosomes; glyoxysomes are found in leaves and seeds; hydrogenosomes are found in anaerobic unicellular ciliates, flagellates, and fungi. The latter two microbody structures belong to lower eukaryotic species, and all these compartments metabolize hydrogen peroxide. According to the endosymbiotic theory, peroxisomes, mitochondria, and chloroplasts may have derived from free-living prokaryotic ancestors. Only mitochondria and chloroplasts are semi-autonomous organelles, containing a DNA genome, which encodes for just some of their proteins.

In the mammalian liver, very-long-chain-fatty acids (VLCFA) are exclusively shortened in peroxisome through a specific β-oxidation system. Then, shortened fatty acids are metabolized by mitochondrial β-oxidation. Peroxisomes also contain the first enzymatic steps of plasmalogen synthesis [1]. In addition, they are involved in maintaining a redox state through the NAD^+/NADH balance, linked to the pyruvate/lactate level. In 1965, a milestone was reached by Hess et al. [2] who described, for the first time, hepatomegaly induced by clofibrate (ethyl 2-(4-chlorophenoxy)-2-methylpropanoate) in rats, subsequently established as a peroxisome proliferator in rodents and an activator of the fatty acid peroxisomal β-oxidation [3]. Later, Isseman and Green [4] identified peroxisome proliferators (PPs) as activator ligands of a special class of nuclear receptors termed Peroxisome Proliferator-Activated Receptor α (PPAR α). Afterward, several PPAR isoforms were characterized as members of the superfamily of the nuclear steroid receptors. It is recognized that the phylogenetic origin of PPARs dates back 200 million years to the fish-mammalian divergence period [5]. PPARs evolved three times faster than other members of the hormone nuclear receptor superfamily, and are represented now in three isoforms (α, β/δ, and γ).

2. Peroxisomal β-Oxidation Systems

In mammalian cells, both mitochondria and peroxisomes can degrade fatty acid chains. Although each organelle harbors its own fatty acid β-oxidation pathway, only the distinct mitochondrial β-oxidation system feeds the oxidative phosphorylation pathway for ATP synthesis, while the peroxisomal β-oxidation pathway participates in cellular thermogenesis [6]. Historically, we owe the first description of the mammalian peroxisomal fatty acid β-oxidation system to Lazarow and de Duve (1976) [7]. Later, a second peroxisomal β-oxidation system was characterized [6]. However, the very-long-chain fatty acids, part of the long-chain class and long-chain dicarboxylic acids, are exclusively processed by the peroxisomal β-oxidation system, whereas other common long-chain fatty acids are oxidized by mitochondria [6,8]. The entry of fatty acids into peroxisome, and activation as acyl-CoAs, depend on ABC membrane transporters (ABCD subfamily) and very-long-chain acyl-CoA synthetases [9,10]. The first β-oxidation system comprises three enzymes: acyl-CoA oxidase 1 (ACOX1), multifunctional protein (L-bifunctional peroxisomal enzyme (L-PBE, also referred to as EHHADH or MFP-1) [11,12], and 3-ketoacyl-CoA thiolases [13] (Figure 1). These three enzymes catalyze four successive reactions, starting with the α,β-dehydrogenation by ACOX1 of the acyl-CoA into 2-trans-enoyl-CoA. L-PBE catalyzes enoyl-CoA hydration into L-3-hydroxacyl-CoA, which is dehydrogenated, giving the 3-ketoacyl-CoA. Then, the 3-ketoacyl-CoA is subjected to a thiolytic cleavage by thiolase to produce one acetyl-CoA molecule and a two-carbon-shortened acyl-CoA [13] (Figure 1).

The second peroxisomal β-oxidation system (Figure 1), converting fatty carboxylates with a 2-methyl branch, such as pristanic acid and bile acid intermediates, includes the 2-methylacyl-CoA-specific oxidases (trihydroxycoprostanoyl-CoA oxidase and pristanoyl-CoA oxidase), the second multifunctional protein (named MFP-2) [11,14,15], and a 58 kDa sterol-carrier protein (SCP-2) containing thiolase activity [6,16] (Figure 1). Although both LBP/MFP-1 and DBP/MFP-2 provide hydratase and dehydrogenase activities, these proteins exhibit opposite stereospecificities. While LBP/MFP-1 hydrates 2-trans-enoyl-CoAs into L-3-hydroxyacyl-CoAs, and dehydrogenates the L-isomers [11,12], the DBP/MFP-2 transforms 2-trans-enoyl-CoAs into D-3-hydroxyacyl-CoA and dehydrogenates the D-isomers [6,11,14,15]. Despite the fact that the MFP enzymes are structurally unrelated to each other, both MFPs can hydrate 2-methyl-enoyl-CoAs [14]. The 3-hydroxy isomers formed by MFP-2 have the same (3R, 2R) configuration, or (24R, 25R) configuration in bile acid intermediates, underlining the role of MFP-2 in both pristanic acid degradation and bile acid synthesis [15]. Recently, it was demonstrated that LBP/MFP-1 is indispensable for the β-oxidation of dicarboxylic acids and the production of their medium-chain derivatives [6,15,17,18].

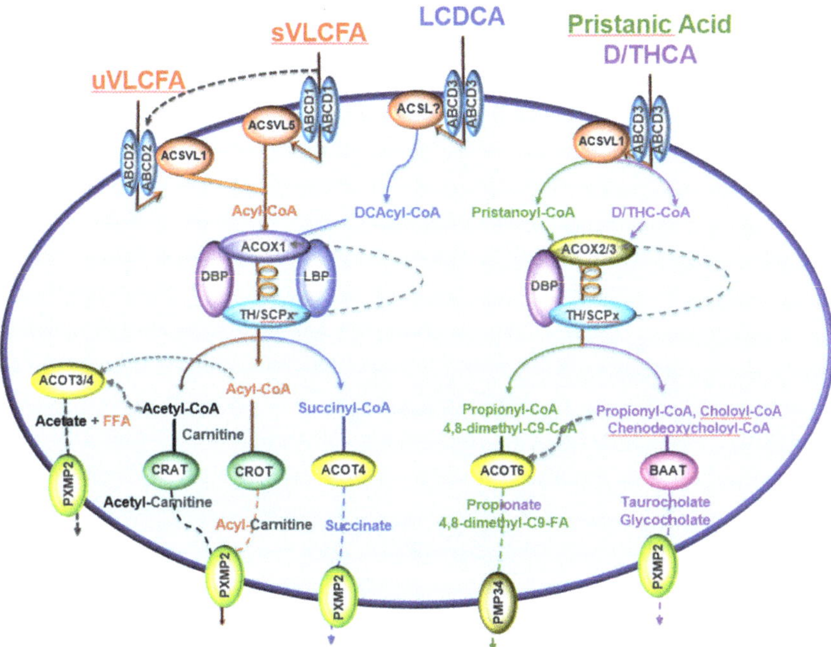

Figure 1. Peroxisomal β-oxidation pathways, including different enzymes and transporters. Saturated or unsaturated very-long-chain fatty acids (sVLCFA or uVLCFA) are imported by the solute ABC-transporters ABCD1 and ABCD2, respectively, into peroxisome, where they are transformed into their acyl-CoAs by one of the peroxisomal acyl-CoA synthetases (ACSVL5 for uVLCFA, and ACSVL1 for uVLCFA). The long-chain dicarboxylic acids (LCDCA), originating from the endoplasmic reticulum ω-oxidation, are imported by ABCD3 and activated to their acyl-CoA thioesters (DCAcyl-CoA) by an unknown acyl-CoA synthetase (ACSL?). The reactions that are catalyzed by ACSVL and ACSL use CoASH and hydrolyze ATP to AMP and pyrophosphate to activate VLCFA or LCDCA molecules, giving acyl-CoA. Acyl-CoA oxidase 1 (ACOX1) is the first flavoenzyme in the straight-chain β-oxidation system, oxidizing sVLCFA, uVLCFA, or LCDCA to their enoyl-CoA derivatives. The second enzyme metabolizing sVLCFA and uVLCFA is the D-bifunctional protein (also called MFP2 or HSD4B17), while the dicarboxylic enoyl-CoA are taken by the L-bifunctional enzyme (also called MFP1 or EHHADH). The thiolytic cleavage is catalyzed by one of the two peroxisomal thiolases (TH: thiolase/ACAA1/2 or SCPx: sterol carrier protein-x). After several rounds of β-oxidation, the peroxisomal system gives shortened acyl-CoA derivatives as hexanoyl- or octanoyl-CoA, and one molecule of acetyl-CoA/round. Both shortened acyl-CoA and acetyl-CoA can be hydrolyzed by acyl-CoA thioesterases 3 or 4 (ACOT3/4) to CoASH, while free fatty acid and acetate are exported by the pore-forming protein PXMP2 (or PMP22) to the cytosol. However, acetyl-CoA and acyl-CoA derivatives can also be transformed to acetyl-carnitine or acyl-carnitine by carnitine acetyl- and carnitine octanoyl transferases (CRAT and CROT), respectively, and then exported by PXMP2 to the cytosol. β-oxidation of DCAcyl-CoAs leads to the production of succinyl-CoA, hydrolyzed to succinate and CoASH by ACOT4, and shipped outside by peroxisome PXMP2. Bile acid intermediates, dihydroxycholestanoic acid (DHCA) and trihydroxycholestanoic acid (THCA), imported by ABCD3 transporter, are beta-oxidized by ACOX2, DBP and SCPx enzymes, leading to the formation of choloyl-CoA and chenodeoxycholoyl-CoA, which are conjugated to glycine or taurine by the bile acid-CoA: amino acid N- acyltransferase (BAAT) and then exported by PXMP2. D/THC-CoA indicate DHCA and THCA co-enzyme A thioesters.

Distinct carnitine transferases and thioesterase enzymes handle products created from the peroxisomal β-oxidation fatty acyl-CoA derivatives. Carnitine moiety is then transferred to the acyl-CoA or the acetyl-CoA by carnitine octanoyltransferase (CROT) or carnitine acetyltransferase (CRAT). On the other hand, a specific peroxisomal thioesterase can hydrolyze the acyl-CoA or the acetyl-CoA, giving a free fatty acid or acetate that can be transported to the cytosol by the peroxisomal membrane solute transporters, such as PXMP2 or PMP34 (Figure 1) [19].

3. Peroxisome Proliferator Response Element, PPRE

PPARα is an ultimate lipid sensor [20] that has the potential to orchestrate and prompt the expression of a plethora of target genes implicated in a broad range of fatty acid metabolism processes [21,22], particularly under conditions of fasting-induced lipolysis and a lipid-rich diet [23–25]. Indeed, PPARα activates many enzymatic pathways involved in fatty acid uptake, intracellular transport [26,27], fatty acid activation and β-oxidation, lipogenesis, ketogenesis and lipoprotein/cholesterol metabolism [28]. As a member of the PPARs family, PPARα regulates the target gene expression in a transcriptional manner through heterodimerization with another transcription factor, the retinoid X receptor (RXR) encoded by the *NR2B1* gene [29,30]. Once activated by a ligand in the ligand-binding domain (LBD), the dimer binds to a specific DNA sequence element, the peroxisome proliferator response element (PPRE), located in the promoter region of target genes, to modulate their expression [31]. It is noteworthy that this regulation can require the recruitment of coregulators [32–36]. The PPARα response element is usually composed of a direct repeat 1 type (DR-1), which means two immediate repetitions of the hexanucleotide AGGTCA consensus sequence, spaced by one nucleotide [37] (Figure 2). PPARα and RXRα bind the first and the second hexamer sequences, respectively. The sequence logo of the PPARα/RXRα PPRE consensus sequence ATGTAGGTCA**A**AGGTCA from the MA1148.1 Jaspar matrix [38], and the associated percentage of the four nucleotides at each position, is presented in Figure 2.

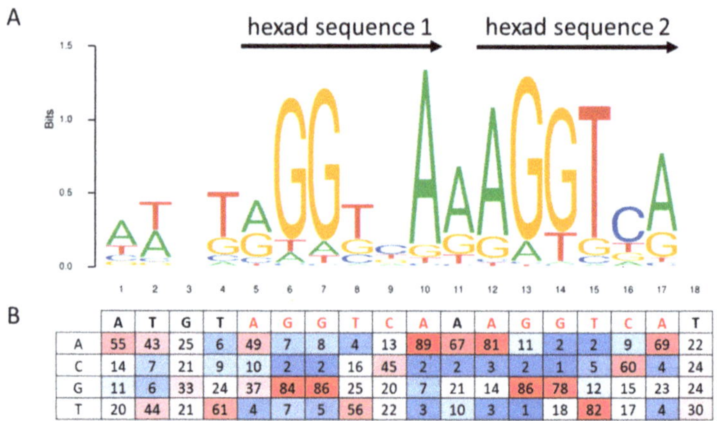

	A	T	G	T	A	G	G	T	C	A	A	A	G	G	T	C	A	T
A	55	43	25	6	49	7	8	4	13	89	67	81	11	2	2	9	69	22
C	14	7	21	9	10	2	2	16	45	2	2	3	2	1	5	60	4	24
G	11	6	33	24	37	84	86	25	20	7	21	14	86	78	12	15	23	24
T	20	44	21	61	4	7	5	56	22	3	10	3	1	18	82	17	4	30

Figure 2. Sequence logo and consensus matrix of the PPARα/RXRα PPRE consensus sequence from MA1148.1 Jaspar matrix [38]. (**A**) Sequence logo of the MA1148.1 Jaspar matrix, presenting the conservation of nucleotides from multiple alignments of 1000 PPARα/RXRα PPRE sequences. Adenosine (A), cytidine (C), guanosine (G), and thymidine (T) nucleotides are respectively green-, blue-, yellow-, and red-colored, and the relative size of the letters represents their frequency in the consensus. The total height of a logo position corresponds to the degree of conservation in the corresponding multiple sequence alignment. (**B**) A table representing the percentage of the four bases for each position of the consensus. The color gradient code highlights the percentage of conservation of bases from blue to red for the whole table.

Among the hundreds of genes known to be regulated by PPARα, eight are encoding enzymes that are commonly localized in the peroxisomal compartment (Figure 3), and belong to the three species: human, mouse, and rat. Table 1 presents validated PPREs sequences for functional genes, of which four (*Acox1*, *Ehhadh*, *Acaa1b*, and *Scp2*) are encoding very-long-chain fatty acid β-oxidation enzymes, and the remaining genes are *Cat*, encoding the catalase enzyme, and *Mlycd* (malonyl-CoA decarboxylase gene), expressing an enzyme with both cytoplasmic and peroxisomal localization. The latter form is believed to be involved in the peroxisomal degradation of malonyl-CoA produced by odd-chain-length dicarboxylic fatty acid β-oxidation [39], and finally, the *Pex11a* gene, which participates particularly in peroxisome biogenesis (Pex11α).

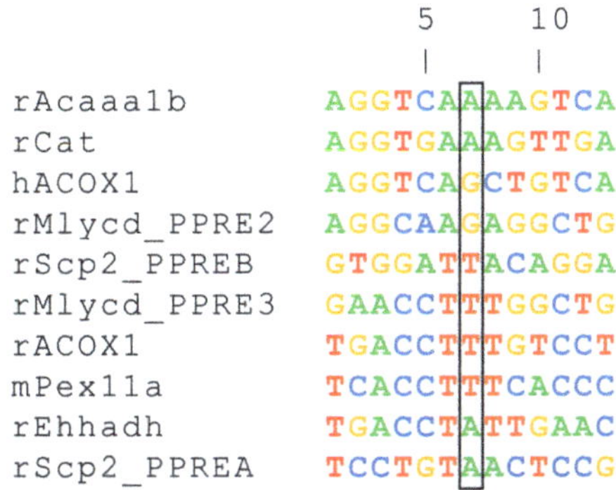

Figure 3. Multiple alignments of PPRE sequences, identified in peroxisomal gene promoters, and experimentally proved to be regulated through PPARα binding. Adenosine (A), cytidine (C), guanosine (G) and thymidine (T) nucleotides are respectively green-, blue-, yellow- and red-colored.

Table 1. Peroxisomal genes experimentally proved to be regulated through the PPARα binding to PPREs.

Gene	Protein	Species	PPRE [a]	PPRE Sequence [b]	Reference
Acaaa1b	acetyl-coenzyme A acyltransferase 1B	rat	PPRE2	AGGTCA**A**AGTCA	[40]
ACOX1	acyl-CoA oxidase 1	human	PPRE	AGGTCA**G**CTGTCA	[41]
Acox1	acyl-CoA oxidase 1	rat	PPRE	TGACCT**T**TGTCCT	[36]
Aldh3a2	aldehyde dehydrogenase family 3, subfamily A2	mouse	PPRE	nd	[42]
Cat	catalase	rat	PPRE	AGGTGA**A**AGTTGA	[43]
Ehhadh	enoyl-CoA hydratase and 3-hydroxyacyl CoA dehydrogenase	rat	PPRE	TGACCT**A**TTGAAC	[44]
Mlycd	malonyl-CoA decarboxylase	rat	PPRE-2	AGGCAA**G**AGGCTG	[45]
Mlycd	malonyl-CoA decarboxylase	rat	PPRE-3	GAACCT**T**TGGCTG	[45]
Pex11a	peroxisomal biogenesis factor 11 alpha	mouse	PPRE	TCACCT**T**TCACCC	[46]
Scp2	sterol carrier protein 2	rat	PPRE-A	TCCTGT**A**ACTCCG	[47]
Scp2	sterol carrier protein 2	rat	PPRE-B	GTGGAT**T**ACAGGA	[47]

[a] corresponds, for each gene, to the PPRE numbering as stated in the corresponding reference. [b] PPREs sequence: PPARα DR-1 sequences are shown with hexads underlined and spacing nucleotides in bold.

4. PPARs and PPARα Structure and Function

Peroxisome proliferator-activated receptors (PPARs) are ligand-regulated transcription factors and belong to a nuclear steroid/thyroid hormone receptor superfamily [48]. Their name originates from their property of peroxisome proliferation [49]. Three PPAR isoforms have been first isolated from the mouse [4], then Xenopus [33,50,51], then rat [52] and human [53], including PPARα (NR1C1), PPARβ/δ (NUC1, NR1C2) and PPARγ (NR1C3). Human PPARα protein consists of 468 amino acid residues, while PPARβ/δ has 441, and PPARγ, 479 aminoacyls long [54]. Each is characterized by a distinct tissue expression profile, a definite ligand binding specificity, and a set of functions implicated in carbohydrate-lipid metabolism, cancer, inflammation, cell proliferation, and differentiation [55–58]. To sustain their protein stability and transcriptional activity, PPARs are subjected to post-translational modifications, such as phosphorylation [59], SUMOylation [60], and ubiquitylation [61]. PPARs act altogether in harmony, to maintain and control cellular and whole-body energy homeostasis by modulating the expression of their specific target genes [57].

The focus here will be on the PPARα isoform. PPARα is a type-II non-steroid ligand-regulated nuclear hormone receptor [4,32,62] transcribed from the human PPARA gene, which spans ~93.2 kb [53] and consists of eight exons [63]. It has been mapped to chromosome 15 in the mouse DNA and to chromosome 22 in humans [31]. The PPARα protein possesses five main functional domains (A–F) embodied in a modular canonical structure [64] (Figure 4). The N-amino terminal end harbors the activation function-1 (AF-1) or A/B domain, which operates autonomously in a ligand-independent manner. The 65 amino acid-long DNA-binding domain (DBD), or C domain, consists of 2 highly conserved zinc finger-like motifs that promote the receptor's binding to the PPRE sequence of the target genes. The D domain or hinge region that bridges the DBD to the ligand-binding domain (LBD) acts as a docking site for cofactors. In the C-terminal region, the LBD, or E/F domain, is responsible for ligand specificity and contains the activation function 2 (AF-2) [28,65]. This latter contains a tyrosine residue on the helix 12, which plays an ultimate role in interacting with the carboxyl group of the ligands [66]. When a ligand enters the LBD pocket of PPARα, the interface of AF-2 stabilizes and facilitates so that PPARα can recruit coactivators [67]. The LBD is still a center of interest in numerous pharmaceutical investigations. Recent publications to date on studies based on X-ray crystallography, referenced in the protein data bank website (PDB; http://www.pdb.org/, accessed on 1 July 2021), provide fascinating, detailed insight into the LBD domain structure, albeit limited to comparing it with other PPAR receptors. It describes a relatively large Y-shaped hydrophobic cavity in the PPARα-LBD pocket volume of 1400 Å^3 [68], which allows PPARα to interact with a broad range of structurally distinct natural and synthetic ligands [67,69].

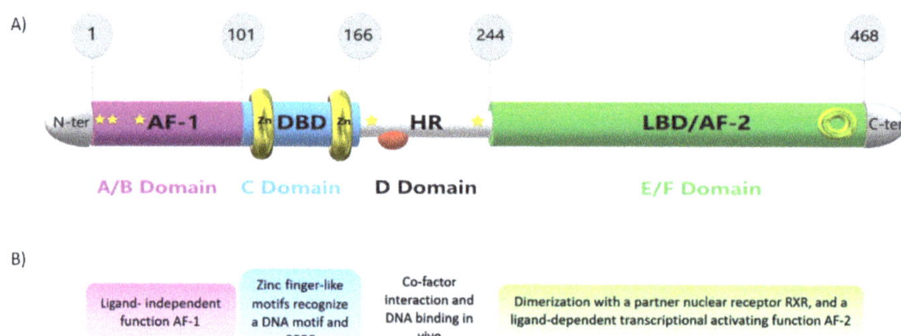

Figure 4. Schematic view of PPARα structure and domain function, with phosphorylation and cofactor binding sites. From the left N-terminus to the right C-terminus of the PPARα protein, (**A**) domain structure of the PPARα protein with the ligand-independent activation function-1 (AF-1) domain or A/B domain shown in purple, the DNA-binding domain (DBD) or C domain shown in blue with two zinc finger-like motifs, the hinge region (HR) or D domain shown in white, and the ligand-binding domain (LBD) or E domain together with the activation function 2 (AF-2) or F domain shown in green. Phosphorylation sites are labeled with yellow stars (6, 12, 21, 179, 230) amino acids, the corepressor site is marked with a red half-sphere, and the coactivator binding site is shown with a green ring. The panels on top show the number of amino acid residues. (**B**) Structural function of A/B, C, D, and E/F domains, respectively.

5. PPARα Ligands

Recently, PPARα-ligands have gained consistent interest in several complex metabolic disease investigations [60,67], such as lipid metabolism disorders. Due to their engagement in physiological and pathophysiological metabolic processes, and their role in activating transcriptional regulatory networks, these ligands are becoming intriguing bona fide treatment opportunities and present a way to unveil many relevant potential roles of PPARα, also known as promising versatile drug targets.

Evidence indicates that a wide variety of lipophilic molecules, the so-called ligands, can activate PPARα, encompassing natural saturated, unsaturated, and polyunsaturated fatty acids (PUFAs) [70,71], and synthetic ligands that are collectively referred to as PPARα-activators [72].

5.1. PPARα Natural Ligands

Natural ligands include endogenous metabolites products derived from the lipid metabolism, such as acyl CoAs [73,74], oxidized fatty acids [63], phospholipids [75], certain nitrated derivatives of fatty acids, eicosanoids [76], endocannabinoid-like molecules [77], and lipoprotein lipolytic products [78]. PPARα natural activators could also originate from an exogenous source that is either found in dietary constituents [65], e.g., dietary ω-3 polyunsaturated fatty acids (docosahexaenoic acid and eicosapentaenoic acid) or issuing from traditionally used medicinal plants (reviewed by Rigano et al. [79]) (Figure 5).

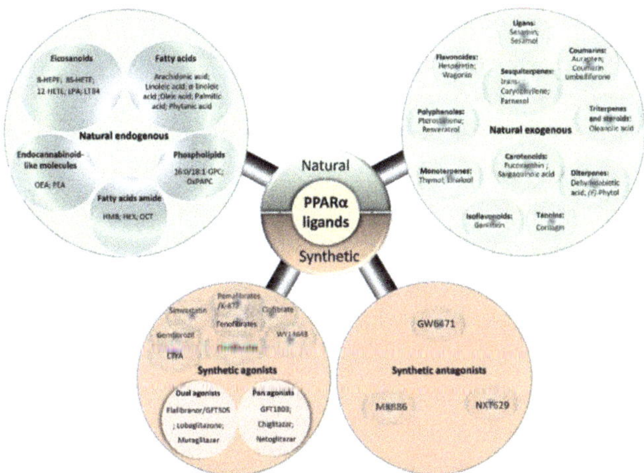

Figure 5. Diagram of different types and classes of PPARα ligands. The natural ligands type encompasses endogenous natural ligands (fatty acids [28,80–82], eicosanoids [31,83–85], phospholipids [75,86], fatty-acid amide [87] and endocannabinoid-like molecules [77], and exogenous natural ligands [30,88] (polyphenol flavonoids, isoflavonoids, monoterpenes, sesquiterpenes, diterpenes, triterpenes and steroids, carotenoids, coumarins, ligans, and tannins). The synthetic ligands type includes various classes of synthetic agonists [28,31,62,67,69,88–91] with various activation and binding modes (single [88,92], dual [28,91,93,94] and pan agonists [28,92]), and synthetic antagonists [89,90,95]. Abbreviations: 8-HEPE: 8-hydroxyeicosapentaenoic acid, 12-HETE: 12-hydroxyeicosatetraenoic acid, 8S-HETE: 8 (S)-hydroxyeicosatetraenoic acids, 16:0/18:1-GPC: phosphatidylcholine(1-palmitoyl-2-oleoyl-sn-glycerol-3-phosphocholine), EPA: eicosapentaenoic acid (20:5), ETYA: eicosatetraynoic acid, HEX: hexadecanamide, HMB: 3-hydroxy-(2,2)-dimethyl butyrate, OCT:9-octadecenamide, OEA: oleoyl-ethanolamide, OxPAPC: oxidized 1-palmitoyl-2-arachidonoyl-sn-glycero-3-phosphocholine, PEA: palmitoyl-ethanolamine.

Numerous findings provide evidence that natural ligands exhibit different binding affinities, which subsequently impact PPARα activation potency. Previous reports showed that omega-3 eicosapentaenoic acid (20:5, ω3) and, to a lesser extent, docosahexaenoic acid (22:6, ω3), are potent ligands [96,97] and consistent activators of PPARα [98–100], while omega-3 PUFA like linolenic acid (C18:3, ω3) and docosapentaenoic (22:5, ω3) acids, and omega-6 PUFA such as linoleic (C18:2, ω6) and arachidonic (C20:4, ω6) acids are weaker PPARα activators [74,99,100]. In addition, experiments performed by Ellinghaus et al. and Zomer et al. [101,102] revealed that phytanic acid (3,7,11,15-tertamethylhexadecanoic acid) is a strong natural physiological ligand for PPARα. These assumptions were then followed by reports from Hostetler et al. [103], showing that PPARα binds the fatty acyl-CoAs (3–20 nM Kds) and branched-chain fatty acyl-CoA (BCFA-CoAs, phytanoyl-CoA, pristanoyl-CoA; Kds near 11 nM) with the highest affinities (i.e., Kd at nM range).

Natural PPARα ligands description studies, using full-length- or chimeric LBD-PPARα constructs, revealed the ability of many saturated and unsaturated fatty acids to activate target gene expression through PPARα modulation. Several PPARα-responsive genes are involved in fatty acid oxidation: (i) mitochondrial β-oxidation pathway (i.e., carnitine palmitoyltransferase 1A) [104]; (ii) microsomal ω-hydroxylation (i.e., CYP4A1-subclass of cytochrome P450 enzymes); and (iii) peroxisomal β-oxidation pathway (i.e., ACOX1; enoyl-CoA hydratase/3-hydroxyacyl-CoA dehydrogenase [105], 3-ketoacyl-CoA Thiolase and SCPx) [7,106,107].

The abovementioned results were recently supported by Chen et al. [108], reporting that feeding animals a diet high in rapeseed oil (rich in erucic acid, a very-long-chain fatty acid) leads to PPARα activation with an adaptive elevation in peroxisomal β-oxidation capacity, which suggested that erucic acid might act as a potential ligand for PPARα. In line with prior communicated data, Maheshwari et al. [109] reported that treating rat Fao cells with a fungal lipid extract rich in monomethyl BCFAs (*Conidiobolus heterosporous*) increases mRNA levels of the PPARα target genes *Acox1*, *Cyp4a1*, *Cpt1A*, and *Slc22A5*, strongly suggesting that BCFAs are similarly potent PPARα activators [109]. Taken together, these relevant results from our laboratory and from others all affirm that the peroxisomal β-oxidation substrates are potent PPARα ligands that modulate the expression of a battery of lipid-metabolizing enzymes to maintain lipid homeostasis and to alleviate the toxic effect of VLCFA and BCFA overload [57].

5.2. PPARα Synthetic Ligands

In the same way, PPARα binds to synthetic ligands termed PPARα activators. Interestingly, PPARα-activators exhibit structural features like a carboxylic acid head and a hydrophobic tail, connected via an aliphatic chain and a central aromatic ring [101]. This group of compounds includes various insecticides (2,4-dichlorophenoxyacetic acid); herbicides (phenoxyacetate derivatives) [110]; surfactants (perfluorooctanoic acid-PFOA); organic chlorinated hydrocarbons solvents such as perchloroethylene and trichloroethylene [111]; food flavors [112]; leukotriene D4 receptor antagonists [113]; phthalate plasticizers, such as di-(2-ethylhexyl)-phthalate and di-(2-ethylhexyl) adipate [114]; and amphipathic carboxylic acids [98]. The latter form the hypolipidemic fibrate class of drugs, acknowledged as the archetypal PPARα agonists, including clofibrates [88,89]; pemafibrates [67,69]; fenofibrates [67], and ciprofibrates [115]. It is notable that certain synthetic ligands are designed to act as dual agonists, like muraglitazar [93], that target both PPARα and PPARγ isotypes; others act as pan-agonists that activate all PPAR receptors like bezafibrates [92]; or as a PPARα partial agonist such as GW9662 [69], known as a PPARγ-selective antagonist (Figure 5). Interestingly, GW9662 displays dual effects by acting as agonist and antagonist against PPARα and also has the ability to enhance agonistic activities of certain less potent fibrates [69], whereas PPARα antagonists like GW6471 [89], MK886 [90], and NXT629 [95] represent the rare range of synthetic ligands that prevent other molecules from binding to this nuclear receptor.

To date, various synthetic single, dual and pan agonists, respectively, are in clinical use as medications to treat dyslipidemia, hyperglycemia in patients with Type 2 diabetes mellitus, hypertriglyceridemia, and cardiovascular disease [28,72,91]. Indeed, potent synthetic ligands could elicit both desirable and undesirable side effects. Studies conducted by Preiss et al. [116] proved that the chronic administration of peroxisome proliferators to rodents causes hepatocellular carcinoma, and it may also increase the risk of gallstones and cause anemia and leukopenia [117]. Much of what we know about PPARα-ligands comes from a collective knowledge primarily derived from rodent studies, via the treatment of mice or rats with synthetic PPARα peroxisome proliferators or by using PPARα null mice [98]. It has been reported that human and mouse PPARα have different binding affinities and physiological effects [118] and are diversely activated by specific ligands, including phthalates and fibrates [119]. Nevertheless, these differences are negligible and do not call into question the tenet of the ultimate role that PPARα plays as a general lipid sensor in both species [98].

To date, tremendous efforts are in progress to develop new, highly PPARα-specific ligands with different activation and binding modes that could more selectively activate PPARα–RXRα transcriptional complex assembly, with tissue-selective and gene-selective activities, to reduce unwanted side effects and assure reasonable safety. In parallel, the "micronutrients" found in food that activate PPAR receptors are gaining increasing interest, as nutritional therapy becomes an unstoppable trend for treating lipid disorders [79].

6. PPARα and Coregulators

The identification of PPAR in the 1990s heralded a new era of biotic and xenobiotic sensing by the liver [4]. The PPAR subfamily of nuclear receptors functions as sensors for fatty acids and fatty acid derivatives and controls critical metabolic pathways involved in lipid and energy metabolism [120,121] and catabolism [122–125]. The transcriptional activation of genes is a complex process that involves the participation of many transcription factors [126]. While the nuclear receptors (NRs) mediated gene-regulation provide the backbone for the transcription factor-specific gene regulation, coregulators provide the much-needed tissue-, cell-, and species-specific differences in the peroxisome proliferator-induced pleiotropic responses of PPARα [127,128]. However, we would like to focus this review section on the role of PPARα and its associated proteins in regulating peroxisomal beta-oxidation genes/pathways. Coregulators are proteins that bind to the nuclear receptor by a specific domain LXXLL, a hallmark for all coregulators [129]. Most coregulators have more than one LXXLL domain and are essential for protein–protein interactions between the nuclear receptor and the coregulator [130] (Figure 6). Moreover, each LXXLL could function in a specific nuclear interaction, suggesting that the coregulators are shared between different NRs.

Coregulators can be broadly classified into subgroups, such as essential vs. non-essential, repressors vs. activators, and DNA binding region (DBD) vs. ligand binding-region (LBD) interacting coregulators [127]. Essential coregulators are proteins deemed critical for the survival of the offspring, and their absence results in embryonic lethality: cAMP-response element-binding protein (CBP); PPAR-interacting protein/activating signal cointegrator 2 (PRIP/ASC2); PPAR-binding protein/mediator complex subunit 1 (PBP/Med1); mediator complex subunit 25 (Med25) [131–133]. Non-essential coregulators are proteins with such critical functional responsibilities that they are usually represented by more than one isoform—steroid receptor coactivators (SRCs) [131,132,134], Asp-Glu-Ala-Asp (DEAD)-box helicases [135–137], sirtuins (SIRT) [96,97], PPARγ coactivators (PGCs) [138–141]—and the loss of one isoform is compensated by others. Repressors that bind to the nuclear receptor PPARα in the absence of/or independent of ligands prevent it from binding to the peroxisomal proliferator response elements (PPRE) of the target genes as nuclear corepressor (NCoR) and silencing mediator of retinoic acid and thyroid hormone (SMRT) [96] (Figure 7A). This group of proteins usually bind to the AF-1 domain of the DNA binding region of the receptor. The ligand-independent coregulators (heat-shock protein-70) could also prevent the PPARα from proteolytic degradation in the cytosol before PPARα could translocate to the nucleus in the activated state [131] and, typically, these proteins bind to the hinge region of the nuclear receptor that interconnects the DNA binding region to the ligand-binding region of the receptor [127]. The activators, on the other hand, could help PPARα zero in onto the specific PPREs of the target genomic region, help attach it to the PPREs with the assistance of nucleosomal-specific functions such as histone methylases (SRC proteins) [142], histone acetyltransferases (CBP/p300) [143,144], DNA-helicases [145], PRIC285 [124], and PRIC320 [146]. The activators would also function by stabilizing the transcriptional complex (PRIP/ASC2 [147]) and potentiate the recruitment of RNA-polymerase complex proteins to the transitional complex (mediator complex, PBP [148,149] (Figure 7B). Additionally, the activators would consist of proteins responsible for separating the transcribed mRNA from the genomic region (protein-L-isoaspartate (D-aspartate) O-methyltransferase (PIMT) [147]. These proteins activate the AF-2 domain of the nuclear receptor and enhance transcription by linking the liganded nuclear receptor to the basal transcription machinery. We have identified almost all these groups of coregulators using either a direct protein–protein interaction assay, such as a yeast two-hybrid assay [150], GST-pull downs [124], and ligand affinity chromatography [141] to identify the PPARα-interacting proteins and a functional transcriptional activation complex [131]. PPARs, like other nuclear receptors, interact with coactivators such as SRC-1 (steroid receptor coactivator-1) or corepressors such as NCoR and SMRT. PPARα-interacting coactivators and corepressors augment or repress, respectively, the PPARα transactivation activity. Since the cloning of SRC-1 twenty-five years ago, over 300 coactivators/coregulators have been identified, with new members still being added to this expanding spectrum. PPARα is known to interact with some of these coregulators [151]. These include CBP/p300-dependent binding complex [152], members of the SRC/p160 superfamily, members of PBP/MED1 complex (PBP/TRAP220/DRIP205/MED1 [133,149,153], members of PRIP/NCoA6 (ASC2/RAP250/TRBP/NRC), members of PRIC complex PRIC285 [124], PRIC295 [141], PRIC320 [146], PPAR gamma-binding proteins, PGC-1α [147,154], and PGC-1β [155,156], as well as coactivator-associated proteins PIMT [131] (NCoA6IP) and coactivator-associated arginine methyltransferase 1 (CARM-1) [157,158]. The PPARα-interacting coregulator (PRIC) complex isolated from rat liver nuclear extracts reveals many coregulators, presumably forming one mega-complex. An almost similar complex was isolated with ciprofibrate as the ligand in affinity chromatography. This diversity raises several issues about the evolutionary importance of the versatility and complexity of coregulatory molecules, their relative abundance in various cell types, and their affinity for a given nuclear receptor in orchestrating transcription in gene-, cell-, and developmental stage-specific transcription. In the absence of a specific ligand, PPARα interacts with the corepressors NCoR and SMRT, but the importance of PPARα action is not well documented, as endogenous ligands could potentially activate PPARα [159]. The homozygous deletion of NCoR or SMRT in mice is embryonically lethal, indicating that they cannot fully compensate for each other during development [160–162]. Furthermore, another corepressor, the receptor-interacting protein 140 (RIP140), which can interact with PPARα, is known to repress the activity of NRs by competing with coactivators and by recruiting downstream effectors such as histone deacetylases (HDACs) [163]. Interestingly, the phenotype of RIP140 knockout mice suggests a role for this corepressor in PPARα signaling, as these mice exhibit resistance to high-fat diet-induced obesity, resulting from the upregulation of genes involved in energy dissipation [163]. Interestingly, hepatic sirtuin 1 (SIRT1) regulates lipid homeostasis by positively regulating PPARα [164,165]. On the other hand, SIRT1 interacts with PPARγ and is regulated by PPARγ in a negative feedback mechanism [166]. SIRT6 binds NCOA2, a PPARα coactivator and part of the SRC family of coactivators; the binding results in the decrease of the acetylation of SRC2/NCOA2 K780 in the liver, thus, interaction with SIRT6 mediates the activation of PPARα and thus the inhibition of SREBP-dependent cholesterol and triglyceride synthesis [167]. The ligand binding to a nuclear receptor triggers a molecular switch that releases corepressors and begins the recruitment of coactivator complexes, such as members of the CBP/p300 family, which exhibit the histone acetyltransferase activity required to facilitate chromatin remodeling. The subsequent recruitment of other coregulators, either singly or as preassembled multi-subunit protein complexes, including mediator complex and RNA polymerase machinery, is facilitated by the interaction of the general basal transcription machinery to enhance the transcription of a specific set of genes [97,168]. As discussed previously, coregulators contain an LXXLL motif that forms two turns of the α-helix and binds to a hydrophobic cleft on the surface of the nuclear receptor. The identification and characterization of coregulators have been derived mostly from in vitro experiments, but there is a paucity of information about individual coactivators in vivo cell- and gene-specific functional roles [131].

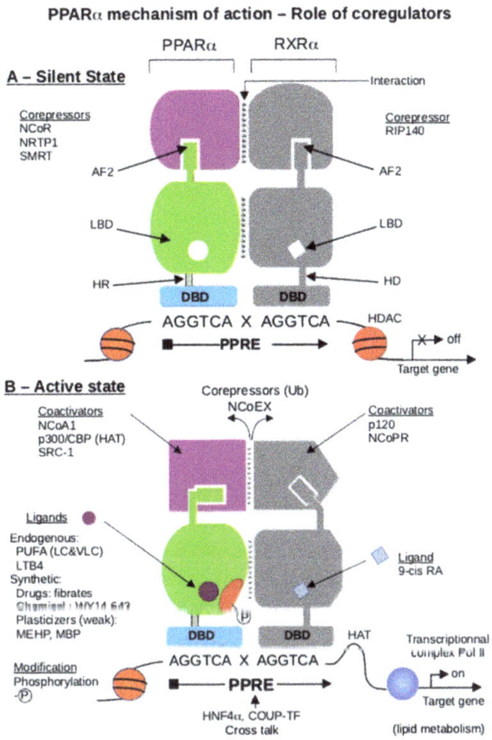

Figure 6. Scheme of heterodimer PPARα/RXRα, located on a PPRE DNA region. The graphic displays two parts: part (**A**), the silent state, and part (**B**), the active state. Part (**A**). In the absence of the ligand, PPARα interacts with transcriptional corepressors (NCoR, SMRT, NRTR-P-1) by recognizing the AF2 region. The same process is at the RXR, where AF2 interacts with RIP140 as RXRα corepressor [169]. Due to the chromatin condensed state, the heterodimer cannot bind the PPRE properly. Part (**B**). In the presence of a PPARα ligand, either a long-chain or very-long-chain polyunsaturated fatty acid, leukotriene LTB4, fibrate or other chemicals [170], and a 9-cis retinoic acid as RXRα ligand, an exchange corepressor/coactivator is made by NCoEX, which suppresses the repressive of corepressor state by ubiquitinylation-inducing degradation by the proteasome system. The fixation of a ligand induces an allosteric LBD conformational change of AF2, allowing the recruitment of coactivators, either NCoR1, p300/CBP, or SRC1 for PPARα, and p120 and NCoPR for RXRα [171]. The CBP-dependent HAT activity induces the remodeling of chromatin and allows the PPARα/RXRα heterodimer to bind to PPRE correctly, then activates the Pol II transcription complex and triggers the transcription of lipid metabolism-encoding genes. Some post-translational

modifications of PPARα regulate its activity [172,173]. For instance, phosphorylation stimulates PPARα transcriptional activity [174]. The HNF4α transcription factor recognizes a similar response element as the PPRE and interplay with PPARα [175]. A comparable mechanism has been reported with the Coup-TF transcription factor. While several works consider PGC-1α [176] as an important coregulator of PPARα, it seems to be more specific for PPAR γ. The 15(S)-HETE, 15-hydroxyicosatetraenoic acid, family of arachidonic acid metabolites; 9-cisRA, retinoic acid cis conformation in carbon 9; AF1, activating domain 1; AF2, activating domain 2; CBP, CREBP binding protein; CoPRs, COPR1 and COPR2 as corepressors of PPAR and RXR, respectively; COUP-TF, chicken ovalbumin upstream promoter transcription factor; CTBP-2, C-terminal binding protein-2; DBD, DNA binding domain; HAT, histone acetyl-transferase; HD, hinge domain; HNF-4α, hepatic nuclear factor 4 α; HDAC, histone de-acetyl transferase; LBD, ligand binding domain; LTB4, leukotrien B4; MBP, mono butyl phthalate; MEHP, mono ethyl hexyl phthalate; NCoA1, nuclear receptor coactivator 1; NCoEX, nuclear receptor corepressor Excit; NCoR1, nuclear receptor corepressor 1; NRTP-1, nuclear repressor transcription factor; p120, protein 120 kDa; p300, protein 300 kDa; Pol II, RNA polymerase class II; PGC-1αPPAR γ co-activator-1α; PPRE, peroxisome proliferator response element; PRIP/RAP250, PPAR interacting-protein methyl transferase; PUFA (LC &VLC), polyunsaturated fatty acids (long-chain or very-long-chain); RIP140 receptor interacting protein corepressor; SMRT, silencing mediator of retinoid and thyroid receptors; SRC1, steroid receptor coactivator-1.

A: Corepressor complex

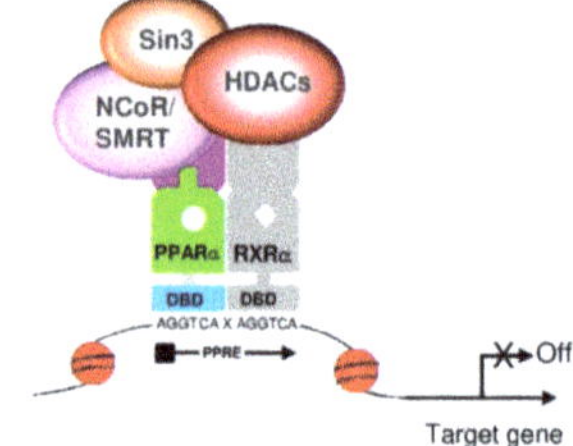

B : Coactivator complex

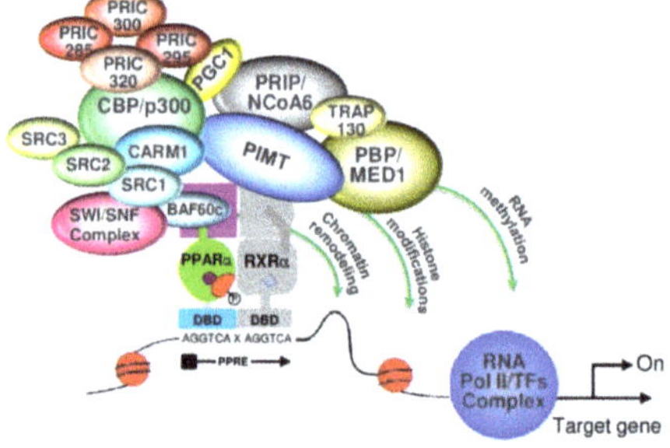

Figure 7. Interaction of PPARα-RXRα heterodimer with corepressor complex (**A**) or coactivator complex (**B**). (**A**) The corepressor complex, including Sin 3, NCoR/SMRT, and HDAC proteins, is recruited to an unliganded PPARα-RXRα heterodimer, so there is no transcription of the PPARα-target genes. (**B**) in the presence of PPARα-ligand, the PPARα-RXRα heterodimer exhibits a conformational change, leading to the dissociation of the corepressor complex, the recruitment of coactivator proteins, and the binding of PPARα to the peroxisome proliferator response element (PPRE). Different subcomplex modules participate in chromatin remodeling, through the acetylation (SRCs, p300) and the methylation (CARM1) of nucleosomes. Mediator components interact with PPARα and promote the recruitment of the basal transcription factors (TFs) to establish a connection with the RNA polymerase II to transcription of PPARα-target genes.

7. Metabolic Regulation of the Peroxisomal β-Oxidation Pathways

The regulation of the peroxisomal pathways is mainly associated with the cellular increase of the peroxisome population, which is highly promoted by several diverse natural and synthetic compounds nominated as peroxisome proliferators (PPs). Such compounds raise a peroxisome number quantitatively, mainly in hepatic parenchymal cells, and provoke delayed pleiotropic responses, including the development of hepatocarcinoma in rats and mice [8,131]. Based on several pieces of experimental evidence, earlier reports from Reddy's group proposed a receptor-mediated mechanism to explain the phenomenon of hepatic peroxisome proliferation. Accordingly, the induction of peroxisomal β-oxidation is a consequence of ligand hepatic overload, leading to lipid metabolism dysregulation, accompanied by an augmentation in the extrahepatic lipolysis and a substantial hepatic influx of free fatty acids [96]. Furthermore, the unique pleiotropic responses raised by structurally unrelated peroxisome proliferators in hepatocytes drive a synchronized transcriptional activation of the peroxisomal β-oxidation genes [13,96,131].

Lazarow and De Duve [7] demonstrated previously that clofibrate administration in rat liver strikingly enhances the peroxisomal β-oxidation activity. A similar observation was reported by Hashimoto and coworkers [177], showing that feeding a diet containing a phthalate ester plasticizer di-(2-ethylhexyl)phthalate, a PPARα activator, leads to a 20-fold increase in the expression of peroxisomal β-oxidation enzymes in rat liver. In addition, a previous study reported that synthetic ligands such as WY-14643 exhibited a high affinity to PPARα, compared to the natural endogenous ligand (16:0/18:1-GPC) in the induction of fatty acid β-oxidation [75]. Moreover, Rogue et al. [93] showed that *Acox1* and *Cpt1A* genes in oleic-acid-overloaded HepaRG cells were significantly upregulated from 1 day, and remained at high levels after 14 days, upon treatment with the dual agonist muraglitazar, which stimulates the fatty acid β-oxidation pathway. These results are in close concordance with previous experiments conducted by Lee et al. [126], showing that after feeding hypolipidemic agents to mice lacking PPARα expression, the mutant animals accumulated lipid droplets in their tissues, which strongly supports the idea that PPARα activators promote the transcription of genes involved in the lipid catabolism process.

Structurally, PPs molecules may be chemically unrelated, including hypolipidemic drugs, such as clofibrate, ciprofibrate, gemfibrozil, and Wy-14,643, as well as some nutritional conditions, especially high-fat diet or vitamin E-deficient diet and leukotriene D4 receptor antagonists. In addition, several herbicides, such as 2,4-dicholophenoxyacetic acid or 4-chloro-2-methylphenoxyacetic acid [8,178] and certain phthalate ester plasticizers, induce a similar liver peroxisome proliferation as do prototypic fibrate derivatives. In addition, the administration to rodents of a C19-steroid, dehydroepiendrosterone, promotes peroxisomal fatty acid β-oxidation and peroxisome proliferation [179]. Though the response to PPs has been demonstrated in several tissues from PPs-treated rodents, the hepatic responsiveness is by far the most powerful, accounting for a 10- to 20-fold induction of peroxisomal fatty acid β-oxidation activities, accompanied by a proliferation of peroxisomes and strong hepatomegaly pathogenesis [8,131].

The description of PPARα-target genes shows that this nuclear hormone receptor largely governs those genes involved in hepatic and cardiac muscle transport, oxidation, and the degradation of lipids. Transcriptionally, PPARα activates several genes, including the lipoprotein lipase gene permitting the release of fatty acids from lipoprotein particles [180], genes encoding fatty acid translocase CD36, and fatty acid-binding protein-facilitating fatty acids capture and transport them through the plasma membrane [8,180]. The acyl-CoA synthetase, activating fatty acids to acyl-CoAs, is another gene-target of PPARα [96,98]. Regarding the genes encoding peroxisomal β-oxidation enzymes, the induction of the peroxisomal fatty acyl-CoA ABC transporter D2 (ALDRP) by PPs was shown to be partially PPARα-dependent in mice hepatocytes [179]; however, the regulation of, e.g., ACOX1, L-PBE and ThB, are entirely reliant on PPARα [8,98,181]. Nevertheless, the regulation of genes implicated in the mitochondrial fatty acid β- oxidation, including the carnitine palmitoyltransferase-1 and the medium chain-acyl-CoA dehydrogenase, is also coordinated by PPARα [98,182,183]. Thus, PPARα arises as a master regulator controlling the hepatic metabolism of free fatty acids. The development of PPARα null mice evidenced the crucial role played by PPARα in the concerted regulation of peroxisome proliferation and expression of its target genes involved in both β- and ω-fatty acid oxidations [181]. By contrast to *Ppara*$^{-/-}$ mice, which exhibit mild hepatic steatosis, *Acox1* null mice develop strong hepatic steatosis, showing a hepatic peroxisome proliferation and the sustained activation of PPARα and expression of its target genes [147,184]. Thus, paradoxically, the defect in ACOX1 activity leads to the hepatic accumulation of ACOX1 substrates, of which some have been shown [147] as efficient endogenous PPARα ligands, mediating the sustained activation of PPARα. On the other hand, the strong PPARα activation of fatty acid β-oxidation genes increases hepatic dicarboxylic acid production and accumulation. Thus, in the absence of ACOX1 activity, these dicarboxylic acids are still unmetabolized and act as firm inhibitors of mitochondrial fatty acid β-oxidation [185]. Moreover, the *Ppara*$^{-/-}$, *Acox1*$^{-/-}$ double-knockout mice exhibit a few periportal clusters of steatotic hepatocytes, and (re-)expression of human *ACOX1* in mice liver results in a substantial reduction in both PPARα activation and hepatic steatosis [8,180]. Peroxisomal fatty acid β-oxidation is induced by starvation in a PPARα-dependent manner, as validated by its impairment in PPARα null mice [8,180]. Accordingly, the deacetylase sirtuin-1 is dispensable to PPARα-inducing peroxisomal fatty acid β-oxidation and needs SIRT1-PPARα interaction, and the deletion of hepatic SIRT1 negatively impacts PPARα signaling [165]. The MAP kinase kinase kinase TGFβ-activated kinase 1 (TAK1) acts upstream to PPARα, and its deletion also impaired the PPARα-dependent induction of peroxisomal fatty acid β-oxidation [186]. PPARα signaling has also been shown to involve the AMPK-SIRT1-PGC-1α axis via the adiponectin receptors [187] (Figure 8). These results strongly highlight the detrimental role of the peroxisomal β-oxidation pathway in the sensing of PPARα activity.

Several peroxisomal β-oxidation substrates display a substantial role as PPARα modulators. It is believed that the activities of (inducible and non-inducible) peroxisomal fatty acid β-oxidation systems are modulated by PPARα [108]. Moreover, several findings provide significant evidence that VLCFA and BCFA, which are considered potentially toxic fatty acids, are potent inducers of PPARα that enhance the transcription of peroxisomal enzymes mediating fatty acid β-oxidation [57,188].

Interestingly, Oleoylethanolamide, a naturally occurring lipid regulating satiety and body weight, exhibited a high-affinity binding to PPARα and the activation of its lipid-metabolizing target genes [189]. Nonetheless, we should consider that most fatty acids are subject to elongation, desaturation, esterification, and β-oxidation, which could modify the availability of PPARα ligands. Accordingly, very-long-chain saturated and unsaturated fatty acids are exclusively metabolized by peroxisomal β-oxidation, which participates in their degradation, synthesis, or retro conversion. One defect in this pathway is associated with the accumulation of VLCFAs and a deficit in certain PUFAs' synthesis, such as DHA. Interestingly, a mouse deficiency of ACOX1, the rate-limiting enzyme in the peroxisomal β-oxidation, leads to the sustained activation of hepatic PPARα and the induction of its target genes [190]. The role of ACOX1 in PPARα lipid sensing was highlighted by *Acox1*$^{-/-}$; *ob/ob* double knockout mice. Thus, the sustained activation of PPARα when linked to the absence of ACOX1 activity attenuates the metabolic consequences of leptin deficiency, due to the *ob/ob* genotype, showing less obesity with the recovery of glucose homeostasis and alleviating insulin resistance [131,147]. Collectively, accumulated data underline the key role of peroxisomal β-oxidation in sensing PPARα-dependent lipid and energy metabolism.

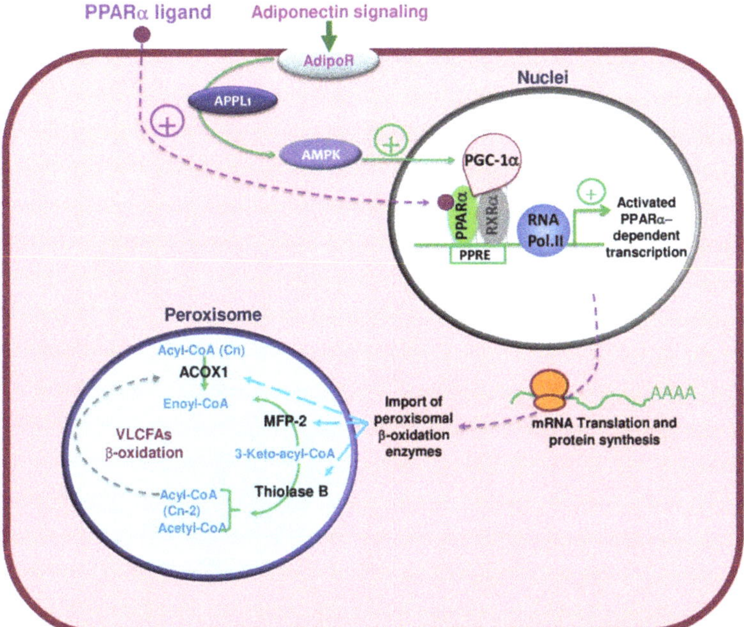

Figure 8. PPARα-dependent regulation of peroxisomal fatty acid b-oxidation in rat liver through adiponectin signaling. Adiponectin: a hormone produced by adipose tissue that plays a role in lipid and glucose metabolism regulation; AdipoR, adiponectin receptor; APPL1, an adaptor protein containing a PH domain, PTB domain, and leucine zipper motif 1, plays a central role as the main contributing factor in the adiponectin and insulin signaling; AMPK, AMP kinase; PGC1- α, PPAR γ coactivator1α; PPRE, peroxisome proliferator-activated receptor; RNA Pol II, RNA-polymerase II; ACOX1, acyl-CoA oxidase 1; MFP2, multifunctional protein 2; thiolase B, 3-ketoacyl-CoA thiolase B; VLCFA, very-long-chain fatty acid.

8. PPARα Expression in Species and Tissue Distribution

8.1. PPARα Expression in Different Species

PPAR is ubiquitous among animal species, i.e., worms [191], insects, fish, frogs [192], reptiles, mammals, including hamsters [193], and humans. A PPARab subtype was detected in zebrafish. This PPARab mutant shows lower expression in liver and visceral mass, which were associated with lipid accumulation [194]. In a jerboa (*Jaculus orientalis*) liver, both active wild-type PPARα (PPARα1 wt) and a truncated PPARα 2 forms were expressed. The availability of active PPARα1 wt is differentially regulated during fasting-associated hibernation [195].

8.2. PPARα Tissue Distribution

PPARα tissue expression is also ubiquitous, although on a different level. PPARα is mainly expressed in tissues with high rates of fatty acid catabolism, i.e., those involved in digestive function (liver, stomach, enterocytes) and muscular activity (heart, skeletal muscle, kidney at proximal tubules). In the nervous system, the expression is moderated (low in retinal, or lacking expression in the central nervous system). Low expression is found in the pancreas and adipose tissue [196], while in the brain, PPARα is found at the highest levels in neurons, followed by astrocytes, and is weakly expressed in microglia [62,197]—more likely, to upregulate the expression of several synaptic related genes coding proteins engaged in excitatory neurotransmission and the neuroprotective mechanism [198–200]. In the immune system, PPARα expression is detected in the spleen, monocytes/macrophages, and neutrophils [201]. In addition, expression is seen in reproductive organs and the epidermis. PPARα is also associated with tumorigenesis in colorectal carcinoma [202]. Concerning the expression in developmental tissue in rats, *Ppara* transcripts are detectable in mouse embryo at 13.5 gestation days, to reach the maximum level at birth [203].

8.3. Lessons from Pparα Knockout

This part of the manuscript provides recent findings from the last five years related to *Ppara* knockout animals, with the intent of disentangling the PPARα's various biological functions in health and disease and to evaluate its engagement in fatty acid catabolism and clearance in liver and heart tissues, where PPARα and FAO are both abundant. A growing body of evidence indicates that PPARα is a crucial regulator of systemic lipid metabolism. PPARα deficiency is considered to be a prime factor that either causes or exacerbates fatty acid metabolism impairment, which leads inevitably to the development of numerous metabolic diseases, to name but a few—obesity [204,205], type 2 diabetes mellitus, insulin resistance, dyslipidemia, myocardial infarction, hepatic steatosis without ethanol consumption, termed non-alcoholic fatty liver disease (NAFLD), which includes severe phenotypes such as non-alcoholic steatohepatitis (NASH), liver fibrosis, and hepatocellular carcinoma [206–210]. Therefore, many investigations were conducted using mainly PPARα knockout mouse models, because of the relative equivalent expression of *Ppara* mRNA between mice and humans in different tissues [98]. Knockout animal models are generated either with the global (*Ppara*$^{-/-}$) or hepatocyte-specific abrogation of the *Ppara* gene, such as *Ppara*$^{Hep-/-}$ (reviewed by Wang et al. [181]). The goal was to identify the pathophysiological mechanisms underlying the abnormal phenotypes associated with PPARα dysfunction and to assess the distinct contribution of hepatic and extrahepatic PPARα to global energy and immune system homeostasis in vivo.

8.4. Lessons from Pparα-KO in the Liver

Hepatic PPARα activation occurs during suckling [211], with a high-fat diet, and during fasting [212–214], boosting fatty acid oxidation (FAO), which participates in the restoration of energy homeostasis and provides energy supply for the extrahepatic tissues. For that reason, most of the studies were focused on hepatic PPARα. Furthermore, hepatic PPARα can protect the liver against fasting/high-fat diet-induced steatosis, by transactivating the genes required for fatty acid catabolism and repressing several inflammatory genes. Thus, during the fasting process, metabolic substrates stored in white adipose tissue are released into the circulation and captured by the liver. Subsequently, this increases β-oxidation and ketogenesis to maintain the energy balance [212]. It was observed that fasted *Ppara*$^{-/-}$ and *Ppara*$^{Hep-/-}$ mice developed hypoketonemia, hypoglycemia, and hypothermia with decreased serum triglycerides. Additionally, the ectopic accumulation of medium-chain fatty acids and long-chain fatty acids in the liver manifests as an increase of hepatic fat mass, termed steatosis, with pronounced oxidative stress and lipid peroxidation compared to wild-type mouse liver. These effects result from the altered mitochondrial and peroxisomal fatty acid β-oxidation pathways in the liver [212,214,215]. Furthermore, mice in which *Ppara* was deleted uniquely in hepatocytes could not modulate bone marrow monocyte egress upon fasting [216], suggesting that PPARα contributes to the regulation of monocyte homeostasis during fasting.

Regarding high-fat diet (HFD)-induced obesity, mice with the hepatocyte-specific deletion of *Ppara* develop steatosis and inflammation [217]. These observations corroborate previous results communicated by Stec and al. [205], showing that *Ppara*$^{Hep-/-}$ mice on HFD had worsened hepatic inflammation associated with steatosis, and exhibited high levels of LDL, which is considered an emerging risk factor for cardiovascular disease in NAFLD. PPARα could also protect against obesity. In *ob/ob* obese mice, the absence of PPARα resulted in increased obesity and led to severe hepatic steatosis [184]. Interestingly, mice lacking only hepatocyte-PPARα developed steatosis spontaneously but without obesity in aging [212,214]. Indeed, extrahepatic PPARα activity blunts and compensates when hepatic PPARα is disrupted, by elevating FAO and lipase activity in other tissues to increase and utilize excess lipid, thus maintaining lipid homeostasis [215]. Likewise, the transcriptome, lipidome, and metabolome results communicated by Régnier et al. and Batatinha et al. [217,218] demonstrate the significant contribution of extrahepatic PPARα activity to the metabolic homeostasis response to HFD consumption. By using double-knockout mice, *Ppara*$^{-/-}$/*Cyp2a5*$^{-/-}$, Chen et al. [108,206] together indicate that PPARα interacts with CYP2a5 (cytochrome P450 2A5) an antioxidant enzyme to protect against steatosis. Fibroblast growth factor 21 (FGF21) acts as a downstream molecule of the PPARα signaling pathway to regulate the liver lipid metabolism and contribute to the CYP2a5 protective effects on alcoholic fatty liver disease [206]. In an experiment conducted by Brocker et al. [219], it was observed that treatment with WY-14643, a PPARα agonist, caused weight loss and severe hepatomegaly in WT and *Ppara*$^{\Delta Mac}$ mice but not in *Ppara*$^{Hep-/-}$ and *Ppara*$^{-/-}$ mice, suggesting that cell proliferation is mediated exclusively by PPARα activation in hepatocytes in response to WY-14643 agonist treatment.

Pparab is one of the two *Ppara* paralogs, highly expressed in zebrafish tissues with high oxidative activity. Li and coworkers generated *Pparab-knockout* in the zebrafish model [194]. *Pparab*-null zebrafish demonstrated a lower expression of critical enzymes involved in FAO, and lower mitochondrial and peroxisomal FAO in the liver and muscle, associated with lipid accumulation in the liver. Furthermore, PPARab deficiency increases glucose oxidation, protein synthesis, and reduced amino acid breakdown, while in rodents, the loss of PPARα increases amino acid breakdown [194].

8.5. Lessons from Ppara-KO in the Heart

PPARα is a crucial regulator of substrate utilization in the heart. Fatty acids are a primary energy source for the heart, and fatty acid β-oxidation provides almost 70% of cardiac ATP; the remainder is obtained primarily from glycolysis and lactate oxidation [220]. Thus, *Ppara* KO mice, in response to chronic pressure overload, exhibit enhanced cardiac dysfunction. In contrast, mild PPARα activation in mice showed a positive effect on myocardial energetic functions, especially during progressive and pressure-overloaded heart failure, revealing the virtue of PPARα-associated FAO modulation as a promising therapeutic strategy for heart failure [221]. In addition, *Ppara* ablation exacerbated myocardial ischemia-reperfusion injury in *Ppara* KO mice models subjected to cardiac ischemia-reperfusion, and interestingly, after the treatment with PEA microparticles (PEA-um®®, 10 mg/Kg), an endogenous PPARα ligand, only *Ppara* WT mice showed the cardioprotective effect of PEA-um®®, but not in *Ppara* KO mice. Although PEA-um®® had a protective and beneficial effect on inflammatory disorders associated with ischemic myocardial failure, it also negatively regulates inflammation through PPARα activation by reducing the activation of the nuclear factor-kB (NF-kB) pathway and production of pro-inflammatory cytokines [222]. Thus, PPARα could augment heart function and cardiac fatty acid oxidation, whereas in the *Ppara* KO mouse model, a more severe sepsis phenotype is observed due to deteriorated cardiac performance and fatty acid oxidation, associated with both a hyperinflammatory cytokine storm as well as immune paralysis [223]. Furthermore, during sepsis, WT hearts showed a decrease in PPARα and other FAO genes' mRNA expression, and this reduction was more dramatic in *Ppara-null* mouse hearts [223]. Taken together, PPARα expression increased fatty acid oxidation and subsequently supported the hyperdynamic cardiac response early during sepsis or pressure-overloaded heart failure, which may prevent morbidity and mortality.

9. PPARα and Micronutrients

As reported above, PPARα is activated by different ligands of both natural and synthetic origins, involved in several signaling and metabolic pathways. Some natural ligands are issued from the lipid metabolism, such as PUFAs and their derivatives. Interestingly, micronutrients, such as minerals, vitamins, phytophenols, and phytosterols are non-energetic compounds with essential signaling activity. Of particular interest, polyphenols, oil products, and some terpenoids and alkaloids impact cell functions through the modulation of PPARα activity.

10. Effect of Polyphenols, Known as Antioxidants and Anti-Aging Compounds

10.1. Resveratrol

Resveratrol, or 3,4′,5-trihydroxystilbene, is a natural polyphenol present in large amounts in Japanese knotweed (*Polygonum cuspidatum*) root. This phytoalexin is produced by a wide variety of plants, some of which are edible for humans, such as grapes, blackberries, blackcurrants, blueberries, and cranberries, to name but a few [224]. However, in the last two decades, the effect of resveratrol on animal models related to several disorders, such as autism spectrum disorder, mitochondrial myopathies, type 2 diabetic nephropathy, or renal lipotoxicity has been increasingly reported.

The effect of resveratrol in the presence of quercetin has been studied on PPARα-mediating uncoupling protein regulation in visceral white adipose tissue from metabolic syndrome rats. Resveratrol treatment leads to a significantly increased expression of both *Ppara* mRNA and protein levels [225]. Remarkably, resveratrol prevents renal lipotoxicity in a high-fat diet-treated mouse model by regulating the PPARα pathway, enhancing the expression of lipolytic genes, and raising the renal PPARα protein level and AMPK phosphorylation level [226]. Due to known dyslipidemia in autism spectrum disorders, PPARs have been proposed as therapeutic targets of resveratrol. Furthermore, in autism, impaired mitochondrial fatty acid oxidation suggests the potential implications for regulating mitochondrial oxidation flux by PPAR activators, especially resveratrol [227].

Numerous natural ligands, including polyphenolic compounds, control the expression of PPAR receptors [228]. They have several health-promoting properties, including antioxidant, anti-inflammatory, and antineoplastic activities. Resveratrol is an active biological modulator of several signaling proteins, including PPARα. Resveratrol activates the AMPK-SIRT1-PGC-1α axis and PPARα via the adiponectin receptors in the renal cortex [187]. Adiponectin has multiple functions, including insulin sensitization and lipid metabolism regulation. Similarly, in mitochondrial myopathy, resveratrol has been shown to potentially target many mitochondrial metabolic pathways comprising fatty acid β-oxidation and oxidative phosphorylation, leading to the up-regulation of the energy supply via AMPkinase-SIRT1-PGC-1α signaling pathways [229].

10.2. Quercetin

Quercetin (2-(3,4-dihydroxyphenyl)-3,5,7-trihydroxy-4H-chromen-4-one) is a flavonoid polyphenol found in plants and a variety of other natural sources—red grape, onion, broccoli, tomatoes and lettuce [224]. PPARα is significantly upregulated and enhances β-oxidation by mulberry-leaf powder containing quercetin [230]. Quercetin-3-*O*-β-D-glucuronide (Q3GA) ameliorates dyslipidemia in fatty livers by modulating the PPARα/sterol regulatory element-binding protein-1c (SREBP-1c) signaling. Q3GA reduced lipogenesis through downregulation of SREBP-1c and fatty acid synthase levels, and raised lipolysis and fatty acid oxidation by increasing the expression of PPARα, carnitine palmitoyl-transferase1 and medium-chain acyl-coenzyme A dehydrogenase, both in vivo and in vitro [231].

10.3. EGCG (Epigallocatechin-3-Gallate)

Epigallocatechin-3-gallate (EGCG) is catechin conjugated with gallic acid. It belongs to the flavonol class and is found abundantly in green tea [232] and cocoa, which have the highest content of catechins, followed by prune juice, broad bean pods, and argan oil [224].

EGCG and green tea polyphenol extract display crosstalk with PPARα. Reported studies in cancer cells revealed that EGCG induced the expression level of PPARα protein in a dose-dependent manner. Clofibrate, a PPARα agonist, blocks heme oxygenase-1 (HO-1) induction and sensitizes cancer cells to EGCG-promoted cell death. Moreover, PPARα interacts with the PPRE of the HO-1 promoter. The activation of PPARα sensitizes cancer cells to epigallocatechin-3-gallate (EGCG) treatment by suppressing HO-1 expression [233].

In rats, green tea polyphenols reduce the renal oxidative stress induced by a high-fat diet through deacetylation of SIRT3 mediated by PPARα upregulation [234].

10.4. Curcumin

Curcumin (1,7-bis-(4-hydroxy-3-methoxyphenyl)-hepta-1,6-diene-3,5-dione) belongs to a chemical class of polyphenols that is extracted from the rhizomes of the turmeric plant (*Curcuma longa*) [224]. Tetrahydro-curcumin improves oleic acid-induced hepatic steatosis and ameliorates insulin resistance in HepG2 cells, likely through downregulation of the expression of the lipogenic proteins, SREBP-1c and PPARγ, and the stimulation of lipolysis by upregulating PPARα and CPT-1a, which are involved in fatty acid β-oxidation [235].

10.5. Anthocyanins

Among berries, blueberries contain higher amounts of anthocyanins. These polyphenols are known to exhibit hypolipidemic properties. Rimando et al. reported that both anthocyanins and catechins do not activate PPARα, while pterostilbene revealed the dose-dependent activation of PPARα in H4IIEC3 hepatocytes [236]. In addition, pterostilbene showed a significant increase in *Ppara* gene expression, but at a lower extent than fenofibrate [236]. Although pterostilbene and resveratrol, as PPARα activators, are under the threshold for effective concentrations in blueberry extract, hepatic mRNA *Ppara* expression has increased in hamsters fed on a diet containing blueberry extract [236].

10.6. Coffee

Coffee consumption has been shown to upregulate mouse hepatic PPARα expression and its target-gene *Acox1*, consequently leading to the induction of liver peroxisomal fatty acid β-oxidation. Such FAO induction, with induced intestinal cholesterol efflux and reduced lipid digestion, prevents the high-fat diet-induced fatty liver through the lipid-sensing modulation of the gut–liver axis [237].

10.7. Edible Oil Products

The effect of polyphenols has been investigated in a rat model of bowel disease by 3 months diet supplementation with extra-virgin olive oil with a high or low phenolic content [238]. The presence of polyphenols in olive oil significantly attenuates the intestinal inflammation associated with hypocholesterolemia and the induction of PPAR-α gene expression in the liver [238]. In a model of insulin resistance of rats fed a high-fat diet, the administration of the major metabolite of oleuropein, hydroxytyrosol, increases the hepatic mRNA levels of *Ppara* and its target genes, i.e., fibroblast growth factor 21 and carnitine palmitoyltransferase 1a [239]. Similarly, mice receiving a high-fat diet develop hepatic steatosis and inflammation, which were attenuated by hydroxytyrosol supplementation through PPARα activation, Nrf2 (nuclear factor, erythroid 2 like 2) mediated-antioxidative pathway, and by the downregulation of NF-κB-associated inflammation [240]. Used as food supplementation, argan oil or olive oil was shown to restore the expression of genes involved in liver mitochondrial and peroxisomal fatty acid β-oxidation and gluconeogenesis in the mice sepsis model when injected with lipopolysaccharides. This preventive effect of argan oil likely involves the hepatic upregulation of PPARα, PGC-1α, and the estrogen-related receptor α [241].

Likewise, ginsenoside Rb3 micronutrients, derived from ginseng, or nuciferine, found in *Nelumbo nucifera* leaves, was shown to activate the PPARα pathway by regulating energy metabolism in cardiomyocytes [242], or hepatic steatosis diabetic streptozocin-induced mice fed a high-fat diet [243], while bilobetin, a biflavonoid, modulates PPARα activity by PKA-dependent phosphorylation. Finally, berberine, an alkaloid, binds PPARα LBD with a hypolipidemic effect and a comparable affinity to fenofibrate [244].

11. Conclusions and Future Directions

In all these tested situations, irrespective of the tissue, animal, or pathological condition, micronutrients appear to have an advantageous effect on *Ppara* expression and activity. Furthermore, almost all these compounds are potent antioxidants and can activate signaling pathways via PGC1-α and AMP kinase. Numerous natural products might modulate PPARα, including terpenes, polyketides, phenylpropanoids, polyphenols, and alkaloids; for instance, the linalool effect is ten times less compared to fenofibrate [88], demonstrating the potential beneficial effects of dietary micro-components to modulate PPARα functions desirably in a population with an ever-increasing high-fat diet consumption. The question is the dietary relevance of these effects, since most of the data were obtained from in vitro studies, and secondly, these micronutrients are often present at very low doses in the diet, except for some polyphenols.

Despite tremendous signs of progress on the critical role of PPARα-dependent regulation in lipid metabolism, the characterization of peroxisomal enzymes and transporters, there are still gaps that need to be filled to fully define the exact role and regulation of PPARα and peroxisomal fatty acid β-oxidation in the cellular homeostasis of lipid metabolism. Particular attention needs to be focused on:

1. The shuttling of substrates and cofactors from and into peroxisome.
2. What is the exact role of peroxisomal β-oxidation in lipid metabolism and cell signaling?
3. How can peroxisome be a mediator and responder of metabolic and environmental stresses?
4. What are the molecular events that are required at the metabolic level?
    (a) Does heterodimerization of PPAR/RXR control the regulation? Is it controlled by coregulators?
    (b) What is the nature of ligands?
    (c) What is the nature of micronutrients? Are they natural agonists or antagonists or their balance?
    (d) Is PPARα the only nuclear receptor governing peroxisomal β-oxidation-related genes?
    (e) How do coregulators play in concert to fine-tune metabolically peroxisomal β-oxidation pathway?

All these as yet unanswered questions deserve our complete focus in the near future.

There is an increasing demand from health institutions and pharmaceutical industries for efficient drugs. PPARα binding pocket-ligand interactions are being increasingly recognized as a source for therapeutic interventions. Bio structural analysis based on X-ray crystallography and ligand structure pharmacophore modeling approaches afford new biophysical and structural parameters that are important in designing and developing novel potent and highly PPARα-specific ligands to preserve human health and safety. However, the overall goal of increasing the peroxisomal fatty acid oxidation and β-oxidation safely, without increasing the lipid peroxidation and free radical-based risk of non-genotoxic carcinogenesis in the high-fat Western diet-fed population, is a challenge that is still unmet and requires continuous exploration of avenues to activate PPARα dependent pathways safely.

Author Contributions: Conceptualization, N.L., M.C.-M. Investigation: M.T.-J., P.A., S.S., M.C.-M., N.L. Writing—original draft: M.T.-J., P.A., S.S., M.C.-M., N.L. Formal analysis: M.T.-J., P.A., S.S., M.C.-M., N.L., B.N. All authors have read and agreed to the published version of the manuscript.

Funding: This research was funded by the Ministère de l'Enseignement et de la Recherche and the CNRST (Mounia Tahri-Joutey, PhD excellence grant number: 17UHP2019, Morocco) and by the Action Intégrée of the Comité Mixte Inter-universitaire Franco-Marocain (n° TBK 19/92 n° Campus France: 41501RJ) from the PHC Toubkal program, Ministère des Affaires Étrangères.

Institutional Review Board Statement: Not applicable.

Data Availability Statement: Not applicable.

Acknowledgments: The authors would like to acknowledge networking support by the COST Action CA 16,112 NutRedOx (Personalized Nutrition in aging society: redox control of major age-related diseases), supported by COST (European Cooperation in Science and Technology). We thank Nathalie Bancod for her helpful contribution in figure conception.

Conflicts of Interest: The authors declare no conflict of interest.

References

1. Latruffe, N.; Vamecq, J. Evolutionary Aspects of Peroxisomes as Cell Organelles, and of Genes Encoding Peroxisomal Proteins. *Biol. Cell* **2000**, *92*, 389–395. [CrossRef]
2. Hess, R.; Stäubli, W.; Riess, W. Nature of the Hepatomegalic Effect Produced by Ethyl-Chlorophenoxy-Isobutyrate in the Rat. *Nature* **1965**, *208*, 856–858. [CrossRef]
3. Lalwani, N.D.; Reddy, M.K.; Qureshi, S.A.; Sirtori, C.R.; Abiko, Y.; Reddy, J.K. Evaluation of Selected Hypolipidemic Agents for the Induction of Peroxisomal Enzymes and Peroxisome Proliferation in the Rat Liver. *Hum. Toxicol.* **1983**, *2*, 27–48. [CrossRef] [PubMed]
4. Issemann, I.; Green, S. Activation of a Member of the Steroid Hormone Receptor Superfamily by Peroxisome Proliferators. *Nature* **1990**, *347*, 645–650. [CrossRef]
5. Zhou, T.; Yan, X.; Wang, G.; Liu, H.; Gan, X.; Zhang, T.; Wang, J.; Li, L. Evolutionary Pattern and Regulation Analysis to Support Why Diversity Functions Existed within PPAR Gene Family Members. *BioMed Res. Int.* **2015**, *2015*, e613910. [CrossRef] [PubMed]
6. Wanders, R.J.; Waterham, H.R.; Ferdinandusse, S. Metabolic Interplay between Peroxisomes and Other Subcellular Organelles Including Mitochondria and the Endoplasmic Reticulum. *Front. Cell Dev. Biol.* **2015**, *3*, 83. [CrossRef] [PubMed]
7. Lazarow, P.B.; De Duve, C. A Fatty Acyl-CoA Oxidizing System in Rat Liver Peroxisomes; Enhancement by Clofibrate, a Hypolipidemic Drug. *Proc. Natl. Acad. Sci. USA* **1976**, *73*, 2043–2046. [CrossRef] [PubMed]
8. Cherkaoui-Malki, M.; Surapureddi, S.; El-Hajj, H.I.; Vamecq, J.; Andreoletti, P. Hepatic Steatosis and Peroxisomal Fatty Acid Beta-Oxidation. *Curr. Drug Metab.* **2012**, *13*, 1412–1421. [CrossRef] [PubMed]
9. Andreoletti, P.; Raas, Q.; Gondcaille, C.; Cherkaoui-Malki, M.; Trompier, D.; Savary, S. Predictive Structure and Topology of Peroxisomal ATP-Binding Cassette (ABC) Transporters. *Int. J. Mol. Sci.* **2017**, *18*, 1593. [CrossRef]
10. Watkins, P.A.; Ellis, J.M. Peroxisomal Acyl-CoA Synthetases. *Biochim. Biophys. Acta* **2012**, *1822*, 1411–1420. [CrossRef] [PubMed]
11. Caira, F.; Clémencet, M.C.; Cherkaoui-Malki, M.; Dieuaide-Noubhani, M.; Pacot, C.; Van Veldhoven, P.P.; Latruffe, N. Differential Regulation by a Peroxisome Proliferator of the Different Multifunctional Proteins in Guinea Pig: CDNA Cloning of the Guinea Pig D-Specific Multifunctional Protein 2. *Biochem. J.* **1998**, *330 Pt 3*, 1361–1368. [CrossRef]
12. Osumi, T.; Ishii, N.; Hijikata, M.; Kamijo, K.; Ozasa, H.; Furuta, S.; Miyazawa, S.; Kondo, K.; Inoue, K.; Kagamiyama, H.; et al. Molecular Cloning and Nucleotide Sequence of the CDNA for Rat Peroxisomal Enoyl-CoA: Hydratase-3-Hydroxyacyl-CoA Dehydrogenase Bifunctional Enzyme. *J. Biol. Chem.* **1985**, *260*, 8905–8910. [CrossRef]
13. Latruffe, N. Human Peroxisomal 3-Ketoacyl-CoA Thiolase: Tissue Expression and Metabolic Regulation: Human Peroxisomal Thiolase. *Adv. Exp. Med. Biol.* **2020**, *1299*, 161–167. [CrossRef] [PubMed]
14. Baes, M.; Van Veldhoven, P.P. Hepatic Dysfunction in Peroxisomal Disorders. *Biochim. Biophys. Acta* **2016**, *1863*, 956–970. [CrossRef] [PubMed]
15. Van Veldhoven, P.P.; De Schryver, E.; Young, S.G.; Zwijsen, A.; Fransen, M.; Espeel, M.; Baes, M.; Van Ael, E. Slc25a17 Gene Trapped Mice: PMP34 Plays a Role in the Peroxisomal Degradation of Phytanic and Pristanic Acid. *Front. Cell Dev. Biol.* **2020**, *8*, 144. [CrossRef] [PubMed]
16. Seedorf, U.; Brysch, P.; Engel, T.; Schrage, K.; Assmann, G. Sterol Carrier Protein X Is Peroxisomal 3-Oxoacyl Coenzyme A Thiolase with Intrinsic Sterol Carrier and Lipid Transfer Activity. *J. Biol. Chem.* **1994**, *269*, 21277–21283. [CrossRef]
17. Ranea-Robles, P.; Violante, S.; Argmann, C.; Dodatko, T.; Bhattacharya, D.; Chen, H.; Yu, C.; Friedman, S.L.; Puchowicz, M.; Houten, S.M. Murine Deficiency of Peroxisomal L-Bifunctional Protein (EHHADH) Causes Medium-Chain 3-Hydroxydicarboxylic Aciduria and Perturbs Hepatic Cholesterol Homeostasis. *Cell. Mol. Life Sci.* **2021**. [CrossRef]
18. Wang, H.; Lu, J.; Chen, X.; Schwalbe, M.; Gorka, J.E.; Mandel, J.A.; Wang, J.; Goetzman, E.S.; Ranganathan, S.; Dobrowolski, S.F.; et al. Acquired Deficiency of Peroxisomal Dicarboxylic Acid Catabolism Is a Metabolic Vulnerability in Hepatoblastoma. *J. Biol. Chem.* **2021**, 100283. [CrossRef]
19. Tillander, V.; Alexson, S.E.H.; Cohen, D.E. Deactivating Fatty Acids: Acyl-CoA Thioesterase-Mediated Control of Lipid Metabolism. *Trends Endocrinol. Metab.* **2017**, *28*, 473–484. [CrossRef] [PubMed]
20. Bowen, K.J.; Kris-Etherton, P.M.; Shearer, G.C.; West, S.G.; Reddivari, L.; Jones, P.J.H. Oleic Acid-Derived Oleoylethanolamide: A Nutritional Science Perspective. *Prog. Lipid Res.* **2017**, *67*, 1–15. [CrossRef] [PubMed]
21. Desvergne, B.; Michalik, L.; Wahli, W. Transcriptional Regulation of Metabolism. *Physiol. Rev.* **2006**, *86*, 465–514. [CrossRef]
22. Varga, T.; Czimmerer, Z.; Nagy, L. PPARs Are a Unique Set of Fatty Acid Regulated Transcription Factors Controlling Both Lipid Metabolism and Inflammation. *Biochim. Biophys. Acta* **2011**, *1812*, 1007–1022. [CrossRef]
23. Mandard, S.; Müller, M.; Kersten, S. Peroxisome Proliferator-Activated Receptor Alpha Target Genes. *Cell. Mol. Life Sci.* **2004**, *61*, 393–416. [CrossRef] [PubMed]

24. More, V.R.; Campos, C.R.; Evans, R.A.; Oliver, K.D.; Chan, G.N.; Miller, D.S.; Cannon, R.E. PPAR-α, a Lipid-Sensing Transcription Factor, Regulates Blood-Brain Barrier Efflux Transporter Expression. *J. Cereb. Blood Flow Metab.* **2017**, *37*, 1199–1212. [CrossRef] [PubMed]
25. Ning, L.-J.; He, A.-Y.; Lu, D.-L.; Li, J.-M.; Qiao, F.; Li, D.-L.; Zhang, M.-L.; Chen, L.-Q.; Du, Z.-Y. Nutritional Background Changes the Hypolipidemic Effects of Fenofibrate in Nile Tilapia (*Oreochromis Niloticus*). *Sci. Rep.* **2017**, *7*, 41706. [CrossRef] [PubMed]
26. Martin, G.; Schoonjans, K.; Lefebvre, A.M.; Staels, B.; Auwerx, J. Coordinate Regulation of the Expression of the Fatty Acid Transport Protein and Acyl-CoA Synthetase Genes by PPARalpha and PPARgamma Activators. *J. Biol. Chem.* **1997**, *272*, 28210–28217. [CrossRef]
27. Motojima, K.; Passilly, P.; Peters, J.M.; Gonzalez, F.J.; Latruffe, N. Expression of Putative Fatty Acid Transporter Genes Are Regulated by Peroxisome Proliferator-Activated Receptor Alpha and Gamma Activators in a Tissue- and Inducer-Specific Manner. *J. Biol. Chem.* **1998**, *273*, 16710–16714. [CrossRef] [PubMed]
28. Han, L.; Shen, W.-J.; Bittner, S.; Kraemer, F.B.; Azhar, S. PPARs: Regulators of Metabolism and as Therapeutic Targets in Cardiovascular Disease. Part I: PPAR-α. *Future Cardiol.* **2017**, *13*, 259–278. [CrossRef]
29. Kandel, B.A.; Thomas, M.; Winter, S.; Damm, G.; Seehofer, D.; Burk, O.; Schwab, M.; Zanger, U.M. Genomewide Comparison of the Inducible Transcriptomes of Nuclear Receptors CAR, PXR and PPARα in Primary Human Hepatocytes. *Biochim. Biophys. Acta* **2016**, *1859*, 1218–1227. [CrossRef]
30. Lefebvre, P.; Benomar, Y.; Staels, B. Retinoid X Receptors: Common Heterodimerization Partners with Distinct Functions. *Trends Endocrinol. Metab.* **2010**, *21*, 676–683. [CrossRef]
31. Wójtowicz, S.; Strosznajder, A.K.; Jeżyna, M.; Strosznajder, J.B. The Novel Role of PPAR Alpha in the Brain: Promising Target in Therapy of Alzheimer's Disease and Other Neurodegenerative Disorders. *Neurochem. Res.* **2020**, *45*, 972. [CrossRef]
32. Corrales, P.; Vidal-Puig, A.; Medina-Gómez, G. PPARs and Metabolic Disorders Associated with Challenged Adipose Tissue Plasticity. *Int. J. Mol. Sci.* **2018**, *19*, 2124. [CrossRef]
33. Dreyer, C.; Krey, G.; Keller, H.; Givel, F.; Helftenbein, G.; Wahli, W. Control of the Peroxisomal Beta-Oxidation Pathway by a Novel Family of Nuclear Hormone Receptors. *Cell* **1992**, *68*, 879–887. [CrossRef]
34. Feige, J.N.; Gelman, L.; Tudor, C.; Engelborghs, Y.; Wahli, W.; Desvergne, B. Fluorescence Imaging Reveals the Nuclear Behavior of Peroxisome Proliferator-Activated Receptor/Retinoid X Receptor Heterodimers in the Absence and Presence of Ligand. *J. Biol. Chem.* **2005**, *280*, 17880–17890. [CrossRef]
35. Kliewer, S.A.; Umesono, K.; Noonan, D.J.; Heyman, R.A.; Evans, R.M. Convergence of 9-Cis Retinoic Acid and Peroxisome Proliferator Signalling Pathways through Heterodimer Formation of Their Receptors. *Nature* **1992**, *358*, 771–774. [CrossRef] [PubMed]
36. Tugwood, J.D.; Issemann, I.; Anderson, R.G.; Bundell, K.R.; McPheat, W.L.; Green, S. The Mouse Peroxisome Proliferator Activated Receptor Recognizes a Response Element in the 5′ Flanking Sequence of the Rat Acyl CoA Oxidase Gene. *EMBO J.* **1992**, *11*, 433–439. [CrossRef] [PubMed]
37. Tzeng, J.; Byun, J.; Park, J.Y.; Yamamoto, T.; Schesing, K.; Tian, B.; Sadoshima, J.; Oka, S. An Ideal PPAR Response Element Bound to and Activated by PPARα. *PLoS ONE* **2015**, *10*, e0134996. [CrossRef]
38. Fornes, O.; Castro-Mondragon, J.A.; Khan, A.; Van Der Lee, R.; Zhang, X.; Richmond, P.A.; Modi, B.P.; Correard, S.; Gheorghe, M.; Baranašić, D.; et al. JASPAR 2020: Update of the Open-Access Database of Transcription Factor Binding Profiles. *Nucleic Acids Res.* **2020**, *48*, D87–D92. [CrossRef] [PubMed]
39. Koch, J.; Pranjic, K.; Huber, A.; Ellinger, A.; Hartig, A.; Kragler, F.; Brocard, C. PEX11 Family Members Are Membrane Elongation Factors That Coordinate Peroxisome Proliferation and Maintenance. *J. Cell Sci.* **2010**, *123*, 3389–3400. [CrossRef]
40. Hansmannel, F.; Clémencet, M.-C.; Le Jossic-Corcos, C.; Osumi, T.; Latruffe, N.; Nicolas-Francés, V. Functional Characterization of a Peroxisome Proliferator Response-Element Located in the Intron 3 of Rat Peroxisomal Thiolase B Gene. *Biochem. Biophys. Res. Commun.* **2003**, *311*, 149–155. [CrossRef] [PubMed]
41. Woodyatt, N.J.; Lambe, K.G.; Myers, K.A.; Tugwood, J.D.; Roberts, R.A. The Peroxisome Proliferator (PP) Response Element Upstream of the Human Acyl CoA Oxidase Gene Is Inactive among a Sample Human Population: Significance for Species Differences in Response to PPs. *Carcinogenesis* **1999**, *20*, 369–372. [CrossRef]
42. Ashibe, B.; Motojima, K. Fatty Aldehyde Dehydrogenase Is Up-Regulated by Polyunsaturated Fatty Acid via Peroxisome Proliferator-Activated Receptor Alpha and Suppresses Polyunsaturated Fatty Acid-Induced Endoplasmic Reticulum Stress. *FEBS J.* **2009**, *276*, 6956–6970. [CrossRef]
43. Girnun, G.D.; Domann, F.E.; Moore, S.A.; Robbins, M.E.C. Identification of a Functional Peroxisome Proliferator-Activated Receptor Response Element in the Rat Catalase Promoter. *Mol. Endocrinol.* **2002**, *16*, 2793–2801. [CrossRef] [PubMed]
44. Bardot, O.; Aldridge, T.C.; Latruffe, N.; Green, S. PPAR-RXR Heterodimer Activates a Peroxisome Proliferator Response Element Upstream of the Bifunctional Enzyme Gene. *Biochem. Biophys. Res. Commun.* **1993**, *192*, 37–45. [CrossRef]
45. Lee, G.Y.; Kim, N.H.; Zhao, Z.-S.; Cha, B.S.; Kim, Y.S. Peroxisomal-Proliferator-Activated Receptor Alpha Activates Transcription of the Rat Hepatic Malonyl-CoA Decarboxylase Gene: A Key Regulation of Malonyl-CoA Level. *Biochem. J.* **2004**, *378*, 983–990. [CrossRef]
46. Shimizu, M.; Yamashita, D.; Yamaguchi, T.; Hirose, F.; Osumi, T. Aspects of the Regulatory Mechanisms of PPAR Functions: Analysis of a Bidirectional Response Element and Regulation by Sumoylation. *Mol. Cell. Biochem.* **2006**, *286*, 33–42. [CrossRef]

47. Lopez, D.; Irby, R.B.; McLean, M.P. Peroxisome Proliferator-Activated Receptor Alpha Induces Rat Sterol Carrier Protein x Promoter Activity through Two Peroxisome Proliferator-Response Elements. *Mol. Cell. Endocrinol.* **2003**, *205*, 169–184. [CrossRef]
48. Evans, R.M.; Barish, G.D.; Wang, Y.-X. PPARs and the Complex Journey to Obesity. *Nat. Med.* **2004**, *10*, 355–361. [CrossRef] [PubMed]
49. Fan, W.; Evans, R. PPARs and ERRs: Molecular Mediators of Mitochondrial Metabolism. *Curr. Opin. Cell Biol.* **2015**, *33*, 49–54. [CrossRef] [PubMed]
50. Green, S.; Wahli, W. Peroxisome Proliferator-Activated Receptors: Finding the Orphan a Home. *Mol. Cell. Endocrinol.* **1994**, *100*, 149–153. [CrossRef]
51. Zhu, Y.; Qi, C.; Korenberg, J.R.; Chen, X.N.; Noya, D.; Rao, M.S.; Reddy, J.K. Structural Organization of Mouse Peroxisome Proliferator-Activated Receptor Gamma (MPPAR Gamma) Gene: Alternative Promoter Use and Different Splicing Yield Two MPPAR Gamma Isoforms. *Proc. Natl. Acad. Sci. USA* **1995**, *92*, 7921–7925. [CrossRef] [PubMed]
52. Göttlicher, M.; Widmark, E.; Li, Q.; Gustafsson, J.A. Fatty Acids Activate a Chimera of the Clofibric Acid-Activated Receptor and the Glucocorticoid Receptor. *Proc. Natl. Acad. Sci. USA* **1992**, *89*, 4653–4657. [CrossRef] [PubMed]
53. Sher, T.; Yi, H.F.; McBride, O.W.; Gonzalez, F.J. CDNA Cloning, Chromosomal Mapping, and Functional Characterization of the Human Peroxisome Proliferator Activated Receptor. *Biochemistry* **1993**, *32*, 5598–5604. [CrossRef] [PubMed]
54. Vamecq, J.; Latruffe, N. Medical Significance of Peroxisome Proliferator-Activated Receptors. *Lancet* **1999**, *354*, 141–148. [CrossRef]
55. Brown, J.D.; Plutzky, J. Peroxisome Proliferator-Activated Receptors as Transcriptional Nodal Points and Therapeutic Targets. *Circulation* **2007**, *115*, 518–533. [CrossRef] [PubMed]
56. Hong, F.; Pan, S.; Guo, Y.; Xu, P.; Zhai, Y. PPARs as Nuclear Receptors for Nutrient and Energy Metabolism. *Molecules* **2019**, *24*, 2545. [CrossRef]
57. Lamichane, S.; Dahal Lamichane, B.; Kwon, S.-M. Pivotal Roles of Peroxisome Proliferator-Activated Receptors (PPARs) and Their Signal Cascade for Cellular and Whole-Body Energy Homeostasis. *Int. J. Mol. Sci.* **2018**, *19*, 949. [CrossRef] [PubMed]
58. Moore, J.T.; Collins, J.L.; Pearce, K.H. The Nuclear Receptor Superfamily and Drug Discovery. *ChemMedChem* **2006**, *1*, 504–523. [CrossRef] [PubMed]
59. Floyd, Z.E.; Stephens, J.M. Controlling a Master Switch of Adipocyte Development and Insulin Sensitivity: Covalent Modifications of PPARγ. *Biochim. Biophys. Acta* **2012**, *1822*, 1090–1095. [CrossRef]
60. Wadosky, K.M.; Willis, M.S. The Story so Far: Post-Translational Regulation of Peroxisome Proliferator-Activated Receptors by Ubiquitination and SUMOylation. *Am. J. Physiol. Heart Circ. Physiol.* **2012**, *302*, H515–H526. [CrossRef] [PubMed]
61. Kim, T.-H.; Kim, M.-Y.; Jo, S.-H.; Park, J.-M.; Ahn, Y.-H. Modulation of the Transcriptional Activity of Peroxisome Proliferator-Activated Receptor Gamma by Protein-Protein Interactions and Post-Translational Modifications. *Yonsei Med. J.* **2013**, *54*, 545–559. [CrossRef] [PubMed]
62. Tufano, M.; Pinna, G. Is There a Future for PPARs in the Treatment of Neuropsychiatric Disorders? *Molecules* **2020**, *25*, 1062. [CrossRef]
63. Bougarne, N.; Weyers, B.; Desmet, S.J.; Deckers, J.; Ray, D.W.; Staels, B.; De Bosscher, K. Molecular Actions of PPARα in Lipid Metabolism and Inflammation. *Endocr. Rev.* **2018**, *39*, 760–802. [CrossRef] [PubMed]
64. Pawlak, M.; Lefebvre, P.; Staels, B. General Molecular Biology and Architecture of Nuclear Receptors. *Curr. Top. Med. Chem.* **2012**, *12*, 486–504. [CrossRef] [PubMed]
65. Lamas Bervejillo, M.; Ferreira, A.M. Understanding Peroxisome Proliferator-Activated Receptors: From the Structure to the Regulatory Actions on Metabolism. *Adv. Exp. Med. Biol.* **2019**, *1127*, 39–57. [CrossRef] [PubMed]
66. Oyama, T.; Toyota, K.; Waku, T.; Hirakawa, Y.; Nagasawa, N.; Kasuga, J.I.; Hashimoto, Y.; Miyachi, H.; Morikawa, K. Adaptability and Selectivity of Human Peroxisome Proliferator-Activated Receptor (PPAR) Pan Agonists Revealed from Crystal Structures. *Acta Crystallogr. D Biol. Crystallogr.* **2009**, *65*, 786–795. [CrossRef] [PubMed]
67. Kawasaki, M.; Kambe, A.; Yamamoto, Y.; Arulmozhiraja, S.; Ito, S.; Nakagawa, Y.; Tokiwa, H.; Nakano, S.; Shimano, H. Elucidation of Molecular Mechanism of a Selective PPARα Modulator, Pemafibrate, through Combinational Approaches of X-Ray Crystallography, Thermodynamic Analysis, and First-Principle Calculations. *Int. J. Mol. Sci.* **2020**, *21*, 361. [CrossRef] [PubMed]
68. Xu, H.E.; Lambert, M.H.; Montana, V.G.; Plunket, K.D.; Moore, L.B.; Collins, J.L.; Oplinger, J.A.; Kliewer, S.A.; Gampe, R.T.; McKee, D.D.; et al. Structural Determinants of Ligand Binding Selectivity between the Peroxisome Proliferator-Activated Receptors. *Proc. Natl. Acad. Sci. USA* **2001**, *98*, 13919–13924. [CrossRef] [PubMed]
69. Kamata, S.; Oyama, T.; Saito, K.; Honda, A.; Yamamoto, Y.; Suda, K.; Ishikawa, R.; Itoh, T.; Watanabe, Y.; Shibata, T.; et al. PPARα Ligand-Binding Domain Structures with Endogenous Fatty Acids and Fibrates. *iScience* **2020**, *23*, 101727. [CrossRef]
70. Forman, B.M.; Chen, J.; Evans, R.M. Hypolipidemic Drugs, Polyunsaturated Fatty Acids, and Eicosanoids Are Ligands for Peroxisome Proliferator-Activated Receptors Alpha and Delta. *Proc. Natl. Acad. Sci. USA* **1997**, *94*, 4312–4317. [CrossRef]
71. Kliewer, S.A.; Sundseth, S.S.; Jones, S.A.; Brown, P.J.; Wisely, G.B.; Koble, C.S.; Devchand, P.; Wahli, W.; Willson, T.M.; Lenhard, J.M.; et al. Fatty Acids and Eicosanoids Regulate Gene Expression through Direct Interactions with Peroxisome Proliferator-Activated Receptors Alpha and Gamma. *Proc. Natl. Acad. Sci. USA* **1997**, *94*, 4318–4323. [CrossRef]
72. Takada, I.; Makishima, M. Peroxisome Proliferator-Activated Receptor Agonists and Antagonists: A Patent Review (2014-Present). *Expert Opin. Ther. Pat.* **2020**, *30*, 1–13. [CrossRef]

73. Elholm, M.; Dam, I.; Jorgensen, C.; Krogsdam, A.M.; Holst, D.; Kratchmarova, I.; Gottlicher, M.; Gustafsson, J.A.; Berge, R.; Flatmark, T.; et al. Acyl-CoA Esters Antagonize the Effects of Ligands on Peroxisome Proliferator-Activated Receptor Alpha Conformation, DNA Binding, and Interaction with Co-Factors. *J. Biol. Chem.* **2001**, *276*, 21410–21416. [CrossRef] [PubMed]
74. Hostetler, H.A.; Kier, A.B.; Schroeder, F. Very-Long-Chain and Branched-Chain Fatty Acyl-CoAs Are High Affinity Ligands for the Peroxisome Proliferator-Activated Receptor Alpha (PPARalpha). *Biochemistry* **2006**, *45*, 7669–7681. [CrossRef]
75. Chakravarthy, M.V.; Lodhi, I.J.; Yin, L.; Malapaka, R.R.V.; Xu, H.E.; Turk, J.; Semenkovich, C.F. Identification of a Physiologically Relevant Endogenous Ligand for PPARalpha in Liver. *Cell* **2009**, *138*, 476–488. [CrossRef]
76. Brown, J.D.; Karimian Azari, E.; Ayala, J.E. Oleoylethanolamide: A Fat Ally in the Fight against Obesity. *Physiol. Behav.* **2017**, *176*, 50–58. [CrossRef] [PubMed]
77. Campolongo, P.; Roozendaal, B.; Trezza, V.; Cuomo, V.; Astarita, G.; Fu, J.; McGaugh, J.L.; Piomelli, D. Fat-Induced Satiety Factor Oleoylethanolamide Enhances Memory Consolidation. *Proc. Natl. Acad. Sci. USA* **2009**, *106*, 8027–8031. [CrossRef]
78. Azhar, S. Peroxisome Proliferator-Activated Receptors, Metabolic Syndrome and Cardiovascular Disease. *Future Cardiol.* **2010**, *6*, 657–691. [CrossRef] [PubMed]
79. Rigano, D.; Sirignano, C.; Taglialatela-Scafati, O. The Potential of Natural Products for Targeting PPARα. *Acta Pharm. Sin. B* **2017**, *7*, 427–438. [CrossRef] [PubMed]
80. Green, S. PPAR: A Mediator of Peroxisome Proliferator Action. *Mutat. Res.* **1995**, *333*, 101–109. [CrossRef]
81. Yu, K.; Bayona, W.; Kallen, C.B.; Harding, H.P.; Ravera, C.P.; McMahon, G.; Brown, M.; Lazar, M.A. Differential Activation of Peroxisome Proliferator-Activated Receptors by Eicosanoids. *J. Biol. Chem.* **1995**, *270*, 23975–23983. [CrossRef]
82. Goto, T.; Takahashi, N.; Kato, S.; Egawa, K.; Ebisu, S.; Moriyama, T.; Fushiki, T.; Kawada, T. Phytol Directly Activates Peroxisome Proliferator-Activated Receptor Alpha (PPARalpha) and Regulates Gene Expression Involved in Lipid Metabolism in PPARalpha-Expressing HepG2 Hepatocytes. *Biochem. Biophys. Res. Commun.* **2005**, *337*, 440–445. [CrossRef]
83. Wahli, W.; Michalik, L. PPARs at the Crossroads of Lipid Signaling and Inflammation. *Trends Endocrinol. Metab.* **2012**, *23*, 351–363. [CrossRef]
84. Narala, V.R.; Adapala, R.K.; Suresh, M.V.; Brock, T.G.; Peters-Golden, M.; Reddy, R.C. Leukotriene B4 Is a Physiologically Relevant Endogenous Peroxisome Proliferator-Activated Receptor-Alpha Agonist. *J. Biol. Chem.* **2010**, *285*, 22067–22074. [CrossRef]
85. Lin, Q.; Ruuska, S.E.; Shaw, N.S.; Dong, D.; Noy, N. Ligand Selectivity of the Peroxisome Proliferator-Activated Receptor Alpha. *Biochemistry* **1999**, *38*, 185–190. [CrossRef]
86. Delerive, P.; Furman, C.; Teissier, E.; Fruchart, J.; Duriez, P.; Staels, B. Oxidized Phospholipids Activate PPARalpha in a Phospholipase A2-Dependent Manner. *FEBS Lett.* **2000**, *471*, 34–38. [CrossRef]
87. Roy, A.; Kundu, M.; Jana, M.; Mishra, R.K.; Yung, Y.; Luan, C.-H.; Gonzalez, F.J.; Pahan, K. Identification and Characterization of PPARα Ligands in the Hippocampus. *Nat. Chem. Biol.* **2016**, *12*, 1075–1083. [CrossRef] [PubMed]
88. Bernardes, A.; Souza, P.C.T.; Muniz, J.R.C.; Ricci, C.G.; Ayers, S.D.; Parekh, N.M.; Godoy, A.S.; Trivella, D.B.B.; Reinach, P.; Webb, P.; et al. Molecular Mechanism of Peroxisome Proliferator-Activated Receptor α Activation by WY14643: A New Mode of Ligand Recognition and Receptor Stabilization. *J. Mol. Biol.* **2013**, *425*, 2878–2893. [CrossRef] [PubMed]
89. Huang, H.-T.; Liao, C.-K.; Chiu, W.-T.; Tzeng, S.-F. Ligands of Peroxisome Proliferator-Activated Receptor-Alpha Promote Glutamate Transporter-1 Endocytosis in Astrocytes. *Int. J. Biochem. Cell Biol.* **2017**, *86*, 42–53. [CrossRef]
90. Moraes, L.A.; Piqueras, L.; Bishop-Bailey, D. Peroxisome Proliferator-Activated Receptors and Inflammation. *Pharmacol. Ther.* **2006**, *110*, 371–385. [CrossRef]
91. Mirza, A.Z.; Althagafi, I.I.; Shamshad, H. Role of PPAR Receptor in Different Diseases and Their Ligands: Physiological Importance and Clinical Implications. *Eur. J. Med. Chem.* **2019**, *166*, 502–513. [CrossRef]
92. Tenenbaum, A.; Motro, M.; Fisman, E.Z. Dual and Pan-Peroxisome Proliferator-Activated Receptors (PPAR) Co-Agonism: The Bezafibrate Lessons. *Cardiovasc. Diabetol.* **2005**, *4*, 14. [CrossRef] [PubMed]
93. Rogue, A.; Anthérieu, S.; Vluggens, A.; Umbdenstock, T.; Claude, N.; de la Moureyre-Spire, C.; Weaver, R.J.; Guillouzo, A. PPAR Agonists Reduce Steatosis in Oleic Acid-Overloaded HepaRG Cells. *Toxicol. Appl. Pharmacol.* **2014**, *276*, 73–81. [CrossRef] [PubMed]
94. Shin, N.-R.; Park, S.-H.; Ko, J.-W.; Cho, Y.-K.; Lee, I.-C.; Kim, J.-C.; Shin, I.-S.; Kim, J.-S. Lobeglitazone Attenuates Airway Inflammation and Mucus Hypersecretion in a Murine Model of Ovalbumin-Induced Asthma. *Front. Pharmacol.* **2018**, *9*, 906. [CrossRef] [PubMed]
95. Stebbins, K.J.; Broadhead, A.R.; Cabrera, G.; Correa, L.D.; Messmer, D.; Bundey, R.; Baccei, C.; Bravo, Y.; Chen, A.; Stock, N.S.; et al. In Vitro and in Vivo Pharmacology of NXT629, a Novel and Selective PPARα Antagonist. *Eur. J. Pharmacol.* **2017**, *809*, 130–140. [CrossRef]
96. Duszka, K.; Gregor, A.; Guillou, H.; König, J.; Wahli, W. Peroxisome Proliferator-Activated Receptors and Caloric Restriction-Common Pathways Affecting Metabolism, Health, and Longevity. *Cells* **2020**, *9*, 1708. [CrossRef]
97. Kosgei, V.J.; Coelho, D.; Gueant-Rodriguez, R.M.; Gueant, J.L. Sirt1-PPARS Cross-Talk in Complex Metabolic Diseases and Inherited Disorders of the One Carbon Metabolism. *Cells* **2020**, *9*, 1882. [CrossRef]
98. Kersten, S.; Stienstra, R. The Role and Regulation of the Peroxisome Proliferator Activated Receptor Alpha in Human Liver. *Biochimie* **2017**, *136*, 75–84. [CrossRef] [PubMed]
99. Laleh, P.; Yaser, K.; Alireza, O. Oleoylethanolamide: A Novel Pharmaceutical Agent in the Management of Obesity-an Updated Review. *J. Cell. Physiol.* **2019**, *234*, 7893–7902. [CrossRef] [PubMed]

100. Pawar, A.; Jump, D.B. Unsaturated Fatty Acid Regulation of Peroxisome Proliferator-Activated Receptor Alpha Activity in Rat Primary Hepatocytes. *J. Biol. Chem.* **2003**, *278*, 35931–35939. [CrossRef] [PubMed]
101. Ellinghaus, P.; Wolfrum, C.; Assmann, G.; Spener, F.; Seedorf, U. Phytanic Acid Activates the Peroxisome Proliferator-Activated Receptor Alpha (PPARalpha) in Sterol Carrier Protein 2-/Sterol Carrier Protein x-Deficient Mice. *J. Biol. Chem.* **1999**, *274*, 2766–2772. [CrossRef] [PubMed]
102. Zomer, A.W.; van Der Burg, B.; Jansen, G.A.; Wanders, R.J.; Poll-The, B.T.; van Der Saag, P.T. Pristanic Acid and Phytanic Acid: Naturally Occurring Ligands for the Nuclear Receptor Peroxisome Proliferator-Activated Receptor Alpha. *J. Lipid Res.* **2000**, *41*, 1801–1807. [CrossRef]
103. Hostetler, H.A.; Petrescu, A.D.; Kier, A.B.; Schroeder, F. Peroxisome Proliferator-Activated Receptor Alpha Interacts with High Affinity and Is Conformationally Responsive to Endogenous Ligands. *J. Biol. Chem.* **2005**, *280*, 18667–18682. [CrossRef] [PubMed]
104. Brady, P.S.; Marine, K.A.; Brady, L.J.; Ramsay, R.R. Co-Ordinate Induction of Hepatic Mitochondrial and Peroxisomal Carnitine Acyltransferase Synthesis by Diet and Drugs. *Biochem. J.* **1989**, *260*, 93–100. [CrossRef] [PubMed]
105. Marcus, S.L.; Miyata, K.S.; Zhang, B.; Subramani, S.; Rachubinski, R.A.; Capone, J.P. Diverse Peroxisome Proliferator-Activated Receptors Bind to the Peroxisome Proliferator-Responsive Elements of the Rat Hydratase/Dehydrogenase and Fatty Acyl-CoA Oxidase Genes but Differentially Induce Expression. *Proc. Natl. Acad. Sci. USA* **1993**, *90*, 5723–5727. [CrossRef]
106. Reddy, J.K.; Mannaerts, G.P. Peroxisomal Lipid Metabolism. *Annu. Rev. Nutr.* **1994**, *14*, 343–370. [CrossRef]
107. Zhang, B.; Marcus, S.L.; Miyata, K.S.; Subramani, S.; Capone, J.P.; Rachubinski, R.A. Characterization of Protein-DNA Interactions within the Peroxisome Proliferator-Responsive Element of the Rat Hydratase-Dehydrogenase Gene. *J. Biol. Chem.* **1993**, *268*, 12939–12945. [CrossRef]
108. Chen, X.; Shang, L.; Deng, S.; Li, P.; Chen, K.; Gao, T.; Zhang, X.; Chen, Z.; Zeng, J. Peroxisomal Oxidation of Erucic Acid Suppresses Mitochondrial Fatty Acid Oxidation by Stimulating Malonyl-CoA Formation in the Rat Liver. *J. Biol. Chem.* **2020**, *295*, 10168–10179. [CrossRef] [PubMed]
109. Maheshwari, G.; Ringseis, R.; Wen, G.; Gessner, D.K.; Rost, J.; Fraatz, M.A.; Zorn, H.; Eder, K. Branched-Chain Fatty Acids as Mediators of the Activation of Hepatic Peroxisome Proliferator-Activated Receptor Alpha by a Fungal Lipid Extract. *Biomolecules* **2020**, *10*, 1259. [CrossRef] [PubMed]
110. Latruffe, N.; Cherkaoui Malki, M.; Nicolas-Frances, V.; Clemencet, M.C.; Jannin, B.; Berlot, J.P. Regulation of the Peroxisomal Beta-Oxidation-Dependent Pathway by Peroxisome Proliferator-Activated Receptor Alpha and Kinases. *Biochem. Pharmacol.* **2000**, *60*, 1027–1032. [CrossRef]
111. Klaunig, J.E.; Babich, M.A.; Baetcke, K.P.; Cook, J.C.; Corton, J.C.; David, R.M.; DeLuca, J.G.; Lai, D.Y.; McKee, R.H.; Peters, J.M.; et al. PPARalpha Agonist-Induced Rodent Tumors: Modes of Action and Human Relevance. *Crit. Rev. Toxicol.* **2003**, *33*, 655–780. [CrossRef] [PubMed]
112. Reddy, J.K.; Lalwai, N.D. Carcinogenesis by Hepatic Peroxisome Proliferators: Evaluation of the Risk of Hypolipidemic Drugs and Industrial Plasticizers to Humans. *Crit. Rev. Toxicol.* **1983**, *12*, 1–58. [CrossRef]
113. Gonzalez, F.J.; Peters, J.M.; Cattley, R.C. Mechanism of Action of the Nongenotoxic Peroxisome Proliferators: Role of the Peroxisome Proliferator-Activator Receptor Alpha. *J. Natl. Cancer Inst.* **1998**, *90*, 1702–1709. [CrossRef]
114. Maloney, E.K.; Waxman, D.J. Trans-Activation of PPARalpha and PPARgamma by Structurally Diverse Environmental Chemicals. *Toxicol. Appl. Pharmacol.* **1999**, *161*, 209–218. [CrossRef] [PubMed]
115. Akbiyik, F.; Cinar, K.; Demirpence, E.; Ozsullu, T.; Tunca, R.; Haziroglu, R.; Yurdaydin, C.; Uzunalimoglu, O.; Bozkaya, H. Ligand-Induced Expression of Peroxisome Proliferator-Activated Receptor Alpha and Activation of Fatty Acid Oxidation Enzymes in Fatty Liver. *Eur. J. Clin. Investig.* **2004**, *34*, 429–435. [CrossRef] [PubMed]
116. Preiss, D.; Tikkanen, M.J.; Welsh, P.; Ford, I.; Lovato, L.C.; Elam, M.B.; LaRosa, J.C.; DeMicco, D.A.; Colhoun, H.M.; Goldenberg, I.; et al. Lipid-Modifying Therapies and Risk of Pancreatitis: A Meta-Analysis. *JAMA* **2012**, *308*, 804–811. [CrossRef] [PubMed]
117. Estrela, G.R.; Arruda, A.C.; Torquato, H.F.V.; Freitas-Lima, L.C.; Perilhão, M.S.; Wasinski, F.; Budu, A.; Fock, R.A.; Paredes-Gamero, E.J.; Araujo, R.C. Gemfibrozil Induces Anemia, Leukopenia and Reduces Hematopoietic Stem Cells via PPAR-α in Mice. *Int. J. Mol. Sci.* **2020**, *21*, 5050. [CrossRef] [PubMed]
118. Oswal, D.P.; Balanarasimha, M.; Loyer, J.K.; Bedi, S.; Soman, F.L.; Rider, S.D.; Hostetler, H.A. Divergence between Human and Murine Peroxisome Proliferator-Activated Receptor Alpha Ligand Specificities. *J. Lipid Res.* **2013**, *54*, 2354–2365. [CrossRef] [PubMed]
119. Oswal, D.P.; Alter, G.M.; Rider, S.D.; Hostetler, H.A. A Single Amino Acid Change Humanizes Long-Chain Fatty Acid Binding and Activation of Mouse Peroxisome Proliferator-Activated Receptor α. *J. Mol. Graph. Model.* **2014**, *51*, 27–36. [CrossRef] [PubMed]
120. Chawla, A.; Repa, J.J.; Evans, R.M.; Mangelsdorf, D.J. Nuclear Receptors and Lipid Physiology: Opening the X-Files. *Science* **2001**, *294*, 1866–1870. [CrossRef] [PubMed]
121. Krey, G.; Braissant, O.; L'Horset, F.; Kalkhoven, E.; Perroud, M.; Parker, M.G.; Wahli, W. Fatty Acids, Eicosanoids, and Hypolipidemic Agents Identified as Ligands of Peroxisome Proliferator-Activated Receptors by Coactivator-Dependent Receptor Ligand Assay. *Mol. Endocrinol.* **1997**, *11*, 779–791. [CrossRef]
122. Cave, M.C.; Clair, H.B.; Hardesty, J.E.; Falkner, K.C.; Feng, W.; Clark, B.J.; Sidey, J.; Shi, H.; Aqel, B.A.; McClain, C.J.; et al. Nuclear Receptors and Nonalcoholic Fatty Liver Disease. *Biochim. Biophys. Acta* **2016**, *1859*, 1083–1099. [CrossRef] [PubMed]

123. Francque, S.; Szabo, G.; Abdelmalek, M.F.; Byrne, C.D.; Cusi, K.; Dufour, J.F.; Roden, M.; Sacks, F.; Tacke, F. Nonalcoholic Steatohepatitis: The Role of Peroxisome Proliferator-Activated Receptors. *Nat. Rev. Gastroenterol. Hepatol.* **2021**, *18*, 24–39. [CrossRef] [PubMed]
124. Sinha, R.A.; Rajak, S.; Singh, B.K.; Yen, P.M. Hepatic Lipid Catabolism via PPARalpha-Lysosomal Crosstalk. *Int. J. Mol. Sci.* **2020**, *21*, 2391. [CrossRef] [PubMed]
125. Wagner, N.; Wagner, K.-D. The Role of PPARs in Disease. *Cells* **2020**, *9*, 2367. [CrossRef] [PubMed]
126. Lee, S.S.; Pineau, T.; Drago, J.; Lee, E.J.; Owens, J.W.; Kroetz, D.L.; Fernandez-Salguero, P.M.; Westphal, H.; Gonzalez, F.J. Targeted Disruption of the Alpha Isoform of the Peroxisome Proliferator-Activated Receptor Gene in Mice Results in Abolishment of the Pleiotropic Effects of Peroxisome Proliferators. *Mol. Cell. Biol.* **1995**, *15*, 3012–3022. [CrossRef] [PubMed]
127. Amber-Vitos, O.; Chaturvedi, N.; Nachliel, E.; Gutman, M.; Tsfadia, Y. The Effect of Regulating Molecules on the Structure of the PPAR-RXR Complex. *Biochim. Biophys. Acta* **2016**, *1861*, 1852–1863. [CrossRef] [PubMed]
128. Surapureddi, S.; Yu, S.; Bu, H.; Hashimoto, T.; Yeldandi, A.V.; Kashireddy, P.; Cherkaoui-Malki, M.; Qi, C.; Zhu, Y.J.; Rao, M.S.; et al. Identification of a Transcriptionally Active Peroxisome Proliferator-Activated Receptor Alpha -Interacting Cofactor Complex in Rat Liver and Characterization of PRIC285 as a Coactivator. *Proc. Natl. Acad. Sci. USA* **2002**, *99*, 11836–11841. [CrossRef] [PubMed]
129. Skowron, K.J.; Booker, K.; Cheng, C.; Creed, S.; David, B.P.; Lazzara, P.R.; Lian, A.; Siddiqui, Z.; Speltz, T.E.; Moore, T.W. Steroid Receptor/Coactivator Binding Inhibitors: An Update. *Mol. Cell. Endocrinol.* **2019**, *493*, 110471. [CrossRef] [PubMed]
130. Surapureddi, S.; Rana, R.; Reddy, J.K.; Goldstein, J.A. Nuclear Receptor Coactivator 6 Mediates the Synergistic Activation of Human Cytochrome P-450 2C9 by the Constitutive Androstane Receptor and Hepatic Nuclear Factor-4alpha. *Mol. Pharmacol.* **2008**, *74*, 913–923. [CrossRef]
131. Misra, P.; Reddy, J.K. Peroxisome Proliferator-Activated Receptor-α Activation and Excess Energy Burning in Hepatocarcinogenesis. *Biochimie* **2014**, *98*, 63–74. [CrossRef]
132. Rana, R.; Surapureddi, S.; Kam, W.; Ferguson, S.; Goldstein, J.A. Med25 Is Required for RNA Polymerase II Recruitment to Specific Promoters, Thus Regulating Xenobiotic and Lipid Metabolism in Human Liver. *Mol. Cell. Biol.* **2011**, *31*, 466–481. [CrossRef] [PubMed]
133. Spitler, K.M.; Ponce, J.M.; Oudit, G.Y.; Hall, D.D.; Grueter, C.E. Cardiac Med1 Deletion Promotes Early Lethality, Cardiac Remodeling, and Transcriptional Reprogramming. *Am. J. Physiol. Heart Circ. Physiol.* **2017**, *312*, H768–H780. [CrossRef] [PubMed]
134. De Vera, I.M.S.; Zheng, J.; Novick, S.; Shang, J.; Hughes, T.S.; Brust, R.; Munoz-Tello, P.; Gardner, W.J., Jr.; Marciano, D.P.; Kong, X.; et al. Synergistic Regulation of Coregulator/Nuclear Receptor Interaction by Ligand and DNA. *Structure* **2017**, *25*, 1506–1518.e4. [CrossRef]
135. Lai, Y.-H.; Choudhary, K.; Cloutier, S.C.; Xing, Z.; Aviran, S.; Tran, E.J. Genome-Wide Discovery of DEAD-Box RNA Helicase Targets Reveals RNA Structural Remodeling in Transcription Termination. *Genetics* **2019**, *212*, 153–174. [CrossRef] [PubMed]
136. Song, C.; Hotz-Wagenblatt, A.; Voit, R.; Grummt, I. SIRT7 and the DEAD-Box Helicase DDX21 Cooperate to Resolve Genomic R Loops and Safeguard Genome Stability. *Genes Dev.* **2017**, *31*, 1370–1381. [CrossRef]
137. Taschuk, F.; Cherry, S. DEAD-Box Helicases: Sensors, Regulators, and Effectors for Antiviral Defense. *Viruses* **2020**, *12*, 181. [CrossRef] [PubMed]
138. Arconzo, M.; Piccinin, E.; Moschetta, A. Increased Risk of Acute Liver Failure by Pain Killer Drugs in NAFLD: Focus on Nuclear Receptors and Their Coactivators. *Dig. Liver Dis.* **2021**, *53*, 26–34. [CrossRef] [PubMed]
139. Fornes, D.; Gomez Ribot, D.; Heinecke, F.; Roberti, S.L.; Capobianco, E.; Jawerbaum, A. Maternal Diets Enriched in Olive Oil Regulate Lipid Metabolism and Levels of PPARs and Their Coactivators in the Fetal Liver in a Rat Model of Gestational Diabetes Mellitus. *J. Nutr. Biochem.* **2020**, *78*, 108334. [CrossRef]
140. Kalliora, C.; Kyriazis, I.D.; Oka, S.-I.; Lieu, M.J.; Yue, Y.; Area-Gomez, E.; Pol, C.J.; Tian, Y.; Mizushima, W.; Chin, A.; et al. Dual Peroxisome-Proliferator-Activated-Receptor-α/γ Activation Inhibits SIRT1-PGC1α Axis and Causes Cardiac Dysfunction. *JCI Insight* **2019**, *5*, 129556. [CrossRef] [PubMed]
141. Luo, C.; Widlund, H.R.; Puigserver, P. PGC-1 Coactivators: Shepherding the Mitochondrial Biogenesis of Tumors. *Trends Cancer* **2016**, *2*, 619–631. [CrossRef] [PubMed]
142. Stallcup, M.R.; Poulard, C. Gene-Specific Actions of Transcriptional Coregulators Facilitate Physiological Plasticity: Evidence for a Physiological Coregulator Code. *Trends Biochem. Sci.* **2020**, *45*, 497–510. [CrossRef] [PubMed]
143. Emmett, M.J.; Lazar, M.A. Integrative Regulation of Physiology by Histone Deacetylase 3. *Nat. Rev. Mol. Cell Biol.* **2019**, *20*, 102–115. [CrossRef]
144. Jaiswal, B.; Gupta, A. Modulation of Nuclear Receptor Function by Chromatin Modifying Factor TIP60. *Endocrinology* **2018**, *159*, 2199–2215. [CrossRef]
145. Jankowsky, E.; Guenther, U.-P. A Helicase Links Upstream ORFs and RNA Structure. *Curr. Genet.* **2019**, *65*, 453–456. [CrossRef] [PubMed]
146. Surapureddi, S.; Viswakarma, N.; Yu, S.; Guo, D.; Rao, M.S.; Reddy, J.K. PRIC320, a Transcription Coactivator, Isolated from Peroxisome Proliferator-Binding Protein Complex. *Biochem. Biophys. Res. Commun.* **2006**, *343*, 535–543. [CrossRef]
147. Jia, Y.; Liu, N.; Viswakarma, N.; Sun, R.; Schipma, M.J.; Shang, M.; Thorp, E.B.; Kanwar, Y.S.; Thimmapaya, B.; Reddy, J.K. PIMT/NCOA6IP Deletion in the Mouse Heart Causes Delayed Cardiomyopathy Attributable to Perturbation in Energy Metabolism. *Int. J. Mol. Sci.* **2018**, *19*, 1485. [CrossRef]

148. Jeronimo, C.; Robert, F. The Mediator Complex: At the Nexus of RNA Polymerase II Transcription. *Trends Cell Biol.* **2017**, *27*, 765–783. [CrossRef]
149. Soutourina, J. Transcription Regulation by the Mediator Complex. *Nat. Rev. Mol. Cell Biol.* **2018**, *19*, 262–274. [CrossRef]
150. Paiano, A.; Margiotta, A.; De Luca, M.; Bucci, C. Yeast Two-Hybrid Assay to Identify Interacting Proteins. *Curr. Protoc. Protein Sci.* **2019**, *95*, e70. [CrossRef]
151. O'Malley, B.W. Origins of the Field of Molecular Endocrinology: A Personal Perspective. *Mol. Endocrinol.* **2016**, *30*, 1015–1018. [CrossRef]
152. Kamei, Y.; Xu, L.; Heinzel, T.; Torchia, J.; Kurokawa, R.; Gloss, B.; Lin, S.C.; Heyman, R.A.; Rose, D.W.; Glass, C.K.; et al. A CBP Integrator Complex Mediates Transcriptional Activation and AP-1 Inhibition by Nuclear Receptors. *Cell* **1996**, *85*, 403–414. [CrossRef]
153. Sabari, B.R.; Dall'Agnese, A.; Boija, A.; Klein, I.A.; Coffey, E.L.; Shrinivas, K.; Abraham, B.J.; Hannett, N.M.; Zamudio, A.V.; Manteiga, J.C.; et al. Coactivator Condensation at Super-Enhancers Links Phase Separation and Gene Control. *Science* **2018**, *361*, eaar3958. [CrossRef] [PubMed]
154. Tan, H.W.S.; Anjum, B.; Shen, H.-M.; Ghosh, S.; Yen, P.M.; Sinha, R.A. Lysosomal Inhibition Attenuates Peroxisomal Gene Transcription via Suppression of PPARA and PPARGC1A Levels. *Autophagy* **2019**, *15*, 1455–1459. [CrossRef]
155. Dumesic, P.A.; Egan, D.F.; Gut, P.; Tran, M.T.; Parisi, A.; Chatterjee, N.; Jedrychowski, M.; Paschini, M.; Kazak, L.; Wilensky, S.E.; et al. An Evolutionarily Conserved UORF Regulates PGC1alpha and Oxidative Metabolism in Mice, Flies, and Bluefin Tuna. *Cell Metab.* **2019**, *30*, 190–200.e6. [CrossRef] [PubMed]
156. Petr, M.; Stastny, P.; Zajac, A.; Tufano, J.J.; Maciejewska-Skrendo, A. The Role of Peroxisome Proliferator-Activated Receptors and Their Transcriptional Coactivators Gene Variations in Human Trainability: A Systematic Review. *Int. J. Mol. Sci.* **2018**, *19*, 1472. [CrossRef] [PubMed]
157. Behera, A.K.; Bhattacharya, A.; Vasudevan, M.; Kundu, T.K. P53 Mediated Regulation of Coactivator Associated Arginine Methyltransferase 1 (CARM1) Expression Is Critical for Suppression of Adipogenesis. *FEBS J.* **2018**, *285*, 1730–1744. [CrossRef] [PubMed]
158. Xu, W.; Chen, H.; Du, K.; Asahara, H.; Tini, M.; Emerson, B.M.; Montminy, M.; Evans, R.M. A Transcriptional Switch Mediated by Cofactor Methylation. *Science* **2001**, *294*, 2507–2511. [CrossRef] [PubMed]
159. Kang, Z.; Fan, R. PPARα and NCOR/SMRT Corepressor Network in Liver Metabolic Regulation. *FASEB J.* **2020**, *34*, 8796–8809. [CrossRef]
160. Ghisletti, S.; Huang, W.; Jepsen, K.; Benner, C.; Hardiman, G.; Rosenfeld, M.G.; Glass, C.K. Cooperative NCoR/SMRT Interactions Establish a Corepressor-Based Strategy for Integration of Inflammatory and Anti-Inflammatory Signaling Pathways. *Genes Dev.* **2009**, *23*, 681–693. [CrossRef]
161. Jepsen, K.; Gleiberman, A.S.; Shi, C.; Simon, D.I.; Rosenfeld, M.G. Cooperative Regulation in Development by SMRT and FOXP1. *Genes Dev.* **2008**, *22*, 740–745. [CrossRef]
162. Kumar, S.; Cunningham, T.J.; Duester, G. Nuclear Receptor Corepressors Ncor1 and Ncor2 (Smrt) Are Required for Retinoic Acid-Dependent Repression of Fgf8 during Somitogenesis. *Dev. Biol.* **2016**, *418*, 204–215. [CrossRef]
163. Duong, V.; Augereau, P.; Badia, E.; Jalaguier, S.; Cavailles, V. Regulation of Hormone Signaling by Nuclear Receptor Interacting Proteins. *Adv. Exp. Med. Biol.* **2008**, *617*, 121–127. [CrossRef]
164. Ogawa, K.; Yagi, T.; Guo, T.; Takeda, K.; Ohguchi, H.; Koyama, H.; Aotani, D.; Imaeda, K.; Kataoka, H.; Tanaka, T. Pemafibrate, a Selective PPARα Modulator, and Fenofibrate Suppress Microglial Activation through Distinct PPARα and SIRT1 Dependent Pathways. *Biochem. Biophys. Res. Commun.* **2020**, *524*, 385–391. [CrossRef]
165. Purushotham, A.; Schug, T.T.; Xu, Q.; Surapureddi, S.; Guo, X.; Li, X. Hepatocyte-Specific Deletion of SIRT1 Alters Fatty Acid Metabolism and Results in Hepatic Steatosis and Inflammation. *Cell Metab.* **2009**, *9*, 327–338. [CrossRef] [PubMed]
166. Han, L.; Zhou, R.; Niu, J.; McNutt, M.A.; Wang, P.; Tong, T. SIRT1 Is Regulated by a PPAR{gamma}-SIRT1 Negative Feedback Loop Associated with Senescence. *Nucleic Acids Res.* **2010**, *38*, 7458–7471. [CrossRef] [PubMed]
167. Naiman, S.; Huynh, F.K.; Gil, R.; Glick, Y.; Shahar, Y.; Touitou, N.; Nahum, L.; Avivi, M.Y.; Roichman, A.; Kanfi, Y.; et al. SIRT6 Promotes Hepatic Beta-Oxidation via Activation of PPARalpha. *Cell Rep.* **2019**, *29*, 4127–4143.e8. [CrossRef] [PubMed]
168. Glass, C.K.; Rosenfeld, M.G. The Coregulator Exchange in Transcriptional Functions of Nuclear Receptors. *Genes Dev.* **2000**, *14*, 121–141. [PubMed]
169. Fritah, A.; Christian, M.; Parker, M.G. The Metabolic Coregulator RIP140: An Update. *Am. J. Physiol. Endocrinol. Metab.* **2010**, *299*, E335–E340. [CrossRef] [PubMed]
170. Venkata, N.G.; Robinson, J.A.; Cabot, P.J.; Davis, B.; Monteith, G.R.; Roberts-Thomson, S.J. Mono(2-Ethylhexyl)Phthalate and Mono-n-Butyl Phthalate Activation of Peroxisome Proliferator Activated-Receptors Alpha and Gamma in Breast. *Toxicol. Lett.* **2006**, *163*, 224–234. [CrossRef] [PubMed]
171. Dawson, M.I.; Xia, Z. The Retinoid X Receptors and Their Ligands. *Biochim. Biophys. Acta* **2012**, *1821*, 21–56. [CrossRef]
172. Brunmeir, R.; Xu, F. Functional Regulation of PPARs through Post-Translational Modifications. *Int. J. Mol. Sci.* **2018**, *19*, 1738. [CrossRef] [PubMed]
173. Iershov, A.; Nemazanyy, I.; Alkhoury, C.; Girard, M.; Barth, E.; Cagnard, N.; Montagner, A.; Chretien, D.; Rugarli, E.I.; Guillou, H.; et al. The Class 3 PI3K Coordinates Autophagy and Mitochondrial Lipid Catabolism by Controlling Nuclear Receptor PPARα. *Nat. Commun.* **2019**, *10*, 1566. [CrossRef] [PubMed]

174. Shalev, A.; Siegrist-Kaiser, C.A.; Yen, P.M.; Wahli, W.; Burger, A.G.; Chin, W.W.; Meier, C.A. The Peroxisome Proliferator-Activated Receptor Alpha Is a Phosphoprotein: Regulation by Insulin. *Endocrinology* **1996**, *137*, 4499–4502. [CrossRef] [PubMed]
175. Chamouton, J.; Latruffe, N. PPARα/HNF4α Interplay on Diversified Responsive Elements. Relevance in the Regulation of Liver Peroxisomal Fatty Acid Catabolism. *Curr. Drug Metab.* **2012**, *13*, 1436–1453. [CrossRef]
176. Scarpulla, R.C. Metabolic Control of Mitochondrial Biogenesis through the PGC-1 Family Regulatory Network. *Biochim. Biophys. Acta* **2011**, *1813*, 1269–1278. [CrossRef]
177. Hashimoto, T. Individual Peroxisomal Beta-Oxidation Enzymes. *Ann. N. Y. Acad. Sci.* **1982**, *386*, 5–12. [CrossRef] [PubMed]
178. Reddy, J.K. Peroxisome Proliferators and Peroxisome Proliferator-Activated Receptor Alpha: Biotic and Xenobiotic Sensing. *Am. J. Pathol.* **2004**, *164*, 2305–2321. [CrossRef]
179. Raas, Q.; Gondcaille, C.; Hamon, Y.; Leoni, V.; Caccia, C.; Menetrier, F.; Lizard, G.; Trompier, D.; Savary, S. CRISPR/Cas9-Mediated Knockout of Abcd1 and Abcd2 Genes in BV-2 Cells: Novel Microglial Models for X-Linked Adrenoleukodystrophy. *Biochim. Biophys. Acta Mol. Cell Biol. Lipids* **2019**, *1864*, 704–714. [CrossRef] [PubMed]
180. Dixon, E.D.; Nardo, A.D.; Claudel, T.; Trauner, M. The Role of Lipid Sensing Nuclear Receptors (PPARs and LXR) and Metabolic Lipases in Obesity, Diabetes and NAFLD. *Genes* **2021**, *12*, 645. [CrossRef] [PubMed]
181. Wang, Y.; Nakajima, T.; Gonzalez, F.J.; Tanaka, N. PPARs as Metabolic Regulators in the Liver: Lessons from Liver-Specific PPAR-Null Mice. *Int. J. Mol. Sci.* **2020**, *21*, 2061. [CrossRef]
182. Haro, D.; Marrero, P.F.; Relat, J. Nutritional Regulation of Gene Expression: Carbohydrate-, Fat- and Amino Acid-Dependent Modulation of Transcriptional Activity. *Int. J. Mol. Sci.* **2019**, *20*, 1386. [CrossRef] [PubMed]
183. Vega, R.B.; Kelly, D.P. Cardiac Nuclear Receptors: Architects of Mitochondrial Structure and Function. *J. Clin. Investig.* **2017**, *127*, 1155–1164. [CrossRef] [PubMed]
184. Gao, Q.; Jia, Y.; Yang, G.; Zhang, X.; Boddu, P.C.; Petersen, B.; Narsingam, S.; Zhu, Y.-J.; Thimmapaya, B.; Kanwar, Y.S.; et al. PPARα-Deficient Ob/Ob Obese Mice Become More Obese and Manifest Severe Hepatic Steatosis Due to Decreased Fatty Acid Oxidation. *Am. J. Pathol.* **2015**, *185*, 1396–1408. [CrossRef] [PubMed]
185. Tonsgard, J.H.; Getz, G.S. Effect of Reye's Syndrome Serum on Isolated Chinchilla Liver Mitochondria. *J. Clin. Investig.* **1985**, *76*, 816–825. [CrossRef] [PubMed]
186. Inokuchi-Shimizu, S.; Park, E.J.; Roh, Y.S.; Yang, L.; Zhang, B.; Song, J.; Liang, S.; Pimienta, M.; Taniguchi, K.; Wu, X.; et al. TAK1-Mediated Autophagy and Fatty Acid Oxidation Prevent Hepatosteatosis and Tumorigenesis. *J. Clin. Investig.* **2014**, *124*, 3566–3578. [CrossRef] [PubMed]
187. Park, H.S.; Lim, J.H.; Kim, M.Y.; Kim, Y.; Hong, Y.A.; Choi, S.R.; Chung, S.; Kim, H.W.; Choi, B.S.; Kim, Y.S.; et al. Resveratrol Increases AdipoR1 and AdipoR2 Expression in Type 2 Diabetic Nephropathy. *J. Transl. Med.* **2016**, *14*, 176. [CrossRef]
188. Wanders, R.J.A.; Ferdinandusse, S.; Brites, P.; Kemp, S. Peroxisomes, Lipid Metabolism and Lipotoxicity. *Biochim. Biophys. Acta* **2010**, *1801*, 272–280. [CrossRef] [PubMed]
189. Pontis, S.; Ribeiro, A.; Sasso, O.; Piomelli, D. Macrophage-Derived Lipid Agonists of PPAR-α as Intrinsic Controllers of Inflammation. *Crit. Rev. Biochem. Mol. Biol.* **2016**, *51*, 7–14. [CrossRef]
190. Vluggens, A.; Andreoletti, P.; Viswakarma, N.; Jia, Y.; Matsumoto, K.; Kulik, W.; Khan, M.; Huang, J.; Guo, D.; Yu, S.; et al. Reversal of Mouse Acyl-CoA Oxidase 1 (ACOX1) Null Phenotype by Human ACOX1b Isoform [Corrected]. *Lab. Investig.* **2010**, *90*, 696–708. [CrossRef]
191. Qi, W.; Gutierrez, G.E.; Gao, X.; Dixon, H.; McDonough, J.A.; Marini, A.M.; Fisher, A.L. The ω-3 Fatty Acid α-Linolenic Acid Extends Caenorhabditis Elegans Lifespan via NHR-49/PPARα and Oxidation to Oxylipins. *Aging Cell* **2017**, *16*, 1125–1135. [CrossRef]
192. Fock, E.; Lavrova, E.; Bachteeva, V.; Nikolaeva, S.; Parnova, R. Suppression of Fatty Acid β-Oxidation and Energy Deficiency as a Cause of Inhibitory Effect of E. Coli Lipopolysaccharide on Osmotic Water Transport in the Frog Urinary Bladder. *Comp. Biochem. Physiol. Part C Toxicol. Pharmacol.* **2019**, *218*, 81–87. [CrossRef]
193. Ling, Y.; Shi, Z.; Yang, X.; Cai, Z.; Wang, L.; Wu, X.; Ye, A.; Jiang, J. Hypolipidemic Effect of Pure Total Flavonoids from Peel of Citrus (PTFC) on Hamsters of Hyperlipidemia and Its Potential Mechanism. *Exp. Gerontol.* **2020**, *130*, 110786. [CrossRef]
194. Li, L.-Y.; Lv, H.-B.; Jiang, Z.-Y.; Qiao, F.; Chen, L.-Q.; Zhang, M.-L.; Du, Z.-Y. Peroxisomal Proliferator-Activated Receptor α-b Deficiency Induces the Reprogramming of Nutrient Metabolism in Zebrafish. *J. Physiol.* **2020**, *598*, 4537–4553. [CrossRef]
195. El Kebbaj, Z.; Andreoletti, P.; Mountassif, D.; Kabine, M.; Schohn, H.; Dauca, M.; Latruffe, N.; El Kebbaj, M.S.; Cherkaoui-Malki, M. Differential Regulation of Peroxisome Proliferator-Activated Receptor (PPAR)-Alpha 1 and Truncated PPAR Alpha 2 as an Adaptive Response to Fasting in the Control of Hepatic Peroxisomal Fatty Acid Beta-Oxidation in the Hibernating Mammal. *Endocrinology* **2008**, *150*, 1192–1201. [CrossRef]
196. Braissant, O.; Foufelle, F.; Scotto, C.; Dauça, M.; Wahli, W. Differential Expression of Peroxisome Proliferator-Activated Receptors (PPARs): Tissue Distribution of PPAR-Alpha, -Beta, and -Gamma in the Adult Rat. *Endocrinology* **1996**, *137*, 354–366. [CrossRef] [PubMed]
197. Warden, A.; Truitt, J.; Merriman, M.; Ponomareva, O.; Jameson, K.; Ferguson, L.B.; Mayfield, R.D.; Harris, R.A. Localization of PPAR Isotypes in the Adult Mouse and Human Brain. *Sci. Rep.* **2016**, *6*, 27618. [CrossRef] [PubMed]
198. Mariani, M.M.; Malm, T.; Lamb, R.; Jay, T.R.; Neilson, L.; Casali, B.; Medarametla, L.; Landreth, G.E. Neuronally-Directed Effects of RXR Activation in a Mouse Model of Alzheimer's Disease. *Sci. Rep.* **2017**, *7*, 42270. [CrossRef] [PubMed]

199. Raso, G.M.; Esposito, E.; Vitiello, S.; Iacono, A.; Santoro, A.; D'Agostino, G.; Sasso, O.; Russo, R.; Piazza, P.V.; Calignano, A.; et al. Palmitoylethanolamide Stimulation Induces Allopregnanolone Synthesis in C6 Cells and Primary Astrocytes: Involvement of Peroxisome-Proliferator Activated Receptor-α. *J. Neuroendocrinol.* **2011**, *23*, 591–600. [CrossRef]
200. Roy, A.; Jana, M.; Corbett, G.T.; Ramaswamy, S.; Kordower, J.H.; Gonzalez, F.J.; Pahan, K. Regulation of Cyclic AMP Response Element Binding and Hippocampal Plasticity-Related Genes by Peroxisome Proliferator-Activated Receptor α. *Cell Rep.* **2013**, *4*, 724–737. [CrossRef]
201. Marx, N.; Duez, H.; Fruchart, J.-C.; Staels, B. Peroxisome Proliferator-Activated Receptors and Atherogenesis: Regulators of Gene Expression in Vascular Cells. *Circ. Res.* **2004**, *94*, 1168–1178. [CrossRef] [PubMed]
202. Morinishi, T.; Tokuhara, Y.; Ohsaki, H.; Ibuki, E.; Kadota, K.; Hirakawa, E. Activation and Expression of Peroxisome Proliferator-Activated Receptor Alpha Are Associated with Tumorigenesis in Colorectal Carcinoma. *PPAR Res.* **2019**, *2019*, 7486727. [CrossRef] [PubMed]
203. Kliewer, S.A.; Forman, B.M.; Blumberg, B.; Ong, E.S.; Borgmeyer, U.; Mangelsdorf, D.J.; Umesono, K.; Evans, R.M. Differential Expression and Activation of a Family of Murine Peroxisome Proliferator-Activated Receptors. *Proc. Natl. Acad. Sci. USA* **1994**, *91*, 7355–7359. [CrossRef] [PubMed]
204. Costet, P.; Legendre, C.; Moré, J.; Edgar, A.; Galtier, P.; Pineau, T. Peroxisome Proliferator-Activated Receptor Alpha-Isoform Deficiency Leads to Progressive Dyslipidemia with Sexually Dimorphic Obesity and Steatosis. *J. Biol. Chem.* **1998**, *273*, 29577–29585. [CrossRef]
205. Stec, D.E.; Gordon, D.M.; Hipp, J.A.; Hong, S.; Mitchell, Z.L.; Franco, N.R.; Robison, J.W.; Anderson, C.D.; Stec, D.F.; Hinds, T.D. Loss of Hepatic PPARα Promotes Inflammation and Serum Hyperlipidemia in Diet-Induced Obesity. *Am. J. Physiol. Regul. Integr. Comp. Physiol.* **2019**, *317*, R733–R745. [CrossRef] [PubMed]
206. Chen, X.; Ward, S.C.; Cederbaum, A.I.; Xiong, H.; Lu, Y. Alcoholic Fatty Liver Is Enhanced in CYP2A5 Knockout Mice: The Role of the PPARα-FGF21 Axis. *Toxicology* **2017**, *379*, 12–21. [CrossRef]
207. Francque, S.; Verrijken, A.; Caron, S.; Prawitt, J.; Paumelle, R.; Derudas, B.; Lefebvre, P.; Taskinen, M.-R.; Van Hul, W.; Mertens, I.; et al. PPARα Gene Expression Correlates with Severity and Histological Treatment Response in Patients with Non-Alcoholic Steatohepatitis. *J. Hepatol.* **2015**, *63*, 164–173. [CrossRef]
208. Ip, E.; Farrell, G.C.; Robertson, G.; Hall, P.; Kirsch, R.; Leclercq, I. Central Role of PPARalpha-Dependent Hepatic Lipid Turnover in Dietary Steatohepatitis in Mice. *Hepatology* **2003**, *38*, 123–132. [CrossRef]
209. Patsouris, D.; Reddy, J.K.; Müller, M.; Kersten, S. Peroxisome Proliferator-Activated Receptor Alpha Mediates the Effects of High-Fat Diet on Hepatic Gene Expression. *Endocrinology* **2006**, *147*, 1508–1516. [CrossRef]
210. Stienstra, R.; Mandard, S.; Patsouris, D.; Maass, C.; Kersten, S.; Müller, M. Peroxisome Proliferator-Activated Receptor Alpha Protects against Obesity-Induced Hepatic Inflammation. *Endocrinology* **2007**, *148*, 2753–2763. [CrossRef]
211. Rando, G.; Tan, C.K.; Khaled, N.; Montagner, A.; Leuenberger, N.; Bertrand-Michel, J.; Paramalingam, E.; Guillou, H.; Wahli, W. Glucocorticoid Receptor-PPARα Axis in Fetal Mouse Liver Prepares Neonates for Milk Lipid Catabolism. *Elife* **2016**, *5*, e11853. [CrossRef]
212. Montagner, A.; Polizzi, A.; Fouché, E.; Ducheix, S.; Lippi, Y.; Lasserre, F.; Barquissau, V.; Régnier, M.; Lukowicz, C.; Benhamed, F.; et al. Liver PPARα Is Crucial for Whole-Body Fatty Acid Homeostasis and Is Protective against NAFLD. *Gut* **2016**, *65*, 1202–1214. [CrossRef] [PubMed]
213. Polizzi, A.; Fouché, E.; Ducheix, S.; Lasserre, F.; Marmugi, A.P.; Mselli-Lakhal, L.; Loiseau, N.; Wahli, W.; Guillou, H.; Montagner, A. Hepatic Fasting-Induced PPARα Activity Does Not Depend on Essential Fatty Acids. *Int. J. Mol. Sci.* **2016**, *17*, 1624. [CrossRef] [PubMed]
214. Régnier, M.; Polizzi, A.; Lippi, Y.; Fouché, E.; Michel, G.; Lukowicz, C.; Smati, S.; Marrot, A.; Lasserre, F.; Naylies, C.; et al. Insights into the Role of Hepatocyte PPARα Activity in Response to Fasting. *Mol. Cell. Endocrinol.* **2018**, *471*, 75–88. [CrossRef] [PubMed]
215. Brocker, C.N.; Patel, D.P.; Velenosi, T.J.; Kim, D.; Yan, T.; Yue, J.; Li, G.; Krausz, K.W.; Gonzalez, F.J. Extrahepatic PPARα Modulates Fatty Acid Oxidation and Attenuates Fasting-Induced Hepatosteatosis in Mice. *J. Lipid Res.* **2018**, *59*, 2140–2152. [CrossRef]
216. Jordan, S.; Tung, N.; Casanova-Acebes, M.; Chang, C.; Cantoni, C.; Zhang, D.; Wirtz, T.H.; Naik, S.; Rose, S.A.; Brocker, C.N.; et al. Dietary Intake Regulates the Circulating Inflammatory Monocyte Pool. *Cell* **2019**, *178*, 1102–1114.e17. [CrossRef] [PubMed]
217. Régnier, M.; Polizzi, A.; Smati, S.; Lukowicz, C.; Fougerat, A.; Lippi, Y.; Fouché, E.; Lasserre, F.; Naylies, C.; Bétoulières, C.; et al. Hepatocyte-Specific Deletion of Pparα Promotes NAFLD in the Context of Obesity. *Sci. Rep.* **2020**, *10*, 6489. [CrossRef] [PubMed]
218. Batatinha, H.A.P.; Lima, E.A.; Teixeira, A.A.S.; Souza, C.O.; Biondo, L.A.; Silveira, L.S.; Lira, F.S.; Rosa Neto, J.C. Association Between Aerobic Exercise and Rosiglitazone Avoided the NAFLD and Liver Inflammation Exacerbated in PPAR-α Knockout Mice. *J. Cell. Physiol.* **2017**, *232*, 1008–1019. [CrossRef] [PubMed]
219. Brocker, C.N.; Yue, J.; Kim, D.; Qu, A.; Bonzo, J.A.; Gonzalez, F.J. Hepatocyte-Specific PPARA Expression Exclusively Promotes Agonist-Induced Cell Proliferation without Influence from Nonparenchymal Cells. *Am. J. Physiol. Gastrointest. Liver Physiol.* **2017**, *312*, G283–G299. [CrossRef] [PubMed]
220. Stanley, W.C.; Recchia, F.A.; Lopaschuk, G.D. Myocardial Substrate Metabolism in the Normal and Failing Heart. *Physiol. Rev.* **2005**, *85*, 1093–1129. [CrossRef] [PubMed]
221. Kaimoto, S.; Hoshino, A.; Ariyoshi, M.; Okawa, Y.; Tateishi, S.; Ono, K.; Uchihashi, M.; Fukai, K.; Iwai-Kanai, E.; Matoba, S. Activation of PPAR-α in the Early Stage of Heart Failure Maintained Myocardial Function and Energetics in Pressure-Overload Heart Failure. *Am. J. Physiol. Heart Circ. Physiol.* **2017**, *312*, H305–H313. [CrossRef] [PubMed]

222. Di Paola, R.; Cordaro, M.; Crupi, R.; Siracusa, R.; Campolo, M.; Bruschetta, G.; Fusco, R.; Pugliatti, P.; Esposito, E.; Cuzzocrea, S. Protective Effects of Ultramicronized Palmitoylethanolamide (PEA-Um) in Myocardial Ischaemia and Reperfusion Injury in VIVO. *Shock* **2016**, *46*, 202–213. [CrossRef] [PubMed]
223. Standage, S.W.; Waworuntu, R.L.; Delaney, M.A.; Maskal, S.M.; Bennion, B.G.; Duffield, J.S.; Parks, W.C.; Liles, W.C.; McGuire, J.K. Nonhematopoietic Peroxisome Proliferator-Activated Receptor-α Protects Against Cardiac Injury and Enhances Survival in Experimental Polymicrobial Sepsis. *Crit. Care Med.* **2016**, *44*, e594–e603. [CrossRef]
224. Yammine, A.; Namsi, A.; Vervandier-Fasseur, D.; Mackrill, J.J.; Lizard, G.; Latruffe, N. Polyphenols of the Mediterranean Diet and Their Metabolites in the Prevention of Colorectal Cancer. *Molecules* **2021**, *26*, 3483. [CrossRef] [PubMed]
225. Castrejón-Tellez, V.; Rodríguez-Pérez, J.M.; Pérez-Torres, I.; Pérez-Hernández, N.; Cruz-Lagunas, A.; Guarner-Lans, V.; Vargas-Alarcón, G.; Rubio-Ruiz, M.E. The Effect of Resveratrol and Quercetin Treatment on PPAR Mediated Uncoupling Protein (UCP-) 1, 2, and 3 Expression in Visceral White Adipose Tissue from Metabolic Syndrome Rats. *Int. J. Mol. Sci.* **2016**, *17*, 1069. [CrossRef]
226. Zhou, Y.; Lin, S.; Zhang, L.; Li, Y. Resveratrol Prevents Renal Lipotoxicity in High-Fat Diet-Treated Mouse Model through Regulating PPAR-α Pathway. *Mol. Cell. Biochem.* **2016**, *411*, 143–150. [CrossRef]
227. Barone, R.; Rizzo, R.; Tabbì, G.; Malaguarnera, M.; Frye, R.E.; Bastin, J. Nuclear Peroxisome Proliferator-Activated Receptors (PPARs) as Therapeutic Targets of Resveratrol for Autism Spectrum Disorder. *Int. J. Mol. Sci.* **2019**, *20*, 1878. [CrossRef] [PubMed]
228. Fantacuzzi, M.; De Filippis, B.; Amoroso, R.; Giampietro, L. PPAR Ligands Containing Stilbene Scaffold. *Mini Rev. Med. Chem.* **2019**, *19*, 1599–1610. [CrossRef] [PubMed]
229. Bastin, J.; Djouadi, F. Resveratrol and Myopathy. *Nutrients* **2016**, *8*, 254. [CrossRef] [PubMed]
230. Sun, X.; Yamasaki, M.; Katsube, T.; Shiwaku, K. Effects of Quercetin Derivatives from Mulberry Leaves: Improved Gene Expression Related Hepatic Lipid and Glucose Metabolism in Short-Term High-Fat Fed Mice. *Nutr. Res. Pract.* **2015**, *9*, 137–143. [CrossRef]
231. Wang, L.L.; Zhang, Z.C.; Hassan, W.; Li, Y.; Liu, J.; Shang, J. Amelioration of Free Fatty Acid-Induced Fatty Liver by Quercetin-3-O-β-D-Glucuronide through Modulation of Peroxisome Proliferator-Activated Receptor-Alpha/Sterol Regulatory Element-Binding Protein-1c Signaling. *Hepatol. Res.* **2016**, *46*, 225–238. [CrossRef]
232. Chen, D.; Daniel, K.G.; Kuhn, D.J.; Kazi, A.; Bhuiyan, M.; Li, L.; Wang, Z.; Wan, S.B.; Lam, W.H.; Chan, T.H.; et al. Green Tea and Tea Polyphenols in Cancer Prevention. *Front. Biosci.* **2004**, *9*, 2618–2631. [CrossRef] [PubMed]
233. Zhang, S.; Yang, X.; Luo, J.; Ge, X.; Sun, W.; Zhu, H.; Zhang, W.; Cao, J.; Hou, Y. PPARα Activation Sensitizes Cancer Cells to Epigallocatechin-3-Gallate (EGCG) Treatment via Suppressing Heme Oxygenase-1. *Nutr. Cancer* **2014**, *66*, 315–324. [CrossRef] [PubMed]
234. Yang, H.; Zuo, X.Z.; Tian, C.; He, D.L.; Yi, W.J.; Chen, Z.; Zhang, P.W.; Ding, S.B.; Ying, C.J. Green Tea Polyphenols Attenuate High-Fat Diet-Induced Renal Oxidative Stress through SIRT3-Dependent Deacetylation. *Biomed. Environ. Sci.* **2015**, *28*, 455–459. [CrossRef] [PubMed]
235. Chen, J.-W.; Kong, Z.-L.; Tsai, M.-L.; Lo, C.-Y.; Ho, C.-T.; Lai, C.-S. Tetrahydrocurcumin Ameliorates Free Fatty Acid-Induced Hepatic Steatosis and Improves Insulin Resistance in HepG2 Cells. *J. Food Drug Anal.* **2018**, *26*, 1075–1085. [CrossRef]
236. Rimando, A.M.; Khan, S.I.; Mizuno, C.S.; Ren, G.; Mathews, S.T.; Kim, H.; Yokoyama, W. Evaluation of PPARα Activation by Known Blueberry Constituents. *J. Sci. Food Agric.* **2016**, *96*, 1666–1671. [CrossRef] [PubMed]
237. Vitaglione, P.; Mazzone, G.; Lembo, V.; D'Argenio, G.; Rossi, A.; Guido, M.; Savoia, M.; Salomone, F.; Mennella, I.; De Filippis, F.; et al. Coffee Prevents Fatty Liver Disease Induced by a High-Fat Diet by Modulating Pathways of the Gut-Liver Axis. *J. Nutr. Sci.* **2019**, *8*, e15. [CrossRef] [PubMed]
238. Bigagli, E.; Toti, S.; Lodovici, M.; Giovannelli, L.; Cinci, L.; D'Ambrosio, M.; Luceri, C. Dietary Extra-Virgin Olive Oil Polyphenols Do Not Attenuate Colon Inflammation in Transgenic HLAB-27 Rats but Exert Hypocholesterolemic Effects through the Modulation of HMGCR and PPAR-α Gene Expression in the Liver. *Lifestyle Genom.* **2018**, *11*, 99–108. [CrossRef] [PubMed]
239. Pirozzi, C.; Lama, A.; Simeoli, R.; Paciello, O.; Pagano, T.B.; Mollica, M.P.; Di Guida, F.; Russo, R.; Magliocca, S.; Canani, R.B.; et al. Hydroxytyrosol Prevents Metabolic Impairment Reducing Hepatic Inflammation and Restoring Duodenal Integrity in a Rat Model of NAFLD. *J. Nutr. Biochem.* **2016**, *30*, 108–115. [CrossRef] [PubMed]
240. Valenzuela, R.; Illesca, P.; Echeverría, F.; Espinosa, A.; Rincón-Cervera, M.Á.; Ortiz, M.; Hernandez-Rodas, M.C.; Valenzuela, A.; Videla, L.A. Molecular Adaptations Underlying the Beneficial Effects of Hydroxytyrosol in the Pathogenic Alterations Induced by a High-Fat Diet in Mouse Liver: PPAR-α and Nrf2 Activation, and NF-KB down-Regulation. *Food Funct.* **2017**, *8*, 1526–1537. [CrossRef]
241. El Kebbaj, R.; Andreoletti, P.; El Hajj, H.I.; El Kharrassi, Y.; Vamecq, J.; Mandard, S.; Saih, F.-E.; Latruffe, N.; El Kebbaj, M.S.; Lizard, G.; et al. Argan Oil Prevents Down-Regulation Induced by Endotoxin on Liver Fatty Acid Oxidation and Gluconeogenesis and on Peroxisome Proliferator-Activated Receptor Gamma Coactivator-1α, (PGC-1α), Peroxisome Proliferator-Activated Receptor α (PPARα) and Estrogen Related Receptor α (ERRα). *Biochim. Open* **2015**, *1*, 51–59. [CrossRef] [PubMed]
242. Chen, X.; Wang, Q.; Shao, M.; Ma, L.; Guo, D.; Wu, Y.; Gao, P.; Wang, X.; Li, W.; Li, C.; et al. Ginsenoside Rb3 Regulates Energy Metabolism and Apoptosis in Cardiomyocytes via Activating PPARα Pathway. *Biomed. Pharmacother.* **2019**, *120*, 109487. [CrossRef] [PubMed]

243. Zhang, C.; Deng, J.; Liu, D.; Tuo, X.; Xiao, L.; Lai, B.; Yao, Q.; Liu, J.; Yang, H.; Wang, N. Nuciferine Ameliorates Hepatic Steatosis in High-Fat Diet/Streptozocin-Induced Diabetic Mice through a PPARα/PPARγ Coactivator-1α Pathway. *Br. J. Pharmacol.* **2018**, *175*, 4218–4228. [CrossRef] [PubMed]
244. Yu, H.; Li, C.; Yang, J.; Zhang, T.; Zhou, Q. Berberine Is a Potent Agonist of Peroxisome Proliferator Activated Receptor Alpha. *Front. Biosci.* **2016**, *21*, 1052–1060. [CrossRef]

International Journal of
Molecular Sciences

Review

Role of Peroxisome Proliferator-Activated Receptors (PPARs) in Energy Homeostasis of Dairy Animals: Exploiting Their Modulation through Nutrigenomic Interventions

Faiz-ul Hassan [1,2,†], Asif Nadeem [3,†], Zhipeng Li [1], Maryam Javed [4], Qingyou Liu [1], Jahanzaib Azhar [3], Muhammad Saif-ur Rehman [2], Kuiqing Cui [1,*] and Saif ur Rehman [1,*]

1 State Key Laboratory for Conservation and Utilization of Subtropical Agro-Bioresources, Guangxi University, Nanning 530005, China; f.hassan@uaf.edu.pk (F.-u.H.); zp.li@gxu.edu.cn (Z.L.); qyliu-gene@gxu.edu.cn (Q.L.)
2 Institute of Animal and Dairy Sciences, Faculty of Animal Husbandry, University of Agriculture, Faisalabad 38040, Pakistan; shsaifurrehman@yahoo.com
3 Department of Biotechnology, Virtual University of Pakistan, Lahore 54000, Pakistan; asif.nadeem@vu.edu.pk (A.N.); jahanzaib.azhar@vu.edu.pk (J.A.)
4 Institute of Biochemistry and Biotechnology, University of Veterinary and Animal Sciences Lahore, Lahore 54000, Pakistan; Maryam.javed@uvas.edu.pk
* Correspondence: kqcui@gxu.edu.cn (K.C.); Saif_ali28@yahoo.com (S.u.R.)
† These authors contributed equally.

Citation: Hassan, F.-u.; Nadeem, A.; Li, Z.; Javed, M.; Liu, Q.; Azhar, J.; Rehman, M.S.-u.; Cui, K.; Rehman, S.u. Role of Peroxisome Proliferator-Activated Receptors (PPARs) in Energy Homeostasis of Dairy Animals: Exploiting Their Modulation through Nutrigenomic Interventions. *Int. J. Mol. Sci.* **2021**, *22*, 12463. https://doi.org/10.3390/ijms222212463

Academic Editors: Walter Wahli and Manuel Vázquez-Carrera

Received: 29 September 2021
Accepted: 16 November 2021
Published: 18 November 2021

Abstract: Peroxisome proliferator-activated receptors (PPARs) are the nuclear receptors that could mediate the nutrient-dependent transcriptional activation and regulate metabolic networks through energy homeostasis. However, these receptors cannot work properly under metabolic stress. PPARs and their subtypes can be modulated by nutrigenomic interventions, particularly under stress conditions to restore cellular homeostasis. Many nutrients such as polyunsaturated fatty acids, vitamins, dietary amino acids and phytochemicals have shown their ability for potential activation or inhibition of PPARs. Thus, through different mechanisms, all these nutrients can modulate PPARs and are ultimately helpful to prevent various metabolic disorders, particularly in transition dairy cows. This review aims to provide insights into the crucial role of PPARs in energy metabolism and their potential modulation through nutrigenomic interventions to improve energy homeostasis in dairy animals.

Keywords: nuclear receptors; PPARs; nutrigenomics; energy homeostasis; dairy animals

1. Introduction

Dairy animals provide milk and dairy products, which are considered some of the most important sources of nutrients for the human diet globally. Dairy production is the key element of sustainable agriculture in the tropics and subtropics. The rapidly increasing human population urges for consolidated efforts to ensure the abundant future availability of milk and dairy products. Therefore, problems and challenges associated with milk production and dairy animal health should be addressed to enhance the production of animals. Dairy animals experience diverse types of stress at different production stages in which the transition period is one of the most stressful stages in the life of dairy cattle. During the transition period and other stressful stages, the metabolic health of the animal is compromised, resulting in enhanced production of non-esterified fatty acids (NEFA) and ketone bodies (kb). Other major conditions associated with these stress conditions include insulin resistance, low blood sugar levels and inflammation [1], which lead to toxicity, fatty liver, ketosis and other metabolic syndromes, ultimately reducing the performance of dairy animals [2].

Nuclear receptors are known to regulate physiological events of metabolism and control the homeostasis of glucose and lipid metabolism. They are also implicated in mediating the long-term effects of early environmental and nutritional experiences on the onset of chronic metabolic disorders in humans and animals [3]. Nuclear receptors belong to a family of ligand-regulated transcription factors that are activated by steroid hormones, such as progesterone, estrogen, and different other lipid-soluble signals such as oxysterols, thyroid hormone, and retinoic acid [4]. In contrast to other messengers, ligands are one of the intercellular messengers that can cross the plasma membrane barrier and directly interact with nuclear receptors instead of interacting with cells surface receptors. These nuclear receptors, once activated, can directly regulate the transcription of respective genes and control many biological processes, including the reproduction, development, proliferation of cells, and metabolism. Despite the fact that the nuclear receptors primarily work as transcription factors, but some have additionally been found to regulate the function of cells inside the cytoplasm [5]. More than 50 nuclear receptors are being reported in human genomes [4,6]. Ligands for these have been discovered, except for a few "orphan receptors" [7]. Major nuclear receptors with more comprehensive experimental data and their ligands are summarized in Table 1.

Table 1. Nuclear receptors along with their ligands [8].

Receptor Name	Abbreviation	Ligand
Progesterone receptor	PR	Progesterone
Estrogen receptor	ER	Estrogen
Liver X receptor	LXR	Oxysterols
Vitamin D3 receptor	VDR	Vitamin D3
Androgen receptor	AR	Testosterone
Glucocorticoid receptor	GR	Cortisol
Thyroid hormone receptor	TR	Thyroid hormone
Retinoic acid-related receptor	RXR	Rexinoids
Mineralocorticoid receptor	MR	Aldosterone
Peroxisome proliferator activated receptor g	PPARγ	Fatty acid Metabolites
Retinoid orphan receptor	ROR	?
Estrogen-related receptor	ERR	?

All of the nuclear receptors have a common structure comprised of the highly variable amino-terminal domain that incorporates a few particular regions of transactivation (the A/B domain, also referred to as AF1 for activation function 1), a central conserved DNA-binding domain that contains two Zn fingers (the C domain), a short region responsible for nuclear localization (the D domain), and a large fairly well-conserved carboxy-terminal ligand-binding domain (the E domain, or LBD) that contributes to interactions of the subset of nuclear receptors that form heterodimers [4]. Further, a highly variable carboxy-terminal tail (the F domain) that in most cases has unknown functions is also present, as shown in Figure 1.

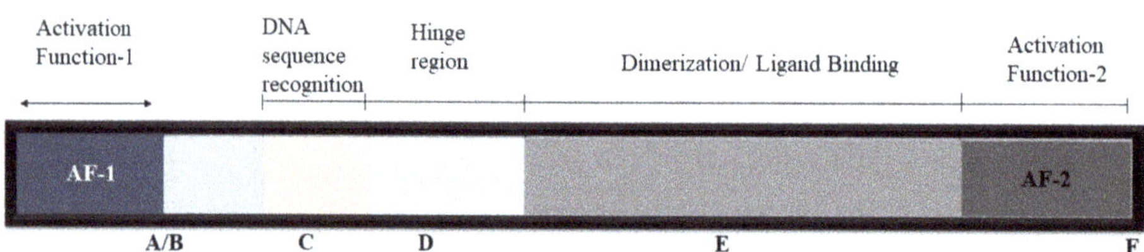

Figure 1. General basic association of atomic receptors [9]. Letters from A to F represented the domains of nuclear receptors from N-end to C-end.

Research studies on metabolic syndromes have identified a close connection between metabolic abnormalities and nuclear receptors, including PPARs, farnesoid X receptors (FXRs), liver X receptors (LXRs) and glucocorticoid receptors (GRs) [3]. PPARs are widely studied nuclear receptors that are known to regulate and control metabolic changes both in humans and animals. Essentially, PPARs were first identified as novel members of the nuclear receptors from *Xenopus* frogs [10] and exhibited to induce the multiplication of peroxisomes in the cells. The PPARα was the first member of these receptors that was identified in mammals during the analysis of molecular targets for liver peroxisome proliferators [11]. The characterization of the PPARA (encoding PPARα) gene in adult mice revealed that PPARα found in humans and dairy animals is abundantly expressed in the liver, heart and kidney. After the discovery of PPARα, the other isotypes were also discovered, including PPARγ and PPARβ/δ [10]. The PPARs form heterodimers and function with the retinoid-X-receptor (RXR). Once a specific ligand binds to receptor dimer, it induces the covalent modification in the structure of PPARs, which activates these nuclear receptors [12]. The activated dimer PPAR/RXR binds to the PPAR response element, which is a specific DNA sequence in the promoter region of target genes, leading to the control of their expression. The PPAR response element is a hexanucleotide (AGGTCA) repeat separated by only a single nucleotide and varies for each PPAR member. All the members of PPARs are activated by the specific ligand concentrations (usually in μM range) both in the case of humans and ruminants [13].

A literature survey showed that information regarding the role of PPARs in lipid metabolism, the regulation of the expression of different genes and proteins and tissue distribution is mainly available in humans compared to dairy animals. However, Bionaz et al. analyzed the relative distribution of PPARs in bovine tissues of dairy cows and bovine cell lines through gene expression analysis by qPCR [14]. Their findings showed that the overall relative distribution of PPARs in dairy animals is quite similar to other species. Later, some studies also showed the relative distribution of PPARs in different organs of dairy animals, including rumen, adipose tissue, liver, kidney, lungs and mammary tissues. The biological and metabolic roles of PPARs have shown that they are the major molecules that regulate energy homeostasis [15], and hence, they are ideal candidates to address metabolic disorders in dairy animals through nutritional interventions.

2. Nuclear Receptors' Mode of Action

The potential mode of action of nuclear receptors is a prerequisite to better understanding the role of PPARs in energy homeostasis. The nuclear receptors can control transcriptional events by exerting a positive effect directly or by repressing regulated promoters. The protein–protein interactions can mediate a repressive effect on other signaling pathways under the regulation of transcription factors such as AP-1, NF-kappa-B, or C/EBP [9]. Figure 2 describes the involved elements and the processes that elicit the biological response.

2.1. Transcriptional Activation

Generally, transcription activation includes ligand-dependent conformational modifications of the chromatin REs-associated nuclear receptors, activating corepressor complex discharge and the successive deployment of coactivator complexes that alter the chromatin structure and facilitate the transcription initiation complex's assembly at the promoter regions. Numerous NR coactivators have been identified, and the repertoire is unique to a few cell types, signals, and genes. Therefore, the agonists binding activates the transfer of corepressors for coactivators critical for the transcription activation process. Moreover, the ligand-dependent interaction of NR corepressors, such as LCoR and RIP140, through LXXLL motifs could hinder the transcription process [16,17].

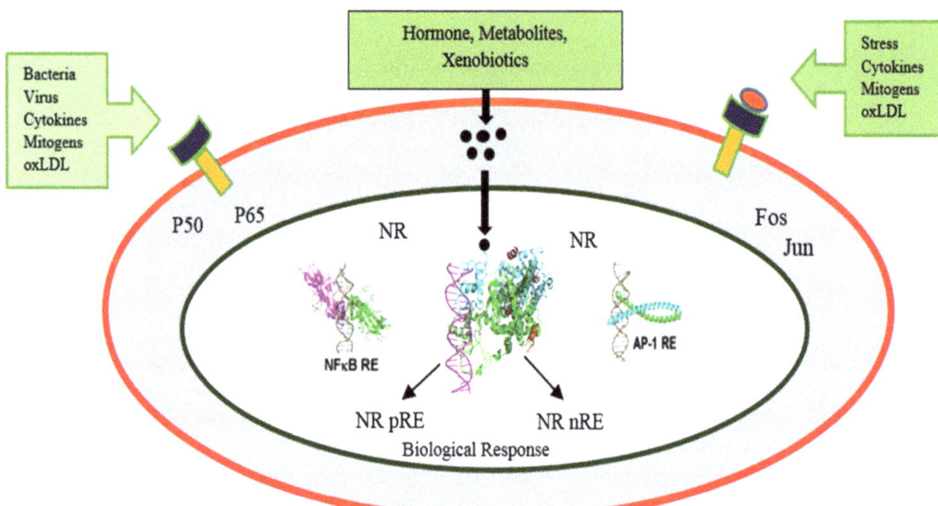

Figure 2. Nuclear receptors work in two apparent manners. Firstly, through the binding of a ligand, these receptors can frame, Heterodimers with RXR that outcomes in their connection with a specific positive response element of gene and, in this manner, can cause transcription of mRNA of genes that are targeted. On the other hand, repressive, negative response elements (nRE) have likewise been observed to interact with these receptors [9].

2.2. Nuclear Receptor Corepressor Binding

In the absence of ligands, the NRs are found to be associated with the corepressor complexes. These complexes consist of a subunit (SMRT/NcoR2 or NCoR1) that interacts directly with the receptor by means of the LXXLL motif that has a consensus sequence (L/I-X-X-I/V-I or LXXXI/LXXXI/L) also referred to as the CoRNR [18,19]. This CoRNR box motif interacts, as the coactivator LXXLL motif, with amino acids from the LBD hydrophobic groove. Part of the CoR binding interface is obscured as the remodeling of the binding of agonist and helix 12 positioning takes place. The availability of CoR binding interfaces as well as new CoRNR boxes indicates the use of alternate methods for the interaction of NR–corepressor [20–22]. The corepressor complexes are also developed around the NCoR or SMRT subunits that have a conserved repression domain and act as a vital point for core repressive machinery (such as GPS2, HDAC3, TBL1/TBLR1B) to assemble. Under certain situations, ligand-binding is adequate to prevent the recruitment of corepressors, such as RXR and TR, but in these cases, the active corepressor complex needs to be eliminated.

2.3. Nuclear Receptor Coactivator Binding

The identification of SRC-1/NCoA1 as a coactivator of the progesterone receptor [23] led to the discovery of more than 350 coactivators up to now. This tremendous volume of polypeptides had their involvement in different enzymatic processes related to the chromatin remodeling, histone modulation, transcription initiation and elongation, mRNA splicing and elongation and nuclear receptor complexes' proteasomal termination [24]. These coactivators are further categorized into two subfamilies. The members of the first subfamily of coactivators are involved in the direct interaction with NR AF-1 and 2 regions such as SRC coactivators, CBP and p300. The members of second subfamily coactivators can interact with primary coactivators such as CoCoA, CARM1 and Fli-I. The primary and secondary activators work in a coordinated manner to regulate the promoters [25].

p160 and p300 Families

Coactivators such as P160, p300 and cAMP-response element-binding protein (CBP) belongs to the p160 family such as NCoA1/SRC-1 and NCoA2/TIF2, commonly referred as SCRP1 or GRIP1, and possess binding affinity with the LBD of NR by means of an Alpha-Helical LXXLL motif [26]. The p160 family coactivators include NCOA 1/SRC-1, NCOA 2/TIF2 (SRC-2 or GRIP1) and NCoA3/RAC3, also known as SRC-3, ACTR, PCIP or TRAM-1. The p300 and CBP coactivators have a histone acetylase transferase (HAT) function, which plays a crucial role in NR-mediated transcriptional regulation [27]. The acetylation of histone H4 N-terminal tail inhibits the interactions of the histone H4 N-terminal with the histone H2A/H2B dimer and disrupts chromatin compaction [28]. Thus, the chromatin is then decondensed causing the initiation complex at the promoter site to be attached.

2.4. *Transcriptional Repression*

2.4.1. Transcriptional Repression by Unliganded Receptors

In the absence of a ligand, some nuclear receptors can effectively downregulate the transcription. Thus, corepressor complexes recruiting is linked to this process. The most widely studied complex is the nuclear receptor corepressor (NCoR) that acts as a silencing mediator of thyroid and retinoid receptors (SMRT), G-protein pathway suppressor 2 (GPS2), histone deacetylase 3 (HDAC3), TBL-1-like related protein (TBLR1) and transducin-α-like 1 (TBL1). A well-characterized function of HDACs in transcriptional repression is to create a condensed, transcriptional inactive chromatin structure by the N-terminal lysine of histone proteins deacetylation. The SMRT and NCoR have been reported to possess a deacetylase-activating domain which can activate the enzymatic activity of HDAC3 [29].

Moreover, some other corepressor complexes, such as the SWI/SNF complex, CoRest and PRC1 and 2 complexes, have been further identified. Similarly, the NCoR/SMRT complex can bind with multiprotein elements of the promoter's site, resulting in covalent histone and DNA changes, accompanied by chromatin contraction and/or DNA masking. The dissociation of the corepressor complex from the DNA-bound receptor is a key step in NR-mediated transcription activation. In vitro experiments along with the data from the crystal structures have shown that agonist-induced conformational changes are adequate for SMRT or NCoR alienation. However, dynamic models of de-repression involving post-translational modifications of corepressor complex subunits leading either to their nuclear exclusion and/or degradation have been described [30].

2.4.2. Direct Trans-Repression by Ligand Activated Receptors

The negative transcription regulation of certain genes can be repressed by ligand-bounded NRs. The mechanistic role of these ligand-bounded NRs has already been described in-depth for the GRs and TRs. These NRs have been suggested to recognize, bind and downregulate particular target genes. Studies have shown that the response elements which negatively regulate the glucocorticoid or rnGREs and thyroid elements or nTREs differ from those of response elements that positively control the activation of transcription process [31,32]. The negative response elements for GR and TR possess the overlapping binding sites that control the response elements transcriptional cis-effect for transcriptional factors such as Oct-1/Pbx, AP-1 and SP1 [33–35]. This indicates that the negative glucocorticoid reaction elements associated with other transcription factors can exert such an action. A recent study has identified a new class of negative glucocorticoid REs, which are arranged as 1 bp spacers inverted repeats and facilitate the glucocorticoids to promote the recruitment of GR–corepressor complexes [36]. For T3-mediated repression of transcription, a similar type of mechanistic principle does not exist. The SMRT corepressor insertion in nTRE enhances the histone deacetylation that has also been reported for the αTSH gene. The SMRT dissociation is associated with histone acetylation and gene suppression after treatment with an agonist [32,37]. Additionally, functional studies have revealed the involvement of SRC-1 in liganded TR transcriptional repression [38,39].

The mechanism involved in the reversal of the transcriptional function is not clear yet, but it can be regulated by means of post-translation changes, including acetylation or SUMOylation of promoter-associated histones, phosphorylation and/coregulatory proteins [24,40]. Therefore, direct repression could happen through distinct receptors and context-dependent pathways. These findings also indicated the versatility of coregulator complexes that either positively or negatively impact the products of the transcription following the stimulation by NR agonists.

2.4.3. Tethered Transrepression by Liganded Receptors

The process called the tethered transrepression contains negative crosstalks of ligand-activated nuclear receptors with other signal-dependent transcription factors, including NF-kappa-B and AP-1. Inflammation in different cells of the central nervous system, the immune system and in the liver, etc., is modulated by this process and also interferes with cell proliferation in many tissues. Various putative mechanisms have also been proposed to explain such repression: (i) the inhibition of PIC assembly on NF-kappa- or AP-1-regulated promoters; (ii) the inhibition of RNA polymerase II change to elongation-competent form; (iii) the upregulation of NK-kappa-B inhibitors [41]; (v) the coactivators exclusion by competitive inhibition [42,43]; (vi) direct physical interaction with AP-1 or NF-kappa-B subunits (p65 commonly) [43], but this mechanism is even more complicated and intricate with several other factors in the cell [44].

Moreover, after being partly affected by PPARγ, GR and LXR agonists, for each receptor, the inhibition was about one-third or half of the gene induced by TLR-3, 4 or 9 active macrophages inflammatory elements. Interestingly, each receptor was partly overlapping with inhibited clusters of genes [45].

The NR structural features unique to transrepression are not well described yet. Research using comprehensive mutagenesis of T3R, RAR, PPARμ, GR and ER has not provided a simple, harmonized model for tethered transrepression [46,47]. Thus, it is evident that the coactivators recruitment through the Domain AF-2, as well as direct DNA connections, is not necessary for this process. Furthermore, it also became apparent that homo- or heterodimerization was not obligatory [47,48]. The unavailability of defined molecular structures for transrepression is the major hindrance in devising screening methods for the detection of dissociated ligands that preferentially induce tethered transrepression in inflammatory diseases.

3. Role of PPARs and Coregulators in Energy Homeostasis

Energy is an absolute necessity to provide subsistence to all the living beings and is usually derived from the metabolism of ingested nutrients. Primarily, in human beings, glucose and long-chain fatty acids derived from food are utilized to produce energy. The cellular requirement for energy is satisfied through the oxidation reactions occurring in mitochondria to metabolize nutrients. Both in normal as well as in induced cells, the transcriptional regulation network controls the demand and supply of diverse cellular physiological states in both normal and induced cells such as during fasting or exercise. PPARs are classified as a part of the superfamily of nuclear receptors within this transcriptional network and control the nutrient-dependent transcription [49–51]. PPARs function as fatty acid sensors in order to control various metabolic processes, and different biological activities such as inflammation, adipogenesis, insulin sensitivity, lipid metabolism, reproduction as well as cell growth and differentiation. PPARs control these functions through the activation of target genes via the attachment of endogenous ligands to the receptors' ligand-binding domain. This binding leads to conformational change, which further enables PPARs to heterodimerize the retinoid X receptor and facilitate the attachment and dissociation of transcription-related important small accessory molecules. This heterodimerized complex formed at the PPAR response elements (PPREs) then transactivates target mitochondrial and peroxisome-related genes. This event cascade controls a protein network concerned with systemic energy homeostasis [51–53]. All the isotypes of PPARs have an indispensable role in lipid and fatty acid metabolism through direct binding or the modulation of the target genes related to fat metabolism [49]. Though all these PPAR isoforms share similar mode of action and function, distinct biological and pharmacological differences exist among them. The PPARβ/δ and PPARα have a metabolic role in the promotion of energy dissipation, but PPARγ stimulates the energy storage. PPARβ/δ improves fatty acids oxidation in different body tissues and also normalizes plasma lipid content. PPARβ/δ and PPARγ boost insulin sensitivity, while PPARα do not. PPARβ/δ-mediated glucose regulation is different from that of PPARγ; however, PPARβ/δ and PPARγ both are implicated in fiber distribution in the skeletal muscle, metabolism of glucose in the liver and controlling pancreatic islet cells function [53]. In lipid catabolism, the PPARα enhances the fatty acid oxidation during situations such as fasting, while PPARγ acts on the adipose tissue during the anabolic process to improve lipogenesis [54,55]. Thus, understanding the role of PPARs in energy homeostasis is important to further investigate the PPARs' role in producing energy in different body parts.

3.1. The ATP-Dependent Remodeling Complex SWI/SNF

In yeast, the SWI/SNF complex is an evolutionarily conserved multi-subunit complex, which uses the energy of ATP hydrolysis to mobilize nucleosomes and remodel chromatin and thereby regulate the transcription of target genes. In ATP-dependent chromatin restructuring [56], the evolutionary conserved SWI/SNF families play a significant role in catalyzing the DNA histone disruption and the nucleosome sliding around the DNA [57]. A multimeric agent of 1.2 MDa is the human homolog BAF complex of BRG1/hBRM, BAF Polypeptides (BAF45a/b/c/d, BAF47, BAF53a/b, BAF57, BAF155/170, BAF60, BAF250a, BAF200, Brd7 and Brd9) and actin. A number of these subunits have LXXLL motifs which have not only been identified in the form of ER, RR [58], RAR [59], FXR [60] and GR [61] coactivators but also as SHP corepressors [62], which also incorporated with corepressor complexes in order to combine SWI/SNF components [63]. Intriguingly, in the mouse liver, the BAF 60a subunit showed a circadian expression, which regulates the expression of clock and metabolic genes by acting as a coregulator of RORα [64].

3.2. The Mediator Complexes

The mediator complex was initially identified in yeast, such as the SWI/SNF complex, and consequently characterized in many other eukaryotic cells. Many studies have identified its role as a catalyst for the transcription pre-initiation complex, abbreviated as PIC, assembly of activated promoters. The mediator plays a major role in RNA II-controlled transcription mechanism by direct association with RNA polymerase II, such as TFIID and TFIIH, and elongation factors [65]. Different studies have reported the function of the mediator in NR and made it clear that complexes linked to mediators are specifically associated with NRs. The mediator consisted of four structural modules and had more than 20 subunits, among which LXXLL motifs [66] were developed from the Med1 subunit. The hepatic steatosis Med1 KO causes PPARα-dependent steatosis [67], in line with the coactivator functions of the liver [68] and PPARα [69].

3.3. PPARs Signaling in Different Body Parts

The liver is the prime body organ involved in energy metabolism to fulfill the body's energy requirements, and PPARα receptors are also distributed in the liver, which controls the uptake and breakdown of fatty acids through ketogenesis and β-oxidation in fasting conditions [70,71]. It has also been described that PPARα knockout in mice causes the suppression of fatty acids uptake and oxidation and the impairment of ketogenesis as well as gluconeogenesis. Furthermore, the function related evidence for PPARβ/δ were also reported as the PPARβ/δ knockout decreases the expression of genes involved in glucose and lipogenesis, while PPARβ/δ overexpression controls genes that are responsible for energy metabolism [49]. In the liver, PPARα is indispensable for glucose homeostasis. The mice deficient in PPARα showed a substantial blood glucose level reduction following

24 h of fasting. The upregulated expression of TRB3 (a positive controller of the cellular reaction to the levels of insulin and Akt/protein kinase B blocker) directly regulates the PPARα transcription that negatively influences insulin signaling [72]. Moreover, PPARα also enhances the production of acetyl-CoA enzyme and fatty acids oxidation through upregulating Acyl-CoA dehydrogenase expression in mitochondria. PPARα controls de novo lipogenesis in the case of positive energy balance to provide fatty acids, deposited in the form of triglycerides, which can also be employed during starvation [73].

Brown adipose tissue (BAT) acts as a caloric storage site, and white adipose tissue (WAT) as lipid storage also holds importance in energy homeostasis. These tissues are involved in endocrinal functions, which also release different types of hormones, including adipokines and cytokines, which subsequently initiate systemic energy metabolism signaling. Through feedback mechanisms, these tissues control energy homeostasis by receiving signals from the metabolic active sites in peripheral tissues and the central nervous system [74–76]. Substantially, PPARγ is expressed in these tissues and plays a lead role in the gene activation required for the uptake and deposition of fatty acids as well as the differentiation of adipose tissue [77]. Non-adipogenic cells are differentiated into adipocytes through the ectopic expression of PPARγ [78]. The PPARγ knockout in embryonic fibroblasts completely disrupts the differentiation process [79]. In vivo studies have revealed the importance of PPARγ for adipocytes production and survival in animals as negative mutations (heterozygous and dominant) in the PPARγ in humans cause lipodystrophy [15,80]. In BAT, PPARα controls the expression of mitochondrial uncoupling protein 1 (UCP1) and PGC1α, but the obliteration of PPARα decreases the protein expression upon exposure to normal and cold conditions while the fatty acids' metabolism is not affected. The enhanced energy metabolism has also been observed in response to the enhanced expression of the FAO gene induced by the activation of PPARα in human and murine adipocytes [49]. Liu et al. reported PPARγ as a positive regulator of milk fat synthesis in dairy cow mammary epithelial cells through improving cell viability, proliferation ability and triacylglycerol secretion [81]. It was also reported that acetic acid and palmitic acid could regulate milk fat synthesis in dairy cow mammary epithelial cells through PPARγ signaling. Shi et al. have cloned the PPARγ gene in the dairy goat mammary gland and explored its function in vitro [82]. It was reported that PPARγ in the goat mammary gland directly controls the synthesis of milk fat through the activation of the transcription regulators, such as sterol regulatory element-binding transcription factor-1 [82,83].

Skeletal body muscles are the significant sites for glucose usage mediated through insulin, lipids metabolism, glycogen storage and oxidation of fatty acid as well as regulation of HDL and cholesterol levels. PPARβ/δ expression is dominant in the skeletal muscles and controls the translation of genes associated with energy metabolism [71,84–86]. Moreover, it also regulates the activity of genes related to triglyceride hydrolysis, lipids uptake, fatty acids oxidation, and uncoupling proteins activation to liberate the energy required by OXPHOS. The protein CPT1 is also programmed to regulate the oxidation of the long-chain fatty acids. PPARβ/T activates the metabolic adaptability of the transcription factor FOXO1 and the pyruvate dehydrogenases kinase 4 (PDK4), which inhibits the complex of pyruvate dehydrogenase. This makes CPT1 a rate-limiting factor for the oxidation of carbohydrates in the muscles. Moreover, PDK4 also controls the regulation of several genes that are involved in lipid efflux, energy usage and increases β-oxidation of fatty acids [84,85]. Furthermore, in PPARβ/δ transgenic mice, metabolism of glucose was greatly amplified [84] as PPARβ could initiate the transcription of lactate dehydrogenase B (LDHB) to regulate the muscle fatty acid metabolism required for glucose oxidation [87]. On the other hand, PPARγ coactivator-1α or PGC-1α, which is a mitochondrial biogenesis regulator, controls the energy metabolism in skeletal muscle through catabolic reactions to produce aerobic ATP. The PPARβ/δ stimulates the expression of PGC-1α to control the skeletal muscles' metabolic activity by enhancing the synthesis of mitochondrial proteins [88–90].

The PPARα and PPARβ/δ are predominantly expressed in the intestines [91,92], and the triglycerides' metabolism in the intestine is crucial for systemic energy homeostasis. Dietary triglycerides are hydrolyzed into free fatty acids in the intestinal lumen and then taken up by epithelial cells of the intestine to the endoplasmic reticulum and again converted into triglycerides [92]. Studies in animals have shown a relationship between energy utilization, intestinal colonization and weight gain that controls the angiopoietin-like protein 4 (ANGPTL4) expression in the intestinal epithelium. ANGPTL4 is a secreted protein that regulates lipid and glucose homeostasis, and its deletion leads to changes in metabolism, reduced oils absorption in the intestine and intestinal mucosa thickening. PPARγ is reported to be involved in the regulation of the fatty acid metabolism via β-oxidation. PPARγ controls ANGPTL4 expression, and short-chain fatty acids stimulate PPARγ and are the major energy resources for colonocytes [93]. Wy-14643 is a PPARα agonist that stimulates the production of the enzymes implicated in fatty acid oxidation and ketogenesis, such as mitochondrial 3-hydroxy-3-methylglutaryl-CoA synthase and CPT1A in the small intestine [94]. PPARα also controls different phase I enzymes and transporters (related to oxidation) as well as uptake of fatty acids. PPARα is activated through the nutritional route and regulates fatty acid oxidation, cholesterol and glucose transporters [95]. PPARα is crucially involved in the regulation of the phage I/II metabolism and also controls the expression of transporter genes in the small intestine [96]. A synthetic agonist (K-877) of PPARα has been shown to control the intestinal fatty acid oxidation and mRNA expression of apo-lipoprotein while reducing plasma levels of triglycerides. The downregulation of Npc1l1 and upregulated expression of Abca1 has been observed in response to treatment with K-877. Npc1l1 is a rate-restricting transporter for absorption of cholesterol in the murine small intestine, while Abca1 is a vital molecule that participates in the production of HDL-C in the small intestine [97].

3.4. Energy Homeostasis through Co-Regulators of PPARs

The energy homeostasis could be controlled through feedback mechanisms involving various types of extraordinarily interconnected pathways. About 320 coregulators and 38 co-modulators for PPARs have been reported in the Nuclear Receptor Signaling Atlas. The direct interaction of PPARs and the crosstalk of PPARs with other pathways contribute to systemic energy homeostasis [98].

Balanced mitochondrial energy production is being regulated through the coordinated effect of coactivators and corepressors where PGC-1α and PPARγ act as co-modulators for the initiation of mitochondrial aerobic metabolism. However, the effect of PGC1α on mitochondria is antagonized via nuclear corepressor 1 (NCOR1). The knocking out of NCOR1 has been shown to imitate the overexpression of PGC-1α phenotypically, which is involved in the transcriptional output of ERR and PPARs. Nuclear receptor interacting protein 1 (NRIP1) interacts with both PPARs and ERR and decreases the target gene expression levels that participate in the consumption of energy. In previous studies, the mice with the deletion of NIRP were slim and presented enhanced insulin sensitivity and glucose tolerance and endurance to diet-induced obesity [50,98,99]. Hes6 protein, hepatocyte nuclear factor α (HNFα) and the PPARs coregulate each other's expression under different nutritional conditions and also control the transcription events during the metabolic reactions [80,100]. PPARγ, along with the transcription factor such as CCAAT/enhancer-binding protein α (C/EBPα), is a crucial regulator in the last phase of adipogenesis. Energy homeostasis by the mediator complex subunit 1 (MED1) through PPARs plays an essential role in a liver-specific knockout of MED1 and demonstrated impaired activities of PPARα and PPARγ in murine models [98].

Since the role of PPARs in different energy homeostasis cascades in various organs has been established, it can be stated that PPARs could be the target for the treatment of disorders, such as inflammation, obesity, diabetes, dyslipidemia, neurodegenerative disorders and cardiac myopathy, when these cascades are disrupted in disease conditions due to metabolic energy imbalance [15,101].

4. Nutritional Modulation of PPARs to Modify Gene Expression and Metabolic Networks

Dietary nutrients can modulate the metabolic networks of PPARs as nutrients, and their products directly control the PPAR activities through acting as natural ligands of PPARs. Diverse nutrients have been shown to affect the action of PPARs, but PPARs depict the greatest inclination for mono-unsaturated and poly-unsaturated fatty acids as demonstrated by different ligand-binding assays [102,103]. It is evident that each type of PPAR triggers a distinct gene network regardless of their overlapping expressions, which indicates the exhibition of ligand-specific properties of PPARs [103,104]. Furthermore, the administration of high-fat diet results in the modulation of PPARα target genes [105]. Comparative nutrigenomic study in mice revealed the influence of several individual dietary fatty acids on hepatic gene expression. These findings concluded that (1) an increase in the chain length of fatty acids and the extent of unsaturation enhanced the total genes that were upregulated and that (2) genes controlled through dietary unsaturated fatty acids do not change in the PPARα knockout murine model depicting PPARα as end target, and the expression levels of same genes were increased in the murine model after the administration with the PPARα activator WY14643 [106].

The modulation of PPAR expression and function through nutrients can be studied by imposing nutrient deprivation conditions on diverse tissues. The properties of all known PPARs are influenced in the fasting state; for example, PPARα signaling in the liver has shown to be upregulated via fasting through increased expression levels of the coactivator PGC-1α, and thus controls hepatic gluconeogenesis and fatty acid oxidation [107,108]. Furthermore, increased expression levels of PPARδ during fasting are affected by plasma fatty acids derived from adipose, hence highlighting a distinct task as a plasma fatty acid sensor in the liver for PPARδ [108].

Several nutrients and their derivatives are being observed for the modulation of PPARs to modify metabolic networks and gene expression through direct and indirect mechanisms. Macronutrients such as nucleotides, fatty acids and their metabolites, amino acids, monosaccharides and micronutrients, such as vitamins, can control the expression of specific genes directly by interacting with transcription factors in the promoter region through cis-regulatory elements. However, many nutrients regulate genes indirectly by modulating the intracellular action/secretion of hormones, such as thyroid hormone, glucocorticoids, glucagon and insulin, which alters the gene expression and thus improves metabolic networks. Many dietary nutrients have been shown to modulate the expression of PPARs in animals (Figure 3), among which some significant factors are described below.

4.1. Poly Unsaturated Fatty Acids (PUFA)

Polyunsaturated fatty acids are categorized as *n-3* and *n-6* fatty acids and could exert opposing effects on receptor signaling. Out of these two classes, *n-3* fatty acids are shown to have an agonistic effect, while *n-6* fatty acids are reported to be inhibitory [109]. PUFAs are shown to bind directly to the PPARα and are involved in the activation of transcription, thus controlling metabolic networks. It has been reported that PUFAs are required in the μM range to bind with PPARα, and these could be derived from dietary nutrients [110]. Interestingly, *n-3* fatty acids are reported to be greater activators of PPARα as compared to *n-6* fatty acids in vivo [111]. Furthermore, many eicosanoids and their derivatives are shown to activate PPARα with a high affinity than other PUFA precursors [112]. Studies have represented that acylethanolamines, including oleoylethanolamide (OEA), palmitoylethanolamide (PEA) and anandamide (AEA) are also PPARα activators [113]. Moreover, PPARα activation by oleoylethanolamide (OEA) leads to appetite and lipolysis suppression, while palmitoylethanolamide (PEA) exerts anti-inflammatory activity when activating the PPARα [114]. The ligands for PPARα are also known to bind PPARβ/δ, but their activation is lower than the PPARα. PUFAs also serve as ligands for PPARγ and are involved in the activation of PPARγ. For example, n-3 fatty acid activates the PPARγ and can result in the prevention of high-fat-diet-induced inflammation in adipose tissues [115]. Collectively, PUFAs are the natural ligands for all the subtypes of PPARs, but their subsequent activation potential varies. These molecules control the PPARs activity in the body and thus have a role in regulating metabolic networks. Although various studies have reported their mechanism of action to activate PPARs, further research is still needed to elucidate the mechanisms of PPARs activation and their distribution.

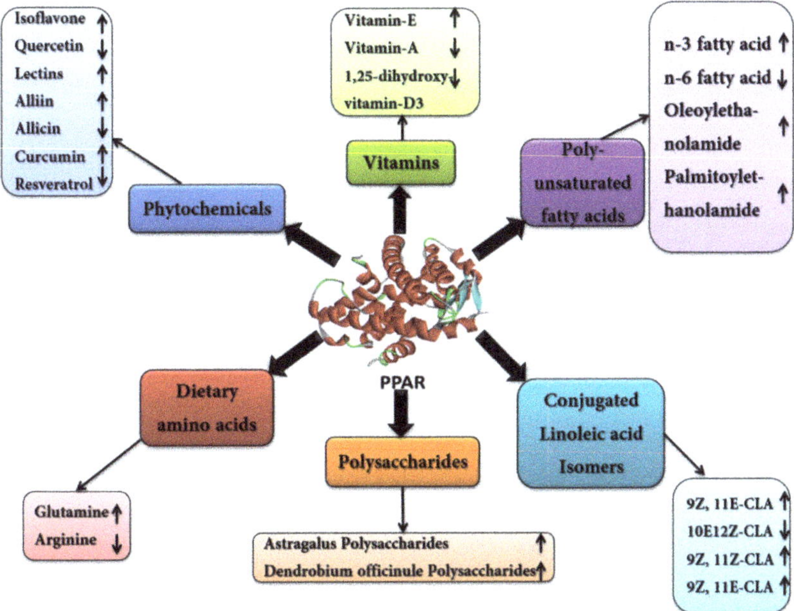

Figure 3. The effect of different nutrients on PPAR. Different nutrients regulate PPAR either by its upregulation or downregulation. The arrow up shows the upregulation of PPAR, while the arrow down shows the downregulation by respective nutrients.

4.2. Conjugated Linoleic-Acids (CLAs)

CLAs are the fatty acids mainly found in foods obtained from ruminant animals [116] and are positional (*cis*- or *trans*-double bond positioning at 7, 9; 8, 10; 9, 11; 10, 12; or 11, 13) and geometrical isomers of the parent linoleic acid molecule (*cis*-9, *cis*-12-18:2, *n*-6). Rumenic acid (9Z, 11*E*-octadecenoic acid, C18:2) is the most abundant natural CLA isomer (over 75–80%) produced through the biohydrogenation of nutritive LAs by ruminant microflora. Because of their numerous health benefits, CLAs are currently being used as nutritional supplements for changing body composition in livestock and humans [117,118], but the mechanisms of the useful properties of CLAs are yet to be explored. CLA isomers serve as ligands for PPARγ, PPARβ/δ and PPARα [119,120], showing differential PPAR activation and health benefits [118,121] (Table 2).

Additionally, a mixture of CLA isomers, i.e., 9Z, 11Z-CLA and 9Z, 11E-CLA, can notably activate the PPARβ/δ in preadipocytes [122]. Therefore, minor structural changes in many CLA isomers can be differentiated by important cellular mechanisms to allow specie and tissue-specific responses [123]. These findings concluded that CLA affects the production of eicosanoids directly or indirectly, abolishes the NF-κB pathway, improvises the activation of PPARγ and decreases proinflammatory cytokines for useful effects on inflammation, ultimately manipulating metabolic syndrome-related conditions, including IR, atherosclerosis and obesity [124]. Therefore, CLAs can directly employ anti-inflammatory properties by modulating the expression of inflammatory mediators through PPARγ-dependent or NF-κB-dependent pathways.

4.3. Dietary Amino Acids

Some of the dietary amino acids have shown the potential to modulate the activity of PPARs, in which glutamine and arginine are the major ones. Glutamine is considered an essential amino acid in situations of metabolic stress and is found to be a special substrate of enterocytes. To date, only a single study has reported the impact of glutamine on PPARγ articulation. Sato et al. examined the impacts of luminal glutamine and arginine on the activity of PPARγ in gut ischemia-reperfusion of a rat model. Luminal glutamine increased the expression of PPARγ, while arginine did not show any significant effect on PPARγ. Furthermore, they also evaluated the effect of a PPARγ antagonist (GW9662) on the action of glutamine. The pre-treatment with GW9662 revokes the impact of glutamine, revealing that glutamine may likewise be a PPARγ agonist, thus signifying its role in metabolic stress [125].

Moreover, the effect of arginine on a gut injury has been investigated, and the supplementation of arginine, which is considered an immune-nutrient, demonstrated a beneficial effect on LPS-induced gut injuries in a pig model [81]. Additionally, upon treatment with arginine, there was a decrease in jejunal TNFa, and an increase in the expression of PPARγ was also observed.

4.4. Vitamins and Minerals

4.4.1. Beta Carotene, Vitamin A, and Its Derivatives

In mammals, beta carotene (BC) is the precursor of apo-carotenoid molecules, i.e., retinoids (vitamin A and its derivatives) [126]. There is an increasing sign that BC and retinoids can affect the physiology of adipocytes as signaling molecules by acting on adiposity in humans [127]. The levels of circulating BC are inversely associated with the risk of human type-2 diabetes [128–130], while the decreased levels of plasma carotenoids, including BC, are usually found in obese children [131].

The BC 15,15′-monooxygenase (*Bcmo1*) is the major contributing enzyme for the production of retinoid, which converts BC into all-*trans*-retinal [132]. Its expression is controlled by PPAR-γ [133,134] induced during the differentiation of adipocyte [135], and *Bcmo1* knockout mice showed an enhanced expression of PPAR-γ genes in fat-deposits and are very susceptible to fat-induced obesity [132]. Retinaldehyde, the primary product of BC cleavage, inhibits the activity of PPAR-γ both in mouse models and adipocyte cell cultures [136]. The role of *Bcmo1* is verified in signaling of the RA receptor (RAR) and the production of Retinoic acid (RA) in adipocytes [135]. Furthermore, BC-derived long-chain apo-carotenoids, such as β-13-apocarotenone, proved to inhibit the activity of retinoid X receptor-alpha (RXRα), while β-apo-149-carotenal hinders the adipogenesis and activity of PPAR-γ in cell culture [137]. BC supplementation can reduce the activity of PPAR-γ and downregulate its target genes, decreasing the adiposity of mice. Thus, BC can significantly control the adiposity in mice, and *Bcmo1* critically regulates the PPAR-γ, which is the key element for the connection between PPAR-γ and RAR-signaling pathways that ultimately control the body adiposity [138].

4.4.2. Vitamin E: Alpha Tocopherols and Tocotrienols

Vitamin E is the fat-soluble vitamin family comprised of 8-lipophilic natural compounds including four tocotrienols with an unsaturated-isoprenoid sidechain designated as α, β, γ, and δ, and four tocopherols with a saturated phytyl-tail [139,140]. Soybean, cottonseed and corn are the commercially produced vegetable oils that have high amounts of most common dietary tocopherols (α- and γ-tocopherols) [141,142]. Both α- and γ-tocopherol shown to activate expression of PPAR-γ and transactivation of cancer cells in the colon, but α-tocopherol modulate PPAR-γ expression better than γ-tocopherol [143,144].

Tocotrienols are non-toxic naturally occurring compounds used as dietary supplements to prevent damage with aging due to dysregulated inflammatory responses. Recently, in vivo anti-inflammatory properties of dietary supplements evaluated in mice and chickens with two natural proteasome-inhibitors, i.e., δ-tocotrienol and quercetin [145,146], revealed decreased levels of nitric oxide [147] and serum tumor necrosis factor-alpha (TNF-α). Furthermore, the direct effect of tocotrienols on lipidic metabolism, with an anti-atherogenic effect on rats, humans and mice, has been also reported.

In vitro studies revealed that tocotrienols inhibit the 3-hydroxy-3-methyl-glutaryl-CoA reductase and consequently decrease cholesterol synthesis. For instance, the body fats in rats were decreased by the oral administration of a tocotrienol-rich fraction (TRF) of palm oil containing γ-tocotrienol, while in an in vitro study, the phosphorylation of Akt in 3T3-L1 preadipocytes and adipocyte differentiation was suppressed by TRF through the reduced expression of insulin-induced PPAR-γ [148]. Tocotrienol can serve as an anti-adipogenic vitamin due to nutrient-mediated regulation of body fat, but further research is required in this regard.

4.4.3. Retinoic Acid and 1,25-Dihydroxy Vitamin D3 (1,25$(OH)_2D_3$)

Some of the properties of RA, such as the deposition of fats [149], adipocyte differentiation [150,151] and the expression of adipokines, such as resistin, leptin and serum retinol binding protein, is facilitated by RAR, which interferes with the activity of PPAR-γ [149,151]. The 1,25-dihydroxy vitamin D3 (1,25 (OH)2 D3) is an active form of vitamin D and has been shown to restrict the adipogenesis in the bone marrow of SAM-P/6 mice associated with decreased expression of PPAR-γ2 [152]. The suppressed expression of PPAR-γ2 by RA and 1,25-dihydroxy vitamin D3 inhibit the differentiation of adipocytes in 3T3-L1 preadipocytes [153].

Table 2. Effect of different nutrients on PPARs modulation.

Nutrients	PPARs Regulation	References
Polyunsaturated fatty acids (PUFA)		
n-3 fatty acids	Activate both PPARα and PPARγ and lead to prevention of inflammation in adipocytes	[109,115]
n-6 fatty acid	Inhibitors of PPAR receptor signalling and regulate metabolic network	[109]
Oleoylethanolamide	Activate PPARα and induce lipolysis	[114]
Palmitoylethanolamide	Activate PPARα and provide anti-inflammatory activity	[114]
Conjugated Linoleic acid (CLA) Isomers		
9Z, 11E-CLA	Enhance PPAR-γ activation and exerts strong anti-cancer effects	[119,154]
10E, 12Z-CLA	Inhibits the PPAR-γ activation causing inflammation, IR and adipocyte delipidation	[155]
9Z, 11Z-CLA and 9Z, 11E-CLA	Enhanced activation of PPARβ/δ in preadipocytes	[122]
Dietary Amino acids		
Glutamine	Increase the expression of PPARγ and prevent metabolic stress	[125]
Arginine	Decrease the jejunal TNFa and increase the expression of PPARγ and beneficial against gut injury	[81]
Vitamins		
Vitamin-A [Beta Carotene (BC)]	BC supplementation can reduce the activity of PPAR-γ	[138]
Vitamin- E (Tocopherols)	α-tocopherol modulate PPAR-γ expression better than γ-tocopherol	[143,144]
1,25-dihydroxy vitamin-D3	Decrease the expression of PPAR-γ2 and regulate lipid metabolism	[152]

Table 2. *Cont.*

Nutrients	PPARs Regulation	References
Phytochemicals		
Isoflavone	Act as a ligand for PPAR to regulate lipid metabolism	[156]
Quercetin	Inhibits the activity of all isoform of PPARs except that of PPAR-γ and prevent accumulation of fat in the liver	[119]
Lectins	Up-regulate the PPAR-γ2 and provide an adipogenic effect on mesenchymal cells	[155]
Alliin	Activates the PPAR-γ and provides a cardioprotective effect	[157]
Allicin	Inhibit the PPAR-γ2 and therefore inhibits the differentiation and inflammation of the human preadipocytes	[158]
Curcumin	Activates the PPAR-γ and confer antioxidant and anti-inflammatory activity	[159]
Resveratrol	Down-regulate PPAR-γ1−3 mRNA expression in humans and provide anti-diabetic and anti-obesity effects	[160]
Triterpenes	Suppress PPAR-γ expression and prevent cancer development	[161]
Polysaccharides	Suppress PPAR-γ expression and exert anti-cancerous activity	[161]

4.5. Phytochemicals

Plants possess biological active chemical compounds that are known as phytochemicals [162]. Importantly, flavonoids, lectins, alliin, allicin, curcumin, triterpenes and resveratrol have been observed to regulate lipid and glucose metabolism through the modulation of PPAR [127].

A soy isoflavone, genistein, regulates lipid metabolism by activating the PPAR-γ or PPAR-γ-independent mechanism [156]. On the contrary, quercetin is a flavonol that inhibits the activity of all isoforms of PPAR except PPAR-γ and prevents fat accumulation in the liver [119]. In literature, chicken feed supplemented with quercetin has prevented dysregulation of inflammatory responses by downregulating NO and TNF-α during aging, while isoflavones have the potential to induce cancer via hormone-dependent regulation [154].

The bioactivity of lectins from vegan sources has been reported several times related to immune responses or gastrointestinal tract during allergens exposure. Moreover, its adipogenic effect has also been reported in humans and animal tissues [155]. Banana, garlic and dietary lectins boosted the hematopoietic stem progenitor cell pool in addition to an adipogenic effect on mesenchymal cells of mice by enhancing PPAR-γ2 expression. Furthermore, these dietary lectins interact with insulin receptors and activate the Mitogen-activated protein kinase (MEK)-dependent Extracellular signal-regulated kinase (ERK) pathway [163].

For more than 5000 years, garlic has been used as a culinary spice and medicinal herb. It has abundant antioxidant and organosulfur compounds that impart antibacterial and anti-infectious properties to it. Alliin and allicin are organosulfurated compounds extracted from garlic that possess a cardioprotective effect and anti-inflammatory effect, respectively [157]. Alliin lowers the TNF-α serum level in humans while allicin lowers or inhibits the expression of the CCAAT-enhancer-binding protein, CCAAT-enhancer-binding protein-alpha (C/EBP)-α and PPAR-γ2 during human preadipocytes differentiation [158].

Curcumin is the principal component of turmeric with potent anti-inflammatory, anti-cancerous, and antioxidant activities. Curcumin can suppress sepsis through PPAR-γ and decrease IFN-γ production in primed lymphocytes and *iNOS* gene expression in infected macrophages [159]. On the other hand, resveratrol, which has strong antioxidant properties, along with anti-obesity, anti-carcinogenic, neuroprotective, anti-aging, anti-diabetic and analgesic activity, targets PPAR-γ [164]. Resveratrol modulates white adipose-tissue metabolism and prevents dysregulation of advanced glycosylation end-products (AGE) via PPAR-γ mediated suppression of receptor for AGE in macrophage. It upregulates SIRT1, FOXO1 and adiponectin and downregulates PPAR-γ1−3 mRNA expression in human visceral adipocytes [160].

Polysaccharides and triterpenes have been used as a treatment for atherosclerosis and inhibit invasive behavior, angiogenesis and proliferation in cancer models. Moreover, they significantly promote adiponectin production and adipocyte differentiation by the downregulation of PPAR-γ, SREBP-1c and C/EBP and suppress the expression of genes involved in lipid transport, synthesis and storage [161].

5. Biological Benefits of PPARs Modulation in Dairy Animals

Overall, the comparison of the function of PPARs in humans, mice and ruminants has revealed that PPAR isotypes have a similar role in the metabolism of lipids in different species, including dairy animals. However, some differences are found in their specific roles across species as PPARs are found to be more specific for unsaturated FAs in monogastric species, while in ruminants, the PPARs are more specific for saturated long-chain FAs [165]. Aside from their role in the metabolism of lipids, the PPARs also influence immune responses through the modulation of immune signaling pathways such as AP-1, STAT-1 and NFkb through protein–protein DNA-independent interactions, a greater yield of milk production and controlling metabolic stress in dairy animals [166].

5.1. Energy Metabolism and Lipid Oxidation in Various Organs

PPARα is known to have an important role in fatty acid catabolism in mitochondria, peroxisomes and microsomes in the liver. PPARs are also believed to have a role in maintaining energy balance in various animals, such as ruminants and goats. Among its energy metabolism roles, carnitine homeostasis is well established in a diverse range of mammalian species, such as chicken, humans, mice, pigs, rats and diary animals [165]. Aside from PPARα, a few studies on the role of PPARβ/δ in energy metabolism have also been conducted. However, one role that is continuously associated with it is lipid metabolism. In the mammalian skeletal and heart muscles, the PPARβ/δ regulates fatty acid catabolism. These PPAR isotypes have also been found to have a crucial role in all of the reproductive tissues that have been investigated so far. In addition to this, research shows that PPARβ/δ has an important function in the metabolism of lipids in goat mammary cells, specifically lipid oxidation and secretion [83].

5.2. Adipogenesis and Milk Production

In ovines and bovines, PPARγ performs a critical role in adipogenesis, and its expression in the adipose tissues of these animals is high. PPARγ appears to be the critical mediator of lipogenesis by reacting quickly to stimuli signals in response to dietary energy intake [165]. For adipogenesis, PPARγ expression is both a requirement and a necessity. In bovines, PPARγ activation in bovine adipose tissue results in the upregulation of the genes that permit pre-adipocytes to differentiate into mature adipocytes or cells capable of storing triacylglycerol (TAG). One of the significant roles it plays in dairy animals from an economic benefit point of view is controlling the synthesis of milk fat in bovines. The PPAR expression increases during the transition state, which prevents the animal from metabolic stress [167]. The PPARγ gene is conserved in bovines, goats, humans and mice, as revealed by homology alignments. The research by Shi et al. confirmed the importance of PPARγ in modulating milk fat production. These findings showed that PPARγ is involved in regulating TAG production and release in the mammary cells of goats, highlighting the PPARγ's functional significance of milk production in mammary cells. PPARγ affects the expression of genes involved in the regulation of milk fat in primary goat mammary epithelial cells as revealed by using a combination of experimental approaches, including gene expression analysis, PPARγ-specific activation, luciferase-PPRE tests and siRNA interference [82].

5.3. Controlling Inflammation

In dairy animals, PPARs activation is also associated with anti-inflammatory effects in dairy cows during their transition period [168]. A Japanese group showed for the first time that PPARγ could perform an anti-inflammatory effect in dairy animals by injecting human recombinant TNF with agonist Thiazolidinedione TZD into dairy steers for 9 days. They discovered that TZD therapy partially restored TNF-induced insulin resistance [169]. The TZD impact was most likely due to increased insulin signaling via PPARs activation, which also counteracted TNF [170]. In dairy animals, the anti-inflammatory impact of PPAR is evoked not just by mitigating the effects of TNF but also by lowering TNF synthesis. Bovine peripheral blood mononuclear cells treated in vitro with 100 M of conjugated linoleic acid isomer (t10, c12-CLA) inhibited the TNF production, thus controlling the inflammation and overexpression of pro-inflammatory cytokines [171]. These pieces of evidence highlight the importance of PPARs modulated through their agonists and antagonists to control inflammatory stress in dairy animals and can be further used as targets to reduce metabolic stresses.

5.4. PPARs and Fatty Liver Syndrome of Dairy Animals

Since PPARs bind to and are activated by long-chain fatty acids and their metabolites, the PPARs play an extremely important role in nutrient metabolism. Due to the industrialization of the dairy industry and extensive farming that aims to maximize profits from each dairy animal, milk and meat production of dairy animals, such as cattle and bovines, has increased. Among many factors, one such factor is the use of energy-rich diets containing high amounts of carbohydrates and lipids. However, the intake of such high caloric diets in combination with the sedentary rearing system in the industrial farms, diseases such as non-alcoholic fatty liver (NAFLD) have become common, which has substantially increased the morbidity and mortality rates of these animals around the world. Among various NAFLD conditions, fatty liver syndrome constitutes one of the most common disorders that affect dairy cows and buffaloes during the perinatal period, and it is triggered by a negative nutritional balance following calving. Dairy cows are a good model for studying pathologies such as fatty liver syndrome and various other NAFLD types and their etiology [172]. A recent piece of research revealed that fatty liver disease in dairy cows during early lactation is linked to poor hepatic mitochondrial activity (like abnormal acetylation of amino acid lysine). PPARγ has a role in NAFLD through regulation of glucose and lipid metabolism and differentiation of adipocytes as well as modulation of inflammatory responses in the liver. PPARγ controls the expression of various target genes in adipocytes, participates in the adipocyte differentiation, influences lipid metabolism and principally controls signal transduction in the pancreatic islet cells, all of which contribute to the onset and progression of NAFLD. PPARγ is important in modulating lipid oxidation and lipogenesis, in addition to its role in adipocyte development [173].

5.5. PPARs Interaction with Gut Microbiome and Animal Health

There is a dearth of information related to the role of PPARs and gut microbiome in dairy animals and how their dysregulation leads to various pathologies related to metabolic stress. A recent review by Hassan et al. has provided interesting information regarding the interaction between gut microbiome and PPARs and their overall effect on the health of humans. As high-calorie diets containing high carbohydrates diminish good bacteria that aid in metabolizing various nutrients, low-energy diets containing high fiber do the opposite. It increases the population of certain bacteria such as *Bifidobacterium*, *Lactococcus* and *Streptococcus* that have been shown to alleviate fatty liver syndrome. This is attributed to the increase in the population of folate-producing bacteria whose metabolic product, folate, induces PPARα, which is involved in lipid oxidation in the liver [174]. Since isotypes such as PPARγ are highly conserved among mammals, some of the information should be implicated in dairy animals to improve their economic output and health.

One of the effects of high-calorie diets in all mammals is the modulation of gut microbiota. High-calorie intake is thought to affect gut microbial balance via the TLR4–PPAR pathway. This can lead to (1) a rise in the number of microorganisms that produce inflammasomes (lipopolysaccharides) and (2) a reduction in the short-chain FAs. As a result, systemic inflammation rises, whereas the synthesis of short-chain FAs falls. Short-chain FAs activate PPARs in adipose tissue to control lipolytic genes, such as adipose triglyceride lipase and hormone-sensitive lipase, as well as lipogenic genes such as glycerol kinase and phosphoenolpyruvate carboxykinase, which aid in the appropriate metabolization and utilization of lipids. PPARs activity, on the other hand, declines as a result of a shortage of short-chain fatty acids, resulting in the formation of extra fat and subsequently its storage and increased inflammation [175].

5.6. Other Benefits

Extra-hepatic signals, including hepatokines such as fibroblast growth factor 21 (FGF21) and angiopoietin-like 4 (ANGPTL4) [176,177], have been described in monogastrics as PPARs targets that perform a key role in bovines related to the adaptation of tissues to low-energy state levels of the body, such as undernutrition, fasting and transition to lactation [178,179].

The downregulation of the glucose transport mechanism in the bovine endothelial cells caused by excessive glucose has been also shown to be regulated by PPAR/δ [180]. It has been previously demonstrated that activated PPARδ suppresses the solute carrier family 2 member 1 (or facilitated glucose transporter GLUT1) expression while, at the same time, increasing the calreticulin expression, a protein that promotes GLUT1 mRNA degradation. Given the low levels of circulating glucose in ruminants (<4 mM in dairy cows) [181] against ca. 5 mM in humans and >6 mM in mice), the condition investigated in the study (high glucose) has presumably minimal significance for ruminants. However, GLUT1 is among the most significant glucose transporters whose expression rises dramatically during the lactation period in mammary tissue of dairy cows [182]. Moreover, the modulation of glucose transport by PPARβ/δ might have ramifications in milk synthesis. As a result, these PPARs isotypes could be crucial in providing glucose for lactose production. Moreover, PPAR β/δ expression is significantly decreased in the mammary glands during lactation [183], which coincides with an upsurge in the expression of numerous glucose transporters, particularly GLUT1. If it is true, it opens up the possibility of employing PPARβ/δ antagonists to boost milk production.

Since PPARs have been identified as promising targets for improving metabolism and general wellbeing through nutritional interventions and various agonists and antagonists, further investigations are essentially required to provide physiological insights into the therapeutic role of PPARs in addressing various metabolic disorders in dairy animals.

6. Conclusions

PPARs are considered to be major nuclear receptors to control energy homeostasis in the body through various mechanisms in different body parts. Various nutrients can act as a ligand for PPARs for their modulation, in which PUFAs, dietary amino acids, vitamins and phytochemicals are the major ones. These nutrients modulate PPARs by regulating their expression and signaling in different body parts and lead to the control of metabolic networks.

Author Contributions: Conceptualization, S.u.R., Z.L. and K.C.; material searching, F.-u.H., A.N., M.J. and Q.L.; resources, S.u.R., Q.L., Z.L. and K.C.; writing—original draft preparation, F.-u.H. and A.N.; writing—review and editing, F.-u.H., A.N., M.J., S.u.R., J.A., M.S.-u.R., Q.L., Z.L. and K.C. All authors have read and agreed to the published version of the manuscript.

Funding: The present study was supported by the National Natural Science Fund (U20A2051, 31760648 and 31860638), Guangxi Distinguished scholars Program (201835) and Guangxi University Post-doctorate Fellowship Research Grant (A3130051019).

Institutional Review Board Statement: Not applicable.

Informed Consent Statement: Not applicable.

Data Availability Statement: Not applicable.

Conflicts of Interest: The authors declare no conflict of interest.

References

1. Trevisi, E.; Amadori, M.; Archetti, I.; Lacetera, N.; Bertoni, G. Inflammatory response and acute phase proteins in the transition period of high-yielding dairy cows. *Acute Phase Proteins Early Non-Specif. Biomark. Hum. Vet. Dis.* **2011**, *15*, 355–379.
2. Sordillo, L.M.; Contreras, G.; Aitken, S.L. Metabolic factors affecting the inflammatory response of periparturient dairy cows. *Anim. Health Res. Rev.* **2009**, *10*, 53–63. [CrossRef] [PubMed]
3. Cai, D.; Liu, H.; Zhao, R. Nuclear Receptors in Hepatic Glucose and Lipid Metabolism During Neonatal and Adult Life. *Curr. Protein Pept. Sci.* **2017**, *18*, 548–561. [CrossRef] [PubMed]
4. Mangelsdorf, D.J.; Thummel, C.; Beato, M.; Herrlich, P.; Schütz, G.; Umesono, K.; Blumberg, B.; Kastner, P.; Mark, M.; Chambon, P. The nuclear receptor superfamily: The second decade. *Cell* **1995**, *83*, 835. [CrossRef]
5. Wu, Q.; Chambliss, K.; Umetani, M.; Mineo, C.; Shaul, P.W. Non-nuclear estrogen receptor signaling in the endothelium. *J. Biol. Chem.* **2011**, *286*, 14737–14743. [CrossRef]
6. Sun, Z.; Xu, Y. Nuclear receptor coactivators (NCOAs) and corepressors (NCORs) in the brain. *Endocrinology* **2020**, *161*, bqaa083. [CrossRef]
7. Burris, T.P.; Busby, S.A.; Griffin, P.R. Targeting orphan nuclear receptors for treatment of metabolic diseases and autoimmunity. *Chem. Biol.* **2012**, *19*, 51–59. [CrossRef]
8. Sever, R.; Glass, C.K. Signaling by nuclear receptors. *Cold Spring Harb. Perspect. Biol.* **2013**, *5*, a016709. [CrossRef]
9. Pawlak, M.; Lefebvre, P.; Staels, B. General molecular biology and architecture of nuclear receptors. *Curr. Top. Med. Chem.* **2012**, *12*, 486–504. [CrossRef]
10. Dreyer, C.; Krey, G.; Keller, H.; Givel, F.; Helftenbein, G.; Wahli, W. Control of the peroxisomal β-oxidation pathway by a novel family of nuclear hormone receptors. *Cell* **1992**, *68*, 879–887. [CrossRef]
11. Issemann, I.; Green, S. Activation of a member of the steroid hormone receptor superfamily by peroxisome proliferators. *Nature* **1990**, *347*, 645–650. [CrossRef]
12. Waku, T.; Shiraki, T.; Oyama, T.; Fujimoto, Y.; Maebara, K.; Kamiya, N.; Jingami, H.; Morikawa, K. Structural insight into PPARγ activation through covalent modification with endogenous fatty acids. *J. Mol. Biol.* **2009**, *385*, 188–199. [CrossRef]
13. Forman, B.M.; Chen, J.; Evans, R.M. The Peroxisome Proliferator-activated Receptors: Ligands and Activators a. *Ann. N. Y. Acad. Sci.* **1996**, *804*, 266–275. [CrossRef]
14. Bionaz, M.; Chen, S.; Khan, M.J.; Loor, J.J. Functional role of PPARs in ruminants: Potential targets for fine-tuning metabolism during growth and lactation. *PPAR Res.* **2013**, *2013*, 684159. [CrossRef]
15. Wang, Y.-X. PPARs: Diverse regulators in energy metabolism and metabolic diseases. *Cell Res.* **2010**, *20*, 124–137. [CrossRef]
16. Lee, C.-H.; Chinpaisal, C.; Wei, L.-N. Cloning and characterization of mouse RIP140, a corepressor for nuclear orphan receptor TR2. *Mol. Cell. Biol.* **1998**, *18*, 6745–6755. [CrossRef]
17. Fernandes, I.; Bastien, Y.; Wai, T.; Nygard, K.; Lin, R.; Cormier, O.; Lee, H.S.; Eng, F.; Bertos, N.R.; Pelletier, N. Ligand-dependent nuclear receptor corepressor LCoR functions by histone deacetylase-dependent and-independent mechanisms. *Mol. Cell* **2003**, *11*, 139–150. [CrossRef]
18. Hu, X.; Li, Y.; Lazar, M.A. Determinants of CoRNR-dependent repression complex assembly on nuclear hormone receptors. *Mol. Cell. Biol.* **2001**, *21*, 1747–1758. [CrossRef]
19. Hu, X.; Lazar, M.A. The CoRNR motif controls the recruitment of corepressors by nuclear hormone receptors. *Nature* **1999**, *402*, 93–96. [CrossRef]
20. Varlakhanova, N.; Snyder, C.; Jose, S.; Hahm, J.B.; Privalsky, M.L. Estrogen receptors recruit SMRT and N-CoR corepressors through newly recognized contacts between the corepressor N terminus and the receptor DNA binding domain. *Mol. Cell. Biol.* **2010**, *30*, 1434–1445. [CrossRef]
21. Le Maire, A.; Teyssier, C.; Erb, C.; Grimaldi, M.; Alvarez, S.; De Lera, A.R.; Balaguer, P.; Gronemeyer, H.; Royer, C.A.; Germain, P. A unique secondary-structure switch controls constitutive gene repression by retinoic acid receptor. *Nat. Struct. Mol. Biol.* **2010**, *17*, 801–807. [CrossRef] [PubMed]
22. Phelan, C.A.; Gampe, R.T., Jr.; Lambert, M.H.; Parks, D.J.; Montana, V.; Bynum, J.; Broderick, T.M.; Hu, X.; Williams, S.P.; Nolte, R.T. Structure of Rev-erbα bound to N-CoR reveals a unique mechanism of nuclear receptor–co-repressor interaction. *Nat. Struct. Mol. Biol.* **2010**, *17*, 808. [CrossRef] [PubMed]
23. Onate, S.A.; Tsai, S.Y.; Tsai, M.-J.; O'Malley, B.W. Sequence and characterization of a coactivator for the steroid hormone receptor superfamily. *Science* **1995**, *270*, 1354–1357. [CrossRef] [PubMed]

24. Han, S.J.; Lonard, D.M.; O'Malley, B.W. Multi-modulation of nuclear receptor coactivators through posttranslational modifications. *Trends Endocrinol. Metab.* **2009**, *20*, 8–15. [CrossRef] [PubMed]
25. Dasgupta, S.; Lonard, D.M.; O'Malley, B.W. Nuclear receptor coactivators: Master regulators of human health and disease. *Annu. Rev. Med.* **2014**, *65*, 279–292. [CrossRef]
26. Chen, D.; Huang, S.-M.; Stallcup, M.R. Synergistic, p160 coactivator-dependent enhancement of estrogen receptor function by CARM1 and p300. *J. Biol. Chem.* **2000**, *275*, 40810–40816. [CrossRef]
27. Roth, S.Y.; Denu, J.M.; Allis, C.D. Histone acetyltransferases. *Annu. Rev. Biochem.* **2001**, *70*, 81–120. [CrossRef]
28. Luger, K.; Mäder, A.W.; Richmond, R.K.; Sargent, D.F.; Richmond, T.J. Crystal structure of the nucleosome core particle at 2.8 Å resolution. *Nature* **1997**, *389*, 251–260. [CrossRef]
29. Li, J.; Wang, J.; Wang, J.; Nawaz, Z.; Liu, J.M.; Qin, J.; Wong, J. Both corepressor proteins SMRT and N-CoR exist in large protein complexes containing HDAC3. *EMBO J.* **2000**, *19*, 4342–4350. [CrossRef]
30. Perissi, V.; Jepsen, K.; Glass, C.K.; Rosenfeld, M.G. Deconstructing repression: Evolving models of co-repressor action. *Nat. Rev. Genet.* **2010**, *11*, 109–123. [CrossRef]
31. Sakai, D.D.; Helms, S.; Carlstedt-Duke, J.; Gustafsson, J.-A.; Rottman, F.M.; Yamamoto, K.R. Hormone-mediated repression: A negative glucocorticoid response element from the bovine prolactin gene. *Genes Dev.* **1988**, *2*, 1144–1154. [CrossRef]
32. Chatterjee, V.; Lee, J.-K.; Rentoumis, A.; Jameson, J.L. Negative regulation of the thyroid-stimulating hormone alpha gene by thyroid hormone: Receptor interaction adjacent to the TATA box. *Proc. Natl. Acad. Sci. USA* **1989**, *86*, 9114–9118. [CrossRef]
33. Diamond, M.I.; Miner, J.N.; Yoshinaga, S.K.; Yamamoto, K.R. Transcription factor interactions: Selectors of positive or negative regulation from a single DNA element. *Science* **1990**, *249*, 1266–1272. [CrossRef]
34. Subramaniam, N.; Cairns, W.; Okret, S. Glucocorticoids repress transcription from a negative glucocorticoid response element recognized by two homeodomain-containing proteins, Pbx and Oct-1. *J. Biol. Chem.* **1998**, *273*, 23567–23574. [CrossRef]
35. Villa, A.; Santiago, J.; Belandia, B.; Pascual, A. A response unit in the first exon of the β-amyloid precursor protein gene containing thyroid hormone receptor and Sp1 binding sites mediates negative regulation by 3, 5, 3′-triiodothyronine. *Mol. Endocrinol.* **2004**, *18*, 863–873. [CrossRef]
36. Surjit, M.; Ganti, K.P.; Mukherji, A.; Ye, T.; Hua, G.; Metzger, D.; Li, M.; Chambon, P. Widespread negative response elements mediate direct repression by agonist-liganded glucocorticoid receptor. *Cell* **2011**, *145*, 224–241. [CrossRef]
37. Berghagen, H.; Ragnhildstveit, E.; Krogsrud, K.; Thuestad, G.; Apriletti, J.; Saatcioglu, F. Corepressor SMRT functions as a coactivator for thyroid hormone receptor T3Rα from a negative hormone response element. *J. Biol. Chem.* **2002**, *277*, 49517–49522. [CrossRef]
38. Takeuchi, Y.; Murata, Y.; Sadow, P.; Hayashi, Y.; Seo, H.; Xu, J.; O'Malley, B.W.; Weiss, R.E.; Refetoff, S. Steroid receptor coactivator-1 deficiency causes variable alterations in the modulation of T3-regulated transcription of genes in vivo. *Endocrinology* **2002**, *143*, 1346–1352. [CrossRef]
39. Weiss, R.E.; Xu, J.; Ning, G.; Pohlenz, J.; O'Malley, B.W.; Refetoff, S. Mice deficient in the steroid receptor co-activator 1 (SRC-1) are resistant to thyroid hormone. *EMBO J.* **1999**, *18*, 1900–1904. [CrossRef]
40. Wang, D.; Xia, X.; Weiss, R.E.; Refetoff, S.; Yen, P.M. Distinct and histone-specific modifications mediate positive versus negative transcriptional regulation of TSHα promoter. *PLoS ONE* **2010**, *5*, e9853. [CrossRef]
41. Auphan, N.; DiDonato, J.A.; Rosette, C.; Helmberg, A.; Karin, M. Immunosuppression by glucocorticoids: Inhibition of NF-κB activity through induction of IκB synthesis. *Science* **1995**, *270*, 286–290. [CrossRef] [PubMed]
42. Saijo, K.; Collier, J.G.; Li, A.C.; Katzenellenbogen, J.A.; Glass, C.K. An ADIOL-ERβ-CtBP transrepression pathway negatively regulates microglia-mediated inflammation. *Cell* **2011**, *145*, 584–595. [CrossRef] [PubMed]
43. Evans, M.J.; Eckert, A.; Lai, K.; Adelman, S.J.; Harnish, D.C. Reciprocal antagonism between estrogen receptor and NF-κB activity in vivo. *Circ. Res.* **2001**, *89*, 823–830. [CrossRef] [PubMed]
44. Beck, I.M.; Vanden Berghe, W.; Vermeulen, L.; Yamamoto, K.R.; Haegeman, G.; De Bosscher, K. Crosstalk in inflammation: The interplay of glucocorticoid receptor-based mechanisms and kinases and phosphatases. *Endocr. Rev.* **2009**, *30*, 830–882. [CrossRef]
45. Ogawa, S.; Lozach, J.; Benner, C.; Pascual, G.; Tangirala, R.K.; Westin, S.; Hoffmann, A.; Subramaniam, S.; David, M.; Rosenfeld, M.G. Molecular determinants of crosstalk between nuclear receptors and toll-like receptors. *Cell* **2005**, *122*, 707–721. [CrossRef]
46. Sanchez-Pacheco, A.; Martínez-Iglesias, O.; Mendez-Pertuz, M.; Aranda, A. Residues K128, 132, and 134 in the thyroid hormone receptor-α are essential for receptor acetylation and activity. *Endocrinology* **2009**, *150*, 5143–5152. [CrossRef]
47. Benkoussa, M.; Brand, C.; Delmotte, M.-H.; Formstecher, P.; Lefebvre, P. Retinoic acid receptors inhibit AP1 activation by regulating extracellular signal-regulated kinase and CBP recruitment to an AP1-responsive promoter. *Mol. Cell. Biol.* **2002**, *22*, 4522–4534. [CrossRef]
48. Rauch, A.; Seitz, S.; Baschant, U.; Schilling, A.F.; Illing, A.; Stride, B.; Kirilov, M.; Takacz, A.; Schmidt-Ullrich, R.; Ostermay, S. Glucocorticoids suppress bone formation by attenuating osteoblast differentiation via the monomeric glucocorticoid receptor. *Cell Metab.* **2010**, *11*, 517–531. [CrossRef]
49. Fan, W.; Evans, R. PPARs and ERRs: Molecular mediators of mitochondrial metabolism. *Curr. Opin. Cell Biol.* **2015**, *33*, 49–54. [CrossRef]
50. Kota, B.P.; Huang, T.H.-W.; Roufogalis, B.D. An overview on biological mechanisms of PPARs. *Pharm. Res.* **2005**, *51*, 85–94. [CrossRef]

51. Ahmed, W.; Ziouzenkova, O.; Brown, J.; Devchand, P.; Francis, S.; Kadakia, M.; Kanda, T.; Orasanu, G.; Sharlach, M.; Zandbergen, F. PPARs and their metabolic modulation: New mechanisms for transcriptional regulation? *J. Intern. Med.* **2007**, *262*, 184–198. [CrossRef]
52. Wu, J.; Chen, L.; Zhang, D.; Huo, M.; Zhang, X.; Pu, D.; Guan, Y. Peroxisome proliferator-activated receptors and renal diseases. *Front. Biosci.* **2009**, *14*, 995–1009. [CrossRef]
53. Dubois, V.; Eeckhoute, J.; Lefebvre, P.; Staels, B. Distinct but complementary contributions of PPAR isotypes to energy homeostasis. *J. Clin. Investig.* **2017**, *127*, 1202–1214. [CrossRef]
54. Kersten, S.; Desvergne, B.; Wahli, W. Roles of PPARs in health and disease. *Nature* **2000**, *405*, 421–424. [CrossRef]
55. Ahmed, I.; Rehman, S.U.; Shahmohamadnejad, S.; Zia, M.A.; Ahmad, M.; Saeed, M.M.; Akram, Z.; Iqbal, H.; Liu, Q. Therapeutic Attributes of Endocannabinoid System against Neuro-Inflammatory Autoimmune Disorders. *Molecules* **2021**, *26*, 3389. [CrossRef]
56. York, B.; O'Malley, B.W. Steroid receptor coactivator (SRC) family: Masters of systems biology. *J. Biol. Chem.* **2010**, *285*, 38743–38750. [CrossRef]
57. Liu, N.; Hayes, J.J. When push comes to shove: SWI/SNF uses a nucleosome to get rid of a nucleosome. *Mol. Cell* **2010**, *38*, 484–486. [CrossRef]
58. Chiba, H.; Muramatsu, M.; Nomoto, A.; Kato, H. Two human homologues of Saccharomyces cerevisiae SWI2/SNF2 and Drosophila brahma are transcriptional coactivators cooperating with the estrogen receptor and the retinoic acid receptor. *Nucleic Acids Res.* **1994**, *22*, 1815–1820. [CrossRef]
59. Link, K.A.; Burd, C.J.; Williams, E.; Marshall, T.; Rosson, G.; Henry, E.; Weissman, B.; Knudsen, K.E. BAF57 governs androgen receptor action and androgen-dependent proliferation through SWI/SNF. *Mol. Cell. Biol.* **2005**, *25*, 2200–2215. [CrossRef]
60. Miao, J.; Fang, S.; Lee, J.; Comstock, C.; Knudsen, K.E.; Kemper, J.K. Functional specificities of Brm and Brg-1 Swi/Snf ATPases in the feedback regulation of hepatic bile acid biosynthesis. *Mol. Cell. Biol.* **2009**, *29*, 6170–6181. [CrossRef]
61. Hsiao, P.-W.; Fryer, C.J.; Trotter, K.W.; Wang, W.; Archer, T.K. BAF60a mediates critical interactions between nuclear receptors and the BRG1 chromatin-remodeling complex for transactivation. *Mol. Cell. Biol.* **2003**, *23*, 6210–6220. [CrossRef]
62. Kemper, J.K.; Kim, H.; Miao, J.; Bhalla, S.; Bae, Y. Role of an mSin3A-Swi/Snf chromatin remodeling complex in the feedback repression of bile acid biosynthesis by SHP. *Mol. Cell. Biol.* **2004**, *24*, 7707–7719. [CrossRef]
63. Underhill, C.; Qutob, M.S.; Yee, S.-P.; Torchia, J. A novel nuclear receptor corepressor complex, N-CoR, contains components of the mammalian SWI/SNF complex and the corepressor KAP-1. *J. Biol. Chem.* **2000**, *275*, 40463–40470. [CrossRef]
64. Tao, W.; Chen, S.; Shi, G.; Guo, J.; Xu, Y.; Liu, C. SWItch/sucrose nonfermentable (SWI/SNF) complex subunit BAF60a integrates hepatic circadian clock and energy metabolism. *Hepatology* **2011**, *54*, 1410–1420. [CrossRef]
65. Conaway, R.C.; Conaway, J.W. Origins and activity of the Mediator complex. *Semin. Cell Dev. Biol.* **2011**, *22*, 729–734. [CrossRef]
66. Taatjes, D.J. The human Mediator complex: A versatile, genome-wide regulator of transcription. *Trends Biochem. Sci.* **2010**, *35*, 315–322. [CrossRef]
67. Bai, L.; Jia, Y.; Viswakarma, N.; Huang, J.; Vluggens, A.; Wolins, N.E.; Jafari, N.; Rao, M.S.; Borensztajn, J.; Yang, G. Transcription coactivator mediator subunit MED1 is required for the development of fatty liver in the mouse. *Hepatology* **2011**, *53*, 1164–1174. [CrossRef]
68. Ge, K.; Cho, Y.-W.; Guo, H.; Hong, T.B.; Guermah, M.; Ito, M.; Yu, H.; Kalkum, M.; Roeder, R.G. Alternative mechanisms by which mediator subunit MED1/TRAP220 regulates peroxisome proliferator-activated receptor γ-stimulated adipogenesis and target gene expression. *Mol. Cell. Biol.* **2008**, *28*, 1081–1091. [CrossRef]
69. Ge, K.; Guermah, M.; Yuan, C.-X.; Ito, M.; Wallberg, A.E.; Spiegelman, B.M.; Roeder, R.G. Transcription coactivator TRAP220 is required for PPARγ2-stimulated adipogenesis. *Nature* **2002**, *417*, 563–567. [CrossRef]
70. Pyper, S.R.; Viswakarma, N.; Yu, S.; Reddy, J.K. PPARα: Energy combustion, hypolipidemia, inflammation and cancer. *Nucl. Recept. Signal.* **2010**, *8*, nrs.08002. [CrossRef]
71. Dressel, U.; Allen, T.L.; Pippal, J.B.; Rohde, P.R.; Lau, P.; Muscat, G.E. The peroxisome proliferator-activated receptor β/δ agonist, GW501516, regulates the expression of genes involved in lipid catabolism and energy uncoupling in skeletal muscle cells. *Mol. Endocrinol.* **2003**, *17*, 2477–2493. [CrossRef] [PubMed]
72. Lefebvre, P.; Chinetti, G.; Fruchart, J.-C.; Staels, B. Sorting out the roles of PPARα in energy metabolism and vascular homeostasis. *J. Clin. Investig.* **2006**, *116*, 571–580. [CrossRef] [PubMed]
73. Feingold, K.R.; Wang, Y.; Moser, A.; Shigenaga, J.K.; Grunfeld, C. LPS decreases fatty acid oxidation and nuclear hormone receptors in the kidney. *J. Lipid Res.* **2008**, *49*, 2179–2187. [CrossRef] [PubMed]
74. Choe, S.S.; Huh, J.Y.; Hwang, I.J.; Kim, J.I.; Kim, J.B. Adipose tissue remodeling: Its role in energy metabolism and metabolic disorders. *Front. Endocrinol.* **2016**, *7*, 30. [CrossRef]
75. Birsoy, K.; Festuccia, W.T.; Laplante, M. A comparative perspective on lipid storage in animals. *J. Cell Sci.* **2013**, *126*, 1541–1552. [CrossRef] [PubMed]
76. Rosen, E.D.; Spiegelman, B.M. Adipocytes as regulators of energy balance and glucose homeostasis. *Nature* **2006**, *444*, 847–853. [CrossRef] [PubMed]
77. Tontonoz, P.; Hu, E.; Graves, R.A.; Budavari, A.I.; Spiegelman, B.M. mPPAR gamma 2: Tissue-specific regulator of an adipocyte enhancer. *Genes Dev.* **1994**, *8*, 1224–1234. [CrossRef] [PubMed]
78. Tontonoz, P.; Hu, E.; Spiegelman, B.M. Stimulation of adipogenesis in fibroblasts by PPARγ2, a lipid-activated transcription factor. *Cell* **1994**, *79*, 1147–1156. [CrossRef]

79. Rosen, E.D.; Hsu, C.-H.; Wang, X.; Sakai, S.; Freeman, M.W.; Gonzalez, F.J.; Spiegelman, B.M. C/EBPα induces adipogenesis through PPARγ: A unified pathway. *Genes Dev.* **2002**, *16*, 22–26. [CrossRef]
80. Mullican, S.E.; DiSpirito, J.R.; Lazar, M.A. The orphan nuclear receptors at their 25-year reunion. *J. Mol. Endocrinol.* **2013**, *51*, T115–T140. [CrossRef]
81. Liu, Y.; Huang, J.; Hou, Y.; Zhu, H.; Zhao, S.; Ding, B.; Yin, Y.; Yi, G.; Shi, J.; Fan, W. Dietary arginine supplementation alleviates intestinal mucosal disruption induced by Escherichia coli lipopolysaccharide in weaned pigs. *Br. J. Nutr.* **2008**, *100*, 552–560. [CrossRef]
82. Shi, H.; Luo, J.; Zhu, J.; Li, J.; Sun, Y.; Lin, X.; Zhang, L.; Yao, D.; Shi, H. PPARγ regulates genes involved in triacylglycerol synthesis and secretion in mammary gland epithelial cells of dairy goats. *PPAR Res.* **2013**, *2013*, 310948. [CrossRef]
83. Shi, H.; Zhang, C.; Zhao, W.; Luo, J.; Loor, J. Peroxisome proliferator-activated receptor delta facilitates lipid secretion and catabolism of fatty acids in dairy goat mammary epithelial cells. *J. Dairy Sci.* **2017**, *100*, 797–806. [CrossRef]
84. Manickam, R.; Wahli, W. Roles of Peroxisome Proliferator-Activated Receptor β/δ in skeletal muscle physiology. *Biochimie* **2017**, *136*, 42–48. [CrossRef]
85. Cho, S.Y.; Jeong, H.W.; Sohn, J.H.; Seo, D.-B.; Kim, W.G.; Lee, S.-J. An ethanol extract of Artemisia iwayomogi activates PPARδ leading to activation of fatty acid oxidation in skeletal muscle. *PLoS ONE* **2012**, *7*, e33815. [CrossRef]
86. Periasamy, M.; Herrera, J.L.; Reis, F.C. Skeletal muscle thermogenesis and its role in whole body energy metabolism. *Diabetes Metab. J.* **2017**, *41*, 327–336. [CrossRef]
87. Gan, Z.; Burkart-Hartman, E.M.; Han, D.-H.; Finck, B.; Leone, T.C.; Smith, E.Y.; Ayala, J.E.; Holloszy, J.; Kelly, D.P. The nuclear receptor PPARβ/δ programs muscle glucose metabolism in cooperation with AMPK and MEF2. *Genes Dev.* **2011**, *25*, 2619–2630. [CrossRef]
88. Lamichane, S.; Lamichane, B.D.; Kwon, S.-M. Pivotal Roles of Peroxisome Proliferator-Activated. *PPARs Cell. Whole Body Energy Metab.* **2019**, *19*, 382.
89. Pérez-Schindler, J.; Svensson, K.; Vargas-Fernández, E.; Santos, G.; Wahli, W.; Handschin, C. The coactivator PGC-1α regulates skeletal muscle oxidative metabolism independently of the nuclear receptor PPARβ/δ in sedentary mice fed a regular chow diet. *Diabetologia* **2014**, *57*, 2405–2412. [CrossRef]
90. Thach, T.T.; Lee, C.-K.; woo Park, H.; Lee, S.-J.; Lee, S.-J. Syringaresinol induces mitochondrial biogenesis through activation of PPARβ pathway in skeletal muscle cells. *Bioorgan. Med. Chem. Lett.* **2016**, *26*, 3978–3983. [CrossRef]
91. Vrins, C.L.; van der Velde, A.E.; van den Oever, K.; Levels, J.H.; Huet, S.; Elferink, R.P.O.; Kuipers, F.; Groen, A.K. Peroxisome proliferator-activated receptor delta activation leads to increased transintestinal cholesterol efflux. *J. Lipid Res.* **2009**, *50*, 2046–2054. [CrossRef]
92. Higashimura, Y.; Naito, Y.; TAkAGI, T.; Uchiyama, K.; Mizushima, K.; YOSHIkAWA, T. Propionate promotes fatty acid oxidation through the up-regulation of peroxisome proliferator-activated receptor α in intestinal epithelial cells. *J. Nutri. Sci. Vitaminol.* **2015**, *61*, 511–515. [CrossRef]
93. Korecka, A.; de Wouters, T.; Cultrone, A.; Lapaque, N.; Pettersson, S.; Doré, J.; Blottière, H.M.; Arulampalam, V. ANGPTL4 expression induced by butyrate and rosiglitazone in human intestinal epithelial cells utilizes independent pathways. *Am. J. Physiol.-Gastrointest. Liver Physiol.* **2013**, *304*, G1025–G1037. [CrossRef]
94. Azari, E.K.; Leitner, C.; Jaggi, T.; Langhans, W.; Mansouri, A. Possible role of intestinal fatty acid oxidation in the eating-inhibitory effect of the PPAR-α agonist Wy-14643 in high-fat diet fed rats. *PLoS ONE* **2013**, *8*, e74869.
95. Bünger, M.; de Groot, P.J.; Bosch-Vermeulen, H.; Hooiveld, G.J.; Muller, M. PPARalpha-mediated effects of dietary lipids on intestinal barrier gene expression. *BMC Genom.* **2008**, *9*, 231.
96. Van den Bosch, H.M.; Bünger, M.; de Groot, P.J.; van der Meijde, J.; Hooiveld, G.J.; Müller, M. Gene expression of transporters and phase I/II metabolic enzymes in murine small intestine during fasting. *BMC Genom.* **2007**, *8*, 267. [CrossRef]
97. Takei, K.; Nakagawa, Y.; Wang, Y.; Han, S.-i.; Satoh, A.; Sekiya, M.; Matsuzaka, T.; Shimano, H. Effects of K-877, a novel selective PPARα modulator, on small intestine contribute to the amelioration of hyperlipidemia in low-density lipoprotein receptor knockout mice. *J. Pharmacol. Sci.* **2017**, *133*, 214–222. [CrossRef]
98. Lempradl, A.; Pospisilik, J.A.; Penninger, J.M. Exploring the emerging complexity in transcriptional regulation of energy homeostasis. *Nat. Rev. Genet.* **2015**, *16*, 665–681. [CrossRef]
99. Yamamoto, H.; Williams, E.G.; Mouchiroud, L.; Canto, C.; Fan, W.; Downes, M.; Héligon, C.; Barish, G.D.; Desvergne, B.; Evans, R.M. NCoR1 is a conserved physiological modulator of muscle mass and oxidative function. *Cell* **2011**, *147*, 827–839. [CrossRef]
100. Martinez-Jimenez, C.P.; Kyrmizi, I.; Cardot, P.; Gonzalez, F.J.; Talianidis, I. Hepatocyte nuclear factor 4α coordinates a transcription factor network regulating hepatic fatty acid metabolism. *Mol. Cell. Biol.* **2010**, *30*, 565–577. [CrossRef]
101. Tyagi, S.; Gupta, P.; Saini, A.S.; Kaushal, C.; Sharma, S. The peroxisome proliferator-activated receptor: A family of nuclear receptors role in various diseases. *J. Adv. Pharm. Technol. Res.* **2011**, *2*, 236. [CrossRef] [PubMed]
102. Krey, G.; Braissant, O.; L'Horset, F.; Kalkhoven, E.; Perroud, M.; Parker, M.G.; Wahli, W. Fatty acids, eicosanoids, and hypolipidemic agents identified as ligands of peroxisome proliferator-activated receptors by coactivator-dependent receptor ligand assay. *Mol. Endocrinol.* **1997**, *11*, 779–791. [CrossRef] [PubMed]
103. Itoh, T.; Fairall, L.; Amin, K.; Inaba, Y.; Szanto, A.; Balint, B.L.; Nagy, L.; Yamamoto, K.; Schwabe, J.W. Structural basis for the activation of PPARγ by oxidized fatty acids. *Nat. Struct. Mol. Biol.* **2008**, *15*, 924–931. [CrossRef] [PubMed]

104. Xu, H.E.; Lambert, M.H.; Montana, V.G.; Parks, D.J.; Blanchard, S.G.; Brown, P.J.; Sternbach, D.D.; Lehmann, J.M.; Wisely, G.B.; Willson, T.M. Molecular recognition of fatty acids by peroxisome proliferator–activated receptors. *Mol. Cell* **1999**, *3*, 397–403. [CrossRef]
105. Cavalieri, D.; Calura, E.; Romualdi, C.; Marchi, E.; Radonjic, M.; Van Ommen, B.; Müller, M. Filling gaps in PPAR-alpha signaling through comparative nutrigenomics analysis. *BMC Genom.* **2009**, *10*, 596. [CrossRef]
106. Sanderson, L.M.; de Groot, P.J.; Hooiveld, G.J.; Koppen, A.; Kalkhoven, E.; Müller, M.; Kersten, S. Effect of synthetic dietary triglycerides: A novel research paradigm for nutrigenomics. *PLoS ONE* **2008**, *3*, e1681. [CrossRef]
107. Yoon, J.C.; Puigserver, P.; Chen, G.; Donovan, J.; Wu, Z.; Rhee, J.; Adelmant, G.; Stafford, J.; Kahn, C.R.; Granner, D.K. Control of hepatic gluconeogenesis through the transcriptional coactivator PGC-1. *Nature* **2001**, *413*, 131–138. [CrossRef]
108. Rakhshandehroo, M.; Knoch, B.; Müller, M.; Kersten, S. Peroxisome proliferator-activated receptor alpha target genes. *PPAR Res.* **2010**, *2010*, 612089. [CrossRef]
109. Schmitz, G.; Ecker, J. The opposing effects of n− 3 and n− 6 fatty acids. *Prog. Lipid Res.* **2008**, *47*, 147–155. [CrossRef]
110. Khan, S.A.; Heuvel, J.P.V. Reviews: Current topicsrole of nuclear receptors in the regulation of gene expression by dietary fatty acids. *J. Nutr. Biochem.* **2003**, *14*, 554–567. [CrossRef]
111. Power, G.W.; Newsholme, E.A. Dietary fatty acids influence the activity and metabolic control of mitochondrial carnitine palmitoyltransferase I in rat heart and skeletal muscle. *J. Nutr.* **1997**, *127*, 2142–2150. [CrossRef]
112. Yu, K.; Bayona, W.; Kallen, C.B.; Harding, H.P.; Ravera, C.P.; McMahon, G.; Brown, M.; Lazar, M.A. Differential activation of peroxisome proliferator-activated receptors by eicosanoids. *J. Biol. Chem.* **1995**, *270*, 23975–23983. [CrossRef]
113. Borrelli, F.; Izzo, A.A. Role of acylethanolamides in the gastrointestinal tract with special reference to food intake and energy balance. *Best Pract. Res. Clin. Endocrinol. Metab.* **2009**, *23*, 33–49. [CrossRef]
114. O'Sullivan, S. Cannabinoids go nuclear: Evidence for activation of peroxisome proliferator-activated receptors. *Br. J. Pharmacol.* **2007**, *152*, 576–582. [CrossRef]
115. Huber, J.; Löffler, M.; Bilban, M.; Reimers, M.; Kadl, A.; Todoric, J.; Zeyda, M.; Geyeregger, R.; Schreiner, M.; Weichhart, T. Prevention of high-fat diet-induced adipose tissue remodeling in obese diabetic mice by n-3 polyunsaturated fatty acids. *Int. J. Obesity* **2007**, *31*, 1004–1013. [CrossRef]
116. Kim, S.; Shin, H.-J.; Kim, S.Y.; Kim, J.H.; Lee, Y.S.; Kim, D.-H.; Lee, M.-O. Genistein enhances expression of genes involved in fatty acid catabolism through activation of PPARα. *Mol. Cell. Endocrinol.* **2004**, *220*, 51–58. [CrossRef]
117. Ricketts, M.-L.; Moore, D.D.; Banz, W.J.; Mezei, O.; Shay, N.F. Molecular mechanisms of action of the soy isoflavones includes activation of promiscuous nuclear receptors. A review. *J. Nutr. Biochem.* **2005**, *16*, 321–330. [CrossRef]
118. Mezei, O.; Li, Y.; Mullen, E.; Ross-Viola, J.S.; Shay, N.F. Dietary isoflavone supplementation modulates lipid metabolism via PPARα-dependent and-independent mechanisms. *Physiol. Genom.* **2006**, *26*, 8–14. [CrossRef]
119. Yeh, S.-L.; Yeh, C.-L.; Chan, S.-T.; Chuang, C.-H. Plasma rich in quercetin metabolites induces G2/M arrest by upregulating PPAR-γ expression in human A549 lung cancer cells. *Planta Med.* **2011**, *77*, 992–998. [CrossRef]
120. Kobori, M.; Masumoto, S.; Akimoto, Y.; Oike, H. Chronic dietary intake of quercetin alleviates hepatic fat accumulation associated with consumption of a Western-style diet in C57/BL6J mice. *Mol. Nutr. Food Res.* **2011**, *55*, 530–540. [CrossRef]
121. McMichael-Phillips, D.F.; Harding, C.; Morton, M.; Roberts, S.A.; Howell, A.; Potten, C.S.; Bundred, N.J. Effects of soy-protein supplementation on epithelial proliferation in the histologically normal human breast. *Am. J. Clin. Nutr.* **1998**, *68*, 1431S–1435S. [CrossRef] [PubMed]
122. Alibin, C.P.; Kopilas, M.A.; Anderson, H.D. Suppression of Cardiac Myocyte Hypertrophy by Conjugated Linoleic Acid Role of Peroxisome Proliferator-Activated Receptors α and γ. *J. Biol. Chem.* **2008**, *283*, 10707–10715. [CrossRef] [PubMed]
123. Hodge, G.; Hodge, S.; Han, P. Allium sativum (garlic) suppresses leukocyte inflammatory cytokine production in vitro: Potential therapeutic use in the treatment of inflammatory bowel disease. *Cytom. J. Int. Soc. Anal. Cytol.* **2002**, *48*, 209–215. [CrossRef] [PubMed]
124. Iciek, M.; Kwiecień, I.; Włodek, L. Biological properties of garlic and garlic-derived organosulfur compounds. *Environ. Mol. Mutagenes.* **2009**, *50*, 247–265. [CrossRef]
125. Sato, N.; Moore, F.; Kone, B.; Zou, L.; Smith, M.; Childs, M.; Moore-Olufemi, S.; Schultz, S.; Kozar, R. Differential induction of PPAR-γ by luminal glutamine and iNOS by luminal arginine in the rodent postischemic small bowel. *Am. J. Physiol.-Gastrointest. Liver Physiol.* **2006**, *290*, G616–G623. [CrossRef]
126. Von Lintig, J. Colors with functions: Elucidating the biochemical and molecular basis of carotenoid metabolism. *Annu. Rev. Nutr.* **2010**, *30*, 35–56. [CrossRef]
127. Ortuño Sahagún, D.; Márquez-Aguirre, A.; Quintero-Fabián, S.; López-Roa, R.; Rojas-Mayorquín, A. Modulation of PPAR-γ by nutraceutics as complementary treatment for obesity-related disorders and inflammatory diseases. *PPAR Res.* **2012**, *2012*, 318613. [CrossRef]
128. Ford, E.S.; Will, J.C.; Bowman, B.A.; Narayan, K.V. Diabetes mellitus and serum carotenoids: Findings from the Third National Health and Nutrition Examination Survey. *Am. J. Epidemiol.* **1999**, *149*, 168–176. [CrossRef]
129. Ylönen, K.; Alfthan, G.; Groop, L.; Saloranta, C.; Aro, A.; Virtanen, S.M.; Group, B.R. Dietary intakes and plasma concentrations of carotenoids and tocopherols in relation to glucose metabolism in subjects at high risk of type 2 diabetes: The Botnia Dietary Study. *Am. J. Clin. Nutr.* **2003**, *77*, 1434–1441. [CrossRef]

130. Coyne, T.; Ibiebele, T.I.; Baade, P.D.; Dobson, A.; McClintock, C.; Dunn, S.; Leonard, D.; Shaw, J. Diabetes mellitus and serum carotenoids: Findings of a population-based study in Queensland, Australia. *Am. J. Clin. Nutr.* **2005**, *82*, 685–693. [CrossRef]
131. Burrows, T.L.; Warren, J.M.; Colyvas, K.; Garg, M.L.; Collins, C.E. Validation of overweight children's fruit and vegetable intake using plasma carotenoids. *Obesity* **2009**, *17*, 162–168. [CrossRef]
132. Hessel, S.; Eichinger, A.; Isken, A.; Amengual, J.; Hunzelmann, S.; Hoeller, U.; Elste, V.; Hunziker, W.; Goralczyk, R.; Oberhauser, V. CMO1 deficiency abolishes vitamin A production from β-carotene and alters lipid metabolism in mice. *J. Biol. Chem.* **2007**, *282*, 33553–33561. [CrossRef]
133. Boulanger, A.; McLemore, P.; Copeland, N.G.; Gilbert, D.J.; Jenkins, N.A.; Yu, S.S.; Gentleman, S.; Redmond, T.M. Identification of beta-carotene 15, 15′-monooxygenase as a peroxisome proliferator-activated receptor target gene. *FASEB J.* **2003**, *17*, 1304–1306. [CrossRef]
134. Gong, X.; Tsai, S.-W.; Yan, B.; Rubin, L.P. Cooperation between MEF2 and PPARγ in human intestinal β, β-carotene 15, 15′-monooxygenase gene expression. *BMC Mol. Biol.* **2006**, *7*, 7. [CrossRef]
135. Lobo, G.P.; Amengual, J.; Li, H.N.M.; Golczak, M.; Bonet, M.L.; Palczewski, K.; Von Lintig, J. β, β-carotene decreases peroxisome proliferator receptor γ activity and reduces lipid storage capacity of adipocytes in a β, β-carotene oxygenase 1-dependent manner. *J. Biol. Chem.* **2010**, *285*, 27891–27899. [CrossRef]
136. Ziouzenkova, O.; Orasanu, G.; Sharlach, M.; Akiyama, T.E.; Berger, J.P.; Viereck, J.; Hamilton, J.A.; Tang, G.; Dolnikowski, G.G.; Vogel, S. Retinaldehyde represses adipogenesis and diet-induced obesity. *Nat. Med.* **2007**, *13*, 695–702. [CrossRef]
137. Eroglu, A.; Hruszkewycz, D.P.; Curley, R.W., Jr.; Harrison, E.H. The eccentric cleavage product of β-carotene, β-apo-13-carotenone, functions as an antagonist of RXRα. *Arch. Biochem. Biophys.* **2010**, *504*, 11–16. [CrossRef]
138. Amengual, J.; Gouranton, E.; van Helden, Y.G.; Hessel, S.; Ribot, J.; Kramer, E.; Kiec-Wilk, B.; Razny, U.; Lietz, G.; Wyss, A. Beta-carotene reduces body adiposity of mice via BCMO1. *PLoS ONE* **2011**, *6*, e20644. [CrossRef]
139. Wang, X.; Quinn, P.J. Vitamin E and its function in membranes. *Prog. Lipid Res.* **1999**, *38*, 309–336. [CrossRef]
140. Constantinou, C.; Papas, A.; Constantinou, A.I. Vitamin E and cancer: An insight into the anticancer activities of vitamin E isomers and analogs. *Int. J. Cancer* **2008**, *123*, 739–752. [CrossRef]
141. Traber, M.G. Vitamin E regulatory mechanisms. *Annu. Rev. Nutr.* **2007**, *27*, 347–362. [CrossRef]
142. Ross, A.C.; Caballero, B.; Cousins, R.J.; Tucker, K.L.; Ziegler, T.R. *Modern Nutrition in Health and Disease*, 11th ed.; Williams & Wilkins: Philadelphia, PA, USA, 2006.
143. Stone, W.L.; Krishnan, K.; Campbell, S.E.; Qui, M.; Whaley, S.G.; Yang, H. Tocopherols and the treatment of colon cancer. *Ann. N. Y. Acad. Sci.* **2004**, *1031*, 223–233. [CrossRef]
144. Campbell, S.E.; Stone, W.L.; Whaley, S.G.; Qui, M.; Krishnan, K. Gamma (γ) tocopherol upregulates peroxisome proliferator activated receptor (PPAR) gamma (γ) expression in SW 480 human colon cancer cell lines. *BMC Cancer* **2003**, *3*, 25. [CrossRef]
145. Qureshi, A.A.; Tan, X.; Reis, J.C.; Badr, M.Z.; Papasian, C.J.; Morrison, D.C.; Qureshi, N. Inhibition of nitric oxide in LPS-stimulated macrophages of young and senescent mice by δ-tocotrienol and quercetin. *Lipids Health Dis.* **2011**, *10*, 239. [CrossRef]
146. Qureshi, A.A.; Reis, J.C.; Qureshi, N.; Papasian, C.J.; Morrison, D.C.; Schaefer, D.M. δ-Tocotrienol and quercetin reduce serum levels of nitric oxide and lipid parameters in female chickens. *Lipids Health Dis.* **2011**, *10*, 39. [CrossRef]
147. Mosca, A.; Paleari, R.; Ivaldi, G.; Galanello, R.; Giordano, P. The role of haemoglobin A2 testing in the diagnosis of thalassaemias and related haemoglobinopathies. *J. Clin. Pathol.* **2009**, *62*, 13–17. [CrossRef]
148. Uto-Kondo, H.; Ohmori, R.; Kiyose, C.; Kishimoto, Y.; Saito, H.; Igarashi, O.; Kondo, K. Tocotrienol suppresses adipocyte differentiation and Akt phosphorylation in 3T3-L1 preadipocytes. *J. Nutr.* **2009**, *139*, 51–57. [CrossRef]
149. Ribot, J.; Felipe, F.; Bonet, M.L.; Palou, A. Changes of adiposity in response to vitamin A status correlate with changes of PPARγ2 expression. *Obesity Res.* **2001**, *9*, 500–509. [CrossRef]
150. Kuri-Harcuch, W. Differentiation of 3T3-F442A cells into adipocytes is inhibited by retinoic acid. *Differentiation* **1982**, *23*, 164–169. [CrossRef] [PubMed]
151. Schwarz, E.J.; Reginato, M.J.; Shao, D.; Krakow, S.L.; Lazar, M.A. Retinoic acid blocks adipogenesis by inhibiting C/EBPbeta-mediated transcription. *Mol. Cell. Biol.* **1997**, *17*, 1552–1561. [CrossRef] [PubMed]
152. Duque, G.; Macoritto, M.; Kremer, R. 1,25(OH)$_2$D$_3$ inhibits bone marrow adipogenesis in senescence accelerated mice (SAM-P/6) by decreasing the expression of peroxisome proliferator-activated receptor gamma 2 (PPARγ2). *Exp. Gerontol.* **2004**, *39*, 333–338. [CrossRef] [PubMed]
153. Hida, Y.; Kawada, T.; Kayahashi, S.; Ishihara, T.; Fushiki, T. Counteraction of retinoic acid and 1, 25-dihydroxyvitamin D3 on up-regulation of adipocyte differentiation with PPARγ ligand, an antidiabetic thiazolidinedione, in 3T3-L1 cells. *Life Sci.* **1998**, *62*, PL205–PL211. [CrossRef]
154. Sirtori, C.R.; Arnoldi, A.; Johnson, S.K. Phytoestrogens: End of a tale? *Ann. Med.* **2005**, *37*, 423–438. [CrossRef] [PubMed]
155. Bajaj, M.; Hinge, A.; Limaye, L.S.; Gupta, R.K.; Surolia, A.; Kale, V.P. Mannose-binding dietary lectins induce adipogenic differentiation of the marrow-derived mesenchymal cells via an active insulin-like signaling mechanism. *Glycobiology* **2011**, *21*, 521–529. [CrossRef]
156. Dang, Z.-C.; Audinot, V.; Papapoulos, S.E.; Boutin, J.A.; Löwik, C.W. Peroxisome proliferator-activated receptor γ (PPARγ) as a molecular target for the soy phytoestrogen genistein. *J. Biol. Chem.* **2003**, *278*, 962–967. [CrossRef]
157. Sangeetha, T.; Quine, S.D. Protective effect of S-allyl cysteine sulphoxide (alliin) on glycoproteins and hematology in isoproterenol induced myocardial infarction in male Wistar rats. *J. Appl. Toxicol.* **2008**, *28*, 710–716. [CrossRef]

158. Keophiphath, M.; Priem, F.; Jacquemond-Collet, I.; Clément, K.; Lacasa, D. 1,2-vinyldithiin from garlic inhibits differentiation and inflammation of human preadipocytes. *J. Nutr.* **2009**, *139*, 2055–2060. [CrossRef]
159. Adapala, N.; Chan, M.M. Long-term use of an antiinflammatory, curcumin, suppressed type 1 immunity and exacerbated visceral leishmaniasis in a chronic experimental model. *Lab. Investig.* **2008**, *88*, 1329–1339. [CrossRef]
160. Dos Santos Costa, C.; Rohden, F.; Hammes, T.O.; Margis, R.; Bortolotto, J.W.; Padoin, A.V.; Mottin, C.C.; Guaragna, R.M. Resveratrol upregulated SIRT1, FOXO1, and adiponectin and downregulated PPARγ1–3 mRNA expression in human visceral adipocytes. *Obesity Surg.* **2011**, *21*, 356–361. [CrossRef]
161. Xu, Z.; Chen, X.; Zhong, Z.; Chen, L.; Wang, Y. Ganoderma lucidum polysaccharides: Immunomodulation and potential anti-tumor activities. *Am. J. Chin. Med.* **2011**, *39*, 15–27. [CrossRef]
162. Sawant, R.; Godghate, A. Qualitative phytochemical screening of rhizomes of Curcuma longa Linn. *Int. J. Sci. Environ.* **2013**, *2*, 634–641.
163. Hinge, A.; Bajaj, M.; Limaye, L.; Surolia, A.; Kale, V. Oral administration of insulin receptor-interacting lectins leads to an enhancement in the hematopoietic stem and progenitor cell pool of mice. *Stem Cells Development* **2010**, *19*, 163–174. [CrossRef]
164. Rayalam, S.; Della-Fera, M.A.; Yang, J.-Y.; Park, H.J.; Ambati, S.; Baile, C.A. Resveratrol potentiates genistein's antiadipogenic and proapoptotic effects in 3T3-L1 adipocytes. *J. Nutr.* **2007**, *137*, 2668–2673. [CrossRef]
165. Bionaz, M.; Hausman, G.J.; Loor, J.J.; Mandard, S. Physiological and nutritional roles of PPAR across species. *PPAR Res.* **2013**, *2013*, 807156. [CrossRef]
166. Oliveira, A.C.P.; Bertollo, C.M.; Rocha, L.T.S.; Nascimento, E.B., Jr.; Costa, K.A.; Coelho, M.M. Antinociceptive and antiedematogenic activities of fenofibrate, an agonist of PPAR alpha, and pioglitazone, an agonist of PPAR gamma. *Eur. J. Pharmacol.* **2007**, *561*, 194–201. [CrossRef]
167. Bionaz, M.; Loor, J.J. ACSL1, AGPAT6, FABP3, LPIN1, and SLC27A6 are the most abundant isoforms in bovine mammary tissue and their expression is affected by stage of lactation. *J. Nutr.* **2008**, *138*, 1019–1024. [CrossRef]
168. Varga, T.; Czimmerer, Z.; Nagy, L. PPARs are a unique set of fatty acid regulated transcription factors controlling both lipid metabolism and inflammation. *Biochim. Biophys. Acta (BBA)—Mol. Basis Dis.* **2011**, *1812*, 1007–1022. [CrossRef]
169. Kushibiki, S.; Hodate, K.; Shingu, H.; Ueda, Y.; Shinoda, M.; Mori, Y.; Itoh, T.; Yokomizo, Y. Insulin resistance induced in dairy steers by tumor necrosis factor alpha is partially reversed by 2, 4-thiazolidinedione. *Domestic Anim. Endocrinol.* **2001**, *21*, 25–37. [CrossRef]
170. Hauner, H. The mode of action of thiazolidinediones. *Diabetes/Metabolism Res. Rev.* **2002**, *18*, S10–S15. [CrossRef]
171. Perdomo, M.C.; Santos, J.E.; Badinga, L. Trans-10, cis-12 conjugated linoleic acid and the PPAR-γ agonist rosiglitazone attenuate lipopolysaccharide-induced TNF-α production by bovine immune cells. *Domest. Anim. Endocrinol.* **2011**, *41*, 118–125. [CrossRef]
172. He, X.; Gao, J.; Hou, H.; Qi, Z.; Chen, H.; Zhang, X.-X. Inhibition of mitochondrial fatty acid oxidation contributes to development of nonalcoholic fatty liver disease induced by environmental cadmium exposure. *Environ. Sci. Technol.* **2019**, *53*, 13992–14000. [CrossRef]
173. Filip-Ciubotaru, F.; Foia, L.; Manciuc, C.; Grigore, C. PPARs: Structure, mechanisms of action and control. Note I. *Revista Medico-Chirurgicala a Societatii de Medici si Naturalisti din Iasi* **2011**, *115*, 477–484.
174. Hasan, A.U.; Rahman, A.; Kobori, H. Interactions between host PPARs and gut microbiota in health and disease. *Int. J. Mol. Sci.* **2019**, *20*, 387. [CrossRef]
175. Hasan, A.U.; Ohmori, K.; Hashimoto, T.; Kamitori, K.; Yamaguchi, F.; Rahman, A.; Tokuda, M.; Kobori, H. PPARγ activation mitigates glucocorticoid receptor-induced excessive lipolysis in adipocytes via homeostatic crosstalk. *J. Cell. Biochem.* **2018**, *119*, 4627–4635. [CrossRef]
176. Kersten, S. Regulation of lipid metabolism via angiopoietin-like proteins. *Biochem. Soc. Trans.* **2005**, *33*, 1059–1062. [CrossRef]
177. Kharitonenkov, A.; Shiyanova, T.L.; Koester, A.; Ford, A.M.; Micanovic, R.; Galbreath, E.J.; Sandusky, G.E.; Hammond, L.J.; Moyers, J.S.; Owens, R.A. FGF-21 as a novel metabolic regulator. *J. Clin. Investig.* **2005**, *115*, 1627–1635. [CrossRef]
178. Loor, J.J.; Everts, R.E.; Bionaz, M.; Dann, H.M.; Morin, D.E.; Oliveira, R.; Rodriguez-Zas, S.L.; Drackley, J.K.; Lewin, H.A. Nutrition-induced ketosis alters metabolic and signaling gene networks in liver of periparturient dairy cows. *Physiol. Genom.* **2007**, *32*, 105–116. [CrossRef]
179. Schoenberg, K.M.; Giesy, S.L.; Harvatine, K.J.; Waldron, M.R.; Cheng, C.; Kharitonenkov, A.; Boisclair, Y.R. Plasma FGF21 is elevated by the intense lipid mobilization of lactation. *Endocrinology* **2011**, *152*, 4652–4661. [CrossRef]
180. Riahi, Y.; Sin-Malia, Y.; Cohen, G.; Alpert, E.; Gruzman, A.; Eckel, J.; Staels, B.; Guichardant, M.; Sasson, S. The Natural Protective Mechanism Against Hyperglycemia in Vascular Endothelial Cells: Roles of the Lipid Peroxidation Product 4-Hydroxydodecadienal and Peroxisome Proliferator–Activated Receptor δ. *Diabetes* **2010**, *59*, 808–818. [CrossRef]
181. Bionaz, M.; Trevisi, E.; Calamari, L.; Librandi, F.; Ferrari, A.; Bertoni, G. Plasma paraoxonase, health, inflammatory conditions, and liver function in transition dairy cows. *J. Dairy Sci.* **2007**, *90*, 1740–1750. [CrossRef]
182. Bionaz, M.; Loor, J.J. Gene networks driving bovine mammary protein synthesis during the lactation cycle. *Bioinform. Biol. Insights* **2011**, *5*, BBI.S7003. [CrossRef] [PubMed]
183. Bionaz, M.; Periasamy, K.; Rodriguez-Zas, S.L.; Everts, R.E.; Lewin, H.A.; Hurley, W.L.; Loor, J.J. Old and new stories: Revelations from functional analysis of the bovine mammary transcriptome during the lactation cycle. *PLoS ONE* **2012**, *7*, e33268. [CrossRef] [PubMed]

International Journal of *Molecular Sciences*

MDPI

Article

Pemafibrate Prevents Retinal Dysfunction in a Mouse Model of Unilateral Common Carotid Artery Occlusion

Deokho Lee [1,2,†], Yohei Tomita [1,2,3,†], Heonuk Jeong [1,2], Yukihiro Miwa [1,2,4], Kazuo Tsubota [5], Kazuno Negishi [2] and Toshihide Kurihara [1,2,*]

1 Laboratory of Photobiology, Keio University School of Medicine, Tokyo 160-8582, Japan; deokholee@keio.jp (D.L.); yohei.tomita@childrens.harvard.edu (Y.T.); jeong.h@keio.jp (H.J.); yukihiro226@gmail.com (Y.M.)
2 Department of Ophthalmology, Keio University School of Medicine, Tokyo 160-8582, Japan; kazunonegishi@keio.jp
3 Department of Ophthalmology, Boston Children's Hospital, Harvard Medical School, Boston, MA 02115, USA
4 Animal Eye Care, Tokyo Animal Eye Clinic, Tokyo 158-0093, Japan
5 Tsubota Laboratory, Inc., Tokyo 160-0016, Japan; tsubota@tsubota-lab.com
* Correspondence: kurihara@z8.keio.jp; Tel.: +81-3-5636-3204
† These authors contributed equally to this work.

Abstract: Cardiovascular diseases lead to retinal ischemia, one of the leading causes of blindness. Retinal ischemia triggers pathological retinal glial responses and functional deficits. Therefore, maintaining retinal neuronal activities and modulating pathological gliosis may prevent loss of vision. Previously, pemafibrate, a selective peroxisome proliferator-activated receptor alpha modulator, was nominated as a promising drug in retinal ischemia. However, a protective role of pemafibrate remains untouched in cardiovascular diseases-mediated retinal ischemia. Therefore, we aimed to unravel systemic and retinal alterations by treating pemafibrate in a new murine model of retinal ischemia caused by cardiovascular diseases. Adult C57BL/6 mice were orally administered pemafibrate (0.5 mg/kg) for 4 days, followed by unilateral common carotid artery occlusion (UCCAO). After UCCAO, pemafibrate was continuously supplied to mice until the end of experiments. Retinal function (a-and b-waves and the oscillatory potentials) was measured using electroretinography on day 5 and 12 after UCCAO. Moreover, the retina, liver, and serum were subjected to qPCR, immunohistochemistry, or ELISA analysis. We found that pemafibrate enhanced liver function, elevated serum levels of fibroblast growth factor 21 (FGF21), one of the neuroprotective molecules in the eye, and protected against UCCAO-induced retinal dysfunction, observed with modulation of retinal gliosis and preservation of oscillatory potentials. Our current data suggest a promising pemafibrate therapy for the suppression of retinal dysfunction in cardiovascular diseases.

Keywords: common carotid artery occlusion; electroretinography; fibroblast growth factor 21; pemafibrate; peroxisome proliferator-activated receptor alpha; retinal ischemia

Citation: Lee, D.; Tomita, Y.; Jeong, H.; Miwa, Y.; Tsubota, K.; Negishi, K.; Kurihara, T. Pemafibrate Prevents Retinal Dysfunction in a Mouse Model of Unilateral Common Carotid Artery Occlusion. *Int. J. Mol. Sci.* **2021**, *22*, 9408. https://doi.org/10.3390/ijms22179408

Academic Editors: Manuel Vázquez-Carrera and Walter Wahli

Received: 28 July 2021
Accepted: 27 August 2021
Published: 30 August 2021

1. Introduction

Ocular ischemic syndrome (OIS) is a vision-threatening disease caused by carotid artery stenosis or occlusion [1]. The first case was reported in 1963 as a disease associated with internal carotid artery occlusion [2]. About 7.5 cases per million are annually diagnosed with OIS [3]. It is most common in old males, and patients with underlying diabetes, hypertension, and hyperlipidemia are more likely to have this disease. Atherosclerosis has also been known to be one of the most common causes for the development of OIS [4]. Besides, carotid artery dissection, giant cell arteritis, and trauma can have high chances to cause OIS [5–10]. Unfortunately, there is no current effective treatment in OIS. Moreover, precise mechanisms of OIS have not been fully unraveled yet.

Experimental murine models of carotid artery occlusion have been applied to study OIS [11–18]. From an anatomical point of view, the retina is supplied with oxygen/blood from the ophthalmic artery, one of the internal carotid artery's branches of the common carotid artery. In this regard, occlusion of the carotid artery can cause retinal ischemia leading to vision loss [19,20]. There have been several ways of developing murine models of carotid artery occlusion depending on the species. Two common carotid arteries could be occluded to induce retinal ischemia in rats. As the circle of Willis in rats is well-structured, the rats which receive bilateral common carotid artery occlusion (BCCAO) could be developed as experimental models of retinal ischemia [11–13,15]. Two common carotid arteries could not be occluded in mice because of a high rate of mouse death (almost 100%) as they may have a lack of posterior communicating arteries in the circle of Willis [17,21,22]. Therefore, bilateral common carotid artery stenosis (BCCAS) has been alternately attempted to induce severe retinal ischemic injuries in mice [14]. However, the concern about a high rate of death during and after BCCAO or BCCAS in rats or mice has not been solved in that the experimental models still die easily. Hence, unilateral common carotid artery occlusion (UCCAO) has been tried and developed in mice for studying retinal ischemia more stably [17,18,23]. Even though several phenotypes for retinal ischemia have been described [17,18,23], a rescue for retinal ischemia has not been considerably studied in this model. In this regard, the development of a cure for retinal ischemia in this model could be intriguing.

Peroxisome proliferator-activated receptor alpha (PPARα) is a well-known drug against hyperlipidemia. This agent can potentially reduce triglyceride levels and increase high-density lipoprotein cholesterol (HDL-C) levels [24]. The Fenofibrate Intervention and Event Lowering in Diabetes (FIELD) and The Action to Control Cardiovascular Risk in Diabetes (ACCORD) eye studies showed that fenofibrate, a well-known PPARα agonist, reduced the need for laser therapy and progression of diabetic retinopathy [25,26]. Thus, this drug was recently approved for preventing diabetic retinopathy in Australia. Several studies have shown that fenofibrate has therapeutic effects on retinal diseases in animal models [27–29]. However, fenofibrate may potentially induce renal dysfunction, and it may cause rhabdomyolysis when administered with a statin. Thus, clinicians needed to take care of this part when they prescribed fenofibrate for diabetic patients with renal dysfunction.

Pemafibrate, a novel selective PPARα modulator (SPPARMα), has been developed as a therapeutic agent against hyperlipidemia to reduce this side effect. Our previous study showed that pemafibrate might prevent pathological neovascularization in a murine model of oxygen-induced ischemic retinopathy and preserve retinal function in a streptozotocin-induced diabetic mouse model [30,31]. Another group showed that pemafibrate might prevent retinal inflammation and vascular leakage in a rat's diabetic model and prevent apoptosis in the ganglion cells damaged by N-methyl-D-aspartate (NMDA)-induced excitotoxicity [32,33]. Taken together, we assumed that pemafibrate could be used for neuroprotection against various retinal ischemic injuries.

In this study, we aimed to investigate the protective effects of pemafibrate in a murine model of retinal ischemia induced by UCCAO, which resembles OIS.

2. Results

2.1. Suppression of Retinal Dysfunction by Pemafibrate Administration in a Mouse Model of UCCAO-Induced Retinal Ischemia

According to our timeline of experiments, pemafibrate was orally administered to adult male mice (0.5 mg/kg/day) for 4 days before UCCAO (Figure A1). The administration of pemafibrate did not significantly change the body weight of adult male mice. After 4 days of oral administration of pemafibrate, retinal ischemia was induced by occlusion of the right common carotid artery which is connected to the internal carotid artery stretched toward the ophthalmic artery (Figure A1). We found that the body weight of adult male mice dramatically decreased 1 day after UCCAO (Figure A1). Pemafibrate (0.5 mg/kg/day) was consecutively supplied to UCCAO-operated mice, and we found that administration of pemafibrate did not dramatically change the body weight of UCCAO-operated mice.

However, there was a slightly decreasing tendency in the body weight of UCCAO-operated mice after continuous oral administration of pemafibrate.

Next, to investigate the protective effects of pemafibrate against retinal dysfunction in UCCAO-operated mice, electroretinography (ERG) was performed (Figures 1 and 2). Before UCCAO, there was no significant difference in the amplitudes of a-and b-waves and the oscillatory potentials (OPs) between PBS-administered and pemafibrate-administered naïve mice (Figure A2). Previously, we demonstrated that retinal dysfunction was started from day 3 to day 7 after UCCAO [23,27]. Therefore, we primarily checked retinal dysfunction 5 days after UCCAO. We found that reduction in the amplitudes of a-and b-waves in UCCAO-operated mice was slightly suppressed by the oral administration of pemafibrate (Figure 1A,B). However, there was no statistical significance between PBS-administered UCCAO-operated mice and pemafibrate-administered UCCAO-operated mice. Next, we found that reduction in the amplitudes of OPs in UCCAO-operated mice was significantly suppressed by the oral administration of pemafibrate (Figure 1C,D).

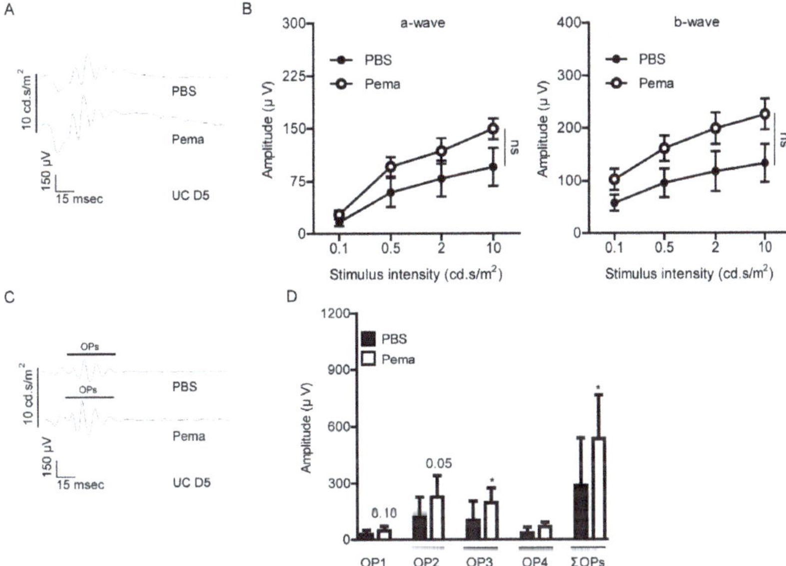

Figure 1. Protective effects of pemafibrate against retinal dysfunction on day 5 after UCCAO. (**A**,**B**) Representative waveforms (10 cd·s/m^2) of a-and b-waves and quantitative analyses (n = 9–10 per group) showed that oral administration of pemafibrate had a slight suppressing tendency in a reduction in the amplitudes of a-wave and b-wave in the UCCAO-operated eye 5 days after UCCAO. The data were analyzed using two-way ANOVA followed by a Bonferroni post hoc test. The data were presented as mean ± standard error of the mean. (**C**,**D**) Representative waveforms (10 cd·s/m^2) of oscillatory potentials (OPs) and quantitative analyses showed that pemafibrate significantly suppressed reduction in the amplitudes of OPs (OP1, OP2, OP3, and ΣOPs) in UCCAO-induced retinal ischemic mice (n = 9–10 per group). * $p < 0.05$. The data were analyzed using two-tailed Student's t-test. The data were presented as mean ± standard deviation. Pema; pemafibrate. UC; unilateral common carotid artery occlusion. ns; not significant.

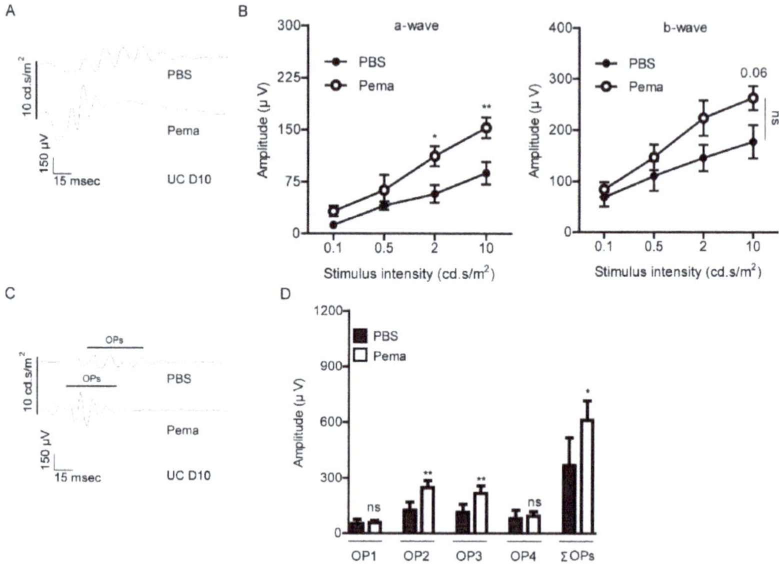

Figure 2. Protective effects of pemafibrate against retinal dysfunction 10 days after UCCAO. (**A**,**B**) Representative waveforms (10 cd·s/m^2) of a-and b-waves and quantitative analyses (n = 5 per group) showed that oral administration of pemafibrate had a suppressing tendency in the reduction in the amplitudes of a-wave and b-wave in the UCCAO-operated eye 10 days after UCCAO with statistical significance. The data were analyzed using two-way ANOVA followed by a Bonferroni post hoc test (a-wave and b-wave). One datum was further analyzed using two-tailed Student's t-test (b-wave; p = 0.06). The data were presented as mean ± standard error of the mean. (**C**,**D**) Representative waveforms (10 cd·s/m^2) of oscillatory potentials (OPs) and quantitative analyses showed that pemafibrate significantly suppressed reduction in the amplitudes of OPs (OP1, OP2, OP3, and ΣOPs) in UCCAO-induced retinal ischemic mice (n = 5 per group). * $p < 0.05$, ** $p < 0.01$. The data were analyzed using two-tailed Student's t-test. The data were presented as mean ± standard deviation. Pema; pemafibrate. UC; unilateral common carotid artery occlusion. ns; not significant.

Furthermore, reduction in the amplitudes of a-and b-waves in UCCAO-operated mice was slightly kept suppressed by the oral administration of pemafibrate 10 days after UCCAO (Figure 2A,B). Finally, we found that reduction in the amplitudes of OPs in UCCAO-operated mice was maintained to be suppressed by oral administration of pemafibrate (Figure 2C,D).

Next, we examined the molecular mechanism underlying pemafibrate-mediated preservation of retinal function against UCCAO. Previously, we found that UCCAO decreased the expression of synaptophysin, one of the well-known synaptic vesicle proteins [27]. This protein is plentifully expressed in inner retinal neuronal cells which are the cellular source for OPs [34–36]. Even though there was no statistical significance, we found a decreasing synaptophysin expression after UCCAO was slightly suppressed in the pemafibrate-administered UCCAO-operated retina (Figure A3).

2.2. *Suppression of Pathological Retinal Gliosis by Pemafibrate Administration in a Mouse Model of UCCAO-Induced Retinal Ischemia*

Reactive gliosis has been used as a responsive marker for retinal ischemic damages [37]. For further investigation of protective roles of pemafibrate against ischemic retinal dysfunction in UCCAO-operated mice, immunohistochemistry (IHC) was performed for detecting pathological reactive gliosis in the retina. Previously, we demonstrated that retinal gliosis was started from day 1 and more clearly seen on day 7 after UCCAO [17,23].

Therefore, we checked retinal gliosis from day 2 to day 5 after UCCAO (Figure 3). We found that slightly increased pathological glial responses on day 2 after UCCAO, observed by morphology scoring, were reduced in pemafibrate-administered UCCAO-operated mice (Figure 3A). Furthermore, as expected, dramatically increased pathological glial responses were seen 5 days after UCCAO, and these responses were significantly reduced in pemafibrate-administered UCCAO-operated mice (Figure 3B).

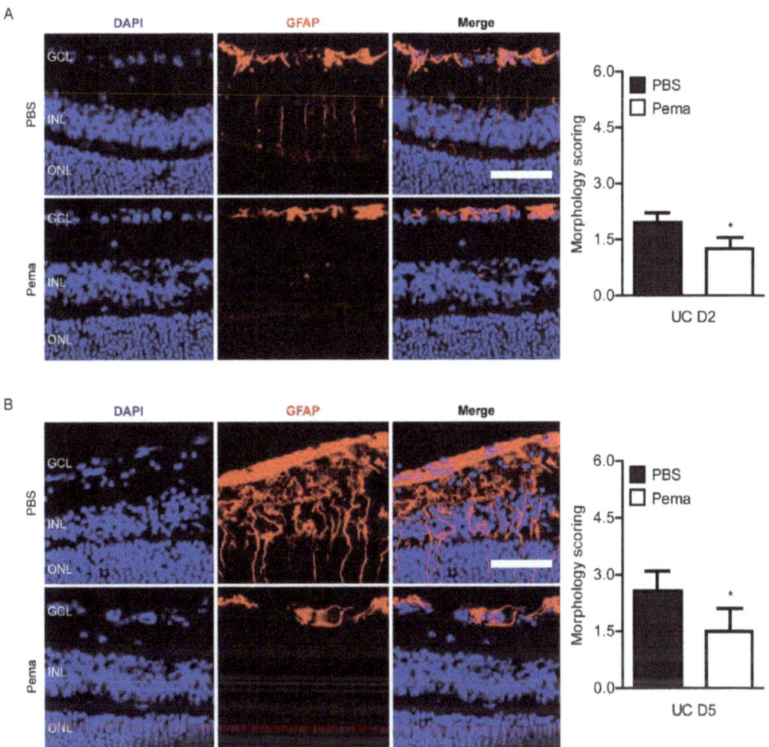

Figure 3. Modulation of pathological reactive gliosis after oral administration of pemafibrate. (**A**) Representative images and quantitative analyses ($n = 4$ per group) showed that slightly increased reactive retinal gliosis stained by GFAP in UCCAO-operated mice was reduced by administration of pemafibrate on day 2 after UCCAO. (**B**) Representative images and quantitative analyses ($n = 4$–5 per group) showed that dramatically increased reactive retinal gliosis stained by GFAP in UCCAO-operated mice were reduced by the administration of pemafibrate on day 5 after UCCAO. Scale bar: 50 μm. The data were analyzed using two-tailed Student's *t*-test. Graphs were presented as mean with ± standard deviation. * $p < 0.05$. GCL: ganglion cell layer; INL: inner nuclear layer; ONL: outer nuclear layer. Pema; pemafibrate. UC; unilateral common carotid artery occlusion.

2.3. Screening of Hypoxia-Ischemia-Related Gene Expressions after Pemafibrate Administration in a Mouse Model of UCCAO-Induced Retinal Ischemia

Previously, it was reported that several hypoxia-ischemia-related gene expressions (*Epo*, *Bnip3*, *Vegfa*, *Ccl2*, *Ccl12*, and *Glut1*) were induced 1 day after UCCAO [17,27]. Therefore, we screened changes in these gene expressions after oral administration of pemafibrate (Figure 4). We found that the expression of *Glut1* significantly increased in the retina of pemafibrate-administered UCCAO-operated mice. Expressions of the other genes were not significantly altered by oral administration of pemafibrate.

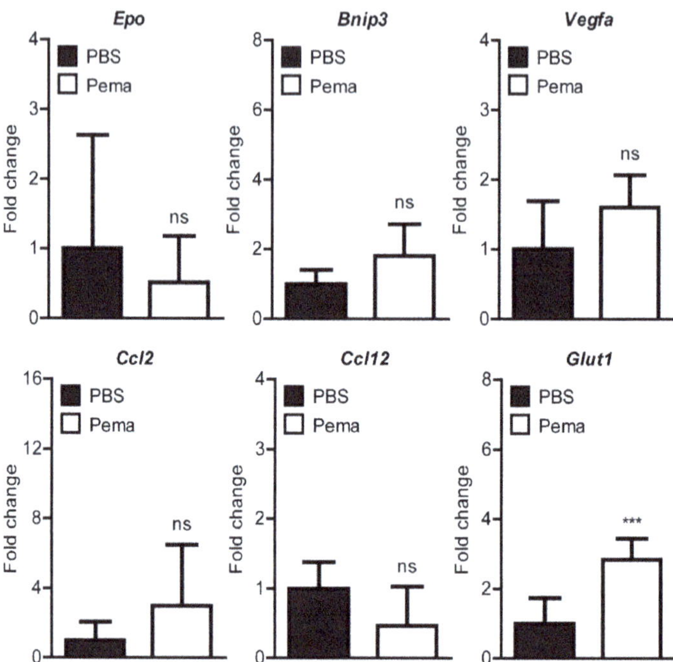

Figure 4. Screening of alterations in retinal hypoxia-ischemia-related gene expressions by oral administration of pemafibrate in UCCAO-operated mice. Primarily, genes reported to be slightly or dramatically altered after UCCAO were selected; *Epo*, *Bnip3*, *Vegfa*, *Ccl2*, *Ccl12*, and *Glut1*. Quantitative analyses (n = 6 per group) showed that oral administration of pemafibrate significantly reduced the expression of *Glut1* in the retina 1 day after UCCAO. However, the other genes' expressions were not changed by oral administration of pemafibrate. *** $p < 0.001$. The data were analyzed using two-tailed Student's t-test and presented as mean ± standard deviation. Pema; pemafibrate. ns; not significant.

2.4. Induction of PPARα Target Genes by Pemafibrate Administration in a Mouse Model of UCCAO-Induced Retinal Ischemia

We examined whether expressions of PPARα downstream genes could be induced by the oral administration of pemafibrate. The retina was targeted as it is our primary region of interest. We could not find any significant change in *Ucp3*, *Fabp4*, *Vldlr*, *Fgf21*, and *Acox1* between the PBS-administered retina and the pemafibrate-administered retina on the day of UCCAO surgery (Figure 5A). Furthermore, we could not find any significant change in *Ucp3*, *Fabp4*, *Fgf21*, and *Acox1* 1 day after UCCAO. Even though a significant increase in *Vldlr* expression was detected, the fold change was not dramatic at all.

Next, the liver was targeted as it has been known as a region for pemafibrate-induced PPARα activation [24,38,39]. The livers in pemafibrate-administered UCCAO-operated mice seemed larger than those in PBS-administered UCCAO-operated mice while we collected the samples. Therefore, the liver weight was calculated with the body weight, and we found that the relative liver weight gradually increased after consecutive administration of pemafibrate with statistical significance, in comparison with that in PBS-administered UCCAO-operated mice (Figure 5B). Furthermore, PPARα downstream genes (*Ucp3*, *Fabp4*, *Vldlr*, *Fgf21*, and *Acox1*) in the liver increased significantly and dramatically, in comparison with those in PBS-administered UCCAO-operated mice (Figure 5C). Especially, two genes (*Ucp3* and *Vldlr*) were gradually increased in a time-dependent manner. Even though there was fluctuation, *Fabp4* was also time-dependently increased after long-term repetitive oral administration of pemafibrate. When it comes to *Fgf21*, its gene expression was dramatically induced at the early stage of repetitive oral administration of pemafibrate and gradually decreased after long-term repetitive administration of pemafibrate. There was a gradual increasing tendency in the expression of *Acox1* until 5 days after UCCAO, and its expression was detected to the basal level on day 10 after UCCAO.

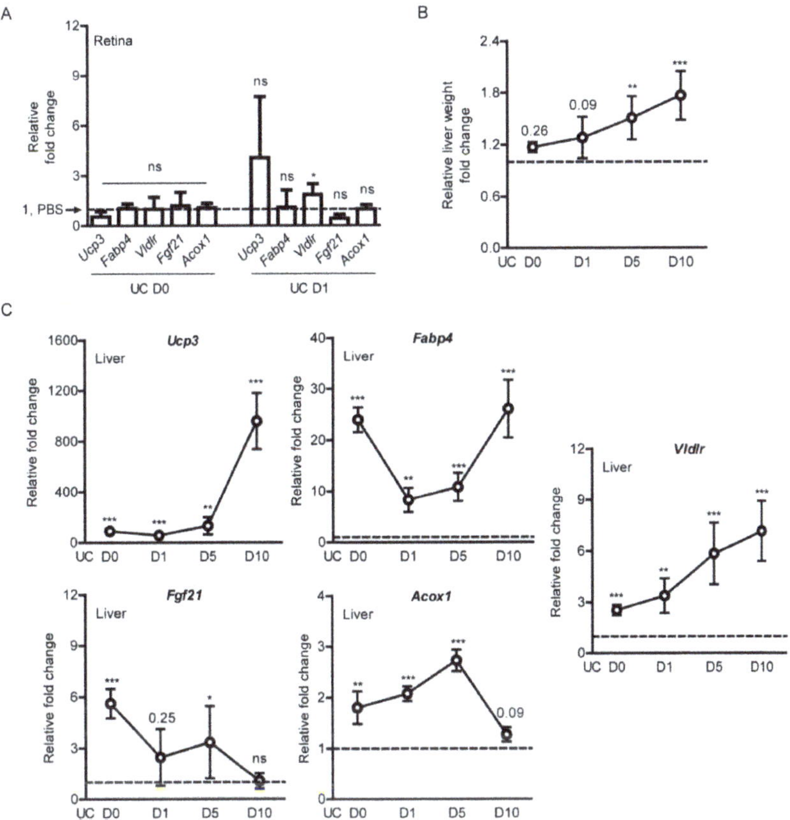

Figure 5. Induction in PPARα downstream gene expressions in the liver by oral administration of pemafibrate in UCCAO-operated mice. (**A**) Quantitative analyses (n = 4–6 per group) showed that oral administration of pemafibrate did not dramatically increase PPARα downstream gene expressions in the retina. The data were analyzed using two-tailed Student's t-test and presented as mean ± standard deviation. (**B**) Quantitative analyses (n = 4–6 per group) showed that the relative liver weight (the liver weight/the body weight) in pemafibrate-administered mice was significantly higher than that in PBS-administered mice. The data were analyzed using two-tailed Student's t-test and were presented as mean ± standard error of the mean. (**C**) Quantitative analyses (n = 4–5 per group) showed that oral administration of pemafibrate significantly increased PPARα downstream gene expressions in the liver. The data were analyzed using two-tailed Student's t-test and presented as mean ± standard deviation. The value for PBS-administered mice was indicated as a dotted line; 1. * $p < 0.05$, ** $p < 0.01$, *** $p < 0.001$. Pema; pemafibrate, UC; unilateral common carotid artery occlusion. ns; not significant.

Moreover, we examined whether pemafibrate is able to increase serum levels of FGF21. Increases in serum FGF21 levels by PPARα agonists have been reported in various experimental models and clinical studies [27,30,31,40–42]. As expected, elevated serum levels of FGF21 were dramatically seen after oral administration of pemafibrate on the day of the UCCAO surgery (Figure 6A). Furthermore, increased serum FGF21 levels were continuously observed in pemafibrate-administered UCCAO-operated mice until the end of experiments, in comparison with PBS-administered UCCAO-operated mice.

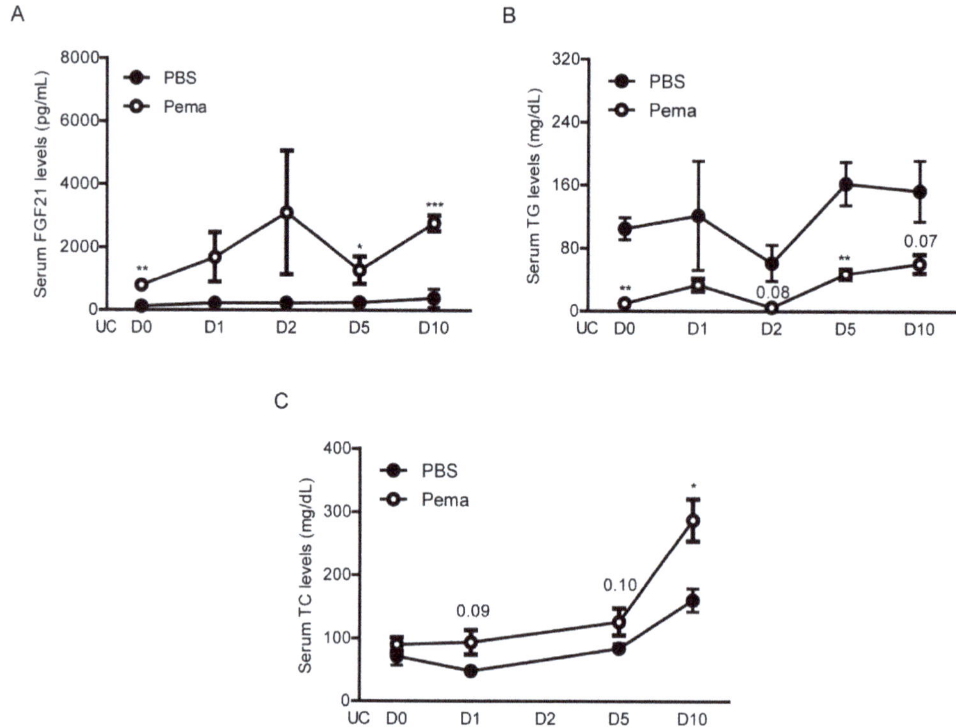

Figure 6. Changes in serum levels of FGF21, TG, and TC by oral administration of pemafibrate in UCCAO-operated mice. (**A**) Quantitative analyses (n = 3–9 per group) showed that oral administration of pemafibrate significantly increased serum levels of FGF21. The data were analyzed using two-tailed Student's t-test and presented as mean ± standard error of the mean. (**B**,**C**) Quantitative analyses (n = 3–8 per group) showed that oral administration of pemafibrate significantly decreased serum levels of TG and increased serum levels of TC. The data were analyzed using two-tailed Student's t-test and presented as mean ± standard error of the mean. * $p < 0.05$, ** $p < 0.01$, *** $p < 0.001$. Pema; pemafibrate, UC; unilateral common carotid artery occlusion, TG; triglyceride, TC; total cholesterol.

Next, triglyceride (TG) and total cholesterol (TC) levels in the serum were examined (Figure 6B,C), as TG and TC levels have also been reported to be changed by administration of pemafibrate in various experimental models and clinical studies [31,43–45]. Expectedly, we detected continuous decreases in serum TG levels and increases in serum TC levels in pemafibrate-administered UCCAO-operated mice in comparison with PBS-administered UCCAO-operated mice.

2.5. Observation of Retinal Thickness Changes by Pemafibrate Administration in a Mouse Model of UCCAO-Induced Retinal Ischemia

Previously, we could not find changes in retinal thickness 14 days after UCCAO [17,23]. In case of unknown effects of pemafibrate on retinal thickness, we examined whether retinal thickness could be changed by the oral administration of pemafibrate using optical coherence tomography (OCT) (Figure A4). Expectedly, we could not find any particular change in the retina and retinal thickness after oral administration of pemafibrate on day 10 after UCCAO.

3. Discussion

We revealed that the consecutive oral administration of pemafibrate, a selective PPARα modulator, suppressed pathological retinal gliosis and functional neuronal deficits in a murine model of retinal ischemia by UCCAO. Furthermore, we found significant increases in PPARα target gene expressions in the liver, not in the retina, reduction in serum levels of TG, and elevation in serum levels of FGF21 and TC. Previous single-cell data demonstrated that PPARα had low expression in the retina [46]. On the other hand, FGFR1, a crucial receptor for FGF21 function, was highly expressed in several types of cells in the retina. That could be the reason that pemafibrate did not activate PPARα extensively in the retina. On the other hand, it is reported that PPARα is a key modulator of hepatic FGF21 [47]. This is consistent with our previous reports and other studies that PPARα agonists (pemafibrate or fenofibrate) increase serum levels of FGF21 as well as boost liver function to exert therapeutic effects in ischemic retinopathies such as diabetic retinopathy or ocular ischemic syndrome [27,30,31,48].

FGF21 comprises 209 amino acids, and its protein regulates critical metabolic pathways [49–52]. FGF21 is produced in various tissues, especially in the liver [52], and improves lipid profiles in patients with type 2 diabetes [53]. So far, several studies have shown FGF21's therapeutic roles in the retina in vitro and in vivo (Table 1). Fu et al. showed that long-acting FGF21 suppressed neovascularization in mice by suppressing TNF-α expression via increasing adiponectin levels [54]. They also showed that long-acting FGF21 preserved retinal function (analyzed using ERG) in streptozotocin-induced diabetic mice and Akita mice which mimic type 1 diabetes [55]. Additionally, they showed that FGF21 suppressed oxidative stress-induced inflammation in 661W cells. Our group showed that long-acting FGF21 reduced retinal vascular leakage in a murine model of retinal vascular leakage and demonstrated that long-acting FGF21 maintained claudin-1 expression in human endothelial cells [56]. We recently reported that long-acting FGF21 improved retinal neuronal function through Müller glial remodeling in P23H mice, studied along with in vitro rat retinal Müller glial cells [57]. On top of that, we showed that pemafibrate showed therapeutic effects against pathological neovascularization in a murine model of oxygen-induced retinopathy and rescued retinal function in streptozotocin-induced diabetic mice through increasing FGF21 levels in the blood [30,31]. Taken together, increases in serum FGF21 levels induced by pemafibrate administration may also have the same protective effects on the UCCAO-induced ischemic retina. However, further studies are needed to see if direct FGF21 injection could exert cellular protection in the ischemic retina.

Table 1. Therapeutic Roles of FGF21 in the Eye (studied using PF-05231023, a long-acting FGF21 analog).

Author	Year of Publication	Journal	In Vitro Cell Type	Effect	In Vivo Experimental Model	Effect
Fu et al. [54]	2017	*Cell Reports*	HRMEC	Promotes cell migration	OIR; VldlrKO; Laser-induced CNV	Suppresses NV via decreasing TNF-α expression
Fu et al. [55]	2018	*Diabetes*	661 W	Inhibits oxidative stress-induced inflammation	STZ; Akita mouse	Rescues retinal morphology and function
Tomita and Ozawa et al. [30]	2019	*IJMS*	661 W	Inhibits a HIF activity	-	-
Tomita et al. [56]	2020	*IJMS*	HRMEC	Prevents vascular permeability	mVEGF164-induced retinal vascular leakage mouse	Preserves an expression of tight junction protein
Tomita and Lee et al. [31]	2020	*IJMS*	PC12D	Increases synaptophysin protein expression	-	-
Fu and Qiu et al. [57]	2021	*iScience*	rMC-1	Increases SRF protein expression	P23H mutation mouse	Modulates retinal glial responses

HRMEC: human retinal microvascular endothelial cell; 661W: cone photoreceptor cell; PC12D: pheochromocytoma 12D neuronal cells; rMC-1: rat retinal Müller glia; HIF: hypoxia-inducible factor; SRF: serum response factor; OIR: oxygen-induced retinopathy; Vldlr KO: very-low-density lipoprotein receptor knock out; CNV: choroidal neovascularization; NV: neovascularization; STZ: streptozotocin-induced diabetes; mVEGF164: mouse vascular endothelial growth factor 164.

Based on our current data, an increase in *Glut1* expression was seen in the pemafibrate-administered UCCAO-induced ischemic retina. Previously, we also demonstrated that

Glut1 expression increased in the retina of the same ischemic murine model after consecutive oral administration of fenofibrate, a well-known PPARα agonist [27]. We assume that elevation in serum levels of FGF21 is one of the reasons for the induction of *Glut1* in the retina. FGF21 has been suggested to exert a therapeutic effect on glucose and lipid metabolisms in mice [58]. Regarding this effect, a clinical trial has been studied using a novel long-acting FGF21 pegbelfermin which may have therapeutic effects on nonalcoholic fatty liver disease and nonalcoholic steatohepatitis [24]. Previously, FGF21 showed a synergistic effect with insulin on glucose absorption associated with an enhancement in *Glut1* expression [59]. FGF21 could regulate glucose and lipid metabolisms through the induction of FGF21 downstream signaling molecules including *Glut1* [60,61]. In adipocytes, an increase in *Glut1* mRNA expression has been along with upregulation of FGF21 [62]. Moreover, cardiac protection by administering FGF21 against ischemia/reperfusion-induced cardiac damages has been explained by the upregulation of GLUT1 [63]. Suppressed *Glut1* expression may impair an entry of glucose into photoreceptors, which results in a lack of lipid and glucose fuel for retinal function [64]. In this regard, elevated serum levels of FGF21 may support the induction of *Glut1*. This effect may bring positive outcomes to the damaging retina under acute hypoperfused states through modulation of glucose metabolism. However, previous reports suggested that the suppression of diabetic retinopathy could be involved with GLUT1 inhibition [65,66]. There may have a discrepancy between experimental models of our OIS and diabetic retinopathy in that blood glucose levels between them are totally different and the duration of diseases are not the same either. In fact, controversial reports on GLUT1 expression in diabetic retinopathy itself already exist. In the retina and its microvessels of streptozotocin-induced diabetes, downregulated GLUT1 expression was detected [67]. As determined by GLUT1 immunogold staining, compensatory downregulation of GLUT1 on the inner blood-retinal barrier was not seen in diabetic rats [68]. Chronic hyperglycemia led to a decrease in GLUT1 protein expression without alteration in its mRNA expression in the retina of diabetic Goto Kakizaki rats and alloxan-treated diabetic rabbits [69]. Taken together, more studies are needed for understanding the potential role of GLUT1 depending on the disease states.

It is reported that gliosis and loss of the amplitudes of OPs could be a hallmark of the early phase in a streptozotocin-induced diabetic mouse model and OIS mouse models [14,23,70]. Based on our preliminary data, there was a high correlation between pathological gliosis and loss of the amplitudes of OPs in the UCCAO-operated eye (Figure A5). This implies that retinal functional deficits may be along with the induction of pathological gliosis. Previously, fenofibrate modulated pathological gliosis and improved ERG abnormalities in *db/db* mice [48,71]. Similarly, pemafibrate modulated pathological gliosis and preserved retinal function in the ischemic retina based on our current data. In fact, we previously reported that pemafibrate maintained the amplitudes of OPs in a murine model of diabetes via maintaining the expression of synaptophysin, a marker of synapse [31]. Even though the expression of synaptophysin had a slight increase by pemafibrate administration, we assume that our current results have a consistency with the results in our previous study [31] and other PPARα studies [48,71].

In our current research, pemafibrate maintained the amplitude of a-wave in the UCCAO model, which indicates that the function of photoreceptors was protected by pemafibrate administration [72,73]. Müller glial cells have an important role in maintaining photoreceptors as well as retinal pigment epithelium [74]. We recently reported that FGF21 preserved photoreceptor function via modulating Müller glial cells in P23H mice which mimic human retinitis pigmentosa [57]. Additionally, FGF21 increased the synapse formation pathway in the retina and induced Müller glial axon development genes. The synaptic connection between the inner retina and the outer retina was also preserved by FGF21 treatment. Taken together, pemafibrate may have the potential to rescue inner and outer retinal cells via modulating Müller glial cells, observed by the preservation of OPs and a-wave. However, we need further studies to clarify the mechanism. On the other hand, pemafibrate did not affect retinal thickness as seen in our current data. In fact, we did not observe dramatic changes in retinal thickness in UCCAO-operated mice in comparison with that in sham-operated mice [23]. Taken together, pemafibrate could primarily influence retinal function, which is consistent with our previous report [31].

In our study, reduced levels of TG were seen after administration of pemafibrate. Furthermore, increased levels of TC (speculated as HDL-C [31,44,75]) were shown after the administration of pemafibrate. High levels of TG are suggested as one of the risk factors in human cardiovascular diseases [76–78]. Furthermore, it was reported that the TG/HDL-C ratio was highly associated with an increased risk of developing retinopathy [79,80]. It has also been reported that the high TG/HDL-C ratio could act on endothelial dysfunction, chronic low-grade inflammation, and coagulation [79,81]. Although the experimental UCCAO mouse model may not have dramatic metabolic stresses systemically, high levels of TG may exacerbate stenosis of CCA in human metabolic cardiovascular disease states, and pemafibrate could have preventive and protective roles on the stenosis of CCA via decreasing high TG levels in the blood. Similarly, metabolic changes could be considered as important factors in the development of retinal diseases [82,83]. Although we did not deeply cover systemic metabolic changes by the administration of pemafibrate, PPARα activation has been suggested as a regulator of β-oxidation [38,84]. WY16463, one of the selective PPARα agonists, showed reducing effects on the number of retinal angiomatous proliferation-like vascular lesions in the *Vldlr−/−* retina via the possible mechanism of enhancement of β-oxidation [64]. Pemafibrate also has been suggested to enhance β-oxidation [85,86]. Taken together, we speculate that more therapeutic effects of pemafibrate could be seen if we develop a new murine model of retinal ischemia by UCCAO in metabolic disorder models (which are more clinically relevant) and treat pemafibrate in those ischemic retinas. This will be further studied.

In the current study, the oral administration of pemafibrate was tested. Even though various methods for drug administration such as intraperitoneal injection, or intravitreal injection could be tested in our UCCAO model, we believe that our current method is patient-friendly (in terms of repetitive administrations of pemafibrate) and pain-free in the eye or body (as it is a non-invasive procedure) [87].

Now, Pemafibrate to Reduce Cardiovascular OutcoMes by Reducing Triglycerides IN patiENts With diabeTes (PROMINENT) study in patients with type 2 diabetes mellitus and dyslipidemia is undergoing all over the world (ClinicalTrials.gov Identifier: NCT03071692, accessed on 20 July 2021). Unfortunately, PROMINENT eye study, which tried to evaluate patient with diabetic retinopathy, was terminated because of a lack of recruited patients. Another clinical trial for pemafibrate has been completed for nonalcoholic fatty liver disease (NAFLD) and clinical scientists are waiting for the results (ClinicalTrials.gov Identifier: NCT03350165, phase 2, accessed on 20 July 2021). If the positive effect of pemafibrate on cardiovascular diseases or NAFLD could be seen, pemafibrate might have chances to be repositioned for retinal diseases in the future.

In conclusion, even though we need more links regarding retinal protection by activating PPARα in the liver, we suggest a promising pemafibrate therapy in carotid artery occlusion-induced ischemic retinopathy, with boosting liver function, regulating serum levels of FGF21, TC, and TG, and suppressing retinal dysfunction (Figure 7).

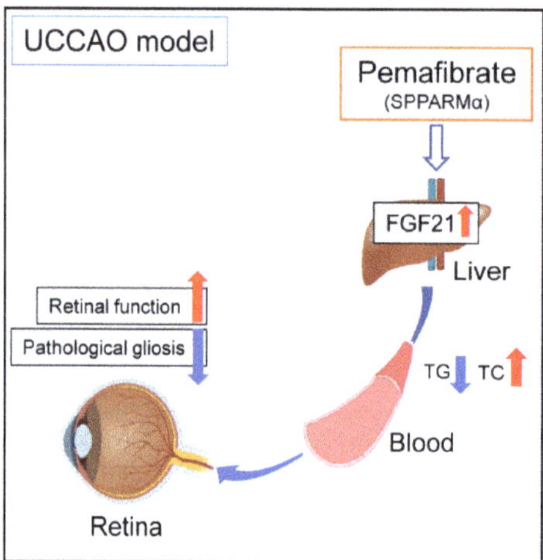

Figure 7. A working hypothesis of the protective mechanism against retinal dysfunction by administering pemafibrate in a murine model of retinal ischemia by UCCAO. The possible mechanisms for suppression of retinal dysfunction induced in cardiovascular diseases are that consecutive administration of systemic selective PPARα modulator (SPPARMα) pemafibrate enhances liver function and upregulates PPARα target genes in the liver, and elevated levels of serum FGF21 (one of the strong neuroprotective agents) modulate pathological gliosis and maintain the amplitudes of OPs. Indirectly, a reduction in levels of TG and an induction in levels of TC may have a risk-decreasing effect on developing retinopathy in humans. TG; triglyceride, TC; total cholesterol.

4. Materials and Methods

4.1. Animal

A number of 6–8 weeks old male C57BL/6 mice were obtained from CLEA Japan (Tokyo, Japan) and supplied freely with food and water under a twelve-hour light-dark cycle in a temperature-managed room. All protocols were permitted by the Ethics Committee on Animal Research of the Keio University School of Medicine (Approved number #16017/2020). All procedures complied with the ARVO Statement for the Use of Animals in Ophthalmic and Vision Research and the international standards of animal care and use, Animal Research: Reporting in Vivo Experiments (ARRIVE) guidelines (accessed on 20 July 2021, http://www.nc3rs.org.uk/arrive-guidelines).

4.2. A Murine Model of UCCAO-Induced Retinal Ischemia and Oral Administration of Pemafibrate

Randomized mice were orally provided 0.5% DMSO-dissolved PBS or pemafibrate (0.5 mg/kg in 0.5% DMSO-dissolved PBS) for four days daily before UCCAO. A mouse model of UCCAO-induced retinal ischemia was induced, as previously described [17]. Briefly, deep anesthesia was induced to mice with a combination of midazolam (40 μg/100 μL; Sandoz, Tokyo, Japan), medetomidine (7.5 μg/100 μL; Orion, Espoo, Finland), and butorphanol tartrate (50 μg/100 μL; Meiji Seika Pharma, Tokyo, Japan) [22]. The mouse neck was incised to observe the common carotid artery. Then, the common carotid artery in the right side was permanently occluded using 6–0 silk sutures. Wounds of the neck were clearly sutured, and the mouse was recovered. Pemafibrate was continuously supplied to mice daily 1 day after UCCAO until the end of experiments. The body weight was measured during the whole experimental period, and the liver weight was measured on the day of sample collection.

4.3. Optical Coherence Tomography (OCT)

OCT (Envisu R4310, Leica, Wetzlar, Germany) was conducted as previously described [22,27]. Briefly, mice were subjected to mydriasis by a combination of 0.5% tropicamide and 0.5% phenylephrine (Santen Pharmaceutical, Osaka, Japan). After 5 min, mice were anesthetized as same as Section 4.2. Anesthetized mice were quickly subjected to OCT analyses. B-scan images were obtained from equatorial slices of en-face scans, and images in 0.2, 0.4, and 0.6 mm from the optic nerve head were taken. Retinal thickness was measured from the outer retina to the inner retina as we described [27].

4.4. Electroretinography (ERG)

ERG was conducted as previously described [27]. Briefly, mice were placed for more than 12 h for dark adaptation. Pupils were dilated as Section 4.3. Mice were anesthetized as Section 4.2 after 5 min incubation. Recording of scotopic ERG responses was processed using a Ganzfeld dome and LED stimulators with an acquisition system (PuREC, MAYO, Inazawa, Japan). The amplitudes of a-wave and b-wave were measured with various light stimuli. Furthermore, the amplitudes of OPs were measured at the four peaks of OPs as previously described [23].

4.5. Immunohistochemistry (IHC)

IHC was performed as previously [23]. Briefly, eyes were fixed with PFA (4%), and O.C.T. Compound (Sakura Tissue-Tek, Tokyo, Japan) was applied to embed the eyes for frozen sectioning. The sagittal sectioning slides using Cryostat (Leica CM3050S, Leica, Wetzlar, Germany) were incubated in a blocking solution (PBS + 0.1% Triton + 0.1% BSA). Then, a primary antibody (GFAP 1:400, Cat #13-0300, Thermo Fisher Scientific, Waltham, MA, USA) was added to the eyes. The eyes were washed with PBS + 0.1% Triton and soaked into a solution of a species-appropriate fluorescence-conjugated secondary antibody (Thermo Fisher Scientific, Waltham, MA, USA) for several hours. After washing with PBS + 0.1% Triton three times, DAPI was shortly incubated. After washing with PBS again, the eyes were mounted and examined via a fluorescence microscope (LSM710, Carl Zeiss, Jena, Germany), as previously described [22]. The fluorescence immunoreactivity was quantified by a morphology score as previously described [12,17,23] with a minor modification: 0 = no signal, 1 = labeled processes in the ganglion cell layer, 2 = weakly labeled processes in the inner retinal layer, including the ganglion cell layer, and 3 = strongly labeled processes in the entire retinal layer including the inner and outer retinas.

4.6. Measurement of Serum FGF21, TC, and TG Levels

After blood collection and serum extraction as previously described [27,31], serum samples were evaluated with an FGF21 ELISA kit (Cat #RD291108200R, BioVendor Laboratory Medicine, Brno, Czech Republic), a TC kit (Cat #STA-384, Cell Biolabs, Inc., San Diego, CA, USA), and a TG kit (Cat #STA-396, Cell Biolabs, Inc., San Diego, CA, USA) following the manufacturer's instructions.

4.7. Quantitative PCR

Quantitative PCR was conducted, as previously described [27]. Briefly, the retina and the liver mRNA were extracted using an RNeasy Plus Mini Kit (Qiagen, Venlo, The Netherlands). RT-PCR was conducted with a ReverTra Ace® qPCR RT Master Mix with gDNA Remover (TOYOBO, Osaka, Japan). Quantitative PCR was conducted using a THUNDERBIRD® SYBR® qPCR Mix (TOYOBO, Osaka, Japan) with the Step One Plus Real-Time PCR system (Applied Biosystems, Waltham, MA, USA). The primers that we used are entered in Table 2. The fold alteration between levels of different transcripts was calculated by the ΔΔCT protocol.

4.8. Western Blotting

Western Blotting was conducted as described in our previous paper [27]. We used anti-synaptophysin (1:1000, Cat #SAB4502906, Sigma, Tokyo, Japan) and anti-β-Actin (1:5000, #3700, Cell Signaling Technology, Danvers, MA, USA). After incubation of primary antibodies, HRP-conjugated secondary antibodies (1:1000 for anti-synaptophysin; 1:5000 for anti-β-Actin, GE Healthcare, Chicago, IL, USA) were put to the membrane. Intensities of the bands were quantified via NIH ImageJ program (National Institutes of Health, Bethesda, MD, USA).

Table 2. Primer list.

Name	Direction	Sequence (5′ → 3′)	Accession Number
Hprt	Forward Reverse	TCAGTCAACGGGGGACATAAA GGGGCTGTACTGCTTAACCAG	NM_013556.2
Epo	Forward Reverse	GGCCATAGAAGTTTGGCAAG CCTCTCCCGTGTACAGCTTC	NM_007942
Bnip3	Forward Reverse	GCTCCCAGACACCACAAGAT TGAGAGTAGCTGTGCGCTTC	NM_009760.4
Vegfa	Forward Reverse	CCTGGTGGACATCTTCCAGGAGTACC GAAGCTCATCTCTCCTATGTGCTGGC	AY707864.1
Glut1	Forward Reverse	CAGTTCGGCTATAACACTGGTG GCCCCCGACAGAGAAGATG	NM_011400.3
Ccl2	Forward Reverse	CCCAATGAGTAGGCTGGAGA TCTGGACCCATTCCTTCTTG	NM_011333.3
Ccl12	Forward Reverse	GCTACAGGAGAATCACAAGCAGC ACGTCTTATCCAAGTGGTTTATGG	NM_011331.3
Ucp3	Forward Reverse	GGAGTCTCACCTGTTTACTGACAACT GCACAGAAGCCAGCTCCAA	NM_009464.3
Fabp4	Forward Reverse	CCGCAGACGACAGGA CTCATGCCCTTTCATAAACT	NM_024406.3
Fgf21	Forward Reverse	AACAGCCATTCACTTTGCCTGAGC GGCAGCTGGAATTGTGTTCTGACT	NM_020013.4
Vldlr	Forward Reverse	GAGCCCCTGAAGGAATGCC CCTATAACTAGGTCTTTGCAGATATGG	NM_001161420.1
Acox1	Forward Reverse	TCTTCTTGAGACAGGGCCCAG GTTCCGACTAGCCAGGCATG	AF006688.1

4.9. Statistical Analysis

Data were analyzed with GraphPad Prism 5 (GraphPad Program, San Diego, CA, USA) and calculated by using a two-way Student's *t*-test or two-way ANOVA followed by a Bonferroni post hoc test depending on the dataset. Any *p*-values of less than 0.05 were regarded as statistically significant.

Author Contributions: Conceptualization, D.L., Y.T. and T.K.; methodology, D.L., H.J. and Y.M.; validation, D.L.; formal analysis, D.L.; investigation, D.L.; resources, T.K.; data curation, D.L. and Y.T.; writing—original draft preparation, D.L. and Y.T.; writing—review and editing, K.T., K.N. and T.K.; visualization, D.L. and Y.T.; supervision, T.K.; project administration, T.K.; funding acquisition, T.K. All authors have read and agreed to the published version of the manuscript.

Funding: This work is supported by Grants-in-Aid for Scientific Research (KAKENHI, number 15K10881 and 18K09424) from the Ministry of Education, Culture, Sports, Science and Technology (MEXT) to T.K.

Institutional Review Board Statement: Protocols using animals were permitted by the Ethics Committee on Animal Research of the Keio University School of Medicine (Approved number #16017/2020). Procedures complied with the ARVO Statement for the Use of Animals in Ophthalmic and Vision Research in accordance with the international standards of animal care and use, ARRIVE (Animal Research: Reporting in Vivo Experiments) guidelines (accessed on 20 July 2021, http://www.nc3rs.org.uk/arrive-guidelines).

Informed Consent Statement: Not applicable.

Data Availability Statement: The data presented in this study are available on request from the corresponding author.

Acknowledgments: We thank K. Kurosaki and A. Kawabata for critical discussions. Furthermore, we thank Kowa Company for providing pemafibrate.

Conflicts of Interest: Yukihiro Miwa is employed by Tokyo Animal Eye Clinic and Kazuo Tsubota is CEO in Tsubota Laboratory, Inc. The remaining authors declare no conflict of interest.

Appendix A

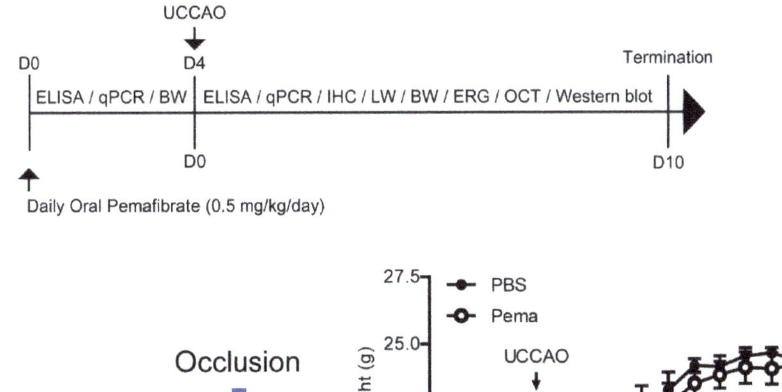

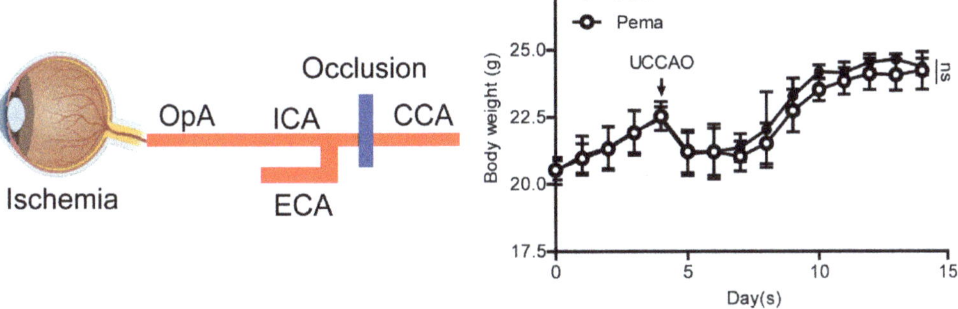

Figure A1. General monitoring for adult mice after consecutive oral administration of pemafibrate. A schematic illustration shows oral administration of pemafibrate (0.5 mg/kg/day) to mice and a time point of the UCCAO surgery and experiments followed. ELISA; enzyme-linked immunosorbent assay, qPCR; quantitative PCR, BW; body weight, IHC; immunohistochemistry, LW; liver weight, ERG; electroretinography, OCT; optical coherence tomography, UCCAO; unilateral common carotid artery occlusion. A schematic illustration of retinal ischemia induction by UCCAO. Retinal ischemia could be induced by occlusion (a blue bar) of the common carotid artery (CCA) as the ophthalmic artery (OpA) is originated from the internal carotid artery (ICA) of CCA. ECA; external carotid artery. Quantitative analyses (n = 5–10 per group) showed that the body weight of mice became lower after UCCAO. There was no dramatic difference in the body weight between pemafibrate-administered mice and PBS-administered mice. However, mice showed a slight decrease in the body weight after consecutive administration of pemafibrate without any statistical significance. $p > 0.05$. The data were analyzed using two-way ANOVA followed by a Bonferroni post hoc test. The data were presented as mean ± standard deviation. Pema; pemafibrate. ns; not significant.

Appendix B

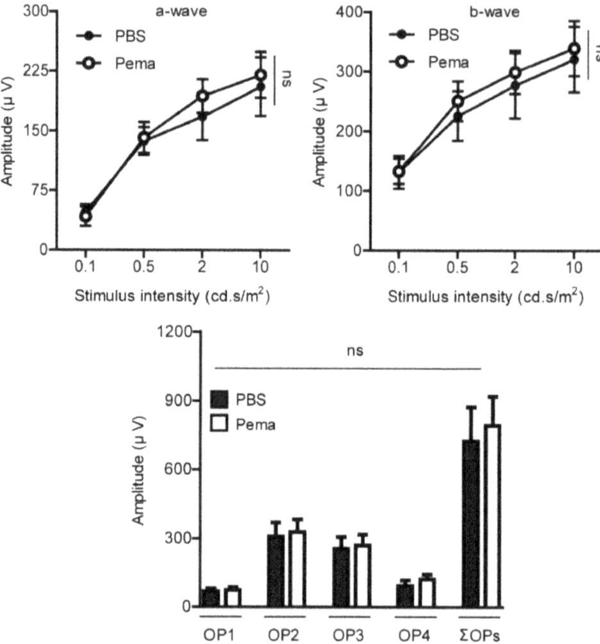

Figure A2. General measurements of retinal function for adult mice after consecutive oral administration of pemafibrate right before UCCAO. Quantitative analyses (n = 5 per group) showed that oral administration of pemafibrate had no effect on retinal function (a-wave, b-wave, and OPs) in adult naïve mice. $p > 0.05$. The data (a-and b-waves) were analyzed using two-way ANOVA followed by a Bonferroni post hoc test and presented as mean ± standard error of the mean. The data (OPs) were analyzed using two-tailed Student's t-test and presented as mean ± standard deviation. Pema; pemafibrate, OPs; oscillatory potentials. ns; not significant.

Appendix C

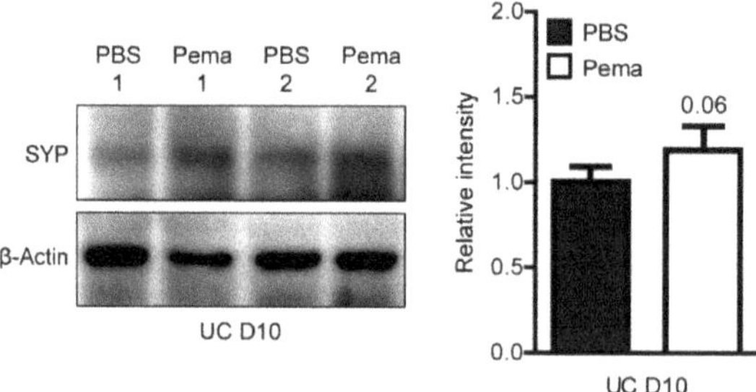

Figure A3. A slight increase in retinal synaptophysin (SYP) expression by oral administration of pemafibrate in UCCAO-operated mice. A representative image and quantitative analysis (n = 4 per group) showed that SYP expression slightly increased by oral administration of pemafibrate 10 days after UCCAO. $p = 0.06$. The data were analyzed using Student's t-test and presented as mean ± standard deviation. Pema; pemafibrate, UC; unilateral common carotid artery occlusion.

Appendix D

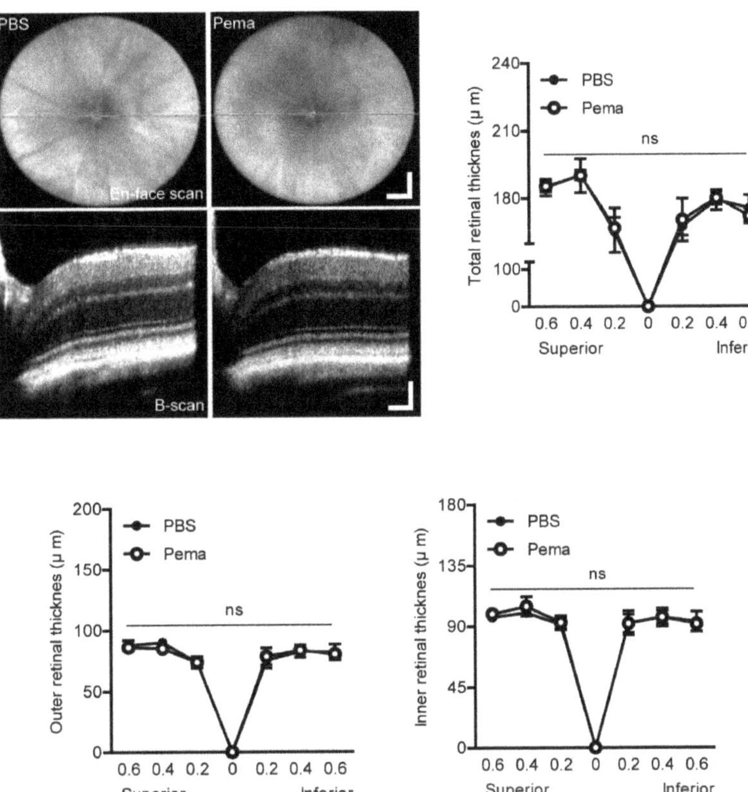

Figure A4. No alteration in retinal thickness by oral administration of pemafibrate in UCCAO-operated mice. Representative OCT images (b-scan) in the PBS-and pemafibrate-administered UCCAO-operated retinas and quantitative analyses (n = 5 per group) showed that there was no change in retinal thickness (total, outer, and inner retinal layers) on day 10 after UCCAO. The values in the horizontal axis of the graph stand for 0.2, 0.4, and 0.6 mm distance from the optic nerve head (0) that was detected by the green line (en-face scan). Representative OCT images were taken at 0.4 mm from the optic nerve head. The data were analyzed using two-way ANOVA followed by a Bonferroni post hoc test and presented as a spider diagram (mean ± standard deviation). $p > 0.05$. Scale bars are 200 (en-face scan; vertical and horizontal bars) and 200 and 100 (b-scan; vertical and horizontal bars) μm, respectively. Pema; pemafibrate, OCT; optical coherence tomography. ns; not significant.

Appendix E

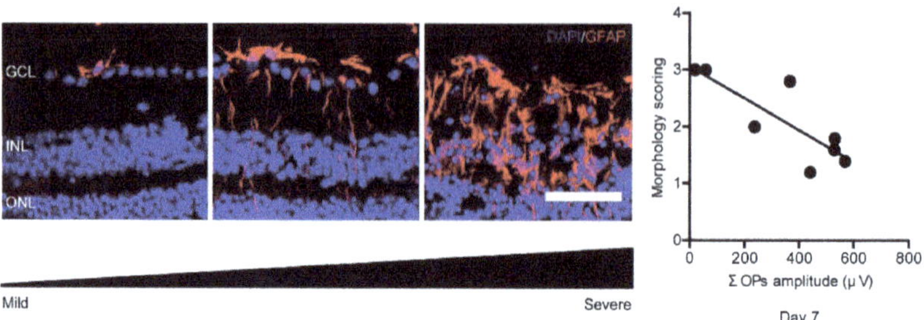

Figure A5. A relationship between pathological retinal gliosis and retinal dysfunction 7 days after UCCAO. Representative images of morphology scoring for pathological retinal gliosis (acquired from our preliminary experiments) and visualization in a correlation between pathological retinal gliosis and the amplitudes of ΣOPs showed that there was a high correlation between pathological gliosis and loss of the amplitudes of ΣOPs in the UCCAO-operated eye, determined by regression analyses. Dots ($n = 8$) represent each morphology scoring and the amplitude of ΣOPs. A line represents the linear fit of the data points. Scale bar: 50 µm. Slope: −0.0028; Y-intercept: 3.06; r^2: 0.68. $p < 0.05$. GCL: ganglion cell layer; INL: inner nuclear layer; ONL: outer nuclear layer, UC; unilateral common carotid artery occlusion.

References

1. Terelak-Borys, B.; Skonieczna, K.; Grabska-Liberek, I. Ocular ischemic syndrome—A systematic review. *Med. Sci. Monit.* **2012**, *18*, RA138–RA144. [CrossRef]
2. Hedges, T.R. Ophthalmoscopic findings in internal carotid artery occlusion. *Am. J. Ophthalmol.* **1963**, *55*, 1007–1012. [CrossRef]
3. Sturrock, G.D.; Mueller, H.R. Chronic ocular ischaemia. *Br. J. Ophthalmol.* **1984**, *68*, 716–723. [CrossRef]
4. Brown, G.C.; Magargal, L.E. The ocular ischemic syndrome. *Int. Ophthalmol.* **1988**, *11*, 239–251. [CrossRef]
5. Duker, J.S.; Belmont, J.B. Ocular ischemic syndrome secondary to carotid artery dissection. *Am. J. Ophthalmol.* **1988**, *106*, 750–752. [CrossRef]
6. Hamed, L.M.; Guy, J.R.; Moster, M.L.; Bosley, T. Giant cell arteritis in the ocular ischemic syndrome. *Am. J. Ophthalmol.* **1992**, *113*, 702–705. [CrossRef]
7. Sadun, A.; Sebag, J.; Bienfang, D.C. Complete bilateral internal carotid artery occlusion in a young man. *J. Clin. Neuro-Ophthalmol.* **1983**, *3*, 63–66.
8. Malhotra, R. Management of ocular ischaemic syndrome. *Br. J. Ophthalmol.* **2000**, *84*, 1428–1431. [CrossRef]
9. Sood, G.; Siddik, A.B. Ocular ischemic syndrome. In *StatPearls*; StatPearls Publishing LLC: Treasure Island, FL, USA, 2021.
10. Sivalingam, A.; Brown, G.C.; Magargal, L.E. The ocular ischemic syndrome—III Visual prognosis and the effect of treatment. *Int. Ophthalmol.* **1991**, *15*, 15–20. [CrossRef]
11. Sun, W.; Geng, Y.; Chen, Y.; Tang, X.-H.; Zhang, Y.; Gu, S.-H.; Xie, J.-J.; Zhang, Z.-A.; Tian, X.-S. Differences of brain pathological changes and cognitive function after bilateral common carotid artery occlusion between Sprague-Dawley and Wistar rats. *Acta Physiol. Sin.* **2019**, *71*, 705–716.
12. Yamamoto, H.; Schmidt-Kastner, R.; Hamasaki, D.I.; Yamamoto, H.; Parel, J.-M. Complex neurodegeneration in retina following moderate ischemia induced by bilateral common carotid artery occlusion in Wistar rats. *Exp. Eye Res.* **2006**, *82*, 767–779. [CrossRef] [PubMed]
13. Qin, Y.; Ji, M.; Deng, T.; Luo, D.; Zi, Y.; Pan, L.; Wang, Z.; Jin, M. Functional and morphologic study of retinal hypoperfusion injury induced by bilateral common carotid artery occlusion in rats. *Sci. Rep.* **2019**, *9*, 1–10. [CrossRef]
14. Crespo-Garcia, S.; Reichhart, N.; Skosyrski, S.; Foddis, M.; Wu, J.; Figura, A.; Herrspiegel, C.; Fuchtemeier, M.; Sassi, C.; Dirnagl, U.; et al. Individual and temporal variability of the retina after chronic bilateral common carotid artery occlusion (BCCAO). *PLoS ONE* **2018**, *13*, e0193961.
15. Lavinsky, D.; Arterni, N.S.; Achaval, M.; Netto, C.A. Chronic bilateral common carotid artery occlusion: A model for ocular ischemic syndrome in the rat. *Graefe's Arch. Clin. Exp. Ophthalmol.* **2005**, *244*, 199–204. [CrossRef] [PubMed]
16. Kalesnykas, G.; Tuulos, T.; Uusitalo, H.; Jolkkonen, J. Neurodegeneration and cellular stress in the retina and optic nerve in rat cerebral ischemia and hypoperfusion models. *Neuroscience* **2008**, *155*, 937–947. [CrossRef] [PubMed]
17. Lee, D.; Kang, H.; Yoon, K.Y.; Chang, Y.Y.; Song, H.B. A mouse model of retinal hypoperfusion injury induced by unilateral common carotid artery occlusion. *Exp. Eye Res.* **2020**, *201*, 108275. [CrossRef]

18. Lee, B.J.; Jun, H.O.; Kim, J.H. Astrocytic cystine/glutamate antiporter is a key regulator of erythropoietin expression in the ischemic retina. *FASEB J.* **2019**, *33*, 6045–6054. [CrossRef]
19. Hayreh, S.S.; Zimmerman, M.B. Ocular arterial occlusive disorders and carotid artery disease. *Ophthalmol. Retin.* **2016**, *1*, 12–18. [CrossRef]
20. Weymouth, W.; Pedersen, C. Central retinal artery occlusion associated with carotid artery occlusion. *Clin. Pr. Cases Emerg. Med.* **2019**, *3*, 233–236. [CrossRef]
21. Yang, G.; Kitagawa, K.; Matsushita, K.; Mabuchi, T.; Yagita, Y.; Yanagihara, T.; Matsumoto, M. C57BL/6 strain is most susceptible to cerebral ischemia following bilateral common carotid occlusion among seven mouse strains: Selective neuronal death in the murine transient forebrain ischemia. *Brain Res.* **1997**, *752*, 209–218. [CrossRef]
22. Lee, D.; Miwa, Y.; Jeong, H.; Ikeda, S.-I.; Katada, Y.; Tsubota, K.; Kurihara, T. A murine model of ischemic retinal injury induced by transient bilateral common carotid artery occlusion. *J. Vis. Exp.* **2020**. [CrossRef] [PubMed]
23. Lee, D.; Jeong, H.; Miwa, Y.; Shinojima, A.; Katada, Y.; Tsubota, K.; Kurihara, T. Retinal dysfunction induced in a mouse model of unilateral common carotid artery occlusion. *PeerJ* **2021**, *9*, e11665. [CrossRef]
24. Tomita, Y.; Lee, D.; Tsubota, K.; Kurihara, T. PPARα agonist oral therapy in diabetic retinopathy. *Biomedicines* **2020**, *8*, 433. [CrossRef]
25. Keech, A.; Mitchell, P.; Summanen, P.; O'Day, J.; Davis, T.; Moffitt, M.; Taskinen, M.-R.; Simes, R.; Tse, D.; Williamson, E.; et al. Effect of fenofibrate on the need for laser treatment for diabetic retinopathy (FIELD study): A randomised controlled trial. *Lancet* **2007**, *370*, 1687–1697. [CrossRef]
26. ACCORD Study Group and ACCORD Eye Study Group. Effects of medical therapies on retinopathy progression in type 2 diabetes. *N. Engl. J. Med.* **2010**, *363*, 233–244. [CrossRef] [PubMed]
27. Lee, D.; Tomita, Y.; Miwa, Y.; Jeong, H.; Mori, K.; Tsubota, K.; Kurihara, T. Fenofibrate protects against retinal dysfunction in a murine model of common carotid artery occlusion-induced ocular ischemia. *Pharmaceuticals* **2021**, *14*, 223. [CrossRef]
28. Noonan, J.E.; Jenkins, A.J.; Ma, J.-X.; Keech, A.C.; Wang, J.J.; Lamoureux, E.L. An update on the molecular actions of fenofibrate and its clinical effects on diabetic retinopathy and other microvascular end points in patients with diabetes. *Diabetes* **2013**, *62*, 3968–3975. [CrossRef]
29. Chen, Y.; Hu, Y.; Lin, M.; Jenkins, A.J.; Keech, A.C.; Mott, R.; Lyons, T.J.; Ma, J.-X. Therapeutic effects of PPARα agonists on diabetic retinopathy in type 1 diabetes models. *Diabetes* **2012**, *62*, 261–272. [CrossRef]
30. Tomita, Y.; Ozawa, N.; Miwa, Y.; Ishida, A.; Ohta, M.; Tsubota, K.; Kurihara, T. Pemafibrate prevents retinal pathological neovascularization by increasing FGF21 level in a murine oxygen-induced retinopathy model. *Int. J. Mol. Sci.* **2019**, *20*, 5878. [CrossRef]
31. Tomita, Y.; Lee, D.; Miwa, Y.; Jiang, X.; Ohta, M.; Tsubota, K.; Kurihara, T. Pemafibrate protects against retinal dysfunction in a murine model of diabetic retinopathy. *Int. J. Mol. Sci.* **2020**, *21*, 6243. [CrossRef]
32. Shiono, A.; Sasaki, H.; Sekine, R.; Abe, Y.; Matsumura, Y.; Inagaki, T.; Tanaka, T.; Kodama, T.; Aburatani, H.; Sakai, J.; et al. PPARα activation directly upregulates thrombomodulin in the diabetic retina. *Sci. Rep.* **2020**, *10*. [CrossRef] [PubMed]
33. Fujita, N.; Sase, K.; Tsukahara, C.; Arizono, I.; Takagi, H.; Kitaoka, Y. Pemafibrate prevents retinal neuronal cell death in NMDA-induced excitotoxicity via inhibition of p-c-Jun expression. *Mol. Biol. Rep.* **2020**, *48*, 195–202. [CrossRef] [PubMed]
34. Wachtmeister, L. Oscillatory potentials in the retina: What do they reveal. *Prog. Retin. Eye Res.* **1998**, *17*, 485–521. [CrossRef]
35. Heynen, H.; Wachtmeister, L.; van Norren, D. Origin of the oscillatory potentials in the primate retina. *Vis. Res.* **1985**, *25*, 1365–1373. [CrossRef]
36. Yonemura, D.; Kawasaki, K. New approaches to ophthalmic electrodiagnosis by retinal oscillatory potential, drug-induced responses from retinal pigment epithelium and cone potential. *Doc. Ophthalmol.* **1979**, *48*, 163–222. [CrossRef]
37. de Hoz, R.; Rojas, B.; Ramírez, A.; Salazar, J.J.; Gallego, B.I.; Triviño, A.; Ramírez, J.M. Retinal macroglial responses in health and disease. *BioMed Res. Int.* **2016**, *2016*, 2954721. [CrossRef]
38. Sasaki, Y.; Raza-Iqbal, S.; Tanaka, T.; Murakami, K.; Anai, M.; Osawa, T.; Matsumura, Y.; Sakai, J.; Kodama, T. Gene expression profiles induced by a novel selective peroxisome proliferator-activated receptor α modulator (SPPARMα) pemafibrate. *Int. J. Mol. Sci.* **2019**, *20*, 5682. [CrossRef]
39. Badman, M.K.; Pissios, P.; Kennedy, A.R.; Koukos, G.; Flier, J.S.; Maratos-Flier, E. Hepatic fibroblast growth factor 21 is regulated by PPARα and is a key mediator of hepatic lipid metabolism in ketotic states. *Cell Metab.* **2007**, *5*, 426–437. [CrossRef]
40. Ong, K.-L.; Januszewski, A.S.; O'Connell, R.; Jenkins, A.; Xu, A.; Sullivan, D.R.; Barter, P.J.; Hung, W.-T.; Scott, R.S.; Taskinen, M.-R.; et al. The relationship of fibroblast growth factor 21 with cardiovascular outcome events in the fenofibrate intervention and event lowering in diabetes study. *Diabetologia* **2014**, *58*, 464–473. [CrossRef]
41. Ong, K.-L.; O'Connell, R.; Januszewski, A.S.; Jenkins, A.J.; Xu, A.; Sullivan, D.R.; Barter, P.J.; Scott, R.S.; Taskinen, M.-R.; Waldman, B.; et al. Baseline circulating FGF21 concentrations and increase after fenofibrate treatment predict more rapid glycemic progression in type 2 diabetes: Results from the FIELD study. *Clin. Chem.* **2017**, *63*, 1261–1270. [CrossRef] [PubMed]
42. Ong, K.L.; Rye, K.-A.; O'Connell, R.; Jenkins, A.J.; Brown, C.; Xu, A.; Sullivan, D.R.; Barter, P.J.; Keech, A.C.; FIELD Study Investigators. Long-term fenofibrate therapy increases fibroblast growth factor 21 and retinol-binding protein 4 in subjects with type 2 diabetes. *J. Clin. Endocrinol. Metab.* **2012**, *97*, 4701–4708. [CrossRef] [PubMed]

43. Sairyo, M.; Kobayashi, T.; Masuda, D.; Kanno, K.; Zhu, Y.; Okada, T.; Koseki, M.; Ohama, T.; Nishida, M.; Sakata, Y.; et al. A novel selective PPARα modulator (SPPARMα), K-877 (pemafibrate), attenuates postprandial hypertriglyceridemia in mice. *J. Atheroscler. Thromb.* **2018**, *25*, 142–152. [CrossRef]
44. Bando, H.; Taneda, S.; Manda, N. Efficacy and safety of low-dose pemafibrate therapy for hypertriglyceridemia in patients with type 2 diabetes. *JMA J.* **2021**, *4*, 135–140. [CrossRef] [PubMed]
45. Komiya, I.; Yamamoto, A.; Sunakawa, S.; Wakugami, T. Pemafibrate decreases triglycerides and small, dense LDL, but increases LDL-C depending on baseline triglycerides and LDL-C in type 2 diabetes patients with hypertriglyceridemia: An observational study. *Lipids Heal. Dis.* **2021**, *20*, 1–11. [CrossRef]
46. Macosko, E.Z.; Basu, A.; Satija, R.; Nemesh, J.; Shekhar, K.; Goldman, M.; Tirosh, I.; Bialas, A.R.; Kamitaki, N.; Martersteck, E.M.; et al. Highly parallel genome-wide expression profiling of individual cells using nanoliter droplets. *Cell* **2015**, *161*, 1202–1214. [CrossRef] [PubMed]
47. Lundåsen, T.; Hunt, M.C.; Nilsson, L.-M.; Sanyal, S.; Angelin, B.; Alexson, S.E.; Rudling, M. PPARα is a key regulator of hepatic FGF21. *Biochem. Biophys. Res. Commun.* **2007**, *360*, 437–440. [CrossRef]
48. Enright, J.M.; Zhang, S.; Thebeau, C.; Siebert, E.; Jin, A.; Gadiraju, V.; Zhang, X.; Chen, S.; Semenkovich, C.F.; Rajagopal, R. Fenofibrate reduces the severity of neuroretinopathy in a type 2 model of diabetes without inducing peroxisome proliferator-activated receptor alpha-dependent retinal gene expression. *J. Clin. Med.* **2020**, *10*, 126. [CrossRef]
49. Eitoh, N. FGF21 as a hepatokine, adipokine, and myokine in metabolism and diseases. *Front. Endocrinol.* **2014**, *5*, 107. [CrossRef]
50. IItoh, N.; Nakayama, Y.; Konishi, M. Roles of FGFs as paracrine or endocrine signals in liver development, health, and disease. *Front. Cell Dev. Biol.* **2016**, *4*, 30. [CrossRef]
51. Nishimura, T.; Nakatake, Y.; Konishi, M.; Itoh, N. Identification of a novel FGF, FGF-21, preferentially expressed in the liver. *Biochim. Biophys. Acta Gene Struct. Expr.* **2000**, *1492*, 203–206. [CrossRef]
52. Tezze, C.; Romanello, V.; Sandri, M. FGF21 as modulator of metabolism in health and disease. *Front. Physiol.* **2019**, *10*, 419. [CrossRef] [PubMed]
53. Talukdar, S.; Zhou, Y.; Li, D.; Rossulek, M.; Dong, J.; Somayaji, V.; Weng, Y.; Clark, R.; Lanba, A.; Owen, B.; et al. A long-acting FGF21 molecule, PF-05231023, decreases body weight and improves lipid profile in non-human primates and type 2 diabetic subjects. *Cell Metab.* **2016**, *23*, 427–440. [CrossRef]
54. Fu, Z.; Gong, Y.; Liegl, R.; Wang, Z.; Liu, C.-H.; Meng, S.S.; Burnim, S.B.; Saba, N.J.; Fredrick, T.W.; Morss-Walton, P.; et al. FGF21 administration suppresses retinal and choroidal neovascularization in mice. *Cell Rep.* **2017**, *18*, 1606–1613. [CrossRef] [PubMed]
55. Fu, Z.; Wang, Z.; Liu, C.-H.; Gong, Y.; Cakir, B.; Liegl, R.; Sun, Y.; Meng, S.S.; Burnim, S.B.; Arellano, I.; et al. Fibroblast growth factor 21 protects photoreceptor function in type 1 diabetic mice. *Diabetes* **2018**, *67*, 974–985. [CrossRef] [PubMed]
56. Tomita, Y.; Fu, Z.; Wang, Z.; Cakir, B.; Cho, S.S.; Britton, W.; Sun, Y.; Hellström, A.; Talukdar, S.; Smith, L.E. Long-acting FGF21 inhibits retinal vascular leakage in in vivo and in vitro models. *Int. J. Mol. Sci.* **2020**, *21*, 1188. [CrossRef]
57. Fu, Z.; Qiu, C.; Cagnone, G.; Tomita, Y.; Huang, S.; Cakir, B.; Kotoda, Y.; Allen, W.; Bull, E.; Akula, J.D.; et al. Retinal glial remodeling by FGF21 preserves retinal function during photoreceptor degeneration. *iScience* **2021**, *24*, 102376. [CrossRef]
58. Staiger, H.; Keuper, M.; Berti, L.; de Angelis, M.H.; Häring, H.-U. Fibroblast growth factor 21—Metabolic role in mice and men. *Endocr. Rev.* **2017**, *38*, 468–488. [CrossRef]
59. Liu, M.; Cao, H.; Hou, Y.; Sun, G.; Li, D.; Wang, W. Liver plays a major role in FGF-21 mediated glucose homeostasis. *Cell. Physiol. Biochem.* **2018**, *45*, 1423–1433. [CrossRef]
60. Ye, X.; Qi, J.; Yu, D.; Wu, Y.; Zhu, S.; Li, S.; Wu, Q.; Ren, G.; Li, D. Pharmacological efficacy of FGF21 analogue, liraglutide and insulin glargine in treatment of type 2 diabetes. *J. Diabetes Complicat.* **2017**, *31*, 726–734. [CrossRef]
61. Yu, D.; Ye, X.; Wu, Q.; Li, S.; Yang, Y.; He, J.; Liu, Y.; Zhang, X.; Yuan, Q.; Liu, M.; et al. Insulin sensitizes FGF21 in glucose and lipid metabolisms via activating common AKT pathway. *Endocrine* **2015**, *52*, 527–540. [CrossRef]
62. Li, K.; Li, L.; Yang, M.; Liu, H.; Boden, G.; Yang, G. The effects of fibroblast growth factor-21 knockdown and over-expression on its signaling pathway and glucose-lipid metabolism in vitro. *Mol. Cell. Endocrinol.* **2012**, *348*, 21–26. [CrossRef]
63. Hu, S.; Cao, S.; Liu, J. Role of angiopoietin-2 in the cardioprotective effect of fibroblast growth factor 21 on ischemia/reperfusion-induced injury in H9c2 cardiomyocytes. *Exp. Ther. Med.* **2017**, *14*, 771–779. [CrossRef]
64. Joyal, J.-S.; Sun, Y.; Gantner, M.L.; Shao, Z.; Evans, L.P.; Saba, N.; Fredrick, T.; Burnim, S.; Kim, J.S.; Patel, G.; et al. Retinal lipid and glucose metabolism dictates angiogenesis through the lipid sensor Ffar1. *Nat. Med.* **2016**, *22*, 439–445. [CrossRef]
65. Mohammad, H.M.; Sami, M.M.; Makary, S.; Toraih, E.A.; Mohamed, A.O.; El-Ghaiesh, S.H. Neuroprotective effect of levetiracetam in mouse diabetic retinopathy: Effect on glucose transporter-1 and GAP43 expression. *Life Sci.* **2019**, *232*, 116588. [CrossRef]
66. You, Z.-P.; Zhang, Y.-L.; Shi, K.; Shi, L.; Zhang, Y.-Z.; Zhou, Y.; Wang, C.-Y. Suppression of diabetic retinopathy with GLUT1 siRNA. *Sci. Rep.* **2017**, *7*, 1–10. [CrossRef] [PubMed]
67. Badr, G.A.; Tang, J.; Ismail-Beigi, F.; Kern, T.S. Diabetes downregulates GLUT1 expression in the retina and its microvessels but not in the cerebral cortex or its microvessels. *Diabetes* **2000**, *49*, 1016–1021. [CrossRef] [PubMed]
68. Fernandes, R.; Suzuki, K.-I.; Kumagai, A.K. Inner blood-retinal barrier GLUT1 in long-term diabetic rats: An immunogold electron microscopic study. *Investig. Opthalmology Vis. Sci.* **2003**, *44*, 3150–3154. [CrossRef] [PubMed]
69. Fernandes, R.; Carvalho, A.L.; Kumagai, A.; Seica, R.; Hosoya, K.-I.; Terasaki, T.; Murta, J.; Pereira, P.; Faro, C. Downregulation of retinal GLUT1 in diabetes by ubiquitinylation. *Mol. Vis.* **2004**, *10*, 618–628.

70. Sergeys, J.; Etienne, I.; Van Hove, I.; Lefevere, E.; Stalmans, I.; Feyen, J.H.M.; Moons, L.; Van Bergen, T. Longitudinal in vivo characterization of the streptozotocin-induced diabetic mouse model: Focus on early inner retinal responses. *Investig. Opthalmology Vis. Sci.* **2019**, *60*, 807–822. [CrossRef]
71. Bogdanov, P.; Hernández, C.; Corraliza, L.; Carvalho, A.R.; Simó, R. Effect of fenofibrate on retinal neurodegeneration in an experimental model of type 2 diabetes. *Acta Diabetol.* **2014**, *52*, 113–122. [CrossRef]
72. Brown, K.T. The electroretinogram: Its components and their origins. *Vis. Res.* **1968**, *8*, 633-IN6. [CrossRef]
73. Robson, J.G.; Saszik, S.M.; Ahmed, J.; Frishman, L.J. Rod and cone contributions to the A-wave of the electroretinogram of the macaque. *J. Physiol.* **2003**, *547*, 509–530. [CrossRef] [PubMed]
74. Vecino, E.; Rodriguez, F.D.; Ruzafa, N.; Pereiro, X.; Sharma, S.C. Glia-neuron interactions in the mammalian retina. *Prog. Retin. Eye Res.* **2016**, *51*, 1–40. [CrossRef]
75. Hennuyer, N.; Duplan, I.; Paquet, C.; Vanhoutte, J.; Woitrain, E.; Touche, V.; Colin, S.; Vallez, E.; Lestavel, S.; Lefebvre, P.; et al. The novel selective PPARα modulator (SPPARMα) pemafibrate improves dyslipidemia, enhances reverse cholesterol transport and decreases inflammation and atherosclerosis. *Atherosclerosis* **2016**, *249*, 200–208. [CrossRef] [PubMed]
76. Singh, A.; Singh, R. Triglyceride and cardiovascular risk: A critical appraisal. *Indian J. Endocrinol. Metab.* **2016**, *20*, 418–428. [CrossRef] [PubMed]
77. Ye, X.; Kong, W.; Zafar, M.I.; Chen, L.-L. Serum triglycerides as a risk factor for cardiovascular diseases in type 2 diabetes mellitus: A systematic review and meta-analysis of prospective studies. *Cardiovasc. Diabetol.* **2019**, *18*, 1–10. [CrossRef] [PubMed]
78. Sarwar, N.; Danesh, J.; Eiriksdottir, G.; Sigurdsson, G.; Wareham, N.; Bingham, S.; Boekholdt, S.M.; Khaw, K.-T.; Gudnason, V. Triglycerides and the risk of coronary heart disease. *Circulation* **2007**, *115*, 450–458. [CrossRef]
79. Zoppini, G.; Negri, C.; Stoico, V.; Casati, S.; Pichiri, I.; Bonora, E. Triglyceride-high-density lipoprotein cholesterol is associated with microvascular complications in type 2 diabetes mellitus. *Metabolism* **2012**, *61*, 22–29. [CrossRef] [PubMed]
80. Chang, Y.-C.; Wu, W.-C. Dyslipidemia and diabetic retinopathy. *Rev. Diabet. Stud.* **2013**, *10*, 121–132. [CrossRef]
81. Mohamed, Q.; Gillies, M.C.; Wong, T.Y. Management of diabetic retinopathy. *JAMA* **2007**, *298*, 902–916. [CrossRef]
82. Fu, Z.; Chen, C.T.; Cagnone, G.; Heckel, E.; Sun, Y.; Cakir, B.; Tomita, Y.; Huang, S.; Li, Q.; Britton, W.; et al. Dyslipidemia in retinal metabolic disorders. *EMBO Mol. Med.* **2019**, *11*, e10473. [CrossRef] [PubMed]
83. Jacob, J.J.; Chopra, R.; Chander, A. Ocular associations of metabolic syndrome. *Indian J. Endocrinol. Metab.* **2012**, *16*, S6–S11. [CrossRef] [PubMed]
84. Naiman, S.; Huynh, F.; Gil, R.; Glick, Y.; Shahar, Y.; Touitou, N.; Nahum, L.; Avivi, M.Y.; Roichman, A.; Kanfi, Y.; et al. SIRT6 promotes hepatic beta-oxidation via activation of PPARα. *Cell Rep.* **2019**, *29*, 4127–4143. [CrossRef] [PubMed]
85. Raza-Iqbal, S.; Tanaka, T.; Anai, M.; Inagaki, T.; Matsumura, Y.; Ikeda, K.; Taguchi, A.; Gonzalez, F.J.; Sakai, J.; Kodama, T. Transcriptome analysis of K-877 (a novel selective PPARα modulator (SPPARMα))-regulated genes in primary human hepatocytes and the mouse liver. *J. Atheroscler. Thromb.* **2015**, *22*, 754–772. [CrossRef] [PubMed]
86. Sasaki, Y.; Asahiyama, M.; Tanaka, T.; Yamamoto, S.; Murakami, K.; Kamiya, W.; Matsumura, Y.; Osawa, T.; Anai, M.; Fruchart, J.-C.; et al. Pemafibrate, a selective PPARα modulator, prevents non-alcoholic steatohepatitis development without reducing the hepatic triglyceride content. *Sci. Rep.* **2020**, *10*, 1–10. [CrossRef] [PubMed]
87. Giovannitti, J.A.; Trapp, L.D. Adult sedation: Oral, rectal, IM, IV. *Anesth. Prog.* **1991**, *38*, 154–171.

International Journal of
Molecular Sciences

MDPI

Review

The Role of PPAR Alpha in the Modulation of Innate Immunity

Maja Grabacka [1,*], Małgorzata Pierzchalska [1], Przemysław M. Płonka [2] and Piotr Pierzchalski [3]

1 Department of Biotechnology and General Technology of Foods, Faculty of Food Technology, University of Agriculture, ul. Balicka 122, 30-149 Cracow, Poland; malgorzata.pierzchalska@urk.edu.pl
2 Department of Biophysics, Faculty of Biochemistry, Biophysics and Biotechnology, Jagiellonian University, ul. Gronostajowa 7, 30-387 Cracow, Poland; przemyslaw.plonka@uj.edu.pl
3 Department of Medical Physiology, Faculty of Health Sciences, Jagiellonian University Medical College, ul. Michałowskiego 12, 31-126 Cracow, Poland; piotr.pierzchalski@uj.edu.pl
* Correspondence: maja.grabacka@urk.edu.pl; Tel.: +48-12-662-4796

Abstract: Peroxisome proliferator-activated receptor α is a potent regulator of systemic and cellular metabolism and energy homeostasis, but it also suppresses various inflammatory reactions. In this review, we focus on its role in the regulation of innate immunity; in particular, we discuss the PPARα interplay with inflammatory transcription factor signaling, pattern-recognition receptor signaling, and the endocannabinoid system. We also present examples of the PPARα-specific immunomodulatory functions during parasitic, bacterial, and viral infections, as well as approach several issues associated with innate immunity processes, such as the production of reactive nitrogen and oxygen species, phagocytosis, and the effector functions of macrophages, innate lymphoid cells, and mast cells. The described phenomena encourage the application of endogenous and pharmacological PPARα agonists to alleviate the disorders of immunological background and the development of new solutions that engage PPARα activation or suppression.

Keywords: pattern-recognition receptors; phagocytosis; nitric oxide synthase; fenofibrate; oleoylethanolamide; palmitoylethanolamide

Citation: Grabacka, M.; Pierzchalska, M.; Płonka, P.M.; Pierzchalski, P. The Role of PPAR Alpha in the Modulation of Innate Immunity. *Int. J. Mol. Sci.* **2021**, *22*, 10545. https://doi.org/10.3390/ijms221910545

Academic Editors: Manuel Vázquez-Carrera and Walter Wahli

Received: 27 August 2021
Accepted: 26 September 2021
Published: 29 September 2021

1. Introduction

Innate immunity comprises a sophisticated set of defensive processes, which are evolutionarily very old and originated concomitantly with the development of multicellular organisms. The defense against invading pathogens is a crucial physiological mechanism that guarantees survival. The development of these mechanisms is a manifestation of a constant race between pathogens (including unicellular pro- and eukaryotic invaders) and host. The biological processes involved in the innate immune response are very complex and tightly regulated on multiple levels, because they may be very harmful when left unsupervised. Recent advances in the elucidation of such a regulation revealed a dense network of connections among immune cell functions, signaling pathways, and cellular metabolism. Peroxisome proliferator-activated receptor α (PPARα) has emerged as an important player in this network, and this review aims to present several aspects of its involvement in the regulation of innate immunity.

2. The New Perspective on Innate Immunity

Innate immunity has evolved to react very rapidly to injury or invasion, and it involves an immediate mobilization of a broad range of inflammatory responses of rather low specificity. Traditionally, the lack of memory was regarded as an intrinsic feature of innate immunity; nevertheless, recent discoveries in this field have led to a thorough revision of this picture and a presentation of the concept of 'innate immune memory' (reviewed in [1]). The innate immune memory differs substantially from its adaptive counterpart, because it lacks somatic gene rearrangement processes and specific epitope-recognizing receptors. Due to the gradual improvement depending on the history of host–pathogen interactions, it is also called 'trained immunity', with genetic recombination events being substituted by the development of epigenetic imprinting and/or changes in miRNA transcriptome. The observations of the innate immune cells' behavior during the exposure to various unrelated pathogens revealed the 'priming' phenomenon, whereby previous contact with one microbial component modulates the response to other pathogenic challenges [1,2]. This modulation can form a certain kind of cross-protection, which is manifested by a nonspecific improved resistance to second infection after an episode of pathogen-associated molecular pattern (PAMP) recognition by pattern-recognition receptors (PRRs) [2]. Such phenomena have been reported in insects (*Tenebrio mollitor* larvae) [3], in planaria (*Schmitdtea mediterranea*) [4], and in Pacific oyster *Crassostrea gigas* [5]. Notably, invertebrates, which lack lymphocyte-based adaptive immunity mechanisms and rely solely on innate responses to fight infections, have developed a high level of sequence diversity and structural complexity of PRRs (e.g., lectins, Toll-like receptors (TLRs), and NOD/NLR-like proteins (see Section 4.4)), as well as soluble or extracellular fibrinogen-related proteins (FREPs) [6,7]. Recognition of PAMPs, such as β-1,3-glucans and peptidoglycan, triggers specific invertebrate antimicrobial effector mechanisms, for instance, activation of prophenoloxidase (and related hemocyanins) that catalyze melanin formation from reactive dihydroxyphenylalanine (DOPA) and DOPAquinone intermediates [8,9].

The three main steps of the innate response are (1) building of a physical and chemical barrier, (2) recognition of foreign invaders and distinguishing from 'self' structural elements, and (3) phagocytosis and production of cytotoxic compounds that help to destroy engulfed particles or are released to damage objects too large to be phagocytosed. For example, various epithelial cells not only form a physical barrier of epithelium protecting the body from the external environment but also secrete hydrolytic enzymes and alarmins such as various antimicrobial peptides (AMPs) [10]. To distinguish between self and foreign molecules and cells, PRRs bind particular molecules characteristic for certain groups of common pathogens of viral, bacterial, or fungal origin, such as nucleic acids and their components (e.g., double-stranded RNA, nonmethylated CpG contacting DNA, nucleotides, and nucleosides), saccharide cell-wall components (e.g., peptidoglycan, lipopolysaccharide, chitin, and zymosan), phospholipids (i.e., cardiolipin of microbial origin), or particular proteins (e.g., formylmethionine-containing peptides and flagellin), usually regarded as PAMPs. The same mechanisms are responsible for the response to disrupted cell contents released during necrosis, which are immunogenic, such as mitochondrial formylated peptides, cardiolipin-containing inner mitochondrial membrane, and ATP (damage-associated molecular patterns, DAMPs) [11,12]. In a localization where invasion or sterile injury take place, phagocytosis leading to the elimination of a danger is triggered. It is carried out by professional phagocytes (polymorphonuclear neutrophils, mononuclear monocytes, and macrophages residing in tissues), para-professional phagocytes (dendritic cells), and nonprofessional phagocytes (epithelial cells and fibroblasts) [13,14].

During phagocytosis, the engulfed particles or microbial cells need to be destroyed intracellularly by a variety of microbicidal molecules stored in cytoplasmic granules, such as antimicrobial peptides (AMPs, e.g., azurocidin and defensins), proteolytic enzymes (e.g., elastase, cathepsin G, collagenase, gelatinase, and metalloproteinases), and reactive oxygen, nitrogen, and halogenated species [15]. Cytotoxic reactive oxygen species are generated during respiratory burst and include the superoxide anion ($O^{\bullet-}$), produced by NADPH oxidase, as well as hydrogen peroxide generated by superoxide dismutase from $O^{\bullet-}$. NADPH oxidase, which is assembled from the transmembrane cytochrome b558, numerous cytosolic phox (phagocyte oxidase) subunits, and small GTPase Rac2, releases $O^{\bullet-}$ directly into the phagosome or the extracellular space [16]. A small fraction of superoxide (about 1%) may give rise to a highly reactive hydroxyl radical in reaction with ferric ions (Fe^{3+}) [16,17]. Neutrophil myeloperoxidase uses hydrogen peroxide and halides to form hypochlorous or hypobromous acids, as well as highly bactericidal chloramines. Mononuclear phagocytes express inducible nitric oxide synthase and produce cytotoxic nitric oxide (NO) from arginine. During the active phase of oxidative burst, NO, which freely diffuses across membranes, reacts with $O^{\bullet -}$, giving rise to peroxynitrite ($ONOO^-$), a strong oxidative agent able to induce nitrative or oxidative damage to proteins and lipids of microbial cells [18]. At later stages of phagocytosis, the phagosome fuses with strongly acidic lysosomes to form phagolysosomes which also contain numerous hydrolytic enzymes, such as proteinases, lipases, and lysozyme.

3. The Main Populations of Innate Immune Cells

Professional phagocytes, such as neutrophils, monocytes/macrophages, or microglia, play a central role in innate immunity, because they perform both regulatory and effector tasks. Macrophages of peripheral tissues belong to the reticuloendothelial system and are known under various customary names according to localization: Kupffer cells (liver), Langerhans cells (skin), osteoclast (bone), etc. Microglia are also skilled phagocytes of myeloid origin that reside exclusively in the central nervous system and share numerous common features with macrophages [19]. The phagocytic capacity of monocytes and monocyte-derived macrophages depends on the expression pattern of specific surface markers, as well as their phenotypic polarization. A recent report [20] showed that M2 macrophages (stimulated with IL-4 and IL-10) presented a twofold higher phagocytic capacity of *E. coli* than M1 macrophages (IFNγ, LPS-stimulated), and the expression level of a surface marker CD209 directly correlated with a high phagocytic capacity. The plethora of stimuli determine which pathway the cell follows, called 'polarization'. M1-polarized macrophages respond to so-called 'classical' activation by typical proinflammatory cytokines, such as IFNγ, secrete other proinflammatory factors (TNFα, IL-1β, IL-6, and IL-12) and chemokines (e.g., CCL1, CCL5, and CXCL10) to recruit other leukocyte populations, and release cytotoxic NO (see below). M2 macrophages represent an opposite, anti-inflammatory phenotype as a result of the so-called 'alternative' activation by IL-4, IL-13, parasitic (helminth, fungal) infections, or immunosuppressing factors, such as IL-10 and glucocorticoids. They express mannose receptor (CD206) and arginase-1, and they secrete the anti-inflammatory IL-10 cytokine, TGF-β, and trophic polyamines (putrescine, spermidine, etc.), collectively contributing to inflammation resolution and tissue regeneration [21,22]. The M1/M2 paradigm was recently broadened and enriched with further details, such as division of the M2 group into more specific M2a, M2b, M2c, and M2d phenotypes [23,24]. However, an opinion currently prevails that, due to macrophage plasticity, there is rather a continuum of phenotypes than distinct, exclusive, and restricted cell profiles [25].

In addition to the aforementioned professional and nonprofessional phagocytes, other cell populations take part in the innate immune defense, namely, innate lymphoid cells (ILCs) from lymphoid lineage and mast cells, eosinophils, basophils, and myeloid-derived suppressor cells from myeloid lineage [25]. Mast cells, secreting heparin and histamine, reside in many tissues and organs, such as connective tissue, skin, lungs, gastrointestinal mucosa, and in proximity to blood vessels [26].

Myeloid-derived suppressor cells (MDSCs) form a heterogenous and plastic population of cells of myeloid origin that inhibit T-cell responses and are able to promote differentiation toward Tregs [25,27]; therefore, they actively contribute to inflammation resolution by being recruited to the site of inflammation by proinflammatory cytokines, such as IL-6.

The last, most recently discovered, and somewhat elusive group of innate immunity effectors comprises so-called innate lymphoid cells (ILCs) [25,28]. They show a common pattern of surface markers ($CD45^+$ $CD127^+$ $CD3^-$ $CD19^-$) and are divided into three main groups (ILC1, ILC2, and ILC3) on the basis of the expression of particular transcription factors and a distinct profile of secreted cytokines [28–30]. Natural killer (NK) cells and large granular lymphocytes (LGLs) belong to ILC1 [31,32], whereas ILC2 and ILC3 cells are mainly associated with mucosal membranes [29,33]. ILC3 cells derived from fetal liver are among the first lymphoid cells that populate gastrointestinal tract, and they play an important role in the development of tolerance to commensal microbiota [34,35]. They secrete

IL-17, IL-22, and lymphoid tissue inducer (LTi), which are critical factors for maintaining mucosal barrier function, sustaining the balance between the inflammatory response to pathogenic microbes, and creating the tolerogenic milieu for probiotic bacteria [28,35]. Collectively, ILC cells are involved in the coordination of various aspects of innate immunity and contribute to immune homeostasis regulation; therefore, they are regarded as an equivalent of Th lymphocytes in adaptive immunity.

4. Peroxisome Proliferator-Activated Receptor alpha (PPARα) and Its Role in Inflammation

Tissue injury and the onset of infection immediately evoke an innate immune response and trigger inflammation. As pointed out by Roman scholar Aulus Cornelius Celsus in the first century, local acute inflammation is manifested by *calor*, *rubor*, *dolor*, and *tumor*, i.e., increased temperature, redness, pain, and edema [36]. These symptoms reflect the action of proinflammatory lipid mediators, histamine, and cytokines released by tissue-infiltrating leukocytes that induce vasodilation and increase endothelial permeability and expression of adhesion molecules on the endothelial surface and in the extracellular matrix underneath. These events lead to extravasation of circulation leukocytes, chemotaxis, and accumulation of interstitial fluid, causing edema (*tumor*). The increased interstitial flow and metabolic activity of proliferating cells generate local heat and flushing (*calor* and *rubor*). Inflammatory pain (*dolor*) is evoked by activation of transient receptor potential cation channel vanilloid subfamily member 1 (TRPV1), which is present on sensory neurons of the peripheral nervous system [37]. The TRPV1 activation leads to an influx of Ca^{2+} and membrane depolarization, followed by the opening of voltage-gated sodium channels and creation of an action potential [37]. TRPV1 receptors are present not only on neurons, but also on immunocompetent cells (T lymphocytes, mast cells), epithelia, keratinocytes, and vascular endothelial cells [38]. TRPV1 channels are activated by various lipid inflammatory mediators, such as COX-2 products (prostaglandins), lipoxygenase 15-LOX products (e.g., 15-hydroperoxyeicosatetraenoic acid, 15-HPETE), and polyamines of molecules released after cell injury, e.g., ATP and adenosine [37]. The links between PPARα and molecular events that spark inflammation and underlie its main symptoms are outlined below (Figure 1).

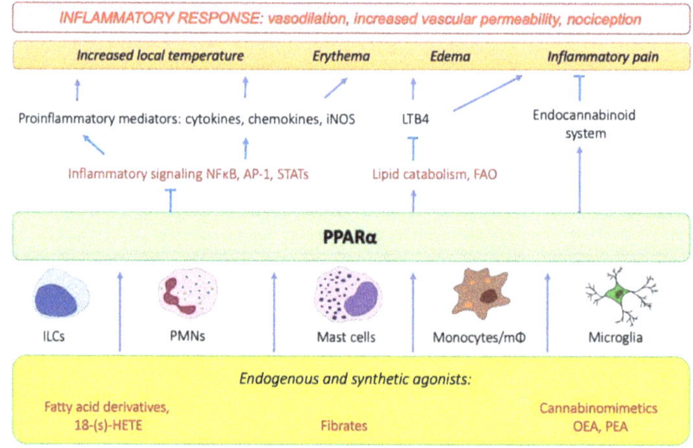

Figure 1. The involvement of PPARα in the modulation of inflammation through interfering with the main inflammatory transcription factors (NF-κB, nuclear factor κB; AP-1, activation protein 1; STATs, signal transducers and activators of transcription) through activating lipid catabolic pathways and participating in the endocannabinoid system (see Section 7). iNOS, inducible nitric oxide synthase; FAO, fatty-acid oxidation; LTB_4, leukotriene B_4; OEA, oleylethanolamide; PEA, palmitoylethanolamide.

4.1. PPARα as a Nuclear Receptor Present in Peripheral Tissues and Immunocompetent Cells

Peroxisome proliferator-activated receptors (PPARs) belong to a family of nuclear receptors that act as transcription factors activated by lipid-soluble ligands. Such ligands are able to cross the plasma membrane directly and bind the intracellular target proteins. PPARs are represented by three isotypes, PPARα, PPARβ/δ, and PPARγ, encoded by separate genes. They show tissue-specific expression patterns and mainly govern lipid, carbohydrate, and amino-acid metabolism, as well as exert other pleiotropic functions, including immunomodulatory activities. All three PPAR isotypes exhibit potent anti-inflammatory properties and have a strong impact on various aspects of the physiology of the immune system. In this review, we focus on peroxisome proliferator-activated receptor alpha (PPARα), which is particularly responsible for the regulation of fatty-acid catabolism and ketogenesis [39,40], also in addition to being deeply involved in the modulation of innate immunity responses. Below, we outline the active participation of PPARα in physiological processes that operate behind all four cardinal symptoms of inflammation, i.e., alleviating edema and pain and contributing to resolution of acute phase.

As a transcription factor, PPARα is involved in the activation of gene transcription, which is carried out by binding the heterodimer of PPARα and the pan-PPAR obligatory partner, retinoid X receptor (RXR), to consensus motifs in the target promoters. The active heterodimer is formed when both partners have their agonists bound. The most potent endogenous PPARα agonists include fatty acids and their derivatives: saturated stearic and palmitic acids, fatty acyl amides such as oleylethanolamide (OEA) and palmitoylethanolamide (PEA), LOX products such as 5-(*S*)-HETE and 8-(*S*)-HETE, and leukotriene B_4 (LTB_4) [41–44]. There is the only one bona fide RXR ligand known so far, which is 9-*cis*-13,14-dihydroretinoic acid, successfully identified after many years of searching, whereas 9-*cis*-retinoic acid, frequently used experimentally, is one of the most potent pharmacological RXR agonists [45,46]. Pharmacological PPARα agonists, such as fibrates, are clinically used to normalize blood lipid profile, particularly to lower concentrations of cholesterol and low-density lipoprotein fractions [47]. Fenofibrate and gemfibrozil are the most widely prescribed drugs from a fibrate group, and they are generally very well tolerated [48]. Nevertheless, some adverse effects have been reported in patients chronically taking fibrates, with myopathy and rhabdomyolysis being the most common problems [49]. The structures of endogenous ligands, as well as the most important synthetic agonists and antagonists, are presented in Table 1.

Interestingly, in addition to the tissues with a high rate of fatty-acid catabolism, such as the liver, cardiac muscle, and kidneys, PPARα is generally expressed in $CD45^+$ leukocytes [50], including numerous innate immune cell populations: basophils [51], eosinophils [52], monocytes and macrophages [30,53–55], Kupffer cells [56], Langerhans cells [57], osteoclasts [58], and microglia [59].

The classical PPARα targets include the genes encoding enzymes from the fatty-acid mitochondrial and peroxisomal β-oxidation (acyl-CoA dehydrogenases, acyl-CoA oxidases), ω-oxidation and ω-hydroxylation (cytochromes P450), and ketogenesis (3-hydroxy-3methylglutaryl-CoA synthase) [60–62]. Importantly, in addition to this canonical mode of action, PPARα is able to transrepress certain genes through at least three mechanisms [63]: (i) initiating protein–protein interactions and sequestration of coactivators that are common to PPARα and other pathways, (ii) cross-coupling of the PPARα/RXR complex with other transcription factors, which leads to mutual cross-inhibition of both participating proteins, and (iii) interference with signal-transducing proteins, i.e., where the PPARα/RXR complex inhibits phosphorylation of MAP-kinase cascade members.

Table 1. Chemical structures of PPARα endogenous agonists, synthetic agonists used in experimental studies, clinically used pharmacological agonists, and synthetic antagonists, including examples of novel *N*-phenylsulfonylamide compounds (the structures of 3- and 10- series according to [64]).

	PPARα Agonists and Antagonists
Natural agonists	8-(S)-HETE; 5-(S)-HETE; 14,15-diHETrE; Leukotriene B_4; OEA; PEA
Synthetic agonists	Wy-14643 (pirinixic acid); GW7647
Agonists applied in clinic: fibrate derivatives	Fenofibrate; Clofibrate; Gemfibrozil; Ciprofibrate
Synthetic antagonists	GW6471; *N*-(phenylsulfonyl)amide derivatives: 3 series; 10 series

4.2. PPARα-Mediated Transrepression of Main Inflammatory Transcription Factors

Transrepressive activity toward nuclear factor κB (NF-κB), activation protein (AP-1), and signal transducers and activators of transcription (STATs) is responsible for PPARα's profound anti-inflammatory action. PPARα physically interacts with the p65 Rel homology domain through its C-terminal fragment and simultaneously binds the JNK-responsive part of c-Jun with its N-terminal fragment (Figure 2a) [65]. Formation of this complex sequesters p65 and c-Jun from binding to the IL-6 promoter and blocks IL-1-induced IL-6 production. The direct inhibitory interaction between PPARα and NF-κB p65 subunit was also reported in cardiomyocytes [66]. In this case, sirtuin 1 (Sirt1) initiated formation of the Sirt1–PPARα–p65 complex, which led to PPARα-dependent p65 inactivation and transrepression of proinflammatory NF-κB-regulated genes, such as monocyte chemoattractant protein 1 (MCP1, Figure 2b) [66]. Sirt1 induced p65 deacetylation, which also had a negative impact on NF-κB activity because acetylation is required for its activity [67]. The deacetylation effect was absent after treatment with PPARα antagonist GW6471 or in PPARα$^{-/-}$ cells, which indicates PPARα involvement [66].

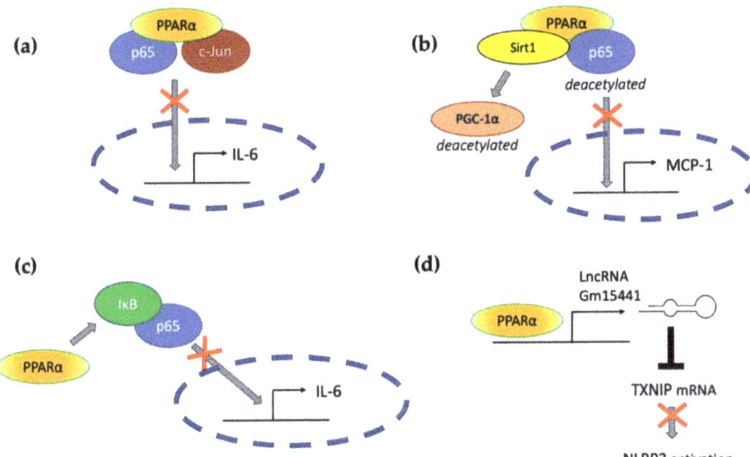

Figure 2. The molecular mechanisms responsible for PPARα-mediated suppression of proinflammatory signaling pathways (see the main text for explanation) (**a**) through a direct interaction with p65 and c-Jun, (**b**) through interaction with Sirt1 and subsequent deacetylation of p65, (**c**) through activation of IκB, and (**d**) through transactivation of long noncoding RNA Gm15441, which interferes with the stability of thioredoxin-interacting protein (TXNIP) mRNA and blocks NLRP3 inflammasome activation.

An additional mechanism responsible for PPARα interference with the NF-κB pathway was also identified in hepatocytes, where PPARα bound and transactivated NF-κB inhibitor alpha (IκBα), which increased the amount of this protein [68]. Accumulated IκBα binds NF-κB, thereby masking its nuclear localization signal, which arrests it in the cytoplasm and blocks its activity as a transcription factor [69]. PPARα was also responsible for the decreased phosphorylation of NF-κB subunits p65 and p50 [68], which was another event with a negative impact on NF-κB activity, because phosphorylation of its subunits is necessary for their optimal function [70]. The interference of PPARα with NF-κB action prevented IL-1 induced IL-6 expression in liver tissues (Figure 2c) [68].

The antagonism between PPARα and NF-κB and AP-1 underlies blocking of the expression of proinflammatory cytokines and effector proteins in various cell and animal models. PPARα ligand K-111 (2,2-dichloro-12-(4-chlorophenyl)-dodecanoic acid) inhibited LPS-induced IL-6 production in Raw 264.7 macrophages on the mRNA and protein level [71]. This effect was exerted through the inhibition of stress-activated protein kinase

(SAPK)/c-Jun N-terminal kinase (JNK), NF-κB p65 phosphorylation, and induction of IκBα protein level [71]. PPARα activation in monocytes was shown to inhibit LPS- or IL-1β-induced expression of tissue factor (TF), a membrane glycoprotein responsible for initiation of coagulation cascade [72,73]. The mechanism involved a previously mentioned blockade of the target gene promoter activity through the antagonism between PPARα and NFκB and AP-1 [72].

Interleukins released by immune cells exert their biological functions through specific cell surface receptors, which transduce signals through the Janus family of kinases (JAK) and phosphorylation STAT transcription factors [74]. Various STAT proteins are negatively regulated by PPARα. For example, a bidirectional cross-inhibitory relationship between PPARα and STAT5b was described [75–77]. STAT5b is responsible for signal transduction from the IL-2 receptor [78]. IL-2 is a very important cytokine, crucial for both innate and adaptive immunity, being indispensable for NK cell proliferation and maturation, as well as promoting the development, differentiation, and proinflammatory response of both Th1 and Th2 cells [78,79].

4.3. PPARα and Inflammatory Lipid Mediators

Another important mechanism of the anti-inflammatory action of PPARα involves the catabolism of lipid mediators, such as leukotriene B_4 (LTB_4). The elegant study by Devchand and colleagues [80] revealed that LTB_4 is a potent and specific PPARα ligand that induces expression of PPARα-transactivated genes of the peroxisomal β-oxidation pathway, namely, acyl-CoA oxidase, which is a rate-limiting enzyme of LTB_4 catabolism. $PPAR\alpha^{-/-}$ mice subjected to a topical application of 5-LOX-inducing inflammatory agent and LTB_4 showed signs of tissue inflammation much longer (by about 30–40%) than wt mice, which were able to clear LTB_4 from circulation much faster [80]. This experiment illustrates the importance of PPARα in the resolution of inflammation. This role of PPARα is necessary for regulation of the innate immune response, because proinflammatory lipid mediators, such as LTB_4, are not only strong chemotactic agents for neutrophils and other leukocytes, but they also facilitate PMNs extravasation and diapedesis at the local site of inflammation and increase vascular permeability in this region [81,82]. By restricting LTB_4 duration, PPARα alleviates three out of four inflammation symptoms (heat, flushing, and edema). Moreover, PMNs are not only recipients of LTB_4 signals, but they are also activated to its production via a positive autocrine feedback loop [83]. Therefore, the PPARα-regulated LTB_4 clearance protects from an overexaggerated inflammatory response and its transition from acute to destructive chronic state. The other eicosanoids, the products of either COX, i.e., prostaglandins PGD_1, PGD_2, PGA_1, and PGA_2, or 5-LOX product 8-(S)-HETE, also activate PPARα [84], which opens the possibility of modulating their impact on the cells with PPARα expression, whether in immunocompetent cells, such as monocytes/macrophages that express high levels of this receptor, or in the inflamed tissue. Such an activity contributes to tissue protection from inflammatory damage and facilitates regeneration.

4.4. PPARα Crosstalk with Pattern Recognition Receptors

Vertebrates take advantage of the PRR functions and employ them to sense all sorts of factors that induce tissue homeostatic imbalance. The PRR receptors are activated by the numerous compounds comprising specific structural entities referred to as the microbial-associated molecular patterns (MAMPs) or the Damage-associated molecular patterns (DAMPs). Several types of PRRs are broadly present in both immune and nonimmune cells, and their activation sparked by contacts with microorganisms, viruses, and some fragments of damaged cells or an alteration in the functioning of cell components (e.g., cytoskeleton or mitochondria malfunction or endoplasmic reticular stress) is the main trigger of the innate immunity response [85]. The PRRs can be divided into four main subfamilies: the Toll-like receptors (TLRs), the nucleotide-binding oligomerization domain (NOD)–leucin-rich repeat (LRR)-containing receptors (NLRs), the retinoic acid-inducible gene 1-like receptors (RLRs), and the C-type lectin receptors (CLRs) [11]. Nevertheless, some other cellular proteins can serve as PRRs in certain situations, e.g., the glycolytic enzyme, hexokinase II, which is able to spot the microbial sugar, *N*-acetylglucosamine, when this building block of peptidoglycan happens to be present in the cytoplasm [86]. In this section, we address the question of how PPARα may be involved in the MAMP and DAMP recognition process in various tissues and cells.

The noteworthy information on TLR and PPARα crosstalk comes from the studies on PPARα knockout (KO) mice and cells derived from these animals. The colonic macrophages from KO mice did not produce the regulatory IL-10, but secreted IL-6, IL-1β, and IL-12, potent inducers of Th1 and Th17 differentiation. Moreover, innate immune ILC3 cells isolated from the colon of PPARα KO mice produce lower levels of IL-22 compared with those from WT mice, which results in the impaired secretion of antimicrobial peptides and commensal dysbiosis. This indicates that PPARα regulates the ILC3 effector functions, which are important for both fighting infections and sustaining tolerance to commensal microbiota. The absence of PPARα affects the species composition of the microbiome and leads to increased representation of segmented filamentous bacteria (SFB). All these facts render the KO mice prone to gut inflammation development and are indirect proof of the critical role of PPARα activation in gut immunological homeostasis [30].

It is well known that interactions between the microbiota and intestinal cells engage Toll-like receptors [87], e.g., SFB regulate the process of Th17 differentiation in the intestine via activation of TLR5 by flagellin [88], and TLR4 ligand LPS from Gram-negative bacteria stimulates Th17 differentiation in vitro [89]. It seems that these events can be modulated by PPARα ligands. Accordingly, it was shown that macrophages from PPARα knockout mice are characterized by higher expression levels of mRNA for proinflammatory cytokines IL1β and IL6, as well as for COX-2 and NF-κB (p65) upon TLR4 ligand stimulation (LPS 50 ng/mL, 5 h), as compared to wild-type cells. It seems that PPARα deficiency speeds up LPS-induced inflammatory responses in murine macrophages [54]. Another study on PPARα KO mice indicated that PPARα was essential for the anti-inflammatory effect of acute exercises. Its absence induced overexpression of proinflammatory cytokines in LPS-treated macrophages isolated from mice 24 h post exercise [90].

TLR ligands can regulate PPARα activity, and PPARα agonists influence the expression of TLRs, as well as proteins involved in signaling from TLRs in various cells of both immune and nonimmune types. Becker et al. studied the involvement of LPS in the regulation of PPARα in murine lungs and showed that 24 h on from a prolonged LPS challenge (daily intranasal administration of 1 μg LPS for 4 consecutive days), a profound inhibition of PPARα mRNA expression took place [91]. LPS, peptidoglycan, and flagellin (ligands of TLR4, TLR1/2, and TLR5, respectively) strongly suppressed PPARα activity in rat astrocytes acting at the mRNA and protein expression level [92]. On the other hand, it was shown that fenofibrate, a pharmacological PPAR agonist, significantly inhibited the TLR4, MYD-88, and NF-κB mRNA expression, as well as TNFα production, in murine melanoma B16F10 LPS-stimulated cells [93]. The strong relationship between TLR4 and the PPARα signaling pathway was also clearly demonstrated in a model of endotoxin-induced uveitis. This study suggested that fenofibrate can also attenuate LPS-induced cytokine production, inhibit NF-κB signaling, and suppress TLR4 expression in retinal pigment epithelial cells. Simultaneously, LPS could act as a direct PPARα antagonist in a PPARα reporter cell line [94]. All these experimental data point to a subtle tuning and complicated interplay between activation of PPARα and the TLR signaling pathway, which is needed for the homeostatic balance between triggering and resolution of the inflammatory response in tissues.

4.5. PPARα and the Regulation of Inflammasomes

The inflammasomes, the complex molecular platforms formed in the cytoplasm (mainly in macrophages, but also in other nonimmune cells, such as endothelial and epithelial cells encountering various DAMPs and MAMPs), are now considered the key element of innate immunity. They are the multiprotein complexes composed of cytoplasmic sensors (mainly NLR family members), adaptive proteins (apoptosis-associated speck-like protein, ASC, or PY-CARD), and effectors (such as cysteine proteinase precursor or pro-caspase-1). In the case of some nonconventional inflammasomes, pro-caspase-1 is substituted by pro-caspase-11 in murine cells and pro-caspase 4/5 in human cells. The complex formation enables the proteolysis of pro-IL1β and pro-IL18 and the release of active cytokines into the cell microenvironment and bloodstream, which drives local or systemic inflammation [95]. Alternatively, the inflammasome formation induces a chain of events leading to pyroptosis—the special type of a programmed cell death connected to an inflammatory state. The molecular mechanisms contributing to inflammasome activity are not yet completely understood, but it is believed that the process of their formation requires two subsequent signals, e.g., LPS binding to TLR4 on the cell membrane as the primary signal and K^+ efflux, cytosolic release of lysosomal cathepsins, or mitochondria-derived factors and reactive oxygen species generation as the secondary signal [96]. The regulation of inflammasome activation can occur at both signals on the post-transcriptional and post-translational levels [97].

It was shown in some animal models that PPARα activation can profoundly suppress the inflammasome-induced tissue injury, thereby contributing to the resolution of inflammation. This can be partially attributed to the downregulation of TLR expression by PPARα and interference with the primary step of inflammasome activation. However, in PPARα KO mice with lung inflammation caused by *Pseudomonas aeruginosa* introduction, a significant increase in expression of NLRP-3, ASC-1, and caspase-1, as compared with infected wt mice, was observed [98]. This indicates that PPARα expression background is also important for the supply of inflammasome building blocks.

Acute liver injury is a disease strongly connected with NLPR3 inflammasome activity. In the context of this pathology, Brocker et al. proposed a mechanism connecting fasting, PPARα, and the reduction in liver inflammation and injury. They showed that the long noncoding RNA gene Gm15441 contained a PPARα-binding site within its promoter, and the Gm15441 RNA expression was activated by PPARα ligand Wy-14643. Gm15441 suppressed its antisense transcript, encoding thioredoxin-interacting protein (TXNIP). This subsequently decreased TXNIP-stimulated NLRP3 inflammasome activation (Figure 2d) [99].

Moreover, it was shown that OEA, an endogenous bioactive lipid and a natural ligand of PPARα, prevented tissue damage in the onset of LPS/D-galactosamine (D-Gal)-induced acute liver injury. OEA administration increased PPARα expression in murine liver subjected to LPS/D-Gal treatment. In turn, the liver protein levels of IL-1β and NLRP3 inflammasome components, NLRP3 protein and pro-caspase-1, were enhanced after LPS/D-Gal injection in mice. The increase in these proteins was alleviated by OEA addition to the diet [100]. The OEA anti-inflammatory effects were also evident in dextran sulfate sodium (DSS)-induced mice colitis, and the effect was mediated by the inhibition of NLRP3, NF-κB, or MyD88-dependent pathways [101].

5. PPARα's Role in the Innate Immunity Effector Processes: ROS/RNS Production

An important component of the innate immunity in animals is generation of active forms of oxygen (mainly superoxide) and active forms of nitrogen, mainly nitric oxide and its derivatives [102]. The form of nitric oxide synthase (NOS) traditionally associated with inflammation is the so-called inducible nitric oxide synthase (iNOS or NOS 2). NOS 2 belongs to the enzymatic family of nitric oxide synthases (NOS), being the evolutionarily most distant member of the family. NOS 2 may be expressed in numerous types of cells and tissues [103]. The other two, NOS 1 and NOS 3, also called 'constitutive' or Ca^{2+}-dependent enzymes, are present constitutively in many tissues and cells of the organism, mainly but not solely in some neurons (NOS 1), as well as endothelial cells (NOS 3) [104]. They generate a lower level of NO than NOS 2, despite their comparable enzymatic activity in vitro [102]. Importantly, under various conditions, all NOS enzymes are a source of active forms of nitrogen and oxygen; in the absence of L-arginine, they simply produce superoxide and may be an important source of oxidative/nitrosative stress [105].

PPARα agonists may downregulate NOS 2 [106,107], while they stimulate both NOS 3 [108], which plays a protective role in the cardiovascular system, and NOS 1 (see [109,110]). NOS 2 is expressed de novo under the influence of proinflammatory factors [102], and, as it is not dependent on calcium, it can only be down regulated by inhibition of the enzymatic activity or proteolytic degradation of the enzyme. NOS activity also depends on competition with the alternate substrate consumer arginase, which produces urea and L-ornithine instead of L-citrulline and nitric oxide [111,112]. The possibility of switching the main path of L-arginine metabolism from the generation of NO and citrulline to the generation of urea and ornithine is a basis for the functional diversification of M1 and M2 macrophages. M1 macrophages, unlike M2 macrophages, generate free radicals and are the proinflammatory type of these cells (as mentioned in Section 3). They contribute to the development of inflammation-driven tumors [107]. PPARα, as an attenuator of inflammation and free-radical production, acts in this case as an antitumor agent. Parallel to tumor progression and diversification of the tumor macrophageal phenotype toward M2, the situation becomes more ambiguous and unpredictable. The actual effect of activation of PPARα clearly depends on the type of tumor and its phase of development [108]. Indeed, fenofibrate inhibited the development of micrometastases of melanoma BHM in Syrian hamster lung, but did not affect the kinetics of the primary tumor growth, nor the progression of macro-metastases [113]. It must be added that, recently, particular attention has been paid to the possibility of manipulation of NOS 2 activity by its selective inhibitors in order to achieve a desirable level of human monocyte physiological response [114].

The second mechanism of innate defense that involves the production of highly reactive small chemical molecules is respiratory (or oxidative) burst carried out by phagocytes. PPARα agonists were shown to increase macrophage microbicidal activity through intensification of ROS production during respiratory burst. This was caused by PPARα-dependent elevated expression of crucial transmembrane (gp91phox) and cytosolic (p47phox and p67phox) components of NADPH oxidase [115]. Interestingly, increased ROS production led to the generation of oxidized low-density lipoproteins (oxLDL), which further stimulated PPARα activation. Activated PPARα downregulated NO production via transrepression of iNOS [115]. This is an example of PPARα differently regulating various innate immunity effector molecules, in this case, ROS and RNS. An unexpectedly interesting transcriptional regulation occurs in the promoter of another gene crucial for the generation of reactive species during respiratory burst, namely, myeloperoxidase (MPO). The human promoter of this gene contains primate-specific Alu elements that are repetitive DNA mobile fragments spread throughout the human genome in about 1 million copies [116]. The Alu fragment in the MPO gene promoter contains four hexamer sequences identical to or closely resembling canonical PPAR response elements (PPREs): AGGTCA, with 2 or 4 bp spacing between them [117]. The third and fourth hexamers serve as PPREs and accommodate PPARα/RXR or PPARγ/RXR heterodimers, which enables transcriptional regulation by PPAR ligands. Surprisingly, MPO expression is regulated by PPARα agonist GW9578 and PPARγ agonist MCC-555 in opposite directions in human macrophages, depending on the differentiation pathway; MPO is significantly downregulated in macrophages derived from MG-CSF-treated monocytes and upregulated in M-CSF differentiated cells [117]. The difference could probably be attributed to the differential utilization of nuclear co-repressors, such as NCoR or silencing mediator of retinoid and thyroid receptors (SMRT), in macrophages differentiated with GM- vs. M-DAMP [117]. Notably, such a mode of regulation is entirely human-specific, because mice do not possess Alu elements in their genome.

6. PPARα as an Immunomodulator during Infections

Truly immunomodulatory action does not lie in the unilateral inhibition or activation of all inflammatory processes, but in selective influence on the chosen aspects of innate immunity. Such an immunomodulatory action of PPARα has been observed in parasitic or microbial infections. One example of such an activity relates to the induction of M2 polarization in macrophages of patients infected with *Trypanosoma cruzi*, a parasitic euglenoid, which is responsible for Chagas disease development. The experiment carried out on the infected mice showed that PPARα agonist Wy-14643 elevated the expression of M2 macrophage markers, arginase-1, mannose receptor (CD206), Ym1, and TGFβ, and decreased the production of proinflammatory molecules characteristic of the M1 phenotype, such as iNOS, NO, IL-1β, IL-6 and TNFα [118]. However, this phenotypic switch was accompanied by a PPARα (but not PPARγ)-dependent increase in phagocytic capacity and efficiency of parasite phagocytosis [118]. These results indicate that PPARα activation might have therapeutic significance, because its immunomodulatory action, on the one hand, strengthens macrophage effector capacity, but, on the other hand, helps to alleviate severe chronic inflammation associated with Chagas disease, which is destructive to various organs.

Similar immunomodulatory activity of PPARα in the context of phagocytosis was described in primary peritoneal macrophage and microglia cultures treated with several PPARα agonists: endogenous cannabinomimetic (see below), PEA, fenofibrate, or palmitic acid [119]. These compounds, particularly PEA, significantly enhanced phagocytosis and intracellular killing of *E. coli* by macrophages and microglial cells. Although PEA pretreatment reduced the levels of proinflammatory cytokines (IL-1β, IL-6, and TNFα) and chemokines (CXCL1) in the tissues of mice subjected to intracerebellar or intraperitoneal *E. coli* infection, it induced a very effective bacterial clearance from blood, spleens, and cerebelli, which translated into improved survival of these animals [119]. These results suggest a prophylactic potential of PPARα activation in the case of bacterial infections.

Another example illustrating that the exaggerated inflammatory response is not beneficial for the host is tuberculosis infection. In this case, PPARα's immunomodulatory and metabolic roles are connected, leading to a better outcome for wt mice infected with mycobacteria (*Bacillus* Calmette–Guerin or *M. tuberculosis*) in comparison with PPARα KO mice [120]. The absence of PPARα resulted in more rapidly increasing intracellular bacterial load in macrophages, heavier bacteremia in the lungs, spleen, and liver, and a significantly higher level of inflammatory cytokines TNFα and IL-6 in the lungs, as compared to wt PPARα mice. The exaggerated inflammatory response was associated with a higher number of granuloma lesions in the lungs of PPARα KO mice. Granuloma lesions are the manifestation of unsuccessful host defense against mycobacteria, because they are full of dead leukocytes, damaged lung tissue multinucleated giant cells, and macrophages converted to foam cells, filled with lipid-containing vesicles, which create a favorable energy source for surviving and proliferating mycobacteria [121]. Pharmacological PPARα agonists GW7647 and Wy-14643 induced phagosomal maturation through activation of transcription factor EB (TFEB) and significantly reduced the survival of intracellular bacteria, which resulted from increased fatty-acid β-oxidation and elimination of lipid-rich bodies [120]. This is an example of the interconnection between PPARα-mediated lipid catabolism and its immunomodulating effects, which support effective antimicrobial innate defense.

Despite a large body of evidence documenting the beneficial outcomes of PPARα activation in various diseases with an inflammatory background, there are also certain conditions in which PPARα-mediated immunomodulation is hazardous. The illustrative example is a situation where, after viral influenza infection, a subsequent bacterial (e.g., staphylococcal) superinfection occurs. Antibiotic-resistant *Staphylococci* are frequent cause of life-threatening nosocomial infections in patients hospitalized due to viral pulmonary infections. Tam and colleagues [122] found out that the presence of PPARα was responsible for a more severe course of superinfection and a higher mortality in wt mice as compared to PPARα KO mice. Viral infection that was induced prior to challenge with *S. aureus* led to increased PPARα expression in lungs. Moreover, the lipidomic analysis of bronchoalveolar lavage fluid from infected mice revealed that superinfection resulted in a significant enrichment of several inflammatory lipid mediators, such as LOX product LTE_4 and CYP450 products 11,12-dihydroxyeicosatrienoic acid (11,12-diHETrE) and 14,15-diHETrE, as compared to single infection, whether viral or bacterial. 14,15-diHETre is a very potent PPARα agonist [123]. The inhibition of NF-κB signaling mediated by activated PPARα led to a blunted proinflammatory response to bacteria and loss of control over bacterial growth, which inflicted higher mortality [122]. Superinfection caused the decreased expression of macrophage inflammatory genes IL-1β, IL-6, CXCL5, and MMP-9, as well as a scavenger receptor Marco, which resulted in less efficient phagocytosis and heavier bacterial burden. Moreover, PPARα activation led to increased necroptosis (a programmed RIPK3 kinase-dependent lytic cell death), which was responsible for lung tissue damage and dramatically worsened the condition of infected animals [122].

The still scarce, but gradually emerging experimental data indicate that PPARα affects the innate host response to viral infections. Such an involvement is beneficial in certain situations, but could be detrimental in other conditions. The overexpression of PPARα homolog in a grouper fish (*Epinephelus coioides*, EcPPARα) blocked interferon- and NF-κB-induced cytokine expression during viral infections, which led to acute cytopathic injuries and heavier multiplicity of infection [124]. The topic of viral infection onset is currently very important due to its relationship with the ongoing COVID-19 pandemic. A study performed on primary human bronchial epithelial cells infected with SARS-CoV-2 revealed severe alterations in the gene transcription pattern that manifested endoplasmic reticular and mitochondrial stress, metabolic reprogramming toward intensive lipid synthesis and accumulation, impaired fatty-acid oxidation, and upregulated aerobic glycolysis via activation of the NF-κB pathway [125]. Such a metabolic signature suggests that infection impairs PPARα signaling. Therefore, the restoration of PPARα activity could be beneficial through reversal of these changes and metabolic 'repair'. Indeed, the treatment of the infected cell cultures with PPARα ligand fenofibrate alleviated the dysregulation of lipid metabolism, blocked infection-induced phospholipid accumulation, and remarkably decreased viral load by 100-fold within 3 days and 1000-fold within 5 days [125]. These results seem to support the hypothesis that fenofibrate treatment could alleviate the acute infection symptoms during COVID-19 by supporting fatty-acid metabolism in alveolar epithelial cells, improving pulmonary endothelial cell function, and calming down the cytokine storm, leading to a better outcome for the patients [126].

7. Interplay between PPARα and the Endocannabinoid System: Implications for Inflamma-Tion, Neuroprotection, and Analgesia

7.1. Analgesic Lipid Mediators as PPARα Agonists

Mechanical tissue damage, hypersensitivity reactions or local infection result in inflammation, which evokes a nociceptive response and pain. Pain signals are elicited by proalgesic lipid mediators, such as lysophospholipids and PDE_2, or hydroxylated derivatives of linoleic acid (e.g., 13-hydroxyoctadecanoic acid, 13-HODE), which increase the excitability of nociceptive neurons [127]. Nevertheless, another group of endogenous lipid mediators possesses opposite, analgesic activity. Acting through cannabinoid receptors CB1 and/or CB2, they mitigate the excitability of sensory nociceptive neurons. This is a part of the so-called endocannabinoid system, which includes the ligands *N*-arachidonoylethanolamine (AEA, anandamide) and 2-arachidonoyl-glycerol (2-AG), which were first discovered, and their receptors, cannabinoid receptors CB1 and CB2 expressed in the CNS and immunocompetent cells, respectively, as well as TRPV1 and endocannabinoid-synthesizing and -degrading enzymes [128,129]. Later, other fatty-acid ethanolamides (FAEs), such as *N*-palmitoylethanolamide (PEA) and *N*-oleoylethanolamide (OEA), were detected in mammalian and invertebrate tissues [130–132]. OEA and PEA are biologically relevant and potent PPARα agonists, with EC_{50} values of 0.12 μM and 3 μM, respectively [44,133], which links PPARα with the endocannabinoid system. Numerous biological hormone-like functions of OEA and PEA are widely known, including analgesic and anti-nociceptive cannabinomimetic activities, although they are not bona fide CB1 or CB2 agonists [134]. Endocannabinoids and cannabinomimetics are synthesized on demand from membrane phospholipids, but can also be accumulated intracellularly in lipid droplets [135,136].

They are abundantly present in the brain, leukocytes, gastrointestinal tract, and other tissues [137–139].

The most common FAE biosynthesis route involves the formation of *N*-acyl-phosphatidylethanolamine from phosphatidylethanolamine by calcium-dependent *N*-acyl-transferase and subsequent conversion to *N*-acyl-ethanolamine by *N*-acyl-phosphatidylethanolamine-hydrolyzing phospholipase D (NAPE-PLD) [140]. Several other biosynthesis pathways that engage other phospholipases and glycerophosphodiesterases are also possible (for a review, see [128]). Endocannabinoids are absorbed by cells and metabolized by intracellular fatty-acid amide hydrolase (FAAH) or *N*-acylethanolamine-hydrolyzing acid amidase (NAAA) [141].

OEA and PEA exert analgesia and reduce nociception in various animal models of inflammatory pain [142,143]. PEA and synthetic PPARα ligands (GW7647, Wy-14634, perfluorooctanoic acid) produce analgesic effects and strongly reduce edema in chemically induced models of inflammation [142,144–146]. Although, in some cases, OEA acted independently of PPARα presence [143], PEA-induced nociception and anti-inflammatory actions were exerted through PPARα [142,145]. Importantly, PEA-mediated activation of PPARα in CNS through intracerebroventricular PEA application was able to reduce peripheral inflammatory response (a paw edema after carrageenan injection) [146]. This demonstrated a distant endocrine action of PEA, despite the molecular mechanism involving inhibition of the NF-κB signaling pathway in CNS tissue [146]. A PPARα involvement was also demonstrated in the experiments with a synthetic PPARα agonist GW7647, which induced synergistic enhancement of AEA analgesic properties in a chemically induced inflammatory pain model [145,147]. The antinociceptive action of GW7647 depended on the activity of large conductance potassium channels, which further supported an involvement of endocannabinoid system [145,147]. The potentiation of endocannabinoid binding to CB1 and CB2 receptors by cognate molecules, which are not agonists themselves, was observed and named 'the entourage effect' [148]. In the case of AEA, PEA, and OEA, such an effect could be explained by FAAH engagement in PEA and OEA hydrolysis, sparing the large pool of AEA from degradation and allowing it to activate CB receptors. Indeed, the entourage effect has been described as an enhanced vasodilation activity of AEA through TRPV1 by PEA and OEA in the endothelium [149]. In summary, all these results indicate that PPARα signaling contributes to inflammatory pain control through cannabinomimetics OEA and PEA (Figure 3) [127].

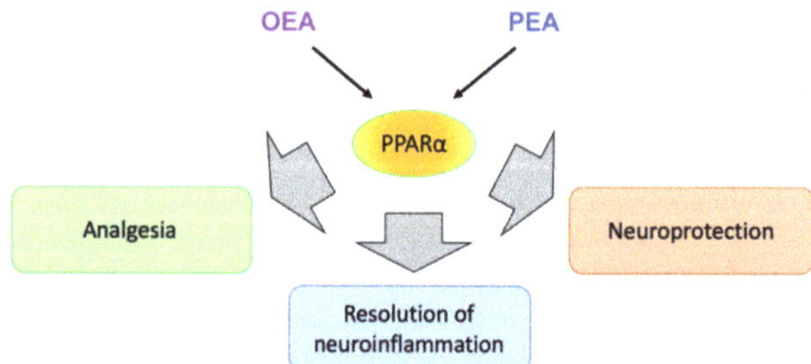

Figure 3. Endocannabinoids OEA and PEA exert analgesic, anti-inflammatory, and neuroprotective actions through PPARα activation. A detailed explanation is provided in the text.

7.2. PPARα Involvement in Resolution of Neuroinflammation

The presence of OEA and PEA in CNS implicates their activity in the physiology of neurons and glial cells. Both compounds were shown to exert beneficial effects by counteracting the glial inflammatory responses and by providing cytoprotection over neuronal cells and their activities in various neuropathic states. Neuroinflammation and exaggerated glial reactivity are associated with numerous neurodegenerative diseases, traumatic injuries, ischemia/reperfusion stress, and neuropathic pain [150–152]. The brain is regarded as 'an immune-privileged' organ, protected from peripheral proinflammatory stimuli by the blood–brain barrier, but microglia, astrocytes, and mast cells are capable of triggering neuroinflammation [153]. Aberrant or chronic activation of these cells in the CNS leads to increased expression of TLRs, cytokines (TNFα, IL-6), chemokines (CXCL6) metalloproteinases, ROS, and RNS, which results in the loss of calcium homeostasis, neuronal damage, or apoptosis [151–153]. The potential of lipid amides, called ALIAmides (autacoid local injury antagonists) to counteract neurogenic inflammation and mast-cell degranulation, was proposed by Rita Levi-Montalcini, a Nobel laureate (1988), for her discoveries in the field of neurobiology [154]. Indeed, numerous studies demonstrated that OEA and PEA, classified as ALIAmides, could provide neuroprotection via downregulation of inflammatory responses in the brain through modulation of glial cell functions. Benito and colleagues discovered that *N*-fatty acylethanolamines (OEA, PEA, AEA) and synthetic agonists of PPARα (Wy-14643) and PPARγ (troglitazone) alleviate the inflammatory response induced by the treatment of astrocytes with β-amyloid peptide fragments [155]. The anti-inflammatory effects were mediated by PPARα, PPARγ, and TRPV1 activity, but not through CB1 or CB2 [155]. The neuroprotective action of PEA and an endocannabinoid 2-AG was observed in an excitatory model of neuronal damage in organotypic hippocampal slice cultures [156]. PEA and 2-AG rescued about 50% of neurons from NMDA-induced cell death, acting on microglial cells, albeit through different and mutually suppressing mechanisms. PEA blocked microglial inflammatory activities, such as NO production and the acquisition of ameboid morphology, characteristic of an activated condition [156]. These effects were associated with PPARα nuclear translocation, which suggests its involvement in the process.

7.3. *PPARα-Mediated Regulation of Microglia and Macrophage Functions*

The glia-directed activity of PEA was studied by Scuderi and coauthors, who, in a series of papers, demonstrated that PEA or synthetic PPARα agonists, in a PPARα-dependent manner, decreased markers of glial inflammation and improved neuronal viability in animal models of Alzheimer's disease, as well as in mixed glio-neuronal cell cultures and organotypic neural cultures [157–159]. The immunomodulatory activity of PEA and the interplay between PPARα and the endocannabinoid system were also analyzed in primary microglial and macrophage cultures [160]. This study revealed that CB2 mRNA and protein levels were significantly increased by the treatment with PEA and a synthetic PPARα agonist GW7647, and this effect was evoked by the PPARα/RXR heterodimer binding to the promoter and transactivation of the gene encoding CB2 [160]. PEA induced microglial effector functions in a PPARα-dependent manner and improved the phagocytosis and killing of *Porphyromonas gingivalis* by microglia and chemotaxis to 2-AG [160]. In addition to the modulation of antimicrobial phagocytosis-based defense, PEA can modulate regenerative functions of macrophages, such as efferocytosis (i.e., phagocytosis and clearance of apoptotic cells) [161]. PEA is produced endogenously by M2c-polarized but not M1-polarized macrophages [161]. Exogenous chronic administration of PEA limited early plaque formation, protected from accumulation of the proinflammatory M1 macrophage within the plaque, and promoted efferocytosis by M2a- and M2c-polarized macrophages, which delayed the onset of arteriosclerosis [161]. These results show that endogenous PPARα ligand PEA is capable of modulating microglia and macrophage biological functions.

7.4. *PPARα's Role in Restoration of Neural Function after Injury or Infection*

Neuroprotective OEA activity was also demonstrated as an inhibition of so-called glial scar (i.e., zones enriched with reactive inflammatory astrocytes, microglia, fibroblasts, and accumulated extracellular matrix components) formation, after focal cerebral ischemia injury [162]. Glial scar is a natural physiological reaction to injury, but it impedes neurite formation, axon regrowth, and recovery after brain stroke. OEA increased PPARα expression in the cerebral cortex and downregulated glial scar markers (S100B, glial fibrillary acidic protein GFAP, metalloproteinases MMP-2, MMP-9, and neurocan) in the ischemic region through a PPARα-dependent mechanism [162]. Importantly, these biological processes translated into a better recovery of motor function in mice after stroke [162]. OEA also decreases the inflammatory response of endothelial cells (such as IL-6, IL-8, ICAM-1, and VCAM expression) evoked by TNFα, in a PPARα- and CB2-dependent manner [163].

The biological activities of OEA and PEA seem similar and sometimes overlap, but are not always identical, as shown in different experimental settings. An intriguing difference between OEA and PEA actions was observed in a study that analyzed functional impairments of neurological functions in an animal model of neonatal anoxia/ischemia-induced brain injury [164]. PEA, but not OEA treatment was capable of limiting hippocampal astrogliosis markers (e.g., ionized calcium-binding adaptor protein Iba-1, GFAP) and restoring PPARα protein expression in anoxia/ischemia-affected brain regions [164]. These effects were associated with improved cognitive abilities and a better recovery of spatial and recognition memory, as compared to control animals subjected to anoxia/ischemia [164]. Nevertheless, OEA was proved effective in ameliorating cognitive deficits and in supporting neurogenesis in ischemia-affected brain regions of rats subjected to middle cerebral artery occlusion [165].

An important immunomodulatory action of OEA and PEA involves TLR3 signaling during the innate response to viral infections. A recent report by Flannery et al. [166] demonstrated that intracerebroventricular administration of a TLR3 ligand, viral mimetic polyinosinic–polycytidynic acid (poly I:C), led to the induction of hypothalamic interferon- and NF-κB-regulated pathways of proinflammatory gene expression and hyperthermia. The treatment with both OEA and PEA attenuated TLR3-mediated hyperthermia, but only OEA (not PEA) was effective in the downregulation of poly I:C-induced inflammatory gene expression, including TNFα, iNOS, IL-1β, COX-2, interferon gamma-induced protein 10 (IP-10), and interferon-regulated factor IRF7. The fact that the PPARα antagonist GW6471 attenuated these effects indicated the PPARα involvement in this regulation [166]. These results have important implications for the current pandemic of SARS-CoV-2 infections, which often cause complications within the CNS, manifested by neurological and mental disorders, such as impaired memory, attention, anxiety, depression, and dementia [167].

7.5. PPARα and Endocannabinoid Involvement in the Regulation of Mast-Cell Functions

Mast cells are important innate immunity cells that, due to their rapid degranulation, can control the onset of inflammation in various tissues. PEA was shown to reduce local accumulation and the activation of mast cells in various inflammatory models: (i) after substance P injection to ear pinna [154], (ii) during chemically induced allergic dermatitis in mice [168], (iii) in myelin basic protein (MBP)-induced neuronal injury in a neuron–glia–mast cell coculture model of multiple sclerosis [169], (iv) in rat mast cell line RBL-2H3 [170], (v) after ischemia/reperfusion inflammatory injury of intestine after splanchnic artery occlusion in mice [171], and (vi) during chemically induced colitis which serves as an animal model of inflammatory bowel disease [172]. In all these experimental models, PEA suppressed a variety of effector reactions produced by mast cells or other leukocytes, such as chemotaxis, degranulation, enzyme release, and induction of proinflammatory cytokines. This suppression of mast-cell activity led to alleviation of inflammatory tissue damage and improved physiological tissue function. A common molecular mechanism could be involved in these effects, because, regardless of the model used, they were mediated, at least partially, by PPARα and CB2 activation [168–170], as well as, in some cases, by GPR55 and TRPV1 [172], which further supports the role of PPARα in the modulation of innate immunity and its connections with the endocannabinoid system.

However, a very intriguing recent discovery has shed new light on the connection among cannobinomimetics, mast cells, and metabolism, namely, ketogenesis. The publication from Daniele Piomelli's group revealed the unexpected role of histamine secreted by mast cells as a mediator necessary to induce ketogenesis in the liver in the state of food deprivation [173]. The mode of metabolic regulation involves an OEA-mediated action on hepatocytes. Routinely, after feeding, OEA is produced in the small intestine from consumed dietary lipids and takes part in food intake control as a satiety mediator via PPARα activation [133,174]. However, during food deprivation, ketogenesis depends on liver-derived OEA. A crucial role in this process is played by a population of mast cells that reside in the gastrointestinal tract and release histamine in the fasting state. Histamine enters the liver through portal circulation and stimulates hepatocytes to OEA secretion via activation of histamine H1 receptors [173]. Furthermore, OEA binding to PPARα in hepatocytes activates transcription of PPARα-target genes that control ketogenesis, including ACAT1, HMGSC2, and Fgf21 [173]. These results provide a novel link between mast cells as innate immunity effectors, cannabinomimetic PPARα ligand OEA, and PPARα-dependent ketogenesis as a metabolic response to fasting.

8. Evolutionary Aspects of PPARα-Mediated Immunomodulation

One of the crucially important features of the innate response is the speed and immediateness of the reaction to menacing invaders. In higher vertebrates, the accurate and prompt launching of the innate mechanisms buys time for the preparation of systemic adaptive immunity. In invertebrates, the effectiveness of innate immunity is a matter of life and death. The precise regulation of the innate responses is a multithreaded process that engages various signaling pathways, including the activity of nuclear receptors, such as PPARs. Such a regulation determines the success in coping with parasitic, viral, and bacterial infections, in addition to providing a hospitable environment for commensal microbiota and restricting inflammation-related tissue damage and injury.

PPARs and NOS serve as an illustrative example of how the elements of innate immunity and their regulatory mechanisms coevolved in the animal kingdom. On the one hand, NOS belongs to a large family of evolutionarily ancient enzymes that includes numerous pro- and eukaryotic flavodoxins [175,176]. There have been several hypotheses of their reciprocal relationship in invertebrates in the function of hemolymph homeostasis maintenance and the destruction of pathogens, i.e., probably unified in hemocytic NOS, as is the case for horseshoe crabs [175,177]. On the other hand, PPARs, despite their origin in the nuclear receptor family that emerged in metazoans, evolved in animals only as late as in the branch of Deuterostomata, whereas, in chordates, their presence dates from the evolution of Branchiostomata [178]. Consequently, they are present in all the vertebrates, but (except for Branchiostomata) absent in invertebrates [178]. Their presence seems to correspond to the evolution of the immune system and adipose tissue, but their tissue specificity does not overlap with their functional diversification. The most basic branch of this family seems to be represented by PPARγ, and the evolution of the whole family comprised two duplications of the genes, the first moving PPARγ apart, and the other dividing the other group into the PPARβ and α subfamilies [179]. This must have taken place on the level of ancient, primitive Teleostei [178,179].

Meanwhile, the diversified NOS family tree must root as deeply as in some Protista, as present in a differentiated side-branch in slime molds, fungi, and practically all Eukaryota including (a loosely related variant) high plants (*Arabidopsis thaliana* [180]). This may explain the engagement of PPARs in the functioning of various NOS in vertebrates. Upon evolution, the diversification of the NOS family has been consistently appreciated, whereas the engagement of PPARs in various aspects of NOS functioning may have been more or less accidental (Figure 4).

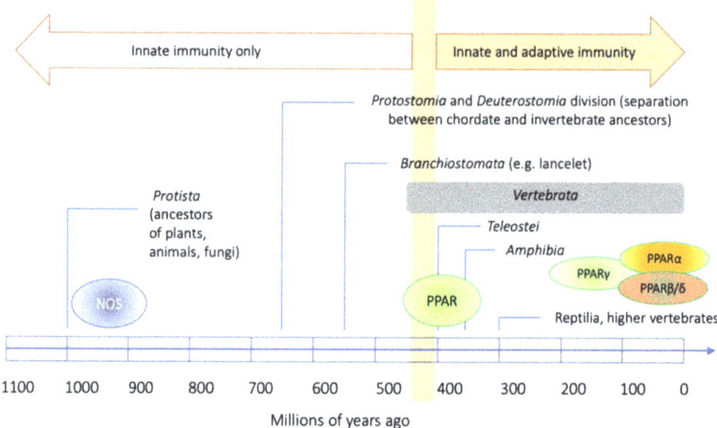

Figure 4. Long evolution of NOS as a background for shorter vertebrate-related evolution of PPARs and its involvement in the immune response in various animal phyla. The time scale is only illustrative and was based on [181].

9. Conclusions and Perspectives

PPARα as a transcription factor exerts a strong impact on cellular metabolism and intracellular signal transduction events, which alters the physiology and behavior of PPARα-expressing cells of both immune and nonimmune provenance. These physiological alterations underlie the immunomodulatory actions of PPARα presented in previous chapters. The broad spectrum of actions of endogenous and pharmacological PPARα agonists directed toward the immune system encourage the development of more commonly used therapeutic application of PPARα-targeted solutions in various infectious diseases and disorders of immunological background. The currently ongoing SARS-CoV-2 pandemic has created a dire need to revise the canonical approaches to the treatment of viral infections and has opened an unexpected possibility for new attempts, such as applying PPARα agonists to calm down the destructive cytokine storm in severe COVID-19 cases.

Author Contributions: Conceptualization, M.G.; literature survey and discussions on the topic, M.G., M.P., P.M.P. and P.P.; writing—original draft preparation, M.G., M.P., P.M.P. and P.P.; writing—review and editing, M.G., M.P., P.M.P. and P.P.; figure preparation, M.G. All authors have read and agreed to the published version of the manuscript.

Funding: This research was funded by N43/DBS/000158 to P.P.

Conflicts of Interest: The authors declare no conflict of interest.

Abbreviations

2-AG, 2-arachidonoyl-glycerol; ACAT1, acetoacetyl-CoA thiolase 1; AEA, *N*-arachidonoylethanolamine; AMPs, antimicrobial peptides; AP-1, activation protein 1; CB, cannabinoid receptors; CLRs, C-type lectin receptors; COX, cyclooxygenase; CSF, colony-stimulating factor; DAMPs, damage-associated molecular patterns; DOPA, dihydroxyphenylalanine; FAAH, atty-acid amide hydrolase; FAEs, fatty-acid ethanolamides; FAO, fatty-acid oxidation; FGF21, fibroblast growth factor 21; FREPs, fibrinogen-related proteins; HETE, hydroxyeicoatetraenoic acid; HMGCS2, 3-hydroxy-3-methylglytaryl-CoA synthetase 2; HPETE, hydroperoxyeicosatetraenoic acid; IDO, indoleamine-2,3-dioxygenase; IL, interleukin; ILCs, innate lymphoid cells; IRF, interferon-regulated factor; JAK, Janus-activated kinase; JNK, c-Jun N-terminal kinase; KO, knockout; LOX, lipoxygenase; LPS, lipopolysaccharide; LT, leukotriene; MAMPs, microbial-associated molecular patterns; MBP, myelin basic protein; MCP1, monocyte chemoattractant protein 1; MDSCs, myeloid-derived suppressor cells; MMP-9, matrix metalloproteinase 9; NAAA, *N*-acylethanolamine-hydrolyzing acid amidase; NAPE-PLD, *N*-acyl-phosphatidylethanolamine-hydrolyzing phospholipase D; NCoR, nuclear receptor co-repressor; NF-κB, nuclear factor κB; NLR, nucleotide-binding oligomerization domain (NOD)–leucin-rich repeat (LRR)-containing receptors; NO, nitric oxide; NOD, nucleotide-binding oligomerization domain; NOS, nitric oxide synthase; OEA, oleylethanolamide; PAMPs, pathogen-associated molecular patterns; PEA, palmitoylethanolamide; PG, prostaglandin; PPAR, peroxisome proliferator-activated receptor; PPRE, peroxisome proliferator response element; PRRs, pattern-recognition receptors; RIG1, retinoic acid inducible gene 1; RLR, retinoic acid inducible gene 1 (RIG1)-like receptors; RNS, reactive nitrogen species; ROR, retinoid orphan receptor; ROS, reactive oxygen species; RXR, retinoid X receptor; SAPK, stress-activated protein kinase; SMRT, silencing mediator of retinoid and thyroid receptors; STAT, signal transducer and activator of transcription; TF, tissue factor; TFEB, transcription factor EB; TGF, transforming growth factor; TLR, Toll-like receptors; TNF, tumor necrosis factor; TRPV1, transient receptor potential cation channel vanilloid subfamily member 1; TXNIP, thioredoxin-interacting protein.

References

1. Boraschi, D.; Italiani, P. Innate immune memory: Time for adopting a correct terminology. *Front. Immunol.* **2018**, *9*, 799. [CrossRef]
2. Netea, M.G.; Quintin, J.; van der Meer, J.W. Trained immunity: A memory for innate host defense. *Cell Host Microbe* **2011**, *9*, 355–361. [CrossRef]
3. Moret, Y.; Siva-Jothy, M.T. Adaptive innate immunity? Responsive-mode prophylaxis in the mealworm beetle, Tenebrio molitor. *Proc. Biol. Sci.* **2003**, *270*, 2475–2480. [CrossRef]
4. Torre, C.; Abnave, P.; Tsoumtsa, L.L.; Mottola, G.; Lepolard, C.; Trouplin, V.; Gimenez, G.; Desrousseaux, J.; Gempp, S.; Levasseur, A.; et al. *Staphylococcus aureus* promotes Smed-PGRP-2/Smed-setd8-1 methyltransferase signalling in planarian neoblasts to sensitize anti-bacterial gene responses during re-infection. *EBioMedicine* **2017**, *20*, 150–160. [CrossRef] [PubMed]
5. Zhang, T.; Qiu, L.; Sun, Z.; Wang, L.; Zhou, Z.; Liu, R.; Yue, F.; Sun, R.; Song, L. The specifically enhanced cellular immune responses in Pacific oyster (*Crassostrea gigas*) against secondary challenge with *Vibrio splendidus*. *Dev. Comp. Immunol.* **2014**, *45*, 141–150. [CrossRef]
6. Wang, X.; Zhao, Q.; Christensen, B.M. Identification and characterization of the fibrinogen-like domain of fibrinogen-related proteins in the mosquito, Anopheles gambiae, and the fruitfly, *Drosophila melanogaster*, genomes. *BMC Genom.* **2005**, *6*, 114. [CrossRef] [PubMed]
7. Melillo, D.; Marino, R.; Italiani, P.; Boraschi, D. Innate immune memory in invertebrate metazoans: A Critical appraisal. *Front. Immunol.* **2018**, *9*, 1915. [CrossRef]
8. Cerenius, L.; Soderhall, K. Immune properties of invertebrate phenoloxidases. *Dev. Comp. Immunol.* **2021**, *122*, 104098. [CrossRef]
9. Coates, C.J.; Nairn, J. Diverse immune functions of hemocyanins. *Dev. Comp. Immunol.* **2014**, *45*, 43–55. [CrossRef] [PubMed]
10. Yang, D.; Han, Z.; Oppenheim, J.J. Alarmins and immunity. *Immunol. Rev.* **2017**, *280*, 41–56. [CrossRef]
11. Amarante-Mendes, G.P.; Adjemian, S.; Branco, L.M.; Zanetti, L.C.; Weinlich, R.; Bortoluci, K.R. Pattern Recognition Receptors and the Host Cell Death Molecular Machinery. *Front. Immunol.* **2018**, *9*, 2379. [CrossRef] [PubMed]
12. Olive, C. Pattern recognition receptors: Sentinels in innate immunity and targets of new vaccine adjuvants. *Expert Rev. Vaccines* **2012**, *11*, 237–256. [CrossRef]
13. Rabinovitch, M. Professional and non-professional phagocytes: An introduction. *Trends Cell Biol.* **1995**, *5*, 85–87. [CrossRef]
14. Lim, J.J.; Grinstein, S.; Roth, Z. Diversity and versatility of phagocytosis: Roles in innate immunity, tissue remodeling, and homeostasis. *Front. Cell Infect. Microbiol.* **2017**, *7*, 191. [CrossRef]
15. Witko-Sarsat, V.; Descamps-Latscha, B. Neutrophil-derived Oxidants and proteinases as immunomodulatory mediators in inflammation. *Mediat. Inflamm.* **1994**, *3*, 257–273. [CrossRef] [PubMed]
16. Nguyen, G.T.; Green, E.R.; Mecsas, J. Neutrophils to the ROScue: Mechanisms of NADPH oxidase activation and bacterial resistance. *Front. Cell Infect. Microbiol.* **2017**, *7*, 373. [CrossRef] [PubMed]
17. Kettle, A.J.; Winterbourn, C.C. Myeloperoxidase: A key regulator of neutrophil oxidant production. *Redox Rep.* **1997**, *3*, 3–15. [CrossRef]
18. Prolo, C.; Alvarez, M.N.; Radi, R. Peroxynitrite, a potent macrophage-derived oxidizing cytotoxin to combat invading pathogens. *Biofactors* **2014**, *40*, 215–225. [CrossRef]
19. Yamasaki, R.; Lu, H.; Butovsky, O.; Ohno, N.; Rietsch, A.M.; Cialic, R.; Wu, P.M.; Doykan, C.E.; Lin, J.; Cotleur, A.C.; et al. Differential roles of microglia and monocytes in the inflamed central nervous system. *J. Exp. Med.* **2014**, *211*, 1533–1549. [CrossRef]
20. Schulz, D.; Severin, Y.; Zanotelli, V.R.T.; Bodenmiller, B. In-Depth characterization of monocyte-derived macrophages using a mass cytometry-based phagocytosis assay. *Sci. Rep.* **2019**, *9*, 1925. [CrossRef]
21. Gerner, E.W.; Meyskens, F.L., Jr. Polyamines and cancer: Old molecules, new understanding. *Nat. Rev. Cancer* **2004**, *4*, 781–792. [CrossRef] [PubMed]

22. Luk, G.D. Essential role of polyamine metabolism in hepatic regeneration. Inhibition of deoxyribonucleic acid and protein synthesis and tissue regeneration by difluoromethylornithine in the rat. *Gastroenterology* **1986**, *90*, 1261–1267. [CrossRef]
23. Murray, P.J. Macrophage polarization. *Annu. Rev. Physiol.* **2017**, *79*, 541–566. [CrossRef]
24. Locati, M.; Curtale, G.; Mantovani, A. Diversity, Mechanisms, and significance of macrophage plasticity. *Annu. Rev. Pathol.* **2020**, *15*, 123–147. [CrossRef] [PubMed]
25. Gasteiger, G.; D'Osualdo, A.; Schubert, D.A.; Weber, A.; Bruscia, E.M.; Hartl, D. Cellular innate immunity: An old game with new players. *J. Innate. Immun.* **2017**, *9*, 111–125. [CrossRef]
26. Erb, K.J.; Holloway, J.W.; Le Gros, G. Mast cells in the front line. Innate immunity. *Curr. Biol.* **1996**, *6*, 941–942. [CrossRef]
27. Sendo, S.; Saegusa, J.; Morinobu, A. Myeloid-derived suppressor cells in non-neoplastic inflamed organs. *Inflamm. Regen.* **2018**, *38*, 19. [CrossRef]
28. Killig, M.; Glatzer, T.; Romagnani, C. Recognition strategies of group 3 innate lymphoid cells. *Front. Immunol.* **2014**, *5*, 142. [CrossRef]
29. Trabanelli, S.; Gomez-Cadena, A.; Salome, B.; Michaud, K.; Mavilio, D.; Landis, B.N.; Jandus, P.; Jandus, C. Human innate lymphoid cells (ILCs): Toward a uniform immune-phenotyping. *Cytom. B Clin. Cytom.* **2018**, *94*, 392–399. [CrossRef]
30. Manoharan, I.; Suryawanshi, A.; Hong, Y.; Ranganathan, P.; Shanmugam, A.; Ahmad, S.; Swafford, D.; Manicassamy, B.; Ramesh, G.; Koni, P.A.; et al. Homeostatic PPARalpha signaling limits inflammatory responses to commensal microbiota in the intestine. *J. Immunol.* **2016**, *196*, 4739–4749. [CrossRef]
31. Sagebiel, A.F.; Steinert, F.; Lunemann, S.; Korner, C.; Schreurs, R.; Altfeld, M.; Perez, D.; Reinshagen, K.; Bunders, M.J. Tissue-resident Eomes(+) NK cells are the major innate lymphoid cell population in human infant intestine. *Nat. Commun.* **2019**, *10*, 975. [CrossRef]
32. Gordon, S.M.; Chaix, J.; Rupp, L.J.; Wu, J.; Madera, S.; Sun, J.C.; Lindsten, T.; Reiner, S.L. The transcription factors T-bet and Eomes control key checkpoints of natural killer cell maturation. *Immunity* **2012**, *36*, 55–67. [CrossRef] [PubMed]
33. Herbert, D.R.; Douglas, B.; Zullo, K. Group 2 innate lymphoid cells (ILC2): Type 2 immunity and helminth immunity. *Int. J. Mol. Sci.* **2019**, *20*, 2276. [CrossRef]
34. Hepworth, M.R.; Fung, T.C.; Masur, S.H.; Kelsen, J.R.; McConnell, F.M.; Dubrot, J.; Withers, D.R.; Hugues, S.; Farrar, M.A.; Reith, W.; et al. Immune tolerance. Group 3 innate lymphoid cells mediate intestinal selection of commensal bacteria-specific CD4(+) T cells. *Science* **2015**, *348*, 1031–1035. [CrossRef] [PubMed]
35. Withers, D.R.; Hepworth, M.R. Group 3 innate lymphoid cells: Communications hubs of the intestinal immune system. *Front. Immunol.* **2017**, *8*, 1298. [CrossRef]
36. Celsus, A.C. *De Medicina*, 1st ed.; Nicolaus Laurentii: Florence, Italy, 1478.
37. Frias, B.; Merighi, A. Capsaicin, nociception and pain. *Molecules* **2016**, *21*, 797. [CrossRef]
38. Nilius, B.; Owsianik, G. The transient receptor potential family of ion channels. *Genome Biol.* **2011**, *12*, 218. [CrossRef] [PubMed]
39. Desvergne, B.; Wahli, W. Peroxisome proliferator-activated receptors: Nuclear control of metabolism. *Endocrinol. Rev.* **1999**, *20*, 649–688. [CrossRef]
40. Sertznig, P.; Seifert, M.; Tilgen, W.; Reichrath, J. Present concepts and future outlook: Function of peroxisome proliferator-activated receptors (PPARs) for pathogenesis, progression, and therapy of cancer. *J. Cell Physiol.* **2007**, *212*, 1–12. [CrossRef]
41. Kamata, S.; Oyama, T.; Saito, K.; Honda, A.; Yamamoto, Y.; Suda, K.; Ishikawa, R.; Itoh, T.; Watanabe, Y.; Shibata, T.; et al. PPARalpha ligand-binding domain structures with endogenous fatty acids and fibrates. *iScience* **2020**, *23*, 101727. [CrossRef]
42. Wahli, W.; Michalik, L. PPARs at the crossroads of lipid signaling and inflammation. *Trends Endocrinol. Metab.* **2012**, *23*, 351–363. [CrossRef]
43. Lo Verme, J.; Fu, J.; Astarita, G.; La Rana, G.; Russo, R.; Calignano, A.; Piomelli, D. The nuclear receptor peroxisome proliferator-activated receptor-alpha mediates the anti-inflammatory actions of palmitoylethanolamide. *Mol. Pharmacol.* **2005**, *67*, 15–19. [CrossRef]
44. LoVerme, J.; La Rana, G.; Russo, R.; Calignano, A.; Piomelli, D. The search for the palmitoylethanolamide receptor. *Life Sci.* **2005**, *77*, 1685–1698. [CrossRef]
45. de Lera, A.R.; Krezel, W.; Ruhl, R. An endogenous mammalian retinoid X receptor ligand, at last! *ChemMedChem* **2016**, *11*, 1027–1037. [CrossRef]
46. Ruhl, R.; Krzyzosiak, A.; Niewiadomska-Cimicka, A.; Rochel, N.; Szeles, L.; Vaz, B.; Wietrzych-Schindler, M.; Alvarez, S.; Szklenar, M.; Nagy, L.; et al. 9-cis-13,14-dihydroretinoic acid is an endogenous retinoid acting as RXR ligand in mice. *PLoS Genet.* **2015**, *11*, e1005213. [CrossRef] [PubMed]
47. Najib, J. Fenofibrate in the treatment of dyslipidemia: A review of the data as they relate to the new suprabioavailable tablet formulation. *Clin. Ther.* **2002**, *24*, 2022–2050. [CrossRef]
48. Blais, J.E.; Tong, G.K.Y.; Pathadka, S.; Mok, M.; Wong, I.C.K.; Chan, E.W. Comparative efficacy and safety of statin and fibrate monotherapy: A systematic review and meta-analysis of head-to-head randomized controlled trials. *PLoS ONE* **2021**, *16*, e0246480. [CrossRef] [PubMed]
49. Alsheikh-Ali, A.A.; Kuvin, J.T.; Karas, R.H. Risk of adverse events with fibrates. *Am. J. Cardiol.* **2004**, *94*, 935–938. [CrossRef] [PubMed]
50. Kaipainen, A.; Kieran, M.W.; Huang, S.; Butterfield, C.; Bielenberg, D.; Mostoslavsky, G.; Mulligan, R.; Folkman, J.; Panigrahy, D. PPARalpha deficiency in inflammatory cells suppresses tumor growth. *PLoS ONE* **2007**, *2*, e260. [CrossRef] [PubMed]

51. Fujimura, Y.; Tachibana, H.; Yamada, K. Peroxisome proliferator-activated receptor ligands negatively regulate the expression of the high-affinity IgE receptor Fc epsilon RI in human basophilic KU812 cells. *Biochem. Biophys. Res. Commun.* **2002**, *297*, 193–201. [CrossRef]
52. Woerly, G.; Honda, K.; Loyens, M.; Papin, J.P.; Auwerx, J.; Staels, B.; Capron, M.; Dombrowicz, D. Peroxisome proliferator-activated receptors alpha and gamma down-regulate allergic inflammation and eosinophil activation. *J. Exp. Med.* **2003**, *198*, 411–421. [CrossRef] [PubMed]
53. Chinetti, G.; Fruchart, J.C.; Staels, B. Peroxisome proliferator-activated receptors: New targets for the pharmacological modulation of macrophage gene expression and function. *Curr. Opin. Lipidol.* **2003**, *14*, 459–468. [CrossRef]
54. Babaev, V.R.; Ishiguro, H.; Ding, L.; Yancey, P.G.; Dove, D.E.; Kovacs, W.J.; Semenkovich, C.F.; Fazio, S.; Linton, M.F. Macrophage expression of peroxisome proliferator-activated receptor-alpha reduces atherosclerosis in low-density lipoprotein receptor-deficient mice. *Circulation* **2007**, *116*, 1404–1412. [CrossRef] [PubMed]
55. Wu, L.; Zhang, X.; Zheng, L.; Zhao, H.; Yan, G.; Zhang, Q.; Zhou, Y.; Lei, J.; Zhang, J.; Wang, J.; et al. RIPK3 orchestrates fatty acid metabolism in tumor-associated macrophages and hepatocarcinogenesis. *Cancer Immunol. Res.* **2020**, *8*, 710–721. [CrossRef] [PubMed]
56. Brocker, C.N.; Yue, J.; Kim, D.; Qu, A.; Bonzo, J.A.; Gonzalez, F.J. Hepatocyte-specific PPARA expression exclusively promotes agonist-induced cell proliferation without influence from nonparenchymal cells. *Am. J. Physiol. Gastrointest. Liver Physiol.* **2017**, *312*, G283–G299. [CrossRef]
57. Dubrac, S.; Stoitzner, P.; Pirkebner, D.; Elentner, A.; Schoonjans, K.; Auwerx, J.; Saeland, S.; Hengster, P.; Fritsch, P.; Romani, N.; et al. Peroxisome proliferator-activated receptor-alpha activation inhibits Langerhans cell function. *J. Immunol.* **2007**, *178*, 4362–4372. [CrossRef]
58. Poulsen, R.C.; Moughan, P.J.; Kruger, M.C. Long-chain polyunsaturated fatty acids and the regulation of bone metabolism. *Exp. Biol. Med.* **2007**, *232*, 1275–1288. [CrossRef]
59. Warden, A.; Truitt, J.; Merriman, M.; Ponomareva, O.; Jameson, K.; Ferguson, L.B.; Mayfield, R.D.; Harris, R.A. Localization of PPAR isotypes in the adult mouse and human brain. *Sci. Rep.* **2016**, *6*, 27618. [CrossRef]
60. Kroetz, D.L.; Yook, P.; Costet, P.; Bianchi, P.; Pineau, T. Peroxisome proliferator-activated receptor alpha controls the hepatic CYP4A induction adaptive response to starvation and diabetes. *J. Biol. Chem.* **1998**, *273*, 31581–31589. [CrossRef]
61. Hashimoto, T.; Fujita, T.; Usuda, N.; Cook, W.; Qi, C.; Peters, J.M.; Gonzalez, F.J.; Yeldandi, A.V.; Rao, M.S.; Reddy, J.K. Peroxisomal and mitochondrial fatty acid beta-oxidation in mice nullizygous for both peroxisome proliferator-activated receptor alpha and peroxisomal fatty acyl-CoA oxidase. Genotype correlation with fatty liver phenotype. *J. Biol. Chem.* **1999**, *274*, 19228–19236. [CrossRef]
62. Vila-Brau, A.; De Sousa-Coelho, A.L.; Mayordomo, C.; Haro, D.; Marrero, P.F. Human HMGCS2 regulates mitochondrial fatty acid oxidation and FGF21 expression in HepG2 cell line. *J. Biol. Chem.* **2011**, *286*, 20423–20430. [CrossRef]
63. Daynes, R.A.; Jones, D.C. Emerging roles of PPARs in inflammation and immunity. *Nat. Rev. Immunol.* **2002**, *2*, 748–759. [CrossRef]
64. Ammazzalorso, A.; Bruno, I.; Florio, R.; De Lellis, L.; Laghezza, A.; Cerchia, C.; De Filippis, B.; Fantacuzzi, M.; Giampietro, L.; Maccallini, C.; et al. Sulfonimide and amide derivatives as novel PPARalpha antagonists: Synthesis, antiproliferative activity, and docking studies. *ACS Med. Chem. Lett.* **2020**, *11*, 624–632. [CrossRef]
65. Delerive, P.; De Bosscher, K.; Besnard, S.; Vanden Berghe, W.; Peters, J.M.; Gonzalez, F.J.; Fruchart, J.C.; Tedgui, A.; Haegeman, G.; Staels, B. Peroxisome proliferator-activated receptor alpha negatively regulates the vascular inflammatory gene response by negative cross-talk with transcription factors NF-kappaB and AP-1. *J. Biol. Chem.* **1999**, *274*, 32048–32054. [CrossRef] [PubMed]
66. Planavila, A.; Iglesias, R.; Giralt, M.; Villarroya, F. Sirt1 acts in association with PPARalpha to protect the heart from hypertrophy, metabolic dysregulation, and inflammation. *Cardiovasc. Res.* **2011**, *90*, 276–284. [CrossRef]
67. Rothgiesser, K.M.; Fey, M.; Hottiger, M.O. Acetylation of p65 at lysine 314 is important for late NF-kappaB-dependent gene expression. *BMC Genom.* **2010**, *11*, 22. [CrossRef] [PubMed]
68. Zhang, N.; Chu, E.S.; Zhang, J.; Li, X.; Liang, Q.; Chen, J.; Chen, M.; Teoh, N.; Farrell, G.; Sung, J.J.; et al. Peroxisome proliferator activated receptor alpha inhibits hepatocarcinogenesis through mediating NF-kappaB signaling pathway. *Oncotarget* **2014**, *5*, 8330–8340. [CrossRef] [PubMed]
69. Jacobs, M.D.; Harrison, S.C. Structure of an IkappaBalpha/NF-kappaB complex. *Cell* **1998**, *95*, 749–758. [CrossRef]
70. Chen, L.F.; Greene, W.C. Shaping the nuclear action of NF-kappaB. *Nat. Rev. Mol. Cell Biol.* **2004**, *5*, 392–401. [CrossRef]
71. Murakami, K.; Bujo, H.; Unoki, H.; Saito, Y. Effect of PPARalpha activation of macrophages on the secretion of inflammatory cytokines in cultured adipocytes. *Eur. J. Pharmacol.* **2007**, *561*, 206–213. [CrossRef]
72. Marx, N.; Mackman, N.; Schonbeck, U.; Yilmaz, N.; Hombach, V.; Libby, P.; Plutzky, J. PPARalpha activators inhibit tissue factor expression and activity in human monocytes. *Circulation* **2001**, *103*, 213–219. [CrossRef]
73. Neve, B.P.; Corseaux, D.; Chinetti, G.; Zawadzki, C.; Fruchart, J.C.; Duriez, P.; Staels, B.; Jude, B. PPARalpha agonists inhibit tissue factor expression in human monocytes and macrophages. *Circulation* **2001**, *103*, 207–212. [CrossRef]
74. Haque, S.J.; Sharma, P. Interleukins and STAT signaling. *Vitam. Horm.* **2006**, *74*, 165–206. [CrossRef]
75. Shipley, J.M.; Waxman, D.J. Down-regulation of STAT5b transcriptional activity by ligand-activated peroxisome proliferator-activated receptor (PPAR) alpha and PPARgamma. *Mol. Pharmacol.* **2003**, *64*, 355–364. [CrossRef] [PubMed]
76. Shipley, J.M.; Waxman, D.J. Simultaneous, bidirectional inhibitory crosstalk between PPAR and STAT5b. *Toxicol. Appl. Pharmacol.* **2004**, *199*, 275–284. [CrossRef] [PubMed]

77. Zhou, Y.C.; Waxman, D.J. STAT5b down-regulates peroxisome proliferator-activated receptor alpha transcription by inhibition of ligand-independent activation function region-1 trans-activation domain. *J. Biol. Chem.* **1999**, *274*, 29874–29882. [CrossRef] [PubMed]
78. Lin, J.X.; Leonard, W.J. The role of Stat5a and Stat5b in signaling by IL-2 family cytokines. *Oncogene* **2000**, *19*, 2566–2576. [CrossRef] [PubMed]
79. Bendickova, K.; Fric, J. Roles of IL-2 in bridging adaptive and innate immunity, and as a tool for cellular immunotherapy. *J. Leukoc. Biol.* **2020**, *108*, 427–437. [CrossRef]
80. Devchand, P.R.; Keller, H.; Peters, J.M.; Vazquez, M.; Gonzalez, F.J.; Wahli, W. The PPARalpha-leukotriene B4 pathway to inflammation control. *Nature* **1996**, *384*, 39–43. [CrossRef]
81. Dahlen, S.E.; Bjork, J.; Hedqvist, P.; Arfors, K.E.; Hammarstrom, S.; Lindgren, J.A.; Samuelsson, B. Leukotrienes promote plasma leakage and leukocyte adhesion in postcapillary venules: In vivo effects with relevance to the acute inflammatory response. *Proc. Natl. Acad. Sci. USA* **1981**, *78*, 3887–3891. [CrossRef]
82. Lindbom, L.; Hedqvist, P.; Dahlen, S.E.; Lindgren, J.A.; Arfors, K.E. Leukotriene B4 induces extravasation and migration of polymorphonuclear leukocytes in vivo. *Acta Physiol. Scand.* **1982**, *116*, 105–108. [CrossRef] [PubMed]
83. Marleau, S.; Fruteau de Laclos, B.; Sanchez, A.B.; Poubelle, P.E.; Borgeat, P. Role of 5-lipoxygenase products in the local accumulation of neutrophils in dermal inflammation in the rabbit. *J. Immunol.* **1999**, *163*, 3449–3458.
84. Yu, K.; Bayona, W.; Kallen, C.B.; Harding, H.P.; Ravera, C.P.; McMahon, G.; Brown, M.; Lazar, M.A. Differential activation of peroxisome proliferator-activated receptors by eicosanoids. *J. Biol. Chem.* **1995**, *270*, 23975–23983. [CrossRef] [PubMed]
85. Christofides, A.; Konstantinidou, E.; Jani, C.; Boussiotis, V.A. The role of peroxisome proliferator-activated receptors (PPAR) in immune responses. *Metabolism* **2021**, *114*, 154338. [CrossRef]
86. Wolf, A.J.; Reyes, C.N.; Liang, W.; Becker, C.; Shimada, K.; Wheeler, M.L.; Cho, H.C.; Popescu, N.I.; Coggeshall, K.M.; Arditi, M.; et al. Hexokinase Is an innate immune receptor for the detection of bacterial peptidoglycan. *Cell* **2016**, *166*, 624–636. [CrossRef]
87. Frosali, S.; Pagliari, D.; Gambassi, G.; Landolfi, R.; Pandolfi, F.; Cianci, R. How the intricate interaction among toll-like receptors, microbiota, and intestinal immunity can influence gastrointestinal pathology. *J. Immunol. Res.* **2015**, *2015*, 489821. [CrossRef]
88. Wang, Y.; Yin, Y.; Chen, X.; Zhao, Y.; Wu, Y.; Li, Y.; Wang, X.; Chen, H.; Xiang, C. Induction of intestinal Th17 cells by flagellins from segmented filamentous bacteria. *Front. Immunol.* **2019**, *10*, 2750. [CrossRef] [PubMed]
89. Park, J.H.; Jeong, S.Y.; Choi, A.J.; Kim, S.J. Lipopolysaccharide directly stimulates Th17 differentiation in vitro modulating phosphorylation of RelB and NF-kappaB1. *Immunol. Lett.* **2015**, *165*, 10–19. [CrossRef]
90. Silveira, L.S.; Pimentel, G.D.; Souza, C.O.; Biondo, L.A.; Teixeira, A.A.S.; Lima, E.A.; Batatinha, H.A.P.; Rosa Neto, J.C.; Lira, F.S. Effect of an acute moderate-exercise session on metabolic and inflammatory profile of PPAR-alpha knockout mice. *Cell Biochem. Funct.* **2017**, *35*, 510–517. [CrossRef]
91. Becker, J.; Delayre-Orthez, C.; Frossard, N.; Pons, F. Regulation of peroxisome proliferator-activated receptor-alpha expression during lung inflammation. *Pulm. Pharmacol. Ther.* **2008**, *21*, 324–330. [CrossRef]
92. Chistyakov, D.V.; Aleshin, S.E.; Astakhova, A.A.; Sergeeva, M.G.; Reiser, G. Regulation of peroxisome proliferator-activated receptors (PPAR) alpha and -gamma of rat brain astrocytes in the course of activation by toll-like receptor agonists. *J. Neurochem.* **2015**, *134*, 113–124. [CrossRef]
93. Dana, N.; Javanmard, H.S.; Vaseghi, G. The effect of fenofibrate, a PPARalpha activator on toll-like receptor-4 signal transduction in melanoma both in vitro and in vivo. *Clin. Transl. Oncol.* **2020**, *22*, 486–494. [CrossRef]
94. Shen, W.; Gao, Y.; Lu, B.; Zhang, Q.; Hu, Y.; Chen, Y. Negatively regulating TLR4/NF-kappaB signaling via PPARalpha in endotoxin-induced uveitis. *Biochim. Biophys. Acta* **2014**, *1842*, 1109–1120. [CrossRef] [PubMed]
95. Lamkanfi, M.; Dixit, V.M. Mechanisms and functions of inflammasomes. *Cell* **2014**, *157*, 1013–1022. [CrossRef] [PubMed]
96. Zheng, D.; Liwinski, T.; Elinav, E. Inflammasome activation and regulation: Toward a better understanding of complex mechanisms. *Cell Discov.* **2020**, *6*, 36. [CrossRef] [PubMed]
97. Seok, J.K.; Kang, H.C.; Cho, Y.Y.; Lee, H.S.; Lee, J.Y. Regulation of the NLRP3 inflammasome by post-translational modifications and small molecules. *Front. Immunol.* **2020**, *11*, 618231. [CrossRef]
98. Gugliandolo, E.; Fusco, R.; Ginestra, G.; D'Amico, R.; Bisignano, C.; Mandalari, G.; Cuzzocrea, S.; Di Paola, R. Involvement of TLR4 and PPAR-alpha receptors in host response and NLRP3 Inflammasome activation, against pulmonary infection with *Pseudomonas aeruginosa*. *Shock* **2019**, *51*, 221–227. [CrossRef] [PubMed]
99. Brocker, C.N.; Kim, D.; Melia, T.; Karri, K.; Velenosi, T.J.; Takahashi, S.; Aibara, D.; Bonzo, J.A.; Levi, M.; Waxman, D.J.; et al. Long non-coding RNA Gm15441 attenuates hepatic inflammasome activation in response to PPARA agonism and fasting. *Nat. Commun.* **2020**, *11*, 5847. [CrossRef]
100. Hu, J.; Zhu, Z.; Ying, H.; Yao, J.; Ma, H.; Li, L.; Zhao, Y. Oleoylethanolamide protects against acute liver injury by regulating Nrf-2/HO-1 and NLRP3 pathways in mice. *Front. Pharmacol.* **2020**, *11*, 605065. [CrossRef]
101. Lama, A.; Provensi, G.; Amoriello, R.; Pirozzi, C.; Rani, B.; Mollica, M.P.; Raso, G.M.; Ballerini, C.; Meli, R.; Passani, M.B. The anti-inflammatory and immune-modulatory effects of OEA limit DSS-induced colitis in mice. *Biomed. Pharmacother.* **2020**, *129*, 110368. [CrossRef]
102. Alderton, W.K.; Cooper, C.E.; Knowles, R.G. Nitric oxide synthases: Structure, function and inhibition. *Biochem. J.* **2001**, *357*, 593–615. [CrossRef]

103. Kleinert, H.; Schwarz, P.M.; Forstermann, U. Regulation of the expression of inducible nitric oxide synthase. *Biol. Chem.* **2003**, *384*, 1343–1364. [CrossRef] [PubMed]
104. Forstermann, U.; Boissel, J.P.; Kleinert, H. Expressional control of the 'constitutive' isoforms of nitric oxide synthase (NOS I and NOS III). *FASEB J.* **1998**, *12*, 773–790. [CrossRef] [PubMed]
105. Pou, S.; Keaton, L.; Surichamorn, W.; Rosen, G.M. Mechanism of superoxide generation by neuronal nitric-oxide synthase. *J. Biol. Chem.* **1999**, *274*, 9573–9580. [CrossRef] [PubMed]
106. Paukkeri, E.L.; Leppanen, T.; Sareila, O.; Vuolteenaho, K.; Kankaanranta, H.; Moilanen, E. PPARalpha agonists inhibit nitric oxide production by enhancing iNOS degradation in LPS-treated macrophages. *Br. J. Pharmacol.* **2007**, *152*, 1081–1091. [CrossRef] [PubMed]
107. Van Ginderachter, J.A.; Movahedi, K.; Van den Bossche, J.; De Baetselier, P. Macrophages, PPARs, and cancer. *PPAR Res.* **2008**, *2008*, 169414. [CrossRef]
108. Maccallini, C.; Mollica, A.; Amoroso, R. The positive regulation of eNOS signaling by PPAR agonists in cardiovascular diseases. *Am. J. Cardiovasc. Drugs* **2017**, *17*, 273–281. [CrossRef]
109. Tanaka, S.; Hosogi, S.; Sawabe, Y.; Shimamoto, C.; Matsumura, H.; Inui, T.; Marunaka, Y.; Nakahari, T. PPARalpha induced NOS1 phosphorylation via PI3K/Akt in guinea pig antral mucous cells: NO-enhancement in Ca(2+)-regulated exocytosis. *Biomed. Res.* **2016**, *37*, 167–178. [CrossRef]
110. Tanaka, S.; Sugiyama, N.; Takahashi, Y.; Mantoku, D.; Sawabe, Y.; Kuwabara, H.; Nakano, T.; Shimamoto, C.; Matsumura, H.; Marunaka, Y.; et al. PPARalpha autocrine regulation of Ca(2)(+)-regulated exocytosis in guinea pig antral mucous cells: NO and cGMP accumulation. *Am. J. Physiol. Gastrointest. Liver Physiol.* **2014**, *307*, G1169–G1179. [CrossRef]
111. Bune, A.J.; Shergill, J.K.; Cammack, R.; Cook, H.T. L-arginine depletion by arginase reduces nitric oxide production in endotoxic shock: An electron paramagnetic resonance study. *FEBS Lett.* **1995**, *366*, 127–130. [CrossRef]
112. Mills, C.D.; Kincaid, K.; Alt, J.M.; Heilman, M.J.; Hill, A.M. M-1/M-2 macrophages and the Th1/Th2 paradigm. *J. Immunol.* **2000**, *164*, 6166–6173. [CrossRef] [PubMed]
113. Grabacka, M.; Placha, W.; Plonka, P.M.; Pajak, S.; Urbanska, K.; Laidler, P.; Slominski, A. Inhibition of melanoma metastases by fenofibrate. *Arch. Dermatol. Res.* **2004**, *296*, 54–58. [CrossRef]
114. Gallorini, M.; Rapino, M.; Schweikl, H.; Cataldi, A.; Amoroso, R.; Maccallini, C. Selective inhibitors of the inducible nitric oxide synthase as modulators of cell responses in LPS-stimulated human monocytes. *Molecules* **2021**, *26*, 4419. [CrossRef]
115. Teissier, E.; Nohara, A.; Chinetti, G.; Paumelle, R.; Cariou, B.; Fruchart, J.C.; Brandes, R.P.; Shah, A.; Staels, B. Peroxisome proliferator-activated receptor alpha induces NADPH oxidase activity in macrophages, leading to the generation of LDL with PPAR-alpha activation properties. *Circ. Res.* **2004**, *95*, 1174–1182. [CrossRef]
116. Deininger, P. Alu elements: Know the SINEs. *Genome Biol.* **2011**, *12*, 236. [CrossRef] [PubMed]
117. Reynolds, W.F.; Kumar, A.P.; Piedrafita, F.J. The human myeloperoxidase gene is regulated by LXR and PPARalpha ligands. *Biochem. Biophys. Res. Commun.* **2006**, *349*, 846–854. [CrossRef]
118. Penas, F.; Mirkin, G.A.; Vera, M.; Cevey, A.; Gonzalez, C.D.; Gomez, M.I.; Sales, M.E.; Goren, N.B. Treatment in vitro with PPARalpha and PPARgamma ligands drives M1-to-M2 polarization of macrophages from T. cruzi-infected mice. *Biochim. Biophys. Acta* **2015**, *1852*, 893–904. [CrossRef] [PubMed]
119. Redlich, S.; Ribes, S.; Schutze, S.; Nau, R. Palmitoylethanolamide stimulates phagocytosis of *Escherichia coli* K1 by macrophages and increases the resistance of mice against infections. *J. Neuroinflamm.* **2014**, *11*, 108. [CrossRef]
120. Kim, Y.S.; Lee, H.M.; Kim, J.K.; Yang, C.S.; Kim, T.S.; Jung, M.; Jin, H.S.; Kim, S.; Jang, J.; Oh, G.T.; et al. PPAR alpha activation mediates innate host defense through induction of TFEB and lipid catabolism. *J. Immunol.* **2017**, *198*, 3283–3295. [CrossRef]
121. Peyron, P.; Vaubourgeix, J.; Poquet, Y.; Levillain, F.; Botanch, C.; Bardou, F.; Daffe, M.; Emile, J.F.; Marchou, B.; Cardona, P.J.; et al. Foamy macrophages from tuberculous patients' granulomas constitute a nutrient-rich reservoir for *M. tuberculosis* persistence. *PLoS Pathog.* **2008**, *4*, e1000204. [CrossRef]
122. Tam, V.C.; Suen, R.; Treuting, P.M.; Armando, A.; Lucarelli, R.; Gorrochotegui-Escalante, N.; Diercks, A.H.; Quehenberger, O.; Dennis, E.A.; Aderem, A.; et al. PPARalpha exacerbates necroptosis, leading to increased mortality in postinfluenza bacterial superinfection. *Proc. Natl. Acad. Sci. USA* **2020**, *117*, 15789–15798. [CrossRef] [PubMed]
123. Ng, V.Y.; Huang, Y.; Reddy, L.M.; Falck, J.R.; Lin, E.T.; Kroetz, D.L. Cytochrome P450 eicosanoids are activators of peroxisome proliferator-activated receptor alpha. *Drug Metab. Dispos.* **2007**, *35*, 1126–1134. [CrossRef]
124. Yang, M.; Wang, Y.; Chen, J.; Wang, Q.; Wei, S.; Wang, S.; Qin, Q. Functional analysis of Epinephelus coioides peroxisome proliferative-activated receptor alpha (PPARalpha): Involvement in response to viral infection. *Fish. Shellfish Immunol.* **2020**, *102*, 257–266. [CrossRef]
125. Ehrlich, A.; Uhl, S.; Ioannidis, K.; Hofree, M.; tenOever, B.R.; Nahmias, Y. The SARS-CoV-2 Transcriptional Metabolic Signature in Lung Epithelium. Available online: https://ssrn.com/abstract=3650499 (accessed on 28 September 2021).
126. Heffernan, K.S.; Ranadive, S.M.; Jae, S.Y. Exercise as medicine for COVID-19: On PPAR with emerging pharmacotherapy. *Med. Hypotheses* **2020**, *143*, 110197. [CrossRef]
127. Piomelli, D.; Hohmann, A.G.; Seybold, V.; Hammock, B.D. A lipid gate for the peripheral control of pain. *J. Neurosci.* **2014**, *34*, 15184–15191. [CrossRef]
128. Chiurchiu, V.; Battistini, L.; Maccarrone, M. Endocannabinoid signalling in innate and adaptive immunity. *Immunology* **2015**, *144*, 352–364. [CrossRef]

129. Guindon, J.; Hohmann, A.G. The endocannabinoid system and pain. *CNS Neurol. Disord. Drug Targets* **2009**, *8*, 403–421. [CrossRef] [PubMed]
130. Bisogno, T.; Ventriglia, M.; Milone, A.; Mosca, M.; Cimino, G.; Di Marzo, V. Occurrence and metabolism of anandamide and related acyl-ethanolamides in ovaries of the sea urchin *Paracentrotus lividus*. *Biochim. Biophys. Acta* **1997**, *1345*, 338–348. [CrossRef]
131. Sepe, N.; De Petrocellis, L.; Montanaro, F.; Cimino, G.; Di Marzo, V. Bioactive long chain N-acylethanolamines in five species of edible bivalve molluscs. Possible implications for mollusc physiology and sea food industry. *Biochim. Biophys. Acta* **1998**, *1389*, 101–111. [CrossRef]
132. Matias, I.; Bisogno, T.; Melck, D.; Vandenbulcke, F.; Verger-Bocquet, M.; De Petrocellis, L.; Sergheraert, C.; Breton, C.; Di Marzo, V.; Salzet, M. Evidence for an endocannabinoid system in the central nervous system of the leech Hirudo medicinalis. *Brain Res. Mol. Brain Res.* **2001**, *87*, 145–159. [CrossRef]
133. Fu, J.; Gaetani, S.; Oveisi, F.; Lo Verme, J.; Serrano, A.; Rodriguez De Fonseca, F.; Rosengarth, A.; Luecke, H.; Di Giacomo, B.; Tarzia, G.; et al. Oleylethanolamide regulates feeding and body weight through activation of the nuclear receptor PPAR-alpha. *Nature* **2003**, *425*, 90–93. [CrossRef]
134. Bradshaw, H.B.; Walker, J.M. The expanding field of cannabimimetic and related lipid mediators. *Br. J. Pharmacol.* **2005**, *144*, 459–465. [CrossRef]
135. Oddi, S.; Fezza, F.; Pasquariello, N.; De Simone, C.; Rapino, C.; Dainese, E.; Finazzi-Agro, A.; Maccarrone, M. Evidence for the intracellular accumulation of anandamide in adiposomes. *Cell Mol. Life Sci.* **2008**, *65*, 840–850. [CrossRef]
136. Kaczocha, M.; Glaser, S.T.; Chae, J.; Brown, D.A.; Deutsch, D.G. Lipid droplets are novel sites of N-acylethanolamine inactivation by fatty acid amide hydrolase-2. *J. Biol. Chem.* **2010**, *285*, 2796–2806. [CrossRef] [PubMed]
137. Bisogno, T.; Maurelli, S.; Melck, D.; De Petrocellis, L.; Di Marzo, V. Biosynthesis, uptake, and degradation of anandamide and palmitoylethanolamide in leukocytes. *J. Biol. Chem.* **1997**, *272*, 3315–3323. [CrossRef] [PubMed]
138. Tsuboi, K.; Ikematsu, N.; Uyama, T.; Deutsch, D.G.; Tokumura, A.; Ueda, N. Biosynthetic pathways of bioactive N-acylethanolamines in brain. *CNS Neurol. Disord. Drug Targets* **2013**, *12*, 7–16. [CrossRef]
139. Fegley, D.; Gaetani, S.; Duranti, A.; Tontini, A.; Mor, M.; Tarzia, G.; Piomelli, D. Characterization of the fatty acid amide hydrolase inhibitor cyclohexyl carbamic acid 3′-carbamoyl-biphenyl-3-yl ester (URB597): Effects on anandamide and oleoylethanolamide deactivation. *J. Pharmacol. Exp. Ther.* **2005**, *313*, 352–358. [CrossRef]
140. Ogura, Y.; Parsons, W.H.; Kamat, S.S.; Cravatt, B.F. A calcium-dependent acyltransferase that produces N-acyl phosphatidylethanolamines. *Nat. Chem. Biol.* **2016**, *12*, 669–671. [CrossRef]
141. Ueda, N.; Tsuboi, K.; Uyama, T. Metabolism of endocannabinoids and related N-acylethanolamines: Canonical and alternative pathways. *FEBS J.* **2013**, *280*, 1874–1894. [CrossRef] [PubMed]
142. LoVerme, J.; Russo, R.; La Rana, G.; Fu, J.; Farthing, J.; Mattace-Raso, G.; Meli, R.; Hohmann, A.; Calignano, A.; Piomelli, D. Rapid broad-spectrum analgesia through activation of peroxisome proliferator-activated receptor-alpha. *J. Pharmacol. Exp. Ther.* **2006**, *319*, 1051–1061. [CrossRef]
143. Suardiaz, M.; Estivill-Torrus, G.; Goicoechea, C.; Bilbao, A.; de Fonseca, R.F. Analgesic properties of oleoylethanolamide (OEA) in visceral and inflammatory pain. *Pain* **2007**, *133*, 99–110. [CrossRef] [PubMed]
144. Taylor, B.K.; Dadia, N.; Yang, C.B.; Krishnan, S.; Badr, M. Peroxisome proliferator-activated receptor agonists inhibit inflammatory edema and hyperalgesia. *Inflammation* **2002**, *26*, 121–127. [CrossRef]
145. Russo, R.; LoVerme, J.; La Rana, G.; D'Agostino, G.; Sasso, O.; Calignano, A.; Piomelli, D. Synergistic antinociception by the cannabinoid receptor agonist anandamide and the PPAR-alpha receptor agonist GW7647. *Eur. J. Pharmacol.* **2007**, *566*, 117–119. [CrossRef]
146. D'Agostino, G.; La Rana, G.; Russo, R.; Sasso, O.; Iacono, A.; Esposito, E.; Raso, G.M.; Cuzzocrea, S.; Lo Verme, J.; Piomelli, D.; et al. Acute intracerebroventricular administration of palmitoylethanolamide, an endogenous peroxisome proliferator-activated receptor-alpha agonist, modulates carrageenan-induced paw edema in mice. *J. Pharmacol. Exp. Ther.* **2007**, *322*, 1137–1143. [CrossRef] [PubMed]
147. Russo, R.; Loverme, J.; La Rana, G.; Compton, T.R.; Parrott, J.; Duranti, A.; Tontini, A.; Mor, M.; Tarzia, G.; Calignano, A.; et al. The fatty acid amide hydrolase inhibitor URB597 (cyclohexylcarbamic acid 3′-carbamoylbiphenyl-3-yl ester) reduces neuropathic pain after oral administration in mice. *J. Pharmacol. Exp. Ther.* **2007**, *322*, 236–242. [CrossRef]
148. Ben-Shabat, S.; Fride, E.; Sheskin, T.; Tamiri, T.; Rhee, M.H.; Vogel, Z.; Bisogno, T.; De Petrocellis, L.; Di Marzo, V.; Mechoulam, R. An entourage effect: Inactive endogenous fatty acid glycerol esters enhance 2-arachidonoyl-glycerol cannabinoid activity. *Eur. J. Pharmacol.* **1998**, *353*, 23–31. [CrossRef]
149. Ho, W.S.; Barrett, D.A.; Randall, M.D. 'Entourage' effects of N-palmitoylethanolamide and N-oleoylethanolamide on vasorelaxation to anandamide occur through TRPV1 receptors. *Br. J. Pharmacol.* **2008**, *155*, 837–846. [CrossRef]
150. Steardo, L., Jr.; Bronzuoli, M.R.; Iacomino, A.; Esposito, G.; Steardo, L.; Scuderi, C. Does neuroinflammation turn on the flame in Alzheimer's disease? Focus on astrocytes. *Front. Neurosci.* **2015**, *9*, 259. [CrossRef]
151. Skaper, S.D.; Facci, L.; Giusti, P. Glia and mast cells as targets for palmitoylethanolamide, an anti-inflammatory and neuroprotective lipid mediator. *Mol. Neurobiol.* **2013**, *48*, 340–352. [CrossRef] [PubMed]
152. Amor, S.; Puentes, F.; Baker, D.; van der Valk, P. Inflammation in neurodegenerative diseases. *Immunology* **2010**, *129*, 154–169. [CrossRef] [PubMed]
153. Colombo, E.; Farina, C. Astrocytes: Key regulators of neuroinflammation. *Trends Immunol.* **2016**, *37*, 608–620. [CrossRef] [PubMed]

154. Aloe, L.; Leon, A.; Levi-Montalcini, R. A proposed autacoid mechanism controlling mastocyte behaviour. *Agents Actions* **1993**, *39*, C145–C147. [CrossRef]
155. Benito, C.; Tolon, R.M.; Castillo, A.I.; Ruiz-Valdepenas, L.; Martinez-Orgado, J.A.; Fernandez-Sanchez, F.J.; Vazquez, C.; Cravatt, B.F.; Romero, J. beta-Amyloid exacerbates inflammation in astrocytes lacking fatty acid amide hydrolase through a mechanism involving PPAR-alpha, PPAR-gamma and TRPV1, but not CB(1) or CB(2) receptors. *Br. J. Pharmacol.* **2012**, *166*, 1474–1489. [CrossRef]
156. Hohmann, U.; Pelzer, M.; Kleine, J.; Hohmann, T.; Ghadban, C.; Dehghani, F. Opposite effects of neuroprotective cannabinoids, palmitoylethanolamide, and 2-arachidonoylglycerol on function and morphology of microglia. *Front. Neurosci.* **2019**, *13*, 1180. [CrossRef] [PubMed]
157. Bronzuoli, M.R.; Facchinetti, R.; Steardo, L., Jr.; Romano, A.; Stecca, C.; Passarella, S.; Steardo, L.; Cassano, T.; Scuderi, C. Palmitoylethanolamide dampens reactive astrogliosis and improves neuronal trophic support in a triple transgenic model of Alzheimer's disease: In vitro and in vivo evidence. *Oxid. Med. Cell. Longev.* **2018**, *2018*, 4720532. [CrossRef]
158. Scuderi, C.; Stecca, C.; Valenza, M.; Ratano, P.; Bronzuoli, M.R.; Bartoli, S.; Steardo, L.; Pompili, E.; Fumagalli, L.; Campolongo, P.; et al. Palmitoylethanolamide controls reactive gliosis and exerts neuroprotective functions in a rat model of Alzheimer's disease. *Cell Death Dis.* **2014**, *5*, e1419. [CrossRef] [PubMed]
159. Scuderi, C.; Valenza, M.; Stecca, C.; Esposito, G.; Carratu, M.R.; Steardo, L. Palmitoylethanolamide exerts neuroprotective effects in mixed neuroglial cultures and organotypic hippocampal slices via peroxisome proliferator-activated receptor-alpha. *J. Neuroinflamm.* **2012**, *9*, 49. [CrossRef]
160. Guida, F.; Luongo, L.; Boccella, S.; Giordano, M.E.; Romano, R.; Bellini, G.; Manzo, I.; Furiano, A.; Rizzo, A.; Imperatore, R.; et al. Palmitoylethanolamide induces microglia changes associated with increased migration and phagocytic activity: Involvement of the CB2 receptor. *Sci. Rep.* **2017**, *7*, 375. [CrossRef]
161. Rinne, P.; Guillamat-Prats, R.; Rami, M.; Bindila, L.; Ring, L.; Lyytikainen, L.P.; Raitoharju, E.; Oksala, N.; Lehtimaki, T.; Weber, C.; et al. Palmitoylethanolamide promotes a proresolving macrophage phenotype and attenuates atherosclerotic plaque formation. *Arterioscler. Thromb. Vasc. Biol.* **2018**, *38*, 2562–2575. [CrossRef]
162. Luo, D.; Zhang, Y.; Yuan, X.; Pan, Y.; Yang, L.; Zhao, Y.; Zhuo, R.; Chen, C.; Peng, L.; Li, W.; et al. Oleoylethanolamide inhibits glial activation via moudulating PPARalpha and promotes motor function recovery after brain ischemia. *Pharmacol. Res.* **2019**, *141*, 530–540. [CrossRef]
163. Xu, X.; Guo, H.; Jing, Z.; Yang, L.; Chen, C.; Peng, L.; Wang, X.; Yan, L.; Ye, R.; Jin, X.; et al. N-oleoylethanolamine reduces inflammatory cytokines and adhesion molecules in TNF-alpha-induced human umbilical vein endothelial cells by activating CB2 and PPAR-alpha. *J. Cardiovasc. Pharmacol.* **2016**, *68*, 280–291. [CrossRef]
164. Holubiec, M.I.; Romero, J.I.; Suarez, J.; Portavella, M.; Fernandez-Espejo, E.; Blanco, E.; Galeano, P.; de Fonseca, F.R. Palmitoylethanolamide prevents neuroinflammation, reduces astrogliosis and preserves recognition and spatial memory following induction of neonatal anoxia-ischemia. *Psychopharmacology* **2018**, *235*, 2929–2945. [CrossRef] [PubMed]
165. Yang, L.C.; Guo, H.; Zhou, H.; Suo, D.Q.; Li, W.J.; Zhou, Y.; Zhao, Y.; Yang, W.S.; Jin, X. Chronic oleoylethanolamide treatment improves spatial cognitive deficits through enhancing hippocampal neurogenesis after transient focal cerebral ischemia. *Biochem. Pharmacol.* **2015**, *94*, 270–281. [CrossRef]
166. Flannery, L.E.; Kerr, D.M.; Hughes, E.M.; Kelly, C.; Costello, J.; Thornton, A.M.; Humphrey, R.M.; Finn, D.P.; Roche, M. N-acylethanolamine regulation of TLR3-induced hyperthermia and neuroinflammatory gene expression: A role for PPARalpha. *J. Neuroimmunol.* **2021**, *358*, 577654. [CrossRef] [PubMed]
167. Taquet, M.; Geddes, J.R.; Husain, M.; Luciano, S.; Harrison, P.J. 6-month neurological and psychiatric outcomes in 236 379 survivors of COVID-19: A retrospective cohort study using electronic health records. *Lancet Psychiatry* **2021**, *8*, 416–427. [CrossRef]
168. Vaia, M.; Petrosino, S.; De Filippis, D.; Negro, L.; Guarino, A.; Carnuccio, R.; Di Marzo, V.; Iuvone, T. Palmitoylethanolamide reduces inflammation and itch in a mouse model of contact allergic dermatitis. *Eur. J. Pharmacol.* **2016**, *791*, 669–674. [CrossRef] [PubMed]
169. Skaper, S.D.; Facci, L.; Romanello, S.; Leon, A. Mast cell activation causes delayed neurodegeneration in mixed hippocampal cultures via the nitric oxide pathway. *J. Neurochem.* **1996**, *66*, 1157–1166. [CrossRef]
170. Facci, L.; Dal Toso, R.; Romanello, S.; Buriani, A.; Skaper, S.D.; Leon, A. Mast cells express a peripheral cannabinoid receptor with differential sensitivity to anandamide and palmitoylethanolamide. *Proc. Natl. Acad. Sci. USA* **1995**, *92*, 3376–3380. [CrossRef]
171. Di Paola, R.; Impellizzeri, D.; Torre, A.; Mazzon, E.; Cappellani, A.; Faggio, C.; Esposito, E.; Trischitta, F.; Cuzzocrea, S. Effects of palmitoylethanolamide on intestinal injury and inflammation caused by ischemia-reperfusion in mice. *J. Leukoc. Biol.* **2012**, *91*, 911–920. [CrossRef] [PubMed]
172. Borrelli, F.; Romano, B.; Petrosino, S.; Pagano, E.; Capasso, R.; Coppola, D.; Battista, G.; Orlando, P.; Di Marzo, V.; Izzo, A.A. Palmitoylethanolamide, a naturally occurring lipid, is an orally effective intestinal anti-inflammatory agent. *Br. J. Pharmacol.* **2015**, *172*, 142–158. [CrossRef]
173. Misto, A.; Provensi, G.; Vozella, V.; Passani, M.B.; Piomelli, D. Mast cell-derived histamine regulates liver ketogenesis via oleoylethanolamide signaling. *Cell Metab.* **2019**, *29*, 91–102e5. [CrossRef] [PubMed]
174. Fu, J.; Kim, J.; Oveisi, F.; Astarita, G.; Piomelli, D. Targeted enhancement of oleoylethanolamide production in proximal small intestine induces across-meal satiety in rats. *Am. J. Physiol. Regul. Integr. Comp. Physiol.* **2008**, *295*, R45–R50. [CrossRef]

175. Golderer, G.; Werner, E.R.; Leitner, S.; Grobner, P.; Werner-Felmayer, G. Nitric oxide synthase is induced in sporulation of Physarum polycephalum. *Genes Dev.* **2001**, *15*, 1299–1309. [CrossRef]
176. Hall, D.A.; Vander Kooi, C.W.; Stasik, C.N.; Stevens, S.Y.; Zuiderweg, E.R.; Matthews, R.G. Mapping the interactions between flavodoxin and its physiological partners flavodoxin reductase and cobalamin-dependent methionine synthase. *Proc. Natl. Acad. Sci. USA* **2001**, *98*, 9521–9526. [CrossRef] [PubMed]
177. Radomski, M.W.; Palmer, R.M.; Moncada, S. Modulation of platelet aggregation by an L-arginine-nitric oxide pathway. *Trends Pharmacol. Sci.* **1991**, *12*, 87–88. [CrossRef]
178. Escriva, H.; Safi, R.; Hanni, C.; Langlois, M.C.; Saumitou-Laprade, P.; Stehelin, D.; Capron, A.; Pierce, R.; Laudet, V. Ligand binding was acquired during evolution of nuclear receptors. *Proc. Natl. Acad. Sci. USA* **1997**, *94*, 6803–6808. [CrossRef]
179. Zhou, T.; Yan, X.; Wang, G.; Liu, H.; Gan, X.; Zhang, T.; Wang, J.; Li, L. Evolutionary pattern and regulation analysis to support why diversity functions existed within PPAR gene family members. *Biomed. Res. Int.* **2015**, *2015*, 613910. [CrossRef]
180. Crawford, N.M. Mechanisms for nitric oxide synthesis in plants. *J. Exp. Bot.* **2006**, *57*, 471–478. [CrossRef]
181. Dohrmann, M.; Worheide, G. Dating early animal evolution using phylogenomic data. *Sci. Rep.* **2017**, *7*, 3599. [CrossRef]

International Journal of *Molecular Sciences*

MDPI

Article

Analysis of *PPARγ* Signaling Activity in Psoriasis

Vladimir Sobolev [1,2,*], Anastasia Nesterova [3], Anna Soboleva [2,4], Alexandre Mezentsev [1,2], Evgenia Dvoriankova [2], Anastas Piruzyan [2], Elena Denisova [2], Olga Melnichenko [5] and Irina Korsunskaya [2]

1 I. Mechnikov Research Institute for Vaccines and Sera RAMS, Russian Federation, Malyy Kazennyy Lane, 5, 105064 Moscow, Russia; mesentsev@yahoo.com

2 Centre of Theoretical Problems of Physico-Chemical Pharmacology, Russian Academy of Sciences, Russian Federation, Srednyaya Kalitnikovskaya Street, 30, 109029 Moscow, Russia; annasobo@mail.ru (A.S.); dvoriankova@mail.ru (E.D.); pirstas2000@hotmail.com (A.P.); evdenissova@rambler.ru (E.D.); marykor@bk.ru (I.K.)

3 Life Science Research and Development Department, Elsevier Inc., Rockville, MD 20850, USA; nesterova.anastasia@gmail.com

4 Scientific Research Institute of Human Morphology, 3 Tsurupa Street, 117418 Moscow, Russia

5 Moscow Scientific and Practical Center of Dermatovenereology and Cosmetology, Russian Federation, 17, Leninskiy Avenue, 119071 Moscow, Russia; dr.melnichenko@gmail.com

* Correspondence: vlsobolew@gmail.com

Abstract: In our previous work, we built the model of *PPARγ* dependent pathways involved in the development of the psoriatic lesions. Peroxisome proliferator-activated receptor gamma (*PPARγ*) is a nuclear receptor and transcription factor which regulates the expression of many proinflammatory genes. We tested the hypothesis that low levels of *PPARγ* expression promote the development of psoriatic lesions triggering the *IL17*-related signaling cascade. Skin samples of normally looking and lesional skin donated by psoriasis patients and psoriatic $CD3^+$ Tcells samples ($n = 23$) and samples of healthy $CD3^+$ T cells donated by volunteers ($n = 10$) were analyzed by real-time PCR, ELISA and immunohistochemistry analysis. We found that the expression of *PPARγ* is downregulated in human psoriatic skin and laser treatment restores the expression. The expression of *IL17*, *STAT3*, *FOXP3*, and *RORC* in psoriatic skin before and after laser treatment were correlated with *PPARγ* expression according to the reconstructed model of *PPARγ* pathway in psoriasis.In conclusion, we report that *PPARγ* weakens the expression of genes that contribute in the development of psoriatic lesion. Our data show that transcriptional regulation of *PPARγ* expression by *FOSL1* and by *STAT3/FOSL1* feedback loop may be central in the psoriatic skin and T-cells.

Keywords: psoriasis; peroxisome proliferator-activated receptor gamma (*PPARγ*); real-time PCR; ELISA; immunohistochemistry; signaling pathway

Citation: Sobolev, V.; Nesterova, A.; Soboleva, A.; Mezentsev, A.; Dvoriankova, E.; Piruzyan, A.; Denisova, E.; Melnichenko, O.; Korsunskaya, I. Analysis of *PPARγ* Signaling Activity in Psoriasis. *Int. J. Mol. Sci.* **2021**, *22*, 8603. https://doi.org/10.3390/ijms22168603

Academic Editors: Manuel Vázquez-Carrera and Walter Wahli

Received: 12 July 2021
Accepted: 3 August 2021
Published: 10 August 2021

1. Introduction

Peroxisome proliferator-activated receptors (*PPARs*) form a group of nucleus receptors that play an important role in the mammalian physiological system and function as a transcription factor [1]. There are three known *PPAR* isoforms, *PPARα*, *PPARβ/δ*, and *PPARγ*, which have significant sequence and structure homology, but exhibit different tissue distribution, selectivity, and sensitivity to ligands, which leads to the regulation of different gene sets by different receptors [2,3].

After binding to the ligand, *PPARs* form a heterodimer with the liver X receptor (LXR), then heterodimerize with the retinoid X receptor (RXR) and bind to the peroxisome proliferator response elements (PPRE) in the promoter regions of target genes [4,5].

PPARγ is the most studied *PPAR* subtype, which is expressed predominantly in the heart, adipose tissue, colon, kidneys, spleen, intestine, skeletal muscle, liver, macrophages, and skin. In the skin, *PPARγ* controls the genetic regulation of gene network expression involved in cell proliferation, differentiation, and inflammatory response [6].

There is an increased expression of *PPARγ* in skin adipocytes, where it plays a critical role in their differentiation [7,8]. *PPARγ* also has an important functional role in the regulation of skin barrier permeability as an inhibitor of keratinocyte cell proliferation and a promoter of terminal differentiation of the epidermis. In addition, being an important regulator of lipid metabolism, it stimulates the production of cholesterol and ceramides in keratinocytes [1,9].

As far as psoriasis is an inflammatory skin disease characterized by epidermal hyperproliferation and abnormal keratinocyte differentiation, proteins involved in *PPARγ* signaling can be considered as potential targets for treatment. Specific *PPARγ* ligands (such as BRL49653/rosiglitazone or pioglitazone) have been shown to inhibit the production of many inflammatory mediators and cytokines in various cell types, including monocytes, lymphocytes, and epithelial cells [10,11]. Studies in a mouse model of hyperproliferative skin disease have shown that local administration of *PPARγ* ligands thiazolidinediones family (ciglitazone and troglitazone) reduces epidermal hyperplasia [12].

Therefore, *PPARγ* can impede the progress of psoriasis, downregulating the expression of proinflammatory genes in a ligand-dependent manner, counteracting the activity of transcription factors.

Previously, we reconstructed several pathway models of molecular mechanisms of psoriasis. Models describe the transition to TH17 cell signaling during the differentiation of psoriatic T cells. In summary, genetic mutations in interleukin receptor (*IL23R*) may cause shift to the TH17 cells production which results in elevated levels of *IL17* and *IL22* expression, which, in turn, activates keratinocytes to release different cytokines and chemokines for attracting neutrophils and other inflammatory cells in the psoriatic lesion [13,14]. In the last work we build the model that describes a hypothesis that low activity of *PPARγ* signaling may promote psoriasis. We applied network analysis to build the model and we used public microarrays data to find statistically significant molecular cascades, cell processes, molecular regulators and expression targets of *PPARγ* [15] (see Supplementary Materials).

In this work, to test the hypothesis of low activity of *PPARγ* signaling in psoriasis, we measured gene expression of *PPARγ* and several key members of the reconstructed model in skin samples and in $CD3^+$ T cells from patients with psoriasis. Additionally, we tested the expression of *PPARγ* signaling in human psoriatic skin before and after laser treatment.

2. Materials and Methods

2.1. Patients and Samples

We analyzed biopsies and peripheral blood samples from patients who were treated in the V G Korolenko Hospital, Moscow Scientific and Practical Centre of Dermatovenerology and Cosmetology. Total were analyzed from 23 patients with plaque-type psoriasis and 10 healthy controls. The age of patients varied from 25 to 56 years (Table 1). Patients were diagnosed with *Psoriasis vulgaris*. The diagnoses were confirmed by the pathomorphological examination of skin biopsies.

Table 1. Clinical parameters of patients with psoriasis (*Psoriasis vulgaris*).

Sex, *n* (%)	Age	PASI	Disease History in Years
Patients			
M/F (*n* = 23)	43.5 ± 8.8	22 ± 6.2	17.4 ± 5.7
M, 10 (43.5%)	42.9 ± 9.9	18.4 ± 5.5	15.9 ± 4.7
F, 13 (56.5%)	44.9 ± 9.2	24.8 ± 4.2	17.6 ± 6.7
Healthy volunteers			
M/F (*n* = 10)	40.9 ± 9	*n*/a	*n*/a
M, 4 (40%)	40.2 ± 7.3	*n*/a	*n*/a
F, 6 (60%)	41.3 ± 10.6	*n*/a	*n*/a

Local anesthesia and dermatological punch (4 mm) were used for the collection of skin samples. Healthy skin samples were taken at a distance of 3 cm from a psoriatic lesion. The research was approved by the Local Ethical Committee at the Center for Theoretical Problems of Physical-Chemical Pharmacology, Russian Academy of Science, and complies with the principles of the Helsinki Declaration. The laser treatment was provided 2–3 times a week. Skin samples were collected before the treatment and one day after the 7th laser seance.

2.2. Cells Isolation

For peripheral blood mononuclear cells (PBMC) isolation from the whole blood density gradient centrifugation was performed. Ficoll isolation method promoted cell extraction. 7 mL of Ficoll solution (density 1.077 g/cm^3, "DIA-M") was placed into a 15 mL Eppendorf conical tube and then carefully overlaid with 7 mL of the whole blood. After that the tube was centrifuged for 25 min at 1200× g at 4 °C. The interphase containing the cellular layer was collected from the tube and placed into a new 15-mL tube for further washing procedure. 15 mL of DPBS buffer (10× without Ca and Mg, with 0.5%Tween 20, pH 7.4) were added to the cell pellet and then centrifuged for 15 min at 400× g at 20 °C. The supernatant was carefully removed and the wash was repeated once with the only difference of the DPBS buffer volume (10 mL). After the last centrifugation and 500 μL of culture media (RPMI) addition, cell count and viability assessment were performed.

Isolation of total CD3$^+$ T cells were obtained from PBMCs of patients and controls using a total CD3$^+$ T cell isolation kit (Miltenyl Biotec, Bergisch Gladbach, Germany).

For better presentation we summarized all methods in one scheme (Figure 1).

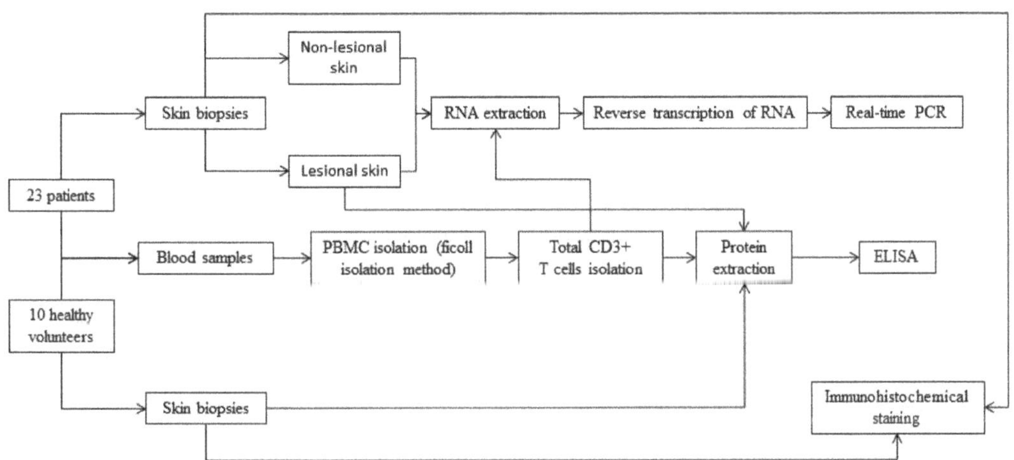

Figure 1. Scheme of experiment procedure.

2.3. PCR

Qiagen spin column and standard RNeasy Mini Kit® for the skin were used for the RNA isolation. Additional treatment of samples with the DNAase (Qiagen®, Germantown, MD, USA) was used to remove DNA traces. RNA concentration was measured with NanoDrop 1000 (Thermo Scientific®, Waltham, MA, USA).

Reverse transcription was done in 200 μL volume; the mixture included the buffer, dNTP, 100 units of reverse transcriptase (M_MLV, Promega®, Madison, WI, USA), 20 units of RNAses inhibitor (RNasin, Promega®, Madison, WI, USA), 500 ng of oligo(dT) primers (DNA-Synthes®, Moscow, Russia), and RNA sample (no more than 100 ng/μL). The mixture was incubated at 37 °C for 1 h.

Real-time PCR was performed in 96-well optical plates using fluorescent dyes SYBR Green (Eurogen®) and custom primers (DNA-Synthesis®). Primer sequences: *PPAR-γ* F: 5′-TCTGGCCCACCAACTTTGGG-3′ R: 5′-CTTCACAAGCATGAACTCCA-3′; *STAT3* F: 5′-ACCAGCAGTATAGCCGCTTC-3′ R: 5′-GCCACAATCCGGGCAATCT-3′; *IL17A* F: 5′-ACAACCGATCCACCTCACCTT-3′ R: 5′- CTTTGCCTCCCAGATCACAGA-3′; *RORC* F: 5′- GTAGAACAGCTGCAGTACAATC-3′ R: 5′-CTTCCAGGTCACTTGGAC-3′; *FOXP3* F: 5′-TCCCAGAGTTCCTCCACAAC-3′ R: 5′-ATTGAGTGTCCGCTGCTTCT-3′. PCR amplifier (Bio-Rad, CFX96™) was used for the amplification with the following program: (1) denaturation at 95 °C for 4 min, (2) denaturation at 94 °C for 15 s, (3–4) annealing and elongation at 60 °C for 30 s, (5) steps 2–4 were repeated 40 times. Levels of the GAPDH gene were used as a control for the expression of targeted genes. Amplification of the GAPDH gene and the studied genes was performed in different test tubes.

To calculate the results, we used numbers from real-time PCR reactions with primer efficiency at least 95%, 0.99 correlation coefficient and the curve (slope) -3.4 ± 0.2. PCR results were analyzed using the $2^{-\Delta\Delta CT}$ method to compare the levels of expressions detected in affected and unaffected samples [16]. Each ΔCt was calculated as ΔCt = Ct (tested gene)−Ct (GAPDH). ΔΔCt was calculated as ΔΔCt = ΔCt (psoriatic skin sample)—ΔCt (health skin sample). The experiments were repeated three times for each sample. Intergroup differences were calculated using the Mann-Whitney U-test.

2.4. ELISA

Human *PPAR-γ* (Peroxisome Proliferator Activated Receptor Gamma) (MBS2503174), Signal Transducer and Activator of Transcription 3 (*STAT3*) (MBS2024094) and Interleukin 17 (*IL17*) (MBS2019491) ELISA Kits (MyBioSource, Inc., San Diego, CA, USA) was applied to detect the *PPARγ* levels in lesional and healthy skin according to the manufacturer's protocol. Briefly, standards and tested samples prepared in assay buffer were loaded on 96-well plate and incubated with immobilized specific antibody for 1 h at 37 °C. After washing with provided solution, the specific antibody conjugated with HRP-streptavidin was added and incubation continued for another 30 min. Then, the presence of antigen was visualized with chromogenic substrate (TMB) and assayed using a microplate reader (RT-2100C, Rayto) at wavelength 450 nm. The antigen was quantified with a standard curve generated with standards of known concentration. The standard curve was constructed by plotting the mean absorbance obtained from each standard against its concentration. The calculation was done using a professional software "Curve Expert 1.4".

2.5. Immunohistochemistry Analysis

Preparation of a paraffin block. To prepare skin micro-sections, a tissue samples up to 5 mm in size was fixed on a substrate to prevent wrinkling. The tissue was fixed in 10% neutral buffered formalin for 24 h at room temperature. Then formalin was washed out of the sample in running water for 6–7 h. Then the tissue was dehydrated in ethyl alcohols of ascending density: 80%-24 h, 96%-24 h, 100%-4 h. To prepare the paraffin block, the sample was kept in a 50/50 ethanol/toluene solution for 40 min at room temperature. Then the skin was kept in 100% toluene for 1 h. The tissue was also kept for 1 day in a 50/50 paraffin/toluene solution at 56 °C for successful penetration of paraffin into the sample. After that, the sample was placed in melted paraffin and kept for 2 days. A paraffin block with the skin sample enclosed in it was prepared with the use of a mold.

Staining of paraffin sections. Paraffin microsections of the human skin samples were obtained with the use of MC-2 sledge microtome. Microsections were placed on positively charged superfrost plus slides. The antigens were visualized by the NOVOLINK imaging system based on the unique compact polymer RE7290-K, designed to visualize mouse immunoglobulins M, G and rabbit immunoglobulins G of primary antibodies. For immunohistochemical staining the sections were dewaxed: toluene for 3 min, 96% ethanol for 3 min, 80% ethanol for 3 min, H_2O for 5 min. Triton X-100 was used to perform antigen unmasking procedure.

2.6. Data Analysis

Literature biomedical network Resnet-2020 and software Pathway Studio were used for enrichment analysis, network analysis and pathway models reconstruction (www.pathwaystudio.com). Resnet—2020 includes interactions between proteins, drugs, diseases, mutations, cells and other biomedical entities and is based on results of text-mining of 3.5 Mln full texts papers and 24 Mln abstracts.

3. Results

3.1. *PPARγ Expression Is Slightly Downregulated in Psoriatic Skin and CD3$^+$ T Cells*

For each of 23 patients, we compared the expression levels of *PPARγ* in the psoriatic skin samples and unlesional skin collected at the distance of 3 cm from the nearest psoriatic plaque. This was necessary to minimize the influence of disease-irrelevant factors on the molecular profile of selected genes [17].

The results of real-time PCR showed that *PPARγ* was downregulated in lesional skin compared to uninvolved skin. The expression level of *PPARγ* in lesional skin was slightly reduced in 1.41 ± 0.27 times (Figure 2). We also found a significant increase in the expression levels of the following genes—*IL17* (42.39 ± 16.68), *STAT3* (4.42 ± 0.90), *RORC* (7.68 ± 1.62, and *FOSL1* (9.72 ± 4.98). In contrast, the expression level of *FOXP3* was decreased in 1.72 ± 0.14 times.

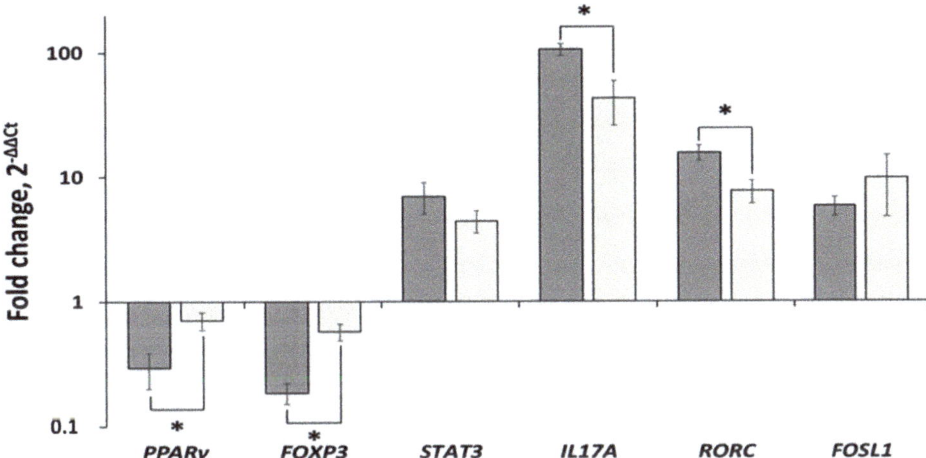

Figure 2. Comparative analysis of changes in the expression levels of *PPARγ*, *STAT3*, *IL17A*, *RORC*, *FOXP3* and *FOSL1* in lesional skin and CD3$^+$ T cells of psoriasis patients. Dark grey bars—lesional vs. uninvolved skin; Light grey bars—CD3$^+$ cells of psoriasis patients vs. same cells of healthy volunteers. The level of gene expression in control group was set to 1. Statistically significant changes in gene expression ($p < 0.05$) are marked with asterisk sign (*).

In the CD3$^+$ T-cells of psoriasis patients, the expression level of *PPARγ* was reduced in 3.4 ± 0.4 times and *FOXP3*—in 5.4 ± 0.16 times compared to healthy volunteers (Figure 2). Moreover, the following genes were upregulated in CD3$^+$ T cells of psoriasis patients—*IL17* (105.2 ± 11.01), *STAT3* (6.98 ± 1.96), *RORC* gene in 15.52 ± 2.18, and *FOSL1* (5.79 ± 0.99).

Since we proposed that the pathogenicity of downregulated *PPARγ*-downstream signalling is different in various types of cells, where this pathway was active, we compared the expression levels of *PPARγ* and the related genes in the CD3$^+$ T cells obtained from the blood of psoriasis patients and lesional skin with similar parameters in CD3$^+$ T cells of healthy volunteers and uninvolved skin, respectively. We found that the expression levels of *PPARγ* and *FOXP3* were decreased in psoriatic CD3$^+$ T cells compared to lesional skin of the same individuals. The observed changes were statistically significant ($p = 0.016$ and >0.001, respectively). In contrast, the expression levels of four other genes were increased.

The changes in the expression levels of *IL17A* and *RORC* were statistically significant ($p = 0.004$ and 0.033, respectively). In the same time, the changes in the expression levels of *STAT3* and *FOSL1* were not significantly different ($p = 0.410$ and 0.278, respectively).

Using an independent method of analysis, we confirmed the differential expression of *PPARγ*, *STAT3* and *IL17* on protein level. The results of ELISA experiments performed on the same group of skin samples (Figure 3, upper panel) discovered significantly higher expression levels of *STAT3* and *IL17* in lesional skin. The expression levels of *STAT3* were 7.91 ± 0.61 and 3.92 ± 0.70 ng/mL ($p < 0.001$) whereas the expression levels of *IL17* were 1.15 ± 0.06 and 0.09 ± 0.01 ng/mL ($p < 0.001$), respectively. In contrast, the expression of *PPARγ* was reduced in lesional skin, compared to healthy skin 2.06 ± 0.20 and 8.02 ± 0.79 ng/mL, $p < 0.001$).

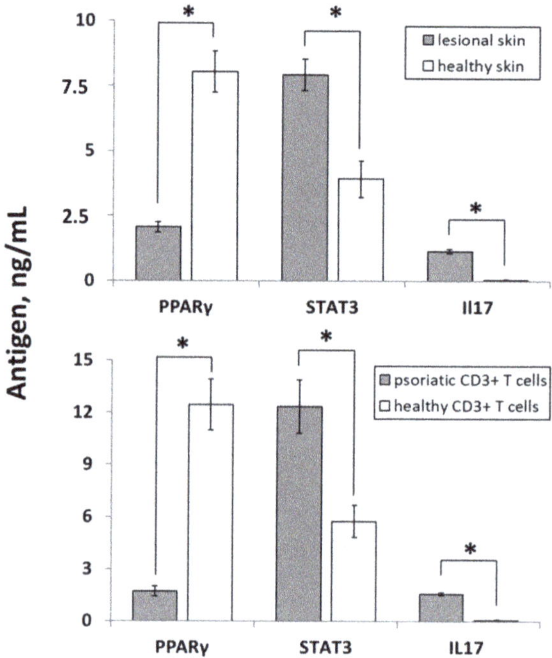

Figure 3. The expression levels of selected proteins in the samples obtained from psoriasis patients ($n = 23$) and healthy volunteers ($n = 10$) assessed by ELISA. The upper panel—lesional vs. healthy skin. The lower panel—samples of $CD3^+$ T cells of psoriasis patients and healthy volunteers. Statistically significant changes in gene expression ($p < 0.05$) are marked with asterisk sign (*).

Expectedly, a similar expression pattern was discovered in $CD3^+$ T-cells (Figure 3, lower panel). The expression levels of *STAT3* were 12.35 ± 1.53 and 5.75 ± 0.95 ng/mL ($p < 0.001$) whereas the expression levels of *IL17* were 1.57 ± 0.09 and 0.07 ± 0.02 ng/mL ($p < 0.001$) in psoriatic and healthy $CD3^+$ T-cells, respectively. In the same time, the expression of *PPARγ* was reduced in psoriatic $CD3^+$ T-cells, compared to same cells of healthy individuals (1.71 ± 0.29 and 12.46 ± 1.47 ng/mL, $p < 0.001$).

To reveal the differences in gene and protein expression between genders we compared the data obtained from male and female patients (Table 2), female patients with and without menopause (Table 3) as well as male and female healthy volunteers (Table 4). The following analysis did not reveal significant gender-specific changes with two exceptions. Firstly, female patients that did not experience menopause had a significantly higher expression of *FOSL1* in $CD3^+$ T-cells (Table 3). Secondarily, healthy female volunteers seemed to have a higher expression of *IL17A* compared to their male counterparts (Table 4).

Table 2. Comparative analysis of gene and protein expression in male (n = 10) and female (n = 13) patients.

Group	*FOSL1*	*PPARγ*	*FOXP3*	*STAT3*	*IL17A*	*RORC*
			(a) qPCR of skin samples, folds;			
males	7.36 ± 3.50	0.66 ± 0.06	0.61 ± 0.07	4.52 ± 0.66	31.27 ± 7.42	6.87 ± 1.26
females	12.29 ± 3.73	0.74 ± 0.10	0.56 ± 0.06	4.39 ± 0.69	47.93 ± 14.02	7.78 ± 1.02
p	0.36	0.52	0.59	0.90	0.35	0.58
			(b) qPCR of CD3⁺ cells, folds;			
males	5.80 ± 1.69	0.27 ± 0.15	0.22 ± 0.06	9.83 ± 3.94	123.99 ± 20.38	13.84 ± 3.65
females	5.79 ± 1.24	0.31 ± 0.13	0.16 ± 0.04	4.79 ± 1.62	90.82 ± 10.69	16.82 ± 2.73
p	1.00	0.86	0.38	0.21	0.14	0.51
			(c) ELISA of skin samples, ng/mL;			
males	N.D.	1.94 ± 0.30	N.D.	7.20 ± 1.01	1.23 ± 0.15	N.D.
females	N.D.	2.16 ± 0.29	N.D.	8.46 ± 0.78	1.08 ± 0.12	N.D.
p		0.61		0.33	0.45	
			(d) ELISA of CD3⁺, ng/mL;			
males	N.D.	2.05 ± 0.55	N.D.	14.33 ± 2.49	1.50 ± 0.13	N.D.
females	N.D.	1.46 ± 0.32	N.D.	10.82 ± 1.97	1.63 ± 0.13	N.D.
p		0.33		0.27	0.49	

Table 3. Comparative analysis of gene and protein expression in female patients have (n = 6) and do not have (n = 7) menopause.

Group	*FOSL1*	*PPARγ*	*FOXP3*	*STAT3*	*IL17A*	*RORC*
			(a) qPCR of skin samples, folds;			
no menopause	7.15 ± 1.39	0.76 ± 0.12	0.57 ± 0.10	4.79 ± 1.10	43.86 ± 10.54	8.71 ± 1.37
menopause	16.69 ± 7.52	0.73 ± 0.16	0.52 ± 0.07	3.83 ± 0.71	59.23 ± 28.35	7.84 ± 1.93
p	0.21	0.88	0.70	0.41	0.60	0.71
			(b) qPCR of CD3⁺ cells, folds;			
no menopause	8.04 ± 1.87	0.24 ± 0.14	0.14 ± 0.04	7.07 ± 2.77	82.59 ± 14.88	16.98 ± 4.68
menopause	3.17 ± 0.75	0.39 ± 0.24	0.18 ± 0.08	2.12 ± 0.60	100.43 ± 15.80	16.64 ± 2.85
p	**0.047**	0.59	0.61	0.13	0.43	0.95
			(c) ELISA of skin samples, ng/mL;			
no menopause	N.D.	2.45 ± 0.46	N.D.	7.69 ± 0.49	0.98 ± 0.16	N.D.
menopause	N.D.	1.81 ± 0.30	N.D.	9.36 ± 1.58	1.21 ± 0.18	N.D.
p		0.29		0.30	0.36	
			(d) ELISA of CD3⁺, ng/mL;			
no menopause	N.D.	1.00 ± 0.42	N.D.	11.41 ± 2.87	1.55 ± 0.21	N.D.
menopause	N.D.	1.05 ± 0.48	N.D.	10.13 ± 2.90	1.72 ± 0.15	N.D.
p		0.26		0.76	0.55	

Table 4. Comparative analysis of gene and protein expression in healthy male (n = 4) and female (n = 6) volunteers.

Group	*FOSL1*	*PPARγ*	*FOXP3*	*STAT3*	*IL17A*	*RORC*
			(a) qPCR of CD3⁺ cells, folds;			
males	1.23 ± 0.52	1.34 ± 0.65	1.26 ± 0.29	1.02 ± 0.55	0.91 ± 0.38	*0.47 ± 0.14*
females	0.85 ± 0.18	0.77 ± 0.13	0.83 ± 0.16	0.99 ± 0.24	1.06 ± 0.23	1.35 ± 0.46
p	0.44	0.41	0.20	0.96	0.73	0.17
			(b) ELISA of skin samples, ng/mL;			
males	N.D.	9.16 ± 1.64	N.D.	3.95 ± 0.73	0.04 ± 0.01	N.D.
females	N.D.	7.27 ± 0.77	N.D.	3.89 ± 1.15	0.07 ± 0.01	N.D.
p		0.27		0.97	**0.049**	
			(c) ELISA of CD3⁺, ng/mL;			
males	N.D.	13.93 ± 1.37	N.D.	6.45 ± 2.09	0.05 ± 0.02	N.D.
females	N.D.	11.47 ± 2.35	N.D.	5.28 ± 0.89	0.09 ± 0.03	N.D.
p		0.46		0.57	0.29	

However, we had several reasons to doubt the significance of these findings. Primarily, the differences reported in Tables 3 and 4 were not confirmed independently. In the first case, the significance of qPCR data was not confirmed by ELISA (Table 3). In the second case, the significance of the findings discovered by ELISA was not confirmed by qPCR (Table 4). Moreover, there we noticed a high data variability within the groups. As we believed, the patients' comorbidities and unreported health issues of volunteers might influence the gene and protein expression. We also have to acknowledge that levels of female sex hormones significantly vary on different stages of the menstrual cycle whereas we drew the blood a day prior discharging the patients and disregarded this matter when we tested healthy volunteers. In addition, the significance of the changes in the expression of *IL17A* could be questioned because of a relatively small sample size (Table 4). Thus, we suggest that there is no association between gene and protein expression and the participants' gender. As we believe, the obtained results do not support the hypothesis that gender could be a risk determinant of psoriasis.

Immunohistochemical skin section profile show that the accumulation of *IL17* is increased in the skin with the development of psoriatic plaque, as compared with the visually unaffected and healthy skin. Also, immunostaining of antibodies against *IL17* showed the staining of the keratinocyte cytoplasm mostly in the suprabasal epidermal layer. In the hyperplastic epidermis, the accumulation of *IL17* is more intense and heterogeneous. At the same time, *IL17* accumulates to a lesser extent in the visually unaffected skin and is only slightly detected in the healthy skin.

In the sections presented, a more intense *PPARy* immunostaining is observed in differentiated suprabasal keratinocytes of the unaffected skin, and less in the tissues of the psoriatic plaque, despite keratinocyte proliferation and hyperplasia development (Figure 4).

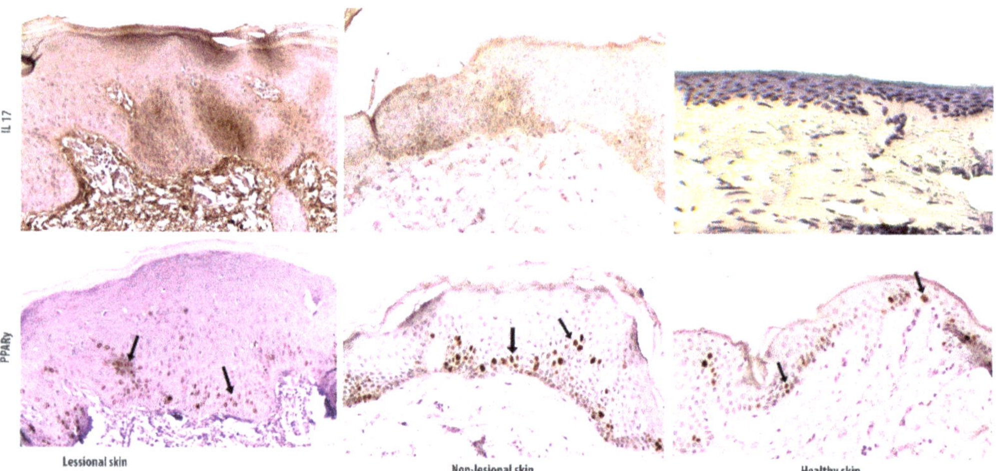

Figure 4. Immunohistochemical staining of antigens in the affected and non-affected psoriatic skin in comparison with the skin of healthy donors. The image was magnified 100×. Black arrows indicate *PPARy* accumulation in the suprabasal layer of epidermal keratinocytes.

3.2. Low Laser Treatment Stabilises PPARγ *Related Signaling in Psoriatic Skin*

For the next step of validation, we studied the expression of *PPARγ*, *STAT3*, *IL17A*, *RORC*, *FOXP3*, and *FOSL1* in human psoriatic skin samples and visually healthy skin samples before and after laser treatment. Patients received low-intensity laser treatment with 1.27 microns wavelength (infrared short waves). Similar to previously published results by different groups of medical researchers, the low-laser treatment had a positive effect on the health of observed patients and reduction of psoriatic skin inflammation was achieved (Figure 5).

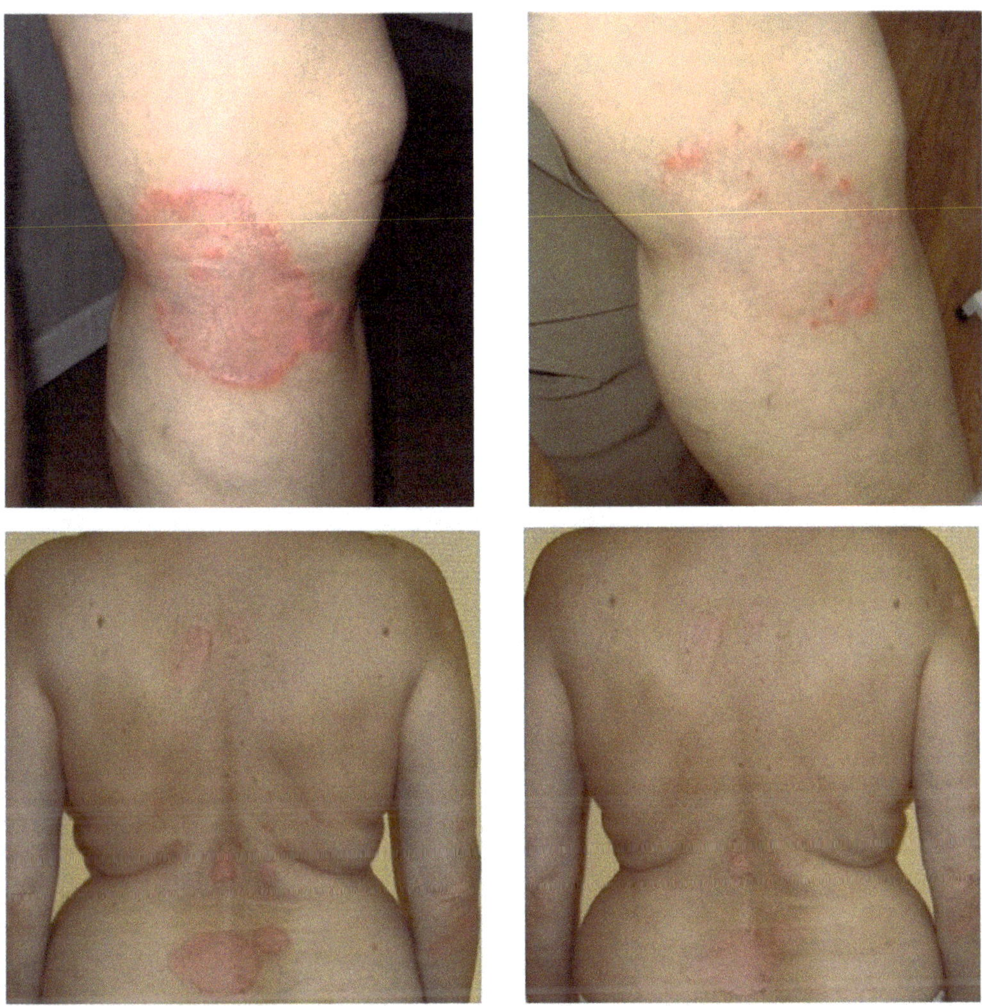

before low-level laser therapy *after low-level laser therapy*

Figure 5. Visual positive effect after low-level laser therapy. Reduction of psoriatic skin inflammation was achieved.

We detected a reliable reduction in the expression of studied *PPARγ*, *STAT3*, *IL17A*, *RORC*, *FOXP3* and *FOSL1* genes after low level (1.27 microns) laser treatment. The level of *STAT3* expression was decreased in 2.08 ± 0.33 times (Figure 6D), *IL17A* in 10.48 ± 3.36 times (Figure 6E), *RORC* in 3.20 ± 0.68 times (Figure 6F) and *FOSL1* in 0.57 ± 0.17 (Figure 6C). The level of the expression of *PPARγ* was increased 2.13 ± 0.47 times (Figure 6A). The level of *FOXP3* was also increased in 2.62 ± 0.39 times (Figure 6B).

Therefore, low laser treatment caused significant growth of the *PPARγ* and *FOXP3* expression while reducing the expression of *STAT3*, *IL17A*, and *RORC*.

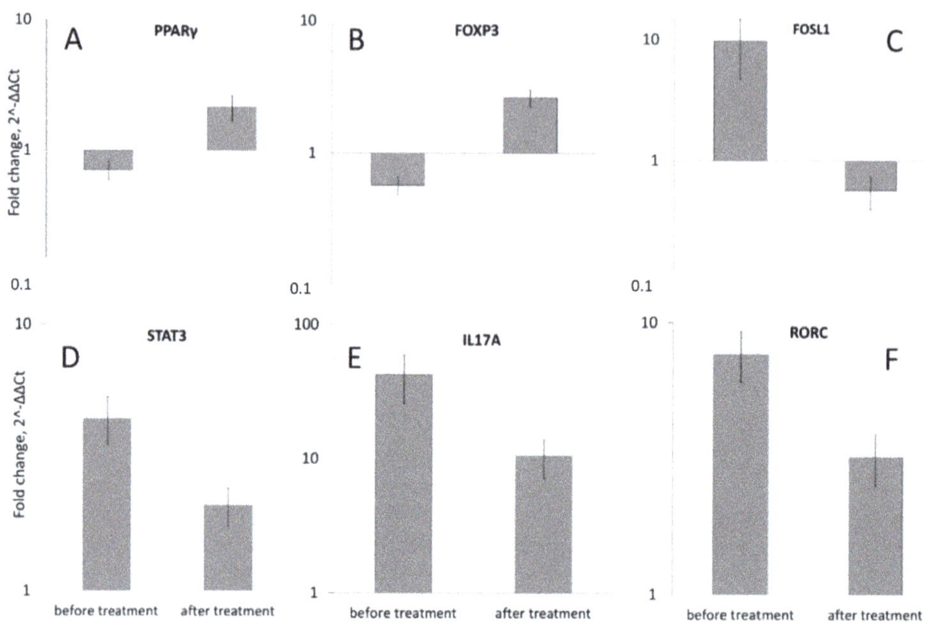

Figure 6. Comparison of *PPARγ* (**A**), *FOXP3* (**B**), *FOSL1* (**C**), *STAT3* (**D**), *IL17A* (**E**) and *RORC* (**F**) genes expression in the skin of 23 patients with psoriasis before and after low-level laser therapy. The levels of mRNA concentration for genes in psoriatic skin samples was calculated in relation to the level of the same genes in unaffected skin samples (which was taken as conditional 1, $p < 0.05$). See supplemental materials for detailed statistics ("PPARG expression file").

4. Discussion

Previously, we built the model of *PPARγ* dependent pathways involved in the development of the psoriatic lesions. The model includes significant molecular cascades such as *IL17* signaling, Toll like receptor and PI3K-AKT pathways from literature network analysis and public microarrays data. In this work we tested the model by measurement mRNA and protein levels of key molecular players in human psoriatic skin and T-cells.

Several key players according to previously reconstructed models of the *PPARγ* signaling were selected for experimental validation of the hypothesis that low levels of *PPARγ* may contribute to the development of psoriatic lesions. There were *IL17A* gene (interleukin 17A), *STAT3* gene (signal transducer and activator of transcription 3), *RORC* gene (retinoid-related orphan receptor-gamma), *FOXP3* gene (forkhead box P3) and *FOSL1* (FOS-like antigen 1) gene (Figure 7).

We detected the repression of *PPARγ* activity in human psoriatic skin and blood immune cells ($CD3^+$ T cells) from 23 patients with real-time PCR method, ELISA and immunohistochemistry analysis. Our results are similar to data from microarray on 58 patients where average *PPARγ* gene expression also was slightly downregulated in psoriatic lesions [18]. Recently, low *PPARγ* expression in $CD4^+$ T cells from 12 psoriasis patients than in healthy controls was reported [19]. Other authors however described the higher level of the *PPARγ* expression in human psoriatic skin compared to healthy skin. But the level of *PPARγ* mRNA was close to the detection limit in their research [20].

In the model we tested in this work *IL17A*, *STATS3*, and *RORC* are statistically significant negative targets of *PPARγ*. We expected that activity of these targets should be higher in psoriatic lesion and slightly decrease after laser treatment. Our experimental results support this idea and they are aligned well with detected low activity of *PPARγ* in psoriatic skin and $CD3^+$ T cells, since *PPARγ* may act as a suppressor of the *IL17* gene transcription by inhibiting his direct transcription factors *RORC* and *STAT3* (Figure 8). In psoriatic cell *STAT3* becomes more active than in healthy cell and, by providing feedback loop via *FOSL1*, further strengthens the downregulation of *PPAR*γ expression (Figure 8).

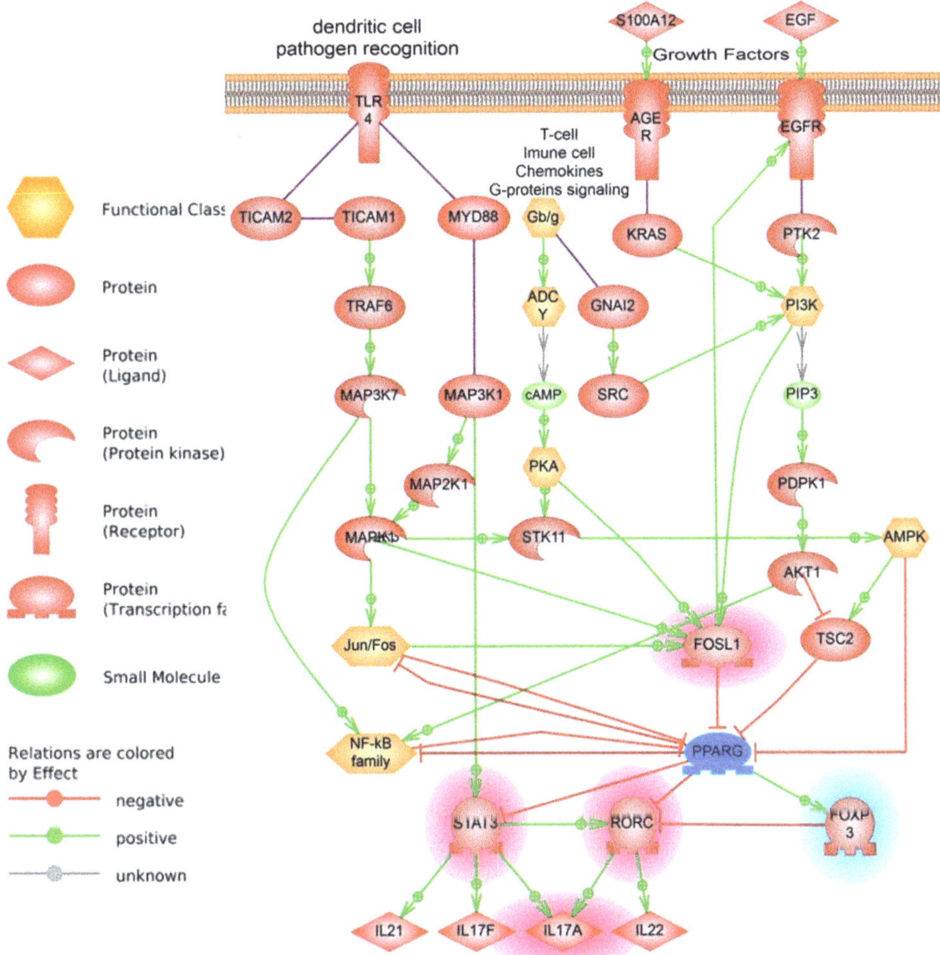

Figure 7. Model of *PPAR*γ pathway in psoriasis (simplified version). Changes in gene expression are highlighted according to results of analysis in red (over-expressed genes) and in blue (down-expressed genes). In psoriatic lesion, cytokines, growth factors, pathogens, apoptotic debris as well as dendritic cells activates TLRs, AGER and EGFR among other receptors on the surface of keratinocytes and T-cells. Activates receptors transfer the signal to their canonical cascades such as G-couple proteins, PI3K, PKA, MAPK1, or calcium (not shown). As a result, several direct inhibitors of *PPARG* protein such as TSC2 or inhibitors of *PPARG* expression (*FOSL1*, Jun/Fos) became over activated. When *PPARG* is inhibited on both RNA expression and protein levels, this causes higher than normal expression of interleukins (such as *IL17A*) via *STAT3* and *RORC* transcription factors. See supplemental materials for references and links to publications that support protein-protein interactions in the model.

While the prominent role of *RORC* in psoriasis as the major controller of Th17 cell differentiation is well described, however, the evidence of *RORC* expression in psoriasis is controversial. In mice T-cells and dendritic cells had increased *STAT3/RORC* expression [18] and patients with psoriasis had elevated level of *RORC* (RORG-t isoform) [20].

In published microarray data, the level of expression of *RORC* was downregulated in most of 58 patients [15,21].

Contrariwise, *FOSL1* may be important for stabilization of psoriatic inflammation. *FOSL1* was reported to have high level of expression in human psoriasis tissues [22,23] and be able to inhibit *PPARγ* directly [24].

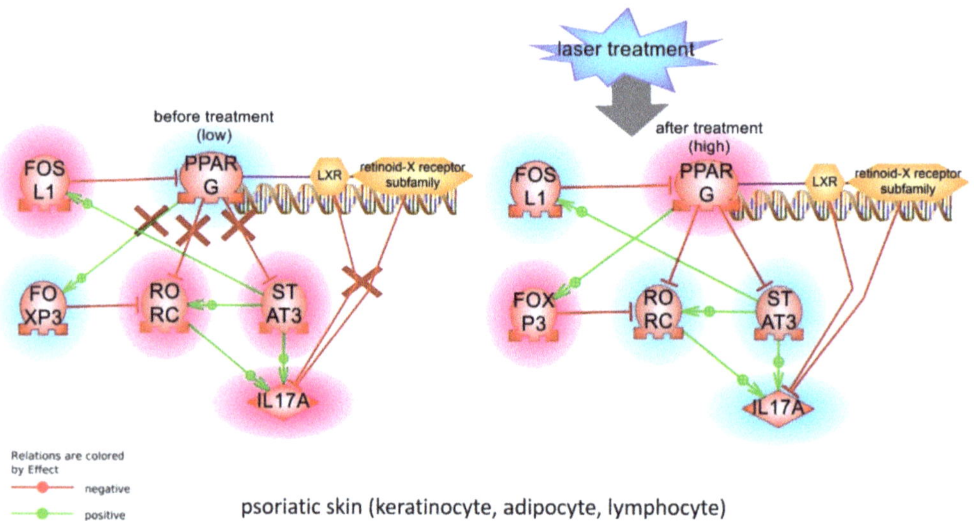

Figure 8. Alignment of tested changes in *PPARγ* signaling before and after laser treatment with the model. Changes in gene expression after treatment are marked with highlights (red—elevated; blue—lowered). Red X symbolise non-functional protein-protein interactions. LXR-RXR complex activity was not tested.

Laser treatment diminishes the *STAT3->FOSL1->PPARγ* feedback loop and restores *PPARγ* activity and slightly reduce *IL17* production (Figure 8). The molecular mechanism of laser treatment is not well understood. Low-intensity laser waves are absorbed by oxygen, CO_2, water molecules switching them into an activated state. Proteins with activated molecules participate in interactions more intensively. There was shown that low laser treatment stimulates Ca^{2+}-related signaling pathways including general membrane reparation and cell proliferation. There are expectations that low-level laser treatment will result in the replacement of "old" cells with new ones thus reducing the inflammation in the psoriatic lesion [25]. Interesting that ozonated autohemotherapy (OAHT) treatment also elevated *PPARγ* expression in $CD4^+$ T cells of patients with psoriasis and decreased patients' PASI scores [19].

Functions of *IL17* and transcription factors we tested in this work are well studied in psoriasis (see more details in [22,26,27], and in our previous publications [28–30]). Other aspects of *PPARγ* related signalling pathways were also studied. For example, interactions between different *PPARs* isoforms are important for their functions. *PPARδ* directly inhibits *PPARγ*, and many pro-inflammatory factors, fatty acid signaling, and "regenerative skin phenotype" pathways (IFNG, TNFA) linked with *PPARδ* stimulation [31]. miRNAs may play important regulator role in *PPARγ* signaling as well [32].

Single nucleotide polymorphisms in *PPARγ* gene are commonly associated with insulin resistance and diabetes. There are no significant associations between mutations in *PPARG* with psoriasis (based on search in OMIM, ClinVAr and Resnet-2020 databases). However, the association between rs1801282 in *PPARγ* and psoriasis, and low level of *PPARγ* expression were reported in Egyptian patients with obesity and metabolic syndrome. Authors concluded that reduced *PPARγ* activity could be the factor responsible for translating the metabolic state among psoriatic patients [33,34].

Can players of *PPARG* signalling be considered as drug targets for psoriasis treatment? We used enrichment analysis with literature biomedical network (Resnet—2020) helps to identify the significant differences in known drugs mechanisms associated with tested model of *PPARγ* signaling. We searched for drugs which were verified in clinical trials or reported in publications as drugs against psoriasis and simultaneously were reported as inhibitors of *IL17, STAT3, FOSL1, RORC* but not *PPARγ* or *FOXP3*. Several substances like corticosteroids and tacrolimus were identified by given criteria. We have found that two other drugs (calcitriol and paclitaxel) that indeed reduce inflammation in psoriasis, however, may not be very effective in psoriasis treatment because they inhibit *PPARγ* or *FOXP3* (Figure 9). Anti-diabetes drugs such as biguanides (metformin) and thiazolidinediones (rosiglitazone and pioglitazone) were studied as additional treatment options for psoriasis. Moreover, it is known that thiazolidinediones act as direct ligand activators of *PPARγ* and it normalizes the histological features of psoriatic skin in vitro [35].

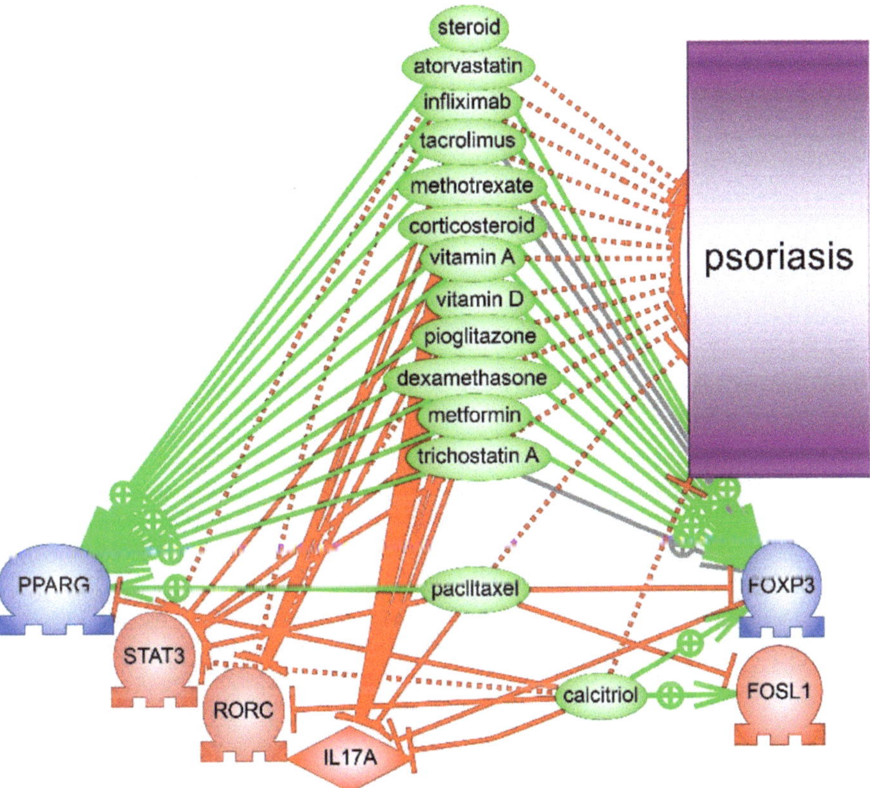

Figure 9. Anti-psoriatic drugs that may stabilize *PPARγ* signaling and supress *IL17* expression. Calcitriol and paclitaxel however have been reported to inhibit *PPARγ* or *FOXP3*. See details of interactions and references in files with models in Supplemental materials.

5. Conclusions

We detected the high level of *RORC* and *STAT3* mRNA in the psoriatic skin of patients which were reduced after laser treatment. Protein level of *STAT3* also were upregulated in psoriatic skin and CD3$^+$ T cells. Also, we report the downregulation of *FOXP3* mRNA expression which is a direct inhibitor of *RORC* and positive target of *PPARγ*. Though, low expression of *PPARγ* as well as high level of *RORC* expression is supported by downregulated *FOXP3* expression and validates reconstructed model. Experimental data we obtained support our model and the hypothesis that in psoriasis low level of *PPARγ* activity stimulates *IL17* synthesis because *STAT3* and *RORC* became less suppressed.

Our research has several limitations. The number of tested samples was relatively small. Not all proteins from the model were tested on protein levels. Also, additional analysis of receptors activation and intermediate cellular cascades may help to evaluate upstream triggers and regulators of *PPARγ*. Finally, the interaction of transcription factors and regulation of gene expression are more complex than tested model. Additional scaffolds proteins such as LXR-RXR complex, histones and chromatin remodelling complexes are involved in gene expression.

In summary, we report that *PPARγ* weakens the expression of genes that contribute in the development of psoriatic lesion. It is not clear what upstream pathway is the most important for *PPARγ/IL17* regulation in psoriasis. Our data show that transcriptional regulation of *PPARγ* expression by *FOSL1* and by *STAT3/FOSL1* feedback loop may be central in the psoriatic skin and T-cells.

Supplementary Materials: The following are available online at https://www.mdpi.com/article/10.3390/ijms22168603/s1.

Author Contributions: Conceptualization, V.S. and A.N.; Methodology, V.S.; Project administration, V.S.; Resources, V.S. and I.K.; Investigation, A.S., E.D. (Evgenia Dvoriankova), A.P., E.D. (Elena Denisova), O.M., V.S.; Writing—original draft, V.S. and A.N.; Writing—review & editing, V.S. and A.N.; Visualization, A.N., A.S., A.M.; Data curation, A.M.; funding acquisition, I.K. All authors have read and agreed to the published version of the manuscript.

Funding: This research received no external funding.

Institutional Review Board Statement: The study was conducted according to the guidelines of the Declaration of Helsinki, and approved by the Ethics Committee (12/02/2020 #2) of Centre of Theoretical Problems of Physico-Chemical Pharmacology, Russian Academy of Sciences, Russian Federation.

Informed Consent Statement: Informed consent was obtained from all subjects involved in the study.

Data Availability Statement: Not applicable.

Conflicts of Interest: The authors declare no conflict of interest.

References

1. Schmuth, M.; Moosbrugger-Martinz, V.; Blunder, S.; Dubrac, S. Role of PPAR, LXR, and PXR in epidermal homeostasis and inflammation. *Biochim. Biophys. Acta* **2014**, *1841*, 463–473. [CrossRef]
2. Sher, T.; Yi, H.F.; McBride, O.W.; Gonzalez, F.J. cDNA cloning, chromosomal mapping, and functional characterization of the human peroxisome proliferator activated receptor. *Biochemistry* **1993**, *32*, 5598–5604. [CrossRef] [PubMed]
3. Sertznig, P.; Reichrath, J. Peroxisome proliferator-activated receptors (PPARs) in dermatology: Challenge and promise. *Dermatoendocrinology* **2011**, *3*, 130–135. [CrossRef]
4. Kliewer, S.A.; Umesono, K.; Noonan, D.J.; Heyman, R.A.; Evans, R.M. Convergence of 9-cis retinoic acid and peroxisome proliferator signalling pathways through heterodimer formation of their receptors. *Nature* **1992**, *358*, 771–774. [CrossRef]
5. Ricote, M.; Glass, C.K. *PPARs* and molecular mechanisms of transrepression. *Biochim. Biophys. Acta* **2007**, *1771*, 926–935. [CrossRef]
6. Jiang, C.; Ting, A.T.; Seed, B. PPAR-gamma agonists inhibit production of monocyte inflammatory cytokines. *Nature* **1998**, *391*, 82–86. [CrossRef]
7. Yessoufou, A.; Wahli, W. Multifaceted roles of peroxisome proliferator-activated receptors (PPARs) at the cellular and whole organism levels. *Swiss Med. Wkly.* **2010**, *140*, w13071. [CrossRef]
8. Nehrenheim, K.; Meyer, I.; Brenden, H.; Vielhaber, G.; Krutmann, J.; Grether-Beck, S. Dihydrodehydrodiisoeugenol enhances adipocyte differentiation and decreases lipolysis in murine and human cells. *Exp. Dermatol.* **2013**, *22*, 638–643. [CrossRef]
9. Adachi, Y.; Hatano, Y.; Sakai, T.; Fujiwara, S. Expressions of peroxisome proliferator-activated receptors (PPARs) are directly influenced by permeability barrier abrogation and inflammatory cytokines and depressed PPARα modulates expressions of chemokines and epidermal differentiation-related molecules in keratinocytes. *Exp. Dermatol.* **2013**, *22*, 606–608. [PubMed]
10. Henson, P. Suppression of macrophage inflammatory responses by PPARs. *Proc. Natl. Acad. Sci. USA* **2003**, *100*, 6295–6296. [CrossRef] [PubMed]

11. Marx, N.; Kehrle, B.; Kohlhammer, K.; Grüb, M.; Koenig, W.; Hombach, V.; Libby, P.; Plutzky, J. *PPAR* activators as antiinflammatory mediators in human T lymphocytes: Implications for atherosclerosis and transplantation-associated arteriosclerosis. *Circ. Res.* **2002**, *90*, 703–710. [CrossRef]
12. Demerjian, M.; Man, M.-Q.; Choi, E.-H.; Brown, B.E.; Crumrine, D.; Chang, S.; Mauro, T.; Elias, P.M.; Feingold, K.R. Topical treatment with thiazolidinediones, activators of peroxisome proliferator-activated receptor-gamma, normalizes epidermal homeostasis in a murine hyperproliferative disease model. *Exp. Dermatol.* **2006**, *15*, 154–160. [CrossRef]
13. Nesterova, A.P.; Klimov, E.A.; Zharkova, M.; Sozin, S.; Sobolev, V.; Ivanikova, N.V.; Shkrob, M.; Yuryev, A. Diseases of the skin and subcutaneous tissue. In *Disease Pathways*; Elsevier: Amsterdam, The Netherlands, 2020; pp. 493–532. ISBN 978-0-12-817086-1. [CrossRef]
14. Nesterova, A.P.; Yuryev, A.; Klimov, E.A.; Zharkova, M.; Shkrob, M.; Ivanikova, N.V.; Sozin, S.; Sobolev, V. *Disease Pathways: An Atlas of Human Disease Signaling Pathways*, 1st ed.; Elsevier: Waltham, MA, USA, 2019; ISBN 978-0-12-817086-1.
15. Sobolev, V.; Nesterova, A.; Soboleva, A.; Dvoriankova, E.; Piruzyan, A.; Mildzikhova, D.; Korsunskaya, I.; Svitich, O. The Model of *PPARγ*-Downregulated Signaling in Psoriasis. *PPAR Res.* **2020**, *2020*, 6529057. [CrossRef] [PubMed]
16. Livak, K.J.; Schmittgen, T.D. Analysis of Relative Gene Expression Data Using Real-Time Quantitative PCR and the $2^{-\Delta\Delta CT}$ Method. *Methods* **2001**, *25*, 402–408. [CrossRef] [PubMed]
17. Yao, Y.; Richman, L.; Morehouse, C.; de los Reyes, M.; Higgs, B.W.; Boutrin, A.; White, B.; Coyle, A.; Krueger, J.; Kiener, P.A.; et al. Type I Interferon: Potential Therapeutic Target for Psoriasis? *PLoS ONE* **2008**, *3*, e2737. [CrossRef] [PubMed]
18. Nadeem, A.; Al-Harbi, N.O.; Ansari, M.A.; Al-Harbi, M.M.; El-Sherbeeny, A.M.; Zoheir, K.M.A.; Attia, S.M.; Hafez, M.M.; Al-Shabanah, O.A.; Ahmad, S.F. Psoriatic inflammation enhances allergic airway inflammation through *IL-23/STAT3* signaling in a murine model. *Biochem. Pharmacol.* **2017**, *124*, 69–82. [CrossRef]
19. Zeng, J.; Tang, Z.; Zhang, Y.; Tong, X.; Dou, J.; Gao, L.; Ding, S.; Lu, J. Ozonated autohemotherapy elevates PPAR-γ expression in CD4+ T cells and serum HDL-C levels, a potential immunomodulatory mechanism for treatment of psoriasis. *Am. J. Transl. Res.* **2021**, *13*, 349–359.
20. Mendoza, G.J.; Almeida, O.; Steinfeld, L. Intermittent fetal bradycardia induced by midpregnancy fetal ultrasonographic study. *Am. J. Obstet. Gynecol.* **1989**, *160*, 1038–1040. [CrossRef]
21. Ding, J.; Gudjonsson, J.E.; Liang, L.; Stuart, P.E.; Li, Y.; Chen, W.; Weichenthal, M.; Ellinghaus, E.; Franke, A.; Cookson, W.; et al. Gene expression in skin and lymphoblastoid cells: Refined statistical method reveals extensive overlap in cis-eQTL signals. *Am. J. Hum. Genet.* **2010**, *87*, 779–789. [CrossRef]
22. Zhu, W.; Li, J.; Su, J.; Li, J.; Li, J.; Deng, B.; Shi, Q.; Zhou, Y.; Chen, X. FOS-like antigen 1 is highly expressed in human psoriasis tissues and promotes the growth of HaCaT cells in vitro. *Mol. Med. Rep.* **2014**, *10*, 2489–2494. [CrossRef]
23. Sobolev, V.V.; Zolotorenko, A.D.; Soboleva, A.G.; Elkin, A.M.; Il'ina, S.A.; Serov, D.N.; Potekaev, N.N.; Tkachenko, S.B.; Minnibaev, M.T.; Piruzyan, A.L. Effects of Expression of Transcriptional Factor AP-1 FOSL1 Gene on Psoriatic Process. *Bull. Exp. Biol. Med.* **2011**, *150*, 632–634. [CrossRef]
24. Hasenfuss, S.C.; Bakiri, L.; Thomsen, M.K.; Williams, E.G.; Auwerx, J.; Wagner, E.F. Regulation of steatohepatitis and *PPARγ* signaling by distinct AP-1 dimers. *Cell Metab.* **2014**, *19*, 84–95. [CrossRef]
25. Avci, P.; Gupta, A.; Sadasivam, M.; Vecchio, D.; Pam, Z.; Pam, N.; Hamblin, M.R. Low-level laser (light) therapy (LLLT) in skin: Stimulating, healing, restoring. *Semin. Cutan. Med. Surg.* **2013**, *32*, 41–52.
26. Petit, R.G.; Cano, A.; Ortiz, A.; Espina, M.; Prat, J.; Muñoz, M.; Severino, P.; Souto, E.B.; García, M.L.; Pujol, M.; et al. Psoriasis: From Pathogenesis to Pharmacological and Nano-Technological-Based Therapeutics. *Int. J. Mol. Sci.* **2021**, *22*, 4983. [CrossRef]
27. Calautti, E.; Avalle, L.; Poli, V. Psoriasis: A STAT3-Centric View. *Int. J. Mol. Sci.* **2018**, *19*, 171. [CrossRef]
28. Sobolev, V.V.; Denisova, E.V.; Korsunskaya, I.M. Alteration of *STAT3* gene expression in psoriasis treatment. *Meditsinskiy Sov. Med. Counc.* **2020**, *12*, 71–74. [CrossRef]
29. Sobolev, V.V.; Soboleva, A.G.; Potekaev, N.N.; Melnichenko, O.O.; Korsunskaya, I.M.; Artemyeva, S.I. *PPARγ* gene expression analysis in psoriasis treatment. *Meditsinskiy Sov. Med. Counc.* **2021**, *8*, 82–87. [CrossRef]
30. Sobolev, V.V.; Mezentsev, A.V.; Ziganshin, R.H.; Soboleva, A.G.; Denieva, M.; Korsunskaya, I.M.; Svitich, O.A. LC-MS/MS analysis of lesional and normally looking psoriatic skin reveals significant changes in protein metabolism and RNA processing. *PLoS ONE* **2021**, *16*, e0240956. [CrossRef] [PubMed]
31. Romanowska, M.; al Yacoub, N.; Seidel, H.; Donandt, S.; Gerken, H.; Phillip, S.; Haritonova, N.; Artuc, M.; Schweiger, S.; Sterry, W.; et al. PPARδ Enhances Keratinocyte Proliferation in Psoriasis and Induces Heparin-Binding EGF-Like Growth Factor. *J. Investig. Dermatol.* **2008**, *128*, 110–124. [CrossRef] [PubMed]
32. Xiuli, Y.; Honglin, W. miRNAs Flowing Up and Down: The Concerto of Psoriasis. *Front. Med.* **2021**, *8*, 646796. [CrossRef] [PubMed]
33. Hegazy, R.A.; Abdel Hay, R.M.; Shaker, O.; Sayed, S.S.; Abdel Halim, D.A. Psoriasis and metabolic syndrome: Is peroxisome proliferator-activated receptor-γ part of the missing link? *Eur. J. Dermatol.* **2012**, *22*, 622–628. [CrossRef] [PubMed]
34. Seleit, I.; Bakry, O.; Abd El Gayed, E.; Ghanem, M. Peroxisome proliferator-activated receptor-γ gene polymorphism in psoriasis and its relation to obesity, metabolic syndrome, and narrowband ultraviolet B response: A case–control study in Egyptian patients. *Indian J. Dermatol.* **2019**, *64*, 192. [PubMed]
35. Ellis, C.N.; Varani, J.; Fisher, G.J.; Zeigler, M.E.; Pershadsingh, H.A.; Benson, S.C.; Chi, Y.; Kurtz, T.W. Troglitazone Improves Psoriasis and Normalizes Models of Proliferative Skin Disease: Ligands for Peroxisome Proliferator-Activated Receptor-γ Inhibit Keratinocyte Proliferation. *Arch. Dermatol.* **2000**, *136*, 609–616. [CrossRef] [PubMed]

International Journal of
Molecular Sciences

MDPI

Review

PPARdelta in Affected Atopic Dermatitis and Psoriasis: A Possible Role in Metabolic Reprograming

Stefan Blunder, Petra Pavel, Deborah Minzaghi and Sandrine Dubrac *

Epidermal Biology Laboratory, Department of Dermatology, Venereology and Allergology, Medical University of Innsbruck, Anichstraße 35, 6020 Innsbruck, Austria; stefan.blunder@i-med.ac.at (S.B.); petra.pavel2@gmail.com (P.P.); deborah.minzaghi@i-med.ac.at (D.M.)

* Correspondence: sandrine.dubrac@i-med.ac.at; Tel.: +43-512-5042-3025; Fax: +43-512-5042-3002

Abstract: Peroxisome proliferator-activated receptors (PPARs) are nuclear hormone receptors expressed in the skin. Three PPAR isotypes, α (NRC1C1), β or δ (NRC1C2) and γ (NRC1C3), have been identified. After activation through ligand binding, PPARs heterodimerize with the 9-cis-retinoic acid receptor (RXR), another nuclear hormone receptor, to bind to specific PPAR-responsive elements in regulatory regions of target genes mainly involved in organogenesis, cell proliferation, cell differentiation, inflammation and metabolism of lipids or carbohydrates. Endogenous PPAR ligands are fatty acids and fatty acid metabolites. In past years, much emphasis has been given to PPARα and γ in skin diseases. PPARβ/δ is the least studied PPAR family member in the skin despite its key role in several important pathways regulating inflammation, keratinocyte proliferation and differentiation, metabolism and the oxidative stress response. This review focuses on the role of PPARβ/δ in keratinocytes and its involvement in psoriasis and atopic dermatitis. Moreover, the relevance of targeting PPARβ/δ to alleviate skin inflammation is discussed.

Keywords: PPAR; atopic dermatitis; psoriasis; metabolic reprograming; glucose; fatty acids

Citation: Blunder, S.; Pavel, P.; Minzaghi, D.; Dubrac, S. PPARdelta in Affected Atopic Dermatitis and Psoriasis: A Possible Role in Metabolic Reprograming. *Int. J. Mol. Sci.* **2021**, *22*, 7354. https://doi.org/10.3390/ijms22147354

Academic Editors: Manuel Vázquez-Carrera and Walter Wahli

Received: 9 June 2021
Accepted: 7 July 2021
Published: 8 July 2021

1. PPARdelta: The Least Studied PPAR Isoform

Peroxisome proliferator-activated receptors (PPARs) are transcription factors belonging to nuclear hormone receptor superfamily. Three PPAR isotypes, α (NRC1C1), β or δ (NRC1C2) and γ (NRC1C3), have been identified in mammals (henceforth, we refer to the β/δ isoform simply as PPARδ). After activation through ligand binding, PPARs heterodimerize with the 9-cis retinoic acid receptor (RXR), another nuclear hormone receptor, to bind to specific PPAR-responsive elements in regulatory regions of target genes, mainly involved in organogenesis, cell proliferation, cell differentiation, inflammation and metabolism of lipids or carbohydrates. Endogenous PPAR ligands are fatty acids and fatty acid metabolites.

PPARδ is ubiquitously expressed in murine tissues with highest expression in liver, muscle, adipose tissue, placenta, small intestine and skin. PPARδ is expressed twofold, 10-fold and 30-fold more in mouse keratinocytes (KCs) compared to mouse liver, quadriceps muscle and thymus, respectively. In most tissues, PPARδ localizes to the nuclear fraction of cells and is hardly detectable in the cytoplasm [1]. In humans, PPARδ mRNA and protein are highly abundant in the thyroid gland and placenta whereas high amounts of mRNA and moderate amounts of protein are detected in the cerebral cortex, skin and esophagus. Of note, inconsistency between protein and RNA levels of PPARδ has been observed in many human tissues and cell types (https://www.proteinatlas.org/ENSG00000112033-PPARD/tissue, accessed on 7 July 2021). There are five human and mouse PPARδ isoforms generated by alternative splicing, which is a mechanism potentially involved in PPARδ regulation, as some PPARδ splice isoforms exhibit reduced translation efficiency [2,3].

The ligand-binding pockets of PPARs have a distinct three-armed T shape, which allows not only straight fatty acids to bind them, but also ligands with multiple branches such as phospholipids and synthetic fibrates. The ligand-binding pocket of PPARδ is smaller than that of PPARγ or PPARα, which limits the binding of large ligands when compared to the other two PPAR isoforms [4]. PPARδ is activated by several endogenous ligands including certain long chain fatty acids (regardless of saturation status), dihomo-γ-linolenic acid, eicosapentaenoic acid, 15(S)-hydroxyeicosatetraenoic acid (HETE), and arachidonic acid, with affinities in the low micromolar range (Table 1). Supraphysiological doses of 8(S)-, 12(S)-, 12(R)-, and 15(S)-HETE efficiently activate PPARδ. 13(S)-hydroxyoctadecadienoic acid (HODE) is considered as weak PPARδ activator [5,6]. Controversial results have been found for prostacyclin (PGI2) and all-trans retinoic acid [7,8]. It has also been reported that 4-hydroxynonenal (4-HNE) and 4-hydroxydodecadienal (4-HDDE), the peroxidation products of polyunsaturated fatty acids, can activate PPARδ, although the mechanism remains unknown [9,10]. Synthetic PPARδ ligands include GW501516, GW0742 and L165041, which preferentially activate PPARδ as compared to PPARα or PPARγ [6]. Recently, 27 new synthetic PPARδ agonists (13 with low nanomolar EC_{50} values) have been discovered [11]. However, it is important to stress that preferential ligand does not mean exclusive ligand and that supraphysiological doses of any of the PPARδ ligands will activate other PPAR isoforms, and the same is true for all PPAR isoforms. For example, bezafibrate, which is known as a PPARα ligand, activates all three PPARs at concentrations ranging from 55 to 110 μM [12]. In the absence of ligand binding, the heterodimer PPARδ-RXR is associated with corepressors and histone deacetylases (HDACs), which inhibit its transcriptional activity. After ligand binding, PPARδ undergoes conformational changes that induce the release of the corepressors and allow it to bind coactivators [7].

The transcriptional activity of PPARδ is modulated by several factors, which are not well characterized but include post-translational modifications such as phosphorylation. Epidermal growth factor receptor (EGFR) has been recently shown to induce PPARδ phosphorylation at Y108 in response to epidermal growth factor (EGF) [13]. Although PPARδ contains several putative phosphorylation sites (Y108, T252, T253, T256), (https://www.phosphosite.org/proteinAction.action?id=24004&showAllSites=true (accessed on 9 May 2021)) [14], little is known about phosphoregulation of PPARδ, in contrast to PPARα and PPARγ. Both cyclic adenosine monophosphate (cAMP) and protein kinase A (PKA) activators increase the ligand-activated and basal activity of PPARδ and could be upstream signals that commit PPARδ to the regulation of glucose and lipid metabolism [14]. In contrast, PPARδ can also be sumoylated at K104, which inhibits its activity [14]. Desumoylation of PPARδ by small ubiquitin-like modifier (SUMO)-specific protease 2 (SENP2) promotes the transcriptional activity of PPARδ, which, in turn, upregulates fatty acid oxidation by enhancing the expression of long-chain-fatty-acid–CoA ligase 1 (ACSL1), carnitine palmitoyltransferase Ib (CPT1b) and mitochondrial uncoupling protein 3 (UCP3) in muscles of mice fed a high fat diet [15]. Moreover, PPARδ contains several ubiquitylation sites, which suggests a potential role of ubiquitin–proteosome degradation in the regulation of its cellular turnover (https://www.phosphosite.org/proteinAction.action?id=24004&showAllSites=true (accessed on 9 May 2021)). Degradation of PPARδ via the proteasome might prevent its accumulation in the nucleus and thereby moderate its cellular activity [16]. In line with this, overexpression of PPARδ in fibroblasts leads to its polyubiquitylation and rapid degradation, a process partially prevented by exposure to the PPARδ synthetic ligand GW501516 [17].

PPARs can also engage in transrepression of other transcription factors. Although transrepression between nuclear factor kappa-light-chain-enhancer of activated B cells (NF-κB), activator protein 1 (AP-1), CCAAT-enhancer-binding protein (C/EBP), signal transducer and activator of transcription (STAT) and nuclear factor of activated T-cells (NF-AT) has been well characterized for PPARα and PPARγ, little is known about transrepression in the context of PPARδ [18,19]. L-165041 is a PPARδ ligand that is less potent and selective than GW501516, yet it promotes the binding of PPARδ to the p65 subunit of NF-κB exerting anti-inflammatory effects [5,20]. Moreover, in the absence of ligand, PPARδ binds directly to the transcription factor B-cell lymphoma 6 (BCL-6), leading to increased expression of proinflammatory cytokines. Indeed, BCL-6 is a transcription factor repressing the expression of various inflammatory genes via direct binding to their promoters or via inhibition of the transcription of nucleotide-binding oligomerization domain-like receptor (NOD)-like receptor family pyrin domain containing 3 (NLRP3) [21,22]. Binding of PPARδ to an agonist disrupts the PPARδ-BCL-6 complex, thus reversing the transcriptional repression of inflammatory genes [23]. Thus, ligand binding to PPARδ alleviates inflammation by enhancing its binding to NF-kB, hence neutralizing the transcriptional activity of NF-kB and/or the release of the anti-inflammatory transcription factor BCL-6. However, PPARδ has also been shown to bind to the N-terminal part of p65 in the absence of exogenous ligand [5]. Therefore, the pro- vs. anti-inflammatory role of PPARδ might be context- and ligand-dependent. Moreover, conformational changes experienced by PPARδ after ligand binding might potentially strengthen or weaken the affinity of PPARδ to p65; however, this has not been studied to date.

Table 1. PPARδ potential endogenous ligands.

Compounds	**Weak Ligands**	**Ligands**
ω3-PUFA	α-Linolenic acid C18:3 γ-Linolenic acid C18:3 Dihomo-γ-linolenic acid DHA C22:6	EPA C20:5
ω6-PUFA	Linoleic acid C18:2 Arachidonic acid C20:4	
ω9-MUFA	Palmitoleic acid C16:1 Elaidic acid C18:1 Erucic acid C22:1 Nervonic acid C24:1	Oleic acid C18:1
Saturated fatty acids	Myristic acid C14:0 Palmitic acid C16:0 Stearic acid C18:0 Behenic acid C22:0	Arachidic acid C20:0
Eicosanoids	5-HpETE 8(S)-HETE 15(S)HpETE 15(S)-HETE 12-HpETE LTA4 9(R)-HODE 12-HpODE 13(S)-HODE 5,15-di-HpETE	5(S)-HETE 15(R)HpETE 15(R)-HETE 12-HETE LTB4 LTC4 9(S)-HODE 5,6-diHETE
Prostaglandins	PGA2 PGB1 PGB2 PGD1 PGD2 PGD3 PGF2α PGF3α PGI2	PGF1α
Lipoxins		LXA4
4-Hydroxyalkenals	4-HDDE	

Adapted from [8]. DHA: docosahexaenoic acid; EPA: eicosapentaenoic acid; 4-HDDE; 4-hydroxydodecadienal; HETE: hydroxyeicosatetraenoic acid; HODE: hydroxyoctadecadienoic acid; LT: leukotriene; LX: lipoxin; PG: prostaglandin.

Although there is likely a set of core effects and target genes of PPARδ common to all cell types and organs, PPARδ has also been shown to exert tissue-specific functions. Moreover, some target genes differ between rodents and humans. Canonical PPARδ target genes are mainly related to lipid metabolism in all cell types [6,19,24–26]. This includes genes involved in fatty acid oxidation (very long-chain specific acyl-CoA dehydrogenase, mitochondrial (*ACADVL*), acyl-CoA oxidase 1 (*ACOX1*), acetyl-CoA acyltransferase 2 (*ACAA2*), catalase (*CAT*), enoyl-CoA hydratase 1 (*ECH1*), pyruvate dehydrogenase kinase 4 (*PDK4*), solute carrier family 25 member 20 (*SLC25A20*), Niemann-Pick C1-like protein 1 (*NPC1L1*), *thiolase B*, *CPT1A*)) or other aspects of lipid metabolism (angiopoietin Like 4 (*ANGPTL4*), fatty acid binding proteins 3-5 (*FABP3-5*), perilipin 2 (*PLIN2*), adipocyte protein 2 (*aP2*)). Other PPARδ target genes exert non-metabolic functions and are involved in immune regulation, such as *CD300A*, *CD52*, LDL receptor related protein 5 (*LRP5*), *NLRC4* and phosphatase and actin regulator 1 (*PHACTR1*) [27]. In muscles, PPARδ controls (i) the entry of long chain fatty acids into cells via *SLC27A1*, *SLC27A3* and *CD36*; (ii) their subsequent activation by forming acyl-CoA via *ACSL3*, *ACSL4*, and acyl-CoA synthetases short chain family member 1 and 2 (*ACSS1-2*); (iii) mitochondrial β-oxidation via *CPT1A*, *CPT1B*, *SLC25A20*, *ACADVL*, and *ACADL*; (iv) peroxisomal β-oxidation via *ACOX1* [28]. In human macrophages, PPARδ regulates the expression of genes involved in lipid metabolism but also electron-transfer-flavoprotein, beta subunit (*ETFB*), electron transfer flavoprotein-ubiquinone oxidoreductase (*ETFDH*) and iron-sulfur cluster assembly 1 (*ISCA1*), which play important roles in electron transfer and iron-sulfur complex assembly and in the immune response via upregulation of *CD1D*, *CD36*, *CD52*, *CD300A*, *LRP5*, *NLRC4* and *PHACTR1* and downregulation of *CCL8*, *CCL13*, *CXCL1*, *IL10*, *IL8* and *TNFA* [27].

The expression of *PPARD* is regulated by various cytokines, hormones, lipid metabolites and other transcription factors. The *PPARD* promoter region contains a vitamin D receptor (VDR) response element [29,30]. Thus, it is likely that there is cross-talk between VDR and the PPARδ pathway, but his has not been investigated in detail despite being of potential pathophysiological interest. AP-1, a transcription factor involved in the inflammatory response, and especially junB, both increase *PPARD* expression [31]. AP-1 mediates the effects of TNF-α, phorbol 12-myristate 13-acetate (TPA) and ceramides on *PPARD/Ppard* expression [32]. Tan et al., in a seminal work, showed that TNF-α promotes the synthesis of ceramides via sphingomyelin hydrolysis, which ultimately activates AP-1 via the mitogen-activated protein kinase kinase kinase 1 (MEKK1) and stress-activated protein kinases (SAPK)/Jun amino-terminal kinases (JNK)/p38 mitogen-activated protein kinases (p38MAPK) pathway [32]. Previous work also showed that *PPARD* can be upregulated by T3-thyroid receptor (TR) [33]. The metabolic regulation of PPARδ has been reviewed elsewhere [7].

2. Metabolic Features of Keratinocytes in Normal Skin

Data on metabolic pathway predominating in keratinocytes is still a controversial topic. Old literature suggests that, to generate ATP, KCs are predominantly committed to glycolysis in the presence of glucose or to mitochondrial respiration in its absence [34]. In suprabasal KCs, limited access to glucose from the dermal vasculature is believed to promote mitochondrial respiration and oxidation of lipids, in contrast to basal KCs, which preferentially use glucose as their main energy substrate [34–36]. In line with this, GLUT1 is the main GLUT isoform in the epidermis and is abundantly expressed in the basal layer, although residual expression can be found in suprabasal layers [37–39]. Recent work showed that decreased glycolysis via inhibition of glucose uptake in KCs promoted cell differentiation, suggesting a major role of glycolysis in KC fate [40]. However, another work proposes a predominating role of mitochondrial-derived ROS in basal KCs as a signal to induce differentiation [41]. This is in line with a recent report showing that NIX, a transcription factor located in mitochondria, controls mitophagy and, in turn, KC differentiation, hence emphasizing the role of mitochondria in KC fate [42]. Thus, further work is required to clarify the relative contribution of glycolysis versus oxidative phosphorylation (OXPHOS) in the control of homeostatic processes in the epidermis.

3. PPARdelta in Psoriasis and Atopic Dermatitis

Atopic dermatitis and psoriasis are two chronic and pruritic inflammatory skin diseases exhibiting pathophysiological commonalities, including impaired epidermal barrier function, immune hyper-responsiveness, and local and systemic symptoms modulated by environmental factors such as the skin microbiome and stress. Moreover, both diseases are associated with a major genetic risk factor, i.e., Filaggrin (*FLG*) loss-of-function mutations in atopic dermatitis and the HLA-Cw0602 allele in psoriasis vulgaris [43,44]. Furthermore, in both atopic dermatitis and psoriasis patients, nonlesional and lesional skin coexists, but the mechanism of transition from the non-affected to the affected condition remains unclear. Atopic dermatitis is one of the most common inflammatory skin diseases worldwide and characterized by skin features such as erythematous and papulovesicular eruptions with oozing, crusting and pruritus as well as associated systemic signs such as food allergies, allergic asthma and rhinitis, anxiety and sleep disorders. At the cellular level, atopic dermatitis is characterized by (a) the complex interplay between impaired epidermal barrier function owing to altered lipid composition of the stratum corneum lipid matrix i.e., a reduction in the chain length of structural lipids (fatty acids and ceramides), (b) a complex Th2-driven inflammation, (c) skin infiltration by eosinophils, basophils and inflammatory dendritic cells, and (d) an altered skin microbiota [43,45–52]. In psoriasis vulgaris, genetic risk factors predominantly affect innate immunity, and to some extent adaptive immunity (IL12p/IL-23R axis, Th1, Th17 cells). Similarly to atopic dermatitis, skin immunological abnormalities in psoriasis are complex and associated with comorbidities (e.g., arthritis and cardiovascular manifestations), pointing to a systemic immune hyper-responsiveness [44,50,53–56].

PPARδ is expressed in all skin cell types, including KCs, fibroblasts, sebocytes, hair follicle cells, melanocytes and Langerhans cells [19,57–59]. PPARδ is the predominant isoform in human KCs and is expressed throughout all epidermal layers [32,60]. Activation of PPARδ with synthetic ligands promotes the expression of human KC differentiation markers such as involucrin (*INV*) and transglutaminase 1 (*TGM1*) [60]. Although there is consensus on the pro-differentiative effects of PPARδ ligands and PPARδ activation in KCs, the effects on KC proliferation are more controversial, with studies showing reduced [60] or enhanced [31] KC proliferation after treatment with the PPARδ ligand L-165041 or GW-501516. Treatment of human KCs with L-165041 gave opposite outcomes in two distinct studies [31,60]. Yet, the use of different treatment regimens of L-165041, i.e., 0.05 μM for 3 days [60] and 1 μM for 7 days [31], might have been responsible for these divergent results, for example by inducing the recruitment of different cofactors and thus engaging PPARδ in different metabolic pathways. Moreover, the direct effects of ligands should not be underestimated because the use of PPARδ siRNA to test the requirement for PPARδ in the cellular response was not carried out in either studies [31,60]. In line with this, L-165041 can activate other PPAR isoforms, i.e., PPARα, PPARγ1 and PPARγ2 at doses as low as 0.05 μM [60]. This underscores that PPAR ligands can exert receptor-independent effects, that metabolic effects might vary with ligand concentrations (e.g., U- or bell-curves), and that the relative contribution of other PPAR isoforms after treatment with ligands might significantly influence experimental results, hence stressing the need for cautious interpretation of data [46]. Human KCs infected with a lentivirus containing an RNAi sequence directed toward PPARδ displayed reduced proliferative capacity, suggesting that PPARδ promotes, rather than dampens, proliferation of human KCs [31]. However, it is also possible that PPARδ exerts both proliferative and differentiative functions according to the cellular context, i.e., basal cells (early KCs, progenitor and stem cells) or suprabasal cells (differentiated cells). As in other cell types, PPARδ is likely a master regulator of fatty acid metabolism in KCs by increasing the uptake of long-chain fatty acids via upregulation of CD36 and fatty acid β-oxidation [60] (Table 2). However, the role of PPARδ in epidermal lipid and glucose metabolism remains under-investigated. Interestingly, the PPARδ target genes in KCs are not identical to those in other organs and cell types (Table 2), suggesting PPARδ has specific cellular functions in the epidermis.

The *PPARD/Ppard* gene is upregulated in lesional skin of patients with psoriasis vulgaris [5,31,61–65] and of mouse models of psoriasis [63,64]. However, although *PPARD* has been identified as a putative pathogenic gene in psoriasis [65], variants at the *PPARD* genomic locus have not been associated with psoriasis. In psoriatic plaques, PPARδ accumulates in KC nuclei in all epidermal layers [5]; however, subcellularly, PPARδ is found both in the cytoplasm and nucleus of KCs in the basal layer and in the stratum spinosum, whereas it is strictly found in nuclei in KCs in the stratum granulosum [5,64]. This suggests that PPARδ is constitutively activated by endogenous ligands in granular KCs of the epidermis in patients with psoriatic lesions [64]. Accordingly, endogenous PPARδ ligands can be produced in psoriatic lesions from the oxidation of arachidonic acid via ALOX8 (mouse) or ALOX12 (mouse and human) [64,66], two enzymes located in the stratum granulosum [66–68]. FABP5 is a fatty acid-binding protein expressed in the epidermis and has been shown to deliver endogenous lipid ligands to PPARδ in KC nuclei and to be a PPARδ target gene [69]. The expression of FABP5 parallels that of PPARδ at both the mRNA and protein levels in psoriasis [5,63]. Thus, in the suprabasal epidermis of psoriatic lesions, it is likely that PPARδ is constitutively activated by endogenous ligands such as arachidonic acid or its derivatives (eicosanoids), which are transported by FABP5 to the nucleus of granular KCs to promote PPARδ–mediated KC terminal differentiation and lipid β-oxidation. Specific overexpression and activation of human PPARδ in suprabasal mouse epidermis has been achieved by generating transgenic mice expressing a Cyp1A1-driven expression of human *PPARD* in KCs followed by topical treatment with the PPARδ agonist GW501516 [62]. Interestingly, these mice developed psoriasis-like inflammation associated with an increased Th17 immune response [62]. In this model, sustained activation of the STAT3 pathway is critically involved in the development of psoriasis-like disease [62]. The constitutive activation of PPARδ in suprabasal epidermis not only promotes terminal KC differentiation but also the production, in KCs, of IL-36 and the pleiotropic pro-inflammatory cytokine IL-1β. The latter can contribute to the activation of skin dendritic cells, which can in turn, skew naïve T cells toward a Th17 phenotype [62]. Moreover, suprabasal mouse KCs overexpressing the constitutively activated human PPARδ probably secrete soluble factors able to trigger the proliferation of basal KCs [62]. In addition, in psoriatic plaques, some PPARδ localize to nuclei in basal KCs to potentially sustain KC proliferation [5,64]. In line with this, previous work suggested that upregulation of PPARδ in the epidermis of psoriatic lesions might contribute to KC hyperproliferation via the upregulation of heparin-binding EGF-like growth factor (HB-EGF) at the mRNA and protein levels [31]. HB-EGF is a ligand that activates EGFR and is expressed in the basal layer of the epidermis, where it has been shown to accelerate wound healing [70]. This might be relevant for psoriasis because disease flares can be induced by physical trauma (the isomorphic or Koebner phenomenon) among other causes. Pioneering work on the pathogenesis of psoriasis showed increased levels of antimicrobial peptides in psoriatic skin breaks the innate tolerance to self-DNA which ultimately drives autoimmunity [71]. Moreover, human genomic DNA fragments enhance *TNFA* and *HBEGF* expression as well as KC proliferation, hence mimicking the KC phenotype in psoriatic skin lesions [72]. Thus, we can speculate that PPARδ in the basal epidermis of psoriatic plaques sustains KC proliferation via mechanisms involving HB-EGF. NF-kB has been shown to inhibit PPARδ-dependent transactivation. However, in lesional psoriasis, p65 NF-kB is sequestered in the cytoplasm of basal KCs, which might allow PPARδ to exert its transcriptional regulation on various genes, including those involved in KC proliferation [5].

PPARδ is upregulated in the epidermis of lesional atopic dermatitis when compared to non-lesional skin but to a lesser extent than in psoriatic lesions [31]. The expression of *FABP5* parallels that of PPARδ in psoriasis and atopic dermatitis [31,73]. Notably, the expression of *Ppard* and *Fabp5* is markedly increased in the epidermis of mouse models of lesional atopic dermatitis [38,74]. Similar to psoriasis, FABP5 is mainly localized to the nuclei of suprabasal KCs, suggesting efficient local generation of PPARδ ligands to sustain the activation of PPARδ [38]. Interestingly, the amounts of arachidonic acid, PGF2α and 5-HETE (PPARδ endogenous ligands) are increased in lesional skin of atopic dermatitis patients when compared to healthy skin [75]. The increased cleavage of membrane phospholipids via cPLA2 in the stratum granulosum can significantly contribute to the accumulation of arachidonic acid and its derivatives in lesional atopic dermatitis skin as well as in psoriatic lesions [76–78]. The role of PPARδ has been less investigated in atopic dermatitis than in psoriasis. However, in both diseases, PPARδ might induce KC hyperproliferation, enhance differentiation and contribute to inflammatory processes.

However, PPARδ can also be envisaged as a key regulator of metabolism, especially in the metabolic shift toward anaerobic glycolysis that has been recently evidenced in psoriatic and atopic lesions [38,79,80]. The production of lactate is largely increased in the epidermis of flaky tail mice and mice treated with MC903, two mouse models of lesional atopic dermatitis [38] and of mice treated with imiquimod, a mouse model of psoriasis [81]. Interestingly, the PPARδ ligand GW610742, when orally administered to *ob/ob* mice, induces lactate accumulation in the liver [82]. Indeed, PPARδ has been shown to regulate the expression of key enzymes involved in glucose metabolism, including in KCs (Table 2) [83–85]. PPARδ can promote anaerobic glycolysis by upregulating PDK, an enzyme that inactivates pyruvate dehydrogenase (PDH) via phosphorylation. PDH is the rate-limiting enzyme involved in pyruvate uptake in mitochondria, which ultimately favors oxidative phosphorylation [86]. Thus, inactivation of PDH by PPARδ-induced PDK inhibits pyruvate uptake in mitochondria, which, in turn, promotes anaerobic glycolysis [87]. In the epidermis of flaky tail mice, there is a shift toward anaerobic glycolysis associated with an enhanced PPARδ pathway including increased PDK1. In line with this, mitochondrial function is not enhanced in the epidermis of flaky tail mice despite a dramatic need for energy to sustain forced KC proliferation and to dampen inflammation [38]. These results are in line with previous work showing that PPARδ antagonism favors mitochondrial function [88].

Table 2. PPARδ target genes and associated pathways in keratinocytes.

	Upregulated	Downregulated
Fatty acid metabolism	FABP5 FABP7 ACADVL ACOX1 CD36 ALOX12B LDLR PLA2G3 ECHB OACT5 BDH1 GDPD3 CRABP2 GM2A	LASS6 GPD1L PRKAB2 CHPT1
Cholesterol metabolism	HMGCS1 HMGCR MVD CYP51 SQLE FDPS LSS FDFT1 DHC7	

Table 2. *Cont.*

	Upregulated	Downregulated
KC proliferation	HB-EGF	EGFR EPS15 EPS8 MCC RBL2 CCNG1 DUSP3 PDGFRA PDGFC CDKN1C
KC differentiation	INV TGM1 TGM3 S100A8 S100A9 S100A16 KRT6B KRT16 KRT17 KRT75 SPRR1B CNFN EHF	DCN KRT15 DUSP3
KC apoptosis	CIDEA	
Inflammation	MMP9 IL1F9 IL1F5 IL1B IL1F6 IL1F8 ILA IL1RA IL18 IL17 IL23A IL22 STAT3	TGFBR2 TGFBR3 LIFR IL1R1
Glucose metabolism	PDK1	PDK4
Oxidative stress	SOD2 CAT ABCC3	
Other	HAS3 GGH UCK2 ATP10B CCNB1 MAPK13 CCNB2 GSPT1 XPC	RBL2 AXL RHOC TTC3 LFNG FXR1 FBLN1 GAB2 PIK3IP1

Table 2. *Cont.*

	Upregulated	Downregulated
Unknown	AKR1B1 ATP12A ACPP MAP4K4 MREG FGFBP1 ARL8B GAS7 CD81 CCDC50 TACC1	SERINC1 EID1 KLF6 RAI14 MTCP1 REEP5 NENF
		OSR2

ABCC3: ATP binding cassette subfamily C member 3; ACAD(V)L: (very) long-chain specific acyl-CoA dehydrogenase, mitochondrial; ACOX1: acyl-CoA oxidase 1; ACPP (ACP3): acid phosphatase 3; AKR1B1: aldo-keto reductase family 1 member B; ALOX: lipoxygenase; ATP10B: ATPase phospholipid transporting 10B; ATP12A: ATPase H+/K+ transporting non-gastric alpha2 subunit; ARL8B: ADP ribosylation factor like GTPase 8B; AXL: AXL receptor tyrosine kinase; BDH1: 3-hydroxybutyrate dehydrogenase 1; CAT: catalase; CCDC50: coiled-coil domain containing 50; CCN: cyclin; CD: cluster of differentiation; CDKN1C: cyclin dependent kinase inhibitor 1C; CHPT1: choline C phosphotransferase 1; CIDEA: cell death inducing DFFA like effector A; CNFN: cornifelin; CRABP2: cellular retinoic acid binding protein 2; CYP51: lanosterol 14α-demethylase; DCN: decorin; DHC7 (DNAH1): dynein axonemal heavy chain 1; DUSP3: dual specificity phosphatase 3; ECHB (HADHB): hydroxyacyl-CoA dehydrogenase trifunctional multienzyme complex subunit beta; EGFR: epidermal growth factor receptor; EHF: ETS homologous factor; EID1: EP300 interacting inhibitor of differentiation 1; EPS: epidermal growth factor receptor pathway substrate; FABP: fatty acid binding protein; FBLN1: fibulin 1; FDFT1: farnesyl-diphosphate farnesyltransferase 1; FDPS: farnesyl diphosphate synthase; FGFBP1: fibroblast growth factor binding protein 1; FXR1: FMR1 autosomal homolog 1; GAB2: GRB2 associated binding protein 2; GAS7: growth arrest specific 7; GDPD3: glycerophosphodiester phosphodiesterase domain containing 3; GGH: gamma-glutamyl hydrolase; GM2A: GM2 ganglioside activator; GPD1L: glycerol-3-phosphate dehydrogenase 1 like; GSPT1: G1 to S phase transition 1; HAS3: hyaluronan synthase 3; HB-EGF: heparin-binding EGF-like growth factor; HMGCR: 3-hydroxy-3-methylglutaryl-CoA reductase; HMGCS1: 3-hydroxy-3-methylglutaryl-CoA synthase 1; IL: interleukin; INV: involucrin; KLF6: kruppel like factor 6; KRT: keratin; LASS6 (CERS6): ceramide synthase 6; LDLR: low density lipoprotein receptor; LFNG: LFNG O-fucosylpeptide 3-beta-N-acetylglucosaminyltransferase; LIFR LIF receptor subunit alpha; LSS: lanosterol synthase; MAP4K4: mitogen-activated protein kinase kinase kinase kinase 4; MAPK13: mitogen-activated protein kinase 13; MCC: MCC regulator of WNT signaling pathway; MMP9: matrix metalloproteinase 9; MREG: melanoregulin; MTCP1: mature T cell proliferation 1; MVD: mevalonate diphosphate decarboxylase; NENF: neudesin neurotrophic factor; OACT5 (LPCAT3): lysophosphatidylcholine acyltransferase 3; OSR2: odd-skipped related transcription factor 2; PDGFC: platelet derived growth factor C; PDGFRA: platelet derived growth factor receptor alpha; PDK: pyruvate dehydrogenase kinase; PIK3IP1: phosphoinositide-3-kinase interacting protein 1; PLA2G3: phospholipase A2 group III; PRKAB2: protein kinase AMP-activated non-catalytic subunit beta 2; RAI14: retinoic acid induced 14; RBL2: RB transcriptional corepressor like 2; REEP5: receptor accessory protein 5; RHOC: ras homolog family member C; S100A: S100 calcium-binding protein A; SERINC1: serine incorporator 1; SOD2: superoxide dismutase 2; SPRR1B: small proline rich protein 1B; SQLE: squalene epoxidase; STAT: signal transducer and activator of transcription; TACC1: transforming acidic coiled-coil containing protein 1; TGFBR: transforming growth factor beta receptor; TGM: transglutaminase; TTC3: tetratricopeptide repeat domain 3; UCK2: uridine-cytidine kinase 2; XPC: XPC complex subunit, DNA damage recognition and repair factor.

PPARδ promotes β-oxidation of fatty acids in all cell types, including KCs (Table 2) [85,89,90]. In flaky tail mice, peroxisomal β-oxidation is upregulated when compared to that of healthy mice, with marked increases in the mRNA, protein and activity levels of ACOX1 [38], a well-known PPARδ downstream target [89,90]. This profile has been observed in another mouse model of lesional atopic dermatitis, i.e., mice topically treated with MC903 [38]. This treatment is associated with decreased proportions of very-long chain fatty acids and ceramides, especially with 24 and 26 carbons [38], as observed in the epidermis of patients with lesional atopic dermatitis [91]. Interestingly, C24 and C26 fatty acids are exclusively oxidized in peroxisomes via ACOX1 [92,93]. Thus, the upregulation of PPARδ in the epidermis of patients with lesional atopic dermatitis might promote peroxisomal β-oxidation of very- and ultra-long-chain fatty acids and ceramides, hence significantly contributing to disease pathogenesis. Indeed, the efficacy of the stratum corneum barrier depends, to a large part, on the lipid composition of the lipid matrix surrounding the corneocytes, which consists of more than 50% fatty acids with 24 and 26 carbons. Interestingly, the proportion of very-long-chain ceramides is also decreased in the epidermis of psoriatic lesions [94] and is associated with increased ACOX1 [38] and PPARδ (see above), thus corroborating the key role of the PPARδ pathway in lipid abnormalities in both lesional atopic dermatitis and psoriasis. In contrast to lesional AD [38], mitochondrial β-oxidation might be increased in psoriasis as suggested by previous work [46] and might further contribute to lipid abnormities.

PPARδ has been shown to be involved in wound healing [95], which might demonstrate relevance in both psoriasis and atopic dermatitis. Indeed, both diseases are characterized by epidermal barrier impairment that can be considered as superficial wounds. In wounded epidermis, PPARδ inhibits KC apoptosis via activation of the phosphoinositide-3-kinase (PI3K)/PKBα/Akt1 pathway and promotes the re-epithelialization of the skin by enhancing KC adhesion and migration [95]. The upstream signal promoting the expression and activation of PPARδ in wounded epidermis is believed to be the accompanying low-grade inflammation, i.e., increased IL-1β and TNF-α, which promotes the synthesis of lipids and the release of bioactive lipids activating PPARδ [95]. In human epidermal equivalents (HEEs) topically treated with sodium dodecyl sulfate (SDS) to inflict epidermal barrier impairment, *PPARD* expression was upregulated at 24 h but not at 6 h post-treatment [96]. This upregulation of *PPARD* requires a rather strong epidermal barrier impairment because a milder epidermal barrier impairment induced by topical treatment of HEEs with acetone, did not result in *PPARD* upregulation [96]. Furthermore, the relatively late upregulation of *PPARD* suggests that it requires the prior synthesis of modulating factors such as lipids and/or cytokines. In line with this, IL-1β but not TNF-α, which are both upregulated after epidermal barrier impairment, is capable of upregulating *PPARD* in KCs [96]. Moreover, epidermal barrier impairment leads to excessive transepidermal water loss, a phenomenon described in both lesional atopic dermatitis and psoriatic plaques as well as in wounded skin. It is thus possible to speculate that, in this context, IL-1β upregulates PPARδ signaling including anaerobic glycolysis via PDK1 and peroxisomal β-oxidation via ACOX1 [38]. In line with these data, placement of occlusive dressing onto the skin of flaky tail mice to reduce transepidermal water loss was found to downregulate ACOX1 [38]. Another candidate upstream of PPARδ in the basal epidermis might be silent mating type information regulation 2 homolog 1 (SIRT1), which is known to promote wound healing [97,98] and enhance PPARδ transcriptional activity [99]. Thus, the chronic epidermal barrier impairment observed in lesional atopic dermatitis and psoriasis might lead to the constitutive activation of a sequential cellular compensatory response aimed at repairing the barrier; this could include upregulation of SIRT1 and production of IL-1β and subsequent release of bioactive lipids to activate PPARδ. This might ultimately lead to uncontrolled inflammation and disruption of epidermal homeostasis. Indeed, PPARδ has been shown to upregulate several genes involved in KC differentiation (e.g., *INV, S100A8, S100A9, TGM3, TGM1*) and proliferation (e.g., *HB-EGF, IL1B, IL17, IL22*) and the inflammatory response (e.g., *IL1B, IL18, IL1A, IL1RA, IL1F, IL17, IL22*) (Table 2).

KC hyper-proliferation, accelerated differentiation and the inflammatory response in psoriatic and atopic lesions require energy that might be provided by enhanced peroxisomal fatty acid β-oxidation and glucose utilization in response to PPARδ activation [38,85]. Anaerobic glycolysis via PPARδ upregulation is an advantageous metabolic pathway to sustain forced KC proliferation because it is a substantial source of ATP, which does not promote oxidative stress, in contrast to mitochondrial metabolism. The side effect of PPARδ upregulation might be the consumption, via ACOX1, of structural lipids, i.e., C24 and C26 fatty acids and ceramides destined to the stratum corneum, thus further compromising the epidermal inside-out barrier. Thus, upregulation of the PPARδ pathway in atopic and psoriatic lesions might be a double-edged sword, by sustaining KC proliferation without worsening oxidative stress but, at the same time, changing the composition of the lipid bilayer in the stratum corneum, resulting in less efficient barrier function. Thus, antagonizing PPARδ to correct metabolic abnormalities in lesional atopic dermatitis and psoriasis plaques might be a new and effective therapeutic strategy to reduce both epidermal hyperplasia and consumption of structural lipids of the stratum corneum lipid matrix.

4. PPARδ as a Therapeutic Target in Atopic Dermatitis and Psoriasis

To date, the therapeutic effects of PPARδ targeting in atopic dermatitis and psoriasis remain underinvestigated. Intriguingly, both PPARδ ligands and antagonists have been proven to dampen skin inflammation. Antagonism of PPARδ by topical application of GSK0660 in transgenic mice expressing Cyp1A1-driven expression of human *PPARD* in KCs and topically treated with the PPARδ agonist GW501516 (mouse model of psoriasis) reduced epidermal thickness, dermal inflammatory infiltrates with $CD4^+$ and $CD8^+$ T lymphocytes and expression of *Il1b* and *Lce3e* but failed to inhibit the expression of *Hb-egf* [100]. However, because the half-life of GSK0660 is only 90 min, this might be a limiting factor for its use as a therapeutic. Consequently, topical treatment with an irreversible PPARδ antagonist would be more appropriate to alleviate psoriasis symptoms. Indeed, a single topical treatment with GSK3787, which covalently binds and permanently inactivates PPARδ showed similar therapeutic efficacy as several topical applications with GSK0660 in mice with psoriasis-like skin inflammation [64,100]. Moreover, GSK3787 reduced the expression of *Il17*, *Il23a*, *Il22* and *Il1b* in these mice [64]. On the other hand, the activation of PPARδ with tetradecylthioacetic acid (TTA) also showed beneficial effects in psoriasis. In a small pilot study, topical treatment of psoriatic plaques with 0.5% TTA reduced the Psoriasis Area and Severity Index (PASI) and skin scaling and inflammation [101]. However, TTA can activate all PPAR isoforms at high doses [60]. Thus, the beneficial effects of TTA are likely the net result of the combined activation of all PPAR isoforms or a direct effect of the molecule. In a mouse model of dermatitis (i.e., mice topically treated with oxazolone, a chemical inducing Th2-predominant inflammation in mouse skin), topical application of GW1514, a PPARδ agonist, reduced epidermal hyperplasia, KC proliferation, transepidermal water loss, skin surface pH, skin infiltration by eosinophils and mast cells, and serum CCL17 [102]. However, it remains to be determined whether these effects are PPARδ-dependent. Topical treatment with GW1514 did not reduce serum IgE levels in oxazolone-treated mice [102], suggesting that this molecule does not reach the blood circulation after topical application. Thus, given the role of PPARδ in psoriasis and atopic dermatitis, PPARδ antagonism, rather than activation, might be the preferred therapeutic approach to treat both diseases. This does not mean that PPARδ ligands would be less advantageous therapeutic options; however, they should be mainly employed for their direct, i.e., PPAR-independent, beneficial effects.

Excessive oxidative stress overtaking the cellular antioxidant response is involved in tumorigenic processes, inflammation and skin aging. Accordingly, both psoriasis and atopic dermatitis are associated with oxidative stress [47,103–105]. The role of PPARδ in the antioxidant response is equivocal. Activation of PPARδ with GW501516 or other agonists has been shown to downregulate the mRNA and protein levels of NF-E2–related factor 2 (NRF2), a master transcription factor controlling the expression of key proteins involved in the cellular detoxification of reactive oxygen species (ROS) [106,107]. In contrast, PPARδ antagonism has been shown to promote the antioxidant response via upregulation of *Nrf2* [88] and to decrease the production of ROS in mitochondria [99]. In line with this, loss of PPARδ in intestinal fibroblasts delayed tumorigenesis, induced NRF2 and reduced oxidative stress [108]. The β-oxidation of very-long-chain fatty acids via ACOX1 produces hydrogen peroxide. In lesional atopic dermatitis and psoriasis, the marked increase in ACOX1 might outstrip the detoxification ability of the cellular antioxidant response and contribute to the epidermal oxidative stress observed in both diseases. Thus, overall, PPARδ might promote oxidative stress in the epidermis. Specifically, PPARδ might promote hydrogen peroxide release by peroxisomes (via ACOX1 activity) and, at the same time, dampen mitochondrial function and, in turn, the production of mitochondria-derived ROS. However, in non-skin cells, PPARδ ligands have been shown to prevent endoplasmic reticulum stress, downregulate NOX4 and reduce ROS production and subsequent inflammation [107,109].

Thus, we can speculate that PPARδ might exert both pro- and antioxidant functions as reported for other transcription factors [46], depending on pathophysiological context, cell type and organelle. Here again PPAR-independent antioxidant effects of PPARδ ligands might be envisaged. Unfortunately, the role of PPARδ in the oxidative response in KCs has never been investigated; PPARδ antagonism might have a potent antioxidant effect via mechanisms that remain to be identified.

Topical treatment with PPARδ agonists or antagonists should be critically evaluated because data on the role of PPARδ in cancer is controversial [19,85,110]. PPARδ has been shown to inhibit non-melanoma skin cancer by enhancing KC terminal differentiation and senescence, blocking KCs in the G2/M phase of the cell cycle, and inhibiting endoplasmic reticulum stress and specific inflammatory pathways [111–113]. However, PPARδ has also been shown to promote KC proliferation via HB-EGF and to contribute to epidermal hyperplasia [38,85]. Moreover, PPARδ can interact with β-catenin, a key mediator in the regulation of the Wnt pathway, which is involved in multiple cellular functions such as embryogenesis and tumorigenesis [114,115]. The overexpression of cytosolic phospholipase A2α (cPLA2α) promotes the binding of PPARδ to β-catenin and, in turn, the binding of the complex to the T-cell factor/lymphoid enhancer factor (TCF/LEF) response element [114,115]. cPLA2α is the rate-limiting enzyme which releases arachidonic acid from membrane phospholipids and, thus, playing a central role in the production of bioactive eicosanoids (including prostaglandins and leukotrienes), some of those are endogenous PPARδ ligands [116]. Thus, activation of PPARδ with endogenous ligands such as arachidonic acid or its derivatives may control cell fate (differentiation vs. proliferation) and malignant cell transformation. It has recently been shown that the PPARδ-β-catenin complex favors the formation of chromatin loops that regulate the transcription of vascular endothelial growth factor A (*VEGFA*), a regulator of angiogenesis during tumorigenesis. Activation of PPARδ via ligand binding releases the loop, which favors the transcription of *VEGFA* [115], and might sustain cancer growth. Furthermore, increased FABP5 is associated with various cancers including skin cancer, by promoting the activation of PPARδ and the upregulation of its oncogenic target genes [19]. It is possible that specific endogenous PPARδ ligands produced during tumorigenic transformation of cells skew PPARδ toward pro-oncogenic functions. The importance of the nature of ligands in driving PPARδ-mediated cellular responses is emphasized by work demonstrating the anti-apoptotic effects of PPARδ after activation with retinoic acid, which was shuttled to KC nuclei by FABP5 [85]. In tumors, this might help cancer cells escape apoptosis. Thus, activation of PPARδ in KCs by specific endogenous ligands might promote tumorigenesis by upregulating oncogenic genes, increasing oxidative stress and favoring a metabolic shift toward anaerobic glycolysis, which might promote non-melanoma skin cancer. Alternatively, competition of synthetic ligands with endogenous ligands to bind to PPARδ might positively intercede in the cellular response in tumors. Although PPARδ is expressed in melanocytes, its role in this cell type has never been investigated, which seems a missed opportunity since ligand-mediated PPARδ activation might protect against melanoma [117,118]. Thus, the role of PPARδ in skin tumorigenesis remains controversial, and the opposing views might owe to the use of different cancer cell lines, patient tissues, cancer staging and progression [7].

An important parameter for the topical utilization of drugs targeting PPARδ to alleviate atopic dermatitis and psoriasis is their transdermal absorption and ability to passage into the bloodstream. Indeed, systemic administration of GW501516 in a mouse model of wound healing showed that PPARδ activation promotes angiogenesis and upregulates matrix metalloproteinase 9 (MMP9) in wounded skin [85,119]. MMP9 is involved in many biological processes and plays roles in tumor progression and invasion, angiogenesis, and determining the composition of the tumor microenvironment [120].

Thus, the competition between endogenous and synthetic ligands/antagonists in a defined pathophysiological context (e.g., inflammation, precancer) might determine the therapeutic versus detrimental outcome of PPARδ targeting. This might also depend on the expression of corepressors/coactivators and other transcription factors engaged in PPARδ transrepression. Due to the therapeutic potential of PPARδ targeting in atopic dermatitis and psoriasis, further studies are necessary to elucidate in depth the role of PPARδ in the skin in various pathophysiological contexts and cell types (e.g., melanocytes) as well as the complex interplay between PPARδ and other transcription factors. Moreover, it is likely that synthetic ligands do not entirely activate PPARδ and that a small fraction of PPARδ remains activated by FABP5-bound endogenous ligands, leading to synergetic or contradictory signals, within cells. This aspect of PPARδ targeting is completely unexplored.

5. Conclusions

Between the years 2000 and 2010, PPARs were thoroughly studied in various organs including skin, but then, enthusiasm significantly waned. Moreover, much of the initial research was focused on PPARα and PPARγ, leaving large gaps in our knowledge of the role of PPARδ in the skin and especially in KCs. Thus, it remains unknown how PPARδ controls KC metabolism or the inflammatory response or the oxidative stress response. Furthermore, PPARδ crosstalk with other receptors such as VDR accentuates its importance in epidermal homeostasis. Therefore, in light of its clear involvement in KC proliferation, differentiation, metabolism, oxidative stress and the inflammatory response (Figure 1), renewed effort should be directed at both basic research and therapeutic strategies targeting PPARδ, including potential local and systemic side effects in psoriasis and atopic dermatitis.

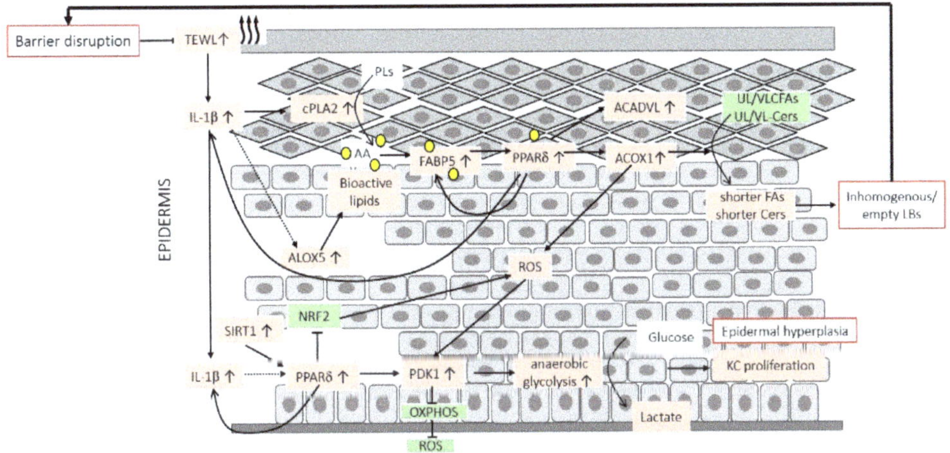

Figure 1. Potential role of PPARδ in keratinocytes in lesional atopic dermatitis and psoriasis: Epidermal barrier impairment, likely originating from (epi)genetic abnormalities, enhances trans-epidermal water loss (TEWL) and the production of IL-1β in granular keratinocytes (KCs), which upregulates $cPLA_2$ involved in the cleavage of membrane phospholipids (PLs) and the release of arachidonic acid (AA). AA and its metabolites, produced by oxidation via ALOX5 into bioactive lipids, are shuttled to the nucleus by FABP5 to activate PPARδ, which, in turn, increases the expression of ACOX1 and ACADVL. Increased ACOX1 consumes ultra- and very-long-chain fatty acids (UL/VLCFAs) and ceramides (Cers), resulting in the improper embedding of stratum corneum lipids into lamellar bodies (LBs), which weakens the efficacy of the stratum corneum barrier, hence perpetuating epidermal barrier impairment. Overactivity of ACOX1 produces excessive hydrogen peroxide, which might signal within granular KCs as well as through all the epidermal layers to cause oxidative stress and metabolic changes. This might be amplified by the downregulation of NRF2 by endogenous ligand-bound PPARδ. In the basal layers, IL-1β, produced either locally or in granular KCs, and SIRT1, which is produced in the lower epidermis, contribute to the activation of PPARδ via unidentified mechanisms. This results in the upregulation of PDK1 and the shift toward anaerobic glycolysis, which circumvents mitochondrial function, including the production of mitochondrial ROS. Anaerobic glycolysis sustains KC hyperproliferation via rapid ATP production.

Author Contributions: Writing—original draft preparation, S.B., D.M., P.P. and S.D.; writing—review and editing, S.D., literature research, D.M., P.P. and S.D.; funding acquisition, S.D. All authors have read and agreed to the published version of the manuscript.

Funding: This work was supported by grants from the Austrian Science Fund and the Tyrol Research Fund (FWF 31662 and FWF 28039) to SD.

Institutional Review Board Statement: Not applicable.

Informed Consent Statement: Not applicable.

Data Availability Statement: Not applicable.

Conflicts of Interest: The authors declare no conflict of interest.

Abbreviations

ACAA2: acetyl-CoA acyltransferase 2; ACAD(V)L: (very) long-chain specific acyl-CoA dehydrogenase, mitochondrial; ACOX1: acyl-CoA oxidase 1; ACSL: long-chain-fatty-acid—CoA ligase; ACSS: acyl-CoA synthetase short chain family member; ALOX: lipoxygenase; ANGPTL4: angiopoietin Like 4; AP-1: activator protein 1; aP2: adipocyte protein 2; ATP: adenosine triphosphate; BCL-6: B-cell lymphoma 6; cAMP: cyclic adenosine monophosphate; CAT: catalase; CCL: CC-chemokine ligand; CD: cluster of differentiation; C/EBP: CCAAT-enhancer-binding protein; cPLA2α: cytosolic phospholipase A2α; CPT1: carnitine palmitoyltransferase I; CXCL: C-X-C motif Chemokine Ligand; CYP1A1: Cytochrome P450, family 1, subfamily A, polypeptide 1; ECH1: enoyl-CoA hydratase 1; EGF: epidermal growth factor; EGFR: epidermal growth factor receptor; ETFB: electron-transfer-flavoprotein, beta subunit; ETFDH: electron transfer flavoprotein-ubiquinone oxidoreductase; FABP: fatty acid binding protein; FLG: filaggrin; HB-EGF: heparin-binding EGF-like growth factor; HDAC: histone deacetylase; HEE: human epidermal equivalent; HETE: hydroxyeicosatetraenoic acid; 4-HDDE: 4-hydroxydodecadienal; HLA: human leukocyte antigen; 4-HNE: 4-hydroxynonenal; HODE: hydroxyoctadecadienoic acid; IG: immunoglobulin; IL: interleukin; INV: involucrin; ISCA1: iron-sulfur cluster assembly 1; JNK: Jun amino-terminal kinases; KC: keratinocyte; LCE3e: late cornified envelope 3e; LRP5: LDL receptor related protein 5; LTB_4: Leukotriene B_4; MEKK1: mitogen-activated protein kinase kinase kinase 1; MMP9: matrix metalloproteinase 9; NCOA/SIRT: nuclear receptor coactivator; NF-AT: nuclear factor of activated T-cells; NF-κB: nuclear factor kappa-light-chain-enhancer of activated B cells; NLRC4: NLR family CARD domain containing 4 protein; NLRP3: nucleotide-binding oligomerization domain-like receptor (NOD)-like receptor family pyrin domain containing 3; NPC1L1: Niemann-Pick C1-like protein 1; NRF2: NF-E2–related factor 2; NOX4: nicotinamide adenine dinucleotide phosphate (NADPH) oxidase 4; OXPHOS: oxidative phosphorylation; PASI: psoriasis area and severity index; p38MAPK: p38 mitogen-activated protein kinase; PDH: pyruvate dehydrogenase; PDK: pyruvate dehydrogenase kinase; PG: prostaglandin; PHACTR1: phosphatase and actin regulator 1; PI3K: phosphoinositide-3-kinase; PK: protein kinase; PLIN2: perilipin 2; PPAR: peroxisome proliferator-activated receptor; ROS: reactive oxygen species; RXR: 9-cis-retinoic acid receptor; S100A: S100 calcium-binding protein A; SAPK: stress-activated protein kinase; SDS: sodium dodecyl sulfate; SENP2: small ubiquitin-like modifier (SUMO)-specific protease 2; SIRT1: silent mating type information regulation 2 homolog 1: SLC: solute carrier family; STAT: signal transducer and activator of transcription; TCF/LEF: T-cell factor/lymphoid enhancer factor; TGM1: transglutaminase 1; Th: T helper; TNF: tumor necrosis factor; TPA: phorbol 12-myristate 13-acetate; TR: T3-thyroid receptor; TTA: tetradecylthioacetic acid; UCP: uncoupling protein; VDR: vitamin D receptor; VEGFA: vascular endothelial growth factor A.

References

1. Girroir, E.E.; Hollingshead, H.E.; He, P.; Zhu, B.; Perdew, G.H.; Peters, J.M. Quantitative expression patterns of peroxisome proliferator-activated receptor-beta/delta (PPARbeta/delta) protein in mice. *Biochem. Biophys. Res. Commun.* **2008**, *371*, 456–461. [CrossRef] [PubMed]
2. Lundell, K.; Thulin, P.; Hamsten, A.; Ehrenborg, E. Alternative splicing of human peroxisome proliferator-activated receptor delta (PPAR delta): Effects on translation efficiency and trans-activation ability. *BMC Mol. Biol.* **2007**, *8*, 70. [CrossRef] [PubMed]

3. Larsen, L.K.; Amri, E.Z.; Mandrup, S.; Pacot, C.; Kristiansen, K. Genomic organization of the mouse peroxisome proliferator-activated receptor beta/delta gene: Alternative promoter usage and splicing yield transcripts exhibiting differential translational efficiency. *Biochem. J.* **2002**, *366*, 767–775. [CrossRef] [PubMed]
4. Harmon, G.S.; Lam, M.T.; Glass, C.K. PPARs and Lipid Ligands in Inflammation and Metabolism. *Chem. Rev.* **2011**, *111*, 6321–6340. [CrossRef] [PubMed]
5. Westergaard, M.; Henningsen, J.; Johansen, C.; Rasmussen, S.; Svendsen, M.L.; Jensen, U.B.; Schrøder, H.D.; Staels, B.; Iversen, L.; Bolund, L.; et al. Expression and localization of peroxisome proliferator-activated receptors and nuclear factor κB in normal and lesional psoriatic skin. *J. Investig. Dermatol.* **2003**, *121*, 1104–1117. [CrossRef]
6. Zarei, M.; Aguilar-Recarte, D.; Palomer, X.; Vázquez-Carrera, M. Revealing the role of peroxisome proliferator-activated receptor β/δ in nonalcoholic fatty liver disease. *Metabolism* **2021**, *114*, 154342. [CrossRef]
7. Liu, Y.; Colby, J.K.; Zuo, X.; Jaoude, J.; Wei, D.; Shureiqi, I. The Role of PPAR-δ in Metabolism, Inflammation, and Cancer: Many Characters of a Critical Transcription Factor. *Int. J. Mol. Sci.* **2018**, *19*, 3339. [CrossRef] [PubMed]
8. Kahremany, S.; Livne, A.; Gruzman, A.; Senderowitz, H.; Sasson, S. Activation of PPARδ: From computer modelling to biological effects. *Br. J. Pharmacol.* **2014**, *172*, 754–770. [CrossRef] [PubMed]
9. Riahi, Y.; Sin-Malia, Y.; Cohen, G.; Alpert, E.; Gruzman, A.; Eckel, J.; Staels, B.; Guichardant, M.; Sasson, S. The natural protective mechanism against hyperglycemia in vascular endothelial cells: Roles of the lipid peroxidation product 4-hydroxydodecadienal and peroxisome proliferator-activated receptor delta. *Diabetes* **2010**, *59*, 808–818. [CrossRef]
10. Coleman, J.D.; Prabhu, K.S.; Thompson, J.T.; Reddy, P.S.; Peters, J.M.; Peterson, B.R.; Reddy, C.C.; Vanden Heuvel, J.P. The oxidative stress mediator 4-hydroxynonenal is an intracellular agonist of the nuclear receptor peroxisome proliferator-activated receptor-β/δ (PPARβ/δ). *Free Radic. Biol. Med.* **2007**, *42*, 1155–1164. [CrossRef]
11. Da'adoosh, B.; Marcus, D.; Rayan, A.; King, F.; Che, J.; Goldblum, A. Discovering highly selective and diverse PPAR-delta agonists by ligand based machine learning and structural modeling. *Sci. Rep.* **2019**, *9*, 1106. [CrossRef] [PubMed]
12. Shearer, B.G.; Hoekstra, W.J. Recent advances in peroxisome proliferator-activated receptor science. *Curr. Med. Chem.* **2003**, *10*, 267–280. [CrossRef]
13. Gou, Q.; Jiang, Y.; Zhang, R.; Xu, Y.; Xu, H.; Zhang, W.; Shi, J.; Hou, Y. PPARδ is a regulator of autophagy by its phosphorylation. *Oncogene* **2020**, *39*, 4844–4853. [CrossRef] [PubMed]
14. Brunmeir, R.; Xu, F. Functional Regulation of PPARs through Post-Translational Modifications. *Int. J. Mol. Sci.* **2018**, *19*, 1738. [CrossRef]
15. Koo, Y.D.; Choi, J.W.; Kim, M.; Chae, S.; Ahn, B.Y.; Kim, M.; Oh, B.C.; Hwang, D.; Seol, J.H.; Kim, Y.-B.; et al. SUMO-Specific Protease 2 (SENP2) Is an Important Regulator of Fatty Acid Metabolism in Skeletal Muscle. *Diabetes* **2015**, *64*, 2420–2431. [CrossRef] [PubMed]
16. Genini, D.; Catapano, C.V. Block of nuclear receptor ubiquitination. A mechanism of ligand-dependent control of peroxisome proliferator-activated receptor delta activity. *J. Biol. Chem.* **2007**, *282*, 11776–11785. [CrossRef]
17. Rieck, M.; Wedeken, L.; Müller-Brüsselbach, S.; Meissner, W.; Müller, R. Expression level and agonist-binding affect the turnover, ubiquitination and complex formation of peroxisome proliferator activated receptor β. *FEBS J.* **2007**, *274*, 5068–5076. [CrossRef]
18. Ricote, M.; Glass, C.K. PPARs and molecular mechanisms of transrepression. *Biochim. Biophys. Acta (BBA) Mol. Cell Biol. Lipids* **2007**, *1771*, 926–935. [CrossRef]
19. Tan, N.S.; Vázquez-Carrera, M.; Montagner, A.; Sng, M.K.; Guillou, H.; Wahli, W. Transcriptional control of physiological and pathological processes by the nuclear receptor PPARβ/δ. *Prog. Lipid Res.* **2016**, *64*, 98–122. [CrossRef] [PubMed]
20. Planavila, A.; Rodríguez-Calvo, R.; Jové, M.; Michalik, L.; Wahli, W.; Laguna, J.C.; Vázquez-Carrera, M. Peroxisome proliferator-activated receptor β/δ activation inhibits hypertrophy in neonatal rat cardiomyocytes. *Cardiovasc. Res.* **2005**, *65*, 832–841. [CrossRef]
21. Chen, D.; Xiong, X.-Q.; Zang, Y.-H.; Tong, Y.; Zhou, B.; Chen, Q.; Li, Y.-H.; Gao, X.-Y.; Kang, Y.-M.; Zhu, G.-Q. BCL6 attenuates renal inflammation via negative regulation of NLRP3 transcription. *Cell Death Dis.* **2017**, *8*, e3156. [CrossRef]
22. Zhang, H.; Qi, X.; Wu, J.; Huang, X.; Zhang, A.; Chen, S.; Ding, X.; Le, S.; Zou, Y.; Xu, H.; et al. BCL6 inhibitor FX1 attenuates inflammatory responses in murine sepsis through strengthening BCL6 binding affinity to downstream target gene promoters. *Int. Immunopharmacol.* **2019**, *75*, 105789. [CrossRef] [PubMed]
23. Lee, C.H.; Chawla, A.; Urbiztondo, N.; Liao, D.; Boisvert, W.A.; Evans, R.M.; Curtiss, L.K. Transcriptional repression of atherogenic inflammation: Modulation by PPARδ. *Science* **2003**, *302*, 453–457. [CrossRef] [PubMed]
24. Zarei, M.; Barroso, E.; Palomer, X.; Escolà-Gil, J.C.; Cedó, L.; Wahli, W.; Vázquez-Carrera, M. Pharmacological PPARβ/δ activation upregulates VLDLR in hepatocytes. *Clin. Investig. Arterioscler.* **2019**, *31*, 111–118.
25. Palomer, X.; Barroso, E.; Pizarro-Delgado, J.; Peña, L.; Botteri, G.; Zarei, M.; Aguilar, D.; Montori-Grau, M.; Vázquez-Carrera, M. PPARβ/δ: A Key Therapeutic Target in Metabolic Disorders. *Int. J. Mol. Sci.* **2018**, *19*, 913. [CrossRef]
26. Zarei, M.; Barroso, E.; Leiva, R.; Barniol-Xicota, M.; Pujol, E.; Escolano, C.; Vázquez, S.; Palomer, X.; Pardo, V.; González-Rodríguez, Á.; et al. Heme-Regulated eIF2α Kinase Modulates Hepatic FGF21 and Is Activated by PPARβ/δ Deficiency. *Diabetes* **2016**, *65*, 3185–3199. [CrossRef]
27. Adhikary, T.; Wortmann, A.; Schumann, T.; Finkernagel, F.; Lieber, S.; Roth, K.; Toth, P.M.; Diederich, W.E.; Nist, A.; Stiewe, T.; et al. The transcriptional PPARβ/δ network in human macrophages defines a unique agonist-induced activation state. *Nucleic Acids Res.* **2015**, *43*, 5033–5051. [CrossRef] [PubMed]

28. Narkar, V.A.; Downes, M.; Yu, R.T.; Embler, E.; Wang, Y.X.; Banayo, E.; Mihaylova, M.M.; Nelson, M.C.; Zou, Y.; Juguilon, H.; et al. AMPK and PPARδ agonists are exercise mimetics. *Cell* **2008**, *134*, 405–415. [CrossRef] [PubMed]
29. Dunlop, T.W.; Väisänen, S.; Frank, C.; Molnár, F.; Sinkkonen, L.; Carlberg, C. The human peroxisome proliferator-activated receptor δ gene is a primary target of 1α,25-dihydroxyvitamin D3 and its nuclear receptor. *J. Mol. Biol.* **2005**, *349*, 248–260. [CrossRef]
30. Magge, S.S.; Guardiola-Diaz, H. Characterization of the mouse peroxisome proliferator-activated receptor δ gene. *Biochem. Biophys. Res. Commun.* **2002**, *290*, 230–235. [CrossRef]
31. Romanowska, M.; al Yacoub, N.; Seidel, H.; Donandt, S.; Gerken, H.; Phillip, S.; Haritonova, N.; Artuc, M.; Schweiger, S.; Sterry, W.; et al. PPARδ enhances keratinocyte proliferation in psoriasis and induces heparin-binding EGF-like growth factor. *J. Investig. Dermatol.* **2008**, *128*, 110–124. [CrossRef]
32. Tan, N.S.; Michalik, L.; Noy, N.; Yasmin, R.; Pacot, C.; Heim, M.; Flühmann, B.; Desvergne, B.; Wahli, W. Critical roles of PPAR β/δ in keratinocyte response to inflammation. *Genes Dev.* **2001**, *15*, 3263–3277. [CrossRef]
33. Lu, C.; Cheng, S.-Y. Thyroid hormone receptors regulate adipogenesis and carcinogenesis via crosstalk signaling with peroxisome proliferator-activated receptors. *J. Mol. Endocrinol.* **2009**, *44*, 143–154. [CrossRef] [PubMed]
34. Decker, R.H. Nature and Regulation of Energy Metabolism in the Epidermis. *J. Investig. Dermatol.* **1971**, *57*, 351–363. [CrossRef]
35. Adachi, K.; Yamasawa, S. Quantitative histochemistry of the primate skin. *I. Hexokinase. J. Investig. Dermatol.* **1966**, *46*, 473–476. [CrossRef]
36. Im, M.J.; Yamasawa, S.; Adachi, K. Quantitative histochemistry of the primate skin. III. Glyceraldehyde-3-phosphate dehydrogenase. *J. Investig. Dermatol.* **1966**, *47*, 35–38. [CrossRef]
37. Bedogni, B.; Powell, M.B. Skin Hypoxia: A Promoting Environmental Factor in Melanomagenesis. *Cell Cycle* **2006**, *5*, 1258–1261. [CrossRef]
38. Pavel, P.; Leman, G.; Hermann, M.; Ploner, C.; Eichmann, T.O.; Minzaghi, D.; Radner, F.P.W.; Del Frari, B.; Gruber, R.; Dubrac, S. Peroxisomal fatty acid oxidation and glycolysis are triggered in mouse models of lesional atopic dermatitis. *JID Innov.* **2021**, 100033, in press. [CrossRef]
39. Cibrian, D.; de la Fuente, H.; Sánchez-Madrid, F. Metabolic Pathways That Control Skin Homeostasis and Inflammation. *Trends Mol. Med.* **2020**, *26*, 975–986. [CrossRef]
40. Sutter, C.H.; Olesen, K.M.; Bhuju, J.; Guo, Z.; Sutter, T.R. AHR Regulates Metabolic Reprogramming to Promote SIRT1-Dependent Keratinocyte Differentiation. *J. Investig. Dermatol.* **2019**, *139*, 818–826. [CrossRef]
41. Hamanaka, R.B.; Chandel, N.S. Mitochondrial metabolism as a regulator of keratinocyte differentiation. *Cell. Logist.* **2013**, *3*, e25456. [CrossRef]
42. Simpson, C.L.; Tokito, M.K.; Uppala, R.; Sarkar, M.K.; Gudjonsson, J.E.; Holzbaur, E.L. NIX initiates mitochondrial fragmentation via DRP1 to drive epidermal differentiation. *Cell Rep.* **2021**, *34*, 108689. [CrossRef]
43. Ständer, S. Atopic Dermatitis. *N. Engl. J. Med.* **2021**, *384*, 1136–1143. [CrossRef]
44. Griffiths, C.E.M.; Armstrong, A.W.; Gudjonsson, J.E.; Barker, J. Psoriasis. *Lancet* **2021**, *397*, 1301–1315. [CrossRef]
45. Moosbrugger-Martinz, V.; Hackl, H.; Gruber, R.; Pilecky, M.; Knabl, L.; Orth-Höller, D.; Dubrac, S. Initial Evidence of Distinguishable Bacterial and Fungal Dysbiosis in the Skin of Patients with Atopic Dermatitis or Netherton Syndrome. *J. Investig. Dermatol.* **2021**, *141*, 114–123. [CrossRef] [PubMed]
46. Minzaghi, D.; Pavel, P.; Dubrac, S. Xenobiotic Receptors and Their Mates in Atopic Dermatitis. *Int. J. Mol. Sci.* **2019**, *20*, 4234. [CrossRef] [PubMed]
47. Blunder, S.; Kõks, S.; Kõks, G.; Reimann, E.; Hackl, H.; Gruber, R.; Moosbrugger-Martinz, V.; Schmuth, M.; Dubrac, S. Enhanced Expression of Genes Related to Xenobiotic Metabolism in the Skin of Patients with Atopic Dermatitis but Not with Ichthyosis Vulgaris. *J. Investig. Dermatol.* **2018**, *138*, 98–108. [CrossRef]
48. Moosbrugger-Martinz, V.; Schmuth, M.; Dubrac, S. A Mouse Model for Atopic Dermatitis Using Topical Application of Vitamin D3 or of Its Analog MC903. *Methods Mol. Biol.* **2017**, *1559*, 91–106. [CrossRef]
49. Moosbrugger-Martinz, V.; Tripp, C.; Clausen, B.; Schmuth, M.; Dubrac, S. Atopic dermatitis induces the expansion of thymus-derived regulatory T cells exhibiting a Th2-like phenotype in mice. *J. Cell. Mol. Med.* **2016**, *20*, 930–938. [CrossRef]
50. Schmuth, M.; Blunder, S.; Dubrac, S.; Gruber, R.; Moosbrugger-Martinz, V. Epidermal barrier in hereditary ichthyoses, atopic dermatitis, and psoriasis. *J. Der Dtsch. Dermatol. Ges.* **2015**, *13*, 1119–1123. [CrossRef] [PubMed]
51. Dubrac, S.; Schmuth, M.; Ebner, S. Atopic dermatitis: The role of Langerhans cells in disease pathogenesis. *Immunol. Cell Biol.* **2010**, *88*, 400–409. [CrossRef] [PubMed]
52. Reynolds, G.; Vegh, P.; Fletcher, J.; Poyner, E.F.M.; Stephenson, E.; Goh, I.; Botting, R.A.; Huang, N.; Olabi, B.; Dubois, A.; et al. Developmental cell programs are co-opted in inflammatory skin disease. *Science* **2021**, *371*, eaba6500. [CrossRef]
53. Yamazaki, F. Psoriasis: Comorbidities. *J. Dermatol.* **2021**, *48*, 732–740. [CrossRef]
54. Zeng, C.; Tsoi, L.C.; Gudjonsson, J.E. Dysregulated epigenetic modifications in psoriasis. *Exp. Dermatol.* **2021**. [CrossRef]
55. Bugaut, H.; Aractingi, S. Major Role of the IL17/23 Axis in Psoriasis Supports the Development of New Targeted Therapies. *Front. Immunol.* **2021**, *12*, 621956. [CrossRef] [PubMed]
56. Kahremany, S.; Hofmann, L.; Harari, M.; Gruzman, A.; Cohen, G. Pruritus in psoriasis and atopic dermatitis: Current treatments and new perspectives. *Pharmacol. Rep.* **2021**, *73*, 443–453. [CrossRef]

57. Billoni, N.; Buan, B.; Gautier, B.; Collin, C.; Gaillard, O.; Mahé, Y.F.; Bernard, B.A. Expression of peroxisome proliferator activated receptors (PPARs) in human hair follicles and PPAR α involvement in hair growth. *Acta Derm. Venereol.* **2000**, *80*, 329–334.
58. Dubrac, S.; Stoitzner, P.; Pirkebner, D.; Elentner, A.; Schoonjans, K.; Auwerx, J.; Saeland, S.; Hengster, P.; Fritsch, P.; Romani, N.; et al. Peroxisome Proliferator-Activated Receptor-α Activation Inhibits Langerhans Cell Function. *J. Immunol.* **2007**, *178*, 4362–4372. [CrossRef]
59. Schmuth, M.; Jiang, Y.J.; Dubrac, S.; Elias, P.M.; Feingold, K.R. Thematic Review Series: Skin Lipids. Peroxisome proliferator-activated receptors and liver X receptors in epidermal biology. *J. Lipid Res.* **2008**, *49*, 499–509. [CrossRef]
60. Westergaard, M.; Henningsen, J.; Kratchmarova, I.; Kristiansen, K.; Svendsen, M.L.; Johansen, C.; Jensen, U.B.; Schrøder, H.D.; Berge, R.K.; Iversen, L.; et al. Modulation of Keratinocyte Gene Expression and Differentiation by PPAR-Selective Ligands and Tetradecylthioacetic Acid. *J. Investig. Dermatol.* **2001**, *116*, 702–712. [CrossRef]
61. Rivier, M.; Safonova, I.; Lebrun, P.; Michel, S.; Griffiths, C.; Ailhaud, G. Differential Expression of Peroxisome Proliferator-Activated Receptor Subtypes During the Differentiation of Human Keratinocytes. *J. Investig. Dermatol.* **1998**, *111*, 1116–1121. [CrossRef]
62. Romanowska, M.; Reilly, L.; Palmer, C.N.; Gustafsson, M.C.; Foerster, J. Activation of PPARβ/δ causes a psoriasis-like skin disease in vivo. *PLoS ONE* **2010**, *5*, e9701. [CrossRef] [PubMed]
63. Chamcheu, J.C.; Chaves-Rodriquez, M.I.; Adhami, V.M.; Siddiqui, I.A.; Wood, G.S.; Longley, B.J.; Mukhtar, H. Upregulation of PI3K/AKT/mTOR, FABP5 and PPARβ/δ in Human Psoriasis and Imiquimod-induced Murine Psoriasiform Dermatitis Model. *Acta. Derm. Venereol.* **2016**, *96*, 854–856. [CrossRef] [PubMed]
64. Wang, X.; Hao, Y.; Wang, X.; Wang, L.; Chen, Y.; Sun, J.; Hu, J. A PPARδ-selective antagonist ameliorates IMQ-induced psoriasis-like inflammation in mice. *Int. Immunopharmacol.* **2016**, *40*, 73–78. [CrossRef]
65. Dou, J.; Zhang, L.; Xie, X.; Ye, L.; Yang, C.; Wen, L.; Shen, C.; Zhu, C.; Zhao, S.; Zhu, Z.; et al. Integrative analyses reveal biological pathways and key genes in psoriasis. *Br. J. Dermatol.* **2017**, *177*, 1349–1357. [CrossRef] [PubMed]
66. Schneider, C.; Strayhorn, W.D.; Brantley, D.M.; Nanney, L.B.; Yull, F.E.; Brash, A.R. Upregulation of 8-lipoxygenase in the dermatitis of IκB-α-deficient mice. *J. Investig. Dermatol.* **2004**, *122*, 691–698. [CrossRef] [PubMed]
67. Funk, C.D.; Keeney, D.S.; Oliw, E.; Boeglin, W.E.; Brash, A.R. Functional Expression and Cellular Localization of a Mouse Epidermal Lipoxygenase. *J. Biol. Chem.* **1996**, *271*, 23338–23344. [CrossRef] [PubMed]
68. Krieg, P.; Fürstenberger, G. The role of lipoxygenases in epidermis. *Biochim. Biophys. Acta (BBA) Mol. Cell Biol. Lipids* **2014**, *1841*, 390–400. [CrossRef]
69. Morgan, E.; Kannan-Thulasiraman, P.; Noy, N. Involvement of Fatty Acid Binding Protein 5 and PPARβ/δ in Prostate Cancer Cell Growth. *PPAR Res.* **2010**, *2010*, 234629. [CrossRef]
70. Werner, S.; Grose, R. Regulation of Wound Healing by Growth Factors and Cytokines. *Physiol. Rev.* **2003**, *83*, 835–870. [CrossRef]
71. Lande, R.; Gregorio, J.; Facchinetti, V.; Chatterjee, B.; Wang, Y.-H.; Homey, B.; Cao, W.; Wang, Y.-H.; Su, B.; Nestle, F.O.; et al. Plasmacytoid dendritic cells sense self-DNA coupled with antimicrobial peptide. *Nature* **2007**, *449*, 564–569. [CrossRef]
72. Luo, Y.; Hara, T.; Kawashima, A.; Ishido, Y.; Suzuki, S.; Ishii, N.; Kambara, T.; Suzuki, K. Pathological role of excessive DNA as a trigger of keratinocyte proliferation in psoriasis. *Clin. Exp. Immunol.* **2020**, *202*, 1–10. [CrossRef] [PubMed]
73. Lee, J.; Kim, B.; Chu, H.; Zhang, K.; Kim, H.; Kim, J.H.; Kim, S.H.; Pan, Y.; Noh, J.Y.; Sun, Z.; et al. FABP5 as a possible biomarker in atopic march: FABP5-induced Th17 polarization, both in mouse model and human samples. *EBioMedicine* **2020**, *58*, 102879. [CrossRef] [PubMed]
74. Gericke, J.; Ittensohn, J.; Mihály, J.; Dubrac, S.; Rühl, R. Allergen-Induced Dermatitis Causes Alterations in Cutaneous Retinoid Mediated Signaling in Mice. *PLoS ONE* **2013**, *8*, e71244. [CrossRef]
75. Töröcsik, D.; Weise, C.; Gericke, J.; Szegedi, A.; Lucas, R.; Mihaly, J.; Worm, M.; Rühl, R. Transcriptomic and lipidomic profiling of eicosanoid/docosanoid signalling in affected and non-affected skin of human atopic dermatitis patients. *Exp. Dermatol.* **2019**, *28*, 177–189. [CrossRef]
76. Chiba, H.; Michibata, H.; Wakimoto, K.; Seishima, M.; Kawasaki, S.; Okubo, K.; Mitsui, H.; Torii, H.; Imai, Y. Cloning of a Gene for a Novel Epithelium-specific Cytosolic Phospholipase A2, cPLA2δ, Induced in Psoriatic Skin. *J. Biol. Chem.* **2004**, *279*, 12890–12897. [CrossRef]
77. Jarrett, R.; Salio, M.; Lloyd-Lavery, A.; Subramaniam, S.; Bourgeois, E.; Archer, C.; Cheung, K.L.; Hardman, C.; Chandler, D.; Salimi, M.; et al. Filaggrin inhibits generation of CD1a neolipid antigens by house dust mite–derived phospholipase. *Sci. Transl. Med.* **2016**, *8*, 325ra18. [CrossRef] [PubMed]
78. Hardman, C.S.; Chen, Y.-L.; Salimi, M.; Jarrett, R.; Johnson, D.; Järvinen, V.J.; Owens, R.J.; Repapi, E.; Cousins, D.J.; Barlow, J.L.; et al. CD1a presentation of endogenous antigens by group 2 innate lymphoid cells. *Sci. Immunol.* **2017**, *2*, eaan5918. [CrossRef]
79. Lian, N.; Shi, L.Q.; Hao, Z.M.; Chen, M. Research progress and perspective in metabolism and metabolomics of psoriasis. *Chin. Med. J.* **2020**, *133*, 2976–2986. [CrossRef] [PubMed]
80. Dutkiewicz, E.P.; Hsieh, K.-T.; Wang, Y.-S.; Chiu, H.-Y.; Urban, P.L. Hydrogel Micropatch and Mass Spectrometry—Assisted Screening for Psoriasis-Related Skin Metabolites. *Clin. Chem.* **2016**, *62*, 1120–1128. [CrossRef]
81. Zhang, Z.; Zi, Z.; Lee, E.E.; Zhao, J.; Contreras, D.C.; South, A.P.; Abel, E.D.; Chong, B.F.; Vandergriff, T.; Hosler, G.A.; et al. Differential glucose requirement in skin homeostasis and injury identifies a therapeutic target for psoriasis. *Nat. Med.* **2018**, *24*, 617–627. [CrossRef]

82. Roberts, L.D.; Hassall, D.G.; Winegar, D.A.; Haselden, J.N.; Nicholls, A.W.; Griffin, J.L. Increased hepatic oxidative metabolism distinguishes the action of Peroxisome proliferator-activated receptor δ from Peroxisome proliferator-activated receptor γ in the ob/ob mouse. *Genome Med.* **2009**, *1*, 115. [CrossRef] [PubMed]
83. Fan, W.; Evans, R. PPARs and ERRs: Molecular mediators of mitochondrial metabolism. *Curr. Opin. Cell Biol.* **2015**, *33*, 49–54. [CrossRef] [PubMed]
84. Lamichane, S.; Lamichane, B.D.; Kwon, S.-M. Pivotal Roles of Peroxisome Proliferator-Activated Receptors (PPARs) and Their Signal Cascade for Cellular and Whole-Body Energy Homeostasis. *Int. J. Mol. Sci.* **2018**, *19*, 949. [CrossRef]
85. Magadum, A.; Engel, F.B. PPARβ/δ: Linking Metabolism to Regeneration. *Int. J. Mol. Sci.* **2018**, *19*, 2013. [CrossRef] [PubMed]
86. Palomer, X.; Salvadó, L.; Barroso, E.; Vázquez-Carrera, M. An overview of the crosstalk between inflammatory processes and metabolic dysregulation during diabetic cardiomyopathy. *Int. J. Cardiol.* **2013**, *168*, 3160–3172. [CrossRef]
87. Kwak, C.-H.; Jin, L.; Han, J.H.; Han, C.W.; Kim, E.; Cho, M.; Chung, T.-W.; Bae, S.-J.; Jang, S.B.; Ha, K.-T. Ilimaquinone Induces the Apoptotic Cell Death of Cancer Cells by Reducing Pyruvate Dehydrogenase Kinase 1 Activity. *Int. J. Mol. Sci.* **2020**, *21*, 6021. [CrossRef] [PubMed]
88. Rubio, B.; Mora, C.; Pintado, C.; Mazuecos, L.; Fernández, A.; López, V.; Andrés, A.; Gallardo, N. The nutrient sensing pathways FoxO1/3 and mTOR in the heart are coordinately regulated by central leptin through PPARβ/δ. Implications in cardiac remodeling. *Metabolism* **2021**, *115*, 154453. [CrossRef] [PubMed]
89. Higgins, L.G.; Garbacz, W.G.; Gustafsson, M.C.; Nainamalai, S.; Ashby, P.R.; Wolf, C.R.; Palmer, C.N. Conditional Expression of Human PPARδ and a Dominant Negative Variant of hPPARδ In Vivo. *PPAR Res.* **2012**, *2012*, 216817. [CrossRef]
90. Luquet, S.; Gaudel, C.; Holst, D.; Lopez-Soriano, J.; Jehl-Pietri, C.; Fredenrich, A.; Grimaldi, P.A. Roles of PPAR delta in lipid absorption and metabolism: A new target for the treatment of type 2 diabetes. *Biochim. Biophys. Acta (BBA) Mol. Basis Dis.* **2005**, *1740*, 313–317. [CrossRef] [PubMed]
91. Berdyshev, E.; Goleva, E.; Bronova, I.; Dyjack, N.; Rios, C.; Jung, J.; Taylor, P.; Jeong, M.; Hall, C.F.; Richers, B.N.; et al. Lipid abnormalities in atopic skin are driven by type 2 cytokines. *JCI Insight* **2018**, *3*, e98006. [CrossRef]
92. Lodhi, I.J.; Semenkovich, C.F. Peroxisomes: A Nexus for Lipid Metabolism and Cellular Signaling. *Cell Metab.* **2014**, *19*, 380–392. [CrossRef]
93. Baldwin, H.A.; Wang, C.; Kanfer, G.; Shah, H.V.; Velayos-Baeza, A.; Dulovic-Mahlow, M.; Brüggemann, N.; Anding, A.; Baehrecke, E.H.; Maric, D.; et al. VPS13D promotes peroxisome biogenesis. *J. Cell Biol.* **2021**, *220*, e202001188. [CrossRef]
94. Li, Q.; Fang, H.; Dang, E.; Wang, G. The role of ceramides in skin homeostasis and inflammatory skin diseases. *J. Dermatol. Sci.* **2020**, *97*, 2–8. [CrossRef]
95. Montagner, A.; Wahli, W.; Tan, N.S. Nuclear receptor peroxisome proliferator activated receptor (PPAR) β/δ in skin wound healing and cancer. *Eur. J. Dermatol.* **2015**, *25*, 4–11.
96. Blunder, S.; Krimbacher, T.; Moosbrugger-Martinz, V.; Gruber, R.; Schmuth, M.; Dubrac, S. Keratinocyte-derived IL-1β induces PPARG downregulation and PPARD upregulation in human reconstructed epidermis following barrier impairment. *Exp. Dermatol.* **2021**. [CrossRef] [PubMed]
97. Qiang, L.; Sample, A.; Liu, H.; Wu, X.; He, Y.-Y. Epidermal SIRT1 regulates inflammation, cell migration, and wound healing. *Sci. Rep.* **2017**, *7*, 14110. [CrossRef] [PubMed]
98. Bielach-Bazyluk, A.; Zbroch, E.; Mysliwiec, H.; Rydzewska-Rosolowska, A.; Kakareko, K.; Flisiak, I.; Hryszko, T. Sirtuin 1 and Skin: Implications in Intrinsic and Extrinsic Aging—A Systematic Review. *Cells* **2021**, *10*, 813. [CrossRef] [PubMed]
99. Cheang, W.S.; Wong, W.T.J.; Wang, L.; Cheng, C.K.; Lau, C.W.; Ma, R.C.W.; Xu, A.; Wang, N.; Huang, Y.; Tian, X.Y. Resveratrol ameliorates endothelial dysfunction in diabetic and obese mice through sirtuin 1 and peroxisome proliferator-activated receptor δ. *Pharmacol. Res.* **2019**, *139*, 384–394. [CrossRef]
100. Hack, K.; Reilly, L.; Palmer, C.; Read, K.D.; Norval, S.; Kime, R.; Booth, K.; Foerster, J. Skin-targeted inhibition of PPAR β/δ by selective antagonists to treat PPAR β/δ-mediated psoriasis-like skin disease in vivo. *PLoS ONE* **2012**, *7*, e37097. [CrossRef] [PubMed]
101. Kuenzli, S.; Saurat, J.-H. Effect of topical PPARβ/δ and PPARγ agonists on plaque psoriasis. A pilot study. *Dermatology* **2003**, *206*, 252–256. [CrossRef]
102. Hatano, Y.; Man, M.Q.; Uchida, Y.; Crumrine, D.; Mauro, T.M.; Feingold, K.R.; Elias, P.M.; Holleran, W.M. Murine atopic dermatitis responds to peroxisome proliferator-activated receptors α and β/δ (but not γ) and liver X receptor activators. *J. Allergy Clin. Immunol.* **2010**, *125*, 160–169.e5. [CrossRef]
103. Bertino, L.; Guarneri, F.; Cannavò, S.P.; Casciaro, M.; Pioggia, G.; Gangemi, S. Oxidative Stress and Atopic Dermatitis. *Antioxidants* **2020**, *9*, 196. [CrossRef] [PubMed]
104. Ji, H.; Li, X.-K. Oxidative Stress in Atopic Dermatitis. *Oxidative Med. Cell. Longev.* **2016**, *2016*, 2721469. [CrossRef] [PubMed]
105. Lin, X.; Huang, T. Oxidative stress in psoriasis and potential therapeutic use of antioxidants. *Free. Radic. Res.* **2016**, *50*, 585–595. [CrossRef] [PubMed]
106. Barroso, E.; Rodriguez-Rodriguez, R.; Chacón, M.R.; Masip, E.M.; Ferrer, L.; Salvadó, L.; Salmerón, E.; Wabistch, M.; Palomer, X.; Vendrell, J.; et al. PPARβ/δ ameliorates fructose-induced insulin resistance in adipocytes by preventing Nrf2 activation. *Biochim. Biophys. Acta (BBA) Mol. Basis Dis.* **2015**, *1852*, 1049–1058. [CrossRef] [PubMed]

107. Jimenez, R.; Toral, M.; Gómez-Guzmán, M.; Romero, M.; Sanchez, M.; Mahmoud, A.; Duarte, J. The Role of Nrf2 Signaling in PPARβ/δ-Mediated Vascular Protection against Hyperglycemia-Induced Oxidative Stress. *Oxidative Med. Cell. Longev.* **2018**, *2018*, 5852706. [CrossRef] [PubMed]
108. Tan, E.H.P.; Sng, M.K.; How, I.S.B.; Chan, J.S.K.; Chen, J.; Tan, C.K.; Wahli, W.; Tan, N.S. ROS release by PPARβ/δ-null fibroblasts reduces tumor load through epithelial antioxidant response. *Oncogene* **2018**, *37*, 2067–2078. [CrossRef]
109. Salvadó, L.; Barroso, E.; Gómez-Foix, A.M.; Palomer, X.; Michalik, L.; Wahli, W.; Vázquez-Carrera, M. PPARβ/δ prevents endoplasmic reticulum stress-associated inflammation and insulin resistance in skeletal muscle cells through an AMPK-dependent mechanism. *Diabetologia* **2014**, *57*, 2126–2135. [CrossRef] [PubMed]
110. Wagner, N.; Wagner, K.-D. The Role of PPARs in Disease. *Cells* **2020**, *9*, 2367. [CrossRef]
111. Peters, J.M.; Kim, D.J.; Bility, M.T.; Borland, M.G.; Zhu, B.; Gonzalez, F.J. Regulatory mechanisms mediated by peroxisome proliferator-activated receptor-β/δ in skin cancer. *Mol. Carcinog.* **2019**, *58*, 1612–1622. [CrossRef]
112. Borland, M.G.; Kehres, E.M.; Lee, C.; Wagner, A.L.; Shannon, B.E.; Albrecht, P.P.; Zhu, B.; Gonzalez, F.J.; Peters, J.M. Inhibition of tumorigenesis by peroxisome proliferator-activated receptor (PPAR)-dependent cell cycle blocks in human skin carcinoma cells. *Toxicology* **2018**, *404–405*, 25–32. [CrossRef] [PubMed]
113. Zhu, B.; Ferry, C.H.; Markell, L.K.; Blazanin, N.; Glick, A.B.; Gonzalez, F.J.; Peters, J.M. The Nuclear Receptor Peroxisome Proliferator-activated Receptor-β/δ (PPARβ/δ) Promotes Oncogene-induced Cellular Senescence through Repression of Endoplasmic Reticulum Stress. *J. Biol. Chem.* **2014**, *289*, 20102–20119. [CrossRef] [PubMed]
114. Han, C.; Lim, K.; Xu, L.; Li, G.; Wu, T. Regulation of Wnt/β-catenin pathway by cPLA2α and PPARδ. *J. Cell. Biochem.* **2008**, *105*, 534–545. [CrossRef]
115. Hwang, I.; Kim, J.; Jeong, S. β-Catenin and Peroxisome Proliferator-activated Receptor-δ Coordinate Dynamic Chromatin Loops for the Transcription of Vascular Endothelial Growth Factor A Gene in Colon Cancer Cells. *J. Biol. Chem.* **2012**, *287*, 41364–41373. [CrossRef]
116. Blunder, S.; Rühl, R.; Moosbrugger-Martinz, V.; Krimmel, C.; Geisler, A.; Zhu, H.; Crumrine, D.; Elias, P.M.; Gruber, R.; Schmuth, M.; et al. Alterations in Epidermal Eicosanoid Metabolism Contribute to Inflammation and Impaired Late Differentiation in FLG-Mutated Atopic Dermatitis. *J. Investig. Dermatol.* **2017**, *137*, 706–715. [CrossRef] [PubMed]
117. Michiels, J.F.; Perrin, C.; Leccia, N.; Massi, D.; Grimaldi, P.; Wagner, N. PPARβ activation inhibits melanoma cell proliferation involving repression of the Wilms' tumour suppressor WT1. *Pflügers Arch.* **2010**, *459*, 689–703. [CrossRef]
118. Lim, J.C.W.; Kwan, Y.P.; Tan, M.S.; Teo, M.H.Y.; Chiba, S.; Wahli, W.; Wang, X. The Role of PPARβ/δ in Melanoma Metastasis. *Int. J. Mol. Sci.* **2018**, *19*, 2860. [CrossRef] [PubMed]
119. Han, J.-K.; Kim, H.-L.; Jeon, K.-H.; Choi, Y.-E.; Lee, H.-S.; Kwon, Y.-W.; Jang, J.-J.; Cho, H.-J.; Kang, H.-J.; Oh, B.-H.; et al. Peroxisome proliferator-activated receptor-δ activates endothelial progenitor cells to induce angio-myogenesis through matrix metallo-proteinase-9-mediated insulin-like growth factor-1 paracrine networks. *Eur. Heart J.* **2013**, *34*, 1755–1765. [CrossRef]
120. Huang, H. Matrix Metalloproteinase-9 (MMP-9) as a Cancer Biomarker and MMP-9 Biosensors: Recent Advances. *Sensors* **2018**, *18*, 3249. [CrossRef]

International Journal of
Molecular Sciences

MDPI

Review

PPARγ and TGFβ—Major Regulators of Metabolism, Inflammation, and Fibrosis in the Lungs and Kidneys

Gábor Kökény [1,2,*], Laurent Calvier [3,4,5,6,*] and Georg Hansmann [5,6,*]

1 Institute of Translational Medicine, Semmelweis University, 1089 Budapest, Hungary
2 International Nephrology Research and Training Center, Semmelweis University, 1089 Budapest, Hungary
3 Department of Molecular Genetics, University of Texas (UT) Southwestern Medical Center, Dallas, TX 75390, USA; calvier.laurent@gmail.com
4 Center for Translational Neurodegeneration Research, University of Texas (UT) Southwestern Medical Center, Dallas, TX 75390, USA
5 Pulmonary Vascular Research Center (PVRC), Hannover Medical School, 30625 Hannover, Germany
6 Department of Pediatric Cardiology and Critical Care, Hannover Medical School, 30625 Hannover, Germany
* Correspondence: kokeny.gabor@med.semmelweis-univ.hu (G.K.); georg.hansmann@gmail.com (G.H.); Tel.: +36-1-210-0100 (G.K.); +49-511-532-9594 (G.H.)

Abstract: Peroxisome proliferator-activated receptor gamma (PPARγ) is a type II nuclear receptor, initially recognized in adipose tissue for its role in fatty acid storage and glucose metabolism. It promotes lipid uptake and adipogenesis by increasing insulin sensitivity and adiponectin release. Later, PPARγ was implicated in cardiac development and in critical conditions such as pulmonary arterial hypertension (PAH) and kidney failure. Recently, a cluster of different papers linked PPARγ signaling with another superfamily, the transforming growth factor beta (TGFβ), and its receptors, all of which play a major role in PAH and kidney failure. TGFβ is a multifunctional cytokine that drives inflammation, fibrosis, and cell differentiation while PPARγ activation reverses these adverse events in many models. Such opposite biological effects emphasize the delicate balance and complex crosstalk between PPARγ and TGFβ. Based on solid experimental and clinical evidence, the present review summarizes connections and their implications for PAH and kidney failure, highlighting the similarities and differences between lung and kidney mechanisms as well as discussing the therapeutic potential of PPARγ agonist pioglitazone.

Keywords: PPARγ; pulmonary arterial hypertension; TGFβ; vascular injury; inflammation; proliferation; kidney fibrosis

Citation: Gábor Kökény, Laurent Calvier and Georg Hansmann PPARγ and TGFβ—Major Regulators of Metabolism, Inflammation, and Fibrosis in the Lungs and Kidneys. *Int. J. Mol. Sci.* **2021**, *22*, 10431. https://doi.org/10.3390/ijms221910431

Academic Editors: Manuel Vázquez-Carrera and Walter Wahli

Received: 31 August 2021
Accepted: 24 September 2021
Published: 28 September 2021

1. Introduction

Peroxisome proliferator-activated receptors (PPARs; α, β/δ, γ) are ligand-activated transcription factors of the nuclear receptor superfamily that regulate metabolic homeostasis of the cell. Among them, PPARγ regulates synthetic metabolism (anabolism) in the adipose tissue and plays an important role in glucose metabolism [1] and cardiac development [2]. The human PPARγ gene contains nine exons spanning over 100 kilobases located on chromosome 3 [3]. The ligand-activated PPARγ regulates target genes by forming a heterodimer with the retinoid X receptor (RXR). Mutations in PPARγ gene have been associated with dysfunctional lipid and glucose homeostasis leading to obesity and type 2 diabetes mellitus (T2DM) [4,5] but also with thyroid cancer [6].

Although PPARγ is predominantly a key regulator of adipocyte homeostasis, it is ubiquitously expressed. Overall, there were predominantly protective effects in the cardiovascular system, including systemic and pulmonary circulation. The diseases and conditions which are positively affected by PPARγ activation in preclinical and/or clinical studies include but are not limited to pulmonary arterial hypertension (PAH), prediabetes/insulin resistance, cardiovascular diseases such as stroke in prediabetes, nephrotic syndrome, kidney, or lung fibrosis, independently of the blood glucose lowering effect [7–12].

Recently, post-transcriptional regulation of PPARγ by microRNAs have been implicated in different diseases [10,13,14]. Protein phosphorylation is another regulatory mechanism that can reduce or increase the transcriptional activity of PPARγ [15].

Since 2007 [16], PPARγ agonists have emerged as promising novel, antiproliferative, anti-inflammatory, insulin-sensitizing, and efficient medications for the treatment of PAH. Still, the results of earlier diabetes studies and their false interpretations, as well as scarce reports on the possible adverse effects, substantially diminished the interest on using pharmacological PPARγ activation for the treatment of cardiovascular diseases, including PAH. However, the recent, very large IRIS trial [17–19] did not confirm any serious adverse effects for the PPARγ agonists pioglitazone when used in patients with insulin resistance/prediabetes—in fact, pioglitazone decreased the risk for stroke and myocardial infarction [17]. The present review summarizes recent experimental and clinical evidences showing how PPARγ participates in the pathogenesis of pulmonary and renal diseases while also highlighting the therapeutic potential of the thiazolidinedione (TZD) class PPARγ agonists (e.g., pioglitazone and rosiglitazone) in these diseases.

2. Role of PPARγ Crosstalk with TGFβ Superfamily Members and microRNAs in Pulmonary Vascular Homeostasis

The pathology of PAH affects not only the pulmonary arteries but also several extrapulmonary organs (heart, skeletal muscle, and adipose tissue) [20–24] that share common metabolic abnormalities (i.e., suppression of mitochondrial glucose oxidation and increased glycolysis, disturbed fatty acid oxidation (FAO), and dyslipidemia/insulin resistance) [16,20,24–26].

PPARγ regulates several target genes that are strongly implicated in the pathobiology of PAH, for instance adiponectin (APN), IL-6, monocyte chemotactic protein-1 (MCP-1/CCL2) or endothelin-1 (ET-1) [25,27]. PPARγ agonists have been proven to exert antiproliferative (on vascular smooth-muscle cells (VSMC)), anti-inflammatory, proangiogenic, and proapoptotic effects in cells, animal models, and patients, emphasizing their therapeutic potential in PAH and other cardiopulmonary diseases, even in the absence of insulin resistance [25].

Bone morphogenetic protein 2 (BMP2) is a ligand of BMPR2 and inhibits VSMC growth. In endothelial cells, however, BMP2 acts as a survival factor and hence may counteract the endothelial cell injury and dysfunction in the early stages of PAH. Loss-of-function mutations in the BMPR2 gene are frequently seen in familial/hereditable (HPAH, 70%, i.e., germline mutations) and idiopathic PAH (IPAH, 10–20%) cases. The recent discovery of an antiproliferative BMP2/BMPR2-PPARγ-ApoE axis [28] in VSMC suggests that dysfunction of BMPR2 reduces endogenous PPARγ activity [28]. Thus, the activation of PPARγ might reverse the PAH phenotype in patients with or without BMPR2 mutations. Pulmonary BMPR2 expression decreases even in the absence of BMPR2 mutations in idiopathic or HPAH and in PAH secondary to connective tissue disease or congenital heart disease [29]. Importantly, PAH patients have reduced pulmonary BMP2 [30], PPARγ [31], and apolipoprotein E (ApoE) mRNA expression [30]. PPARγ inhibits cell growth in hypoxia-exposed human pulmonary arterial smooth-muscle cells (HPASMC) through the suppression of miR-21, and its activation cancels programmed cell death protein 4 (PDCD4) repression, thus facilitating the apoptosis of HPASMC [32]. SCUBE1, a proposed BMP co-receptor has been recently identified as a novel factor in the pathogenesis of PAH. In cultured PAECs, BMPR2 knockdown induced SCUBE1 downregulation, and both plasma and lung biopsy samples of PAH patients demonstrated reduced SCUBE1 expression that correlated with disease severity [33].

The calcineurin inhibitor tacrolimus (FK506) used in picomolar concentrations binds to the BMP signaling repressor FK-binding protein-12 (FKBP12). Low-dose FK506 treatment of floxed endothelial cell-specific Bmpr2$^{-/-}$ mice prevented the development of hypoxia-induced pulmonary arterial muscularization and normalized RVSP. Additionally, a 3 week FK506 treatment was able to reverse established PAH in the SU5416 (VEGFR2 inhibitor)/hypoxia (SuHx) rat model via the activation of apelin that suppresses PASMC proliferation. In human PAECs obtained from iPAH patients, low-dose FK506 reduced endothelial dysfunction [34].

We identified PPARγ as a missing link and a key regulator of the functional antagonism between BMP2 and TGFβ1 pathways in human and murine VSMC [10,14]. In HPASMC, PPARγ activation with pioglitazone inhibited a novel noncanonical TGFβ1-pSTAT3-pFoxO1 pathway, in addition to the inhibition of the canonical TGFβ1-pSmad3/4 axis [10,35]. Additionally, pioglitazone treatment of TGFβ1-overexpressing mice reversed PAH and pulmonary vascular remodeling [10] (Figure 1). Recently, the alleviation of disrupted PPARγ-p53 axis in PAEC from BMPR2 mutant patients emerged as a possible therapeutic potential for PAH [36]. Even in the absence of other possible injuries the cell-specific deficiency of PPARγ in VSMCs was demonstrated to increase pulmonary vascular muscularization in mice, independently of a low-fat or high-fat diet [37].

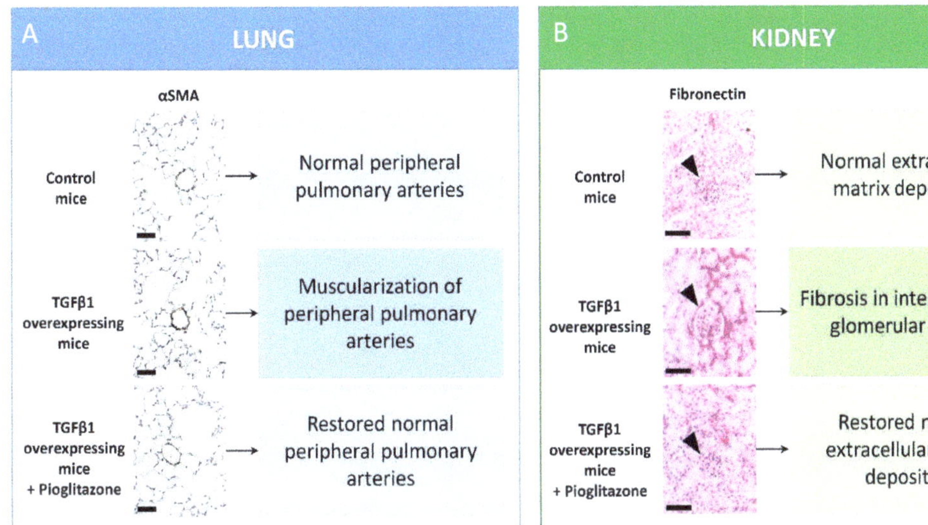

Figure 1. Representative photomicrographs of lung and kidney immunohistochemistry in TGFβ overexpressing mice treated with pioglitazone. Lungs stained for αSMA depicted significant muscularization of peripheral pulmonary arteries in untreated TGFβ overexpressing mice as compared to controls but restored arterial wall morphology upon pioglitazone treatment (**A**). Scale bar: 50 μm. Fibronectin staining of the kidneys in untreated TGFβ overexpressing mice depicts increased tubulointerstitial and glomerular production (arrowhead points on glomeruli) but restored fibronectin content after chronic pioglitazone treatment (**B**). Scale bar: 50 μm.

It has been shown that the miR-130/-301 family promotes pulmonary hypertension through systemic regulation of miRNA networks [38–40], where PPARγ plays a key role as a direct target of this miRNA family. For instance, pulmonary arteries from IPAH patients demonstrated increased miR-130a/-301b expression as compared to controls [10]. Additionally, TGFβ1 stimulation of HPASMC reduces PPARγ-mRNA via miR-130a/-301b, hence suppressing the BMP2/BMPR2-PPARγ axis. Recently, new miRNAs upregulated by the BMP2/PPARγ axis have been identified. In HPASMC, BMP2 induces miR-331-5p, which downregulates the mRNA expression of the platelet isoform of phosphofructokinase (PFKP), a rate-limiting enzyme of glycolysis and pro-proliferative factor that is highly expressed in situ in pulmonary arteries of IPAH patients vs. controls [10]. Activation of the BMP2/BMPR2-PPARγ axis upregulates miR-331-5p and miR-148a (suspected to repress cell proliferation), thus inhibiting proliferation and glucose metabolism in VSMC [10,14].

Heat-shock protein 90 (Hsp90) is a molecular chaperone involved in many cellular protein interactions, and abnormal Hsp90 expression has been recently attributed to

PAH [41,42]. Increased expression levels of cytosolic Hsp90 have been found in PASMCs of PAH patients, and a Hsp90-inhibitor suppressed PASMC proliferation [42]. Targeted inhibition of mitochondrial Hsp90 reversed pulmonary arterial remodeling in the monocrotaline rat model of PAH and in PAH-PASMC in vitro [41]. Hsp90 might also have a strong cellular interplay with PPARγ. Interestingly, Hsp90 stabilized PPARγ in both liver cells [43] and adipocytes [44], and Hsp90 inhibition lowered PPARγ levels, while Hsp90 overexpression diminished PPARγ degradation [43] in liver cells. However, the reduced Hsp90/eNOS signaling and endothelial dysfunction in PAH has been attributed to reduced PPARγ levels, modulated by miR-27b overexpression in HPAECs and also in monocrotaline-induced rat model of PAH [45]. The exposure of ovine PAECs to TGFβ1 resulted in reduced PPARγ expression, mitochondrial dysfunction, and disrupted Hsp90/eNOS signaling [46]. These studies suggest that dysfunctional, boosted TGFβ1 results in suppression of the PPARγ/Hsp90/eNOS signaling, contributing to endothelial dysfunction and PASMC proliferation in PAH.

LRP1 is a recognized vasoprotective receptor that interacts with several ligands, such as growth factors, cytokines, lipoproteins, and extracellular matrix components. LRP1 serves as a co-receptor for TGFBRs inhibiting the growth effect of TGFβ by interacting with Smad2/3 signaling [47]. Reduced vascular LRP1 expression was recently demonstrated in human PAH, and LRP1 in VSMC was found to protect from PAH in vivo [48]. Importantly, the activation of PPARγ by pioglitazone reversed PAH caused by LRP1 deficiency in murine VSMC, inhibiting Smad3, Nox4, and CTGF [48]. Hence, PPARγ activation can normalize TGFβ1/BMP2 homeostasis via regulation of both canonical and non-canonical TGFβ1 pathways and the expression of key miRNAs involved in cell proliferation and glucose/lipid metabolism (summarized in Figure 2).

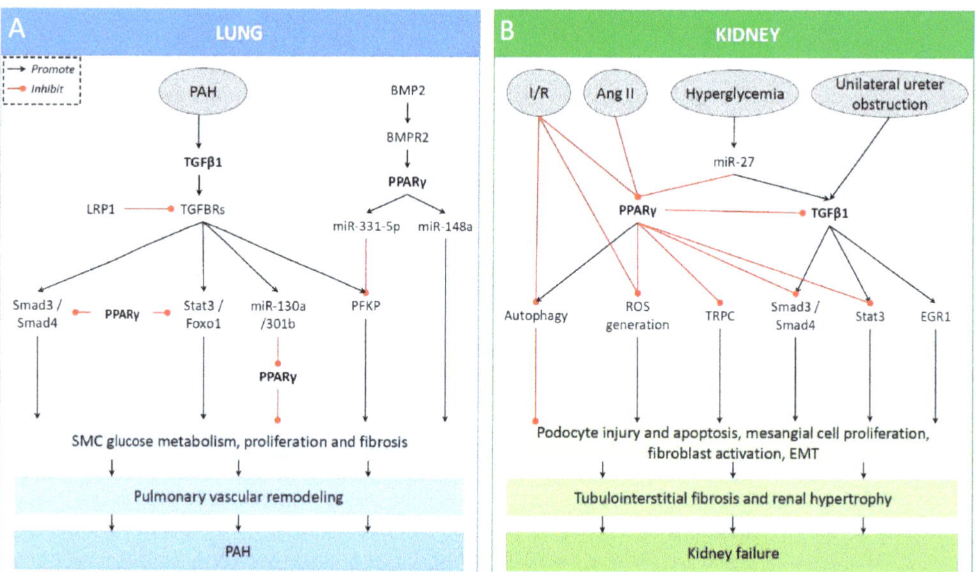

Figure 2. Summary of PPARγ actions in pulmonary arterial hypertension (**A**) and kidney disease models (**B**).

3. Dysregulation of Metabolic Pathways and PPARγ Dysfunction in PAH

The role of dysfunctional PPARγ in the pathogenesis of metabolic disturbances has been demonstrated both in human PAH and in experimental models. In patients suffering from idiopathic pulmonary arterial hypertension, PPARγ mRNA expression was found to be markedly reduced in the failing RV [49]. Knockdown of PPARγ in cultured HPASMC has been associated with reduced PGC1α and with stimulating mitochondrial fragmentation and superoxide production and inducing proliferation [50].

The elevated TG/HDL ratio in PAH patients is the manifestation of lipid and lipoprotein homeostasis alterations due to insulin resistance [20,24]. Decreased fatty acidy oxidation (FAO) can directly cause myocardial lipid accumulation (lipotoxicity) [51], and this occurs in end-stage human PAH-RVs [42] as well as in the SU5416 (VEGFR2 inhibitor)/hypoxia (SuHx) PAH rat model [49]. In addition, the targeted deletion of PPARγ in cardiomyocytes of mice induces biventricular systolic dysfunction even in the absence of PAH [49]. Oral treatment in the SuHx rat model with PPARγ agonist pioglitazone reverses PAH and prevents RV failure through regulating mRNA and miRNA networks that restore mitochondrial fatty acid oxidation (FAO) and prevent lipotoxicity [49]. Studies in cardiomyocytes identified a direct link between miR-197 and miR-146b overexpression and the suppression of genes that drive FAO. PPARγ activation downregulated miR-197 and miR-146b that were upregulated in the SuHx-RV but were also found to be upregulated in the pressure-overloaded failing human RV in end-stage idiopathic PAH [49]. Thus, PPARγ activation could prevent lipotoxicity by normalizing transcriptional and post-transcriptional regulation of the disturbed lipid metabolism and mitochondrial function.

BOLA3 (BolA Family Member 3) is a member of mitochondrial iron-sulfur cluster assembly system. BOLA3 deficiency has been recently attributed to PAH via the activation of glycolysis and fatty acid oxidation, inhibiting glycine catabolism and increasing mitochondrial respiration in PAEC [52]. In cultured PAECs but not PASMCs, hypoxia downregulated BOLA3 expression. In addition, BOLA3 was found to be repressed in lungs of hypoxic C57Bl/6 mice and the SuHx rat PAH model, and in lung biopsies of PH patients. Importantly, orotracheal administration of adeno-associated virus carrying BOLA3 transgene was able to prevent hypoxia-induced PH in mice [52]. In human brown adipose tissue, BOLA3 gene expression was found to be positively correlated to PPARG expression [53].

Tribbles homolog 3 (TRIB3), a pseudokinase of the Tribbles family that inhibits AKT phosphorylation, is involved in several metabolic cellular events in the liver or adipose tissue [54,55]. TRIB3 also plays a role in the development of skeletal muscle insulin resistance and cellular glucotoxicity in diabetes [56]. Recently, TRIB3 was recognized to participate in the pathogenesis of pulmonary hypertension by reducing PPARγ activity [57]. In cultured PAECs, lentiviral overexpression of TRIB3 upregulated ERK1/2 and downregulated PPARγ and eNOS activity under both normoxia and hypoxia. Knockdown of TRIB3 by 50% in hypoxic PAECs reduced ERK1/2 and increased eNOS phosphorylation. The early pioglitazone treatment of rats with hypoxia-induced pulmonary hypertension (HPH) partially ameliorated PH and vascular insulin resistance through reduction of TRIB3 and ERK1/2 activity; pioglitazone also restored eNOS [57]. These findings suggest that hypoxia-induced TRIB3 and insulin resistance in PAECs contributes to PH that can be inhibited by early activation of PPARγ.

4. PPARγ in Renal Glomerular and Epithelial Cell Metabolism

PPARγ protein is expressed in several regions of the kidney, including different renal tubule segments [58], interstitial cells, the juxtaglomerular apparatus, podocytes, mesangial cells, and renal microvascular endothelial cells [59]. Since multiple renal cells show endogenous PPARγ expression and activity, PPARγ might play an important role in maintaining normal homeostasis and function of the kidney. Several studies on synthetic PPARγ agonists showed renoprotective effects of such compounds in both diabetic and nondiabetic kidney diseases and models of renal fibrosis [11,60–62]. PPARγ agonists of the thiazolidinedione class ("TZDs"), such as pioglitazone and rosiglitazone, have been demonstrated to induce PPARγ mRNA and protein expression in podocytes and tubular epithelial cells, in association with the amelioration of aging-related progressive renal injury [63,64]. The effects of PPARγ activation on experimental kidney disease models are summarized in Figure 2.

In the renal glomerulus, glucose and free fatty acids (FFA) are freely filtrated. Approximately 70% of filtrated FFAs are reabsorbed and then metabolized by β-oxidation within mitochondria in the proximal tubules, providing a significant energy source (in form of ATP); on the other hand, high amounts of intracellular fatty acids might limit ammonia production [65]. Mice deficient of PPARγ (having disrupted exon B1 of PPARγ2) and leptin develop metabolic syndrome with dyslipidemia, as well as renal hypertrophy and increased expression of the profibrotic TGFβ in the kidney [66]. Similar to FFA, filtrated glucose is also reabsorbed in the proximal tubules, using sodium-dependent glucose cotransporters (SGLT2 located in segment S1 and SGLT1 in segment S3). Hyperglycemia results in dysfunction of the SGLT-mediated glucose reabsorption in proximal tubular cells and promotes the profibrotic epithelial-to-mesenchymal transition (EMT). Such hyperglycemia-induced EMT can be reversed by PPARγ agonists that restore the SGLT-mediated glucose reabsorption [67].

In a recent study, PPARγ was shown to regulate proximal tubule cell metabolism by suppressing glycolysis and EGF degradation. Indeed, inhibition of PPARγ with GW9662 resulted in proximal tubule cell dysfunction in vitro, and in C57Bl6 mice it caused tubular hypertrophy, increased interstitial collagen deposition, and expression of kidney injury molecule-1 (KIM-1) [68]. These findings implicate that PPARγ agonists might exhibit their antifibrotic effect in the kidneys—at least partly—via modulation of tubular epithelial cell metabolism.

Podocytes play a principal role in the glomerular filtration and also express PPARγ [69]. Fatty acid treatment of podocytes (as a lipotoxicity model) tended to reduce PPARγ expression and led to inflammatory and apoptotic cellular events [70]. Several animal models of podocyte injury revealed the protective effect of PPARγ in podocytes. For instance, in puromycin aminoglycoside (PAN)-induced podocyte damage (that leads to nephrotic syndrome), pioglitazone treatment reduces proteinuria to the same extent as high dose glucocorticoid treatment and effectively attenuates podocyte damage [8]. The protective effect of PPARγ activation in podocytes is attributed to the reduction of profibrotic TGFβ expression and inhibition of apoptosis [64], restoring podocyte synaptopodin expression and ameliorating podocyte foot process effacement [62]. Rosiglitazone reduced aldosterone-induced podocyte damage by restoring nephrin expression and slit diaphragm integrity, as well as by reducing the amount of oxidative radicals [71]. Rosiglitazone also ameliorated the stretch-induced decrease in nephrin expression of podocytes in vitro [72]. Recently, fibroblast growth factor-1 (FGF1) has been demonstrated to reduce TGFβ expression via the induction of PPARγ, which resulted in EMT inhibition on cultured mouse podocytes and diabetic mouse model, reducing fibrosis and proteinuria [73].

One of the mechanisms of how PPARγ can reduce proteinuria and glomerular disease has been demonstrated lately by Sonneveld and colleagues. Transient receptor potential channel C6 (TRPC6) is a nonspecific calcium (Ca^{2+})—conducting ion channel and a transcriptional target of PPARγ and reduced TRPC6—mediated Ca^{2+} influx into podocytes leads to podocyte injury in glomerular disease. Cultured mouse podocytes were treated with pioglitazone or rosiglitazone, which inhibited PAN and adriamycin-induced TRPC6 overexpression and significantly inhibited TRPC6 promoter activity. In vivo, rats treated with pioglitazone developed less podocyte damage and milder albuminuria in an adriamycin-induced nephropathy model [74]. Thus, the activation of PPARγ can restore glomerular function by reducing podocyte damage. Apart from the damaged podocytes, glomerular mesangial cells also play pivotal role in the pathogenesis of glomerular sclerosis and function loss. PPARγ activation in cultured rat glomerular mesangial cells decreased AngII-induced Ca^{2+} influx via reducing TRPC activity, inhibiting mesangial cell proliferation, one of the hallmarks of glomerulosclerosis [75].

Of note, activation of pioglitazone as additional treatment over immunosuppression in a child with refractory nephrotic syndrome reduced proteinuria and increased eGFR, while less immunosuppression was needed to maintain renal function [8]. These studies emphasize the critical role of PPARγ in the regulation of renal epithelial, mesangial cell, and podocyte metabolism and homeostasis.

5. PPARγ in Kidney Fibrosis

Fibroproliferative diseases are estimated to account for up to 45% of mortality worldwide [76], resulting in high demand for new therapies fighting tissue fibrosis. PPARγ agonists emerged in the last decade as such new therapies: reduced albuminuria and nephropathy were observed in T2DM patients treated with TZD-class PPARγ agonists [77].

Epiblast-specific systemic deletion of the PPARγ gene in mice leads to the spontaneous development of T2DM and renal fibrosis in aging mice with glomerular hypertrophy, significant proteinuria and collagen deposition. Interestingly, this is associated with antiphospholipid syndrome, glomerular immune complex deposition, and macrophage infiltration [78]. On the other hand, hyperglycemia was shown to decrease PPARγ activity, associated with the upregulation of miR-27a [79]. MiR-27a represses PPARγ and activates TGFβ/Smad3 signaling leading to tubulointerstitial fibrosis, and both in diabetic rats and patients, the elevated plasma miR-27a was associated with poor renal function [80]. Inhibition of miR-27a both in cultured rat mesangial cells and in streptozotocin-induced diabetic rats (a T1DM model) abrogated the reduction of PPARγ and in vivo decreased renal ECM accumulation and podocyte injury [79]. Pioglitazone treatment of ZDF rats, a model of human T2DM, ameliorated diabetic kidney disease and reduced blood pressure as well as interstitial collagen-I and TGFβ production, which was associated with lower renal expression of Twist-1, an evolutionarily conserved protein that can accelerate renal epithelial-to-mesenchymal transition (EMT) and interstitial fibrosis [81].

Furthermore, several experimental studies show that PPARγ agonists bear antifibrotic effects independent of glycemic control. For instance, in the lung fibrosis model induced by silica exposure in mice, a PPARγ agonist inhibited both the reduction of pulmonary PPARγ and LXRa as well as the increase in TGFβ, fibronectin, and collagen-I expression [82]. Further, PPARγ agonist treatment prevented interstitial fibrosis and inflammation in unilateral ureter obstruction (UUO) mouse model of kidney fibrosis through reduction of renal TGFβ expression [9]. It was recently demonstrated that PPARγ activation in TGFβ transgenic mice inhibits the TGFβ-STAT3 and TGFβ-EGR1 transcriptional activation pathways, thus preventing renal fibrosis induced by elevated circulating TGFβ [11] (Figure 1). In kidney fibrosis, the elevated angiotensin-II levels also reduce renal PPARγ expression both in vivo and in vitro, while the angiotensin-II receptor blocker losartan exerts its renoprotective effects partly via the upregulation of PPARγ [83]. Repression of the TGFβ/Smad signaling by PPARγ agonist treatment was recently demonstrated in the hyperuricemia-induced rat model of renal fibrosis, associated with reduced proteinuria, serum creatinine, and BUN levels as well as interstitial ECM accumulation [84]. Another in vivo study where massive glomerular damage and renal fibrosis has been induced with subtotal nephrectomy in rats has implicated the beneficial effect of combined pioglitazone and angiotensin receptor blocker treatment over monotherapies in preserving podocytes, reducing glomerular macrophage infiltration and tubulointerstitial fibrosis. Intriguingly, pioglitazone—even in monotherapy—was able to reduce glomerulosclerosis [85].

Several in vivo and in vitro models emphasize the antifibrotic, TGFβ1-antagonizing effect of BMP7/ALK3 (activin-like kinase-3). For instance, administration of human recombinant BMP7 to rats subjected to UUO or mice with chronic glomerulonephritis reversed the fibrotic process and tubular damage via increased Smad1/5 signaling and reduced Smad2/3 phosphorylation, counteracting the canonical TGFβ1 signaling [86,87]. The induction of BMP signaling via ALK3 activation also inhibits renal fibrosis and tubular epithelial damage in mouse models of renal ischemia-reperfusion, UUO, or glomerulonephritis [88]. In a recent study, the administration of low-dose FK506 inhibited UUO-induced renal fibrosis in mice and activated ALK3 via ARNT transcription factor in cultured tubular epithelial cells, suggesting the antifibrotic role of FKBP12/ARNT/ALK3/BMP7 signaling [89]. Additionally, BMP7 increased both PPARγ expression and activity in cultured human mesangial cells, and the PPARγ agonist rosiglitazone reduced TNFα induced mesangial cell damage in vitro [90].

Fibroblast activation and proliferation is a key step in kidney fibrosis. PPARγ agonist treatment of primary mouse renal fibroblast suppressed PDGF-induced proliferation by inhibiting AKT phosphorylation and subsequent skp2 expression, which regulates cell proliferation via inhibition of p21/p27 effects blocking cell cycle progression [91]. Recently, it has been demonstrated that PPARγ-HGF production in renal fibroblasts regulates tubular epithelial cell survival. Pioglitazone treatment of cultured fibroblasts induced HGF expression, and conditioned media of these fibroblasts significantly attenuated staurosporine-induced acute epithelial cell injury and apoptosis in vitro, but this effect was abrogated by inhibition of downstream HGF signaling [92].

PPARγ activity has been attributed to a healthy epithelial phenotype of proximal tubular epithelial cells, inhibiting EMT and fibrogenesis. The induction of EMT and interstitial collagen production due to unilateral ureter obstruction (UUO) in mice could be attenuated by PPARγ agonist rosiglitazone, which preserved the proximal tubular cell phenotype [93]. In a recent study, the beneficial effect of PPARγ activation was attributed to increased renal Klotho expression and reduced oxidative stress, which effectively ameliorated the age-related nephrosclerosis in ApoE-null mice [94]. Interestingly, mice with Klotho gene loss of function mutations (kl/kl mice) develop cardiac hypertrophy associated with increased cardiac TGFβ protein expression [95].

6. PPARγ in Renal Inflammation and Cardiovascular Disease

In hyperoxaluric mouse model, pioglitazone suppressed renal calcium-oxalate (CaOx) crystal formation and inflammatory injury by enhancing the PPAR-γ mediated expression of miR-23, which dampened macrophage polarization to inflammatory (M1) phenotype but induced the anti-inflammatory M2 phenotype [96]. In a different model, distal tubules of rats that were treated with ethylene glycol to induce CaOx formation, rosiglitazone reduced CaOx crystal formation, oxidative stress, and TGFβ signaling. Similar results were obtained in vitro, using canine distal tubule cells that were induced with oxalate [97].

Interestingly, mice having a macrophage-specific deletion of PPARγ or RXRa develop lupus-like autoimmune glomerulonephritis and antinuclear antibodies [98]. The anti-inflammatory effect of PPARγ raises the therapeutic potential of PPARγ agonists such as pioglitazone in the prevention of chronic rejection after kidney transplantation (see below). The possible role of PPARγ in the development, severity, or progression of glomerulonephritis has been confirmed by another study using a different approach: When podocyte-specific PPARγ-deficient mice were challenged with anti-GBM nephrotoxic serum, they developed more severe glomerulonephritis with mononuclear cell infiltration as compared to wild-type mice treated with same nephrotoxin. Additionally, human kidney biopsies from patients with rapid progressing glomerulonephritis (RPGN) depicted the absence of PPARγ in the nuclei of cells in affected glomeruli [99].

Cardiovascular disease due to arterial calcification is a major complication in chronic kidney disease patients. One of the leading pathomechanism is hyperphosphatemia-induced arterial calcification and differentiation of VSMC into osteoblasts [100]. Hyperphosphatemia reduced PPARγ and Klotho expression in bovine aortic VSMCs, which were reversed by rosiglitazone treatment [101]. Decreased PPARγ expression was recently associated with hyperphosphatemia-induced osteogenic VSMC differentiation in CKD patients, too, and also in mouse VSMC cell line, where reduced BMP2 expression accompanied reduced PPARγ. Here, rosiglitazone inhibited calcification in vitro and also inhibited the hyperphosphatemia-induced vascular calcification in a mouse model of CKD, and this effect was Klotho dependent [102]. Thus, the PPARγ-Klotho axis plays an important role in the hyperphosphatemia-induced ossification of arterial VSMCs. In addition, recent experimental data suggest that PPARγ also plays a protective vascular role against atherosclerosis development by maintaining vascular homeostasis and reducing vascular inflammation. The long-term pioglitazone treatment of ApoE-null mice (a known model for advanced atherosclerosis) markedly reduced the total atherosclerotic lesion area in the aorta, which was accompanied by lower hepatic expression of proinflammatory cytokines as well as increased plasma superoxide dismutase activity [94]. These important roles of PPARγ and ApoE as key players within the antiproliferative BMP2/BMPR2-PPARγ-ApoE axis were first demonstrated in HPASMC [28].

7. PPARγ in Renal Ischemia Reperfusion Injury

One of the main reasons of acute kidney injury (AKI) is renal ischemia reperfusion injury (IRI), leading to the overproduction of reactive oxygen species (ROS) early during reperfusion. Pioglitazone-pretreated rats subjected to 40 min renal IRI had a minimal decline in renal function and almost normalized fractionated sodium excretion (FENa) and proteinuria, as compared to nontreated IRI rats. This renoprotective effect was accompanied by PPARγ-mediated inhibition of NMDA receptor function [103]. In the most sensitive proximal tubular epithelial cells, ROS triggers apoptosis. PPARγ was shown to reduce ROS generation in kidney epithelial cells after hypoxia in vitro and pioglitazone pretreatment of mice for one week before renal IR reduced AKI. The protective effect of the PPARγ activation was associated with the upregulation of uncoupling protein-1 (UCP1, member of the mitochondrial anion carrier protein family expressed in the mitochondrial inner membrane) in renal epithelia [104]. During renal ischemia/reperfusion, autophagy modulates the extent of kidney injury [105]. Pioglitazone pretreatment of NRK rat kidney cells substantially reduced hypoxia-/reoxygenation-induced apoptosis, via activation of autophagy through the AMPK-mTOR regulatory axis [106].

8. The Role of PPARγ in Transplanted Kidneys

Despite the improved immunosuppressive therapies in the past decades leading to a good control of acute rejection and improving short-term graft survivals, chronic rejection of kidney transplants attributed to chronic allograft nephropathy did not improve significantly. Chronic allograft nephropathy (CAN) is mainly caused by excessive inflammation and fibrosis. Biopsies of transplanted kidneys with chronic allograft nephropathy depict increased vascular and tubulointerstitial PAI-1 (plasminogen activator inhibitor-1, a strong profibrotic molecule) expression that is closely associated with fibrosis severity [107]. In a rat model of glomerulosclerosis induced by subtotal nephrectomy, PPARγ activation reduced PAI-1 expression and ameliorated fibrosis, suggesting that PPARγ exerts a protective role in glomerulosclerotic kidneys by downregulating PAI-1 [108]. Interestingly, PPARγ was found to be upregulated in the same kidney areas where PAI-1 was expressed in human biopsies with CAN, and interstitial macrophages were also PPARγ positive in the fibrotic kidneys. This suggests that PPARγ could be induced as counter-acting response to injury in these kidneys [107].

The potential immunosuppressive and antifibrotic effect of PPARγ was also demonstrated in experimental models of allogenic kidney transplantation. Pharmacological activation of PPARγ preserved kidney function of allografts as well as reducing fibrosis, tubular atrophy, and inflammation [109,110]. Furthermore, PPARγ agonist decreased migration and proliferation of both fibroblasts and macrophages [109].

Still, the long-term survival of allografts following renal transplantation highly depends on development of chronic allograft dysfunction. Using the classical Fisher-to-Lewis renal allograft transplantation model, PPARγ activation by rosiglitazone reduced proteinuria by 30% and also decreased interstitial collagen deposition and expression of profibrotic TGFβ. This was accompanied by the reduced expression of renal inflammatory molecules, reduced NF-kB activity, and also attenuated Smad3 phosphorylation [110].

One of the challenges after organ transplantation is the avoidance of immunosuppressive side effects while inhibiting the rejection of grafts. Side effects of immunosuppression can also include deterioration of renal function, so that the use of the potent immunosuppressant Cyclosporin-A (CsA) is sometimes limited due to its known nephrotoxic side effect. Treatment of rats with PPARγ agonist rosiglitazone appear to protect kidneys from

CsA toxicity, associated with a reduction of oxidative stress, renal TGFβ expression, and tubular mitochondrial damage [111].

9. Resurrection of the PPARγ Agonist Pioglitazone

The TZD class drug rosiglitazone was presumed to increase cardiovascular mortality, but the FDA dropped this assumption in recent years, after evaluation of the RECORD (Rosiglitazone Evaluated for Cardiac Outcomes and Regulation of Glycemia in Diabetes) trial [112].

Pioglitazone improves the systolic and diastolic LV function in rodents and in patients with [113] and without [114] diabetes. Pioglitazone has fewer off-target effects and a better side-effect profile as compared to rosiglitazone. Of note, genetic variation determines PPARγ function and the antidiabetic drug response in vivo [115]. Certain single-nucleotide polymorphisms modify binding of the transcription factor PPARγ to its target genes, influencing the antidiabetic drug response in mice and affecting the individual risk for metabolic disease in humans [115]. Therefore, natural genetic variations modifying the PPARγ function affect the individual disease risk and drug response.

10. Summary and Future Directions

Recent studies using PPARγ agonists—and especially pioglitazone—shed light on multiple pathways that can inhibit or even reverse the pathomechanisms at play in PAH and chronic fibroproliferative kidney diseases. These ways of PPARγ actions are either dependent on or independent of the regulation of cell metabolism. In the lungs for instance, PPARγ activation inhibits canonical TGFβ/Smad3 and noncanonical TGFβ/pSTAT3/pFoxO1 pathways in HPASMC, counteracts BMPR2 dysfunction, and induces the antiproliferative PPARγ/apoE axis. PPARγ activation also improves mitochondrial dysfunction and decreases superoxide production. In the kidneys, pioglitazone ameliorates experimental renal fibrosis by repressing TGFβ/pSTAT3 and TGFβ/EGR1 pathways, reducing podocyte injury and apoptosis—partly through restoration of TRPC6—mediated Ca^{2+} influx. The repression of renal TGFβ/Smad signaling by PPARγ activation inhibits interstitial extracellular matrix (ECM) accumulation and epithelial-to-mesenchymal transition (EMT) in both podocytes and tubular epithelium. Additionally, PPARγ activation reduces inflammation and chronic allograft rejection after experimental kidney transplantation. Recent randomized controlled clinical trials show that PPARγ activation with pioglitazone has beneficial effects in cardiovascular patients without significant adverse effects. The experimental and clinical studies suggest that pioglitazone and other, newly developed PPARγ agonists could become a valuable treatment for PAH and kidney fibrosis.

Author Contributions: G.K. drafted the manuscript, L.C. revised the draft and prepared the figures, and G.H. drafted and revised the manuscript. All authors have read and agreed to the published version of the manuscript.

Funding: This work was supported by the German Research Foundation (DFG HA4348/6-2 KFO311 and HA4348/2-2 to G.H.) and the European Pediatric Pulmonary Vascular Disease Network (www.pvdnetwork.org, accessed on 27 September 2021). Dr. Hansmann receives additional funding from the Federal Ministry of Education and Research (BMBF ViP+ program 03VP08053; BMBF 01KC2001B). Dr. Kökény received financial support from the Hungarian Society for Hypertension Scientific Grant, STIA-OTKA 137266/TMI/2020 of the Semmelweis University Innovation Center, Bolyai Scholarship of the Hungarian Academy of Sciences and the ÚNKP Bolyai+ Scholarship (UNKP-20-5-SE-3).

Institutional Review Board Statement: Not applicable.

Informed Consent Statement: Not applicable.

Conflicts of Interest: The authors declare no conflict of interest.

Abbreviations

ApoE	apolipoprotein-E
BMP2	bone morphogenetic protein 2
BMPR2	bone morphogenetic protein receptor 2
CAN	chronic allograft nephropathy
CKD	chronic kidney disease
CTGF	connective tissue growth factor
ECM	extracellular matrix
EGF	endothelial growth factor
EMT	epithelial-to-mesenchymal transition
FGF1	fibroblast growth factor-1
HPASMC	human pulmonary arterial smooth-muscle cell
LRP1	low-density lipoprotein receptor-related protein 1 (TGFβ receptor 5/ApoE receptor)
IPAH	idiopathic pulmonary arterial hypertension
LV	left ventricle
PAEC	pulmonary endothelial cell
PAH	pulmonary arterial hypertension
PPARγ	peroxisome proliferator-activated receptor gamma
ROS	reactive oxygen species
RV	right ventricle
TGFβ	transforming growth factor-β
UUO	unilateral ureter obstruction
VSMC	vascular smooth-muscle cell

References

1. Dubois, V.; Eeckhoute, J.; Lefebvre, P.; Staels, B. Distinct but complementary contributions of PPAR isotypes to energy homeostasis. *J. Clin. Investig.* **2017**, *127*, 1202–1214. [CrossRef]
2. Barak, Y.; Nelson, M.C.; Ong, E.S.; Jones, Y.Z.; Ruiz-Lozano, P.; Chien, K.R.; Koder, A.; Evans, R.M. PPAR gamma is required for placental, cardiac, and adipose tissue development. *Mol. Cell* **1999**, *4*, 585–595. [CrossRef]
3. Beamer, B.A.; Negri, C.; Yen, C.J.; Gavrilova, O.; Rumberger, J.M.; Durcan, M.J.; Yarnall, D.P.; Hawkins, A.L.; Griffin, C.A.; Burns, D.K.; et al. Chromosomal localization and partial genomic structure of the human peroxisome proliferator activated receptor-gamma (hPPAR gamma) gene. *Biochem. Biophys. Res. Commun.* **1997**, *233*, 756–759. [CrossRef]
4. Ristow, M.; Muller-Wieland, D.; Pfeiffer, A.; Krone, W.; Kahn, C.R. Obesity associated with a mutation in a genetic regulator of adipocyte differentiation. *N. Engl. J. Med.* **1998**, *339*, 953–959. [CrossRef]
5. Barroso, I.; Gurnell, M.; Crowley, V.E.; Agostini, M.; Schwabe, J.W.; Soos, M.A.; Maslen, G.L.; Williams, T.D.; Lewis, H.; Schafer, A.J.; et al. Dominant negative mutations in human PPARgamma associated with severe insulin resistance, diabetes mellitus and hypertension. *Nature* **1999**, *402*, 880–883. [CrossRef]
6. Nikiforova, M.N.; Lynch, R.A.; Biddinger, P.W.; Alexander, E.K.; Dorn, G.W., II; Tallini, G.; Kroll, T.G.; Nikiforov, Y.E. RAS point mutations and PAX8-PPAR gamma rearrangement in thyroid tumors: Evidence for distinct molecular pathways in thyroid follicular carcinoma. *J. Clin. Endocrinol. Metab.* **2003**, *88*, 2318–2326. [CrossRef]
7. Sugawara, A.; Uruno, A.; Kudo, M.; Matsuda, K.; Yang, C.W.; Ito, S. Effects of PPARgamma on hypertension, atherosclerosis, and chronic kidney disease. *Endocr. J.* **2010**, *57*, 847–852. [CrossRef]
8. Agrawal, S.; Chanley, M.A.; Westbrook, D.; Nie, X.; Kitao, T.; Guess, A.J.; Benndorf, R.; Hidalgo, G.; Smoyer, W.E. Pioglitazone Enhances the Beneficial Effects of Glucocorticoids in Experimental Nephrotic Syndrome. *Sci. Rep.* **2016**, *6*, 24392. [CrossRef] [PubMed]
9. Kawai, T.; Masaki, T.; Doi, S.; Arakawa, T.; Yokoyama, Y.; Doi, T.; Kohno, N.; Yorioka, N. PPAR-gamma agonist attenuates renal interstitial fibrosis and inflammation through reduction of TGF-beta. *Lab. Investig.* **2009**, *89*, 47–58. [CrossRef] [PubMed]
10. Calvier, L.; Chouvarine, P.; Legchenko, E.; Hoffmann, N.; Geldner, J.; Borchert, P.; Jonigk, D.; Mozes, M.M.; Hansmann, G. PPARgamma Links BMP2 and TGFbeta1 Pathways in Vascular Smooth Muscle Cells, Regulating Cell Proliferation and Glucose Metabolism. *Cell Metab.* **2017**, *25*, 1118–1134.e7. [CrossRef] [PubMed]
11. Nemeth, A.; Mozes, M.M.; Calvier, L.; Hansmann, G.; Kokeny, G. The PPARgamma agonist pioglitazone prevents TGF-beta induced renal fibrosis by repressing EGR-1 and STAT3. *BMC Nephrol.* **2019**, *20*, 245. [CrossRef]
12. Kokeny, G.; Calvier, L.; Legchenko, E.; Chouvarine, P.; Mozes, M.M.; Hansmann, G. PPARgamma is a gatekeeper for extracellular matrix and vascular cell homeostasis: Beneficial role in pulmonary hypertension and renal/cardiac/pulmonary fibrosis. *Curr. Opin. Nephrol. Hypertens.* **2020**, *29*, 171–179. [CrossRef] [PubMed]
13. Peyrou, M.; Ramadori, P.; Bourgoin, L.; Foti, M. PPARs in Liver Diseases and Cancer: Epigenetic Regulation by MicroRNAs. *PPAR Res.* **2012**, *2012*, 757803. [CrossRef]

14. Calvier, L.; Chouvarine, P.; Legchenko, E.; Hansmann, G. Transforming Growth Factor beta1- and Bone Morphogenetic Protein 2/PPARgamma-regulated MicroRNAs in Pulmonary Arterial Hypertension. *Am. J. Respir. Crit. Care Med.* **2017**, *196*, 1227–1228. [CrossRef]
15. Ahmadian, M.; Suh, J.M.; Hah, N.; Liddle, C.; Atkins, A.R.; Downes, M.; Evans, R.M. PPARgamma signaling and metabolism: The good, the bad and the future. *Nat. Med.* **2013**, *19*, 557–566. [CrossRef]
16. Hansmann, G.; Wagner, R.A.; Schellong, S.; Perez, V.A.; Urashima, T.; Wang, L.; Sheikh, A.Y.; Suen, R.S.; Stewart, D.J.; Rabinovitch, M. Pulmonary arterial hypertension is linked to insulin resistance and reversed by peroxisome proliferator-activated receptor-gamma activation. *Circulation* **2007**, *115*, 1275–1284. [CrossRef]
17. Kernan, W.N.; Viscoli, C.M.; Furie, K.L.; Young, L.H.; Inzucchi, S.E.; Gorman, M.; Guarino, P.D.; Lovejoy, A.M.; Peduzzi, P.N.; Conwit, R.; et al. Pioglitazone after Ischemic Stroke or Transient Ischemic Attack. *N. Engl. J. Med.* **2016**, *374*, 1321–1331. [CrossRef]
18. Young, L.H.; Viscoli, C.M.; Curtis, J.P.; Inzucchi, S.E.; Schwartz, G.G.; Lovejoy, A.M.; Furie, K.L.; Gorman, M.J.; Conwit, R.; Abbott, J.D.; et al. Cardiac Outcomes After Ischemic Stroke or Transient Ischemic Attack: Effects of Pioglitazone in Patients With Insulin Resistance Without Diabetes Mellitus. *Circulation* **2017**, *135*, 1882–1893. [CrossRef] [PubMed]
19. Spence, J.D.; Viscoli, C.M.; Inzucchi, S.E.; Dearborn-Tomazos, J.; Ford, G.A.; Gorman, M.; Furie, K.L.; Lovejoy, A.M.; Young, L.H.; Kernan, W.N.; et al. Pioglitazone Therapy in Patients With Stroke and Prediabetes: A Post Hoc Analysis of the IRIS Randomized Clinical Trial. *JAMA Neurol.* **2019**, *76*, 526–535. [CrossRef]
20. Zamanian, R.T.; Hansmann, G.; Snook, S.; Lilienfeld, D.; Rappaport, K.M.; Reaven, G.M.; Rabinovitch, M.; Doyle, R.L. Insulin resistance in pulmonary arterial hypertension. *Eur. Respir. J.* **2009**, *33*, 318–324. [CrossRef] [PubMed]
21. Malenfant, S.; Potus, F.; Fournier, F.; Breuils-Bonnet, S.; Pflieger, A.; Bourassa, S.; Tremblay, E.; Nehme, B.; Droit, A.; Bonnet, S.; et al. Skeletal muscle proteomic signature and metabolic impairment in pulmonary hypertension. *J. Mol. Med.* **2015**, *93*, 573–584. [CrossRef]
22. Jafri, S.; Ormiston, M.L. Immune regulation of systemic hypertension, pulmonary arterial hypertension, and preeclampsia: Shared disease mechanisms and translational opportunities. *Am. J. Physiol. Regul. Integr. Comp. Physiol.* **2017**, *313*, R693–R705. [CrossRef]
23. Culley, M.K.; Chan, S.Y. Mitochondrial metabolism in pulmonary hypertension: Beyond mountains there are mountains. *J. Clin. Investig.* **2018**, *128*, 3704–3715. [CrossRef] [PubMed]
24. Hemnes, A.R.; Luther, J.M.; Rhodes, C.J.; Burgess, J.P.; Carlson, J.; Fan, R.; Fessel, J.P.; Fortune, N.; Gerszten, R.E.; Halliday, S.J.; et al. Human PAH is characterized by a pattern of lipid-related insulin resistance. *JCI Insight* **2019**, *4*, e123611. [CrossRef] [PubMed]
25. Hansmann, G.; Zamanian, R.T. PPARgamma activation: A potential treatment for pulmonary hypertension. *Sci. Transl. Med.* **2009**, *1*, 12ps14. [CrossRef] [PubMed]
26. Bertero, T.; Oldham, W.M.; Cottrill, K.A.; Pisano, S.; Vanderpool, R.R.; Yu, Q.; Zhao, J.; Tai, Y.; Tang, Y.; Zhang, Y.Y.; et al. Vascular stiffness mechanoactivates YAP/TAZ-dependent glutaminolysis to drive pulmonary hypertension. *J. Clin. Investig.* **2016**, *126*, 3313–3335. [CrossRef]
27. Humbert, M.; Guignabert, C.; Bonnet, S.; Dorfmuller, P.; Klinger, J.R.; Nicolls, M.R.; Olschewski, A.J.; Pullamsetti, S.S.; Schermuly, R.T.; Stenmark, K.R.; et al. Pathology and pathobiology of pulmonary hypertension: State of the art and research perspectives. *Eur. Respir. J.* **2019**, *53*, 1801887. [CrossRef]
28. Hansmann, G.; de Jesus Perez, V.A.; Alastalo, T.P.; Alvira, C.M.; Guignabert, C.; Bekker, J.M.; Schellong, S.; Urashima, T.; Wang, L.; Morrell, N.W.; et al. An antiproliferative BMP-2/PPARgamma/apoE axis in human and murine SMCs and its role in pulmonary hypertension. *J. Clin. Investig.* **2008**, *118*, 1846–1857. [CrossRef]
29. Atkinson, C.; Stewart, S.; Upton, P.D.; Machado, R.; Thomson, J.R.; Trembath, R.C.; Morrell, N.W. Primary pulmonary hypertension is associated with reduced pulmonary vascular expression of type II bone morphogenetic protein receptor. *Circulation* **2002**, *105*, 1672–1678. [CrossRef]
30. Geraci, M.W.; Moore, M.; Gesell, T.; Yeager, M.E.; Alger, L.; Golpon, H.; Gao, B.; Loyd, J.E.; Tuder, R.M.; Voelkel, N.F. Gene expression patterns in the lungs of patients with primary pulmonary hypertension: A gene microarray analysis. *Circ. Res.* **2001**, *88*, 555–562. [CrossRef]
31. Ameshima, S.; Golpon, H.; Cool, C.D.; Chan, D.; Vandivier, R.W.; Gardai, S.J.; Wick, M.; Nemenoff, R.A.; Geraci, M.W.; Voelkel, N.F. Peroxisome proliferator-activated receptor gamma (PPARgamma) expression is decreased in pulmonary hypertension and affects endothelial cell growth. *Circ. Res.* **2003**, *92*, 1162–1169. [CrossRef]
32. Green, D.E.; Murphy, T.C.; Kang, B.Y.; Bedi, B.; Yuan, Z.; Sadikot, R.T.; Hart, C.M. Peroxisome proliferator-activated receptor-gamma enhances human pulmonary artery smooth muscle cell apoptosis through microRNA-21 and programmed cell death 4. *Am. J. Physiol. Lung Cell. Mol. Physiol.* **2017**, *313*, L371–L383. [CrossRef]
33. Sun, W.; Tang, Y.; Tai, Y.Y.; Handen, A.; Zhao, J.; Speyer, G.; Al Aaraj, Y.; Watson, A.; Romanelli, M.E.; Sembrat, J.; et al. SCUBE1 Controls BMPR2-Relevant Pulmonary Endothelial Function: Implications for Diagnostic Marker Development in Pulmonary Arterial Hypertension. *JACC Basic Transl. Sci.* **2020**, *5*, 1073–1092. [CrossRef]
34. Spiekerkoetter, E.; Tian, X.; Cai, J.; Hopper, R.K.; Sudheendra, D.; Li, C.G.; El-Bizri, N.; Sawada, H.; Haghighat, R.; Chan, R.; et al. FK506 activates BMPR2, rescues endothelial dysfunction, and reverses pulmonary hypertension. *J. Clin. Investig.* **2013**, *123*, 3600–3613. [CrossRef]

35. Calvier, L.; Chouvarine, P.; Legchenko, E.; Kokeny, G.; Mozes, M.M.; Hansmann, G. Chronic TGF-beta1 Signaling in Pulmonary Arterial Hypertension Induces Sustained Canonical Smad3 Pathways in Vascular Smooth Muscle Cells. *Am. J. Respir. cell Mol. Biol.* **2019**, *61*, 121–123. [CrossRef] [PubMed]
36. Hennigs, J.K.; Cao, A.; Li, C.G.; Shi, M.; Mienert, J.; Miyagawa, K.; Korbelin, J.; Marciano, D.P.; Chen, P.I.; Roughley, M.; et al. PPARgamma-p53-Mediated Vasculoregenerative Program to Reverse Pulmonary Hypertension. *Circ. Res.* **2021**, *128*, 401–418. [CrossRef] [PubMed]
37. Caglayan, E.; Trappiel, M.; Behringer, A.; Berghausen, E.M.; Odenthal, M.; Wellnhofer, E.; Kappert, K. Pulmonary arterial remodelling by deficiency of peroxisome proliferator-activated receptor-gamma in murine vascular smooth muscle cells occurs independently of obesity-related pulmonary hypertension. *Respir. Res.* **2019**, *20*, 42. [CrossRef] [PubMed]
38. Bertero, T.; Lu, Y.; Annis, S.; Hale, A.; Bhat, B.; Saggar, R.; Saggar, R.; Wallace, W.D.; Ross, D.J.; Vargas, S.O.; et al. Systems-level regulation of microRNA networks by miR-130/301 promotes pulmonary hypertension. *J. Clin. Investig.* **2014**, *124*, 3514–3528. [CrossRef]
39. Bertero, T.; Cottrill, K.; Krauszman, A.; Lu, Y.; Annis, S.; Hale, A.; Bhat, B.; Waxman, A.B.; Chau, B.N.; Kuebler, W.M.; et al. The microRNA-130/301 family controls vasoconstriction in pulmonary hypertension. *J. Biol. Chem.* **2015**, *290*, 2069–2085. [CrossRef] [PubMed]
40. Bertero, T.; Cottrill, K.A.; Lu, Y.; Haeger, C.M.; Dieffenbach, P.; Annis, S.; Hale, A.; Bhat, B.; Kaimal, V.; Zhang, Y.Y.; et al. Matrix Remodeling Promotes Pulmonary Hypertension through Feedback Mechanoactivation of the YAP/TAZ-miR-130/301 Circuit. *Cell Rep.* **2015**, *13*, 1016–1032. [CrossRef]
41. Boucherat, O.; Peterlini, T.; Bourgeois, A.; Nadeau, V.; Breuils-Bonnet, S.; Boilet-Molez, S.; Potus, F.; Meloche, J.; Chabot, S.; Lambert, C.; et al. Mitochondrial HSP90 Accumulation Promotes Vascular Remodeling in Pulmonary Arterial Hypertension. *Am. J. Respir. Crit. Care Med.* **2018**, *198*, 90–103. [CrossRef]
42. Wang, G.K.; Li, S.H.; Zhao, Z.M.; Liu, S.X.; Zhang, G.X.; Yang, F.; Wang, Y.; Wu, F.; Zhao, X.X.; Xu, Z.Y. Inhibition of heat shock protein 90 improves pulmonary arteriole remodeling in pulmonary arterial hypertension. *Oncotarget* **2016**, *7*, 54263–54273. [CrossRef] [PubMed]
43. Wheeler, M.C.; Gekakis, N. Hsp90 modulates PPARgamma activity in a mouse model of nonalcoholic fatty liver disease. *J. Lipid Res.* **2014**, *55*, 1702–1710. [CrossRef] [PubMed]
44. Nguyen, M.T.; Csermely, P.; Soti, C. Hsp90 chaperones PPARgamma and regulates differentiation and survival of 3T3-L1 adipocytes. *Cell Death Differ.* **2013**, *20*, 1654–1663. [CrossRef] [PubMed]
45. Bi, R.; Bao, C.; Jiang, L.; Liu, H.; Yang, Y.; Mei, J.; Ding, F. MicroRNA-27b plays a role in pulmonary arterial hypertension by modulating peroxisome proliferator-activated receptor gamma dependent Hsp90-eNOS signaling and nitric oxide production. *Biochem. Biophys. Res. Commun.* **2015**, *460*, 469–475. [CrossRef]
46. Sun, X.; Lu, Q.; Yegambaram, M.; Kumar, S.; Qu, N.; Srivastava, A.; Wang, T.; Fineman, J.R.; Black, S.M. TGF-beta1 attenuates mitochondrial bioenergetics in pulmonary arterial endothelial cells via the disruption of carnitine homeostasis. *Redox Biol.* **2020**, *36*, 101593. [CrossRef]
47. Huang, S.S.; Ling, T.Y.; Tseng, W.F.; Huang, Y.H.; Tang, F.M.; Leal, S.M.; Huang, J.S. Cellular growth inhibition by IGFBP-3 and TGF-beta1 requires LRP-1. *FASEB J. Off. Publ. Fed. Am. Soc. Exp. Biol.* **2003**, *17*, 2068–2081.
48. Calvier, L.; Boucher, P.; Herz, J.; Hansmann, G. LRP1 Deficiency in Vascular SMC Leads to Pulmonary Arterial Hypertension That Is Reversed by PPARgamma Activation. *Circ. Res.* **2019**, *124*, 1778–1785. [CrossRef]
49. Legchenko, E.; Chouvarine, P.; Borchert, P.; Fernandez-Gonzalez, A.; Snay, E.; Meier, M.; Maegel, L.; Mitsialis, S.A.; Rog-Zielinska, E.A.; Kourembanas, S.; et al. PPARgamma agonist pioglitazone reverses pulmonary hypertension and prevents right heart failure via fatty acid oxidation. *Sci. Transl. Med.* **2018**, *10*, eaao0303. [CrossRef]
50. Yeligar, S.M.; Kang, B.Y.; Bijli, K.M.; Kleinhenz, J.M.; Murphy, T.C.; Torres, G.; San Martin, A.; Sutliff, R.L.; Hart, C.M. PPARgamma Regulates Mitochondrial Structure and Function and Human Pulmonary Artery Smooth Muscle Cell Proliferation. *Am. J. Respir. Cell Mol. Biol.* **2018**, *58*, 648–657. [CrossRef]
51. Brittain, E.L.; Talati, M.; Fessel, J.P.; Zhu, H.; Penner, N.; Calcutt, M.W.; West, J.D.; Funke, M.; Lewis, G.D.; Gerszten, R.E.; et al. Fatty Acid Metabolic Defects and Right Ventricular Lipotoxicity in Human Pulmonary Arterial Hypertension. *Circulation* **2016**, *133*, 1936–1944. [CrossRef]
52. Yu, Q.; Tai, Y.Y.; Tang, Y.; Zhao, J.; Negi, V.; Culley, M.K.; Pilli, J.; Sun, W.; Brugger, K.; Mayr, J.; et al. BOLA (BolA Family Member 3) Deficiency Controls Endothelial Metabolism and Glycine Homeostasis in Pulmonary Hypertension. *Circulation* **2019**, *139*, 2238–2255. [CrossRef]
53. Bai, N.; Ma, J.; Alimujiang, M.; Xu, J.; Hu, F.; Xu, Y.; Leng, Q.; Chen, S.; Li, X.; Han, J.; et al. Bola3 Regulates Beige Adipocyte Thermogenesis via Maintaining Mitochondrial Homeostasis and Lipolysis. *Front. Endocrinol.* **2020**, *11*, 592154. [CrossRef]
54. Qi, L.; Heredia, J.E.; Altarejos, J.Y.; Screaton, R.; Goebel, N.; Niessen, S.; Macleod, I.X.; Liew, C.W.; Kulkarni, R.N.; Bain, J.; et al. TRB3 links the E3 ubiquitin ligase COP1 to lipid metabolism. *Science* **2006**, *312*, 1763–1766. [CrossRef] [PubMed]
55. Du, K.; Herzig, S.; Kulkarni, R.N.; Montminy, M. TRB3: A tribbles homolog that inhibits Akt/PKB activation by insulin in liver. *Science* **2003**, *300*, 1574–1577. [CrossRef] [PubMed]
56. Zhang, W.; Wu, M.; Kim, T.; Jariwala, R.H.; Garvey, W.J.; Luo, N.; Kang, M.; Ma, E.; Tian, L.; Steverson, D.; et al. Skeletal Muscle TRIB3 Mediates Glucose Toxicity in Diabetes and High- Fat Diet-Induced Insulin Resistance. *Diabetes* **2016**, *65*, 2380–2391. [CrossRef] [PubMed]

57. Fan, F.; He, J.; Su, H.; Zhang, H.; Wang, H.; Dong, Q.; Zeng, M.; Xing, W.; Sun, X. Tribbles Homolog 3-Mediated Vascular Insulin Resistance Contributes to Hypoxic Pulmonary Hypertension in Intermittent Hypoxia Rat Model. *Front. Physiol.* **2020**, *11*, 542146. [CrossRef]
58. Yang, T.; Michele, D.E.; Park, J.; Smart, A.M.; Lin, Z.; Brosius, F.C., III; Schnermann, J.B.; Briggs, J.P. Expression of peroxisomal proliferator-activated receptors and retinoid X receptors in the kidney. *Am. J. Physiol.* **1999**, *277*, F966–F973. [CrossRef]
59. Kiss-Toth, E.; Roszer, T. PPARgamma in Kidney Physiology and Pathophysiology. *PPAR Res.* **2008**, *2008*, 183108. [CrossRef]
60. Sarafidis, P.A.; Bakris, G.L. Protection of the kidney by thiazolidinediones: An assessment from bench to bedside. *Kidney Int.* **2006**, *70*, 1223–1233. [CrossRef]
61. Sarafidis, P.A.; Stafylas, P.C.; Georgianos, P.I.; Saratzis, A.N.; Lasaridis, A.N. Effect of thiazolidinediones on albuminuria and proteinuria in diabetes: A meta-analysis. *Am. J. Kidney Dis.* **2010**, *55*, 835–847. [CrossRef]
62. Zuo, Y.; Yang, H.C.; Potthoff, S.A.; Najafian, B.; Kon, V.; Ma, L.J.; Fogo, A.B. Protective effects of PPARgamma agonist in acute nephrotic syndrome. *Nephrol. Dial. Transplant.* **2012**, *27*, 174–181. [CrossRef]
63. Yang, H.C.; Deleuze, S.; Zuo, Y.; Potthoff, S.A.; Ma, L.J.; Fogo, A.B. The PPARgamma agonist pioglitazone ameliorates aging-related progressive renal injury. *J. Am. Soc. Nephrol.* **2009**, *20*, 2380–2388. [CrossRef]
64. Kanjanabuch, T.; Ma, L.J.; Chen, J.; Pozzi, A.; Guan, Y.; Mundel, P.; Fogo, A.B. PPAR-gamma agonist protects podocytes from injury. *Kidney Int.* **2007**, *71*, 1232–1239. [CrossRef] [PubMed]
65. Bobulescu, I.A.; Lotan, Y.; Zhang, J.; Rosenthal, T.R.; Rogers, J.T.; Adams-Huet, B.; Sakhaee, K.; Moe, O.W. Triglycerides in the human kidney cortex: Relationship with body size. *PLoS ONE* **2014**, *9*, e101285. [CrossRef] [PubMed]
66. Martinez-Garcia, C.; Izquierdo, A.; Velagapudi, V.; Vivas, Y.; Velasco, I.; Campbell, M.; Burling, K.; Cava, F.; Ros, M.; Oresic, M.; et al. Accelerated renal disease is associated with the development of metabolic syndrome in a glucolipotoxic mouse model. *Dis. Models Mech.* **2012**, *5*, 636–648. [CrossRef]
67. Lee, Y.J.; Han, H.J. Troglitazone ameliorates high glucose-induced EMT and dysfunction of SGLTs through PI3K/Akt, GSK-3beta, Snail1, and beta-catenin in renal proximal tubule cells. *Am. J. Physiol. Renal. Physiol.* **2010**, *298*, F1263–F1275. [CrossRef]
68. Lyu, Z.; Mao, Z.; Li, Q.; Xia, Y.; Liu, Y.; He, Q.; Wang, Y.; Zhao, H.; Lu, Z.; Zhou, Q. PPARgamma maintains the metabolic heterogeneity and homeostasis of renal tubules. *EBioMedicine* **2018**, *38*, 178–190. [CrossRef]
69. Boerries, M.; Grahammer, F.; Eiselein, S.; Buck, M.; Meyer, C.; Goedel, M.; Bechtel, W.; Zschiedrich, S.; Pfeifer, D.; Laloe, D.; et al. Molecular fingerprinting of the podocyte reveals novel gene and protein regulatory networks. *Kidney Int.* **2013**, *83*, 1052–1064. [CrossRef] [PubMed]
70. Martinez-Garcia, C.; Izquierdo-Lahuerta, A.; Vivas, Y.; Velasco, I.; Yeo, T.K.; Chen, S.; Medina-Gomez, G. Renal Lipotoxicity-Associated Inflammation and Insulin Resistance Affects Actin Cytoskeleton Organization in Podocytes. *PLoS ONE* **2015**, *10*, e0142291.
71. Zhu, C.; Huang, S.; Yuan, Y.; Ding, G.; Chen, R.; Liu, B.; Yang, T.; Zhang, A. Mitochondrial dysfunction mediates aldosterone-induced podocyte damage: A therapeutic target of PPARgamma. *Am. J. Pathol.* **2011**, *178*, 2020–2031. [CrossRef]
72. Miceli, I.; Burt, D.; Tarabra, E.; Camussi, G.; Perin, P.C.; Gruden, G. Stretch reduces nephrin expression via an angiotensin II-AT(1)-dependent mechanism in human podocytes: Effect of rosiglitazone. *Am. J. Physiol. Renal. Physiol.* **2010**, *298*, F381–F390. [CrossRef] [PubMed]
73. Wang, D.; Zhao, T.; Zhao, Y.; Yin, Y.; Huang, Y.; Cheng, Z.; Wang, B.; Liu, S.; Pan, M.; Sun, D.; et al. PPARgamma Mediates the Anti-Epithelial-Mesenchymal Transition Effects of FGF1(DeltaHBS) in Chronic Kidney Diseases via Inhibition of TGF-beta1/SMAD3 Signaling. *Front. Pharmacol.* **2021**, *12*, 690535. [CrossRef] [PubMed]
74. Sonneveld, R.; Hoenderop, J.G.; Isidori, A.M.; Henique, C.; Dijkman, H.B.; Berden, J.H.; Tharaux, P.L.; van der Vlag, J.; Nijenhuis, T. Sildenafil Prevents Podocyte Injury via PPAR-gamma-Mediated TRPC6 Inhibition. *J. Am. Soc. Nephrol.* **2017**, *28*, 1491–1505. [CrossRef]
75. Wei, L.; Mao, J.; Lu, J.; Gao, J.; Zhu, D.; Tian, L.; Chen, Z.; Jia, L.; Wang, L.; Fu, R. Rosiglitazone Inhibits Angiotensin II-Induced Proliferation of Glomerular Mesangial Cells via the Galphaq/Plcbeta4/TRPC Signaling Pathway. *Cell. Physiol. Biochem.* **2017**, *44*, 2228–2242. [CrossRef] [PubMed]
76. Wynn, T.A. Common and unique mechanisms regulate fibrosis in various fibroproliferative diseases. *J. Clin. Investig.* **2007**, *117*, 524–529. [CrossRef] [PubMed]
77. Pistrosch, F.; Passauer, J.; Herbrig, K.; Schwanebeck, U.; Gross, P.; Bornstein, S.R. Effect of thiazolidinedione treatment on proteinuria and renal hemodynamic in type 2 diabetic patients with overt nephropathy. *Horm. Metab. Res.* **2012**, *44*, 914–918. [CrossRef]
78. Toffoli, B.; Gilardi, F.; Winkler, C.; Soderberg, M.; Kowalczuk, L.; Arsenijevic, Y.; Bamberg, K.; Bonny, O.; Desvergne, B. Nephropathy in Pparg-null mice highlights PPARgamma systemic activities in metabolism and in the immune system. *PLoS ONE* **2017**, *12*, e0171474. [CrossRef]
79. Wu, L.; Wang, Q.; Guo, F.; Ma, X.; Ji, H.; Liu, F.; Zhao, Y.; Qin, G. MicroRNA-27a Induces Mesangial Cell Injury by Targeting of PPARgamma, and its In Vivo Knockdown Prevents Progression of Diabetic Nephropathy. *Sci. Rep.* **2016**, *6*, 26072. [CrossRef]
80. Hou, X.; Tian, J.; Geng, J.; Li, X.; Tang, X.; Zhang, J.; Bai, X. MicroRNA-27a promotes renal tubulointerstitial fibrosis via suppressing PPARgamma pathway in diabetic nephropathy. *Oncotarget* **2016**, *7*, 47760–47776. [CrossRef]
81. Wang, Z.; Liu, Q.; Dai, W.; Hua, B.; Li, H.; Li, W. Pioglitazone downregulates Twist-1 expression in the kidney and protects renal function of Zucker diabetic fatty rats. *Biomed. Pharmacother.* **2019**, *118*, 109346. [CrossRef]

82. Yao, W.; Yang, P.; Qi, Y.; Jin, L.; Zhao, A.; Ding, M.; Wang, D.; Li, Y.; Hao, C. Transcriptome analysis reveals a protective role of liver X receptor alpha against silica particle-induced experimental silicosis. *Sci. Total. Environ.* **2020**, *747*, 141531. [CrossRef] [PubMed]
83. Maquigussa, E.; Paterno, J.C.; de Oliveira Pokorny, G.H.; da Silva Perez, M.; Varela, V.A.; da Silva Novaes, A.; Schor, N.; Boim, M.A. Klotho and PPAR Gamma Activation Mediate the Renoprotective Effect of Losartan in the 5/6 Nephrectomy Model. *Front. Physiol.* **2018**, *9*, 1033. [CrossRef]
84. Wang, X.; Deng, J.; Xiong, C.; Chen, H.; Zhou, Q.; Xia, Y.; Shao, X.; Zou, H. Treatment with a PPAR-gamma Agonist Protects Against Hyperuricemic Nephropathy in a Rat Model. *Drug Des. Dev. Ther.* **2020**, *14*, 2221–2233. [CrossRef] [PubMed]
85. Matsushita, K.; Yang, H.C.; Mysore, M.M.; Zhong, J.; Shyr, Y.; Ma, L.J.; Fogo, A.B. Effects of combination PPARgamma agonist and angiotensin receptor blocker on glomerulosclerosis. *Lab. Investig.* **2016**, *96*, 602–609. [CrossRef] [PubMed]
86. Zeisberg, M.; Hanai, J.; Sugimoto, H.; Mammoto, T.; Charytan, D.; Strutz, F.; Kalluri, R. BMP-7 counteracts TGF-beta1-induced epithelial-to-mesenchymal transition and reverses chronic renal injury. *Nat. Med.* **2003**, *9*, 964–968. [CrossRef] [PubMed]
87. Hruska, K.A.; Guo, G.; Wozniak, M.; Martin, D.; Miller, S.; Liapis, H.; Loveday, K.; Klahr, S.; Sampath, T.K.; Morrissey, J. Osteogenic protein-1 prevents renal fibrogenesis associated with ureteral obstruction. *Am. J. Physiol. Renal. Physiol.* **2000**, *279*, F130–F143. [CrossRef] [PubMed]
88. Sugimoto, H.; LeBleu, V.S.; Bosukonda, D.; Keck, P.; Taduri, G.; Bechtel, W.; Okada, H.; Carlson, W., Jr.; Bey, P.; Rusckowski, M.; et al. Activin-like kinase 3 is important for kidney regeneration and reversal of fibrosis. *Nat. Med.* **2012**, *18*, 396–404. [CrossRef]
89. Tampe, B.; Tampe, D.; Nyamsuren, G.; Klopper, F.; Rapp, G.; Kauffels, A.; Lorf, T.; Zeisberg, E.M.; Muller, G.A.; Kalluri, R.; et al. Pharmacological induction of hypoxia-inducible transcription factor ARNT attenuates chronic kidney failure. *J. Clin. Investig.* **2018**, *128*, 3053–3070. [CrossRef]
90. Chan, W.L.; Leung, J.C.; Chan, L.Y.; Tam, K.Y.; Tang, S.C.; Lai, K.N. BMP-7 protects mesangial cells from injury by polymeric IgA. *Kidney Int.* **2008**, *74*, 1026–1039. [CrossRef]
91. Lu, J.; Shi, J.; Gui, B.; Yao, G.; Wang, L.; Ou, Y.; Zhu, D.; Ma, L.; Ge, H.; Fu, R. Activation of PPAR-gamma inhibits PDGF-induced proliferation of mouse renal fibroblasts. *Eur. J. Pharmacol.* **2016**, *789*, 222–228. [CrossRef] [PubMed]
92. Gui, Y.; Lu, Q.; Gu, M.; Wang, M.; Liang, Y.; Zhu, X.; Xue, X.; Sun, X.; He, W.; Yang, J.; et al. Fibroblast mTOR/PPARgamma/HGF axis protects against tubular cell death and acute kidney injury. *Cell Death Differ.* **2019**, *26*, 2774–2789. [CrossRef]
93. Zhao, M.; Chen, Y.; Ding, G.; Xu, Y.; Bai, M.; Zhang, Y.; Jia, Z.; Huang, S.; Zhang, A. Renal tubular epithelium-targeted peroxisome proliferator-activated receptor-gamma maintains the epithelial phenotype and antagonizes renal fibrogenesis. *Oncotarget* **2016**, *7*, 64690–64701. [CrossRef]
94. Shen, D.; Li, H.; Zhou, R.; Liu, M.J.; Yu, H.; Wu, D.F. Pioglitazone attenuates aging-related disorders in aged apolipoprotein E deficient mice. *Exp. Gerontol.* **2018**, *102*, 101–108. [CrossRef]
95. Leifheit-Nestler, M.; Richter, B.; Basaran, M.; Nespor, J.; Vogt, I.; Alesutan, I.; Voelkl, J.; Lang, F.; Heineke, J.; Krick, S.; et al. Impact of Altered Mineral Metabolism on Pathological Cardiac Remodeling in Elevated Fibroblast Growth Factor 23. *Front. Endocrinol.* **2018**, *9*, 333. [CrossRef]
96. Chen, Z.; Yuan, P.; Sun, X.; Tang, K.; Liu, H.; Han, S.; Ye, T.; Liu, X.; Yang, X.; Zeng, J.; et al. Pioglitazone decreased renal calcium oxalate crystal formation by suppressing M1 macrophage polarization via the PPAR-gamma-miR-23 axis. *Am. J. Physiol. Renal. Physiol.* **2019**, *317*, F137–F151. [CrossRef]
97. Liu, Y.D.; Yu, S.L.; Wang, R.; Liu, J.N.; Jin, Y.S.; Li, Y.F.; An, R.H. Rosiglitazone Suppresses Calcium Oxalate Crystal Binding and Oxalate-Induced Oxidative Stress in Renal Epithelial Cells by Promoting PPAR-gamma Activation and Subsequent Regulation of TGF-beta1 and HGF Expression. *Oxidative Med. Cell. Longev.* **2019**, *2019*, 4826525. [CrossRef] [PubMed]
98. Roszer, T.; Menendez-Gutierrez, M.P.; Lefterova, M.I.; Alameda, D.; Nunez, V.; Lazar, M.A.; Fischer, T.; Ricote, M. Autoimmune kidney disease and impaired engulfment of apoptotic cells in mice with macrophage peroxisome proliferator-activated receptor gamma or retinoid X receptor alpha deficiency. *J. Immunol.* **2011**, *186*, 621–631. [CrossRef]
99. Henique, C.; Bollee, G.; Lenoir, O.; Dhaun, N.; Camus, M.; Chipont, A.; Flosseau, K.; Mandet, C.; Yamamoto, M.; Karras, A.; et al. Nuclear Factor Erythroid 2-Related Factor 2 Drives Podocyte-Specific Expression of Peroxisome Proliferator-Activated Receptor gamma Essential for Resistance to Crescentic GN. *J. Am. Soc. Nephrol.* **2016**, *27*, 172–188. [CrossRef]
100. Fang, Y.; Ginsberg, C.; Sugatani, T.; Monier-Faugere, M.C.; Malluche, H.; Hruska, K.A. Early chronic kidney disease-mineral bone disorder stimulates vascular calcification. *Kidney Int.* **2014**, *85*, 142–150. [CrossRef] [PubMed]
101. Cheng, L.; Zhang, L.; Yang, J.; Hao, L. Activation of peroxisome proliferator-activated receptor gamma inhibits vascular calcification by upregulating Klotho. *Exp. Ther. Med.* **2017**, *13*, 467–474. [CrossRef] [PubMed]
102. Liu, L.; Liu, Y.; Zhang, Y.; Bi, X.; Nie, L.; Liu, C.; Xiong, J.; He, T.; Xu, X.; Yu, Y.; et al. High phosphate-induced downregulation of PPARgamma contributes to CKD-associated vascular calcification. *J. Mol. Cell. Cardiol.* **2018**, *114*, 264–275. [CrossRef]
103. Singh, A.P.; Singh, N.; Bedi, P.M. Pioglitazone ameliorates renal ischemia reperfusion injury through NMDA receptor antagonism in rats. *Mol. Cell. Biochem.* **2016**, *417*, 111–118. [CrossRef] [PubMed]
104. Jia, P.; Wu, X.; Pan, T.; Xu, S.; Hu, J.; Ding, X. Uncoupling protein 1 inhibits mitochondrial reactive oxygen species generation and alleviates acute kidney injury. *EBioMedicine* **2019**, *49*, 331–340. [CrossRef] [PubMed]
105. Ling, H.; Chen, H.; Wei, M.; Meng, X.; Yu, Y.; Xie, K. The Effect of Autophagy on Inflammation Cytokines in Renal Ischemia/Reperfusion Injury. *Inflammation* **2016**, *39*, 347–356. [CrossRef] [PubMed]

106. Xi, X.; Zou, C.; Ye, Z.; Huang, Y.; Chen, T.; Hu, H. Pioglitazone protects tubular cells against hypoxia/reoxygenation injury through enhancing autophagy via AMPK-mTOR signaling pathway. *Eur. J. Pharmacol.* **2019**, *863*, 172695. [CrossRef] [PubMed]
107. Revelo, M.P.; Federspiel, C.; Helderman, H.; Fogo, A.B. Chronic allograft nephropathy: Expression and localization of PAI-1 and PPAR-gamma. *Nephrol. Dial. Transplant.* **2005**, *20*, 2812–2819. [CrossRef]
108. Ma, L.J.; Marcantoni, C.; Linton, M.F.; Fazio, S.; Fogo, A.B. Peroxisome proliferator-activated receptor-gamma agonist troglitazone protects against nondiabetic glomerulosclerosis in rats. *Kidney Int.* **2001**, *59*, 1899–1910. [CrossRef]
109. Kiss, E.; Popovic, Z.V.; Bedke, J.; Adams, J.; Bonrouhi, M.; Babelova, A.; Schmidt, C.; Edenhofer, F.; Zschiedrich, I.; Domhan, S.; et al. Peroxisome proliferator-activated receptor (PPAR)gamma can inhibit chronic renal allograft damage. *Am. J. Pathol.* **2010**, *176*, 2150–2162. [CrossRef]
110. Deng, J.; Xia, Y.; Zhou, Q.; Wang, X.; Xiong, C.; Shao, X.; Shao, M.; Zou, H. Protective effect of rosiglitazone on chronic renal allograft dysfunction in rats. *Transpl. Immunol.* **2019**, *54*, 20–28. [CrossRef]
111. Korolczuk, A.; Maciejewski, M.; Smolen, A.; Dudka, J.; Czechowska, G.; Widelska, I. The role of peroxisome-proliferator-activating receptor gamma agonists: Rosiglitazone and 15-deoxy-delta12,14-prostaglandin J2 in chronic experimental cyclosporine A-induced nephrotoxicity. *J. Physiol. Pharmacol.* **2014**, *65*, 867–876. [PubMed]
112. Lazar, M.A. Reversing the curse on PPARgamma. *J. Clin. Investig.* **2018**, *128*, 2202–2204. [CrossRef]
113. Hughes, A.D.; Park, C.; March, K.; Coady, E.; Khir, A.; Chaturvedi, N.; Thom, S.A. A randomized placebo controlled double blind crossover study of pioglitazone on left ventricular diastolic function in type 2 diabetes. *Int. J. Cardiol.* **2013**, *167*, 1329–1332. [CrossRef] [PubMed]
114. Horio, T.; Suzuki, M.; Suzuki, K.; Takamisawa, I.; Hiuge, A.; Kamide, K.; Takiuchi, S.; Iwashima, Y.; Kihara, S.; Funahashi, T.; et al. Pioglitazone improves left ventricular diastolic function in patients with essential hypertension. *Am. J. Hypertens.* **2005**, *18*, 949–957. [CrossRef] [PubMed]
115. Soccio, R.E.; Chen, E.R.; Rajapurkar, S.R.; Safabakhsh, P.; Marinis, J.M.; Dispirito, J.R.; Emmett, M.J.; Briggs, E.R.; Fang, B.; Everett, L.J.; et al. Genetic Variation Determines PPARgamma Function and Anti-diabetic Drug Response In Vivo. *Cell* **2015**, *162*, 33–44. [CrossRef]

International Journal of
Molecular Sciences

MDPI

Review

The Regulatory Roles of PPARs in Skeletal Muscle Fuel Metabolism and Inflammation: Impact of PPAR Agonism on Muscle in Chronic Disease, Contraction and Sepsis

Hannah Crossland [1,2], Dumitru Constantin-Teodosiu [1] and Paul L. Greenhaff [1,2,*]

1 MRC/Versus Arthritis Centre for Musculoskeletal Ageing Research, Division of Physiology, Pharmacology and Neuroscience, School of Life Sciences, Queen's Medical Centre, University of Nottingham, Nottingham NG7 2UH, UK; mbzhc@exmail.nottingham.ac.uk (H.C.); tim.constantin@nottingham.ac.uk (D.C.-T.)
2 National Institute for Health Research Nottingham Biomedical Research Centre, Queen's Medical Centre, University of Nottingham, Nottingham NG7 2UH, UK
* Correspondence: Paul.Greenhaff@nottingham.ac.uk

Abstract: The peroxisome proliferator-activated receptor (PPAR) family of transcription factors has been demonstrated to play critical roles in regulating fuel selection, energy expenditure and inflammation in skeletal muscle and other tissues. Activation of PPARs, through endogenous fatty acids and fatty acid metabolites or synthetic compounds, has been demonstrated to have lipid-lowering and anti-diabetic actions. This review will aim to provide a comprehensive overview of the functions of PPARs in energy homeostasis, with a focus on the impacts of PPAR agonism on muscle metabolism and function. The dysregulation of energy homeostasis in skeletal muscle is a frequent underlying characteristic of inflammation-related conditions such as sepsis. However, the potential benefits of PPAR agonism on skeletal muscle protein and fuel metabolism under these conditions remains under-investigated and is an area of research opportunity. Thus, the effects of PPARγ agonism on muscle inflammation and protein and carbohydrate metabolism will be highlighted, particularly with its potential relevance in sepsis-related metabolic dysfunction. The impact of PPARδ agonism on muscle mitochondrial function, substrate metabolism and contractile function will also be described.

Keywords: skeletal muscle; inflammation; PPARs; substrate metabolism

Citation: Crossland, H.; Constantin-Teodosiu, D.; Greenhaff, P.L. The Regulatory Roles of PPARs in Skeletal Muscle Fuel Metabolism and Inflammation: Impact of PPAR Agonism on Muscle in Chronic Disease, Contraction and Sepsis. *Int. J. Mol. Sci.* **2021**, *22*, 9775. https://doi.org/10.3390/ijms22189775

Academic Editors: Manuel Vázquez-Carrera and Walter Wahli

Received: 11 August 2021
Accepted: 8 September 2021
Published: 10 September 2021

1. Introduction

Peroxisome proliferator-activated receptors (PPARs) are a group of transcription factors implicated in wide-ranging cellular functions, including lipid metabolism, inflammatory responses and cell proliferation and differentiation [1]. Three PPAR subtypes exist (PPARα, PPARδ and PPARγ). They are activated in vivo by endogenous fatty acids and their metabolites and synthetic compounds developed for their lipid-lowering and anti-diabetic actions. Skeletal muscle is a tissue that displays high metabolic flexibility, comprising different fibre types that vary according to their contractile and metabolic properties [2]. For example, slow-twitch type I fibres have a relatively high capillary density, are rich in mitochondria and possess a relatively high capacity for oxidative metabolism during contraction. In contrast, fast-twitch type IIx fibres have a lower capillary density and a high capacity for energy delivery from non-mitochondrial routes during contraction. Disturbances in skeletal muscle energy homeostasis play a key part in the pathogenesis of several chronic non-communicable disease conditions, including type 2 diabetes (T2D) and chronic lung disease. The dysregulation of skeletal muscle energy homeostasis is also a frequent underlying characteristic of acute inflammation-related conditions, such as sepsis [3] and surgical trauma. The role of PPAR agonism in modulating skeletal muscle protein and fuel metabolism in these conditions is relatively poorly understood. Still, the potential of such an approach will be addressed in this article. Specifically, this review will aim to provide an overview of the metabolic regulatory roles of PPARs in energy homeostasis, with a focus on the impacts of PPARδ agonism on skeletal muscle metabolism and contractile function, primarily highlighting studies that have involved in vivo/ex vivo animal models or human volunteers. Furthermore, the focus will be directed towards the potential role of PPARγ agonism in alleviating muscle inflammation and metabolic disturbances during sepsis.

2. Metabolic Functions of PPARs and Their Actions in Skeletal Muscle

Peroxisome proliferator-activated receptors (PPARs) are a group of proteins that belong to the nuclear hormone receptor superfamily of ligand-activated transcription factors. PPAR transcriptional activity is mediated by heterodimers of PPARs with retinoid X receptor (RXR), which subsequently bind to DNA sequence elements (PPREs) in regulatory regions of target genes [4]. Three PPAR subtypes have been identified (PPARα, PPARδ (also known as PPARβ) and PPARγ [5]. Through their interactions with endogenous lipids and lipid metabolites, PPARs have been reported to regulate many metabolic processes, including lipid and glucose homeostasis, cell proliferation and inflammation [1]. Several endogenous compounds, including n-3 and n-6 fatty acids, eicosanoids and phospholipids, have been identified as natural ligands of PPARs. In addition to this, activation of PPAR activity by pharmacological agonists has been identified as a promising treatment strategy for conditions related to insulin resistance and dyslipidaemia, in part through increased fatty acid oxidation in skeletal muscle, thereby decreasing overall body fat content [6].

Each PPAR subtype has been attributed to different tissue-specific expression levels and functions. For example, PPARα is highly expressed in tissue types that undergo significant fatty acid catabolism, such as brown adipose tissue, heart and liver [7]. Activated by polyunsaturated fatty acids (PUFA) and leukotriene, PPARα has an important function in fatty acid catabolism and carbohydrate metabolism [8,9]. Synthetic compounds that act as agonists of PPARα are known as fibrates, whose actions are important in lipid-lowering activities and cardio-protection [10,11]. PPARγ, on the other hand, is variably expressed in adipocytes, macrophages, placenta and other tissues and is activated by specific endogenous fatty acid metabolites (such as 15-deoxy-prostaglandin J2) as well as by a class of insulin sensitisers known as thiazolidinediones (TZDs) [12]. TZDs have been proven to be important in the treatment of T2D. While early TZDs (e.g., Troglitazone) were related to severe hepatic side effects, other newer available TZDs (Rosiglitazone, Pioglitazone) are not toxic to the liver. PPARγ plays a central role in adipogenesis, whereby the insulin-sensitising effect of TZDs may be due to new adipose cell recruitment, enabling increased lipid storage capacity and adipokine secretion [13]. PPARγ activation also regulates the transcription of genes that promote the synthesis of triglycerides [13]. In patients with T2D, administration of TZDs successfully improved insulin-stimulated glucose disposal under euglycaemic-hyperinsulinaemic clamp conditions [14,15], where skeletal muscle plays a central glucose-lowering role. One mechanism by which TZDs exert their insulin sensitising actions on skeletal muscle is through the modulation of adipose secretory factors, such as adiponectin. Increased secretion of adiponectin has been suggested to act as an insulin sensitiser for liver and skeletal muscle, and this occurs through the activation of PPARγ [16].

The role of PPARδ has remained relatively unclear until recently, where it has been associated with a wide range of metabolic functions in vivo [17,18]. It has broad expression across tissues and is activated by various ligands, such as long-chain fatty acids. A developmental regulatory role has been identified for PPARδ, as well as regulation of lipid metabolism [17,18]. It is the predominant isotype in skeletal muscle, where it has been linked to fuel metabolism, energy expenditure, inflammation, and fibre type switching through physical exercise [17,19]. Both PPARα and PPARδ have been demonstrated to regulate genes for proteins involved in fatty acid uptake and oxidation, including lipoprotein lipase (LPL), fatty acid-binding protein 3 (FABP), stearoyl-Coenzyme A desaturase

(SCD)-1 and cluster of differentiation 36 (CD36) [20,21]. During fasting, PPARδ expression is upregulated in rodent skeletal muscles, which is important in regulating the cellular uptake and oxidation of free fatty acids (FFA) as an energy source for ATP production [22].

Through their importance in metabolic regulation, the role of all three PPAR subtypes in skeletal muscle metabolism has been established. For example, one link between PPARs and metabolic regulation in skeletal muscle appears to be through the upregulation of pyruvate dehydrogenase kinase 4 (PDK4), a key regulator of the pyruvate dehydrogenase complex (PDC). The PDC activation status is regulated by various competing PDKs and pyruvate dehydrogenase phosphatase (PDP) proteins [23]. These covalent processes ultimately determine the extent of PDC phosphorylation (i.e., activation). There are four isoforms of PDK (PDK1-4) and two isoforms of PDP (PDP1 and 2) [24,25]. While PDK1 and PDK3 appear to be mainly expressed in the heart, pancreatic islet cells and kidney, PDK2 and PDK4 are expressed in most tissues, including heart and skeletal muscle [24]. Selective PDK4 upregulation has been demonstrated in response to starvation conditions and pathologies such as T2D [26,27], which is thought to be due to changes in FFA availability in skeletal muscle. An increase in fatty acid oxidation via PPARδ agonism [6], and starvation [28], is believed to be responsible for the PDK4 transcriptional activation, thereby inactivating PDC (the rate-limiting enzyme in mitochondrial carbohydrate oxidation). It should be noted, however, that a lack of association between increases in plasma FFA levels and muscle PDK4 expression has been reported during fasting in humans [29], with no observable changes in muscle PPARα expression, indicating that other factors could also be important in PDK4 upregulation. One such factor could be the Forkhead box class O (FOXO) family of transcription factors, which has been linked to promoter binding of the PDK4 gene as a result of FFA-mediated nuclear translocation [30].

In addition to increased availability of endogenous fatty acids and their metabolites being associated with PPAR activation, inflammation has been proven to be a major site of PPAR regulation, which can occur through both direct and indirect mechanisms [31]. As mentioned, PPARs have emerged as targets of drugs used to treat various aspects of the metabolic syndrome, of which inflammation is an underlying key factor. All three PPAR isotypes have been shown to exert anti-inflammatory effects during conditions of chronic low-grade inflammation, characterised by increased circulatory cytokines and acute-phase proteins [32,33]. PPARα was shown to upregulate the expression of IkB, a factor that suppresses the nuclear translocation and transcriptional activity of the pro-inflammatory nuclear factor kappa-light-chain-enhancer of activated B cells (NF-kB) [34]. PPARγ has also been shown to reduce activation of NF-kB, as well as inhibit pro-inflammatory cytokine production in T lymphocytes and induction of anti-inflammatory regulatory molecules of the innate immune system [35].

In summary, all three PPAR subtypes have distinct yet overlapping roles in regulating metabolic function and inflammation (see Table 1), and synthetic compounds aimed at activating the PPARs have been developed for their lipid-lowering and anti-diabetic actions. In skeletal muscle, PPAR activation appears important in the upregulation of PDK4, thereby demonstrating its essential role in regulating carbohydrate oxidation and energy homeostasis. The following section of this review will focus in more detail on the impact of PPARδ agonism on muscle metabolism and contractile function and PPARγ agonism on muscle metabolism and inflammation.

Table 1. Regulation of lipid and carbohydrate metabolism by PPARs in skeletal muscle, adipose tissue and liver.

	Skeletal Muscle	Liver	Adipose
PPARδ	+ FA oxidation −carbohydrate oxidation	+ FA oxidation − lipogenesis	+ FA oxidation
PPARγ	+ FA oxidation + glucose uptake	+ lipogenesis + lipid storage − glucose production	+ adipogenesis + lipogenesis + lipid storage + adipokine production
PPARα	+ FA oxidation	+ FA oxidation − lipid storage	

3. PPARδ Agonism and Skeletal Muscle Metabolism, Contractile Function and Inflammation

Several in vivo animal studies have been performed with the aim of determining the impact of PPARδ agonism on skeletal muscle metabolism and function. We previously demonstrated in our laboratory that 6 days of administration of the PPARδ agonist, GW610742 [36], resulted in increased activity of β-hydroxy acyl-CoA dehydrogenase (β-HAD) in resting rat soleus muscle, which is a key step in β-oxidation in the mitochondria. Compared with control animals, these changes were paralleled by increased expression of muscle PDK2 and PDK4 mRNA and PDK4 protein expression. Thus, evidence points towards PPAR activation in skeletal muscle being, in part, important in mediating FFA-induced PDK4 upregulation in skeletal muscle, thereby contributing to PDC inhibition, suppressing PDC-regulated carbohydrate oxidation, and switching fuel selection towards fat oxidation in skeletal muscle (Figure 1). We also measured the impact of GW610742 on muscle growth-related pathways since FOXO1, which plays a part in PDK4 upregulation, has also been suggested to increase transcription of MAFbx and MuRF1, thereby activating ubiquitin-proteasome mediated muscle proteolysis [37]. In keeping with this, administration of the PPARδ agonist resulted in increases in muscle mRNA and protein expression of MAFbx and MuRF1, suggesting that potentially the induction of muscle atrophy signalling is another consequence of PPARδ agonism. Collectively, the findings pointed to PPARδ agonism being involved in the regulation of muscle fuel selection and the induction of a muscle atrophy programme via a single common signalling pathway. It should be stated, however, there was no evidence of soleus muscle atrophy based on the muscle protein:DNA ratio after 6 days of GW610742 administration compared with control.

In line with the above findings relating to a PPARδ agonism induced switch in muscle fuel selection away from carbohydrate to increased fat oxidation, in another study, mice treated with the PPARδ agonist GW501516 exhibited increased PGC-1α levels, and improved prolonged low-intensity wheel-running performance. They also saw hypertrophy of oxidative slow-twitch myofibres, which are rich in mitochondria, perhaps suggesting increased reliance on the catabolism of FA through mitochondrial beta-oxidation [38]. However, we further reported that when muscle contraction was increased to an intensity that necessitates carbohydrate to become an obligate fuel for contraction, PPARδ agonism negatively affected contractile function in rats [39]. Specifically, male Wistar rats received the PPARδ agonist GW610742X (or vehicle) for 6 days. The gastrocnemius–soleus–plantaris muscle group was isolated and subjected to submaximal electrically evoked contraction using a perfused hindlimb model. The contraction intensity was fixed to guarantee carbohydrate become an essential fuel, and PDC activity was increased, ensuring pyruvate derived acetyl group delivery to the mitochondrion [40]. We observed that PDC activity during contraction was significantly less with the PPARδ agonist than control, while anaerobic metabolism (reflected by phosphocreatine hydrolysis and lactate accumulation) was greater. We proposed that this collectively accounted for the observed impaired contractile function with GW610742X agonist, indicating that PPARδ agonism can impair the contractile muscle function by inhibiting carbohydrate oxidation during muscle contraction where carbohydrate is an obligate fuel.

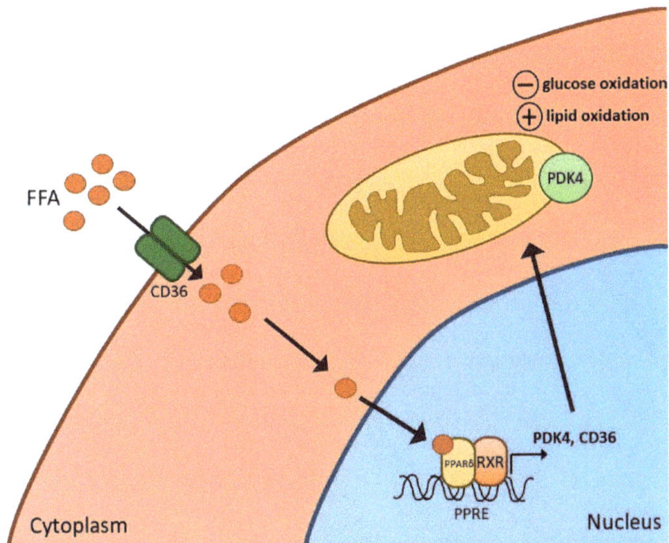

Figure 1. Activation of PPARδ in skeletal muscle. Increased free fatty acids (FFA) and their metabolites enter skeletal muscle via the FFA transporter CD36, resulting in the formation of a heterodimer of PPARδ and retinoid X receptor (RXR), and subsequent activation of PPARδ-dependent genes, such as pyruvate dehydrogenase kinase 4 (PDK4) (and CD36 itself). Activation of PDK4 can result in reduced rates of glucose oxidation as well as increased fatty acid oxidation in mitochondria.

We have also sought to determine whether PPAR transcription factors may be necessary for the high-fat feeding induced inhibition of PDC activation and carbohydrate oxidation during submaximal exercise in humans [41,42]. Healthy male volunteers were given a control diet or an iso-caloric high-fat diet (HFD). They underwent 60 min of submaximal exercise at an intensity equivalent to 75% maximal oxygen uptake. There was a relative increase in expression of PDK4 in muscle with HFD compared to control, alongside reduced PDC activation in muscle. Exercise increased PDC activity and carbohydrate utilisation with both diets, but these measures were diminished with the HFD. In terms of PPAR expression, there was no effect of the high-fat diet on the mRNA expression of PPARδ. However, PPARγ and PPARα were increased at rest, though this increase was not apparent during exercise. Of note, the expression of PPARα mRNA was lower in another group of volunteers that underwent a HFD but were also treated with dichloroacetate (DCA), a potent inhibitor of PDK2 and PDK4 and, therefore, a stimulator of PDC activity, which restored carbohydrate oxidation during exercise in this group. Thus, in humans, these results appear to suggest there may not be significant involvement of PPARs in increasing muscle PDK4 expression, although muscle protein levels of each PPAR were not measured.

Following these findings, further work from our laboratory studied the role of PPARδ (and FOXO1) in palmitate-induced PDC inhibition and carbohydrate use using a skeletal muscle cell model [43]. Myotubes were treated with palmitate for 16 hrs in the presence or absence of continuous electrical pulse stimulation, the latter having been shown to increase glucose uptake and carbohydrate oxidation in muscle cells [44,45], and therefore potentially having the capacity to reverse palmitate-mediated inhibition of PDC. It was observed that palmitate reduced glucose uptake, PDC activity and maximal rates of palmitate derived mitochondrial ATP production whilst also increasing PDK4, PPARδ and PPARα proteins. There was also a significant reduction in the magnitude of FOXO1 phosphorylation, indicating its nuclear translocation and subsequent activation. Electrical pulse stimulation reversed many palmitate-induced changes to carbohydrate oxidation and was associated with reduced PDK4 protein and reduced PPARδ (but not PPARα) protein content. Collectively, while more work is required to elucidate the relative importance of PPARδ and FOXO1 transcription factors in mediating PDK4 transcription, particularly in humans, these findings indicate their potential roles in palmitate-induced impairments in PDC activity and carbohydrate oxidation in skeletal muscle.

As discussed earlier, all three PPAR subtypes have the potential for treating inflammatory states, with their anti-inflammatory effects well characterised [31,32]. Based upon this, targeting PPARs could potentially represent an attractive avenue for alleviating inflammation and metabolic disturbances during conditions such as sepsis. Fibrates are proven to be beneficial in treating dyslipidaemia. They may lower circulating triglyceride levels in the blood by inducing hepatic fatty acid oxidation and increasing levels of high-density lipoproteins [10,11]. TZDs have also proven antidiabetic effects, though their clinical use and development has been limited due to adverse side effects (increased risk of congestive heart failure, weight gain, increased risk of bone fracture). However, inflammation associated with the pathophysiology of T2D is typically chronic and low-grade, while sepsis is an acute, high-grade inflammatory condition, which may increase their utility in this scenario. Sepsis is characterised by an uncontrolled host response to an infection and remains a major cause of morbidity (including muscle atrophy and insulin resistance) and mortality worldwide [46]. Ongoing work aims to understand its pathophysiology and develop novel therapeutics since current available therapeutic strategies remain limited. In patients with sepsis, severe metabolic dysregulation occurs, which contributes to sepsis pathophysiology and resultant organ failure. It has been observed that early and rapid skeletal muscle wasting can occur with critical illness, which can play a major role in causing the increased length of hospital stay and delayed recovery [3]. In conjunction with this, trials aimed at attenuating muscle wasting and improving physical function through either nutritional support or exercise approaches have proved inconsistent.

It is unclear what causes the decline in muscle protein synthesis early in critical illness and could feasibly be related to impaired mitochondrial function or tissue content [47]. Decreased substrate utilisation, including both carbohydrate and fatty acids, is a consequence of critical illness and could result in impaired metabolic function in muscle [48]. Infusion or injection of the bacterial endotoxin lipopolysaccharide (LPS) can reproduce many metabolic changes seen in sepsis [49]. The endotoxemia model, therefore, represents a relevant physiological model of sepsis. In a rat model of LPS-induced septic shock, tissue protein expression (renal and cardiac) of cytosolic and nuclear PPARα, PPARδ and PPARγ and nuclear translocation of these proteins were decreased with LPS [50]. In a rodent model of septic shock, early administration of a selective RXR agonist (bexarotene) was shown to prevent LPS-induced decrease in mean arterial pressure, as well as LPS-induced decreases in tissue PPARα/δ/γ-RXRα heterodimer formation [51]. The concurrent decline in circulating iNOS and LDH levels led the authors to conclude that activation of PPARα/δ/γ-RXRα heterodimers contributes to the beneficial effect of bexarotene to prevent the hypotension associated with inflammation and tissue injury during rat endotoxemia.

In terms of muscle-specific effects of PPARδ agonism on metabolism and inflammation, there have been few studies to date. In a model using cultured muscle cells, one group tested the hypothesis that PPARδ upregulates FOXO1 activity in muscle, thereby upregulating MAFbx and MuRF1 expression during sepsis and glucocorticoid treatment [52]. Activation of PPARδ in myotubes resulted in increased atrophy along with protein degradation and increased FOXO1 activity. Similar changes induced by dexamethasone (used as an agent to cause atrophy) were prevented by treatment with a PPARδ inhibitor. Furthermore, a PPARδ inhibitor given to dexamethasone-treated or septic rats prevented muscle wasting. These findings appear to support the suggestion that PPARδ may regulate activation of a FOXO1 linked atrophy programme in sepsis-induced muscle wasting.

Cardiac failure and decreased uptake and oxidation of fatty acids in the heart are common features of severe sepsis [53]. In a mouse model of sepsis [54], LPS administration rapidly caused downregulation of PPARα, PPARδ, as well as isoforms of thyroid hormone receptor (TR) and RXR (which are required for PPAR transcriptional activity) in the heart. There were also concurrent decreases in the expression of key fatty acid transporter/oxidation genes with LPS treatment. Thus, it is possible that these rapid decreases in the expression of key genes, including PPARα and PPARδ, are important in driving the reductions in cardiac fatty oxidation and myocardial dysfunction in sepsis. The potential protective effects of PPARδ agonism have been studied in relation to changes in LPS-induced apoptosis. In cultured rat cardio-myoblast cells, pre-treatment with the PPARδ agonist GW501516 inhibited increased rates of apoptosis induced by LPS, decreased activity of caspase-3 and increased nuclear translocation of NF-kB [55]. Furthermore, GW501516 increased protein expression of haem oxygenase-1 (HO-1), while inhibition of HO-1 reversed the effects of GW501516 on LPS-induced NF-kB activation. Thus, PPARδ also has anti-apoptotic effects during an LPS challenge in cardiac cells, potentially through suppressing NF-kB activation and via HO-1. Whether these observed effects of PPARδ are relevant to skeletal muscle in terms of inflammation and substrate oxidation during sepsis and related conditions remains to be determined.

To summarise, the effects of PPARδ agonism on skeletal muscle appear to be predominantly related to the switching of fuel utilisation towards increased oxidation of fatty acids, as well as declined carbohydrate oxidation. This can result in impaired function during prolonged muscle contraction where carbohydrate is an obligate fuel. Activation of PPARδ may also induce atrophy-related programmes in skeletal muscle. During sepsis, however, declines in PPAR activity may underlie some of the declines in FFA oxidation in various tissues, indicating that there may be some benefit to PPARδ agonism during these conditions.

4. PPARγ Agonism and Skeletal Muscle Metabolism and Inflammation

As described earlier, PPARγ plays a central role in adipogenesis, and TZDs have been shown to increase insulin-stimulated glucose disposal in T2D patients effectively. The effects of PPARγ agonism on skeletal muscle metabolic regulation remains poorly understood. One study assessed the impact of the PPARγ agonist, Rosiglitazone, on fatty acid transport and oxidation in rat muscle [56]. Seven days of rosiglitazone infusion (1 mg/day) did not alter the rate of fatty acid transport into muscle, but did increase rates of fatty acid oxidation in subsarcolemmal and intermyofibrillar mitochondria. This was accompanied by increases in mitochondrial FAT/CD36 protein, with no changes in citrate synthase or β-HAD activity. The effects of PPARγ activation on lipid metabolism were also studied in human skeletal muscle in vivo [57]. Long-chain fatty acid composition and stearoyl-CoA desaturase 1 (SCD1) were examined following 8 weeks of Rosiglitazone treatment in men with impaired glucose tolerance, with muscle biopsies and hyperinsulinaemic-euglycaemic clamps being carried out before and after rosiglitazone administration. Alongside an increase in insulin sensitivity, SCD1 expression was increased in muscle samples with rosiglitazone treatment. In addition, there was a shift in lipid composition from saturated long-chain fatty acids to unsaturated fatty acids in muscle. These findings clearly indicate a role for PPARγ activation in modulating lipid metabolism in skeletal muscle in vivo.

The impact of TZDs on skeletal muscle lipid and carbohydrate metabolism has been predominantly studied in relation to animal models of T2D [58–60], and patients with T2D [61–63]. In one model of obese Zucker rats [59], 6 weeks of rosiglitazone administration improved glucose tolerance in obese rats, while intramuscular triglyceride content, which was higher in obese compared with lean animals, was further increased following rosiglitazone treatment. There were also increases in skeletal muscle diacylglycerol and ceramide with rosiglitazone treatment, indicating that under these conditions, Rosiglitazone increased insulin sensitivity in obese rats, but this was not through reduced fatty acid accumulation in muscle.

In a study with T2D patients, the effects of three months of rosiglitazone treatment on insulin sensitivity and lipid metabolism were examined during a hyperinsulinaemic-euglycaemic clamp [61]. Insulin-stimulated glucose disposal was improved following rosiglitazone treatment, and reduced plasma fatty acid concentrations and increased extramyocellular lipid levels were observed. Rosiglitazone also promoted increased insulin sensitivity in peripheral adipocytes, indicating that enhanced insulin sensitivity through PPARγ agonism in humans may occur predominantly via improving adipocyte insulin sensitivity, leading to lipid redistribution from insulin-sensitive organs to peripheral adipocytes. In another study with T2D patients, further insight into the mechanisms by which TZDs improved insulin sensitivity in T2D was examined. Pioglitazone treatment for 6 months improved insulin-stimulated glucose disposal in T2D patients. In muscle tissue, there were increases in AMPK and acetyl-CoA carboxylase (ACC) phosphorylation with pioglitazone and increased expression of genes important in fat oxidation and mitochondrial function. These findings suggest some of the mechanisms by which TZDs improve skeletal muscle insulin sensitivity may involve stimulation of AMPK signalling and fat oxidation.

As described in a previous section, more work is required to determine whether different types of PPAR drug targets may have any potential benefit in alleviating inflammation in sepsis, thereby potentially preventing and/or improving certain deleterious consequences associated with the condition. In relation to specifically PPARγ agonists impacting on targeting inflammation, we previously assessed the impact of Rosiglitazone on muscle carbohydrate and protein metabolism in a rat model of LPS-induced endotoxaemia [64]. Initial work from our laboratory [65–67] demonstrated that dysregulation of the Akt/FOXO signalling pathway was important in mediating the development of muscle atrophy during LPS-induced endotoxaemia, specifically through activation of ubiquitin ligases MAFbx and MuRF1. We also proposed that the Akt/FOXO signalling pathway represents a site of molecular crosstalk between insulin and atrophy-related signalling processes during endotoxaemia through FOXO-mediated upregulation of PDK4 and reduced activity of PDC.

To assess the effects of PPARγ agonism on muscle protein and carbohydrate metabolism during endotoxaemia, rats were fed standard chow containing Rosiglitazone (8.5 ± 0.1 mg·kg^{-1}·day^{-1}) for 2 weeks before and during 24 h continuous intravenous infusion of LPS (15 μg·kg^{-1}·h^{-1}) or saline [64]. In terms of muscle inflammation, Rosiglitazone blunted LPS-induced increases in TNF-α and IL-6 mRNA expression. We also examined the subsequent impact of Rosiglitazone on LPS-induced changes in muscle protein degradation pathways, specifically, ubiquitin-proteasome-mediated protein breakdown. Increased expression of key proteolytic regulators (MAFbx and MurF1 mRNA), and activity of the 20S proteasome, were suppressed in the rosiglitazone-treated group of animals in the presence of endotoxaemia. In carbohydrate oxidation, LPS-induced increases in PDK4 gene expression and muscle lactate content were also suppressed with rosiglitazone administration. Collectively, these findings indicated that there were metabolic benefits of rosiglitazone pre-treatment in this LPS model of endotoxaemia, reflected by blunted muscle cytokine accumulation, muscle protein loss and lactate accumulation (Figure 2).

While few studies have assessed the impact of PPARγ agonists on skeletal muscle inflammation and metabolism during inflammatory disorders such as sepsis, there has been some work surrounding the protective effects of PPARγ on myocardial dysfunction in sepsis. One study in mice (using LPS administration as a sepsis model) investigated whether reduced fatty acid oxidation is the underlying cause for cardiac dysfunction in sepsis [68]. LPS administered to mice rapidly decreased cardiac fatty acid oxidation in conjunction with inducing cardiac dysfunction, while gene expression of PPARγ was downregulated. Moreover, activation of PPARγ using a transgenic mouse model (cardiomyocyte-specific PPARγ expression induced by the alpha-myosin heavy chain promoter) protected against cardiac dysfunction induced by LPS, while fatty acid oxidation was not reduced with LPS exposure in these animals. Interestingly, the expression of inflammation-related genes (IL-1α, IL-1β, IL-6 and TNF-α) in response to LPS treatment was similar to wild-type mice.

Rosiglitazone administration in wild-type mice similarly increased fatty acid oxidation, improved cardiac function after treatment with LPS and improved survival, despite not suppressing the expression of cardiac markers of inflammation. These findings are, therefore, promising in terms of the use of PPARγ agonists in sepsis treatment. Still, more work should be done on their mechanism of action and delineating whether their beneficial effects occur through their anti-inflammatory actions.

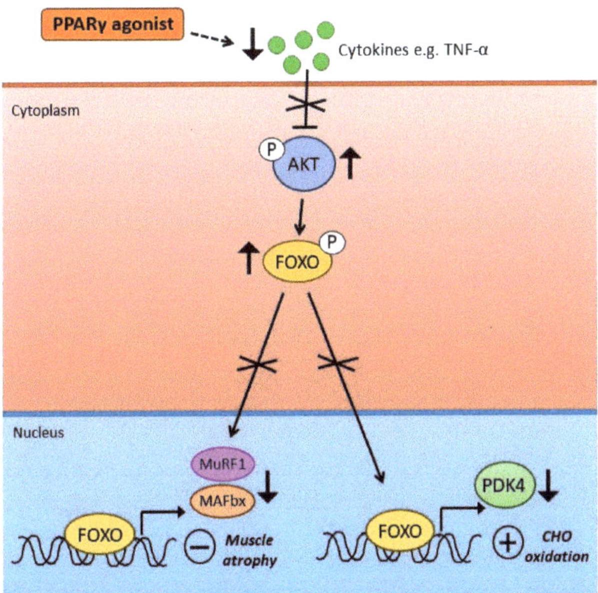

Figure 2. Impact of PPARγ agonism in skeletal muscle during LPS-induced endotoxaemia. Treatment with PPARγ agonists during endotoxaemia suppresses production of pro-inflammatory cytokines (e.g., tumour necrosis factor α: TNF-α). This results in reduced suppression of muscle AKT, and reduced transcriptional activity of Forkhead Box O (FOXO) transcription factors. Reduced activity of FOXO leads to suppression of factors important in increased muscle atrophy (MAFbx and MuRF1) as well as PDK4, a key protein in PDC inhibition.

In relation to myocardial dysfunction, a separate study in rats assessed the mechanism of PPARγ-mediated cardiac protective effects in sepsis [69]. Using rats subjected to caecal ligation and puncture (CLP), a PPARγ agonist (Rosiglitazone) and antagonist (T0070907) were used. The model of sepsis used resulted in significant impairments in cardiac function, with evidence of tissue apoptosis, necrosis and upregulated proinflammatory cytokines. Activation of PPARγ prevented these changes while blocking its activity exacerbated the differences and further reduced survival rates. These results provide other evidence that Rosiglitazone can exert beneficial effects related to reduced cardiac inflammation and cell death. A separate recent study further examined the effects of Rosiglitazone on sepsis-induced myocardial dysfunction in relation to the NF-κB pathway [70]. Here, a model of sepsis was established using female Sprague-Dawley rats, after which one group was administered 3 mg/mg rosiglitazone (daily, for 3 days). Rosiglitazone successfully decreased the number of apoptotic cells in septic animals, while in myocardial tissues, Rosiglitazone lowered TNF-α expression and activity of NF-κB.

To summarise, agonism of PPARγ in vivo appears to improve insulin sensitivity and may also increase fatty acid oxidation in skeletal muscle. However, improved insulin sensitivity may not be related to reductions in fatty acid accumulation in muscle tissue under certain conditions. In relation to inflammatory conditions such as sepsis, PPARγ agonists effectively suppress pro-inflammatory cytokine production and appear to be beneficial in alleviating organ injury and dysfunction, which has promising potential for therapeutic development.

5. Conclusions and Future Perspectives

To conclude, the PPAR family of transcription factors have wide-ranging critical regulatory roles in skeletal muscle and other tissues, from inflammation to fuel selection and contractile function. In terms of PPARδ, evidence suggests that the agonism of PPARδ appears to be primarily related to the switching of substrate utilisation towards increasing the use of fatty acids. In contrast, PPARδ agonism can impair muscle contractile function by inhibiting carbohydrate oxidation during muscle contraction, where carbohydrate is an obligate fuel. Conversely, there may be benefits to PPARδ agonism during certain inflammatory conditions, such as sepsis, since declines in PPAR activity may underlie some of the reductions in FFA oxidation. Similarly, agonism of PPARγ in vivo appears to be an effective anti-inflammatory strategy during sepsis and has proven beneficial in improving organ/tissue function in pre-clinical models. Blunting muscle cytokine accumulation during endotoxaemia in rodents has also been demonstrated to result in metabolic benefits via reduced muscle wasting and lactate accumulation.

Moving forwards, more studies will be required to better define the mechanistic roles of all the PPARs in different physiological and pathophysiological conditions. The potential benefits of PPAR agonism on skeletal muscle protein and carbohydrate/lipid metabolism during sepsis and other inflammatory conditions remains under-investigated and is, therefore, a promising area of research opportunity. Studies using dual or pan-PPAR agonists could also widen the therapeutic potential of these compounds, while cross-tissue studies will be important in evaluating potential off-target effects. Nevertheless, with ongoing drug developments and a greater understanding of the wide-ranging functions of PPARs, these transcription factors will undoubtedly remain critical therapeutic targets for a multitude of metabolic and inflammatory conditions.

Author Contributions: H.C. wrote the first draft of the review manuscript, and H.C., D.C.-T. and P.L.G. added to, edited and reviewed the manuscript. All authors approved the final version of the article. All persons designated as authors qualify for authorship, and all those who are eligible for authorship are listed. All authors have read and agreed to the published version of the manuscript.

Funding: This research was supported by the Medical Research Council [grant number MR/K00414X/1] and Arthritis Research UK [grant number 19891]. The National Institute of Health Research (NIHR) Nottingham Biomedical Research Centre and Nottingham University Hospitals Charities also supported the work.

Institutional Review Board Statement: Not applicable.

Informed Consent Statement: Not applicable.

Data Availability Statement: Not applicable.

Conflicts of Interest: The authors declare no conflict of interest.

References

1. Berger, J.; Wagner, J.A. Physiological and Therapeutic Roles of Peroxisome Proliferator-Activated Receptors. *Diabetes Technol. Ther.* **2002**, *4*, 163–174. [CrossRef] [PubMed]
2. Mukund, K.; Subramaniam, S. Skeletal muscle: A review of molecular structure and function, in health and disease. *Wiley Interdiscip. Rev. Syst. Biol. Med.* **2020**, *12*, e1462. [CrossRef] [PubMed]
3. Puthucheary, Z.A.; Rawal, J.; McPhail, M.; Connolly, B.; Ratnayake, G.; Chan, P.; Hopkinson, N.S.; Padhke, R.; Dew, T.; Sidhu, P.S.; et al. Acute Skeletal Muscle Wasting in Critical Illness. *JAMA* **2013**, *310*, 1591–1600. [CrossRef] [PubMed]
4. Varga, T.; Czimmerer, Z.; Nagy, L. PPARs are a unique set of fatty acid regulated transcription factors controlling both lipid metabolism and inflammation. *Biochim. Biophys. Acta (BBA)-Mol. Basis Dis.* **2011**, *1812*, 1007–1022. [CrossRef] [PubMed]
5. Lee, C.-H.; Olson, P.; Evans, R.M. Minireview: Lipid Metabolism, Metabolic Diseases, and Peroxisome Proliferator-Activated Receptors. *Endocrinology* **2003**, *144*, 2201–2207. [CrossRef] [PubMed]

6. Tanaka, T.; Yamamoto, J.; Iwasaki, S.; Asaba, H.; Hamura, H.; Ikeda, Y.; Watanabe, M.; Magoori, K.; Ioka, R.X.; Tachibana, K.; et al. Activation of peroxisome proliferator-activated receptor δ induces fatty acid β-oxidation in skeletal muscle and attenuates metabolic syndrome. *Proc. Natl. Acad. Sci. USA* **2003**, *100*, 15924–15929. [CrossRef] [PubMed]
7. Abbott, B.D. Review of the expression of peroxisome proliferator-activated receptors alpha (PPARα), beta (PPARβ), and gamma (PPARγ) in rodent and human development. *Reprod. Toxicol.* **2009**, *27*, 246–257. [CrossRef] [PubMed]
8. Baes, M.; Peeters, A. Role of PPARα in hepatic carbohydrate metabolism. *PPAR Res.* **2010**, *2010*, 572405.
9. Muoio, D.M.; MacLean, P.S.; Lang, D.B.; Li, S.; Houmard, J.A.; Way, J.M.; Winegar, D.A.; Corton, J.C.; Dohm, G.L.; Kraus, W.E. Fatty Acid Homeostasis and Induction of Lipid Regulatory Genes in Skeletal Muscles of Peroxisome Proliferator-activated Receptor (PPAR) α Knock-out Mice. Evidence for compensatory regulation by PPARδ. *J. Biol. Chem.* **2002**, *277*, 26089–26097. [CrossRef] [PubMed]
10. Ueno, H.; Saitoh, Y.; Mizuta, M.; Shiiya, T.; Noma, K.; Mashiba, S.; Kojima, S.; Nakazato, M. Fenofibrate ameliorates insulin resistance, hypertension and novel oxidative stress markers in patients with metabolic syndrome. *Obes. Res. Clin. Pr.* **2011**, *5*, e335–e340. [CrossRef]
11. Koh, K.K.; Han, S.H.; Quon, M.J.; Ahn, J.Y.; Shin, E.K. Beneficial Effects of Fenofibrate to Improve Endothelial Dysfunction and Raise Adiponectin Levels in Patients With Primary Hypertriglyceridemia. *Diabetes Care* **2005**, *28*, 1419–1424. [CrossRef]
12. Fajas, L.; Auboeuf, D.; Raspé, E.; Schoonjans, K.; Lefebvre, A.-M.; Saladin, R.; Najib, J.; Laville, M.; Fruchart, J.-C.; Deeb, S.; et al. The Organization, Promoter Analysis, and Expression of the Human PPARγ Gene. *J. Biol. Chem.* **1997**, *272*, 18779–18789. [CrossRef] [PubMed]
13. El Akoum, S. PPAR Gamma at the Crossroads of Health and Disease: A Masterchef in Metabolic Homeostasis. *Endocrinol. Metab. Syndr.* **2014**, *3*, 2161–1017. [CrossRef]
14. Miyazaki, Y.; He, H.; Mandarino, L.J.; DeFronzo, R.A. Rosiglitazone improves downstream insulin receptor signaling in type 2 diabetic patients. *Diabetes* **2003**, *52*, 1943–1950. [CrossRef] [PubMed]
15. Gastaldelli, A.; Ferrannini, E.; Miyazaki, Y.; Matsuda, M.; Mari, A.; DeFronzo, R.A. Thiazolidinediones improve β-cell function in type 2 diabetic patients. *Am. J. Physiol. Metab.* **2007**, *292*, E871–E883. [CrossRef] [PubMed]
16. Yamauchi, T.; Kamon, J.; Waki, H.; Terauchi, Y.; Kubota, N.; Hara, K.; Mori, Y.; Ide, T.; Murakami, K.; Tsuboyama-Kasaoka, N.; et al. The fat-derived hormone adiponectin reverses insulin re-sistance associated with both lipoatrophy and obesity. *Nat. Med.* **2001**, *7*, 941–946. [CrossRef] [PubMed]
17. Ehrenborg, E.; Krook, A. Regulation of Skeletal Muscle Physiology and Metabolism by Peroxisome Proliferator-Activated Receptor δ. *Pharmacol. Rev.* **2009**, *61*, 373–393. [CrossRef]
18. Tan, N.S.; Vázquez-Carrera, M.; Montagner, A.; Sng, M.K.; Guillou, H.; Wahli, W. Transcriptional control of physiological and pathological processes by the nuclear receptor PPARβ/δ. *Prog. Lipid Res.* **2016**, *64*, 98–122. [CrossRef] [PubMed]
19. Phua, W.W.T.; Wong, M.X.Y.; Liao, Z.; Tan, N.S. An apparent functional consequence in skeletal muscle physiology via pe-roxisome proliferator-activated receptors. *Int. J. Mol. Sci.* **2018**, *19*, 1425. [CrossRef]
20. Pawlak, M.; Lefebvre, P.; Staels, B. Molecular mechanism of PPARα action and its impact on lipid metabolism, inflammation and fibrosis in non-alcoholic fatty liver disease. *J. Hepatol.* **2015**, *62*, 720–733. [CrossRef] [PubMed]
21. Dressel, U.; Allen, T.L.; Pippal, J.B.; Rohde, P.R.; Lau, P.; Muscat, G.E.O. The Peroxisome Proliferator-Activated Receptor β/δ Agonist, GW501516, Regulates the Expression of Genes Involved in Lipid Catabolism and Energy Uncoupling in Skeletal Muscle Cells. *Mol. Endocrinol.* **2003**, *17*, 2477–2493. [CrossRef] [PubMed]
22. Peters, S.J.; Harris, R.A.; Heigenhauser, G.J.F.; Spriet, L.L. Muscle fiber type comparison of PDH kinase activity and isoform expression in fed and fasted rats. *Am. J. Physiol. Integr. Comp. Physiol.* **2001**, *280*, R661–R668. [CrossRef] [PubMed]
23. Wieland, O.H. The mammalian pyruvate dehydrogenase complex: Structure and regulation. *Rev. Physiol. Biochem. Pharmacol.* **1983**, *96*, 123–170. [CrossRef]
24. Bowker-Kinley, M.M.; Davis, I.W.; Wu, P.; Harris, A.R.; Popov, M.K. Evidence for existence of tissue-specific regulation of the mammalian pyruvate dehydrogenase complex. *Biochem. J.* **1998**, *329*, 191–196. [CrossRef] [PubMed]
25. Huang, B.; Gudi, R.; Wu, P.; Harris, R.A.; Hamilton, J.; Popov, K.M. Isoenzymes of Pyruvate Dehydrogenase Phosphatase. DNA-derived amino acid sequences, expression, and regulation. *J. Biol. Chem.* **1998**, *273*, 17680–17688. [CrossRef]
26. Sugden, M.C.; Kraus, A.; Harris, R.A.; Holness, M.J. Fibre-type specific modification of the activity and regulation of skeletal muscle pyruvate dehydrogenase kinase (PDK) by prolonged starvation and refeeding is associated with targeted regulation of PDK isoenzyme 4 expression. *Biochem. J.* **2000**, *346*, 651–657. [CrossRef]
27. Wu, P.; Inskeep, K.; Bowker-Kinley, M.M.; Popov, K.M.; Harris, R.A. Mechanism responsible for inactivation of skeletal muscle pyruvate dehydrogenase complex in starvation and diabetes. *Diabetes* **1999**, *48*, 1593–1599. [CrossRef] [PubMed]
28. Tsintzas, K.; Jewell, K.; Kamran, M.; Laithwaite, D.; Boonsong, T.; Littlewood, J.; Macdonald, I.; Bennett, A. Differential regulation of metabolic genes in skeletal muscle during starvation and refeeding in humans. *J. Physiol.* **2006**, *575*, 291–303. [CrossRef]
29. Spriet, L.L.; Tunstall, R.J.; Watt, M.J.; Mehan, K.A.; Hargreaves, M.; Cameron-Smith, D. Pyruvate dehydrogenase activation and kinase expression in human skeletal muscle during fasting. *J. Appl. Physiol.* **2004**, *96*, 2082–2087. [CrossRef]
30. Furuyama, T.; Kitayama, K.; Yamashita, H.; Mori, N. Forkhead transcription factor FOXO1 (FKHR)-dependent induction of PDK4 gene expression in skeletal muscle during energy deprivation. *Biochem. J.* **2003**, *375*, 365–371. [CrossRef]
31. Wahli, W.; Michalik, L. PPARs at the crossroads of lipid signaling and inflammation. *Trends Endocrinol. Metab.* **2012**, *23*, 351–363. [CrossRef] [PubMed]

32. Michalik, L.; Wahli, W. PPARs Mediate Lipid Signaling in Inflammation and Cancer. *PPAR Res.* **2008**, *2008*, 134059. [CrossRef] [PubMed]
33. Bishop-Bailey, D.; Bystrom, J. Emerging roles of peroxisome proliferator-activated receptor-β/δ in inflammation. *Pharmacol. Ther.* **2009**, *124*, 141–150. [CrossRef] [PubMed]
34. Delerive, P.; De Bosscher, K.; Berghe, W.V.; Fruchart, J.-C.; Haegeman, G.; Staels, B. DNA Binding-Independent Induction of IκBα Gene Transcription by PPARα. *Mol. Endocrinol.* **2002**, *16*, 1029–1039. [CrossRef]
35. Huang, W.; Glass, C.K. Nuclear receptors and inflammation control: Molecular mechanisms and pathophysiological rele-vance. *Arterioscler. Thromb. Vasc. Biol.* **2010**, *30*, 1542–1549. [CrossRef] [PubMed]
36. Constantin, D.; Constantin-Teodosiu, D.; Layfield, R.; Tsintzas, K.; Bennett, A.J.; Greenhaff, P.L. PPARδ agonism induces a change in fuel metabolism and activation of an atrophy programme, but does not impair mitochondrial function in rat skeletal muscle. *J. Physiol.* **2007**, *583*, 381–390. [CrossRef] [PubMed]
37. Léger, B.; Cartoni, R.; Praz, M.; Lamon, S.; Dériaz, O.; Crettenand, A.; Gobelet, C.; Rohmer, P.; Konzelmann, M.; Luthi, F.; et al. Akt signalling through GSK-3β, mTOR and Foxo1 is involved in human skeletal muscle hypertrophy and atrophy. *J. Physiol.* **2006**, *576*, 923–933. [CrossRef]
38. Chen, W.; Gao, R.; Xie, X.; Zheng, Z.; Li, H.; Li, S.; Dong, F.; Wang, L. A metabolomic study of the PPARδ agonist GW501516 for enhancing running endurance in Kunming mice. *Sci. Rep.* **2015**, *5*, 9884. [CrossRef]
39. Constantin-Teodosiu, D.; Baker, D.J.; Constantin, D.; Greenhaff, P.L. PPARδ agonism inhibits skeletal muscle PDC activity, mitochondrial ATP production and force generation during prolonged contraction. *J. Physiol.* **2009**, *587*, 231–239. [CrossRef]
40. Constantin-Teodosiu, D.; Cederblad, G.; Hultman, E. PDC activity and acetyl group accumulation in skeletal muscle during isometric contraction. *J. Appl. Physiol.* **1993**, *74*, 1712–1718. [CrossRef]
41. Constantin-Teodosiu, D.; Constantin, D.; Stephens, F.; Laithwaite, D.; Greenhaff, P.L. The Role of FOXO and PPAR Transcription Factors in Diet-Mediated Inhibition of PDC Activation and Carbohydrate Oxidation During Exercise in Humans and the Role of Pharmacological Activation of PDC in Overriding These Changes. *Diabetes* **2012**, *61*, 1017–1024. [CrossRef] [PubMed]
42. Constantin-Teodosiu, D. Regulation of Muscle Pyruvate Dehydrogenase Complex in Insulin Resistance: Effects of Exercise and Dichloroacetate. *Diabetes Metab. J.* **2013**, *37*, 301–314. [CrossRef] [PubMed]
43. Chien, H.-C.; Greenhaff, P.L.; Constantin-Teodosiu, D. PPARδ and FOXO1 Mediate Palmitate-Induced Inhibition of Muscle Pyruvate Dehydrogenase Complex and CHO Oxidation, Events Reversed by Electrical Pulse Stimulation. *Int. J. Mol. Sci.* **2020**, *21*, 5942. [CrossRef]
44. Grosset, J.-F.; Crowe, L.; de Vito, G.; O'Shea, D.; Caulfield, B. Comparative effect of a 1 h session of electrical muscle stimulation and walking activity on energy expenditure and substrate oxidation in obese subjects. *Appl. Physiol. Nutr. Metab.* **2013**, *38*, 57–65. [CrossRef]
45. Nieuwoudt, S.; Mulya, A.; Fealy, C.E.; Martelli, E.; Dasarathy, S.; Prasad, S.V.N.; Kirwan, J.P. In vitro contraction protects against palmitate-induced insulin resistance in C2C12 myotubes. *Am. J. Physiol. Physiol.* **2017**, *313*, C575–C583. [CrossRef]
46. Hotchkiss, R.S.; Moldawer, L.L.; Opal, S.M.; Reinhart, K.; Turnbull, I.R.; Vincent, J.L. Sepsis and septic shock. *Nat. Rev. Dis. Prim.* **2016**, *2*, 16045. [CrossRef] [PubMed]
47. Singer, M. The role of mitochondrial dysfunction in sepsis-induced multi-organ failure. *Virulence* **2014**, *5*, 66–72. [CrossRef] [PubMed]
48. Chiolèro, R.; Revelly, J.P.; Tappy, L. Energy metabolism in sepsis and injury. *Nutrition* **1997**, *13*, 45–51. [CrossRef]
49. Kamisoglu, K.; Haimovich, B.; Calvano, S.E.; Coyle, S.M.; Corbett, S.A.; Langley, R.J.; Kingsmore, S.F.; Androulakis, I.P. Human metabolic response to systemic inflammation: Assessment of the concordance between experimental endotoxemia and clinical cases of sepsis/SIRS. *Crit. Care* **2015**, *19*, 71. [CrossRef] [PubMed]
50. Senol, S.P.; Temiz, M.; Guden, D.S.; Cecen, P.; Sari, A.N.; Sahan-Firat, S.; Falck, J.R.; Dakarapu, R.; Malik, K.U.; Tunctan, B. Contribution of PPARα/β/γ, AP-1, importin-α3, and RXRα to the protective effect of 5,14-HEDGE, a 20-HETE mimetic, against hypotension, tachycardia, and inflammation in a rat model of septic shock. *Inflamm. Res.* **2016**, *65*, 367–387. [CrossRef] [PubMed]
51. Tunctan, B.; Kucukkavruk, S.P.; Temiz-Resitoglu, M.; Guden, D.S.; Sari, A.N.; Sahan-Firat, S. Bexarotene, a Selective RXRα Agonist, Reverses Hypotension Associated with Inflammation and Tissue Injury in a Rat Model of Septic Shock. *Inflammation* **2017**, *41*, 337–355. [CrossRef]
52. Castillero, E.; Alamdari, N.; Aversa, Z.; Gurav, A.; Hasselgren, P.-O. PPARβ/δ Regulates Glucocorticoid- and Sepsis-Induced FOXO1 Activation and Muscle Wasting. *PLoS ONE* **2013**, *8*, e59726. [CrossRef] [PubMed]
53. Drosatos, K.; Lymperopoulos, A.; Kennel, P.J.; Pollak, N.; Schulze, P.C.; Goldberg, I.J. Pathophysiology of Sepsis-Related Car-diac Dysfunction: Driven by Inflammation, Energy Mismanagement, or Both? *Curr. Heart Fail. Rep.* **2015**, *12*, 130–140. [CrossRef] [PubMed]
54. Feingold, K.; Kim, M.S.; Shigenaga, J.; Moser, A.; Grunfeld, C. Altered expression of nuclear hormone receptors and coactivators in mouse heart during the acute-phase response. *Am. J. Physiol. Metab.* **2004**, *286*, E201–E207. [CrossRef]
55. Shi, Y.; Jiang, H.; Yang, X. PPARδ Activation protects H9c2 cardiomyoblasts from LPS-induced apoptosis through the heme oxygenase-1-mediated suppression of NF-κB activation. *Mol. Med. Rep.* **2017**, *15*, 3775–3780. [CrossRef] [PubMed]
56. Benton, C.R.; Holloway, G.P.; Campbell, S.E.; Yoshida, Y.; Tandon, N.N.; Glatz, J.F.; Luiken, J.J.; Spriet, L.L.; Bonen, A. Rosiglitazone increases fatty acid oxidation and fatty acid translocase (FAT/CD36) but not carnitine palmitoyltransferase I in rat muscle mitochondria. *J. Physiol.* **2008**, *586*, 1755–1766. [CrossRef] [PubMed]

57. Mai, K.; Andres, J.; Bobbert, T.; Assmann, A.; Biedasek, K.; Diederich, S.; Graham, I.; Larson, T.R.; Pfeiffer, A.F.; Spranger, J. Rosiglitazone increases fatty acid Δ9-desaturation and decreases elongase activity index in human skeletal muscle in vivo. *Metabolism* **2012**, *61*, 108–116. [CrossRef] [PubMed]
58. Muurling, M.; Mensink, R.P.; Pijl, H.; Romijn, J.A.; Havekes, L.M.; Voshol, P.J. Rosiglitazone improves muscle insulin sensitivity, irrespective of increased triglyceride content, in ob/ob mice. *Metabolism* **2003**, *52*, 1078–1083. [CrossRef]
59. Lessard, S.J.; Giudice, S.L.; Lau, W.; Reid, J.J.; Turner, N.; Febbraio, M.A.; Hawley, J.A.; Watt, M.J. Rosiglitazone Enhances Glucose Tolerance by Mechanisms Other than Reduction of Fatty Acid Accumulation within Skeletal Muscle. *Endocrinology* **2004**, *145*, 5665–5670. [CrossRef]
60. Tan, L.; Song, A.; Ren, L.; Wang, C.; Song, G. Effect of pioglitazone on skeletal muscle lipid deposition in the insulin resistance rat model induced by high fructose diet under AMPK signaling pathway. *Saudi J. Biol. Sci.* **2020**, *27*, 1317–1323. [CrossRef]
61. Mayerson, A.B.; Hundal, R.S.; Dufour, S.; Lebon, V.; Befroy, U.; Cline, G.W.; Enocksson, S.; Inzucchi, S.E.; Shulman, G.I.; Petersen, K.F. The effects of rosiglitazone on insulin sensitivity, lipolysis, and hepatic and skeletal muscle triglyceride content in patients with type 2 diabetes. *Diabetes* **2002**, *51*, 797–802. [CrossRef]
62. Coletta, D.K.; Sriwijitkamol, A.; Wajcberg, E.; Tantiwong, P.; Li, M.; Prentki, M.; Madiraju, M.; Jenkinson, C.P.; Cersosimo, E.; Musi, N.; et al. Pioglitazone stimulates AMP-activated protein kinase signalling and increases the expression of genes involved in adiponectin signalling, mitochondrial function and fat oxidation in human skeletal muscle in vivo: A randomised trial. *Diabetologia* **2009**, *52*, 723–732. [CrossRef] [PubMed]
63. Mensink, M.; Hesselink, M.K.C.; Russell, A.P.; Schaart, G.; Sels, J.-P.; Schrauwen, P. Improved skeletal muscle oxidative enzyme activity and restoration of PGC-1α and PPARβ/δ gene expression upon rosiglitazone treatment in obese patients with type 2 diabetes mellitus. *Int. J. Obes.* **2007**, *31*, 1302–1310. [CrossRef]
64. Crossland, H.; Constantin-Teodosiu, D.; Gardiner, S.M.; Greenhaff, P.L. Peroxisome proliferator-activated receptor γ agonism attenuates endotoxaemia-induced muscle protein loss and lactate accumulation in rats. *Clin. Sci.* **2017**, *131*, 1437–1447. [CrossRef]
65. Crossland, H.; Constantin-Teodosiu, D.; Gardiner, S.M.; Constantin, D.; Greenhaff, P.L. A potential role for Akt/FOXO sig-nalling in both protein loss and the impairment of muscle carbohydrate oxidation during sepsis in rodent skeletal muscle. *J. Physiol.* **2008**, *586*, 5589–5600. [CrossRef] [PubMed]
66. Alamdari, N.; Constantin-Teodosiu, D.; Murton, A.J.; Gardiner, S.M.; Bennett, T.; Layfield, R.; Greenhaff, P.L. Temporal changes in the involvement of pyruvate dehydrogenase complex in muscle lactate accumulation during lipopolysaccharide infusion in rats. *J. Physiol.* **2008**, *586*, 1767–1775. [CrossRef]
67. Murton, A.J.; Alamdari, N.; Gardiner, S.M.; Constantin-Teodosiu, D.; Layfield, R.; Bennett, T.; Greenhaff, P.L. Effects of Endotoxaemia on Protein Metabolism in Rat Fast-Twitch Skeletal Muscle and Myocardium. *PLoS ONE* **2009**, *4*, e6945. [CrossRef] [PubMed]
68. Drosatos, K.; Khan, R.S.; Trent, C.M.; Jiang, H.; Son, N.-H.; Blaner, W.S.; Homma, S.; Schulze, P.C.; Goldberg, I.J. Peroxisome Proliferator–Activated Receptor-γ Activation Prevents Sepsis-Related Cardiac Dysfunction and Mortality In Mice. *Circ. Hear. Fail.* **2013**, *6*, 550–562. [CrossRef]
69. Peng, S.; Xu, J.; Ruan, W.; Li, S.; Xiao, F. PPAR-γ Activation Prevents Septic Cardiac Dysfunction via Inhibition of Apoptosis and Necroptosis. *Oxidative Med. Cell. Longev.* **2017**, *2017*, 8326749. [CrossRef] [PubMed]
70. Zhang, S.M.; Cai, X.F.; Ma, Y.L.; Lu, Q. Effect of rosiglitazone on myocardial injury in septic rats through NF-κB pathway. *Eur. Rev. Med. Pharmacol. Sci.* **2020**, *24*, 452–460. [CrossRef] [PubMed]

International Journal of
Molecular Sciences

Article

Gene Doping with Peroxisome-Proliferator-Activated Receptor Beta/Delta Agonists Alters Immunity but Exercise Training Mitigates the Detection of Effects in Blood Samples

Brigitte Sibille [1], Isabelle Mothe-Satney [1], Gwenaëlle Le Menn [1], Doriane Lepouse [1], Sébastien Le Garf [1], Elodie Baudoin [1], Joseph Murdaca [1], Claudine Moratal [1], Noura Lamghari [1], Giulia Chinetti [2], Jaap G. Neels [1,*,†] and Anne-Sophie Rousseau [1,†]

1 INSERM, Université Côte d'Azur, C3M, 06204 Nice, France; Brigitte.SIBILLE@univ-cotedazur.fr (B.S.); Isabelle.SATNEY@univ-cotedazur.fr (I.M.-S.); gwenaelle.lemenn@gmail.com (G.L.M.); dlepouse@gmail.com (D.L.); seb.legarf@hotmail.fr (S.L.G.); elodiebaudoin68@gmail.com (E.B.); joseph.murdaca@univ-cotedazur.fr (J.M.); Claudine.MORATAL@univ-cotedazur.fr (C.M.); nouritalamghari@gmail.com (N.L.); Anne-Sophie.ROUSSEAU@univ-cotedazur.fr (A.-S.R.)
2 CHU, INSERM, Université Côte d'Azur, C3M, 06204 Nice, France; Giulia.CHINETTI@univ-cotedazur.fr
* Correspondence: jaap.neels@univ-cotedazur.fr; Tel.: +33-(0)-4-89-15-38-40
† These authors shared last authorship.

Citation: Sibille, B.; Mothe-Satney, I.; Le Menn, G.; Lepouse, D.; Le Garf, S.; Baudoin, E.; Murdaca, J.; Moratal, C.; Lamghari, N.; Chinetti, G.; et al. Gene Doping with Peroxisome-Proliferator-Activated Receptor Beta/Delta Agonists Alters Immunity but Exercise Training Mitigates the Detection of Effects in Blood Samples. *Int. J. Mol. Sci.* **2021**, 22, 11497. https://doi.org/10.3390/ijms222111497

Academic Editors: Manuel Vázquez-Carrera and Walter Wahli

Received: 24 September 2021
Accepted: 23 October 2021
Published: 25 October 2021

Publisher's Note: MDPI stays neutral with regard to jurisdictional claims in published maps and institutional affiliations.

Abstract: Synthetic ligands of peroxisome-proliferator-activated receptor beta/delta (PPARβ/δ) are being used as performance-enhancing drugs by athletes. Since we previously showed that PPARβ/δ activation affects T cell biology, we wanted to investigate whether a specific blood T cell signature could be employed as a method to detect the use of PPARβ/δ agonists. We analyzed in primary human T cells the in vitro effect of PPARβ/δ activation on fatty acid oxidation (FAO) and on their differentiation into regulatory T cells (Tregs). Furthermore, we conducted studies in mice assigned to groups according to an 8-week exercise training program and/or a 6-week treatment with 3 mg/kg/day of GW0742, a PPARβ/δ agonist, in order to (1) determine the immune impact of the treatment on secondary lymphoid organs and to (2) validate a blood signature. Our results show that PPARβ/δ activation increases FAO potential in human and mouse T cells and mouse secondary lymphoid organs. This was accompanied by increased Treg polarization of human primary T cells. Moreover, Treg prevalence in mouse lymph nodes was increased when PPARβ/δ activation was combined with exercise training. Lastly, PPARβ/δ activation increased FAO potential in mouse blood T cells. Unfortunately, this signature was masked by training in mice. In conclusion, beyond the fact that it is unlikely that this signature could be used as a doping-control strategy, our results suggest that the use of PPARβ/δ agonists could have potential detrimental immune effects that may not be detectable in blood samples.

Keywords: peroxisome-proliferator-activated receptor; fatty acid oxidation; doping control; regulatory T cells; inflammation; exercise

1. Introduction

The nuclear receptor peroxisome-proliferator-activated receptor beta/delta (PPARβ/δ) plays an important role in muscle physiology [1]. This transcription factor can be activated by endogenous natural ligands, such as certain lipid metabolites, or synthetic ligands, such as GW501516 and GW0742 [1]. The latter substances have also been called "exercise pills" or "exercise mimetics", since they were shown to affect the expression of endurance-related genes and metabolic pathways leading to increased exercise endurance [2]. While synthetic PPARβ/δ agonists so far have not been approved for clinical purposes for treating diseases such as dyslipidemia due to the discovery of carcinogenic properties in preclinical studies on animals, these substances are being abused for performance-enhancing purposes in both humans and horses [3,4]. Therefore, since 2009, the list of prohibited substances and methods of doping, as established by the World Anti-Doping Agency, includes PPARβ/δ agonists. Methods to detect PPARβ/δ agonists are mostly focused on GW501516, and tests were developed for both blood and urine samples [5]. However, the emergence of new substances of this class means that new methods need to be developed that will allow the detection of any PPARβ/δ agonist. One method could be to identify a blood signature that would be specific for PPARβ/δ activation. In this respect, our laboratory has previously published several studies demonstrating that PPARβ/δ activation induces significant immunometabolic changes in T cells. We showed that in vitro and in vivo treatment with GW0742 led to an increase in the mRNA levels of three genes involved in fatty acid oxidation (i.e., *Acaa2*, *Acadvl*, and *Cpt1a*) in isolated mouse primary T cells and lymph nodes, respectively, resulting in increased fatty acid oxidation (FAO) in these cells [6]. Furthermore, increased PPARβ/δ activity had an impact on T cell development in the thymus, resulting in reduced production of αβ-T cells, while γδ-T cell production was unaffected. This led to a decrease in the αβ/γδ T cell ratio in peripheral tissues, including blood.

Regulatory T cells (Tregs) are a subset of T cells important for maintaining self-tolerance by downregulating the immune response, and they do so by secreting immunoregulatory cytokines, such as TGF-β and IL-10, which act to suppress the activity and function of immune effector cells (e.g., CD4+ and CD8+ T cells, monocytes/macrophages, natural killer cells, and dendritic cells) [7]. While mouse Tregs have a certain flexibility in metabolic fuel choice, they have a preference for FAO [8]. It was therefore not unexpected, given our above-mentioned result showing an increased FAO in T cells following activation of PPARβ/δ, that we observed an increased prevalence of CD4+FOXP3+ Tregs in mouse lymph nodes after in vivo GW0742 treatment [9].

Whether these observed PPARβ/δ-induced changes in T cell parameters can be used as a blood signature for detection of the use of PPARβ/δ agonists depends on whether these changes are specific to PPARβ/δ activation and are not also potentially induced by other factors such as acute or chronic exercise. Exercise-induced immune changes have been described previously [10]. Many of them affect the Treg population or the ability of T cells to produce pro/anti-inflammatory cytokines [10,11]. However, the ability of physical fitness or exercise to directly modify the metabolism of immune cells is unproven [12] but could be involved in these changes.

Our objective in this study was to confirm, in human T cells, our previous published observations in mouse T cells, that PPARβ/δ activation leads to an increase in FAO. Likewise, we also wanted to determine the effect of PPARβ/δ activation on the induction of Treg polarization in human T cells. Furthermore, we investigated whether exercise training would interfere with the effects of PPARβ/δ activation on FAO gene expression, T cell ratios, and Treg polarization.

2. Results

2.1. In Vitro Treatment of Human T Cells with GW0742 Increases Their FAO Potential and Polarization in Tregs

We hypothesized that GW0742, a PPARβ/δ agonist, preconditions human T cells and favors their polarization toward the Treg subtype. To validate this hypothesis, we isolated PBMCs (peripheral blood mononuclear cells) from human buffy coats and performed monocyte depletion by adhesion. The remaining cells, enriched in lymphocytes, were placed in culture, activated with beads coated with αCD3 and αCD28 antibodies and IL-2, and treated with 1 μM GW0742 or left untreated for 6 days. We showed (Figure 1A) a significant fourfold induction by GW0742 treatment of carnitine palmitoyl transferase 1a (CPT1a) mRNA that encodes the enzyme limiting the entry of fatty acids into the mitochondria, leading to a 2.6-fold increase in palmitate oxidation (Figure 1B). However, no difference in PPARβ/δ mRNA level was observed (Figure 1A). As already demonstrated in mouse T cells from secondary lymphoid organs [6], activation of the PPARβ/δ pathway in human blood T cells induces the expression of genes encoding FAO proteins (CPT1a) and increases FAO. As Treg cells are very dependent on FAO, we studied the impact of

PPARβ/δ pathway activation on human T cell polarization toward Tregs. In this objective, we cultured CD4+ T cells selected from monocyte-depleted human buffy coats and treated in vitro with TGF-β (5 μg/mL) to induce Treg polarization in the presence of DMSO (TREG) or GW0742 (TREG GW). We used the gating strategy presented in Figure 1C, namely a flow cytometry analysis of CD25+ FOXP3+ cells (Tregs) in CD3+CD4+ human T cells. The presence of TGF-β in the culture medium of CD4+ T cells (Figure 1D, TREG) permitted to almost double (1.94 ± 0.09, # $p < 0.0001$) the percentage of CD25+ FOXP3+ cells (Tregs) in CD3+CD4+ human T cells compared to the condition without TGF-β (Th0) and induced a slight but significant increase in FOXP3 mean fluorescent intensity (MFI) (Figure 1E, TREG, 1.104 ± 0.04, # $p < 0.05$), considered to represent the mean content level of FOXP3 protein in cells. The activation of the PPARβ/δ pathway by GW0742 (Figure 1D, TREG GW) significantly favored Treg polarization in 14 independent experiments, as reflected by the increase in prevalence of CD25+ FOXP3+ cells in CD3+CD4+ human T cells (mean increase 14.9% ± 3.1%) without change in FOXP3 MFI. We, therefore, showed that activation of PPARβ/δ pathway in human blood T cells leads to a change in T cell metabolism, favoring FAO that is accompanied by an increase in Treg polarization. These changes could perhaps be used as a blood signature of the abuse of PPARβ/δ agonists by athletes.

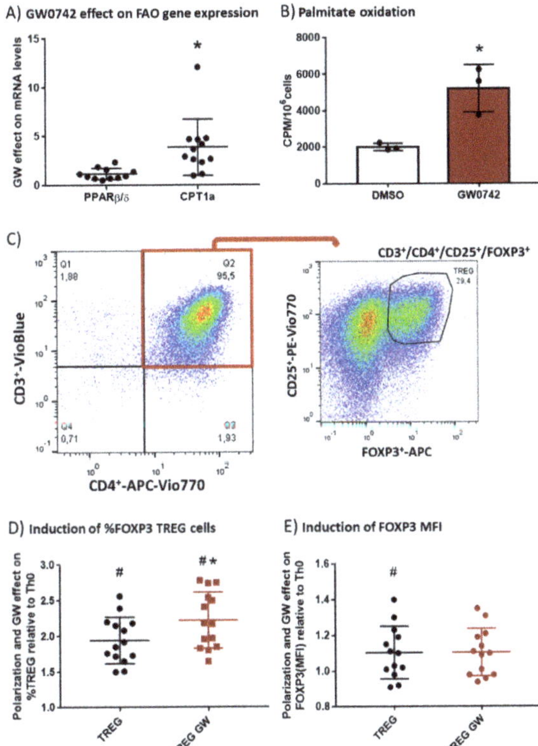

Figure 1. In vitro treatment of human T cells with GW0742 increases their FAO potential and their polarization in Tregs. (**A**) GW0742 1 μM effect (compared to DMSO, $n = 8$) on PPARβ/δ and CPT1a mRNA level reported to RPL27 mRNA level used as housekeeping mRNA on monocyte-depleted human buffy coats activated with αCD3 and αCD28 antibody-coated beads and cultured for 6 days with human IL-2 (20 ng/mL). (**B**) Palmitate oxidation in isolated human CD4+ T cells. FAO was measured as ^{3}H-palmitate conversion to ^3H_2O and quantified as CPM/10^6 cells in in vitro–activated

CD4+ cells treated with 1 μM GW0742 or DMSO (n = 3). (**C**) Gating strategy of flow cytometry analysis of CD25+ FOXP3+ cells (Tregs) in CD3+CD4+ human T cells. (**D**) Fold induction of prevalence (%) of CD25+ FOXP3+ cells (Tregs) in enriched CD4+ T cells (n = 14) derived from monocyte-depleted human buffy coats treated in vitro with TGF-β (5 μg/mL) to induce Treg polarization in the presence of DMSO (TREG) or 1 μM GW0742 (TREG GW) relative to Th0 cells (nonpolarized cells). (**E**) Fold induction of FOXP3+ MFI (mean fluorescent intensity) in CD4+ T cells (n = 14) derived from monocyte-depleted human buffy coats treated in vitro with TGF-β (5 μg/mL) to induce Treg polarization in the presence of DMSO (TREG) or 1 μM GW0742 (TREG GW) relative to Th0 cells. Data are shown as mean ± SD. * $p < 0.05$, GW effect; # $p < 0.05$, TREG vs. Th0 cells (univariate t-test).

2.2. GW0742 Treatment Increases FAO Potential and Leads to Differential Changes in Treg Prevalence in Mouse Secondary Lymphoid Organs Depending on Training Status

We first validated whether the potential signature that was detectable in human cells (i.e., increased FAO potential and Treg polarization) is specific for PPARβ/δ activation. We have previously demonstrated [6] that treatment of murine T cells with GW0742 increased palmitate oxidation and this effect was lost when the cells were co-treated with etomoxir, an inhibitor of CPT1a. We isolated CD4+ T cells from secondary lymphoid organs (SLO) of controls (Cre) or mice invalidated for PPARβ/δ in T cells (KO-T-PPARβ/δ), treated the cells with 1 μM of GW0742 for 6 days and studied the consequences on PPARβ/δ and CPT1a mRNA level. We showed, as seen in Figure 2A, that the treatment of control cells (CreTh0) with GW0742 did not alter the mRNA level of PPARβ/δ. However, GW0742 treatment increased the mRNA level of CPT1a by a factor of 5. This GW0742 effect seemed specific to its action on PPARβ/δ, since the induction of CPT1a mRNA was markedly reduced in cells isolated from KO-T-PPARβ/δ mice (KOTh0).

To discriminate between the effects induced by chronic (training) bouts of exercise and PPARβ/δ activation, mice were trained on treadmills for 8 weeks and they were, or were not, treated with GW0742 mixed with food for 6 weeks (3 mg/kg BW/day). At the end, the spleen and lymph nodes were harvested. GW0742 treatment led to an increase in PPARβ/δ mRNA levels in the lymph nodes but not in the spleen (Figure 2B,D). Moreover, the treatment of control mice with GW0742 for 6 weeks induced a significant increase in CPT1a mRNA by a factor of 3.3 ± 1.5 in the spleen (Figure 2C) and 3.8 ± 1.7 in the lymph nodes (Figure 2E). We did not detect a significant effect of training on PPARβ/δ or CPT1a mRNA levels in the SLO, nor did training alter the effects of GW0742 (Figure 2B–E). Regarding Treg prevalence, GW0742 treatment of mice did not significantly affect the percentage of FOXP3+ T cells (Tregs) in the lymph nodes of mice (Figure 2F), but exercise training independently of GW0742 treatment significantly increased this percentage and decreased the MFI level of FOXP3 (Figure 2F,G). However, there was a significant combined effect of GW0742 treatment and training, significantly increasing the proportion of Treg cells in lymph nodes (Figure 2F). It is noteworthy that the GW0742 treatment did not affect the FOXP3 MFI level of trained mice (Figure 2G). Thus, the GW0742 treatment of mice leads to immunometabolic changes promoting FAO potential and an increased proportion of Treg cells in exercise-trained mouse secondary lymphoid organs.

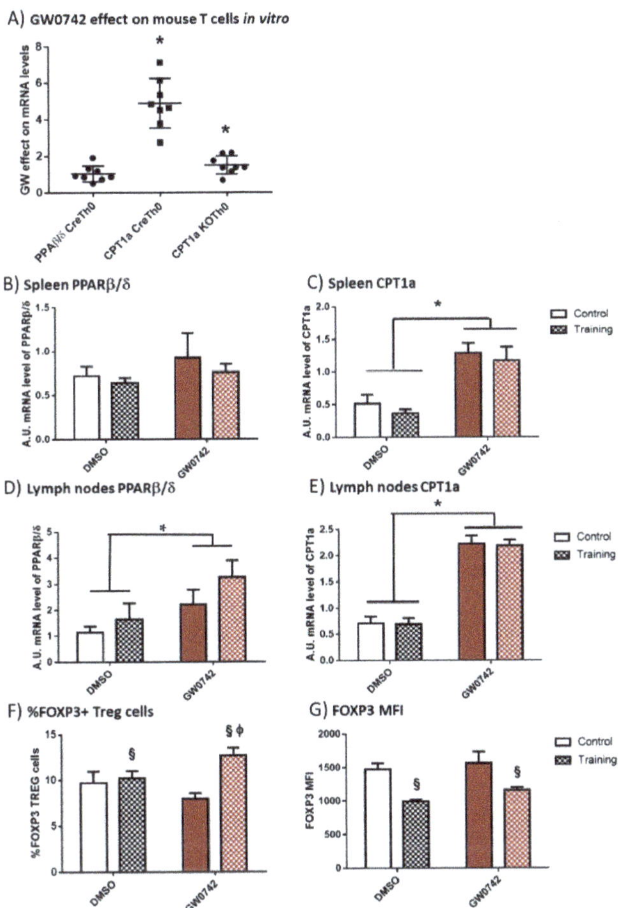

Figure 2. In vitro treatment of mouse T cells with GW0742 increases their FAO potential, and in vivo GW0742 treatment of mice leads to differential changes in FAO potential and Treg profile in the lymph nodes and spleen in trained mice. (**A**) The effect of 1 μM GW0742 (compared to that of DMSO, $n = 6$) on PPARβ/δ and CPT1a mRNA levels normalized to 36B4 mRNA level used as housekeeping mRNA in CD4+ T cells from Lck-Cre (Cre) or KO-T-PPARβ/δ mice (KO) activated with αCD3 and αCD28 antibodies coated-beads and cultured for 6 days with mouse IL-2 (20 ng/mL). (**B**) PPARβ/δ mRNA level and (**C**) CPT1a mRNA level in spleen; (**D**) PPARβ/δ mRNA level and (**E**) CPT1a mRNA level in lymph nodes, from control or trained mice (8 weeks, $n = 6$ per group) treated, or not treated, for 6 weeks with GW0742 (3 mg/kg BW/day). (**F**) Prevalence (%) of FOXP3+ cells (Tregs) in cells extracted from lymph nodes from control or trained mice (8 weeks, $n = 6$ per group) treated, or not treated, for 6 weeks with GW0742 (3 mg/kg BW/day). (**G**) FOXP3+ MFI (mean fluorescent intensity) in cells extracted from lymph nodes from control or trained mice (8 weeks, $n = 6$ per group) treated, or not treated, for 6 weeks with GW0742 (3 mg/kg BW/day). Data are shown as mean ± SD. * $p < 0.05$, GW0742 effect; § $p < 0.05$, training effect; and ϕ $p < 0.05$, interaction effect between training and GW0742 (two-way ANOVA).

2.3. The Detection of GW0742 Effect on FAO Potential Is Masked in the Blood of Trained Mice

In mice, we showed that increased PPARβ/δ activity leads to a defect in T cell development in the thymus with subsequent consequences on T cell populations in peripheral lymphoid organs, characterized by a decrease in the αβ/γδ T cell ratio. This was accompanied by an increase in FAO potential and a concomitant increase in CPT1a mRNA levels in lymphoid organs [6,9]. To definitively validate this signature of increased PPARβ/δ activity, which could be detected in athletes' blood, its effects must be discriminated from those induced by acute and chronic (training) bouts of exhaustive exercise. We submitted mice to acute exercise, training (8 weeks), or long-term treatment (6 weeks) with GW0742 and studied the evolution in the blood of some signature markers (αβ/γδ T cell ratio, CPT1a mRNA levels). We also wanted to check whether the signature of the GW0742 use could be distinguished from that of physical training. We showed that the CD4+/CD8+ T cell ratio (Figure 3A) and the αβ/γδ T cell ratio (Figure 3B) were not altered in the blood either by acute exercise, training, and GW0742 treatment, or by the combination of both GW0742 and training. We found (Figure 3C) that mRNA levels of CPT1a in the blood were not impaired by training but, in contrast, were significantly largely increased by treatment with GW0742 (4.04 ± 3.03-fold increase). This measurement in blood cells of the CPT1a mRNA levels could, thus, constitute a signature of the use of the GW0742. However, and very surprisingly, we can see that the effect of GW0742 on CPT1a mRNA levels was largely and significantly decreased when GW0742 intake was combined with endurance training. Thus, the signature of the use of GW0742 is masked by endurance training, and the measurement of blood cell CPT1a mRNA levels will not be a reliable marker for the use of GW0742.

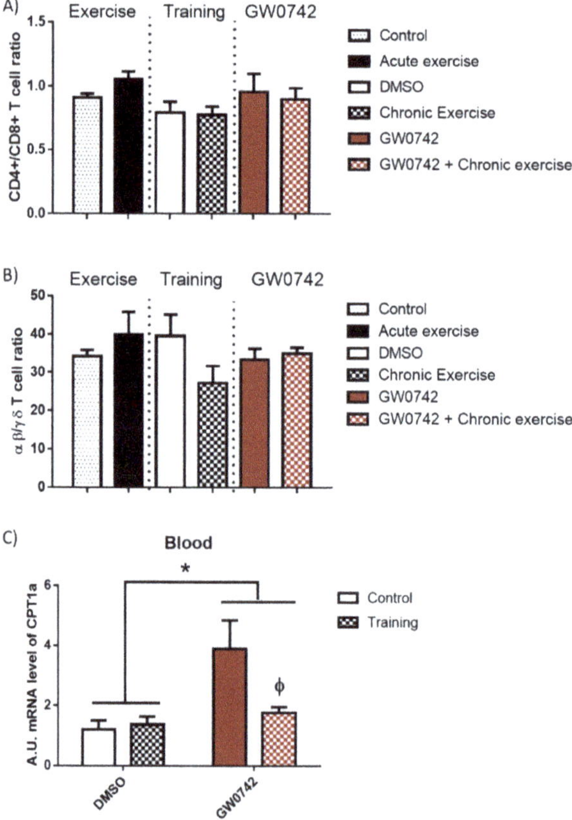

Figure 3. The T cell profile (CD4+/CD8+ T cell ratio, αβ/γδ T cell ratio) is unchanged in the blood by GW0742 treatment, exercise, or training. However, the detection in whole blood of GW0742's effects on the FAO potential is reduced by training. Mice were either subjected to or not given (control, n = 10) acute exercise on a treadmill with a slope of 5° (n = 8), the speed of the treadmill increased by

5 cm/s every 15 min until mouse exhaustion. Another cohort of mice (n = 6 per group) was trained (chronic exercise) on a treadmill for 8 weeks, or not trained, and were then treated, or not treated (given DMSO instead), for 6 weeks with GW0742 (3 mg/kg BW/day). Blood mononuclear cells were isolated using Ficoll gradient, stained with fluorescence-conjugated antibodies, and analyzed with a BD FACS Canto II flow cytometer. (**A**) The CD4+/CD8+ T cell ratio was calculated; (**B**) the $\alpha\beta/\gamma\delta$ T cell ratio was calculated; (**C**) the CPT1a mRNA level in blood cells was normalized by 36B4. Data are shown as mean ± SD. * $p < 0.05$, GW0742 effect, and ϕ $p < 0.05$, interaction effect between training and GW0742 (two-way ANOVA).

3. Discussion

T cells' function is intimately linked to their metabolic programs [13,14]. While Tregs rely heavily on FAO, they have found ways to adapt to different tissue types, such as tumors, to survive in competitive environments [15]. Mouse Treg cells generated through in vitro polarization of CD4^{+} T cells preferentially use FAO [16,17]. However, it is still controversial whether human Treg cell differentiation is dependent on FAO. We show here that the use of substances that activate PPARβ/δ can increase FAO in human T cells in vitro, and as a result increase the prevalence of Tregs. This result is important and new. Human Tregs are metabolically distinct from their mouse counterparts. It is known that ex vivo human Tregs are more glycolytic than ex vivo mouse Tregs [18]. This baseline difference may account for the magnitude of detectable metabolic changes that could be induced by either an endogenous or exogenous modulator of PPARβ/δ activity. Thus, we can assume that the whole-body GW0742 effect on immunometabolism would be more potent in humans compared to that in mice. Since our in vivo studies were conducted in mice, it is plausible that the effects observed in mice will be stronger in humans.

Whether exercise can modulate immune function by metabolic changes remains an underexplored area of research, and the ability of physical fitness or exercise to directly modify the metabolism of immune cells is unproven [12]. In obese mice, metabolic changes induced by exercise training were characterized by an increase in AMPK activity, both in lymphoid tissues and in skeletal muscle [9]. In both tissues, GW0742 treatment had complementary effects to exercise training on the decrease in inflammatory markers [9]. In the present study, in secondary lymphoid tissues, the induction of CPT1a expression was independent of exercise and was characteristic of the GW0742 effect on increasing FAO potential. The magnitude of CPT1a induction was high and suggests that the metabolism of immune cells (mainly T cells) was altered by GW0742 treatment. Furthermore, PPARβ/δ expression was also increased in lymph nodes by GW0742 treatment. However, the prevalence of Tregs was unchanged in the lymph nodes of sedentary mice treated with GW0742. Therefore, we can conclude that, even though GW0742 increased PPARβ/δ and CPT1a expression, it did not increase Treg prevalence in mice lymph nodes. Exercise training significantly decreased the MFI level of FOXP3 but interacted with GW0742, leading to an increase in the prevalence of Tregs. This increase appeared despite an absence of effect on CPT1a mRNA level in secondary lymphoid tissues. Together, these findings suggest that, at least in mice, CPT1a expression levels are disconnected from Treg prevalence. These results are entirely in line with the work of Raud et al. [19] that showed, using a mouse genetic model in which CPT1a was abrogated in T cells, that the ACC2/CPT1a axis is dispensable for Treg cell formation.

Despite an absence of effect on CPT1a mRNA level of exercise in SLOs, it is known that exercise impairs aspects of cellular immune function, probably due to the higher energy cost of exercise and metabolic perturbations in endurance athletes [20]. Indeed, a rapid metabolite turnover can be detected in seconds after an acute bout of endurance exercise, whereas it takes minutes to hours for transcriptomic and proteomic responses accounting for training adaptation [21]. An increase of about 75% of the maximum rate of fat oxidation (whole body measure), which is already high in elite endurance athletes, is observed after a 2 h recovery in a fasting condition from an endurance exercise session [22–24]. As both exercise and GW0742 alter fatty acid availability, we considered it important to choose the most appropriate experimental conditions in mice that allowed discriminating GW0742 effects from those induced by exercise, considering that the signature of an increase in FAO in T cells would be interpreted in an individual athletes' biological passport [25] as a doping signature. Information is available on the internet regarding the oral doses of GW0742 used by athletes for the purpose of doping. The oral dosages used comprise between 10 to 50 mg per day for 4 to 8 weeks, which in terms of availability would correspond to a dose of 1–10 mg/kg administered in mice. Notably, the plasma concentration of the ligand at the 1 mg/kg dose in mice is shown to specifically activate PPARβ/δ [26]. We used a dosage/treatment period in our mouse studies that is quite close to the doping protocol used by athletes by administering a dose of 3 mg/kg persistently in food for 6 weeks. We used blood samples from trained mice to detect interactions between GW0742 and exercise training effects. GW0742 induced an increase in CPT1a mRNA, but surprisingly, this induction was masked by the training status of mice. This questioned the relevance of this signature for doping-control strategies. Another suggested alternative is based on our previous study that proposed the αβ/γδ T cell ratio as a T cell signature to detect activity of the PPARβ/δ pathway [6]. We showed here that neither acute or chronic exercise nor GW0742 treatment changed this αβ/γδ T cell ratio in mouse blood. Perhaps the 6-week GW0742 treatment was not long enough for alterations in T cell development in the thymus to be reflected in the blood (our previous study examined transgenic mice that overexpressed PPARβ/δ in T cells constitutively).

Outside of the potential to use the latter observations to develop novel methods to detect the use of substances that activate the PPARβ/δ pathway, it should be noted that these novel discoveries also suggest that athletes who take PPARβ/δ agonists might seriously disturb their T cell homeostasis, thereby endangering the effectiveness of their immune system. Forcing FAO in CD4+ T cells would result in an increase in metabolic inflexibility. Depending on the (patho)physiological context, this could have either beneficial or deleterious consequences. The immunomodulatory effects of exercise might be mediated by the ability of exercise to adjust and improve Treg number and function [27]. An increase in Tregs would augment immune tolerance, thereby decreasing the risk of development of autoimmune diseases [28]. It should be noted in this context that physical exercise is known to decrease the risk of developing and is beneficial to the management of autoimmune disease [29]. Adipose tissue Tregs have been shown to play a beneficial role in decreasing insulin resistance associated with diet-induced obesity but a deleterious role in age-associated insulin resistance [30,31]. In the context of cancer, Tregs suppress anticancer immunity and, by doing so, hinder protective immunosurveillance of tumors and hamper effective antitumor immune responses [32]. A recent publication demonstrated that PPARβ/δ plays an important role in Treg survival and function in tumors [33]. It was observed that intratumoral Tregs displayed increased expression of multiple PPARβ/δ target genes compared to Tregs from spleen and lymph nodes. Knocking out PPARβ/δ specifically in Tregs led to a reduction in intratumoral Treg accumulation accompanied by decreased tumor growth. Taken together, these data suggest that there is a real possibility that abuse of PPARβ/δ agonists for performance-enhancing purposes might lead to an increased cancer risk and/or a worse outcome when a tumor develops.

To conclude, we show here that the use of substances that activate PPARβ/δ can increase FAO in human T cells and as a result increase the prevalence of Tregs. It is unlikely that this signature could be used as a doping-control strategy in athlete's blood, since these immunometabolic changes are masked in mice by training status. Moreover, our study alerts on the risks of immune surveillance alterations with the use of PPARβ/δ activators in order to improve physical performance.

4. Materials and Methods

4.1. Animal Experiments

4.1.1. Acute Treadmill Exercise

Twelve-week-old wild-type mice (n = 18) purchased from Charles River (Ecully, France) were accustomed to the treadmill (five-lane motorized treadmill, LE8710 M, Bioseb) a week before the running test was performed with a slope of 5° (n = 18). During a warm-up phase, the speed of the treadmill was progressively increased every 2 min for 10 min (5 to 25 cm/s). This phase was followed by an acute exercise phase where the speed of the treadmill was increased by 5 cm/s every 15 min (30 to 40 cm/s) until the mice exhibited signs of exhaustion. The rear of the treadmill was equipped with a low-voltage electric stimulating bar to encourage each mouse to run. The bar was set to deliver 0.2 mA at a frequency of 0.25 Hz, which caused an uncomfortable shock but did not injure the animal. The number of shocks was recorded, and the electric delivery was stopped if 50 shocks were reached.

4.1.2. Physical Training and GW0742 Treatment of Mice

We used 7-week-old C57Bl/6J wild-type mice purchased from Charles River (Ecully, France). Animals were maintained in a 12 h light, 12 h dark cycle and received food (A04 from UAR (Usine d'Alimentation Rationnelle), Villemoisson sur Orge, France) and water ad libitum. The mice were trained (8 weeks) on the five-lane treadmill. The training protocol was divided into three phases (Figure 4). The acclimatation phase lasted 4 weeks, during which the mice were trained in three sessions per week. The overload phase lasted 3 weeks, during which the mice were trained in five sessions per week. Finally, the tapering phase lasted 1 week, during which the mice were trained in three sessions. A training session lasted between 20 and 40 min, the treadmill speed varied between 20 and 40 cm/s, and the belt was positively inclined at 5°. To encourage the mice to run, electrical (0.2 mA–160 kΩ) and mechanical stimulation were used.

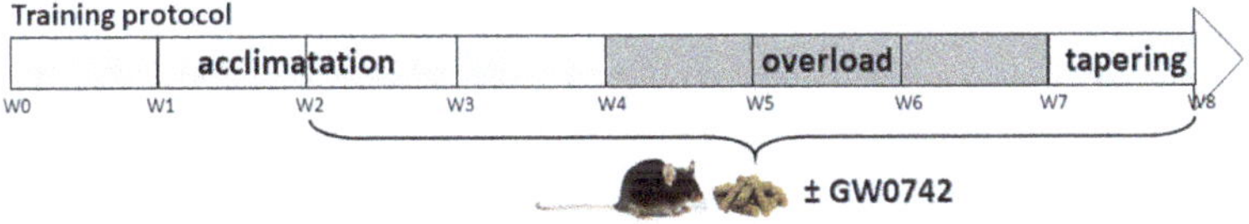

Figure 4. Training and GW0742 mouse treatment procedure.

After 2 weeks of acclimatation to training, mice received a normal chow diet (standard chow diet (A04)) administered ad libitum, supplemented with GW0742 (3 mg/kg BW/day) or with the vehicle (dimethyl sulfoxide, DMSO, 1%). Food was reconstituted as described previously [9,34]. Twice a week, the food was refreshed, and animals were weighed.

Animals were sacrificed (90 min after acute exercise or 24 h after the last training session) by a lethal dose of intraperitoneal ketamine/xylazine (100/16 mg/kg). Blood samples were obtained by cardiac puncture.

4.2. Mouse and Human T Cell Isolation and Treg Polarization

Mouse CD4+ cells from control (Lck-Cre) or KO-T-PPARβ/δ mice [35] were positively selected from secondary lymphoid organs (SLOs, consisting of the inguinal, brachial, and cervical lymph nodes and the spleen). CD4+ T cells were grown as previously described [36] at 4 × 10^5 cells/well in a 48-well plate in RPMI medium supplemented with 1 mM sodium pyruvate, nonessential amino-acid (1×), 1% penicillin/streptomycin, 10% fetal calf serum (FCS), and 50 μM β-mercaptoethanol. Activation beads, covalently bound to αCD3 and αCD28 antibodies, as well as mouse IL-2 (20 ng/mL), were added to the culture medium for the Th0 conditions, as well as PPARβ/δ agonist GW0742 1 μM or DMSO (0.1%). Cell medium was complemented with an equal volume of fresh medium every three days.

Human buffy coats from healthy donors (Établissement Français du Sang, Marseille, France) were used to collect PBMCs by Ficoll density gradient centrifugation. Monocytes were depleted by adherence to Primaria plates for 2 h. The cells of the supernatant (enriched in T cells), or CD4+ T cells isolated by negative selection using a Miltenyi Biotec system (#130-091-155), were plated at 4×10^5 cells/well in a 48-well plate. Activation beads coated with αCD3 and αCD28 antibodies, as well as human IL-2 (20 ng/mL), were added to the culture medium in presence or absence of GW0742 (1 μM) or DMSO (0.1%). After 6 days of activation, the cultures comprised over 95% T cells (data not shown). For Treg polarization experiments, TGF-β (5 μg/mL) was added to the culture medium for 6 days.

4.3. Measurement of β-Oxidation Using ^{3}H-Labeled Palmitate

The isolated human CD4+ cells were cultured at 4×10^5 cells/well in a 48-well plate in RPMI and were activated with anti-CD3/anti-CD28-coated beads. The palmitate β-oxidation was evaluated as previously described in [6,35,36]. Briefly, after 5 days of activation, for the last 24 h, we added in the wells a mix of radioactive and nonradioactive palmitate coupled to BSA (2:1 ratio; 30 μM Na-palmitate, 15 μM fatty-acid-free BSA and 10 μCi (0.83 μM 9,10-^{3}H-palmitic acid (Perkin Elmer)). After a 24 h additional incubation, 100% trichloroacetic acid (10% final) was added to the cell suspensions, and proteins were precipitated. After centrifugation, NaOH (final concentration 0.75 M) was added to the supernatant to increase pH to 12. Subsequently, 400 μL of supernatant was applied to ion-exchange columns (Dowex 1 × 2–400 resin), and ^3H_2O was recovered by eluting with 4 mL of H_2O. A 0.75 mL aliquot was then used for scintillation counting. Results were expressed as CPM (counts per minute) per 10^6 cells.

4.4. Cell Preparation and Flow Cytometry Analysis

Human CD4+ in culture, mouse lymph node cell suspension (1 million), or blood mononuclear cells isolated using Ficoll gradient were stained with fluorescence-conjugated antibodies (αCD3-VioBlue or αCD3-FITC, αCD4-APC-Vio770 or αCD4-APC, αCD25-PE-Vio, αFOXP3-APC, αTCRβ-PEcy7, αTCRγδ-PE), and stained cells were analyzed with a BD FACS Canto II flow cytometer (BD Biosciences, Franklin Lakes, NJ, USA) using Miltenyi Biotec (Paris, France) antibodies. The extracellular labeling (αCD3, αCD4, αTCRβ, αTCRγ, and αCD25) was done at 4 °C for 20 min. After two washes with PBS 0.5% FCS plus 2 mM EDTA, cells were permeabilized and fixed following the manufacturer's protocol (Miltenyi Biotec Kit). The intracellular staining was performed with αFOXP3 after the extracellular labeling. To determine the percentage of Tregs in a cell population, we discriminated CD3+CD4+ cells, and in this population, we gated FOXP3+ cells for mice cells and CD25+FOXP3+ cells for human cells, as shown in Figure 1C. Data were analyzed using FlowJo software.

4.5. RNA Extraction and Quantitative Real-Time PCR

Total RNA was extracted from cells or tissues with Trizol reagent (Invitrogen). For isolating RNA from blood, we used the kit Mouse RiboPure–Blood RNA Isolation (Applied Biosystems) following manufacturer procedure. Then, 1 μg of RNA was reverse-transcribed using a QuantiTect Reverse Transcription Kit (Qiagen) on a Q-cycler II. Quantitative PCR was done using SYBR Premix Ex Taq (Tli RNase H Plus) (Ozyme) on a StepOne machine (Life Technologies). The relative amount of all mRNAs was calculated using the comparative ΔΔCT method, and either 36B4 (for mice) or RPL27 (for humans) was used as the housekeeping gene. Primer sequences are available upon request.

4.6. Statistical Analyses

For each dependent variable under consideration, and according to assumptions for statistical analysis (i.e., normal distribution, equal variance), we performed the following: (1) nonparametric Mann–Whitney (in vitro study); (2) one-way ANOVA analysis (acute exercise effect); (3) two-way ANOVA analyses to investigate independent effects of GW0742 treatment and exercise training, and the interaction effects between GW0742 and training. Statistical significance was accepted at $p < 0.05$. The results are presented as means ± standard deviations. All data were analyzed using StatView and GraphPad Prism v 5.0 software (San Diego, CA, USA).

Author Contributions: Conceptualization, B.S., A.-S.R. and J.G.N.; methodology, B.S., I.M.-S., G.L.M., D.L., S.L.G., E.B., J.M., C.M., N.L., J.G.N. and A.-S.R.; formal analysis, B.S. and A.-S.R.; investigation, B.S., A.-S.R., I.M.-S., G.L.M., D.L., S.L.G., E.B., J.M., C.M. and N.L.; data curation, B.S. and A.-S.R.; writing—original draft preparation, B.S., I.M.-S., J.G.N. and A.-S.R.; writing—review and editing, B.S., I.M.-S., G.L.M., D.L., S.L.G., E.B., J.M., C.M., N.L., G.C., J.G.N. and A.-S.R.; supervision, A.-S.R. and J.G.N.; project administration, A.-S.R. and J.G.N.; funding acquisition, A.-S.R., G.C. and J.G.N. All authors have read and agreed to the published version of the manuscript.

Funding: This work was funded by INSERM, the Université Côte d'Azur, the Fondation pour la Recherche Médicale (FRM, grant DRM20101220437) and the Agence Française de Lutte contre le Dopage (AFLD, grant R17020AA).

Institutional Review Board Statement: All mice experimental procedures were conducted at C3M according to French legislation, following the EU Directive 2010/63 for animal experiments, and were approved by the Institutional Ethic Committee for the Use of Laboratory Animals (CIEPAL-AZUR no. C2EA-28, N-2018110914193037).

Informed Consent Statement: Human blood samples form volunteers were obtained from the Etablissement Français du Sang (EFS) through authorization 2018-00131 and written consent was obtained by EFS for use of the blood for research purposes.

Data Availability Statement: The data presented in this study are available on request from the corresponding author.

Acknowledgments: The authors thank Véronique Corcelle and the animal facility staff (Unit1065, C3M, Institut National de la Santé et de la Recherche Médicale (INSERM)), for their excellent care of mice. The authors gratefully thank W. Wahli for sharing the B6.Ppard^{TM1Mtz} mice (that possess loxP sites up- and downstream of PPARβ/δ exon 4).

Conflicts of Interest: The authors declare no conflict of interest.

References

1. Neels, J.G.; Grimaldi, P.A. Physiological functions of peroxisome proliferator-activated receptor beta. *Physiol. Rev.* **2014**, *94*, 795–858. [CrossRef] [PubMed]
2. Guerrieri, D.; Moon, H.Y.; van Praag, H. Exercise in a Pill: The Latest on Exercise-Mimetics. *Brain Plast.* **2017**, *2*, 153–169. [CrossRef] [PubMed]
3. Van der Gronde, T.; de Hon, O.; Haisma, H.J.; Pieters, T. Gene doping: An overview and current implications for athletes. *Br. J. Sports Med.* **2013**, *47*, 670–678. [CrossRef] [PubMed]
4. Trevisiol, S.; Moulard, Y.; Delcourt, V.; Jaubert, M.; Boyer, S.; Tendon, S.; Haryouli, H.; Taleb, W.; Caroff, M.; Chabot, B.; et al. Comprehensive characterization of the peroxisome proliferator activated receptor-delta agonist GW501516 for horse doping control analysis. *Drug Test. Anal.* **2021**, *13*, 1191–1202. [CrossRef]
5. Pokrywka, A.; Cholbinski, P.; Kaliszewski, P.; Kowalczyk, K.; Konczak, D.; Zembron-Lacny, A. Metabolic modulators of the exercise response: Doping control analysis of an agonist of the peroxisome proliferator-activated receptor delta (GW501516) and 5-aminoimidazole-4-carboxamide ribonucleotide (AICAR). *J. Physiol. Pharmacol.* **2014**, *65*, 469–476.
6. Mothe-Satney, I.; Murdaca, J.; Sibille, B.; Rousseau, A.S.; Squillace, R.; Le Menn, G.; Rekima, A.; Larbret, F.; Pele, J.; Verhasselt, V.; et al. A role for Peroxisome Proliferator-Activated Receptor Beta in T cell development. *Sci. Rep.* **2016**, *6*, 34317. [CrossRef] [PubMed]
7. Sakaguchi, S.; Yamaguchi, T.; Nomura, T.; Ono, M. Regulatory T cells and immune tolerance. *Cell* **2008**, *133*, 775–787. [CrossRef]
8. Newton, R.; Priyadharshini, B.; Turka, L.A. Immunometabolism of regulatory T cells. *Nat. Immunol.* **2016**, *17*, 618–625. [CrossRef] [PubMed]
9. Le Garf, S.; Murdaca, J.; Mothe-Satney, I.; Sibille, B.; Le Menn, G.; Chinetti, G.; Neels, J.G.; Rousseau, A.S. Complementary Immunometabolic Effects of Exercise and PPARbeta/delta Agonist in the Context of Diet-Induced Weight Loss in Obese Female Mice. *Int. J. Mol. Sci.* **2019**, *20*, 5182. [CrossRef]
10. Shaw, D.M.; Merien, F.; Braakhuis, A.; Dulson, D. T-cells and their cytokine production: The anti-inflammatory and immunosuppressive effects of strenuous exercise. *Cytokine* **2018**, *104*, 136–142. [CrossRef] [PubMed]

11. Gleeson, M.; Bishop, N.C.; Stensel, D.J.; Lindley, M.R.; Mastana, S.S.; Nimmo, M.A. The anti-inflammatory effects of exercise: Mechanisms and implications for the prevention and treatment of disease. *Nat. Rev. Immunol.* **2011**, *11*, 607–615. [CrossRef]
12. Padilha, C.S.; Figueiredo, C.; Minuzzi, L.G.; Chimin, P.; Deminice, R.; Kruger, K.; Rosa-Neto, J.C.; Lira, F.S. Immunometabolic responses according to physical fitness status and lifelong exercise during aging: New roads for exercise immunology. *Ageing Res. Rev.* **2021**, *68*, 101341. [CrossRef]
13. Pearce, E.L.; Poffenberger, M.C.; Chang, C.H.; Jones, R.G. Fueling immunity: Insights into metabolism and lymphocyte function. *Science* **2013**, *342*, 1242454. [CrossRef] [PubMed]
14. Blagih, J.; Coulombe, F.; Vincent, E.E.; Dupuy, F.; Galicia-Vazquez, G.; Yurchenko, E.; Raissi, T.C.; van der Windt, G.J.; Viollet, B.; Pearce, E.L.; et al. The energy sensor AMPK regulates T cell metabolic adaptation and effector responses in vivo. *Immunity* **2015**, *42*, 41–54. [CrossRef] [PubMed]
15. Blagih, J.; Hennequart, M.; Zani, F. Tissue Nutrient Environments and Their Effect on Regulatory T Cell Biology. *Front. Immunol.* **2021**, *12*, 637960. [CrossRef]
16. Michalek, R.D.; Gerriets, V.A.; Jacobs, S.R.; Macintyre, A.N.; MacIver, N.J.; Mason, E.F.; Sullivan, S.A.; Nichols, A.G.; Rathmell, J.C. Cutting edge: Distinct glycolytic and lipid oxidative metabolic programs are essential for effector and regulatory CD4+ T cell subsets. *J. Immunol.* **2011**, *186*, 3299–3303. [CrossRef] [PubMed]
17. Gerriets, V.A.; Kishton, R.J.; Nichols, A.G.; Macintyre, A.N.; Inoue, M.; Ilkayeva, O.; Winter, P.S.; Liu, X.; Priyadharshini, B.; Slawinska, M.E.; et al. Metabolic programming and PDHK1 control CD4+ T cell subsets and inflammation. *J. Clin. Investig.* **2015**, *125*, 194–207. [CrossRef] [PubMed]
18. Procaccini, C.; Carbone, F.; Di Silvestre, D.; Brambilla, F.; De Rosa, V.; Galgani, M.; Faicchia, D.; Marone, G.; Tramontano, D.; Corona, M.; et al. The Proteomic Landscape of Human Ex Vivo Regulatory and Conventional T Cells Reveals Specific Metabolic Requirements. *Immunity* **2016**, *44*, 712. [CrossRef]
19. Raud, B.; Roy, D.G.; Divakaruni, A.S.; Tarasenko, T.N.; Franke, R.; Ma, E.H.; Samborska, B.; Hsieh, W.Y.; Wong, A.H.; Stuve, P.; et al. Etomoxir Actions on Regulatory and Memory T Cells Are Independent of Cpt1a-Mediated Fatty Acid Oxidation. *Cell Metab.* **2018**, *28*, 504–515.e7. [CrossRef] [PubMed]
20. Simpson, R.J.; Campbell, J.P.; Gleeson, M.; Kruger, K.; Nieman, D.C.; Pyne, D.B.; Turner, J.E.; Walsh, N.P. Can exercise affect immune function to increase susceptibility to infection? *Exerc. Immunol. Rev.* **2020**, *26*, 8–22. [PubMed]
21. Belhaj, M.R.; Lawler, N.G.; Hoffman, N.J. Metabolomics and Lipidomics: Expanding the Molecular Landscape of Exercise Biology. *Metabolites* **2021**, *11*, 151. [CrossRef] [PubMed]
22. Frandsen, J.; Poggi, A.I.; Ritz, C.; Larsen, S.; Dela, F.; Helge, J.W. Peak Fat Oxidation Rate Is Closely Associated With Plasma Free Fatty Acid Concentrations in Women; Similar to Men. *Front. Physiol.* **2021**, *12*, 696261. [CrossRef] [PubMed]
23. Frandsen, J.; Vest, S.D.; Ritz, C.; Larsen, S.; Dela, F.; Helge, J.W. Plasma free fatty acid concentration is closely tied to whole body peak fat oxidation rate during repeated exercise. *J. Appl. Physiol.* **2019**, *126*, 1563–1571. [CrossRef] [PubMed]
24. Andersson Hall, U.; Edin, F.; Pedersen, A.; Madsen, K. Whole-body fat oxidation increases more by prior exercise than overnight fasting in elite endurance athletes. *Appl. Physiol. Nutr. Metab.* **2016**, *41*, 430–437. [CrossRef]
25. Astolfi, T.; Crettaz von Roten, F.; Kayser, B.; Saugy, M.; Faiss, R. The Influence of Training Load on Hematological Athlete Biological Passport Variables in Elite Cyclists. *Front. Sports Act. Living* **2021**, *3*, 618285. [CrossRef] [PubMed]
26. Takata, Y.; Liu, J.; Yin, F.; Collins, A.R.; Lyon, C.J.; Lee, C.H.; Atkins, A.R.; Downes, M.; Barish, G.D.; Evans, R.M.; et al. PPARdelta-mediated antiinflammatory mechanisms inhibit angiotensin II-accelerated atherosclerosis. *Proc. Natl. Acad. Sci. USA* **2008**, *105*, 4277–4282. [CrossRef]
27. Dorneles, G.P.; Dos Passos, A.A.Z.; Romao, P.R.T.; Peres, A. New Insights about Regulatory T Cells Distribution and Function with Exercise: The Role of Immunometabolism. *Curr. Pharm. Des.* **2020**, *26*, 979–990. [CrossRef]
28. Dominguez-Villar, M.; Hafler, D.A. Regulatory T cells in autoimmune disease. *Nat. Immunol.* **2018**, *19*, 665–673. [CrossRef] [PubMed]
29. Sharif, K.; Watad, A.; Bragazzi, N.L.; Lichtbroun, M.; Amital, H.; Shoenfeld, Y. Physical activity and autoimmune diseases: Get moving and manage the disease. *Autoimmun. Rev.* **2018**, *17*, 53–72. [CrossRef]
30. Feuerer, M.; Herrero, L.; Cipolletta, D.; Naaz, A.; Wong, J.; Nayer, A.; Lee, J.; Goldfine, A.B.; Benoist, C.; Shoelson, S.; et al. Lean, but not obese, fat is enriched for a unique population of regulatory T cells that affect metabolic parameters. *Nat. Med.* **2009**, *15*, 930–939. [CrossRef]
31. Bapat, S.P.; Myoung Suh, J.; Fang, S.; Liu, S.; Zhang, Y.; Cheng, A.; Zhou, C.; Liang, Y.; LeBlanc, M.; Liddle, C.; et al. Depletion of fat-resident Treg cells prevents age-associated insulin resistance. *Nature* **2015**, *528*, 137–141. [CrossRef] [PubMed]
32. Li, C.; Jiang, P.; Wei, S.; Xu, X.; Wang, J. Regulatory T cells in tumor microenvironment: New mechanisms, potential therapeutic strategies and future prospects. *Mol. Cancer* **2020**, *19*, 116. [CrossRef] [PubMed]
33. Wang, H.; Franco, F.; Tsui, Y.C.; Xie, X.; Trefny, M.P.; Zappasodi, R.; Mohmood, S.R.; Fernandez-Garcia, J.; Tsai, C.H.; Schulze, I.; et al. CD36-mediated metabolic adaptation supports regulatory T cell survival and function in tumors. *Nat. Immunol.* **2020**, *21*, 298–308. [CrossRef] [PubMed]
34. Mothe-Satney, I.; Piquet, J.; Murdaca, J.; Sibille, B.; Grimaldi, P.A.; Neels, J.G.; Rousseau, A.S. Peroxisome Proliferator Activated Receptor Beta (PPARbeta) activity increases the immune response and shortens the early phases of skeletal muscle regeneration. *Biochimie* **2017**, *136*, 33–41. [CrossRef] [PubMed]

35. Rousseau, A.S.; Murdaca, J.; Le Menn, G.; Sibille, B.; Wahli, W.; Le Garf, S.; Chinetti, G.; Neels, J.G.; Mothe-Satney, I. Invalidation of the Transcriptional Modulator of Lipid Metabolism PPARbeta/delta in T Cells Prevents Age-Related Alteration of Body Composition and Loss of Endurance Capacity. *Front. Physiol.* **2021**, *12*, 587753. [CrossRef] [PubMed]
36. Le Garf, S.; Sibille, B.; Mothe-Satney, I.; Eininger, C.; Fauque, P.; Murdaca, J.; Chinetti, G.; Neels, J.G.; Rousseau, A.S. Alpha-lipoic acid supplementation increases the efficacy of exercise- and diet-induced obesity treatment and induces immunometabolic changes in female mice and women. *FASEB J.* **2021**, *35*, e21312. [CrossRef]

International Journal of
Molecular Sciences

MDPI

Review

PPARs-Orchestrated Metabolic Homeostasis in the Adipose Tissue

Chen Sun [1,2], Shuyu Mao [2], Siyu Chen [2], Wenxiang Zhang [2] and Chang Liu [1,2,*]

1 College of Pharmacy, Xinjiang Medical University, Urumqi 830054, China; 3119030134@stu.cpu.edu.cn
2 State Key Laboratory of Natural Medicines and School of Life Science and Technology, China Pharmaceutical University, Nanjing 211198, China; 3220030438@stu.cpu.edu.cn (S.M.); siyuchen@cpu.edu.cn (S.C.); wenxiangzhang@cpu.edu.cn (W.Z.)
* Correspondence: changliu@cpu.edu.cn; Tel./Fax: +86-25-86185645

Abstract: It has been more than three decades since peroxisome proliferator-activated receptors (PPARs) were first discovered. Many investigations have revealed the central regulators of PPARs in lipid and glucose homeostasis in response to different nutrient conditions. PPARs have attracted much attention due to their ability to improve metabolic syndromes, and they have also been proposed as classical drug targets for the treatment of hyperlipidemia and type 2 diabetes (T2D) mellitus. In parallel, adipose tissue is known to play a unique role in the pathogenesis of insulin resistance and metabolic syndromes due to its ability to "safely" store lipids and secrete cytokines that regulate whole-body metabolism. Adipose tissue relies on a complex and subtle network of transcription factors to maintain its normal physiological function, by coordinating various molecular events, among which PPARs play distinctive and indispensable roles in adipocyte differentiation, lipid metabolism, adipokine secretion, and insulin sensitivity. In this review, we discuss the characteristics of PPARs with special emphasis on the roles of the different isotypes in adipocyte biology.

Keywords: adipose tissue; PPAR; browning; lipid metabolism

Citation: Sun, C.; Mao, S.; Chen, S.; Zhang,W.; Liu, C. PPARs-Orchestrated Metabolic Homeostasis in the Adipose Tissue. *Int. J. Mol. Sci.* **2021**, 22, 8974. https://doi.org/10.3390/ijms22168974

Academic Editors: Manuel Vázquez-Carrera and Walter Wahli

Received: 26 July 2021
Accepted: 17 August 2021
Published: 20 August 2021

Publisher's Note: MDPI stays neutral with regard to jurisdictional claims in published maps and institutional affiliations.

1. Introduction

Adipose tissue is an essential component of healthy energy homeostasis. Conversely, adipose tissue dysfunction promotes a pro-inflammatory, hyperlipidemic, and insulin-resistant environment that contributes to the pathogenesis of T2D and metabolic syndromes [1]. On the other hand, despite their obesity, some individuals appear to have a healthy metabolism. Moreover, lipodystrophy also contributes to insulin resistance and metabolic syndromes [2]. These diametrically opposite conditions illustrate the complex interplay between adipose tissue and metabolic homeostasis.

PPARs are fatty acid-activated nuclear receptors that belong to the subfamily 1 of the nuclear hormone receptor superfamily of transcription factors, and they have three subtypes: PPARα (also called NR1C1), PPARβ/δ (also called NR1C2), and PPARγ (also called NR1C3) [3]. Like other nuclear receptors, PPARs are composed of several distinct functional domains. PPARs are activated by ligands through the ligand-binding pocket in the C-terminal ligand-binding domain (LBD), which contains a ligand-dependent transactivation function (AF2), and they bind target genes through a highly conserved DNA-binding domain (DBD). In addition, the N-terminal domain (NTD, A/B domain) of PPARs contains a ligand independent activation function (AF1) that can recruit coregulatory proteins to regulate the expression of target genes. After being activated by endogenous ligands, PPARs recruit coregulator proteins with chromatin-remodeling capabilities through AF2, thereby regulating the expression of target genes [4]. The subsequent DNA binding requires dimerization with retinoid X receptor (RXR), and then the PPAR-RXR heterodimer binds to a specific DNA response element called the PPAR response element (PPRE), activating the transactivation of target genes. Meanwhile, conformational changes in PPARs, induced by diverse ligand binding, cause differential recruitment of cofactors and changes in the PPARs' activity, thereby regulating unique physiological processes [5].

In fact, these three PPAR isoforms have some discrepancies in their functions, tissue distributions, and ligand sensitivities, in vivo. PPARα, the first rodent PPAR isoform to be identified and cloned, is expressed predominantly in the tissues that exhibit high capacity for fatty acid catabolism, such as kidneys, brown adipose tissue (BAT), liver, and skeletal muscle [6]. In these tissues, PPARα regulates the adaptive response to nutritional changes by controlling fatty acid metabolism, resulting in energy dissipation. PPARα is activated by hypolipidemic fibrates, which reduce plasma triglycerides by inhibiting the synthesis of very-low-density lipoprotein (VLDL) and increasing fatty acid oxidation in the liver [7]. PPARβ/δ was subsequently cloned from mice after the discovery of PPARα [8]. PPARβ/δ shows a relatively broader expression pattern, which is ubiquitously expressed in the heart, kidneys, skeletal muscle, fat, skin, and gastrointestinal tract, and it plays a crucial role in fatty acid and glucose metabolism [9]. PPARγ, the third member of the PPAR family, is most highly expressed in both white adipose tissue (WAT) and BAT. Due to alternative splicing and differential promoter usage, PPARγ exists as two isoforms, PPARγ1 and PPARγ2, with the former lacking the first 30 amino acids at the N-terminus, and it is expressed in a broad variety of tissues, whereas the latter is highly abundant in adipose tissue. PPARγ is mainly responsible for regulating adipocyte differentiation and lipid metabolism [10]. Thiazolidinediones (TZDs) are synthetic PPARγ ligands with robust insulin-sensitizing activities, and they are used in the treatment of type 2 diabetes [5]. Compared with the other two subtypes, PPARγ seems to play a more important role in the regulation of the biology of adipose tissue.

In this review, we highlight the roles of three PPAR isoforms in maintaining the metabolic homeostasis of adipose tissue and discuss the new findings about PPARs in adipose tissue.

2. Adipose Tissue Classification and Function

Adipose tissue, as a central metabolic organ, is distributed throughout the body and is composed of individual fat depots with diversity in terms of their embryology, topology, morphology, function, and gene expression profile. In mammals, WAT and BAT are the two principal types of adipose tissue. WAT is responsible for the storage and release of fat, and therefore maintains systemic energy balance and plays a role in thermal insulation, as well as in protection from mechanical damage. WAT uptakes fats and carbohydrates from the circulation and converts them into triacylglycerides (TGs) via lipogenesis. During starvation, TGs are hydrolyzed into free fatty acids (FFAs) and glycerol, which are released into the circulation to supply substrates for other tissues. On the other hand, WAT is composed of many different types of cells that secrete a variety of cytokines, chemokines, and hormones; therefore, WAT is described as an important endocrine organ in controlling the systemic energy metabolism. Adipocyte dysfunction is due to excessive lipid load causes alterations in adipokine secretion, tissue inflammation, and ectopic fat accumulation in other tissues, which subsequently cause peripheral metabolic dysfunctions, such as insulin resistance and glucose intolerance; this may explain the many adverse effects of obese states. In addition, according to its location, WAT can be roughly divided into subcutaneous WAT (sWAT) and visceral WAT (vWAT). Different lipid turnovers between sWAT and vWAT may cause distinct metabolic changes in obese states [11]. Lipid accumulation in vWAT is associated with insulin resistance and increased risk of metabolic disease, whereas lipid accumulation in sWAT may even be protective against metabolic syndromes, explaining why some people are metabolically healthy in spite of their obesity [2].

Unlike white adipocytes, which contain a large unilocular lipid droplet that fills the cytoplasm, brown adipocytes contain multilocular lipid droplets and large numbers of mitochondria for the dissipation of energy via uncoupled mitochondrial respiration. In humans, BAT can be estimated in the cervical, axillary, and paraspinal regions by using

PET/CT with 2-deoxy-2-[18F] fluoroglucose [12]. BAT plays an active role in thermoregulation by converting chemical energy into heat. Cold-induced norepinephrine release stimulates lipolysis and β-oxidation in BAT. Thermogenesis is regulated by uncoupling protein 1 (UCP1), which is localized on the inner membrane of mitochondria and uncouples mitochondrial respiration from ATP synthesis. In brief, BAT is a metabolically active tissue that can clear circulating glucose and lipids; therefore, increased BAT activity is associated with several metabolic benefits, such as increased weight loss and improved glucose metabolism and insulin sensitivity.

In rodents, prolonged cold exposure leads not only to the recruitment of brown fat, but also to the appearance of white adipocytes with multilocular fat droplets and UCP1 expression, which is called "browning" [13]. These brown-like adipocytes are termed beige/brite adipocytes—the third classification of adipose tissue—and appear within classical WAT. Although beige/brite adipocytes share characteristics of brown adipocytes and express most brown-adipocyte-specific genes, such as UCP1, cell-death-inducing DNA fragmentation factor alpha subunit-like effector A (Cidea), and peroxisome proliferative activated receptor gamma coactivator 1 alpha (PGC1α), beige/brite adipocytes appear to develop from distinct populations of embryonic precursors and have distinct gene expression signatures [14].

3. PPARγ

PPARγ was first described as a factor induced during adipocyte differentiation, and was subsequently identified as a master regulator of adipocyte differentiation as early as 1994. These early studies indicate that PPARγ is induced and involved in adipogenesis [15,16]. In vivo studies showed that, due to placental defects, embryonic death was caused in whole-body PPARγ knockout mice. In addition, the mice that were chimeric for wild-type and PPARγ-null cells showed little or no contribution from null cells to the development of adipose tissue [17,18]. Tissue-specific gene knockout, mediated by the Cre/loxP strategy, permitted further investigation. Both adipocyte protein 2 promoter-driven Cre (aP2-Cre) and adiponectin-driven Cre (Adipoq–Cre) mouse lines were used to probe into the adipose-specific functions of PPARγ [19,20]. In these models of knockout mice, adipose tissue-specific loss of PPARγ led to critical atrophy of adipose tissue and was accompanied by significant impairment of adipokine secretion. Mechanically, the activation of the transcription factor CCAAT/enhancer binding protein (C/EBP) is one of the most important downstream effects of PPARγ during adipocyte differentiation [21]. Adipogenic transcriptional cooperation between PPARγ and C/EBP is essential in order to fully activate the programming of mature adipocytes. More than 90% of the DNA binding sites of PPARγ are also bound by C/EBP, and PPARγ relies on the induction of proteins of the C/EBP family for the complete activation of the gene transcription that is expressed in mature adipocytes (Figure 1) [22,23]. Moreover, the contributions of two PPARγ isoforms—PPARγ1 and PPARγ2—in adipogenesis are obviously different in vitro. Because the regulatory function of PPARγ2 in adipogenesis cannot be achieved by PPARγ1 in the absence of an exogenous ligand, PPARγ2 is considered the more adipogenic isoform of PPARγ [24]. In fact, the PPARγ1 isoform is sufficient for supporting development of adipose tissue and the fat deposition requirements of a lean mouse model, but the expandability of adipose tissue mainly relies on the PPARγ2 isoform under energy-excess conditions [25]. Once sufficient adipocytes are formed, mature adipocytes—along with infiltrated immune cells—secrete IL-6 and other cytokines, which, by inducing AT-rich interactive domain 5A (Arid5a), further limit the differentiation of adipocytes. Arid5a binds to the PPARγ2 promoter and prevents the activation of PPARγ2. Collectively, the feedback regulation of Arid5a and PPARγ2 maintains the homeostasis of adipose tissue. To effective adipogenesis, inhibition of Arid5a is accomplished by PPARγ2. In contrast, to limit excess adipogenesis, a check of PPARγ2 is accomplished by Arid5a [26]. In addition to its role in adipose tissue development and total storage capacity, PPARγ2 has also been identified as a crucial regulator of the lipid storage rate in adipose tissue. Mice that lack PPARγ2 cope when fat storage demands are low, but acute overfeeding overwhelms the adipose tissue, and lipids are redirected to the muscle, causing insulin resistance [27]. As already mentioned, PPARγ is essential for adipogenesis, and other adipogenic factors must act (at least in part) by activating the expression or activity of PPARγ (no transcriptional regulator that promotes adipocyte differentiation in the absence of PPARγ has been discovered).

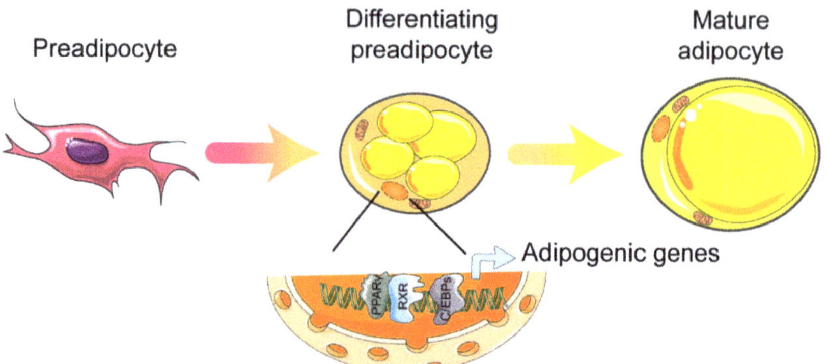

Figure 1. Control of white adipocyte differentiation by PPARγ. Cooperation between PPARγ and C/EBP is essential in order to fully activate the programming of mature adipocytes. Abbreviations: PPARγ, peroxisome proliferator-activated receptor γ; RXR, retinoid X receptor; C/EBPs, CCAAT/enhancer binding proteins. Figure was created using SMART–Servier Medical Art (https://smart.servier.com, the last accessed date is 29 July 2021).

The discovery of PPARγ mutants in human subjects also supports the important role of PPARγ in adipose tissue development [28]. In general, most of the above subjects with PPARγ mutations suffered from partial lipodystrophy, insulin resistance, and dyslipidemia. Significantly, the subcutaneous fat of limbs and the gluteal region was preferentially lost, while the visceral abdominal fat tissue was relatively preserved. In turn, treatment with PPARγ agonists in humans also resulted in redistribution of WAT [29]. In summary, PPARγ also plays a role in determining WAT distribution.

In addition to its critical role in adipogenesis, PPARγ is also indispensable for the state of mature adipocytes. Using the tamoxifen-dependent Cre-ERT2 recombination system, PPARγ was selectively ablated in adipocytes of adult mice, which resulted in the death of PPARγ-ablated adipocytes and formation of newly PPARγ-positive differentiated adipocytes within a few days [30]. Due to the compensatory effect, the remaining adipocytes were hypertrophic and more susceptible to apoptosis, which further gave rise to the presence of inflammation (such as macrophage infiltration and fibrosis) in the adipose tissue [31]. In white adipocytes, PPARγ plays a role in energy storage and adiposity. Both supraphysiological activation of PPARγ by thiazolidinediones (TZDs) and heterozygous PPARγ deficiency prevent adipocyte hypertrophy, but via different mechanisms. TZDs induce adipocyte differentiation and apoptosis, thereby increasing the number of small adipocytes, whereas the reduction of PPARγ decreases lipogenesis and promotes leptin expression in WAT [32]. On the other hand, the loss of the differentiated cell state caused by cell plasticity can result in the inability of the tissue to perform its functions. PPARγ blocks TGF-β signal transduction, thereby inhibiting the loss of adipocyte status [33].

White adipose tissue, as an important energy storage organ, strongly responds changes in nutritional signals and dynamically regulates fat storage; unsurprisingly, PPARγ also contributes to this physiological process. In adipose tissue, PPARγ expression is down-regulated by fasting and insulin-deficient diabetes but induced by exposure to a high-fat diet and insulin [34,35]. The activation of PPARγ in adipocytes promotes the expression of the genes involved in the release of FFA from lipoproteins, FFA uptake, intracellular FFA transport, FFA activation, and FFA esterification [36]. Specifically, adipocytes' lipid uptake and transport are partially regulated by lipoprotein lipase (LPL), differentiation cluster 36 (CD36), and adipocyte protein 2 (Ap2), all of which are upregulated by the response of PPARγ to TZDs treatment [37]. With the uptake of FFAs by adipocytes, PPARγ upregulates phosphoenolpyruvate carboxykinase (PEPCK), which provides a skeleton for the esterification of FFA, promotes the formation of intracellular lipid vesicles, and protects against FFA-induced lipotoxicity [38]. Furthermore, PPARγ promotes efficient storage of triglycerides in unilocular lipid droplets by regulating several lipid-droplet-associated proteins [39]. The results of the ChIP-seq experiments on the differentiated 3T3-L1 adipocytes showed that PPARγ-binding sites were found on the promoters of *Plin1*, *Plin2*, *Plin4*, *Plin5*, *Abhd5*, *Pnpla2*, *G0s2*, *Cidea*, and *Cidec* [40]. Under conditions of nutritional deficiency, PPARγ, as a fatty acid sensor, also activates lipolysis and releases FFA in order to provide maintain the balance of energy metabolism. It has been reported that the activation of PPARγ with rosiglitazone stimulates lipolysis and increases expression of adipose triglyceride lipase (ATGL) and monoacylglycerol lipase (MGL) in rat subcutaneous and visceral WAT [41]. Adipose tissue lipolysis is also stimulated by natriuretic peptides (NPs), which play a key role in maintaining blood pressure and fluid volume. Under overnutrition conditions, PPARγ upregulates high-fat diet (HFD)-dependent NP receptor C (Nprc) expression in adipocytes through long-range distal transcriptional regulation, and thereby attenuates adipocyte NP signaling in obesity [42]. Mitochondrial activity plays an important role in the health and function of adipose tissue. PPARγ induces E3 ubiquitin ligase membrane-associated RING-CH-type finger 5 (March5), which is known as an outer mitochondrial membrane protein, to regulate mitochondrial morphology and dynamics in adipocytes by controlling mitochondrial fusion. The inhibition of PPARγ expression in hypertrophic adipocytes has been observed during obesity, which may explain the decrease in mitochondrial gene expression, including that of March5 [43].

In addition to regulating lipid metabolism in WAT, PPARγ also influences the production of various signal molecules (adipokines) in white adipocytes, including adiponectin, FGF21,TNF-α, MCP-1, and resistin [44]. Adiponectin, an important adipokine, plays a cardinal role in improving obesity and metabolic diseases, and it is induced during adipocyte differentiation. PPARγ is the main regulator of adiponectin expression and processing [45]. Recent studies showed that PPARγ promotes the transport of vesicles containing adiponectin by activating reptin, which has both ATPase and DNA helicase activities. Then, upregulated transport accelerates polymerization and secretion of adiponectin, which facilitate pre-adipocyte differentiation [46]. Leptin is an adipocyte hormone that controls the mass and function of adipose tissue. By using the assay for transposase-accessible chromatin with high throughput (ATAC-seq), the functional requirement of the PPARγ-RXRα complex for the quantitative transcriptional regulation of leptin by binding to leptin regulatory element 1 (LepRE1) was confirmed. This underappreciated role of the PPARγ-RXRα complex is responsible for the quantitative control of leptin expression but does not affect its fat-specific expression [47].

Although PPARγ has been widely studied in the differentiation of WAT, it is also indispensable for the development and function of BAT. Compared with WAT, PPARγ has higher expression in both adult and embryonic BAT [48]. It was observed that PPARγ expression was already high in undifferentiated brown pre-adipocytes in vitro, and it increased further during differentiation [49,50]. Furthermore, PPARγ agonists drive BAT formation, both in vivo and in vitro [51,52]. Certainly, PPARγ is a mediator in the process of recruitment of BAT, whether by itself or in combination with other factors [53]. However, unlike in the case of WAT, C/EBPα is not a necessary factor for the gene expression of PPARγ during brown adipocyte differentiation [54]. For brown adipocytes to acquire their identity and thermogenic capacity, PPARγ recruits PR (PRD1-BF1-RIZ1 homologous) domain containing 16 (PRDM16), histone-lysine N-methyltransferase (EHMT1), and early B-cell factor (EBF2) to form a transcription complex that coordinates the transcriptional circuits toward the brown lineage. PPARγ and PRDM16 form the core part of the transcription complex, and the other two factors, EHMT1 and EBF2, are incorporated into the PPARγ-PRDM16 complex and advance its function in brown adipocytes. In detail, EHMT1, a unique methyltransferase that is specifically purified with PRDM16 by using a mass spectrum, induces the inhibitory H3K9me2 and H3K9me3 at promoter regions of the PRDM16-resident gene, which promotes precursors toward mature brown adipocytes [55]. In the same light, PPARγ recruits EBF2 to its brown-selective binding site and activates the expression of related genes, such as UCP1 [56].

Following the formation of BAT, the PPARγ-PRDM16 complex recruits a different set of cofactors in order to maintain the function of brown fat in adaptive thermogenesis and energy balance, among which PGC1α plays a central role. In brown adipocytes, PGC1α at least partially coactivates PPARγ to promote the expression of genes related to mitogenesis and thermogenesis, including *Cidea*, *Elovl3* and *Ucp1* [57]. Indeed, the PPARγ-PRDM16-PGC1α thermogenic transcription complex fine-tunes the thermogenesis and energy homeostasis by recruiting other cofactors, or it undergoes multiple modifications.

The thermogenic capacity of brown adipose tissue is directly related to intracellular triglyceride storage. The hydrolysis of triglyceride provides the FFA needed for allosteric activation of UCP1, as well as for mitochondrial oxidation, which releases energy in the form of heat during thermogenesis. Interestingly, triglyceride synthesis in BAT is also significantly increased upon cold exposure [58]. Like cold exposure, pharmacological PPARγ activation significantly accelerates triglyceride synthesis, promotes hypertrophy in brown adipocytes (Figure 2), and increases BAT mass. This process is associated with upregulated absorption of fatty acids from circulating triacylglycerol via lipoprotein lipase (LPL), increased generation of glycerol 3-phosphate via glyceroneogenesis and glycerokinase (GK), and elevated esterification of fatty acids via glycerol-3-phosphate acyltransferase (GPA) and diacylglycerol acyltransferase (DGAT), which catalyze the first and last acylation of glycerol-3-phosphate, respectively [59–61]. On the other hand, pharmacological PPARγ activation also upregulates lipolytic genes, such as ATGL and its partner, abhydrolase domain containing 5 (Abdh5), and MGL [62]. However, the release of lipolysis-derived FFA is counteracted by its intracellular recycling and re-esterification back to TAG (Figure 2). Therefore, these higher lipase levels are not translated into higher functional lipolytic rates due to the impairment of sympathetic activity and thyroid status in this condition [63]. In addition to fatty acids, glucose is another important metabolic substrate in supporting BAT thermogenesis, which is explained by the large amount of glucose stored in brown adipocytes in the form of glycogen and the significant increase in glucose uptake caused by sympathetic nerve-mediated thermogenic activation [64]. Unlike cold exposure, pharmacological PPARγ activation dramatically reduces glucose uptake and glycogen contents, which is explained by the impairment of sympathetic activity [65]. Overall, pharmacological PPARγ activation seems to hamper the thermogenetic ability of BAT through other systemic alterations. Another option is that PPARγ is needed for β-adrenergic signaling-mediated induction of brown adipocytes, and GK is, at least in part, required for mediating PPARγ function in BAT [66]. Furthermore, it was reported that pharmacological PPARγ activation enhanced the ability of normal mice to defend against cold-induced hypothermia by switching the fuel preference of BAT from carbohydrates to lipids under cold conditions [67]. Further investigation is required in order to elucidate the mechanism of this shift.

BAT contains large numbers of mitochondria and oxidases, which are used to oxidize fatty acids and glucose in order to dissipate energy. In vivo studies showed that the activation of PPARγ by rosiglitazone was not related to the number of BAT mitochondria or the expression of PGC1α [62]. In addition, rosiglitazone did not affect the expression of PGC1α in brown adipocytes that were cultured in vitro but increased the number of mitochondria and the expression of carnitine palmitoyl transferase 1 (CPT1) [52]. However, this change increases oxygen consumption only in the presence of norepinephrine. In the other words, PPARγ cannot enhance mitochondrial function in brown adipocytes independently of the activation of the sympathetic nervous system. Interestingly, the truncated form of PPARγ2 (52 kDa), but not the full-length PPARγ2, is highly enriched in brown adipocyte mitochondria, and it regulates mtDNA-encoded ETC gene expression [68]. This unexpected regulation may provide an additional level of control for mitochondrial respiration in brown adipocytes.

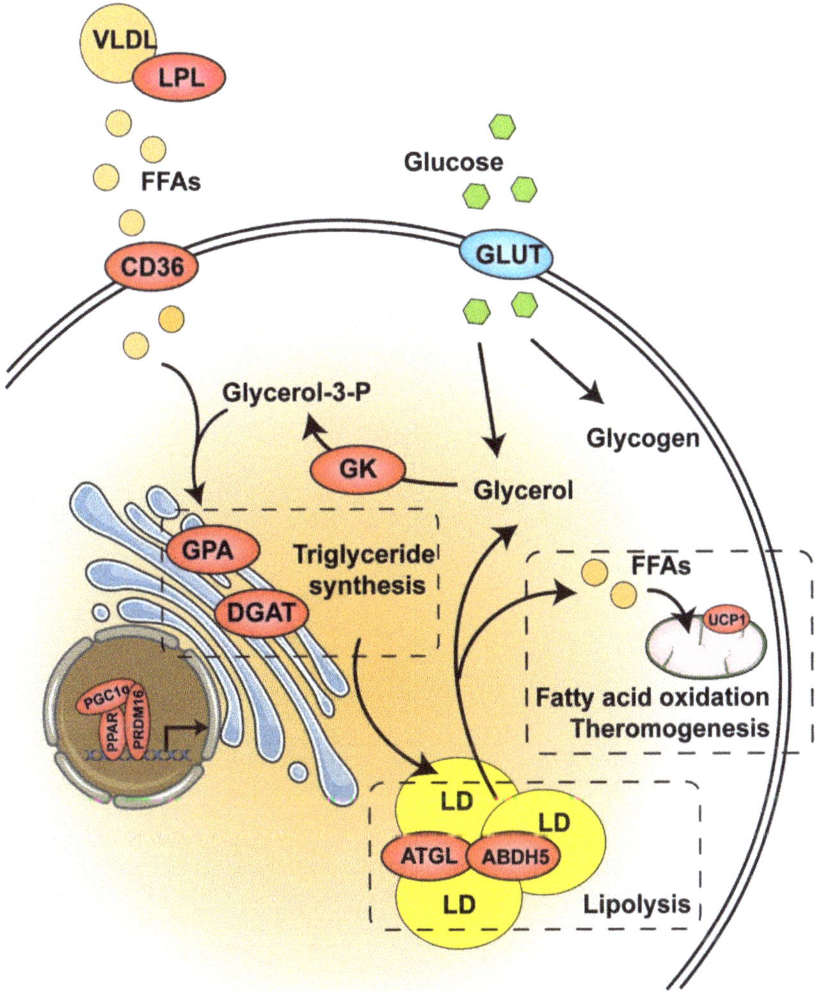

Figure 2. Control of glucose and lipid metabolism by PPAR in brown or beige/brite adipose tissue. Enzymes in red are activated by PPAR. The enzyme in blue remains unchanged. Abbreviations: VLDL, very-low-density lipoprotein; LPL, lipoprotein lipase; FFAs, free fatty acids; CD36, differentiation cluster 36; GPA, glycerol-3-phosphate acyltransferase; DGAT, diacylglycerol acyltransferase; GLUT, glucose transporters; GK, glycerokinase; LD, lipid droplet; ATGL, adipose triglyceride lipase; ABDH5, abhydrolase domain containing 5; UCP1, uncoupling protein 1. Figure was created using SMART–Servier Medical Art (https://smart.servier.com, the last accessed date is 29 July 2021).

The brown-like cells that are recruited through cold exposure and that arise from progenitors expressing TMEM26 and CD137 on the cell surface are referred to as beige adipocytes [69]. The activation of PPARγ by synthetic agonists induces brown fat-gene transcription in white adipocytes, both in vivo and in vitro, and these brown-like cells are referred to as brite adipocytes [70]. The significant morphological differences between brite adipocytes and beige adipocytes have been observed—the former are paucilocular, while the latter are multilocular [71]. PPARγ full agonists, such as classical TZDs, induce a brown fat phenotype in subcutaneous WAT. On the other hand, PPARγ ligands with weak or partial agonism, such as MRL24, nTZDpa, Mbx-102, and BVT.13, exhibit little or no browning effects [72]. Specifically, chronic treatment of TZDs induces activation of the PGC-1α expression [73] and stimulates a powerful stabilization of the PRDM16 Protein [72]. In vitro, brown adipocyte-like cells, which have numerous mitochondria and the presence of UCP1 protein, emerge in TZDs-treated white adipocyte cultures [70]. These cells have increased expression of not only PGC1α and UCP1, but also other brown adipocyte-specific genes, such as carnitine palmitoyltransferase 1b (CPT1B), ELOVL fatty acid elongase 3 (Elovl3), and Cidea [70]. Li and colleagues further demonstrate that PPARγ-induced WAT browning is mediated by SIRT1, PRDM16, C/EBPα and PGC1α. The transcriptional program of BAT is triggered via an SIRT1–PPARγ–PRDM16 cascade, in which PPARγ is deacetylated by SIRT1 on K268 and K293 and then recruits PRDM16 to increase the expression of BAT genes such as *Ucp1* and *Cidea* [74]. Moreover, the CDK inhibitor prevents S273 phosphorylation of PPARγ and promotes the formation of brite adipocytes in WAT [71]. In human adipocytes, kruppel-like factor 11 (KLF11) is directly induced by PPARγ and appears to cooperate with PPARγ in a feed-forward manner to activate and maintain the brite-selective gene program during long-term exposure to rosiglitazone [75]. In addition to the induction of BAT genes, the browning process also involves the repression of WAT genes. The mutation of critical amino acids within helix 7 of the PPARγ LBD suppresses TZD-mediated inhibition of WAT genes, including resistin and angiotensinogen [76]. On the molecular level, the repression of the WAT genes involves recruitment of two members of the carboxy-terminal binding protein family, CtBP1 and CtBP2, which, directed by C/EBPα, to the minimal promoter of the corresponding genes in response to the treatment of TZDs [76]. Therefore, PPARγ depends on its post-translational modifications and cofactor recruitment profiles in order to modulate its ability that activating the distinct genes subsets. However, TZD-induced browning is not associated with increased energy expenditure or weight loss in vivo, although UCP1-mediated uncoupled respiration is enhanced in adipocytes. Hence, it is important to uncouple TZDs' benefits from their adverse effects. Further research showed that the constitutively deacetylated PPARγ (K268R/K293R) mutant mice resisted HFD-induced obesity by increasing brown remodeling in WAT, and they maintained the insulin-sensitizing response to TZD while displaying few adverse effects on fat deposition [77]. Thus, PPARγ deacetylation may dissociate the metabolic benefits of the PPARγ agonist from its adverse effects. Interestingly, the ablation of PPARγ in the sWAT of 12-month-old mice revealed PPARγ preferential regulation of brown fat gene expression for the maintenance of browning programs during aging [78].

4. PPARα

Unlike PPARγ, PPARα is mainly expressed in the liver, and the expression level of PPARα is low in both human and rodent WAT [79]. However, expression of PPARα in human subcutaneous and omental adipose tissue has been reported to be negatively correlated with body mass index (BMI) [80]. PPARα expression is also decreased in the WAT of mice with genetically or HFD-induced obesity, and PPARα agonists can reduce adiposity and improve insulin resistance in such obese mouse models by stimulating both differentiation and fatty acid oxidation in adipocytes [81]. Similarly, the activation of PPARα by GW7647 also stimulates differentiation and fatty acid oxidation in human adipocytes [82]. In terms of adipogenesis, the effect of PPARα seems to be partially shared with PPARγ. On the other hand, PPARα may promote a futile cycle of lipolysis and fatty acid re-esterification through the induction of GK in human white adipocytes [83]. Other studies further clarified the capacity of PPARα to promote lipolysis through several mechanisms in white adipocytes. Firstly, PPARα activation increases the expression of

ATGL and HSL, which catalyze the first two important steps of lipolysis [84]. Secondly, PPARα agonists increase Ap2a2 expression, which facilitates the efficient endocytosis of β-adrenergic receptors (β-ARs) and thereby allows the avoidance of desensitization and internalization of β-ARs caused by prolonged exposure to agonists [85]. Moreover, the activation of PPARα by Wy14,643 upregulates the gene expression of adiponectin receptors (*Adipor1* and *Adipor2*) in the WAT of obese diabetic KKAγ mice [86]. In addition, PPARα has been shown to have a potent anti-inflammatory effect in white adipocytes, by inhibiting CD40 expression via upregulation of SIRT1 expression through the AMPK pathway in TNFα-treated 3T3-L1 cells [87]. All of these actions of PPARα in the WAT can enhance energy consumption and improve adipocyte hypertrophy, as well as obesity-induced insulin resistance. In addition, other studies showed that PPARα has a key role in regulating the crosstalk between the ER and mitochondria. In response to an adiponectin signal, PPARα binds to the activating transcription factor-2 (ATF2) promoter region, resulting in the inhibition of ATF2 transcription, thereby alleviating ER stress and apoptosis in adipocytes [88].

As PPARα is the key regulator of cellular fatty acid uptake and oxidation (Figure 2), it is not surprising that PPARα is highly expressed in BAT. PPARα-deficient mice, despite having a normal BAT morphology, exhibited a thermogenesis-associated disorder in response to cold exposure. [89] Furthermore, compared with PPARα-deficient mice, liver-specific PPARα-null mice had more severe hypothermia after 24h-fasting, indicating that extra-hepatic PPARα is necessary for maintaining whole-body temperature [90]. The possible explanation includes, but is not limited to, the role of PPARα in BAT. PPARα is activated by endogenous ligands that are derived from cold-induced lipolysis, and it upregulates the fatty acid oxidation and thermogenic genes through a cooperative mechanism with PGC1α. This regulation is initiated by the positive feedback loop of PPARα-induced PGC1α expression. In addition, PPARα regulates the expression of PRDM16, which binds to the PPARα-binding site of the PGC1α gene promoter and enhances PGC1α expression [91]. The energy source used for ATP production or thermogenesis is mainly supplied by glucose or fatty acids in BAT. Pyruvate dehydrogenases (PDH) and pyruvate dehydrogenase kinases (PDK) play a key role in the process of supplying energy from glucose. The activated PPARα competes with hepatocyte nuclear factor 4 alpha (HNF4α), a positive regulator, to bind the PDHβ promoter region, thereby suppressing PDHβ expression during cold exposure [92]. This finding agrees with the above-mentioned switching of fuel preference in BAT under cold conditions. Moreover, PPARα activation resulted in a reversal of whitening, with the favored thermogenesis being sustained by enhanced β3-adrenergic stimulation, lipolysis, and β-oxidation in BAT of the mice with HFD-induced obesity [93]. However, recent studies have suggested that PPARα is dispensable in thermogenesis. Even though PPARα agonists enhance the function of BAT, PPARα depletion in BAT did not affect the expression of classic BAT markers (such as *Ucp1*, *Cidea* and *Cox7a1*) [66]. In addition, WT and PPARα-null mice had no differences in BAT function during CL316,243 (β3-adrenergic agonist) treatment [94]. Furthermore, another study confirmed that the redundancy of PPARα with PPARγ is because PPARα binds to a subset of PPARγ sites [95].

Indeed, in vitro PPARα activation in human white adipocytes induces the expression of brown adipocyte-selective genes, such as *PGC1α, PRDM16, UCP1* and *DIO2* [91]. In addition, the activation of PPARα by chronic fenofibrate administration increases the gene expression of PGC1α and irisin, and it yields UCP-1-positive beige cells in the sWAT of the mice with HFD-induced obesity [96]. Furthermore, PPARα-null mice displayed a disrupted induction of thermogenic and brite markers in the inguinal WAT upon β3-adrenergic agonist treatment, which was associated with lower PDK4 expression [94]. Taken together, these data indicate that PPARα activation can promote the browning of WAT. Surprisingly, recent research showed that PPARα, despite its marked upregulation by cold in inguinal WAT, is completely dispensable for cold-induced browning in mice because cold-induced changes in gene expression in inguinal WAT are fully maintained in the absence of PPARα [97]. The reason why PPARα is induced by cold exposure, if it is not involved in regulating gene expression in inguinal WAT, remains unknown. The possible explanation is that PPARα activation is a sufficient and unnecessary condition during WAT browning.

5. PPARβ/δ

Similar to PPARγ, the expression of PPARβ/δ is also upregulated during adipocyte differentiation, and the difference is that PPARγ is expressed at the late stage of differentiation whereas PPARβ/δ is expressed in the initial stage of differentiation [98]. In vitro, PPARβ/δ promotes preadipocyte differentiation by inducing the expression of adipogenesis-related genes, such as PPARγ and fatty acid transporter (FAT) [99]. Another study showed that PPARβ/δ plays an important role in the proliferation of adipocyte precursor cells, and it has only a minor impact on terminal adipocyte differentiation [100]. This finding is consistent with the observation that PPARβ/δ knockout mice had impaired gonadal adipose stores [101]. On the other hand, the activation of PPARβ/δ promotes fatty acid oxidation and energy uncoupling in vivo and in vitro [102]. Moreover, PPARβ/δ prevents angiotensin-II-induced adipocyte hypertrophy and stimulates adipocyte remodeling with smaller adipocytes, while decreasing inflammation and increasing adiponectin secretion via the activation of haem oxygenase-1 (HO-1) expression and the Wnt-canonical pathway [103]. The activation of PPARβ/δ by GW501516 arrested IL-6–dependent reduction in insulin-stimulated Akt phosphorylation and glucose uptake by inhibiting ERK1/2 and inhibiting the activation of the signal transducer and activator of transcription-3 (STAT3) and the upregulation of the suppressor of cytokine signaling 3 (SOCS3) [104]. Otherwise, in adipose tissue-resident macrophages, PPARβ/δ is upregulated by IL-13 to control the polarization of macrophages toward alternative activation, thereby improving insulin sensitivity [105]. Furthermore, PPARβ/δ ablation impairs macrophage M2 polarization, which, in turn, causes inflammation and results in the stimulation of lipolysis and insulin resistance in adipocytes.

In BAT, PPARβ/δ activation induces the expression of genes related to fatty acid oxidation and thermogenesis (Figure 2) [102]. Furthermore, BAT-specific PPARβ/δ knockout mice are compromised with respect to maintaining body temperature during cold exposure, because PGC1α no longer binds to the UCP1 promoter in the absence of PPARβ/δ. Interestingly, PPARβ/δ not only mediates the actions of PGC1α, but also regulates the expression of twist family BHLH transcription factor 1 (twist-1), which inhibits histone H3 acetylation on the promoters of PGC1α target genes, suggesting a negative-feedback regulatory mechanism [106].

As a nutritional signal sensor, PPARβ/δ has been described as a candidate for the induction of adiposal browning [102]. However, in mice with HFD-induced obesity, pharmacological PPARβ/δ activation tackles glucose intolerance and reduces adipocyte size, but not positive UCP1 beige adipocytes were not shown in the sWAT, which may have been because of the enhanced Cidea gene expression, which inhibited the activity of UCP1 by forming a complex [107]. On the other hand, a recent study showed that PPARβ/δ mediates leptin-induced FGF21 expression in the crosstalk of brain–visceral adipose tissue, therefore contributing to the white-to-beige cell transition in WAT via autocrine/paracrine mechanisms [108].

6. Conclusions

Numerous studies support the crucial role of PPARs in maintaining metabolic homeostasis in adipose tissue (Figure 3). PPARγ plays a key role in adipocyte differentiation, and lipid storage, PPARα and PPARβ/δ are primarily involved in adaptive thermogenesis and lipid utilization in adipose tissue. Selective and potent PPARα or PPARγ agonists enhance the activity of BAT and induce the "browning "of WAT (Figure 2). However, over the last decades, market withdrawal and the failure of drug development programs have made people doubt the clinical value of compounds with PPAR-activation functions. Meanwhile, the PPARγ agonist rosiglitazone and dual PPAR agonists displayed ineluctable adverse effects that led to restricted use or halted development. Nevertheless, most of these side effects were either caused by nonspecific and off-target effects of the drugs or excessive PPARγ activation. With the development of targeted therapy, PPAR targeted therapy will regain its brilliance.

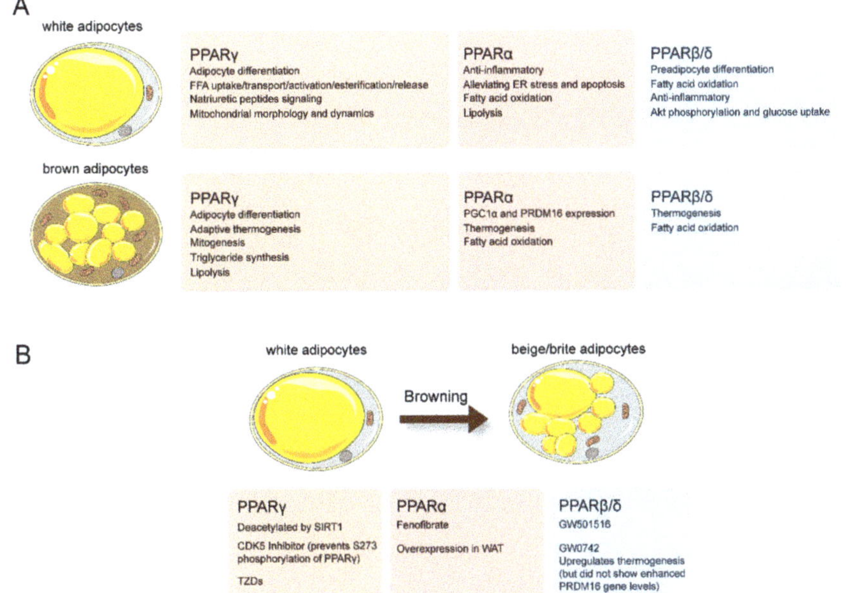

Figure 3. (**A**) Effects of physiologic and pharmacologic PPAR activation on white and brown adipocyte biology. (**B**) Physiologic and pharmacologic PPAR activation promote the "browning" of WAT. Figure was created using SMART–Servier Medical Art (https://smart.servier.com, the last accessed date is 29 July 2021).

At present, significant advances have been made in understanding the sophistication of the function of adipose tissue and the role of adipose tissue in controlling systemic energy balance. Obviously, targeting adipose tissue is an effective strategy for the treatment of obesity, insulin resistance, T2D, and another metabolic syndromes. In adipose tissue, PPARs represent how various metabolic signaling networks converge into a single nuclear factor, and the transcriptional activity of PPARs is controlled by multiple regulative layers of alternative splicing, post-translational modification, and coactivator/suppressor interaction, thereby resulting in its time- and tissue-specific responses. The study of alternative splicing and post-translational modification of PPARs in adipose tissue under diverse physiological and pathological conditions still needs further research. Altogether, we are convinced that the targeting of adipose PPARs in metabolic disorders remains a valuable and promising approach, with a future ahead of it.

Author Contributions: Conceptualization, C.L., S.C. and W.Z.; writing-original draft preparation, C.S.; literature searching, C.S. and S.M.; figures preparation, C.S.; writing-review and editing, all authors; supervision, C.L.; funding acquisition, C.L., S.C., W.Z. and C.S. All authors have read and agreed to the published version of the manuscript.

Funding: This work was financially supported by grants from the National Natural Science Foundation of China (grant no. 92057112 and 31771298 to C.L., 31800992 to S.Y.C., 81800512 to W.X.Z.), the Project of State Key Laboratory of Natural Medicines, China Pharmaceutical University (grant no. SKLNMZZRC202005 to C.L.), and the Priority Academic Program Development of Jiangsu Higher

Education Institutions (PAPD, to C.L. and S.Y.C.), the Postgraduate Research & Practice Innovation Program of Jiangsu Province (grant no. KYCX21_0648 to C.S.).

Institutional Review Board Statement: Not applicable.

Informed Consent Statement: Not applicable.

Data Availability Statement: Not applicable.

Conflicts of Interest: The authors declare no conflict of interest.

References

1. Chait, A.; de Hartigh, L.J. Adipose Tissue Distribution, Inflammation and Its Metabolic Consequences, Including Diabetes and Cardiovascular Disease. *Front. Cardiovasc. Med.* **2020**, *7*, 22. [CrossRef]
2. Kahn, C.R.; Wang, G.; Lee, K.Y. Altered adipose tissue and adipocyte function in the pathogenesis of metabolic syndrome. *J. Clin. Investig.* **2019**, *129*, 3990–4000. [CrossRef]
3. Brunmeir, R.; Xu, F. Functional Regulation of PPARs through Post-Translational Modifications. *Int. J. Mol. Sci.* **2018**, *19*, 1738. [CrossRef]
4. Cataldi, S.; Costa, V.; Ciccodicola, A.; Aprile, M. PPARγ and Diabetes: Beyond the Genome and Towards Personalized Medicine. *Curr. Diabetes Rep.* **2021**, *21*, 1–15. [CrossRef]
5. Ahmadian, M.; Suh, J.M.; Hah, N.; Liddle, C.; Atkins, A.R.; Downes, M.; Evans, R.M. PPARγ signaling and metabolism: The good, the bad and the future. *Nat. Med.* **2013**, *19*, 557–566. [CrossRef] [PubMed]
6. Issemann, I.; Green, S. Activation of a member of the steroid hormone receptor superfamily by peroxisome proliferators. *Nature* **1990**, *347*, 645–650. [CrossRef]
7. Okopień, B.; Bułdak, Ł.; Bołdys, A. Benefits and risks of the treatment with fibrates—A comprehensive summary. *Expert Rev. Clin. Pharmacol.* **2018**, *11*, 1099–1112. [CrossRef]
8. Kliewer, S.A.; Forman, B.M.; Blumberg, B.; Ong, E.S.; Borgmeyer, U.; Mangelsdorf, D.; Umesono, K.; Evans, R. Differential expression and activation of a family of murine peroxisome proliferator-activated receptors. *Proc. Natl. Acad. Sci. USA* **1994**, *91*, 7355–7359. [CrossRef] [PubMed]
9. Bedu, E.; Wahli, W.; Desvergne, B. Peroxisome proliferator-activated receptor β/δ as a therapeutic target for metabolic diseases. *Expert Opin. Ther. Targets* **2005**, *9*, 861–873. [CrossRef] [PubMed]
10. Corrales, P.; Vidal-Puig, A.; Medina-Gómez, G. PPARs and Metabolic Disorders Associated with Challenged Adipose Tissue Plasticity. *Int. J. Mol. Sci.* **2018**, *19*, 2124. [CrossRef] [PubMed]
11. Spalding, K.L.; Bernard, S.; Näslund, E.; Salehpour, M.; Possnert, G.; Appelsved, L.; Fu, K.-Y.; Alkass, K.; Druid, H.; Thorell, A.; et al. Impact of fat mass and distribution on lipid turnover in human adipose tissue. *Nat. Commun.* **2017**, *8*, 15253. [CrossRef] [PubMed]
12. Bhatt, P.S.; Dhillo, W.S.; Salem, V. Human brown adipose tissue—Function and therapeutic potential in metabolic disease. *Curr. Opin. Pharmacol.* **2017**, *37*, 1–9. [CrossRef]
13. Bartelt, A.; Heeren, J. Adipose tissue browning and metabolic health. *Nat. Rev. Endocrinol.* **2014**, *10*, 24–36. [CrossRef]
14. Schoettl, T.; Fischer, I.P.; Ussar, S. Heterogeneity of adipose tissue in development and metabolic function. *J. Exp. Biol.* **2018**, *221*, jeb162958. [CrossRef] [PubMed]
15. Chawla, A.; Schwarz, E.J.; Dimaculangan, D.D.; Lazar, M.A. Peroxisome proliferator-activated receptor (PPAR) gamma: Adipose-predominant expression and induction early in adipocyte differentiation. *Endocrinology* **1994**, *135*, 798–800. [CrossRef]
16. Tontonoz, P.; Hu, E.; Graves, R.A.; Budavari, A.I.; Spiegelman, B.M. mPPAR gamma 2: Tissue-specific regulator of an adipocyte enhancer. *Genes Dev.* **1994**, *8*, 1224–1234. [CrossRef]
17. Rosen, E.D.; Sarraf, P.; Troy, A.E.; Bradwin, G.; Moore, K.; Milstone, D.S.; Spiegelman, B.M.; Mortensen, R.M. PPAR gamma is required for the differentiation of adipose tissue in vivo and in vitro. *Mol. Cell* **1999**, *4*, 611–617. [CrossRef]
18. Barak, Y.; Nelson, M.C.; Ong, E.S.; Jones, Y.Z.; Ruiz-Lozano, P.; Chien, K.R.; Koder, A.; Evans, R.M. PPAR gamma is required for placental, cardiac, and adipose tissue development. *Mol. Cell* **1999**, *4*, 585–595. [CrossRef]
19. Jones, J.R.; Barrick, C.; Kim, K.-A.; Lindner, J.; Blondeau, B.; Fujimoto, Y.; Shiota, M.; Kesterson, R.A.; Kahn, B.B.; Magnuson, M.A. Deletion of PPARgamma in adipose tissues of mice protects against high fat diet-induced obesity and insulin resistance. *Proc. Natl. Acad. Sci. USA* **2005**, *102*, 6207–6212. [CrossRef]
20. Wang, F.; Mullican, S.E.; Dispirito, J.R.; Peed, L.C.; Lazar, M.A. Lipoatrophy and severe metabolic disturbance in mice with fat-specific deletion of PPARγ. *Proc. Natl. Acad. Sci. USA* **2013**, *110*, 18656–18661. [CrossRef]
21. Ghaben, A.L.; Scherer, P.E. Adipogenesis and metabolic health. *Nat. Rev. Mol. Cell Biol.* **2019**, *20*, 242–258. [CrossRef] [PubMed]
22. Lee, J.-E.; Schmidt, H.; Lai, B.; Ge, K. Transcriptional and Epigenomic Regulation of Adipogenesis. *Mol. Cell. Biol.* **2019**, *39*, e00601–e00618. [CrossRef] [PubMed]
23. Lefterova, M.I.; Zhang, Y.; Steger, D.J.; Schupp, M.; Schug, J.; Cristancho, A.; Feng, D.; Zhuo, D.; Stoeckert, C.J., Jr.; Liu, X.S.; et al. PPARγ and C/EBP factors orchestrate adipocyte biology via adjacent binding on a genome-wide scale. *Genes Dev.* **2008**, *22*, 2941–2952. [CrossRef] [PubMed]

24. Ren, D.; Collingwood, T.N.; Rebar, E.J.; Wolffe, A.P.; Camp, H.S. PPARγ knockdown by engineered transcription factors: Exogenous PPARγ2 but not PPARγ1 reactivates adipogenesis. *Genes Dev.* **2002**, *16*, 27–32. [CrossRef] [PubMed]
25. Medina-Gomez, G.; Gray, S.L.; Yetukuri, L.; Shimomura, K.; Virtue, S.; Campbell, M.; Curtis, R.K.; Jimenez-Linan, M.; Blount, M.; Yeo, G.S.H.; et al. PPAR gamma 2 Prevents Lipotoxicity by Controlling Adipose Tissue Expandability and Peripheral Lipid Metabolism. *PLoS Genet.* **2007**, *3*, e64. [CrossRef] [PubMed]
26. Chalise, J.P.; Hashimoto, S.; Parajuli, G.; Kang, S.; Singh, S.K.; Gemechu, Y.; Metwally, H.; Nyati, K.K.; Dubey, P.K.; Zaman, M.M.-U.; et al. Feedback regulation of Arid5a and Ppar-γ2 maintains adipose tissue homeostasis. *Proc. Natl. Acad. Sci. USA* **2019**, *116*, 15128–15133. [CrossRef]
27. Virtue, S.; Petkevicius, K.; Moreno-Navarrete, J.M.; Jenkins, B.; Hart, D.; Dale, M.; Koulman, A.; Fernández-Real, J.M.; Vidal-Puig, A. Peroxisome Proliferator-Activated Receptor γ2 Controls the Rate of Adipose Tissue Lipid Storage and Determines Metabolic Flexibility. *Cell Rep.* **2018**, *24*, 2005–2012.e7. [CrossRef]
28. Mann, J.P.; Savage, D.B. What lipodystrophies teach us about the metabolic syndrome. *J. Clin. Investig.* **2019**, *129*, 4009–4021. [CrossRef]
29. Semple, R.K.; Chatterjee, V.K.; O'Rahilly, S. PPAR gamma and human metabolic disease. *J. Clin. Investig.* **2006**, *116*, 581–589. [CrossRef]
30. Imai, T.; Takakuwa, R.; Marchand, S.; Dentz, E.; Bornert, J.-M.; Messaddeq, N.; Wendling, O.; Mark, M.; Desvergne, B.; Wahli, W.; et al. Peroxisome proliferator-activated receptor γ is required in mature white and brown adipocytes for their survival in the mouse. *Proc. Natl. Acad. Sci. USA* **2004**, *101*, 4543–4547. [CrossRef]
31. He, W.; Barak, Y.; Havener, A.; Olson, P.; Liao, D.; Le, J.; Nelson, M.; Ong, E.; Olefsky, J.M.; Evans, R.M. Adipose-specific peroxisome proliferator-activated receptor γ knockout causes insulin resistance in fat and liver but not in muscle. *Proc. Natl. Acad. Sci. USA* **2003**, *100*, 15712–15717. [CrossRef]
32. Yamauchi, T.; Kamon, J.; Waki, H.; Murakami, K.; Motojima, K.; Komeda, K.; Ide, T.; Kubota, N.; Terauchi, Y.; Tobe, K.; et al. The Mechanisms by Which Both Heterozygous Peroxisome Proliferator-activated Receptor γ (PPARγ) Deficiency and PPARγ Agonist Improve Insulin Resistance. *J. Biol. Chem.* **2001**, *276*, 41245–41254. [CrossRef]
33. Taylor, B.; Shah, A.; Bielczyk-Maczyńska, E. TGF-β is insufficient to induce adipocyte state loss without concurrent PPARγ downregulation. *Sci. Rep.* **2020**, *10*, 1–13. [CrossRef] [PubMed]
34. Rieusset, J.; Andreelli, F.; Auboeuf, D.; Roques, M.; Vallier, P.; Riou, J.P.; Auwerx, J.; Laville, M.; Vidal, H. Insulin acutely regulates the expression of the peroxisome proliferator-activated receptor-gamma in human adipocytes. *Diabetes* **1999**, *48*, 699–705. [CrossRef] [PubMed]
35. Vidal-Puig, A.; Jimenez-Liñan, M.; Lowell, B.B.; Hamann, A.; Hu, E.; Spiegelman, B.; Flier, J.S.; Moller, D. Regulation of PPAR gamma gene expression by nutrition and obesity in rodents. *J. Clin. Investig.* **1996**, *97*, 2553–2561. [CrossRef] [PubMed]
36. Tontonoz, P.; Spiegelman, B.M. Fat and Beyond: The Diverse Biology of PPARgamma. *Annu. Rev. Biochem.* **2008**, *77*, 289–312. [CrossRef]
37. Skat-Rordam, J.; Hojland Ipsen, D.; Lykkesfeldt, J.; Tveden-Nyborg, P. A role of peroxisome proliferator-activated receptor gamma in non-alcoholic fatty liver disease. *Basic Clin. Pharmacol. Toxicol.* **2019**, *124*, 528–537. [CrossRef]
38. Li, J.; Liu, Y.-P. The roles of PPARs in human diseases. *Nucleosides Nucleotides Nucleic Acids* **2018**, *37*, 361–382. [CrossRef]
39. Rodriguez, M.A.D.L.R.; Kersten, S. Regulation of lipid droplet-associated proteins by peroxisome proliferator-activated receptors. *Biochim. Biophys. Acta BBA—Mol. Cell Biol. Lipids* **2017**, *1862*, 1212–1220. [CrossRef]
40. Christian, M. Nuclear receptor-mediated regulation of lipid droplet-associated protein gene expression in adipose tissue. *Horm. Mol. Biol. Clin. Investig.* **2013**, *14*, 87–97. [CrossRef]
41. Festuccia, W.T.; Laplante, M.; Berthiaume, M.; Gelinas, Y.; Deshaies, Y. PPARgamma agonism increases rat adipose tissue lipolysis, expression of glyceride lipases, and the response of lipolysis to hormonal control. *Diabetologia* **2006**, *49*, 2427–2436. [CrossRef] [PubMed]
42. Shi, F.; Simandi, Z.; Nagy, L.; Collins, S. Diet-dependent natriuretic peptide receptor C expression in adipose tissue is mediated by PPARγ via long-range distal enhancers. *J. Biol. Chem.* **2021**, *297*, 1–23. [CrossRef]
43. Bond, S.; Moody, S.; Liu, Y.; Civelek, M.; Villanueva, C.; Gregorevic, P.; Kingwell, B.A.; Hevener, A.L.; Lusis, A.J.; Henstridge, D.C.; et al. The E3 ligase MARCH5 is a PPARγ target gene that regulates mitochondria and metabolism in adipocytes. *Am. J. Physiol. Endocrinol. Metab.* **2019**, *316*, E293–E304. [CrossRef] [PubMed]
44. Muise, E.S.; Azzolina, B.; Kuo, D.W.; El-Sherbeini, M.; Tan, Y.; Yuan, X.; Mu, J.; Thompson, J.R.; Berger, J.P.; Wong, K.K. Adipose Fibroblast Growth Factor 21 Is Up-Regulated by Peroxisome Proliferator-Activated Receptor γ and Altered Metabolic States. *Mol. Pharmacol.* **2008**, *74*, 403–412. [CrossRef]
45. Astapova, O.; Leff, T. Adiponectin and PPARγ: Cooperative and interdependent actions of two key regulators of metabolism. *Vitam. Horm.* **2012**, *90*, 143–162.
46. Zhu, D.; Xu, L.; Wei, X.; Xia, B.; Gong, Y.; Li, Q.; Chen, X. PPARγ enhanced Adiponectin polymerization and trafficking by promoting RUVBL2 expression during adipogenic differentiation. *Gene* **2020**, *764*, 145100. [CrossRef]
47. Zhang, Y.; Dallner, O.S.; Nakadai, T.; Fayzikhodjaeva, G.; Friedman, J.M. A non-canonical-PPARγ/RXRα-binding sequence regulates leptin expression in response to changes in adipose tissue mass. *Proc. Natl. Acad. Sci. USA* **2018**, *115*, E6039–E6047. [CrossRef]

48. Tymciw, T. Hormonal and Temporal Regulation of Adipogenic Genes in Classical Brown Adipocytes. Master's Thesis, Stockholm University, Stockholm, Sweden, 2018.
49. Lindgren, E.M.; Nielsen, R.; Petrovic, N.; Jacobsson, A.; Mandrup, S.; Cannon, B.; Nedergaard, J. Noradrenaline represses PPAR (peroxisome-proliferator-activated receptor) γ2 gene expression in brown adipocytes: Intracellular signalling and effects on PPARγ2 and PPARγ1 protein levels. *Biochem. J.* **2004**, *382*, 597–606. [CrossRef]
50. Valmaseda, A.; Carmona, M.; Barberá, M.; Viñas, O.; Mampel, T.; Iglesias, R.; Villarroya, F.; Giralt, M. Opposite regulation of PPAR-α and -γ gene expression by both their ligands and retinoic acid in brown adipocytes. *Mol. Cell. Endocrinol.* **1999**, *154*, 101–109. [CrossRef]
51. Tai, T.-A.C.; Jennermann, C.; Brown, K.K.; Oliver, B.B.; MacGinnitie, M.A.; Wilkison, W.O.; Brown, H.R.; Lehmann, J.M.; Kliewer, S.A.; Morris, D.C.; et al. Activation of the Nuclear Receptor Peroxisome Proliferator-activated Receptor γ Promotes Brown Adipocyte Differentiation. *J. Biol. Chem.* **1996**, *271*, 29909–29914. [CrossRef] [PubMed]
52. Petrovic, N.; Shabalina, I.; Timmons, J.A.; Cannon, B.; Nedergaard, J. Thermogenically competent nonadrenergic recruitment in brown preadipocytes by a PPARγ agonist. *Am. J. Physiol. Endocrinol. Metab.* **2008**, *295*, E287–E296. [CrossRef]
53. Oelkrug, R.; Polymeropoulos, E.T.; Jastroch, M. Brown adipose tissue: Physiological function and evolutionary significance. *J. Comp. Physiol. B Biochem. Syst. Environ. Physiol.* **2015**, *185*, 587–606. [CrossRef]
54. Linhart, H.G.; Ishimura-Oka, K.; DeMayo, F.; Kibe, T.; Repka, D.; Poindexter, B.; Bick, R.J.; Darlington, G.J. C/EBPα is required for differentiation of white, but not brown, adipose tissue. *Proc. Natl. Acad. Sci. USA* **2001**, *98*, 12532–12537. [CrossRef]
55. Nagano, G.; Ohno, H.; Oki, K.; Kobuke, K.; Shiwa, T.; Yoneda, M.; Kohno, N. Activation of Classical Brown Adipocytes in the Adult Human Perirenal Depot Is Highly Correlated with PRDM16–EHMT1 Complex Expression. *PLoS ONE* **2015**, *10*, e0122584. [CrossRef]
56. Rajakumari, S.; Wu, J.; Ishibashi, J.; Lim, H.-W.; Giang, A.-H.; Won, K.J.; Reed, R.R.; Seale, P. EBF2 Determines and Maintains Brown Adipocyte Identity. *Cell Metab.* **2013**, *17*, 562–574. [CrossRef]
57. Spiegelman, B.M.; Puigserver, P.; Wu, Z. Regulation of adipogenesis and energy balance by PPARgamma and PGC-1. *Int. J. Obes.* **2000**, *24* (Suppl. 4), S8–S10. [CrossRef]
58. Moura, M.A.; Festuccia, W.T.L.; Kawashita, N.H.; Garofalo, M.A.R.; Brito, S.R.C.; Kettelhut, I.C.; Migliorini, R.H. Brown adipose tissue glyceroneogenesis is activated in rats exposed to cold. *Pflügers Arch.* **2005**, *449*, 463–469. [CrossRef]
59. Festuccia, W.T.; Deshaies, Y. Depot specificities of PPARγ ligand actions on lipid and glucose metabolism and their implication in PPARγ-mediated body fat redistribution. *Clin. Lipidol.* **2009**, *4*, 633–642. [CrossRef]
60. Festuccia, W.T.; Blanchard, P.-G.; Turcotte, V.; Laplante, M.; Sariahmetoglu, M.; Brindley, D.N.; Richard, D.; Deshaies, Y. The PPARγ agonist rosiglitazone enhances rat brown adipose tissue lipogenesis from glucose without altering glucose uptake. *Am. J. Physiol. Regul. Integr. Comp. Physiol.* **2009**, *296*, R1327–R1335. [CrossRef]
61. Laplante, M.; Festuccia, W.T.; Soucy, G.; Blanchard, P.-G.; Renaud, A.; Berger, J.P.; Olivecrona, G.; Deshaies, Y. Tissue-specific postprandial clearance is the major determinant of PPARγ-induced triglyceride lowering in the rat. *Am. J. Physiol. Regul. Integr. Comp. Physiol.* **2009**, *296*, R57–R66. [CrossRef]
62. Festuccia, W.T.; Blanchard, P.-G.; Richard, D.; Deshaies, Y. Basal adrenergic tone is required for maximal stimulation of rat brown adipose tissue UCP1 expression by chronic PPAR-γ activation. *Am. J. Physiol. Regul. Integr. Comp. Physiol.* **2010**, *299*, R159–R167. [CrossRef]
63. Festuccia, W.T.; Öztezcan, S.; Laplante, M.; Berthiaume, M.; Michel, C.; Dohgu, S.; Denis, R.G.; Brito, M.N.; Brito, N.A.; Miller, D.S.; et al. Peroxisome Proliferator-Activated Receptor-γ-Mediated Positive Energy Balance in the Rat Is Associated with Reduced Sympathetic Drive to Adipose Tissues and Thyroid Status. *Endocrinology* **2008**, *149*, 2121–2130. [CrossRef]
64. Yau, W.W.; Yen, P.M. Thermogenesis in Adipose Tissue Activated by Thyroid Hormone. *Int. J. Mol. Sci.* **2020**, *21*, 3020. [CrossRef] [PubMed]
65. Festuccia, W.T.; Blanchard, P.G.; Deshaies, Y. Control of Brown Adipose Tissue Glucose and Lipid Metabolism by PPARgamma. *Front. Endocrinol.* **2011**, *2*, 84. [CrossRef]
66. Lasar, D.; Rosenwald, M.; Kiehlmann, E.; Balaz, M.; Tall, B.; Opitz, L.; Lidell, M.E.; Zamboni, N.; Krznar, P.; Sun, W.; et al. Peroxisome Proliferator Activated Receptor Gamma Controls Mature Brown Adipocyte Inducibility through Glycerol Kinase. *Cell Rep.* **2018**, *22*, 760–773. [CrossRef]
67. Gao, R.; Chen, W.; Yan, H.; Xie, X.; Liu, D.; Wu, C.; Zhu, Z.; Li, H.; Dong, F.; Wang, L. PPARγ agonist rosiglitazone switches fuel preference to lipids in promoting thermogenesis under cold exposure in C57BL/6 mice. *J. Proteom.* **2018**, *176*, 24–36. [CrossRef] [PubMed]
68. Chang, J.S.; Ha, K. A truncated PPAR gamma 2 localizes to mitochondria and regulates mitochondrial respiration in brown adipocytes. *PLoS ONE* **2018**, *13*, e0195007. [CrossRef] [PubMed]
69. Wu, J.; Boström, P.; Sparks, L.M.; Ye, L.; Choi, J.H.; Giang, A.-H.; Khandekar, M.; Virtanen, K.A.; Nuutila, P.; Schaart, G.; et al. Beige Adipocytes Are a Distinct Type of Thermogenic Fat Cell in Mouse and Human. *Cell* **2012**, *150*, 366–376. [CrossRef]
70. Petrovic, N.; Walden, T.B.; Shabalina, I.; Timmons, J.A.; Cannon, B.; Nedergaard, J. Chronic Peroxisome Proliferator-activated Receptor γ (PPARγ) Activation of Epididymally Derived White Adipocyte Cultures Reveals a Population of Thermogenically Competent, UCP1-containing Adipocytes Molecularly Distinct from Classic Brown Adipocytes. *J. Biol. Chem.* **2010**, *285*, 7153–7164. [CrossRef] [PubMed]

71. Wang, H.; Liu, L.; Lin, J.Z.; Aprahamian, T.; Farmer, S.R. Browning of White Adipose Tissue with Roscovitine Induces a Distinct Population of UCP1 + Adipocytes. *Cell Metab.* **2016**, *24*, 835–847. [CrossRef]
72. Ohno, H.; Shinoda, K.; Spiegelman, B.M.; Kajimura, S. PPARγ agonists Induce a White-to-Brown Fat Conversion through Stabilization of PRDM16 Protein. *Cell Metab.* **2012**, *15*, 395–404. [CrossRef]
73. Wilson-Fritch, L.; Nicoloro, S.; Chouinard, M.; Lazar, M.A.; Chui, P.C.; Leszyk, J.; Straubhaar, J.; Czech, M.P.; Corvera, S. Mitochondrial remodeling in adipose tissue associated with obesity and treatment with rosiglitazone. *J. Clin. Investig.* **2004**, *114*, 1281–1289. [CrossRef]
74. Qiang, L.; Wang, L.; Kon, N.; Zhao, W.; Lee, S.; Zhang, Y.; Rosenbaum, M.; Zhao, Y.; Gu, W.; Farmer, S.; et al. Brown Remodeling of White Adipose Tissue by SirT1-Dependent Deacetylation of Pparγ. *Cell* **2012**, *150*, 620–632. [CrossRef]
75. Loft, A.; Forss, I.; Siersbæk, M.S.; Schmidt, S.F.; Larsen, A.-S.B.; Madsen, J.G.S.; Pisani, D.F.; Nielsen, R.; Aagaard, M.M.; Mathison, A.; et al. Browning of human adipocytes requires KLF11 and reprogramming of PPARγ superenhancers. *Genes Dev.* **2015**, *29*, 7–22. [CrossRef]
76. Vernochet, C.; Peres, S.B.; Davis, K.E.; McDonald, M.E.; Qiang, L.; Wang, H.; Scherer, P.E.; Farmer, S.R. C/EBPα and the Corepressors CtBP1 and CtBP2 Regulate Repression of Select Visceral White Adipose Genes during Induction of the Brown Phenotype in White Adipocytes by Peroxisome Proliferator-Activated Receptor γ Agonists. *Mol. Cell. Biol.* **2009**, *29*, 4714–4728. [CrossRef] [PubMed]
77. Kraakman, M.J.; Liu, Q.; Postigo-Fernandez, J.; Ji, R.; Kon, N.; Larrea, D.; Namwanje, M.; Fan, L.; Chan, M.; Area-Gomez, E.; et al. PPARγ deacetylation dissociates thiazolidinedione's metabolic benefits from its adverse effects. *J. Clin. Investig.* **2018**, *128*, 2600–2612. [CrossRef] [PubMed]
78. Xu, L.; Ma, X.; Verma, N.K.; Wang, D.; Gavrilova, O.; Proia, R.L.; Finkel, T.; Mueller, E. Ablation of PPAR γ in subcutaneous fat exacerbates age-associated obesity and metabolic decline. *Aging Cell* **2018**, *17*, e12721. [CrossRef]
79. Auboeuf, D.; Rieusset, J.; Fajas, L.; Vallier, P.; Frering, V.; Riou, J.P.; Staels, B.; Auwerx, J.; Laville, M.; Vidal, H. Tissue distribution and quantification of the expression of mRNAs of peroxisome proliferator–activated receptors and liver X receptor-α in humans: No alteration in adipose tissue of obese and NIDDM patients. *Diabetes* **1997**, *46*, 1319–1327. [CrossRef]
80. MacLaren, R.; Cui, W.; Simard, S.; Cianflone, K. Influence of obesity and insulin sensitivity on insulin signaling genes in human omental and subcutaneous adipose tissues. *J. Lipid Res.* **2008**, *49*, 308–323. [CrossRef]
81. Goto, T.; Lee, J.-Y.; Teraminami, A.; Kim, Y.-I.; Hirai, S.; Uemura, T.; Inoue, H.; Takahashi, N.; Kawada, T. Activation of peroxisome proliferator-activated receptor-alpha stimulates both differentiation and fatty acid oxidation in adipocytes. *J. Lipid Res.* **2011**, *52*, 873–884. [CrossRef]
82. Lee, J.-Y.; Hashizaki, H.; Goto, T.; Sakamoto, T.; Takahashi, N.; Kawada, T. Activation of peroxisome proliferator-activated receptor-α enhances fatty acid oxidation in human adipocytes. *Biochem. Biophys. Res. Commun.* **2011**, *407*, 818–822. [CrossRef]
83. Mazzucotelli, A.; Vigueri, N.; Tiraby, C.; Annicotte, J.-S.; Mairal, A.; Klimcakova, E.; Lepin, E.; Delmar, P.; Dejean, S.; Tavernier, G.; et al. The transcriptional coactivator peroxisome proliferator–activated receptor (PPAR) γ coactivator-1α and the nuclear receptor PPARα control the expression of glycerol kinase and metabolism genes independently of PPARγ activation in human white adipocytes. *Diabetes* **2007**, *56*, 2467–2475. [CrossRef] [PubMed]
84. Miranda, J.; Lasa, A.; Fernández-Quintela, A.; García-Marzo, C.; Ayo, J.; Dentin, R.; Portillo, M.P. *cis*-9, *trans*-11, *cis*-15 and *cis*-9, *trans*-13, *cis*-15 CLNA Mixture Activates PPARα in HEK293 and Reduces Triacylglycerols in 3T3-L1 cells. *Lipids* **2011**, *46*, 1005–1012. [CrossRef]
85. Montgomery, M.K.; Bayliss, J.; Keenan, S.; Rhost, S.; Ting, S.B.; Watt, M.J. The role of Ap2a2 in PPARα-mediated regulation of lipolysis in adipose tissue. *FASEB J.* **2019**, *33*, 13267–13279. [CrossRef] [PubMed]
86. Tsuchida, A.; Yamauchi, T.; Takekawa, S.; Hada, Y.; Ito, Y.; Maki, T.; Kadowaki, T. Peroxisome proliferator–activated receptor (PPAR) α activation increases adiponectin receptors and reduces obesity-related inflammation in adipose tissue: Comparison of activation of PPARα, PPARγ, and their combination. *Diabetes* **2005**, *54*, 3358–3370. [CrossRef]
87. Wang, W.; Lin, Q.; Lin, R.; Zhang, J.; Ren, F.; Zhang, J.; Ji, M.; Li, Y. PPARα agonist fenofibrate attenuates TNF-α-induced CD40 expression in 3T3-L1 adipocytes via the SIRT1-dependent signaling pathway. *Exp. Cell Res.* **2013**, *319*, 1523–1533. [CrossRef]
88. Liu, Z.; Gan, L.; Wu, T.; Feng, F.; Luo, D.; Gu, H.; Liu, S.; Sun, C. Adiponectin reduces ER stress-induced apoptosis through PPAR α transcriptional regulation of ATF2 in mouse adipose. *Cell Death Dis.* **2016**, *7*, e2487. [CrossRef] [PubMed]
89. Tong, Y.; Hara, A.; Komatsu, M.; Tanaka, N.; Kamijo, Y.; Gonzalez, F.J.; Aoyama, T. Suppression of expression of muscle-associated proteins by PPARα in brown adipose tissue. *Biochem. Biophys. Res. Commun.* **2005**, *336*, 76–83. [CrossRef] [PubMed]
90. Montagner, A.; Polizzi, A.; Fouché, E.; Ducheix, S.; Lippi, Y.; Lasserre, F.; Barquissau, V.; Regnier, M.; Lukowicz, C.; Benhamed, F.; et al. Liver PPARα is crucial for whole-body fatty acid homeostasis and is protective against NAFLD. *Gut* **2016**, *65*, 1202–1214. [CrossRef]
91. Hondares, E.; Rosell, M.; Diaz-Delfin, J.; Olmos, Y.; Monsalve, M.; Iglesias, R.; Villarroya, F.; Giralt, M. Peroxisome proliferator-activated receptor α (PPARα) induces PPARγ coactivator 1α (PGC-1α) gene expression and contributes to thermogenic activation of brown fat: Involvement of PRDM16. *J. Biol. Chem.* **2011**, *286*, 43112–43122. [CrossRef] [PubMed]
92. Komatsu, M.; Tong, Y.; Li, Y.; Nakajima, T.; Li, G.; Hu, R.; Sugiyama, E.; Kamijo, Y.; Tanaka, N.; Hara, A.; et al. Multiple roles of PPARα in brown adipose tissue under constitutive and cold conditions. *Genes Cells* **2010**, *15*, 91–100. [CrossRef]

93. Miranda, C.S.; Silva-Veiga, F.; Martins, F.F.; Rachid, T.L.; Mandarim-De-Lacerda, C.A.; Souza-Mello, V. PPAR-α activation counters brown adipose tissue whitening: A comparative study between high-fat–and high-fructose–fed mice. *Nutrition* **2020**, *78*, 110791. [CrossRef]
94. Barquissau, V.; Beuzelin, D.; Pisani, D.; Beranger, G.; Mairal, A.; Montagner, A.; Roussel, B.; Tavernier, G.; Marques, M.-A.; Moro, C.; et al. White-to-brite conversion in human adipocytes promotes metabolic reprogramming towards fatty acid anabolic and catabolic pathways. *Mol. Metab.* **2016**, *5*, 352–365. [CrossRef]
95. Shen, Y.; Su, Y.; Silva, F.J.; Weller, A.H.; Sostre-Colon, J.; Titchenell, P.M.; Steger, D.J.; Seale, P.; Soccio, R.E. Shared PPARα/γ target genes regulate brown adipocyte thermogenic function. *Cell Rep.* **2020**, *30*, 3079–3091.e5. [CrossRef]
96. Rachid, T.L.; Penna-de-Carvalho, A.; Bringhenti, I.; Aguila, M.B.; Mandarim-de-Lacerda, C.A.; Souza-Mello, V. Fenofibrate (PPARalpha agonist) induces beige cell formation in subcutaneous white adipose tissue from diet-induced male obese mice. *Mol. Cell. Endocrinol.* **2015**, *402*, 86–94. [CrossRef]
97. Defour, M.; Dijk, W.; Ruppert, P.; Nascimento, E.; Schrauwen, P.; Kersten, S. The Peroxisome Proliferator-Activated Receptor α is dispensable for cold-induced adipose tissue browning in mice. *Mol. Metab.* **2018**, *10*, 39–54. [CrossRef] [PubMed]
98. Vosper, H.; Khoudoli, G.A.; Na Palmer, C. The peroxisome proliferator activated receptor δ is required for the differentiation of THP-1 monocytic cells by phorbol ester. *Nucl. Recept.* **2003**, *1*, 1–10. [CrossRef] [PubMed]
99. Bastie, C.; Holst, D.; Gaillard, D.; Jehl-Pietri, C.; Grimaldi, P.A. Expression of Peroxisome Proliferator-activated Receptor PPARδ Promotes Induction of PPARγ and Adipocyte Differentiation in 3T3C2 Fibroblasts. *J. Biol. Chem.* **1999**, *274*, 21920–21925. [CrossRef] [PubMed]
100. Hansen, J.; Zhang, H.; Rasmussen, T.H.; Petersen, R.K.; Flindt, E.; Kristiansen, K. Peroxisome Proliferator-activated Receptor δ (PPARδ)-mediated Regulation of Preadipocyte Proliferation and Gene Expression Is Dependent on cAMP Signaling. *J. Biol. Chem.* **2001**, *276*, 3175–3182. [CrossRef] [PubMed]
101. Peters, J.M.; Lee, S.S.T.; Li, W.; Ward, J.M.; Gavrilova, O.; Everett, C.; Reitman, M.; Hudson, L.D.; Gonzalez, F.J. Growth, Adipose, Brain, and Skin Alterations Resulting from Targeted Disruption of the Mouse Peroxisome Proliferator-Activated Receptor β(δ). *Mol. Cell. Biol.* **2000**, *20*, 5119–5128. [CrossRef]
102. Wang, Y.-X.; Lee, C.-H.; Tiep, S.; Yu, R.T.; Ham, J.; Kang, H.; Evans, R. Peroxisome-Proliferator-Activated Receptor δ Activates Fat Metabolism to Prevent Obesity. *Cell* **2003**, *113*, 159–170. [CrossRef]
103. Sodhi, K.; Puri, N.; Hyun, K.D.; Hinds, T.D., Jr.; Stechschulte, L.A.; Favero, G.; Rodella, L.; Shapiro, J.I.; Jude, D.; Abraham, N.G. PPAR-delta binding to heme oxygenase 1 promoter prevents angiotensin II induced adipocyte dysfunction in goldblatt hypertensive rats. *Int. J. Obes.* **2014**, *38*, 456–465. [CrossRef] [PubMed]
104. Serrano-Marco, L.; Rodriguez-Calvo, R.; El Kochairi, I.; Palomer, X.; Michalik, L.; Wahli, W.; Vazquez-Cerrera, M. Activation of Peroxisome Proliferator-Activated Receptor-β/-δ (PPAR-β/-δ) Ameliorates Insulin Signaling and Reduces SOCS3 Levels by Inhibiting STAT3 in Interleukin-6-Stimulated Adipocytes. *Diabetes* **2011**, *60*, 1990–1999. [CrossRef] [PubMed]
105. Kang, K.; Reilly, S.; Karabacak, V.; Gangl, M.R.; Fitzgerald, K.; Hatano, B.; Lee, C.-H. Adipocyte-Derived Th2 Cytokines and Myeloid PPARδ Regulate Macrophage Polarization and Insulin Sensitivity. *Cell Metab.* **2008**, *7*, 485–495. [CrossRef]
106. Pan, D.; Fujimoto, M.; Lopes, A.; Wang, Y.-X. Twist-1 Is a PPARδ-Inducible, Negative-Feedback Regulator of PGC-1α in Brown Fat Metabolism. *Cell* **2009**, *137*, 73–86. [CrossRef] [PubMed]
107. Lima, R.T.; Silva-Veiga, F.M.; Graus-Nunes, F.; Bringhenti, I.; Mandarim-de-Lacerda, C.A.; Souza-Mello, V. Differential actions of PPAR-α and PPAR-β/δ on beige adipocyte formation: A study in the subcutaneous white adipose tissue of obese male mice. *PLoS ONE* **2018**, *13*, e0191365.
108. Mazuecos, L.; Pintado, C.; Rubio, B.; Guisantes-Batán, E.; Andrés, A.; Gallardo, N. Leptin, Acting at Central Level, Increases FGF21 Expression in White Adipose Tissue via PPARβ/δ. *Int. J. Mol. Sci.* **2021**, *22*, 4624. [CrossRef]

International Journal of
Molecular Sciences

MDPI

Review

PPARs as Metabolic Sensors and Therapeutic Targets in Liver Diseases

Hugo Christian Monroy-Ramirez [1], Marina Galicia-Moreno [1], Ana Sandoval-Rodriguez [1], Alejandra Meza-Rios [2], Arturo Santos [2] and Juan Armendariz-Borunda [1,2,*]

1 Instituto de Biologia Molecular en Medicina, Centro Universitario de Ciencias de la Salud, Universidad de Guadalajara, Guadalajara 44340, Jalisco, Mexico; hugo.monroyram@academicos.udg.mx (H.C.M.-R.); marina.galicia@academicos.udg.mx (M.G.-M.); anasol44@hotmail.com (A.S.-R.)

2 Tecnologico de Monterrey, Escuela de Medicina y Ciencias de la Salud, Zapopan 45138, Jalisco, Mexico; alejandramezarios@yahoo.com.mx (A.M.-R.); arturo.santos@tec.mx (A.S.)

* Correspondence: armdbo@gmail.com

Abstract: Carbohydrates and lipids are two components of the diet that provide the necessary energy to carry out various physiological processes to help maintain homeostasis in the body. However, when the metabolism of both biomolecules is altered, development of various liver diseases takes place; such as metabolic-associated fatty liver diseases (MAFLD), hepatitis B and C virus infections, alcoholic liver disease (ALD), and in more severe cases, hepatocelular carcinoma (HCC). On the other hand, PPARs are a family of ligand-dependent transcription factors with an important role in the regulation of metabolic processes to hepatic level as well as in other organs. After interaction with specific ligands, PPARs are translocated to the nucleus, undergoing structural changes to regulate gene transcription involved in lipid metabolism, adipogenesis, inflammation and metabolic homeostasis. This review aims to provide updated data about PPARs' critical role in liver metabolic regulation, and their involvement triggering the genesis of several liver diseases. Information is provided about their molecular characteristics, cell signal pathways, and the main pharmacological therapies that modulate their function, currently engaged in the clinic scenario, or in pharmacological development.

Keywords: metabolic alterations; hepatic damage; nuclear factors; pharmacological targets

Citation: Monroy-Ramirez, H.C.; Galicia-Moreno, M.; Sandoval-Rodriguez, A.; Meza-Rios, A.; Santos, A.; Armendariz-Borunda, J. PPARs as Metabolic Sensors and Therapeutic Targets in Liver Diseases. *Int. J. Mol. Sci.* **2021**, 22, 8298. https://doi.org/10.3390/ijms22158298

Academic Editors: Manuel Vázquez-Carrera and Walter Wahli

Received: 28 June 2021
Accepted: 29 July 2021
Published: 2 August 2021

Publisher's Note: MDPI stays neutral with regard to jurisdictional claims in published maps and institutional affiliations.

1. Introduction

The liver is the main organ responsible for biochemical metabolism in the human body, compounds absorbed by the intestine such as nutrients or drugs, first pass through the liver, where they are processed into simpler products, maintaining and regulating their levels in the bloodstream [1]. Carbohydrates and lipids are two components of the diet that are metabolized by the liver to generate the necessary energy, leading to several physiological processes that help maintain body homeostasis [2]. However, a dysfunction in hepatic metabolism can result in the genesis of several hepatic diseases such as_MAFLD, ALD, fibrosis/cirrhosis, viral hepatitis by hepatitis B (HBV) or hepatitis C(HCV) infection, or in some cases HCC [2,3].

Peroxisome proliferator-activated receptors (PPARs) are a family of ligand-dependent transcription factors that regulate essential metabolic processes in the liver and other organs where they are activated by endogenous ligands such as fatty acids and similar compounds. Three isoforms of PPARs are known: PPARα, PPARβ/δ, and PPARγ, all of them with different distribution, affinity and specificity for their agonists, and the ability to modulate lipid metabolism and energy homeostasis in mammals [4]. All the changes that occur during liver injury alter metabolic functionality, aggravate liver damage, and make PPARs important therapeutic targets for the treatment of these diseases [5].

2. Overviews of PPARs α, β/δ and γ

As we previously mentioned, PPARs are transcription factors of nuclear hormone receptor, a family composed by three subtypes: PPARα, PPARβ/δ, and PPARγ; each encoded by a different gene located in different chromosomes and characterized by different distribution patterns and specific ligands [6–8]. In this section, we describe the molecular characteristics and functions of each subtype.

2.1. Structure and Molecular Characteristics

Structurally, PPARs are similar to steroid and thyroid hormone receptors, and they can be stimulated by small lipophilic ligands [9]. In general, the three-dimensional structure of PPARs consists of a canonical domain that is shared with other nuclear receptors, including the amino-terminal AF-1 trans activation domain (A/B domain); a DNA-binding domain (DBD or C domain) in their N-terminus containing two highly conserved zinc finger motifs with globular structure; and a dimerization and ligand-binding domain (LBD or E/F domain) with a ligand-dependent transactivation function AF-2 (promotes the recruitment of co-activators) at the carboxy-terminal region, which is responsible for ligand specificity and PPAR activation binding to the peroxisome proliferator response elements (PPRE) [7–11] (Figure 1A). The LBD is characterized by its size, which is larger than other nuclear receptors; this feature allows a wide range of unsaturated fatty acids to bind [7,10,12]. In addition, PPARs contain a hinge region functioning as a docking site for cofactors (D domain) [8,11]. PPARs' subtype structures are illustrated in Figure 1B.

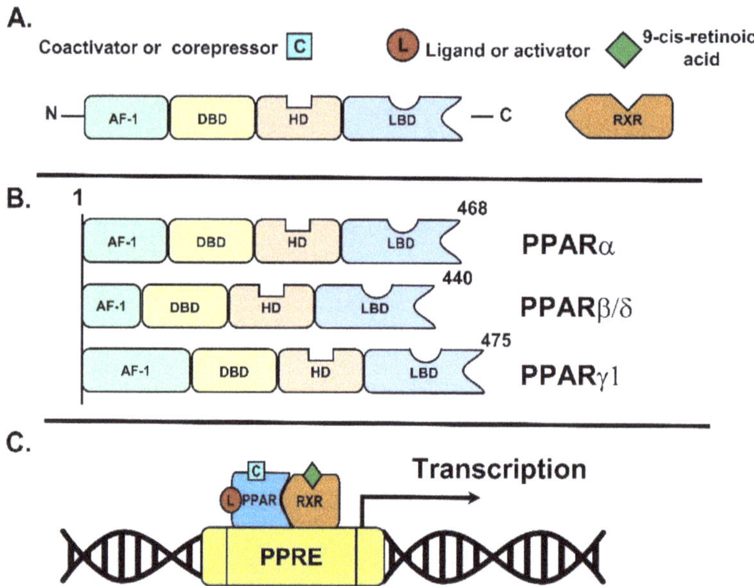

Figure 1. Representation of PPARs' structure and their molecular function. (**A**) The universal structure of PPARs is represented by AF-1, DBD, HD and LBD domains. The active form of PPARs is heterodimerizing with RXR in conjunction with ligands, co-activators and co-repressors that modulate their function. (**B**) PPARs' subtypes representations. (**C**) Transcription of PPARs target genes start upon the union of PPAR-RXR and the ligands and co-activators into PPRE sequences.

2.2. Mechanisms of Action

After interaction with the specific ligands, PPARs are translocated to the nucleus, and heterodimerizes with another nuclear receptor; the retinoid X receptor (RXR), which binds to PPAR through two zinc fingers in the DBD, specifically PPREs present in the vicinity of PPAR-responsive genes promoters, subsequently altering co-activator/co-repressor dynamics to modulate transcription [7,8,10]. PPREs generally have a direct repeat of hexanucleotide core recognition elements (5′AGGTCA-3′) spaced by one or two bp [11]. In addition, the hexanucleotide core has an extension 5′-AACT that provides polarity for heterodimer binding with RXR [12]. Once activated, the heterodimer PPARs/RXR recruit different nuclear receptor co-factors and gene transcription initiates [11] (Figure 1C). In addition, PPAR co-activators are cAMP response element-binding protein, steroid receptor coactivator-1, and the PPARγ coactivator 1α. In addition, co-repressors comprise the nuclear receptor co-repressor, and silence the mediator of the retinoid and thyroid hormone receptor [8,11].

The activation of PPARα and PPARβ/δ mostly facilitates energy combustion and the activation of PPARγ contributes to energy storage [10]. Table 1 summaries the main characteristics of each PPAR subtype.

Table 1. Main characteristics of PPAR subtypes.

PPAR Subtype	PPARα	PPAR β/δ	PPARγ
Gene location	Human chromosome 22q12.2–13.1	Human chromosome 6p21.1–21.2	Human chromosome 3p25
Isoforms	None	None	PPARγ1, PPARγ2, PPARγ3
Tissue distribution	Liver, heart, skeletal muscle (tissues with high fatty acid oxidation rates); brown adipose tissue, kidney, adrenal gland.	Liver, kidney, skeletal and cardiac muscle, adipose tissue, brain, colon, vasculature, esophagus, gut. Ubiquitous.	Mainly in adipose tissue (white and brown). Other tissues such as liver, gut, kidney, retina, immunologic system, muscles, spleen, urinary bladder, heart, lung, brain, vasculature.
Endogenous Ligands	Unsaturated and saturated fatty acids and their derivatives (8-S-hydroxyeicosatetraenoic acid, arachidonic acid lipoxygenase metabolite LTB4, arachidonate monooxygenase metabolite epoxyeicosatrienoic acids), leukotriene derivatives, VLDL hydrolysis products.	Unsaturated fatty acids, arachidonic acid cyclooxygenase metabolite prostacyclin, the linoleic acid 15-lipoxygenase-1 product 13-S-hydroxyoctadecadienoic acid, carbaprostacyclin, components of VLDL.	Polyunsaturated fatty acids, prostanoids (15-deoxy-Δ12, 14-prostaglandin J2 (15-dPGJ2)), 13-hydroxyeicosatetraenoic acid, components of oxidized LDL, eicosanoids, oxidized alkyl phospholipids
Functions	Major regulator of the mitochondrial and peroxisomal β-oxidation (fatty acid metabolism), lowers lipid levels, anti-inflammatory activities.	Increase lipid catabolism, improves the plasma HDL-cholesterol levels and insulin resistance, induce cell proliferation and differentiation, anti-inflammatory activities.	Regulate adipocyte differentiation, lipid storage, and glucose metabolism (improves insulin sensitivity), main regulator of metabolic genes, increase fatty acid oxidation, HDL and uncoupling protein, decrease triglycerides, improves vascular integrity, energy balance, anti-inflammatory activities.

Table 1. *Cont.*

PPAR Subtype	PPARα	PPAR β/δ	PPARγ
Target genes	CYP8B1, FATP, FAT/CD36, liver cytosolic FABP, LPL. Lipid/hormone transport genes. (LEPR, SLC27A2, SLC27A4). Acyl-CoA metabolism. (ACOT12, ACSL3, ACSL3, ACSL5, ACSL1, ACSM3, FABP1, FABP3). β-oxidation. (ACAA2, ACADM, ACADS, ACADVL, CPT1A, CPT2, ETFDH, HADHA, HADHB, SLC25A20, SLC22A5, TXNIP). Ketogenesis/ketolysis genes. (FGF21, HMGCS2), Peroxisomal β-oxidation (ABCD2, ABCD3, ACAA1A, ACOX1, ECH1, HSD17B4). Lipogenesis. (ACACB, AGPAT2, ELOVL5, ELOVL6 FADS1, GPAM, MLYCD, MOD1). Lipases and lipid droplet proteins. (ADFP, CIDEC, PNPLA2, S3-12). Lipoprotein metabolism. (ANGPTL4, APOA1, APOA2, APOA5, APOCIII, LIPC, PCTP, VLDLR). Cholesterol and bile metabolism. (ABCA1, ABCB4, CYP7A1, FXR, LXR)	Genes related with lipid uptake, represses genes that participated in lipid metabolism and efflux. LPL, PGAR, IDK, PDK-1, Ubiquitin C, CPT1, AOX, LCAD, UCP1, UCP3, PGC1-alpha. Tumor angiogenesis (Pdgfrβ, Pdgfb, c-kit)	AP2, CAP, IRS2, GLUT4, GLUT2, adiponectin, ACS, PCK2, LPL, FAT/CD36, FABP, GYK fatty acid transport, acyl-CoA synthetase, glucokinase, leptin, perilipin, GK PEPCK, UCP 1, UCP-2, UCP-3, LXR-alpha, TNF-alpha, IL-6.
References	[8,10–15]	[8,10–14,16,17]	[8,10–14,17–20]

HDL, High density lipoprotein; LDL, low density lipoprotein; VLDL, Very low-density lipoprotein.

Signal Pathways

PPARs activated different signal pathways, mainly via endogenous ligands products of the metabolic pathways of fatty acids, which is the reason why they are called lipid sensors [10]. The main signal transduction pathways related with different PPARs subtypes are recapitulated in Figure 2.

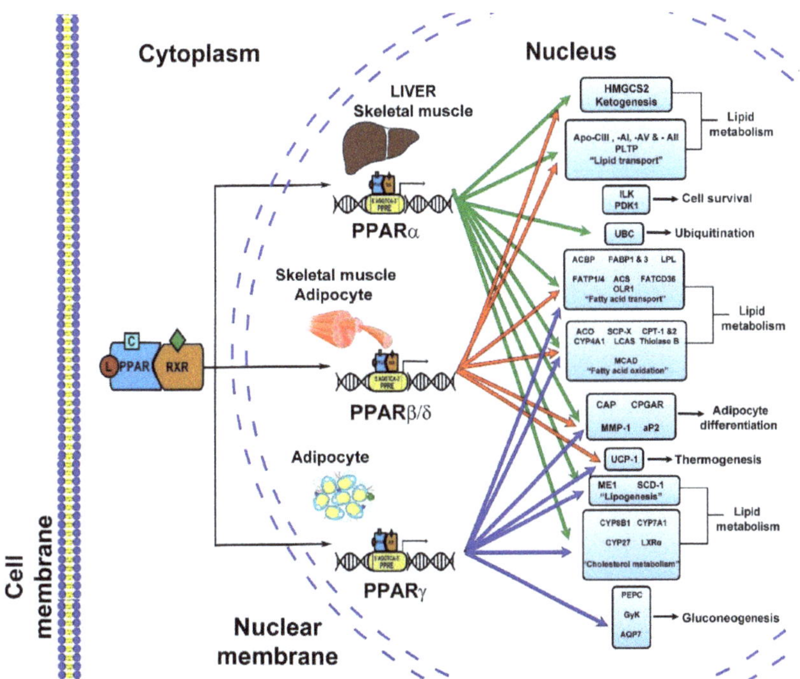

Figure 2. PPARs' signal transduction. The signal pathways where PPARs are involved affects gluconeogenesis, lipid metabolism, thermogenesis, adipocyte differentiation, ubiquitination and cell survival, depending on the target gene and tissues where they are activated.

3. Role of PPARs in Liver Diseases

The most shocking liver diseases worldwide are MAFLD, fibrosis, HCC, HBV and (HCV infections, along with ALD. Several of them, such as HCC, represent a major cause of mortality in the world. Because of this, it is necessary to understand PPARs role as metabolic regulators in the development of these pathologies, and design new effective ligands able to modulate the activity of these receptors, minimizing side effects. The main functions of PPARs in each one of the aforesaid diseases will be described below

3.1. Gene Expression Alterationof PPARs in MAFLD

MAFLD, formerly named non-alcoholic fatty liver disease (NAFLD), affects 20–30% of adult population in western countries [21]. This damage is characterized by hepatic steatosis accompanied by one of three features: overweight or obesity, T2DM, or lean or normal weight with evidence of metabolic dysregulation [22]. During MAFLD, elevated level of circulating free fatty acids increases fat influx into hepatocytes, causing an augmented fatty acid oxidation in mitochondria and peroxisomes leading to ROS generation, causing oxidative stress [23]. Imbalance between ROS generation and antioxidant mechanisms leads to mitochondrial and peroxisome dysfunction; and eventually to apoptosis of hepatocytes exacerbating the proinflammatory events of non-alcoholic steatohepatitis (NASH). Peroxisomes and mitochondria jointly perform various metabolic roles including O_2 and lipid metabolism; these organelles are indispensable in a healthy liver for the breakdown of long-chain fatty acids, very-long-chain fatty acids, and branched-chain fatty acids through α and β-oxidation. Subsequently, they prevent the accumulation of fatty acids (FAs) in the liver. In addition, Acyl-coenzyme A oxidase enzyme (ACOX1) is a rate-limiting enzyme of peroxisomal beta-oxidation of long chain fatty acids exclusive of peroxisomes, alterations in ACOX1 results in hepatic steatosis [24].

3.1.1. PPARα

Of all PPARs; PPARα is the most relevant one to NASH pathogenesis; since it is a metabolic sensor upregulated by fasting and responsible for transcriptional upregulation of β-oxidation genes [25], then altered expression of this transcription factor induces lipogenesis. Therefore, PPARα agonists are potential targets for NASH treatment. Peroxisome biogenesis and proliferation are also regulated by PPARα. After the activation this nuclear transcription factor, the expression of several genes encoding for peroxisomal proteins and genes controlling beta oxidation, fatty acid uptake, triglyceride turnover, bile acid synthesis, adipogenesis, ketogenesis, glucose metabolism and adipocyte differentiation are induced [10,26–28]. Additionally, PPARα exerts anti-inflammatory effects through a negative crosstalk with NF-κB and AP-1 (Activator Protein 1) [29]. In normal conditions, hepatocytes have a high expression of PPARα. In NASH patients, hepatic expression of PPARα decreases and negatively correlates with the severity of the disease [30]. Correspondingly, several authors have reported that expression of PPARα is reduced in NASH models [31,32]. PPARα reduction is believed to be due to an increased expression/activation of Rho-associated protein kinase (ROCK) and to a reduction in peroxisomes number caused by elevated ROS in NASH [33,34]. Hepatic decrease in PPARα expression causes a deficiency in the transcription of its target gene carnitine palmitoyl transferase 1 (CPT-1A) and excessive FAs tend to accumulate in the form of triglycerides, since they cannot go through the inner mitochondrial membrane and reach the mitochondrial matrix for further metabolism [35].

Levels of PPARα are recovered in NASH with statins [31] due to a reduced RhoA cell membrane translocation. Additionally, PPARα increases as a result of lifestyle change or bariatric surgery with the improvement of histological NAFLD score [30].

Intriguingly, some PPARα properties, such as increased DNA synthesis and peroxisome pro-liberation, are observed only in mice and rats, but not in humans. This could be due to the ten-fold higher expression of PPARα in the liver of rodents, that can also partially explain differences in the efficacy of PPARα agonists in experimental models against human studies. Interestingly, it has been suggested that increased hepatic expression of PPARα and target genes involved in fatty acid oxidation is a protective response to high fat diet [36–38].

Mice deficient in PPARα are susceptible to dietary fat-mediated NASH, oxidative stress, cell death and hepatic inflammation [36,39–41]. Pharmacological activation of PPARα in the methionine–choline-deficient diet (MCDD) NASH model, reduces lipid peroxidation and TG content in the liver, and reverses steatohepatitis and fibrosis [42,43]. Additionally, PPARα agonists prevent dietary steatohepatitis by a direct effect on inflammation, independent of its effect on lipid accumulation in hepatocytes and independent of PPARα binding to PPREs [44]. Additionally, Pirfenidone seem to be a PPARα agonist that improves NASH through SIRT1/LKB1/pAMPK signaling [32].

3.1.2. PPARβ/δ

PPARβ/δ is well expressed in hepatocytes, Kupffer cells (KCs) and hepatic stellate cells (HSCs). Most of the work to study MAFLD has been conducted with PPARβ/δ agonist. In MCDD-fed mice, the treatment with PPARδ agonist GW501516 increased hepatic expression of ACOX1, CPT-1A and FABP1 (liver fatty acid binding protein); and decreased hepatic triglyceride content [45]. In obese monkeys, GW501516 normalized serum insulin and TAG concentrations, decreased low-density/lipoprotein cholesterol, and increased high-density lipoprotein cholesterol [46]. Additionally, PPARβ/δ agonist has been shown in high fat diet-fed mice to favor a slender phenotype, improved insulin sensitivity, and prevent hepatic lipid accumulation, due to higher rates of energy expenditure [47]. It also favors an upregulation of Adfp and Cpt1a and enhanced FA oxidation, as well as activation of AMPK and inhibition of sterol regulatory element-binding protein 1c (SREBP-1c), reducing hepatic lipogenesis using GW501516 [48].

In 2008, Riserus et al. published a paper confirming that GW501516 reduced liver fat content, and TG, LDL, ApoB and insulin in plasma in moderate obese men and muscle expression of CPT1b was also significantly increased [49]. Other PPARβ/δ agonists tested in overweight subjects with dyslipidemia demonstrated diminution in GGT and favorable lipid profiles [50]. Adenovirus mediated upregulation of PPARβ/δ into db/db mice resulted in the activation of SREBP-1c, upregulation of lipase, and improved liver steatosis [51]. Similarly, Liu et al. found increased hepatic expression of ACC1, FA uptake and beta oxidation [52]. Quite the opposite, PPARδ-null mice had lower metabolic activity and glucose intolerance [53]. Use of hepatocyte-specific PPARδ null mice identified that hepatic PPARβ/δ augments FA muscle utilization and improves dyslipidemia through a metabolic network between hepatic PPARβ/δ and muscle PPARα. Up to now, there is not enough evidence that PPARβ/δ clinical intervention can be effective for the treatment of MALFD, and carcinogenesis remain a concern.

3.1.3. PPARγ

PPARγ is mostly known by its role in regulation of adipocyte differentiation and high expression of adipose tissue and macrophages. However, hepatic PPARγ expression is robustly induced in NAFLD patients and experimental models [54–57]. Increased activation of PPARγ downregulates the expression of pro-inflammatory cytokines, such as TNF-α, IFN-γ, IL-2, IL-1β, IL-6, MCP-1, and MIP-1β [57] and results in a decreased activation of TLR-4 pathway. In contrast, activation of TLR-4 pathway leads to the downregulation of PPARγ by negatively interfering with NF-kB in macrophages [58]. Additionally, it polarizes macrophages into anti-inflammatory M2 phenotype [59].

PPARγ upregulates proteins associated with lipid uptake, TAG storage, and the formation of lipid droplets, such as FABP4, fat-specific protein 27 (FSP27)/Cidec, CD36, monacylglycerol O-acyltransferase 1, and perilipin 2; then PPARγ hepatic expression promotes steatosis. Overexpression of PPARγ2 in hepatocytes increased steatosis; on the contrary, in mice hepatocyte-specific PPARγ-deficient (Pparγ-DHEP) hepatosteatosis was decreased [60–62]. In Pparγ-DHEP mice, liver expression of genes associated with adipogenesis, and FA uptake were downregulated, but systemic insulin resistance, adiposity, and hyperlipidemia were aggravated [56]. In HFD-fed animals, Pparγ and Srebp1c, CD36 and FAS are overexpressed. Even though activation of PPARγ is steatogenic, treatment with PPARγ ligands to genetically obese or diet-induced NASH mice decreases hepatic TAG due to adiponectic-mediated glucose uptake and AMPK activation, thereby improving FA oxidation in hepatocytes [62]. Thiazolidinediones (TZD), also called glitazones, are PPARγ-ligands and in the absence of adipose tissue, the liver is the primary target for TZD action. Clinical trials utilizing TZDs showed significant improvement in hepatic steatosis and inflammation. However, weight gain concerns and other side effects remain.

3.1.4. Clinical Trials of PPAR-Related Drugs in NASH

Elafibranor is a well-known dual PPARα/δ agonist. A Phase-2b Golden-505 study has demonstrated that, in NASH patients without cirrhosis treated with 120 mg daily for 52 weeks, insulin, sensitivity, glucose homeostasis, and lipid metabolism were improved and inflammation reduced [63]. However, in phase III study, Genfit announced interim results after 72 weeks in RESOLVE-IT study which showed that the trial did not meet histological improvement or NASH resolution without worsening of fibrosis in the ITT intention to treat (ITT) population of 1070 patients with nonalcoholic steatohepatitis (NASH) and fibrosis (https://www.natap.org/2020/AASLD/AASLD_162.htm) (accessed on 30 July 2021).

FXR agonists are used to treat non-alcoholic fatty liver disease (NAFLD), in part because they reduce hepatic lipids.

Obeticholic acid (OCA) is a selective and potent agonists for the farnesoid X receptor (FXR). FXR-PPARγ cascade has demonstrated clinical efficacy in NASH. In the phase 3 double-blind, randomized, placebo-controlled, multicenter, REGENERATE study, it was demonstrated that after 18 months OCA significantly improved fibrosis in 1218 noncirrhotic NASH patients using 25 mg. Additionally, Nonalcoholic Fatty Liver Disease Activity Score decreased (by ≥2 points), and quality of life was impaired, or NASH resolution had greater patient-reported outcomes (PROs) improvements in some domains (ClinicalTrials.gov, Number NCT02548351, https://doi.org/10.1016/j.cgh.2021.07.020) (accessed on 30 July 2021).

The thiazolidinedione class of insulin-sensitizing drugs, including rosiglitazone and pioglitazone, are potent pharmacologic PPARγ agonists. Thiazolidinediones increase plasma adiponectin levels in DM2 and NASH patients [64,65] and levels became similar to the values observed in control subjects. Pioglitazone treatment increases adiponectin concentrations, and improves hepatic insulin resistance and liver histology in NASH [66]. In NASH patients treated with pioglitazone, a reduction was observed in hepatic steatosis but also necroinflammation and fibrosis [67]. It has been suggested that adiponectin may play an important role in mediating the beneficial effects of pioglitazone in NASH patients, inhibiting hepatic fatty acid synthesis, gluconeogenesis and de novo lipogenesis, via activation of AMPK. It also activates PPARα with the stimulation of fatty acid oxidation [68].

Ianifibranor (IVA337) is a next-generation pan-PPAR agonist addressing the pathophysiology of NASH: metabolic, inflammatory and fibrotic. A phase 2b study aiming to evaluate the efficacy and the safety of two doses of IVA337 (800 mg, 1200 mg) per day for 24 weeks versus placebo in adult NASH patients with liver steatosis and moderate to severe necroinflammation without cirrhosis demonstrated that SAF Activity Score (SAF-A) decrease at least 2 points (SAF histological score, calculated as the sum of lobular inflammation score and balloning score) with stable or decreases CRN Fibrosis Score (CRN-F) (Clinical Trial NCT03008070).

3.2. PPARs Expression in Liver Fibrosis

Hepatic fibrosis results from a chronic inflammatory process that affects hepatocytes or biliary cells. Inflammation leads to the activation of effector cells, which results in the accumulation of extracellular matrix components, such as collagens. In liver, HSCs appear to be the primary source of extracellular matrix. These cells change its normal function as a retinol storage cell to a proliferative, contractile and myofibroblastic-like phenotype [69]. Persistent fibrosis leads to cirrhosis—a pathology with an ominous parenchymal lesion and many clinical complications—and even to HCC. Numerous molecular pathways are involved in fibrosis development, but one mainly important is TGF-β pathway. TGF-β is a pleitrotropic cytokine involved as the dominant stimuli for HSCs to produce extracellular matrix (ECM) wound-components and is increased in experimental and clinical fibrosis and its expression is regulated mostly through Smads signaling [70]. A pathway that seems exclusive to liver fibrosis comprises Toll-like receptor 4 (TLR4). TLR4 is activated on HSCs surface by lipopolysaccharides derived from translocated intestinal bacteria, triggering cell activation and fibrogenesis.

3.2.1. PPARα

PPARα is not expressed in rodent or human HSCs [71]. PPARα is poorly expressed in macrophages due to the high levels of IL-1b; but its activation reduces liver inflammation by directly targeting IL-1r antagonist [72]. Oleoylethanolamide, an endocannabinoid-like molecule ameliorated thioacetamide-induced hepatic fibrosis blocking the activation of HSCs inhibiting the expression fibrosis markers, and genes involved in inflammation and extracellular matrix remodeling. These improvements could not be observed in PPARα knockout mice [73].

3.2.2. PPARβ/δ

Contrary to PPARγ role in HSC activation; PPARβ/δ is highly expressed in activated HSCs. In liver injury, PPARβ/δ activation facilitates HSC proliferation by activating p38–JNK–MAPK in CCl4-induced liver fibrosis [74] and augments fibrotic markers expression such as collagen I, α-SMA, TIMPs, and MMPs [75]. PPARβ/δ agonist L165041 and GW501516 increased hepatic expression of fibrosis markers in carbon tetrachloride (CCl4)-injected mice [75,76] and L-165041 increased hepatotoxicity due to HSC activation. Therefore, suppressing PPARβ/δ would be a promising way to avoid fibrosis. PPAR β/δ possesses anti-inflammatory effects in the liver due to direct binding to NF-kB p65 subunit; however, high expression of hepatic proinflammatory factor MCP-1 in CCl4-induced liver disease is associated with PPARβ/δ [77]. PPARβ/δ inhibition reduce liver inflammation through regulation of LPS-mediated TLR4 signaling pathway in cultured hepatocyte cells [78]. PPARβ/δ activation inhibits macrophage activation, showing anti-inflammatory effects [79] and adenoviral over expression of PPARβ/δ in mice decrease JNK signaling and inflammatory markers [80]. Lastly, it has been shown that PPARβ/δ has hepatoprotective effects modulating NF-κB signaling, consequently attenuating CCl_4 hepatotoxicity [81].

3.2.3. PPARγ

PPARγ is a key factor in HSCs activation and fibrosis pathogenesis. PPARγ can regulate the TGF-β/Smads pathway and binds directly to Smad3 and inhibits TGF-β-induced CTGF and α-SMA expression in smooth muscle cells [82]. Several molecules that upregulate PPARγ can inhibit TGF-β production during fibrosis in different tissues [83,84]. PPARγ is involved in HSC transdifferentiation and fibroblast transformation, PPARγ2 is highly expressed in quiescent HSC, and PPARγ is downregulated during HSC activation [85]. Accordingly, PPARγ restoration prompts the change in activated HSC to quiescent HSC and suppresses activity of AP1 [86]. Most PPARγ agonists can reduce hepatic fibrosis by restraining HSC proliferation and driving activated HSC to apoptosis [87]. In addition, PPARγ can reduce the overexpression of α-SMA, type I collagen, and hydroxyproline and thereby inhibit liver fibrosis [88]. PPARγ ameliorate liver fibrosis and inhibit HSC proliferation regulating many transcription factors, such as CCAAT/enhancer binding protein (C/EBP), LXRα and SREBP-1c, which are depleted when HSCs are activated [89]. PPARγ overexpression could directly reverse liver fibrosis in mice fed with a methionine–choline-deficient (MCD) diet by reducing the expression of α-SMA and tissue inhibitors of metalloproteinases (TIMPs) and increasing HSCs cell apoptosis [90]. Capillarization is a term used to describe when liver sinusoidal endothelial cells (LSECs) lack fenestration and develop an organized basement membrane, which is permissive for HSC activation and is a preamble to fibrosis [91]. PPARγ agonist ameliorates LSECs activation and inflammation [92]; while PPARβ/δ and PPARα agonists induce ICAM-1 expression in non-stimulated ECs playing an important role in liver fibrosis [93].

On the other hand, monocyte-derived macrophages and bone-marrow derived macrophages highly express PPARγ [94]. This nuclear factor induces macrophage M2 polarization, and in consequence an anti-inflammatory liver response. In a CCl_4-induced liver damage model, null mice for PPARγ in macrophages and HSC showed aggravated liver necroinflammation and fibrosis compared to Ppary-DHEP mice and control mice demonstrating the important role of alterations in macrophages and HSCs in liver fibrosis [95]. Furthermore, in mice subjected to bile duct ligation rosiglitazone inhibited NF-kB activation and hepatic fibrosis, but these changes disappeared in Ppary-DHEP mice [96]. Crosstalk was observed between PPARγ and FXR in HSC cells, which was involved in regulating inflammation, contributing to the antifibrotic activity of FXR ligands in rodent liver cirrhosis models [97].

In conclusion, the knowledge of PPARs' relationship with HSC activation and inflammation will provide a therapeutic strategy for liver fibrosis.

3.2.4. Clinical Trials of PPAR-Related Drugs in Liver Fibrosis

PPARs play an important role in liver fibrosis, by regulating downstream targeted pathways, such as TGF-β, MAPKs, and NF-κB p65. However, no direct clinical trial is registered in in the official database of the U.S. National Library of Medicine (https:

//clinicaltrials.gov/ct2/home; acceded on 30 July 2021) regarding liver fibrosis, only as part of NASH outcomes.

3.3. PPARs in Hepatocellular Carcinoma

HCC is the most common malignant tumor of the liver, and it is the third leading cause of cancer deaths worldwide [98]. However, the survival of patients with late diagnosis is still limited, as many of the therapies are no longer efficient. Therefore, it is important to search for new therapeutic alternatives that, in conjunction with those mentioned above, might help reduce the incidence of HCC. In the following section the role of PPARs in the development of HCC will be described:

3.3.1. PPARα

The role of PPARα has been debated in the past decade. On one hand, several studies postulate that activation of this transcription factor is fundamental for the development of HCC in a wide variety of experimental animal models and in human HCC cells [99–101]. However, Xiao YB et al. demonstrated, in a total of 804 samples of human HCC, lower expression of PPARα in the nucleus than in those of normal liver tissue; on the other hand, high expression both in nucleus and in cytoplasm of PPARα correlated with a longer survival time of patients with HCC [102].

The differential localization in the nucleus or cytoplasm may be the answer to the pleiotropic effect of PPARα; however, Thomas et al. postulated that a variant transcript of human PPARα lacks full exon 6 due to alternative splicing, generating a truncated PPARα-tr protein lacking ligand binding domain that cannot binds to PPRE, but is capable of autonomously regulating proliferative and proinflammatory genes [103].

In recent years, increasing obesity and diabetes were related to increase in HCC, yet the molecular mechanism correlating both pathologies has not been elucidated. Senni et al. demonstrated that catabolism of fatty acids through β-oxidation is the main mechanism that allows the use of fatty acids in proliferation through the regulation of β-catenin on PPARα [104].

3.3.2. PPARβ/δ

As mentioned above, the functions of PPARβ/δ overlap with those of PPARα in peripheral tissues, while in the liver its functions are more related to the processes regulated by PPARγ. Liu S et al. showed that the overexpression of PPARβ/δ protects the liver of mice from fatty acid overload; in addition, the inflammatory pathways are also decreased, so the risk of developing HCC is probably reduced [80]. On the other hand, Vacca et al. studied the role of this nuclear factor in the modulation of liver proliferation, confirming the low expression of PPARβ/δ in human HCC and the reduced expression of target genes such as Cpt-1 and TGFβ1. They also verified that the PPARβ/δ agonist GW501516 reduces the proliferative potential of Hepa1-6 hepatoma cells [105].

On the other hand, Kim et al. reported metabolic reprogramming in sorafenib-resistant HCC identifying PPARβ/δ as a key regulator of glutamine metabolism and reductive carboxylation, consequently inhibition of PPARβ/δ activity reversed metabolic reprogramming in HCC cells and sensitized them to sorafenib, suggesting PPARβ/δ as a potential therapeutic target [106].

3.3.3. PPARγ

The expression and activation of PPARγ in HCC has been a controversial issue; however, in recent years, Yu et al. demonstrated that PPARγ expression is significantly reduced in tumor tissue compared to non-tumor liver tissue, particularly in early tumors. Lately, this same research group demonstrated that PPARγ activation suppresses migration and invasion of HCC cells, and can inhibit metastasis in an orthotopic HCC model in vivo [99,107].

Recently, Zuo et al. showed that low levels of expression of peroxisome proliferator-activated receptor gamma coactivator 1α (PGC1α) were associated with a poor prognosis in HCC and revealed the molecular mechanism of PGC1α in the metabolism and progression of HCC. PGC1α suppresses HCC cell metastasis by inhibiting the Warburg effect through regulation of the WNT/β-catenin/PDK1 axis, concluding that the tumor suppressor activity of PGC1α depends on PPARγ [108].

Several co-therapies have been developed aimed at modulating the activity of PPARγ Wang et al., proposed that flavonoid avicularin inhibit cell proliferation, migration and invasion, and changes in cell apoptosis and cell cycle, through positive regulation for PPARγ [109]. For its part. telmisartan can modulate the ERK1/2, TAK1 and NF-κB signaling axis, such as agonist of PPARγ, exerting antitumor effects, and increasing tumor sensitivity to sorafenib [110]. Additionally, Abd-El Baset et al. indicated that β-ionone (βI), a cyclic isoprenoid, can regulate the expression of PPARγ, through ofRXR, proposing β-ionone such as a potential chemotherapeutic agent in combination with sorafenib [111]. Finally, it has been shown that simvastatin can inhibit the HIF-1α/PPARγ/PKM2 axis, suppressing PKM2-mediated glycolysis, decreasing cell proliferation, and increasing the expression of apoptotic markers in HCC cells, sensitizing of them to sorafenib treatment [112].

In conclusion, even though the activation and expression of PPARs in HCC development continues to be controversial, in recent years, complementary therapies have been developed that mainly involve PPARα and PPARγ-activation, sensitizing tumor cells to traditional anticancer treatments used in HCC.

3.3.4. Clinical Trials of PPAR-Related Drugs in HCC

Metronomic chemotherapy is a new modality of drug administration in which there is an administration of conventional chemotherapeutic agents at very low doses target activated endothelial cells in tumor, without the risk of developing adverse effects [113]. A prospective one-arm, multicenter phase II clinical trial evaluated the progression-free survival, safety and tolerability of a metronomic chemotherapy, which included capecitabine, rofecoxib (PPARγ agonist, and PPARβ antagonist) and pioglitazone, a PPARγ agonist in 38 patients with non-curative HCC, and the median progression-free survival was 2.7 months, the median overall survival was 6.7 months [114]. As regards side effects, the most common adverse event was edema grade 3, in 66% of patients [114]. This trial offers interesting results about the efficacy of a biotherapy that includes minimal doses of agonists that modulate the response of PPARs, and its safety, in patients with an advanced stage of HCC. Unfortunately, it is one of the few registered clinical trials where this type of therapy is evaluated in HCC patients.

3.4. PPARs in HBV and HCV Infections

Infection with HBV or HCV represents one of the main causes of chronic liver disease in the world. However, in endemic areas, a considerable number of patients are infected with both viruses, mainly as a result of common routes of transmission. Several studies have shown that dual-infected patients have an increased risk of advanced liver disease, fibrosis-cirrhosis, and HCC compared to monoinfected patients. Currently, little is known about the role that PPARs play in the development of the infection [115].

3.4.1. PPARα

PPARα overexpression is characteristic both in the acute and chronic phases of HBV disease; this due to the cccDNA of HBV which has binding sites for global and liver-specific transcription factors such as CCAAT enhancer-binding protein (C/EBP), retinoid X receptor (RXR), and PPARα that bind to enhancer regions I (ENI), Core and Pre-Surface2/Surface promoter proteins [116]. In this manner, after HBV infection there is a PPARα over-expression in hepatocytes characteristic in the G2/M phases of the cellular cycle [117]. Additionally, in the same study performed by Xia et al., they found a negative regulation of TGF-β2 in primary human hepatocytes with HBV infection and TGF-β2 treatment, the levels of PPARα, RXRα, CEBPB mRNA and viral replication decreased significantly [117].

In 2017, Du et al. showed that PPAR agonists such as bezafibrate, fenofibrate and rosiglitazone increase HBV replication, which shows that it is important to analyze viral load in HBV infected patients [118]. Moreover, natural agonists such as resveratrol have a direct effect on Sirtuin-1, promoting PGC1a deacetylation, and this, in turn, supports the transcriptional activity of PPARα, which, according to in vivo and in vitro models, allowed HBV replication [119]. Data of real-time PCR demonstrated that mice knockdown of PPARα or RXRα abolished RSV-induced HBV replication; such a mechanism is clearly dependent on PPARα [119].

HCV infection, through HCV core protein activity, affects the expression and activity of PPARγ in hepatocytes. A decreased expression of this protein is related to the accumulation of lipid droplets in the liver and the eventual development of fatty liver disease [120]. The mechanism is mediated by the HCV core protein, which localizes in the membrane of lipid vesicles and induces hepatic fat accumulation by activating SREBP-1c [120]. One of the mechanisms that explains this effect is through a miRNA. In a study carried out by Shirazaki et al., they infected Huh-7.5 hepatoma cells with a JFH1 strain derived from HCV, finding that this procedure induces the expression of miR-27a [121]. This miRNA targets PPARγ directly, reducing lipid synthesis and increasing lipid secretion; two processes that possibly promote HCV replication and virion efflux [122].

3.4.2. Clinical Trials of PPAR-Related Drugs in Infection HBV/HCV

Despite therapeutic potential of PPARs on HBV/HCV infection, no direct clinical trial is registered in in the official database of the U.S. National Library of Medicine (https://clinicaltrials.gov/ct2/home; acceded on 30 July 2021).

3.5. *PPARs and Their Role in the Development of ALD*

Alcohol is the most socially accepted addictive substance, and its excessive consumption is related to serious health problems [123]. ALD is one of the main causes of death worldwide [124]. This injury is produced by a chronic or binge consumption of ethanol, that is, by ingestion of >40 g or higher of alcohol per day over a prolonged period, or consumption of five standard drinks, 70 g of alcohol in less than 2 h approximately [125]. ALD has a broad spectrum that begins with simple disorders, until more severe forms of liver injury develop. Accumulation of fat in the liver, induced by alcohol consumption (AFL), or steatosis, is the earliest response, and 80–90% of chronic alcohol drinkers develop this damage process; this injury can be reversible through exercise, a low fat-calorie diet consumption, or alcohol withdrawal [125,126]. If noxious stimuli continue, liver damage progresses to an inflammatory lesion, known as alcoholic hepatitis, where only 20–40% of chronic consumers develop it, and can slowly progress to steatohepatitis, fibrosis, cirrhosis, and eventually to HCC [127]. Several risk factors have been identified such as gender, where women are more likely to develop ALD, since there are lower levels of gastric alcohol dehydrogenase, in addition to the presence of a higher proportion of body fat [128,129]. Genetic variants are other risk factors that allow ALD progression, studies demonstrate that variations in patatin-like phospholipase domain-containing protein 3 (PNPLA3), transmembrane 6 superfamily member 2 (TM6SF2) and membrane-bound O-acyltransferase domain-containing protein 7 (MBOAT7) are important genetic determinants of risk and severity of ALD. Although their mechanisms and responses are not entirely clear, mutations in these genetic variants seem to be related to lipid metabolism [125,130,131]. Finally, a co-infection with hepatitis virus B or C can accelerate progression of ALD to liver fibrosis, cirrhosis, or HCC [132].

3.5.1. PPARα

The first alteration that occurs after excessive alcohol is an increase in the proportion of reduced nicotinamide adenine dinucleotide (NADH) and oxidized nicotinamide adenine dinucleotide (NAD+) in hepatocytes [133]. Ingested ethanol is metabolized through the activity of the cytosolic alcohol dehydrogenase enzyme in acetaldehyde, and subsequently in acetate through the participation of the mitochondrial aldehyde dehydrogenase enzyme. Both enzymes use NAD+ as a co-factor, and in response NADH is produced in both steps [134] An increase in NADH levels results in a disruption of mitochondrial β-oxidation of fatty acids, an alteration in energy supply and an increase in fatty acid formation, allowing AFL development [133,134].

Currently, several key molecular mechanisms have been identified as triggers for the AFL development after excessive alcohol intake; one of them is regulated by an increase in the expression of SREBP-1c, and on the other hand, by the decrease in the expression of PPARα. The latter allows ALF generation via fatty acid synthesis induction, and fatty acid-β-oxidation inhibition [135].

Acetaldehyde can inhibit PPARα activity through its covalent binding to the transcription factor, consequently, the binding of PPARα to a specific DNA sequence is also inhibited [136]. On the other hand, alcohol can indirectly block PPARα activation by oxidative response generated by upregulation of cytochrome P450 2E1 activity [137].

In a study carried out by Nakajima et al., it was observed that PPARα knockout mice administered with a liquid diet containing 4% ethanol, exhibited hepatomegaly, inflammation, apoptosis, and fibrosis [138]. RXR function is also affected by the consumption of ethanol. Feeding mice with ethanol showed a decrease in the levels of RXRα protein, which in turn did not allow binding of PPARα/RXRα to DNA, decreasing mRNA for several genes regulated by PPARα, and therefore, the development of steatosis was favored [139].

Information demonstrating that PPARα agonist administration improves hepatic injury induced by alcohol consumption has been generated. In experimental studies with C57BL/6 mice fed with ethanol and treated with WY14643, a PPARα agonist, fat accumulation in the liver was prevented [139,140]. Recently, the hepatoprotective effect of Danshen, a traditional Chinese medicine compound, was evaluated in an experimental model of alcoholic liver damage using male C57BL/6 mice, and in an in vitro model. Danshen was effective in preventing ALD through activation of PPARα and reducing 4-hydroxynonenal levels [141]. Other natural compound that has been shown to be effective in preventing alcoholic liver damage is *Solanum muricatum Ait* (pepino fruit), a common plant cultivated in Taiwan. In an animal model of alcoholic liver damage this compound was effective in improving lipid accumulation induced by ethanol, and the molecular evaluation showed that this response is mediated through induction of hepatic levels of p-AMPK and PPARα; also, this compound reduced SREBP-1c expression, an important hepatic lipogenic enzyme [142].

3.5.2. Clinical Information about PPARs Activity in ALD

Fibrates are PPARα agonists used to treat problems such as dyslipidemia and hypercholesterolemia; however, there is various evidence that demonstrate its effectiveness in reducing alcohol consumption in mice and rats [143,144]. On the other hand, Muñoz et al., in 2020, demonstrated that treatment with Fenofibrate (100 mg/kg) was effective in producing an increase in the expression and activity of the protein alcohol dehydrogenase 1, showing an additional pharmacological mechanism of action to counteract liver damage due to alcohol consumption [145].

Other agonists of these nuclear receptors such as pioglitazone, rosiglitazone, and ciglitazone also have beneficial effects in reducing alcohol consumption [146]; nevertheless, none of the available studies have focused on elucidating the mechanisms by which these agonists can improve liver functionality after chronic damage due to alcohol consumption in humans. Regarding clinical trials, at the present there are no trials registered in in the official database of the U.S. National Library of Medicine (https://clinicaltrials.gov/ct2/home; acceded on 30 July 2021) related with PPARs agonists and ALD patients. This represents an opportunity area to explore the efficacy and safety of PPARα agonist drugs in patients with ALD.

In conclusion, ALD is a pathology responsible for the morbidity and mortality of millions of people around the world. The first harmful response that occurs is steatosis, which occurs in more than 80% of people who consume alcohol in a chronic way. In this phase, the role played by PPARα has allowed the understanding of mechanisms of damage that occurs after alcoholic intake. Agonists of PPARα have demonstrated efficacy at the preclinical level to prevent development of alcoholic liver disease in its most advanced stages; however, it is necessary to continue studying their effects and safety in clinical studies.

4. Conclusions and Perspectives

Liver disease continues to be a challenge to health systems worldwide. In previous years, the search for new therapeutic strategies was focused on the study of fibrogenic processes, and the role of HSC. However, in recent years the efforts of liver disease professionals have focused on the study of early stages of the disease, where accumulation of lipids and metabolic alterations are key processes for the development of these diseases. Figure 3 and Table 2 summarize the role of PPARs as metabolic sensors in different liver diseases.

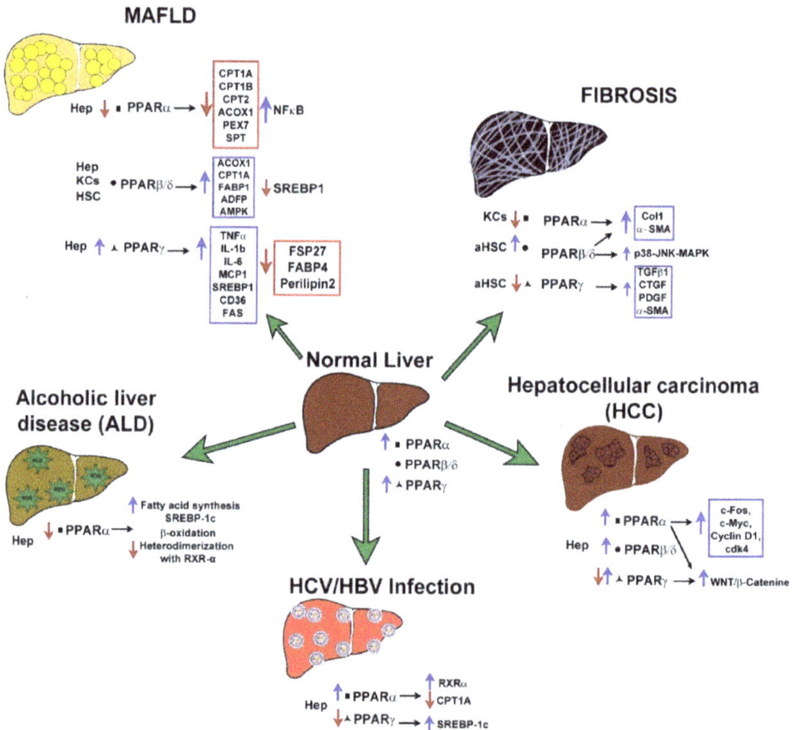

Figure 3. Main molecular targets regulated by PPARs in liver diseases addressed in this review: metabolism-associated fatty liver disease (MAFLD), alcoholic liver disease (ALD), fibrosis, HBV viral hepatitis or HCV infection and hepatocellular carcinoma (HCC). Blue arrows indicate over-activation, while red arrows indicate downregulation of triggered responses. Hep, hepatocytes; KCs, Kupffer Cells; HSCs, Hepatic Stellate Cells.

Table 2. Role of PPARs in liver diseases.

Liver Disease	Expression		Function	Mutation	Reference
MAFLD	Hepatocytes Kupffer Cells Hepatic Stellate Cells	↓ PPARα ↑ PARβ/δ ↑ PPARγ ↑ PARβ/δ ↑ PPARγ	PPARα: Induces lipogenesis PPARβ/δ: Augments liver fat content and decreases insulin sensitivity PPARγ: Promotes steatosis	PPARA: CM003689 association with elevated plasma lipid concentration in diabetes CM025499 CM025500 associated with diabetes PPARG: CM981614, CM981615, CM1617313 associated with Obesity CM066185, CM066187, CM066186, CM066188, CD066392, CX022192 Associated with IR CR032439 association with increased height/lipid metabolism CR057908 association with increased body weight	[25,47,62] The Human Gene Mutation Database, consulted July 2021
Fibrosis	Kupffer Cells Hepatic Stellate Cells (activated)	↓ PPARα ↑ PPARβ/δ ↓ PPARγ	PPARα: Increases oxidative stress and inflammation PPARβ/δ: Facilitates HSC activation PPARγ: key factor in HSCs activation and regulation of inflammation	No mutations associated with liver fibrosis	[73,74,90]
Hepatocellular carcinoma	Hepatocytes	↑ PPARα ↑ PARβ/δ ↓↑PPARγ	PPARα: Regulates expression of B-catenin, c-Fos, c-Myc, Cyclin D1. PPARγ: Modulates activity of p53, ERK1/2, TAK1 and NF-κB	PPARG: R280C, C285Y, Q286P, F287Y, R288C, R288H S289C mutations are potential loss offunction mutations in various aspects including ligand binding for PPARγ activation	[104–106,147]
HBV and HCV infections	Hepatocytes	↑ PPARα ↓ PPARγ	PPARα: Increase fatty acid Synthesis PPARγ: Decrease β-oxidation	No mutations associated with HBV, or HCV infection	[116–119]
Alcoholic liver disease	Hepatocytes	↑ PPARα	PPARα: Increase fatty acid Synthesis Decrease β-oxidation	No mutations associated with alcoholic liver disease	[134–137]

Since the first description of PPARs [148], our knowledge about these nuclear factors has been increasing. At first, PPARs were only considered as regulators of lipid metabolism; however, currently they are considered the main hepatic metabolic mediators, having an important role in various processes such as: cell survival, regulation of ubiquitination, adipocyte differentiation, regulation of thermogenesis and gluconeogenesis mediators. Taking the above into account, the design and study of new pharmacological therapies for the treatment of liver diseases should be aimed at modulating the activity of these transcription factors.

Finally, the use of liver-specific PPAR-null mice has opened the possibility of studying other important mechanisms in which PPARs are involved [149], mainly as mediators of epigenetic regulation mechanisms through their interaction with enzymes such as Sirtuin-1 [32,150], the regulation of PPAR promoters, through DNA methyltransferases (DNMTs) [151], and the regulation of their expression through a variety of microRNAs [152].

Author Contributions: H.C.M.-R. contributed to planning, bibliographic revision, writing of the manuscript, and to figures design; M.G.-M., A.S.-R., and A.S. contributed to the writing of the manuscript and literature review; A.M.-R. contributed to figures design and to the writing of the manuscript; J.A.-B. was responsible for the manuscript planning and revising. All authors have read and agreed to the published version of the manuscript.

Funding: This work was supported in part by Fondo de Desarrollo Cientifico de Jalisco, FODECIJAL, 7941-2019 awarded to J.A.-B. and PRODEP-SEP. Apoyo a la Incorporacion de Nuevos Profesores de Tiempo Completo, Number PTC-1565 to H.C.M-R.

Institutional Review Board Statement: Not applicable.

Informed Consent Statement: Not applicable.

Data Availability Statement: The data used to support this study are included within article as references.

Conflicts of Interest: The authors declare no conflict of interest.

References

1. Trefts, E.; Gannon, M.; Wasserman, D.H. The liver. *Curr. Biol.* **2017**, *27*, R1147–R1151. [CrossRef]
2. Jones, J.G. Hepatic glucose and lipid metabolism. *Diabetology* **2016**, *59*, 1098–1103. [CrossRef]
3. Ding, H.-R.; Wang, J.-L.; Ren, H.-Z.; Shi, X.-L. Lipometabolism and Glycometabolism in Liver Diseases. *BioMed Res. Int.* **2018**, *2018*, 1287127. [CrossRef]
4. Tanaka, N.; Aoyama, T.; Kimura, S.; Gonzalez, F.J. Targeting nuclear receptors for the treatment of fatty liver disease. *Pharmacol. Ther.* **2017**, *179*, 142–157. [CrossRef]
5. Tailleux, A.; Wouters, K.; Staels, B. Roles of PPARs in NAFLD: Potential therapeutic targets. *Biochim. Biophys. Acta (BBA) Mol. Cell Biol. Lipids* **2012**, *1821*, 809–818. [CrossRef] [PubMed]
6. Chinetti, G.; Fruchart, J.-C.; Staels, B. Peroxisome proliferator-activated receptors (PPARs): Nuclear receptors at the crossroads between lipid metabolism and inflammation. *Inflamm. Res.* **2000**, *49*, 497–505. [CrossRef]
7. Darbre, P.D. Disruption of Other Receptor Systems: Progesterone and Glucocorticoid Receptors, Peroxisome Proliferator-Activated Receptors, Pregnane X Receptor, and Aryl Hydrocarbon Receptor. In *Endocrine Disruption and Human Health*, 1st ed.; Philippa, D., Ed.; Academic Press: London, UK, 2015; pp. 111–122. [CrossRef]
8. Lee, W.-S.; Kim, J. Peroxisome Proliferator-Activated Receptors and the Heart: Lessons from the Past and Future Directions. *PPAR Res.* **2015**, *2015*, 1–18. [CrossRef]
9. Tyagi, S.; Gupta, P.; Saini, A.S.; Kaushal, C.; Sharma, S. The peroxisome proliferator-activated receptor: A family of nuclear receptors role in various diseases. *J. Adv. Pharm. Technol. Res.* **2011**, *2*, 236–240. [CrossRef]
10. Grygiel-Górniak, B. Peroxisome proliferator-activated receptors and their ligands: Nutritional and clinical implications—A review. *Nutr. J.* **2014**, *13*, 17. [CrossRef] [PubMed]
11. Guan, Y. Peroxisome Proliferator-Activated Receptor Family and Its Relationship to Renal Complications of the Metabolic Syndrome. *J. Am. Soc. Nephrol.* **2004**, *15*, 2801–2815. [CrossRef] [PubMed]
12. Berger, J.; Moller, D.E. The Mechanisms of Action of PPARs. *Annu. Rev. Med.* **2002**, *53*, 409–435. [CrossRef]
13. Usuda, D.; Kanda, T. Peroxisome proliferator-activated receptors for hypertension. *World J. Cardiol.* **2014**, *6*, 744–754. [CrossRef] [PubMed]
14. Korbecki, J.; Bobiński, R.; Dutka, M. Self-regulation of the inflammatory response by peroxisome proliferator-activated receptors. *Inflamm. Res.* **2019**, *68*, 443–458. [CrossRef] [PubMed]
15. Rakhshandehroo, M.; Knoch, B.; Muller, M.; Kersten, S. Peroxisome Proliferator-Activated Receptor Alpha Target Genes. *PPAR Res.* **2010**, *2010*, 612089. [CrossRef] [PubMed]
16. Reilly, S.; Lee, C.-H. PPARδ as a therapeutic target in metabolic disease. *FEBS Lett.* **2008**, *582*, 26–31. [CrossRef] [PubMed]
17. Wagner, N.; Wagner, K.D. PPAR Beta/Delta and the Hallmarks of Cancer. *Cells* **2020**, *9*, 1133. [CrossRef] [PubMed]
18. Heming, M.; Gran, S.; Jauch, S.-L.; Fischer-Riepe, L.; Russo, A.; Klotz, L.; Hermann, S.; Schäfers, M.; Roth, J.; Barczyk-Kahlert, K. Peroxisome Proliferator-Activated Receptor-γ Modulates the Response of Macrophages to Lipopolysaccharide and Glucocorticoids. *Front. Immunol.* **2018**, *9*, 893–909. [CrossRef] [PubMed]
19. Hernandez-Quiles, M.; Broekema, M.F.; Kalkhoven, E. PPARgamma in Metabolism, Immunity, and Cancer: Unified and Diverse Mechanisms of Action. *Front. Endocrinol.* **2021**, *12*, 624112. [CrossRef]

20. Kim, T.-H.; Kim, M.-Y.; Jo, S.-H.; Park, J.-M.; Ahn, Y.-H. Modulation of the Transcriptional Activity of Peroxisome Proliferator-Activated Receptor Gamma by Protein-Protein Interactions and Post-Translational Modifications. *Yonsei Med. J.* **2013**, *54*, 545–559. [CrossRef]
21. Sayiner, M.; Koenig, A.; Henry, L.; Younossi, Z.M. Epidemiology of Nonalcoholic Fatty Liver Disease and Nonalcoholic Steatohepatitis in the United States and the Rest of the World. *Clin. Liver Dis.* **2016**, *20*, 205–214. [CrossRef]
22. Eslam, M.; Newsome, P.N.; Sarin, S.K.; Anstee, Q.M.; Targher, G.; Romero-Gomez, M.; Zelber-Sagi, S.; Wong, V.W.-S.; Dufour, J.-F.; Schattenberg, J.M.; et al. A new definition for metabolic dysfunction-associated fatty liver disease: An international expert consensus statement. *J. Hepatol.* **2020**, *73*, 202–209. [CrossRef] [PubMed]
23. Masarone, M.; Rosato, V.; Dallio, M.; Gravina, A.G.; Aglitti, A.; Loguercio, C.; Federico, A.; Persico, M. Role of Oxidative Stress in Pathophysiology of Nonalcoholic Fatty Liver Disease. *Oxid. Med. Cell. Longev.* **2018**, *2018*, 1–14. [CrossRef] [PubMed]
24. Terlecky, S.R.; Terlecky, L.J.; Giordano, C.R. Peroxisomes, oxidative stress, and inflammation. *World J. Biol. Chem.* **2012**, *3*, 93–97. [CrossRef]
25. Cave, M.C.; Clair, H.B.; Hardesty, J.E.; Falkner, K.C.; Feng, W.; Clark, B.J.; Sidey, J.; Shi, H.; Aqel, B.A.; McClain, C.J.; et al. Nuclear receptors and nonalcoholic fatty liver disease. *Biochim. Biophys. Acta (BBA) Gene Regul. Mech.* **2016**, *1859*, 1083–1099. [CrossRef] [PubMed]
26. Qi, C.; Zhu, Y.; Reddy, J.K. Peroxisome Proliferator-Activated Receptors, Coactivators, and Downstream Targets. *Cell Biochem. Biophys.* **2000**, *32*, 187–204. [CrossRef]
27. Pawlak, M.; Baugé, E.; Bourguet, W.; De Bosscher, K.; Lalloyer, F.; Tailleux, A.; Lebherz, C.; Lefebvre, P.; Staels, B. The transrepressive activity of peroxisome proliferator-activated receptor alpha is necessary and sufficient to prevent liver fibrosis in mice. *Hepatology* **2014**, *60*, 1593–1606. [CrossRef]
28. Evans, R.M.; Barrish, G.D.; Wang, Y.-X. PPARs and the complex journey to obesity. *Nat. Med.* **2004**, *10*, 355–361. [CrossRef] [PubMed]
29. Delerive, P.; De Bosscher, K.; Besnard, S.; Berghe, W.V.; Peters, J.M.; Gonzalez, F.J.; Fruchart, J.-C.; Tedgui, A.; Haegeman, G.; Staels, B. Peroxisome Proliferator-activated Receptor α Negatively Regulates the Vascular Inflammatory Gene Response by Negative Cross-talk with Transcription Factors NF-κB and AP-1. *J. Biol. Chem.* **1999**, *274*, 32048–32054. [CrossRef] [PubMed]
30. Francque, S.; Verrijken, A.; Caron, S.; Prawitt, J.; Paumelle, R.; Derudas, B.; Lefebvre, P.; Taskinen, M.-R.; Van Hul, W.; Mertens, I.; et al. PPARα gene expression correlates with severity and histological treatment response in patients with non-alcoholic steatohepatitis. *J. Hepatol.* **2015**, *63*, 164–173. [CrossRef]
31. Park, H.-S.; Jang, J.E.; Ko, M.S.; Woo, S.H.; Kim, B.J.; Kim, H.S.; Park, H.S.; Park, I.-S.; Koh, E.H.; Lee, K.-U. Statins Increase Mitochondrial and Peroxisomal Fatty Acid Oxidation in the Liver and Prevent Non-Alcoholic Steatohepatitis in Mice. *Diabetes Metab. J.* **2016**, *40*, 376–385. [CrossRef]
32. Sandoval-Rodriguez, A.; Monroy-Ramirez, H.C.; Meza-Rios, A.; Garcia-Bañuelos, J.; Vera-Cruz, J.; Gutiérrez-Cuevas, J.; Silva-Gomez, J.; Staels, B.; Dominguez-Rosales, J.; Galicia-Moreno, M.; et al. Pirfenidone Is an Agonistic Ligand for PPARα and Improves NASH by Activation of SIRT1/LKB1/pAMPK. *Hepatol. Commun.* **2020**, *4*, 434–449. [CrossRef]
33. Paintlia, A.S.; Paintlia, M.K.; Singh, A.K.; Singh, I. Modulation of Rho-Rock signaling pathway protects oligodendrocytes against cytokine toxicity via PPAR-α-dependent mechanism. *Glia* **2013**, *61*, 1500–1517. [CrossRef]
34. Sumida, Y.; Niki, E.; Naito, Y.; Yoshikawa, T. Involvement of free radicals and oxidative stress in NAFLD/NASH. *Free. Radic. Res.* **2013**, *47*, 869–880. [CrossRef]
35. Brandt, J.M.; Djouadi, F.; Kelly, D.P. Fatty Acids Activate Transcription of the Muscle Carnitine Palmitoyltransferase I Gene in Cardiac Myocytes via the Peroxisome Proliferator-activated Receptor α. *J. Biol. Chem.* **1998**, *273*, 23786–23792. [CrossRef]
36. Patsouris, D.; Reddy, J.K.; Muller, M.; Kersten, S. Peroxisome Proliferator-Activated Receptor α Mediates the Effects of High-Fat Diet on Hepatic Gene Expression. *Endocrinology* **2006**, *147*, 1508–1516. [CrossRef] [PubMed]
37. Kim, S.; Sohn, I.; Ahn, J.-I.; Lee, K.-H.; Lee, Y.S.; Lee, Y.S. Hepatic gene expression profiles in a long-term high-fat diet-induced obesity mouse model. *Gene* **2004**, *340*, 99–109. [CrossRef] [PubMed]
38. Redonnet, A.; Groubet, R.; L-Suberville, C.N.; Bonilla, S.; Martínez, A.; Higueret, P. Exposure to an obesity-inducing diet early affects the pattern of expression of peroxisome proliferator, retinoic acid, and triiodothyronine nuclear receptors in the rat. *Metabolism* **2001**, *50*, 1161–1167. [CrossRef] [PubMed]
39. Abdelmegeed, M.A.; Yoo, S.-H.; Henderson, L.E.; Gonzalez, F.J.; Woodcroft, K.J.; Song, B.-J. PPAR Expression Protects Male Mice from High Fat–Induced Nonalcoholic Fatty Liver. *J. Nutr.* **2011**, *141*, 603–610. [CrossRef] [PubMed]
40. Lalloyer, F.; Wouters, K.; Baron, M.; Caron, S.; Vallez, E.; Vanhoutte, J.; Baugé, E.; Shiri-Sverdlov, R.; Hofker, M.; Staels, B.; et al. Peroxisome Proliferator–Activated Receptor-α Gene Level Differently Affects Lipid Metabolism and Inflammation in Apolipoprotein E2 Knock-In Mice. *Arterioscler. Thromb. Vasc. Biol.* **2011**, *31*, 1573–1579. [CrossRef] [PubMed]
41. Stienstra, R.; Mandard, S.; Patsouris, D.; Maass, C.; Kersten, S.; Muller, M. Peroxisome Proliferator-Activated Receptor α Protects against Obesity-Induced Hepatic Inflammation. *Endocrinology* **2007**, *148*, 2753–2763. [CrossRef]
42. Ip, E.; Farrell, G.; Hall, P.; Robertson, G.; Leclercq, I. Administration of the potent PPAR? agonist, Wy-14,643, reverses nutritional fibrosis and steatohepatitis in mice. *Hepatology* **2004**, *39*, 1286–1296. [CrossRef] [PubMed]
43. Shiri-Sverdlov, R.; Wouters, K.; van Gorp, P.; Gijbels, M.J.; Noel, B.; Buffat, L.; Staels, B.; Maeda, N.; van Bilsen, M.; Hofker, M.H. Early diet-induced non-alcoholic steatohepatitis in APOE2 knock-in mice and its prevention by fibrates. *J. Hepatol.* **2006**, *44*, 732–741. [CrossRef]

44. Pawlak, M.; Lefebvre, P.; Staels, B. General molecular biology and architecture of nuclear receptors. *Curr. Top. Med. Chem.* **2012**, *12*, 486–504. [CrossRef] [PubMed]
45. Nagasawa, T.; Inada, Y.; Nakano, S.; Tamura, T.; Takahashi, T.; Maruyama, K.; Yamazaki, Y.; Kuroda, J.; Shibata, N. Effects of bezafibrate, PPAR pan-agonist, and GW501516, PPARδ agonist, on development of steatohepatitis in mice fed a methionine- and choline-deficient diet. *Eur. J. Pharmacol.* **2006**, *536*, 182–191. [CrossRef]
46. Oliver, W.R., Jr.; Shenk, J.L.; Snaith, M.R.; Russell, C.S.; Plunket, K.D.; Bodkin, N.L.; Lewis, M.C.; Winegar, D.A.; Sznaidman, M.L.; Lambert, M.H.; et al. A selective peroxisome proliferator-activated receptor agonist promotes reverse cholesterol transport. *Proc. Natl. Acad. Sci. USA* **2001**, *98*, 5306–5311. [CrossRef]
47. Tanaka, T.; Yamamoto, J.; Iwasaki, S.; Asaba, H.; Hamura, H.; Ikeda, Y.; Watanabe, M.; Magoori, K.; Ioka, R.X.; Tachibana, K.; et al. Activation of peroxisome proliferator-activated receptor induces fatty acid -oxidation in skeletal muscle and attenuates metabolic syndrome. *Proc. Natl. Acad. Sci. USA* **2003**, *100*, 15924–15929. [CrossRef]
48. Bojic, L.A.; Telford, D.E.; Fullerton, M.D.; Ford, R.J.; Sutherland, B.G.; Edwards, J.Y.; Sawyez, C.G.; Gros, R.; Kemp, B.; Steinberg, G.; et al. PPARδ activation attenuates hepatic steatosis in Ldlr mice by enhanced fat oxidation, reduced lipogenesis, and improved insulin sensitivity. *J. Lipid Res.* **2014**, *55*, 1254–1266. [CrossRef] [PubMed]
49. Risérus, U.; Sprecher, D.; Johnson, T.; Olson, E.; Hirschberg, S.; Liu, A.; Fang, Z.; Hegde, P.; Richards, D.; Sarov-Blat, L.; et al. Activation of Peroxisome Proliferator-Activated Receptor (PPAR) Promotes Reversal of Multiple Metabolic Abnormalities, Reduces Oxidative Stress, and Increases Fatty Acid Oxidation in Moderately Obese Men. *Diabetes* **2008**, *57*, 332–339. [CrossRef] [PubMed]
50. Bays, H.E.; Schwartz, S.; Littlejohn, T., 3rd; Kerzner, B.; Krauss, R.M.; Karpf, D.B.; Choi, Y.-J.; Wang, X.; Naim, S.; Roberts, B.K. MBX-8025, A Novel Peroxisome Proliferator Receptor-δ Agonist: Lipid and Other Metabolic Effects in Dyslipidemic Overweight Patients Treated with and without Atorvastatin. *J. Clin. Endocrinol. Metab.* **2011**, *96*, 2889–2897. [CrossRef]
51. Qin, X.; Xie, X.; Fan, Y.; Tian, J.; Guan, Y.; Wang, X.; Zhu, Y.; Wang, N. Peroxisome proliferator-activated receptor-δ induces insulin-induced gene-1 and suppresses hepatic lipogenesis in obese diabetic mice. *Hepatology* **2008**, *48*, 432–441. [CrossRef]
52. Liu, S.; Brown, J.D.; Stanya, K.; Homan, E.A.; Leidl, M.; Inouye, K.; Bhargava, P.; Gangl, M.R.; Dai, L.; Hatano, B.; et al. A diurnal serum lipid integrates hepatic lipogenesis and peripheral fatty acid use. *Nature* **2013**, *502*, 550–554. [CrossRef]
53. Lee, C.-H.; Olson, P.; Hevener, A.; Mehl, I.; Chong, L.-W.; Olefsky, J.M.; Gonzalez, F.J.; Ham, J.; Kang, H.; Peters, J.; et al. PPAR regulates glucose metabolism and insulin sensitivity. *Proc. Natl. Acad. Sci. USA* **2006**, *103*, 3444–3449. [CrossRef]
54. Pettinelli, P.; Videla, L.A. Up-Regulation of PPAR-γ mRNA Expression in the Liver of Obese Patients: An Additional Reinforcing Lipogenic Mechanism to SREBP-1c Induction. *J. Clin. Endocrinol. Metab.* **2011**, *96*, 1424–1430. [CrossRef]
55. Nakamuta, M.; Kohjima, M.; Morizono, S.; Kotoh, K.; Yoshimoto, T.; Miyagi, I.; Enjoji, M. Evaluation of fatty acid metabo-lism-related gene expression in nonalcoholic fatty liver disease. *Int. J. Mol. Med.* **2005**, *16*, 631–635. [CrossRef] [PubMed]
56. Gavrilova, O.; Haluzik, M.; Matsusue, K.; Cutson, J.J.; Johnson, L.K.; Dietz, K.; Nicol, C.J.; Vinson, C.; Gonzalez, F.J.; Reitman, M. Liver Peroxisome Proliferator-activated Receptor γ Contributes to Hepatic Steatosis, Triglyceride Clearance, and Regulation of Body Fat Mass. *J. Biol. Chem.* **2003**, *278*, 34268–34276. [CrossRef] [PubMed]
57. Ying, S.; Xiao, X.; Chen, T.; Lou, J. PPAR Ligands Function as Suppressors That Target Biological Actions of HMGB1. *PPAR Res.* **2016**, *2016*, 2612743. [CrossRef] [PubMed]
58. Necela, B.M.; Su, W.; Thompson, E.A. Toll-like receptor 4 mediates cross-talk between peroxisome proliferator-activated receptor γ and nuclear factor-κB in macrophages. *Immunology* **2008**, *125*, 344–358. [CrossRef]
59. Bouhlel, M.A.; Derudas, B.; Rigamonti, E.; Dièvart, R.; Brozek, J.; Haulon, S.; Zawadzki, C.; Jude, B.; Torpier, G.; Marx, N.; et al. PPARγ Activation Primes Human Monocytes into Alternative M2 Macrophages with Anti-inflammatory Properties. *Cell Metab.* **2007**, *6*, 137–143. [CrossRef]
60. Matsusue, K.; Haluzik, M.; Lambert, G.; Yim, S.H.; Gavrilova, O.; Ward, J.M.; Brewer, B., Jr.; Reitman, M.L.; Gonzalez, F.J. Li-ver-specific disruption of PPARgamma in leptin-deficient mice improves fatty liver but aggravates diabetic phenotypes. *J. Clin. Investig.* **2003**, *111*, 737–747. [CrossRef] [PubMed]
61. Yu, S.; Matsusue, K.; Kashireddy, P.; Cao, W.-Q.; Yeldandi, V.; Yeldandi, A.V.; Rao, M.S.; Gonzalez, F.J.; Reddy, J.K. Adipocyte-specific Gene Expression and Adipogenic Steatosis in the Mouse Liver Due to Peroxisome Proliferator-activated Receptor γ1 (PPARγ1) Overexpression. *J. Biol. Chem.* **2003**, *278*, 498–505. [CrossRef]
62. Ye, R.; Scherer, P.E. Adiponectin, driver or passenger on the road to insulin sensitivity? *Mol. Metab.* **2013**, *2*, 133–141. [CrossRef] [PubMed]
63. Ratziu, V.; Harrison, S.A.; Francque, S.; Bedossa, P.; Lehert, P.; Serfaty, L.; Romero-Gomez, M.; Boursier, J.; Abdelmalek, M.; Caldwell, S.; et al. Elafibranor, an Agonist of the Peroxisome Proliferator−Activated Receptor−α and −δ, Induces Resolution of Nonalcoholic Steatohepatitis without Fibrosis Worsening. *Gastroenterology* **2016**, *150*, 1147–1159.e5. [CrossRef]
64. Gastaldelli, A.; Miyazaki, Y.; Mahankali, A.; Berria, R.; Pettiti, M.; Buzzigoli, E.; Ferrannini, E.; DeFronzo, R.A. The Effect of Pioglitazone on the Liver: Role of adiponectin. *Diabetes Care* **2006**, *29*, 2275–2281. [CrossRef] [PubMed]
65. Caldwell, S.H.; Argo, C.K.; Al-Osaimi, A.M. Therapy of NAFLD: Insulin sensitizing agents. *J. Clin. Gastroenterol.* **2006**, *40* (Suppl. 1), S61–S66. [PubMed]
66. Aithal, G.P.; Thomas, J.; Kaye, P.V.; Lawson, A.; Ryder, S.D.; Spendlove, I.; Austin, A.S.; Freeman, J.G.; Morgan, L.; Webber, J. Randomized, Placebo-Controlled Trial of Pioglitazone in Nondiabetic Subjects with Nonalcoholic Steatohepatitis. *Gastroenterology* **2008**, *135*, 1176–1184. [CrossRef]

67. Gastaldelli, A.; Harrison, S.; Belfort-Aguiar, R.; Hardies, J.; Balas, B.; Schenker, S.; Cusi, K. Pioglitazone in the treatment of NASH: The role of adiponectin. *Aliment. Pharmacol. Ther.* **2010**, *32*, 769–775. [CrossRef] [PubMed]
68. Viollet, B.; Foretz, M.; Guigas, B.; Horman, S.; Dentin, R.; Bertrand, L.; Hue, L.; Andreelli, F. Activation of AMP-activated protein kinase in the liver: A new strategy for the mana-gement of metabolic hepatic disorders. *J. Physiol.* **2006**, *574*, 41–53. [CrossRef] [PubMed]
69. Higashi, T.; Friedman, S.L.; Hoshida, Y. Hepatic stellate cells as key target in liver fibrosis. *Adv. Drug Deliv. Rev.* **2017**, *121*, 27–42. [CrossRef]
70. Massagué, J.; Chen, Y.G. Controlling TGF-beta signaling. *Genes Dev.* **2000**, *14*, 627–644. [CrossRef]
71. Mandard, S.; Müller, M.; Kersten, S. Peroxisome proliferator-activated receptor a target genes. *Cell. Mol. Life Sci.* **2004**, *61*, 393–416. [CrossRef]
72. Stienstra, R.; Saudale, F.; Duval, C.; Keshtkar, S.; Groener, J.E.M.; Van Rooijen, N.; Staels, B.; Kersten, S.; Muller, M. Kupffer cells promote hepatic steatosis via interleukin-1β-dependent suppression of peroxisome proliferator-activated receptor α activity. *Hepatology* **2009**, *51*, 511–522. [CrossRef]
73. Chen, L.; Li, L.; Chen, J.; Li, L.; Zheng, Z.; Ren, J.; Qiu, Y. Oleoylethanolamide, an endogenous PPAR-α ligand, atte-nuates liver fibrosis targeting hepatic stellate cells. *Oncotarget* **2015**, *6*, 42530–42540. [CrossRef] [PubMed]
74. Kostadinova, R.; Montagner, A.; Gouranton, E.; Fleury, S.; Guillou, H.; Dombrowicz, D.; Desreumaux, P.; Wahli, W. GW501516-activated PPARβ/δ promotes liver fibrosis via p38-JNK MAPK-induced hepatic stellate cell proliferation. *Cell Biosci.* **2012**, *2*, 34. [CrossRef]
75. Hellemans, K.; Michalik, L.; Dittie, A.; Knorr, A.; Rombouts, K.; de Jong, J.; Heirman, C.; Quartier, E.; Schuit, F.; Wahli, W.; et al. Peroxisome proliferator-activated receptor-β signaling contributes to enhanced proliferation of hepatic stellate cells. *Gastroenterology* **2003**, *124*, 184–201. [CrossRef] [PubMed]
76. Iwaisako, K.; Haimerl, M.; Paik, Y.-H.; Taura, K.; Kodama, Y.; Sirlin, C.; Yu, E.; Yu, R.T.; Downes, M.; Evans, R.M.; et al. Protection from liver fibrosis by a peroxisome proliferator-activated receptor agonist. *Proc. Natl. Acad. Sci. USA* **2012**, *109*, E1369–E1376. [CrossRef]
77. Marsillach, J.; Camps, J.; Ferré, N.; Beltran, R.; Rull, A.; Mackness, B.; Mackness, M.; Joven, J. Paraoxonase-1 is related to inflammation, fibrosis and PPAR delta in experimental liver disease. *BMC Gastroenterol.* **2009**, *9*, 3. [CrossRef]
78. Li, Y.; Wang, C.; Lu, J.; Huang, K.; Han, Y.; Chen, J.; Yang, Y.; Liu, B. PPAR δ inhibition protects against palmitic acid-LPS induced lipidosis and injury in cultured hepatocyte L02 cell. *Int. J. Med. Sci.* **2019**, *16*, 1593–1603. [CrossRef]
79. Kang, K.; Reilly, S.; Karabacak, V.; Gangl, M.R.; Fitzgerald, K.; Hatano, B.; Lee, C.-H. Adipocyte-Derived Th2 Cytokines and Myeloid PPARδ Regulate Macrophage Polarization and Insulin Sensitivity. *Cell Metab.* **2008**, *7*, 485–495. [CrossRef] [PubMed]
80. Liu, S.; Hatano, B.; Zhao, M.; Yen, C.-C.; Kang, K.; Reilly, S.; Gangl, M.R.; Gorgun, C.; Balschi, J.A.; Ntambi, J.M.; et al. Role of Peroxisome Proliferator-activated Receptor δ/β in Hepatic Metabolic Regulation. *J. Biol. Chem.* **2011**, *286*, 1237–1247. [CrossRef] [PubMed]
81. Shan, W.; Palkar, P.S.; Murray, I.A.; McDevitt, E.I.; Kennett, M.J.; Kang, B.H.; Isom, H.C.; Perdew, G.H.; Gonzalez, F.J.; Peters, J.M. Ligand Activation of Peroxisome Proliferator–Activated Receptor β/δ (PPARβ/δ) Attenuates Carbon Tetrachloride Hepatotoxicity by Downregulating Proinflammatory Gene Expression. *Toxicol. Sci.* **2008**, *105*, 418–428. [CrossRef]
82. Fu, M.; Zhang, J.; Zhu, X.; Myles, D.E.; Willson, T.M.; Liu, X.; Chen, Y.E. Peroxisome Proliferator-activated Receptor γ Inhibits Transforming Growth Factor β-induced Connective Tissue Growth Factor Expression in Human Aortic Smooth Muscle Cells by Interfering with Smad3. *J. Biol. Chem.* **2001**, *276*, 45888–45894. [CrossRef]
83. Saidi, A.; Kasabova, M.; Vanderlynden, L.; Wartenberg, M.; Kara-Ali, G.H.; Marc, D.; Lecaille, F.; Lalmanach, G. Curcumin inhibits the TGF-β1-dependent differentiation of lung fibroblasts via PPARγ-driven upregulation of cathepsins B and L. *Sci. Rep.* **2019**, *9*, 491. [CrossRef]
84. Xia, Y.; Li, J.; Chen, K.; Feng, J.; Guo, C. Bergenin Attenuates Hepatic Fibrosis by Regulating Autophagy Mediated by the PPAR-γ/TGF-β Pathway. *PPAR Res.* **2020**, *2020*, 1–13. [CrossRef]
85. Zhang, F.; Kong, D.; Lu, Y.; Zheng, S. Peroxisome proliferator-activated receptor-γ as a therapeutic target for hepatic fibrosis: From bench to bedside. *Cell. Mol. Life Sci.* **2013**, *70*, 259–276. [CrossRef]
86. Hazra, S.; Xiong, S.; Wang, J.; Rippe, R.A.; Krishna, V.; Chatterjee, K.; Tsukamoto, H. Peroxisome Proliferator-activated Receptor γ Induces a Phenotypic Switch from Activated to Quiescent Hepatic Stellate Cells. *J. Biol. Chem.* **2004**, *279*, 11392–11401. [CrossRef]
87. He, J.; Hong, B.; Bian, M.; Jin, H.; Chen, J.; Shao, J.; Zhang, F.; Zheng, S. Docosahexaenoic acid inhibits hepatic stellate cell activation to attenuate liver fibrosis in a PPARγ-dependent manner. *Int. Immunopharmacol.* **2019**, *75*, 105816. [CrossRef]
88. Panebianco, C.; Oben, J.A.; Vinciguerra, M.; Pazienza, V. Senescence in hepatic stellate cells as a mechanism of liver fibrosis reversal: A putative synergy between retinoic acid and PPAR-gamma signalings. *Clin. Exp. Med.* **2017**, *17*, 269–280. [CrossRef]
89. She, H.; Xiong, S.; Hazra, S.; Tsukamoto, H. Adipogenic Transcriptional Regulation of Hepatic Stellate Cells. *J. Biol. Chem.* **2005**, *280*, 4959–4967. [CrossRef]
90. Yu, J.; Zhang, S.; Chu, E.S.; Go, M.Y.; Lau, R.H.; Zhao, J.; Wu, C.-W.; Tong, L.; Zhao, J.; Poon, T.C.; et al. Peroxisome proliferator-activated receptors gamma reverses hepatic nutritional fibrosis in mice and suppresses activation of hepatic stellate cells in vitro. *Int. J. Biochem. Cell Biol.* **2010**, *42*, 948–957. [CrossRef]
91. Deleve, L.D. Liver sinusoidal endothelial cells in hepatic fibrosis. *Hepatology* **2015**, *61*, 1740–1746. [CrossRef] [PubMed]

92. Duan, S.Z.; Usher, M.G.; Mortensen, R.M. Peroxisome Proliferator-Activated Receptor-γ–Mediated Effects in the Vasculature. *Circ. Res.* **2008**, *102*, 283–294. [CrossRef]
93. Naidenow, J.; Hrgovic, I.; Doll, M.; Hailemariam-Jahn, T.; Lang, V.; Kleemann, J.; Kippenberger, S.; Kaufmann, R.; Zöller, N.; Meissner, M. Peroxisome proliferator-activated receptor (PPAR) α and δ activators induce ICAM-1 expression in quiescent non stimulated endothelial cells. *J. Inflamm.* **2016**, *13*, 27–36. [CrossRef] [PubMed]
94. Lefere, S.; Puengel, T.; Hundertmark, J.; Penners, C.; Frank, A.K.; Guillot, A.; de Muynck, K.; Heymann, F.; Adarbes, V.; Defrêne, E.; et al. Differential effects of selective- and pan-PPAR agonists on experimental steatohepatitis and hepatic macrophages. *J. Hepatol.* **2020**, *73*, 757–770. [CrossRef]
95. Morán-Salvador, E.; López-Parra, M.; García-Alonso, V.; Titos, E.; Martínez-Clemente, M.; González-Périz, A.; López-Vicario, C.; Barak, Y.; Arroyo, V.; Clària, J. Role for PPARγ in obesity-induced hepatic steatosis as determined by hepatocyte- and macrophage-specific conditional knockouts. *FASEB J.* **2011**, *25*, 2538–2550. [CrossRef]
96. Wei, Z.; Zhao, D.; Zhang, Y.; Chen, Y.; Zhang, S.; Li, Q.; Zeng, P.; Li, X.; Zhang, W.; Duan, Y.; et al. Rosiglitazone ameliorates bile duct ligation-induced liver fibrosis by down-regulating NF-κB-TNF-α signaling pathway in a PPARγ-dependent manner. *Biochem. Biophys. Res. Commun.* **2019**, *519*, 854–860. [CrossRef]
97. Fiorucci, S.; Rizzo, G.; Antonelli, E.; Renga, B.; Mencarelli, A.; Riccardi, L.; Morelli, A.; Pruzanski, M.; Pellicciari, R. Cross-Talk between Farnesoid-X-Receptor (FXR) and Peroxisome Proliferator-Activated Receptor γ Contributes to the Antifibrotic Activity of FXR Ligands in Rodent Models of Liver Cirrhosis. *J. Pharmacol. Exp. Ther.* **2005**, *315*, 58–68. [CrossRef]
98. Ko, K.-L.; Mak, L.-Y.; Cheung, K.-S.; Yuen, M.-F. Hepatocellular carcinoma: Recent advances and emerging medical therapies. *F1000Research* **2020**, *9*, 620. [CrossRef] [PubMed]
99. Shen, B.; Chu, E.S.H.; Zhao, G.; Man, K.; Wu, C.-W.; Cheng, J.T.Y.; Li, G.; Nie, Y.; Lo, C.M.; Teoh, N.; et al. PPARgamma inhibits hepatocellular carcinoma metastases in vitro and in mice. *Br. J. Cancer* **2012**, *106*, 1486–1494. [CrossRef]
100. Bo, Q.-F.; Sun, X.-M.; Liu, J.; Sui, X.-M.; Li, G.-X. Antitumor action of the peroxisome proliferator-activated receptor-γ agonist rosiglitazone in hepatocellular carcinoma. *Oncol. Lett.* **2015**, *10*, 1979–1984. [CrossRef]
101. Misra, P.; Viswakarma, N.; Reddy, J.K. Peroxisome Proliferator-Activated Receptor-α Signaling in Hepatocarcinogenesis. *Subcell. Biochem.* **2013**, *69*, 77–99. [CrossRef]
102. Xiao, Y.-B.; Cai, S.-H.; Liu, L.-L.; Yang, X.; Yun, J.-P. Decreased expression of peroxisome proliferator-activated receptor alpha indicates unfavorable outcomes in hepatocellular carcinoma. *Cancer Manag. Res.* **2018**, *10*, 1781–1789. [CrossRef]
103. Thomas, M.; Bayha, C.; Klein, K.; Müller, S.; Weiss, T.S.; Schwab, M.; Zanger, U.M. The truncated splice variant of peroxisome proliferator-activated receptor alpha, PPARα-tr, autonomously regulates proliferative and pro-inflammatory genes. *BMC Cancer* **2015**, *15*, 488–503. [CrossRef]
104. Senni, N.; Savall, M.; Granados, D.C.; Alves-Guerra, M.-C.; Sartor, C.; Lagoutte, I.; Gougelet, A.; Terris, B.; Gilgenkrantz, H.; Perret, C.; et al. β-catenin-activated hepatocellular carcinomas are addicted to fatty acids. *Gut* **2019**, *68*, 322–334. [CrossRef]
105. Vacca, M.; D'Amore, S.; Graziano, G.; D'Orazio, A.; Cariello, M.; Massafra, V.; Salvatore, L.; Martelli, N.; Murzilli, S.; Sasso, G.L.; et al. Clustering Nuclear Receptors in Liver Regeneration Identifies Candidate Modulators of Hepatocyte Proliferation and Hepatocarcinoma. *PLoS ONE* **2014**, *9*, e104449. [CrossRef]
106. Kim, M.-J.; Choi, Y.-K.; Park, S.Y.; Jang, S.Y.; Lee, J.Y.; Ham, H.J.; Kim, B.-G.; Jeon, H.-J.; Kim, J.-H.; Kim, J.-G.; et al. PPARδ Reprograms Glutamine Metabolism in Sorafenib-Resistant HCC. *Mol. Cancer Res.* **2017**, *15*, 1230–1242. [CrossRef]
107. Yu, J.; Shen, B.; Chu, E.S.H.; Teoh, N.; Cheung, K.-F.; Wu, C.-W.; Wang, S.; Lam, C.N.Y.; Feng, H.; Zhao, J.; et al. Inhibitory role of peroxisome proliferator-activated receptor gamma in hepatocarcinogenesis in mice and in vitro. *Hepatology* **2010**, *51*, 2008–2019. [CrossRef]
108. Zuo, Q.; He, J.; Zhang, S.; Wang, H.; Jin, G.; Jin, H.; Cheng, Z.; Tao, X.; Yu, C.; Li, B.; et al. PPARγ Coactivator-1α Suppresses Metastasis of Hepatocellular Carcinoma by Inhibiting Warburg Effect by PPARγ–Dependent WNT/β-Catenin/Pyruvate Dehydrogenase Kinase Isozyme 1 Axis. *Hepatology* **2021**, *73*, 644–660. [CrossRef] [PubMed]
109. Wang, Z.; Li, F.; Quan, Y.; Shen, J. Avicularin ameliorates human hepatocellular carcinoma via the regulation of NF-κB/COX-2/PPAR-γ activities. *Mol. Med. Rep.* **2019**, *19*, 5417–5423. [CrossRef]
110. Saber, S.; Khodir, A.E.; Soliman, W.E.; Salama, M.M.; Abdo, W.S.; Elsaeed, B.; Nader, K.; Abdelnasser, A.; Megahed, N.; Basuony, M.; et al. Telmisartan attenuates N-nitrosodiethylamine-induced hepatocellular carcinoma in mice by modulating the NF-κB-TAK1-ERK1/2 axis in the context of PPARγ agonistic activity. *Naunyn-Schmiedeberg's Arch. Pharmacol.* **2019**, *392*, 1591–1604. [CrossRef] [PubMed]
111. Abd-Elbaset, M.; Mansour, A.M.; Ahmed, O.M.; Abo-Youssef, A.M. The potential chemotherapeutic effect of β-ionone and/or sorafenib against hepatocellular carcinoma via its antioxidant effect, PPAR-γ, FOXO-1, Ki-67, Bax, and Bcl-2 signaling pathways. *Naunyn-Schmiedeberg's Arch. Pharmacol.* **2020**, *393*, 1611–1624. [CrossRef] [PubMed]
112. Feng, J.; Dai, W.; Mao, Y.; Wu, L.; Li, J.; Chen, K.; Yu, Q.; Kong, R.; Li, S.; Zhang, J.; et al. Simvastatin re-sensitizes hepatocellular carcinoma cells to sorafenib by inhibiting HIF-1α/PPAR-γ/PKM2-mediated glycolysis. *J. Exp. Clin. Cancer Res.* **2020**, *39*, 24. [CrossRef]
113. Maiti, R. Metronomic chemotherapy. *J. Pharmacol. Pharmacother.* **2014**, *5*, 186–192. [CrossRef] [PubMed]
114. Walter, I.; Schulz, U.; Vogelhuber, M.; Wiedmann, K.; Endlicher, E.; Klebl, F.; Andreesen, R.; Herr, W.; Ghibelli, L.; Hackl, C.; et al. Communicative reprogramming non-curative hepatocellular carcinoma with low-dose metronomic chemotherapy, COX-2 inhibitor and PPAR-gamma agonist: A phase II trial. *Med. Oncol.* **2017**, *34*, 1–10. [CrossRef]

115. Gao, W.; Hu, J. Formation of Hepatitis B Virus Covalently Closed Circular DNA: Removal of Genome-Linked Protein. *J. Virol.* **2007**, *81*, 6164–6174. [CrossRef] [PubMed]
116. Turton, K.L.; Meier-Stephenson, V.; Badmalia, M.D.; Coffin, C.S.; Patel, T.R. Host Transcription Factors in Hepatitis B Virus RNA Synthesis. *Viruses* **2020**, *12*, 160. [CrossRef] [PubMed]
117. Xia, Y.; Cheng, X.; Li, Y.; Valdez, K.; Chen, W.; Liang, T.J. Hepatitis B Virus Deregulates the Cell Cycle To Promote Viral Replication and a Premalignant Phenotype. *J. Virol.* **2018**, *92*, e00722-18. [CrossRef]
118. Du, L.; Ma, Y.; Liu, M.; Yan, L.; Tang, H. Peroxisome Proliferators Activated Receptor (PPAR) agonists activate hepatitis B virus replication in vivo. *Virol. J.* **2017**, *14*, 96–105. [CrossRef]
119. Shi, Y.; Li, Y.; Huang, C.; Ying, L.; Xue, J.; Wu, H.; Chen, Z.; Yang, Z. Resveratrol enhances HBV replication through activating Sirt1-PGC-1α-PPARα pathway. *Sci. Rep.* **2016**, *6*, 24744–25756. [CrossRef] [PubMed]
120. Chang, M.-L. Metabolic alterations and hepatitis C: From bench to bedside. *World J. Gastroenterol.* **2016**, *22*, 1461–1476. [CrossRef]
121. Shirasaki, T.; Honda, M.; Shimakami, T.; Horii, R.; Yamashita, T.; Sakai, Y.; Sakai, A.; Okada, H.; Watanabe, R.; Murakami, S.; et al. MicroRNA-27a Regulates Lipid Metabolism and Inhibits Hepatitis C Virus Replication in Human Hepatoma Cells. *J. Virol.* **2013**, *87*, 5270–5286. [CrossRef]
122. Portius, D.; Sobolewski, C.; Foti, M. MicroRNAs-Dependent Regulation of PPARs in Metabolic Diseases and Cancers. *PPAR Res.* **2017**, *2017*, 7058424. [CrossRef]
123. Moss, H.B. The Impact of Alcohol on Society: A Brief Overview. *Soc. Work. Public Health* **2013**, *28*, 175–177. [CrossRef]
124. World Health Organization. *Global Status Report on Alcohol and Health 2018*; Poznyak, V., Rekve, D., Eds.; World Health Organization: Geneva, Switzerland, 2018; ISBN 978-92-4-156563-9.
125. Seitz, H.K.; Bataller, R.; Cortez-Pinto, H.; Gao, B.; Gual, A.; Lackner, C.; Mathurin, P.; Mueller, S.; Szabo, G.; Tsukamoto, H. Alcoholic liver disease. *Nat. Rev. Dis. Primers* **2018**, *4*, 16. [CrossRef]
126. Liu, J. Ethanol and liver: Recent insights into the mechanisms of ethanol-induced fatty liver. *World J. Gastroenterol.* **2014**, *20*, 14672–14685. [CrossRef] [PubMed]
127. Kong, L.-Z.; Chandimali, N.; Han, Y.-H.; Lee, D.S.; Kim, J.-S.; Kim, S.-U.; Kim, T.-D.; Jeong, D.K.; Sun, H.-N.; Kwon, T. Pathogenesis, Early Diagnosis, and Therapeutic Management of Alcoholic Liver Disease. *Int. J. Mol. Sci.* **2019**, *20*, 2712. [CrossRef]
128. Stickel, F.; Datz, C.; Hampe, J.; Bataller, R. Pathophysiology and Management of Alcoholic Liver Disease: Update 2016. *Gut Liver* **2017**, *11*, 173–188. [CrossRef] [PubMed]
129. Gitto, S.; Micco, L.; Conti, F.; Andreone, P.; Bernardi, M. Alcohol and viral hepatitis: A mini-review. *Dig. Liver Dis.* **2009**, *41*, 67–70. [CrossRef] [PubMed]
130. Salameh, H.; Raff, E.; Erwin, A.; Seth, D.; Nischalke, H.D.; Falleti, E.; Burza, M.A.; Leathert, J.; Romeo, S.; Molinaro, A.; et al. PNPLA3 Gene Polymorphism Is Associated with Predisposition to and Severity of Alcoholic Liver Disease. *Am. J. Gastroenterol.* **2015**, *110*, 846–856. [CrossRef] [PubMed]
131. Bucher, S.S.; Stickel, F.; Trépo, E.; Way, M.M.; Herrmann, A.; Nischalke, H.D.; Brosch, M.M.; Rosendahl, J.J.; Berg, T.; Ridinger, M.M.; et al. A genome-wide association study confirms PNPLA3 and identifies TM6SF2 and MBOAT7 as risk loci for alcohol-related cirrhosis. *Nat. Genet.* **2015**, *47*, 1443–1448. [CrossRef]
132. Gao, B.; Bataller, R. Alcoholic Liver Disease: Pathogenesis and New Therapeutic Targets. *Gastroenterology* **2011**, *141*, 1572–1585. [CrossRef]
133. Osna, N.A.; Donohue, T.M., Jr.; Kharbanda, K.K. Alcoholic Liver Disease: Pathogenesis and Current Management. *Alcohol. Res.* **2017**, *38*, 147–161.
134. Donohue, T.D., Jr. Alcohol-induced steatosis in liver cells. *World J. Gastroenterol.* **2007**, *13*, 4974–4978. [CrossRef] [PubMed]
135. Purohit, V.; Gao, B.; Song, B.-J. Molecular Mechanisms of Alcoholic Fatty Liver. *Alcohol. Clin. Exp. Res.* **2009**, *33*, 191–205. [CrossRef]
136. Galli, A.; Pinaire, J.; Fischer, M.; Dorris, R.; Crabb, D.W. The Transcriptional and DNA Binding Activity of Peroxisome Proliferator-activated Receptor α Is Inhibited by Ethanol Metabolism. A novel mechanism for the de-velopment of ethanol-induced fatty liver. *J. Biol. Chem.* **2001**, *276*, 68–75. [CrossRef] [PubMed]
137. Meng, F.-G.; Zhang, X.-N.; Liu, S.-X.; Wang, Y.-R.; Zeng, T. Roles of peroxisome proliferator-activated receptor α in the pathogenesis of ethanol-induced liver disease. *Chem. Biol. Interact.* **2020**, *327*, 109176. [CrossRef]
138. Nakajima, T.; Kamijo, Y.; Tanaka, N.; Sugiyama, E.; Tanaka, E.; Kiyosawa, K.; Fukushima, Y.; Peters, J.M.; Gonzalez, F.J.; Aoyama, T. Peroxisome proliferator-activated receptor? protects against alcohol-induced liver damage. *Hepatology* **2004**, *40*, 972–980. [CrossRef]
139. Fischer, M.; You, M.; Matsumoto, M.; Crabb, D.W. Peroxisome Proliferator-activated Receptor α (PPARα) Agonist Treatment Reverses PPARα Dysfunction and Abnormalities in Hepatic Lipid Metabolism in Ethanol-fed Mice. *J. Biol. Chem.* **2003**, *278*, 27997–28004. [CrossRef] [PubMed]
140. Kong, L.; Ren, W.; Li, W.; Zhao, S.; Mi, H.; Wang, R.; Zhang, Y.; Wu, W.; Nan, Y.; Yu, J. Activation of peroxisome proliferator activated receptor alpha ameliorates ethanol induced steatohepatitis in mice. *Lipids Health Dis.* **2011**, *10*, 246–255. [CrossRef] [PubMed]
141. Ding, L.; Wo, L.; Du, Z.; Tang, L.; Song, Z.; Dou, X. Danshen protects against early-stage alcoholic liver disease in mice via inducing PPARα activation and subsequent 4-HNE degradation. *PLoS ONE* **2017**, *12*, e0186357. [CrossRef]

142. Hsu, J.-Y.; Lin, H.-H.; Hsu, C.-C.; Chen, B.-C.; Chen, J.-H. Aqueous Extract of Pepino (Solanum muriactum Ait) Leaves Ameliorate Lipid Accumulation and Oxidative Stress in Alcoholic Fatty Liver Disease. *Nutrients* **2018**, *10*, 931. [CrossRef]
143. Blednov, Y.A.; Black, M.; Benavidez, J.M.; Stamatakis, E.E.; Harris, R.A. PPAR Agonists: II. Fenofibrate and Tesaglitazar Alter Behaviors Related to Voluntary Alcohol Consumption. *Alcohol. Clin. Exp. Res.* **2016**, *40*, 563–571. [CrossRef]
144. Ferguson, L.B.; Most, D.; Blednov, Y.A.; Harris, R.A. PPAR agonists regulate brain gene expression: Relationship to their effects on ethanol consumption. *Neuropharmacology* **2014**, *86*, 397–407. [CrossRef] [PubMed]
145. Muñoz, D.; Rivera-Meza, M.; Flores-Bastías, O.; Quintanilla, M.E.; Karahanian, E. Fenofibrate-a PPARα agonist-increases alcohol dehydrogenase levels in the liver: Implications for its possible use as an ethanol-aversive drug. *Adicciones* **2020**, *32*, 208–215. [CrossRef]
146. Le Foll, B.; Di Ciano, P.; Panlilio, L.V.; Goldberg, S.R.; Ciccocioppo, R. Peroxisome proliferator-activated receptor (PPAR) agonists as promising new medications for drug addiction: Preclinical evidence. *Curr. Drug Targets* **2013**, *14*, 768–776. [CrossRef] [PubMed]
147. Jang, D.M.; Jang, J.Y.; Kim, H.-J.; Han, B.W. Differential Effects of Cancer-Associated Mutations Enriched in Helix H3 of PPARγ. *Cancers* **2020**, *12*, 3580. [CrossRef]
148. Issemann, I.; Prince, R.A.; Tugwood, J.D.; Green, S. The peroxisome proliferator-activated receptor:retinoid X receptor heterodimer is activated by fatty acids and fibrate hypolipidaemic drugs. *J. Mol. Endocrinol.* **1993**, *11*, 37–47. [CrossRef]
149. Wang, Y.; Nakajima, T.; Gonzalez, F.J.; Tanaka, N. PPARs as Metabolic Regulators in the Liver: Lessons from Liver-Specific PPAR-Null Mice. *Int. J. Mol. Sci.* **2020**, *21*, 2061. [CrossRef]
150. Kosgei, V.J.; Coelho, D.; Guéant-Rodriguez, R.-M.; Guéant, J.-L. Sirt1-PPARS Cross-Talk in Complex Metabolic Diseases and Inherited Disorders of the One Carbon Metabolism. *Cells* **2020**, *9*, 1882. [CrossRef]
151. Hajri, T.; Zaiou, M.; Fungwe, T.; Ouguerram, K.; Besong, S. Epigenetic Regulation of Peroxisome Proliferator-Activated Receptor Gamma Mediates High-Fat Diet-Induced Non-Alcoholic Fatty Liver Disease. *Cells* **2021**, *10*, 1355. [CrossRef]
152. Peyrou, M.; Ramadori, P.; Bourgoin, L.; Foti, M. PPARs in Liver Diseases and Cancer: Epigenetic Regulation by MicroRNAs. *PPAR Res.* **2012**, *2012*, 757803. [CrossRef]

International Journal of
Molecular Sciences

Article

Resveratrol and Quercetin as Regulators of Inflammatory and Purinergic Receptors to Attenuate Liver Damage Associated to Metabolic Syndrome

Agustina Cano-Martínez [1], Rocío Bautista-Pérez [2], Vicente Castrejón-Téllez [1], Elizabeth Carreón-Torres [2], Israel Pérez-Torres [3], Eulises Díaz-Díaz [4], Javier Flores-Estrada [5], Verónica Guarner-Lans [1] and María Esther Rubio-Ruíz [1,*]

1 Department of Physiology, Instituto Nacional de Cardiología Ignacio Chávez, Juan Badiano 1, Sección XVI, Tlalpan, Mexico City 14080, Mexico; agustina.cano@cardiologia.org.mx (A.C.-M.); vicente.castrejon@cardiologia.org.mx (V.C.-T.); veronica.guarner@cardiologia.org.mx (V.G.-L.)
2 Department of Molecular Biology, Instituto Nacional de Cardología Ignacio Chávez, Juan Badiano 1, Sección XVI, Tlalpan, Mexico City 14080, Mexico; rociobtst@yahoo.com (R.B.-P.); juana.carreon@cardiologia.org.mx (E.C.-T.)
3 Department of Cardiovascular Biomedicine, Instituto Nacional de Cardiología Ignacio Chávez, Juan Badiano 1, Sección XVI, Tlalpan, Mexico City 14080, Mexico; israel.perez@cardiologia.org.mx
4 Department of Reproductive Biology, Instituto Nacional de Ciencias Médicas y Nutrición "Salvador Zubirán", Vasco de Quiroga 15, Sección XVI, Tlalpan, Mexico City 14080, Mexico; eulisesd@yahoo.com
5 División de Investigación, Hospital Juárez de México, Av. Instituto Politécnico Nacional 5160, Magdalena de las Salinas, Gustavo A. Madero, Mexico City 07760, Mexico; jose.floresc@salud.gob.mx
* Correspondence: esther_rubio_ruiz@yahoo.com

Citation: Cano-Martínez, A.; Bautista-Pérez, R.; Castrejón-Téllez, V.; Carreón-Torres, E.; Pérez-Torres, I.; Díaz-Díaz, E.; Flores-Estrada, J.; Guarner-Lans, V.; Rubio-Ruíz, M.E. Resveratrol and Quercetin as Regulators of Inflammatory and Purinergic Receptors to Attenuate Liver Damage Associated to Metabolic Syndrome. *Int. J. Mol. Sci.* **2021**, 22, 8939. https://doi.org/10.3390/ijms22168939

Academic Editor: Manuel Vázquez-Carrera

Received: 18 July 2021
Accepted: 17 August 2021
Published: 19 August 2021

Abstract: Nonalcoholic fatty liver disease (NAFLD) is considered a manifestation of metabolic syndrome (MS) and is characterized by the accumulation of triglycerides and a varying degree of hepatic injury, inflammation, and repair. Moreover, peroxisome-proliferator-activated receptors (PPARs) play a critical role in the pathophysiological processes in the liver. There is extensive evidence of the beneficial effect of polyphenols such as resveratrol (RSV) and quercetin (QRC) on the treatment of liver pathology; however, the mechanisms underlying their beneficial effects have not been fully elucidated. In this work, we show that the mechanisms underlying the beneficial effects of RSV and QRC against inflammation in liver damage in our MS model are due to the activation of novel pathways which have not been previously described such as the downregulation of the expression of toll-like receptor 4 (TLR4), neutrophil elastase (NE) and purinergic receptor P2Y2. This downregulation leads to a decrease in apoptosis and hepatic fibrosis with no changes in hepatocyte proliferation. In addition, PPAR alpha and gamma expression were altered in MS but their expression was not affected by the treatment with the natural compounds. The improvement of liver damage by the administration of polyphenols was reflected in the normalization of serum transaminase activities.

Keywords: inflammation; liver damage; toll-like receptor 4; P2Y2 receptor; metabolic syndrome; resveratrol; quercetin

1. Introduction

An increase in the intake of sugars (sucrose and fructose), a lack of physical activity, and genetic predisposition predict the development of metabolic syndrome (MS), independently from obesity and the prevalence of this disease is increasing dramatically in Western and developing countries [1]. MS is a cluster of cardiovascular risk factors associated with obesity and insulin resistance (IR) and is strongly linked to an increase in the level of systemic inflammation markers such as C-reactive protein (CRP), interleukin 6 (IL-6), and tumor necrosis factor-alpha (TNF-α) and an increase in the free fatty acid (FFA) concentration [2]. This disorder is not only associated with a higher risk of appearance of type 2 diabetes and cardiovascular events, but it also impacts the liver in different ways. Nonalcoholic fatty liver disease (NAFLD) is considered the hepatic manifestation of MS and is characterized by triglyceride accumulation and a variable degree of hepatic injury, inflammation, and repair [3,4]. Moreover, some reports suggest a link between liver inflammation and IR [3–5].

FFA, such as saturated fatty acids, activate toll-like receptors (TLR), which are a family of surface receptors that are present in all cells and are typically involved in the innate immune responses [2]. Particularly, TLR2 and TLR4 play a key role in obesity-related inflammation, IR, and vascular dysfunction [6,7]. In the liver, TLR expression was observed on a variety of cells and plays an important part in multiple liver diseases [8]. However, the direct role of TLR4 in these processes in the liver tissue is unclear [9].

When exposed to inflammatory stimuli, neutrophils release a large group of serine proteases, among which neutrophil elastase (NE) is the most important [10]. Obesity is associated with an increase in NE activity and NE is also implicated in IR by inhibiting hepatic Insulin receptor substrate 1. This effect is dependent on the activation of TLR4 [11,12].

After an inflammatory signal, adenosine 5′-triphosphate (ATP) is released into the extracellular space. ATP is known for its important role in intracellular cell metabolic pathways; however, this nucleotide can also act as a danger signal on the purinergic receptors (P2X or P2Y) which are diffusely expressed in various organs including the liver [13]. These receptors are essential regulators of physiological functions and serve as danger signals that trigger inflammation after injury [14]. Purinergic signaling, by P2Y, interacts with other signaling molecules to form a complex network, regulating numerous cellular processes including phagocytosis, chemotaxis, cytokine production, proliferation, differentiation, and death [13,15]. P2Y2 receptors are also associated with fat accumulation, hepatic steatosis, IR, metabolic complications, and inflammation [16,17].

On another hand, peroxisome-proliferator-activated receptors (PPARs) are activated through endogenous agonists (fatty acids and their derivatives) or exogenous agonists and regulate transcriptional activity [18]. Each PPAR isotype possesses specific functional characteristics to control a whole spectrum of physiological functions in the liver, including oxidative stress, lipid, and glucose metabolism, inflammatory responses, regenerative mechanisms, and cell differentiation and proliferation [19].

PPARγ is generally increased in livers with steatosis of both animal models of obesity and NAFLD patients. As opposed to PPARγ, PPARα plays a critical role in the regulation of fatty acid uptake, beta-oxidation, ketogenesis, synthesis of bile acid, and turnover of triglycerides to prevent hepatic steatosis [18]. In addition to its role in the regulation of metabolism, PPARα also has anti-inflammatory effects by inhibiting TLR4 expression and by inhibiting the NF-κB signaling pathway [20]. PPARγ has emerged as a potential target for the treatment of inflammatory diseases such as ulcerative colitis, atherosclerosis, asthma, and rheumatoid arthritis [21]. However, as far as we know, there are no reports on the association of PPAR expression and P2Y2 protein levels in the liver.

In recent years polyphenols such as resveratrol (RSV) and quercetin (QRC), which are present in fruits and vegetables, have gained interest by researchers for preventing and treating diseases, including obesity and obesity-related metabolic diseases [22]. These molecules are available as pills or capsules and people take these nutritional supplements. Although there are studies that demonstrate the antihypertensive, antioxidant, and anti-inflammatory properties in different human and animal models, the mechanisms underlying the beneficial effects of RSV and QRC have not been fully elucidated [22–24].

There is also little evidence of the effect of polyphenols on liver disorders associated with inflammatory and metabolic signaling through TLR4 and purinergic receptors [25]. Although some authors have identified flavonoid derivatives such as potent P2Y2 receptor antagonists, little is known about the effect of flavonoids on purinergic receptor expression [26,27]. Hence, the present study aimed to evaluate the effect of RSV and QRC mixture on the expression of TLR4, NE, and P2Y2 receptors and their association with the expression of PPARs. In addition, we assessed whether the expression of these elements was associated with fibrosis, apoptosis, and proliferation in an MS rat model.

2. Results

2.1. Metabolic Syndrome

The characterization of the MS model was done by analyzing the animal body weight, blood pressure, and intra-abdominal fat and by the serum biochemical analysis. As shown in Table 1, MS animals had an increased body weight and they developed central obesity, hypertension, dyslipidemia (high levels of triglycerides), hyperinsulinemia, and IR (HOMA-IR). Serum adipokine concentrations are higher in the MS group when compared to the Control group.

Table 1. The effects of the administration of RSV + QRC on body characteristics and biochemical parameters in Control and Metabolic syndrome (MS) rats.

	Control	Control/RSV + QRC	MS	MS/RSV + QRC
Weight (g)	492.7 ± 11.2	507.3 ± 15.7	583.3 ± 12.9 ^{a}	441.3 ± 9.3 ^{b}
Central adiposity (g)	5.2 ± 0.7	6.1 ± 0.9	13.1 ± 0.5 ^{a}	8.1 ± 1.4 ^{b}
Blood pressure (mm Hg)	101.7 ± 2.5	106.7 ± 2.4	143.6 ± 1.0 ^{a}	113.2 ± 1.3 ^{b}
Glucose (mg/dL)	92.1 ± 0.9	91.7 ± 0.8	96.5 ± 1.9	93.3 ± 1.0
Total Cholesterol (mg/dL)	56.2 ± 1.7	59.4 ± 3.2	64.1 ± 1.4	62.1 ± 0.8
Triglycerides (mg/dL)	83.6 ± 6.7	78.9 ± 4.2	145.2 ± 6.2 ^{a}	98.3 ± 5.2 ^{b}
Insulin (ng/mL)	0.15 ± 0.04	0.13 ± 0.02	0.48 ± 0.05 ^{a}	0.17 ± 0.02 ^{b}
HOMA index	0.91 ± 0.2	0.62 ± 0.12	2.41 ± 0.3 ^{a}	0.7 ± 0.06 ^{b}
Leptin (ng/dL)	2.6 ± 0.3	2.3 ± 0.1	5.3 ± 0.4 ^{a}	4.1 ± 0.8
Adiponectin (μg/mL)	4.2 ± 0.3	3.8 ± 0.1	6.5 ± 0.5 ^{a}	5.9 ± 0.3

Values are mean ± SEM. $n = 6$ in each group; ^{a} $p < 0.01$ MS without treatment vs. Control without treatment; ^{b} $p < 0.01$ vs. MS group. Abbreviations: MS, metabolic syndrome; HOMA-IR, Homeostatic model assessment of insulin resistance.

As expected, the treatment with RSV + QRC significantly decreased body weight, central adiposity, hypertension, hypertriglyceridemia, and restored IR in the MS group. In the Control group, polyphenol-administration did not affect the body or serum parameters.

2.2. TLR4 Expression

Figure 1 shows the expression of TLR4 in the liver from Control and MS rats. The presence of a label for TLR4 was located in hepatocytes around the central vein. The proportion of TLR4 in the MS group was 2.8 times higher compared to the Control. RSV + QRC administration significantly diminished TLR4 expression in both Control and MS groups although this effect was more evident in the MS rats (53% vs. 87%, respectively) (Figure 1C,D).

2.3. Neutrophil Elastase (NE) Expression

Due to the association of NE with TRL4 expression, we studied if the administration of natural compounds exerts an effect on this enzyme. The presence of the label for NE was located in regions away from the lumen of the vessels. The proportion of NE located in the liver of rats with MS was 5 times higher than that detected in the Control (Figure 2A,B). RSV + QRC treatment in Control and MS animals reverses the proportion of NE by 80% (Figure 2C,D).

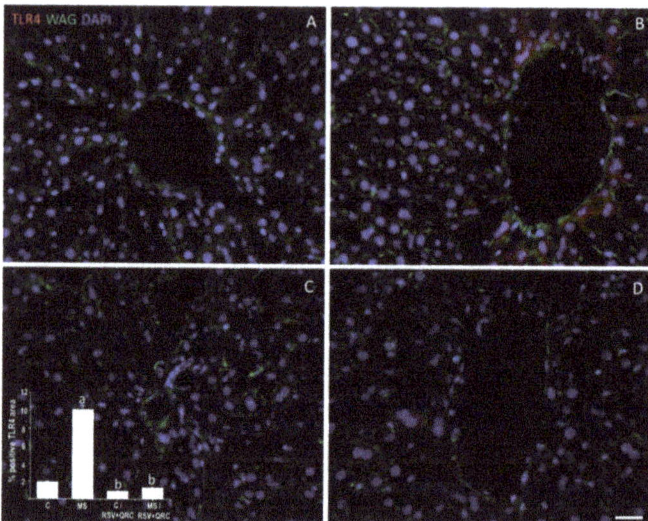

Figure 1. RSV + QRC administration decreased TLR4 expression in the liver from Control and MS rats. The detection of TLR4 immunostaining (red) was located in hepatocytes around of central vein. Wheat germ agglutinin (WAG) labeled with Oregon Green® 488 was used to label the membranes and 2-[4-(Aminoiminomethyl) phenyl]-1H-Indole-6-carboximidamide hydrochloride (DAPI) for nuclei. The graph with the values of the percentage of positive TLR4 area is in the lower-left corner. [a] $p < 0.05$ vs. Control: [b] $p < 0.01$ vs. MS group. Panel (**A**) = Control group, Panel (**B**) = metabolic syndrome (MS) group, Panel (**C**) = Control/RSV + QRC group, Panel (**D**) = MS/RSV + QRC group. Bar = 50 μm.

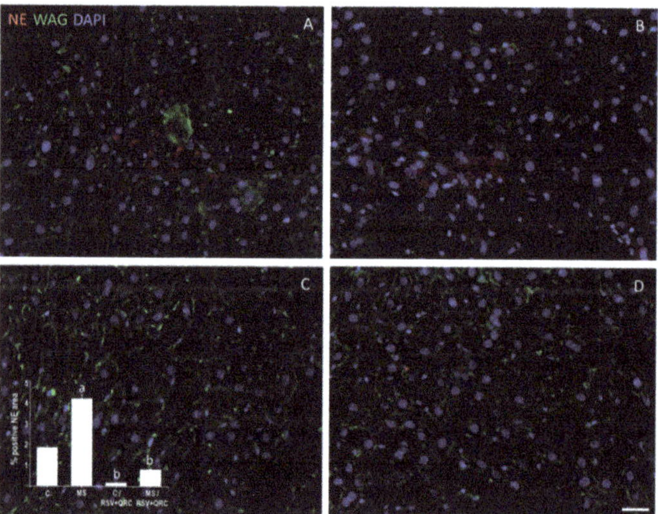

Figure 2. Effect of the administration of natural compounds on Neutrophil elastase (NE) immunodetection. The detection of NE immunostaining (red) was located in regions away from the lumen of the vessels. Wheat germ agglutinin (WAG) labeled with Oregon Green® 488 was used to label the membranes and DAPI for nuclei. The graph with the values of the percentage of positive NE area is in the lower-left corner. [a] $p < 0.05$ vs. Control: [b] $p < 0.01$ vs. MS group. Panel (**A**) = Control group, Panel (**B**) = metabolic syndrome (MS) group, Panel (**C**) = Control/RSV + QRC group, Panel (**D**) = MS/RSV + QRC group. Bar = 50 μm.

2.4. P2Y2 Expression

Figure 3 revealed the differences in the expression of P2Y2 in livers from the Control and MS groups. The presence of labels for P2Y2 receptors was located mainly in hepatocytes around the central vein. The proportion of P2Y2 in the MS group was 32% higher compared to the Control. However, the treatment with natural compounds significantly diminished the P2Y2 expression in the same proportion (50% approximately) in both, Control and MS animals (Figure 3C,D).

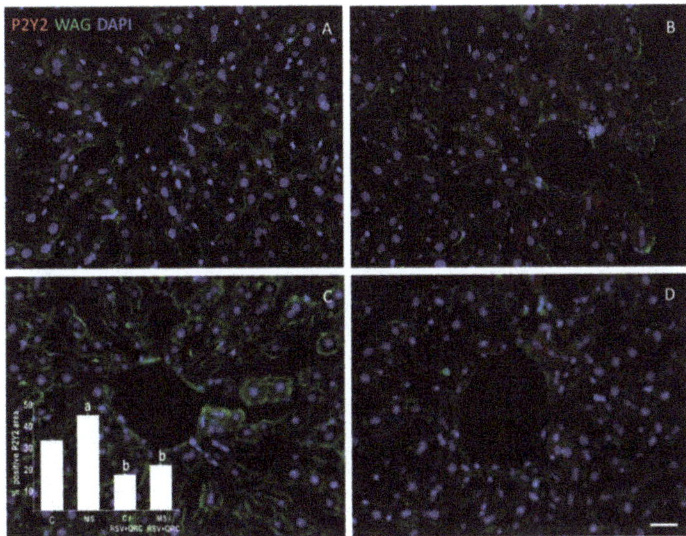

Figure 3. Effect of administration of RSV + QRC on the expression of the P2Y2 receptor in the liver from Control and MS rats. The detection of P2Y2 immunostaining (red) was located in hepatocytes around of central vein. Wheat germ agglutinin (WAG) labeled with Oregon Green® 488 was used to label the membranes and DAPI for nuclei. The graph with the values of the percentage of positive P2Y2 area is in the lower-left corner. ^{a} $p < 0.05$ vs. Control: ^{b} $p < 0.01$ vs. MS group. Panel (**A**) = Control group, Panel (**B**) = metabolic syndrome (MS) group, Panel (**C**) = Control/RSV + QRC group, Panel (**D**) = MS/RSV + QRC group. Bar = 50 µm.

2.5. Fibrosis

Because fibrosis is considered an indicator of liver damage, we analyze this parameter in the liver from the experimental groups. The liver of rats with MS presented mainly perivascular fibrosis (PVF), with indications of interstitial fibrosis (IF) and replacement fibrosis (RF) in the region surrounding the vessels, including both the central vein (CV) and the intralobular vein (ILBV) in the triad. Fibrosis was increased in the tissues from MS animals that were damaged similarly as was found with Masson's trichrome staining (MT) (Figure 4) for total collagen deposits and confirmed with Sirius Red (SR) staining (Figure 5) for collagen I y III. The proportion of total collagen deposits in MS was 3 times more than that observed in the Control tissue. The RSV + QRC administration reduced deposition of fibrosis in the MS group almost reaching Control values (Figure 4).

Collagen I and III accumulation was confirmed by SR analysis. Livers from MS rats had 122% more collagen deposition compared to the livers from Control rats (Figure 5A,B). When Control and MS animals were treated with RSV plus QRC, both groups presented less collagen I and III depositions, although the decrease is more evident in MS animals (59% vs. 85%, respectively) (Figure 5C,D).

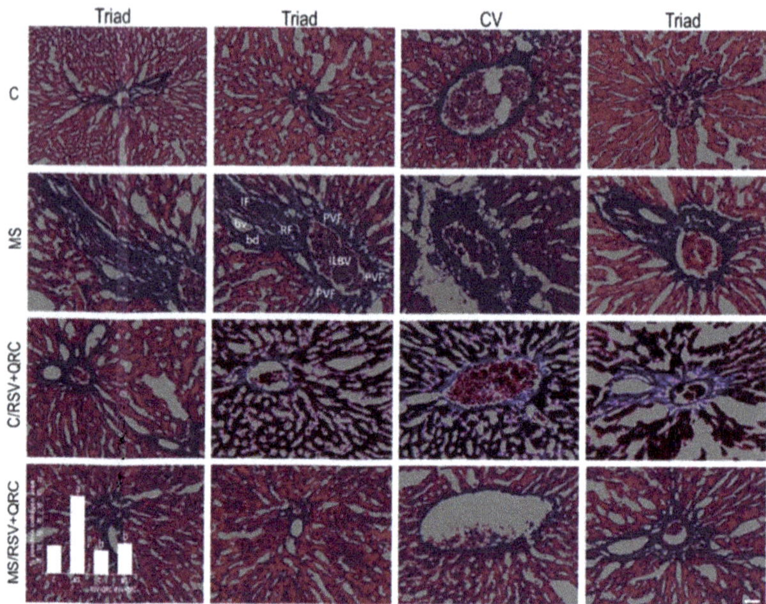

Figure 4. Resveratrol and quercetin administration attenuates liver fibrosis in the liver from MS rats. Representative images of Masson's Trichrome staining; the triad and central vein (CV) in each condition are presented. The proportion of collagen deposits is greater in MS in the perivascular region (PVF), between (IF) and within the hepatocytes (RF) surrounding the vessels, both in the CV and in the triad. In the lower-left corner, the graph with the % total positive collagen area in each group is presented. ^{a} $p < 0.05$ vs. Control: ^{b} $p < 0.01$ vs. MS group. Abbreviations: C = Control, MS = metabolic syndrome; RSV + QRC = resveratrol plus quercetin, ILBV = interlobular vein, bd = bile duct, bv = blood vessel, PVF = perivascular fibrosis; IF = interstitial fibrosis, RF = replacement fibrosis. Bar = 50 μm.

2.6. Apoptosis and Proliferation

Afterward, we researched if the treatment with polyphenols was able to prevent apoptosis using the Terminal deoxynucleotidyl transferase dUTP nick end labeling (TUNEL) assay (Figure 6). Our results showed that livers from rats with MS presents 3.34 times more cells in apoptosis compared to the Control animals (Figure 6A,B). The cells in apoptosis were located towards the lumen of the vessels as well as in the hepatocytes around the vessels, mainly in the central vein. RSV plus QRC treatment significantly reduced apoptosis (72%) in livers from MS rats; while the percentage did not change significantly in the Control group (Figure 6C,D).

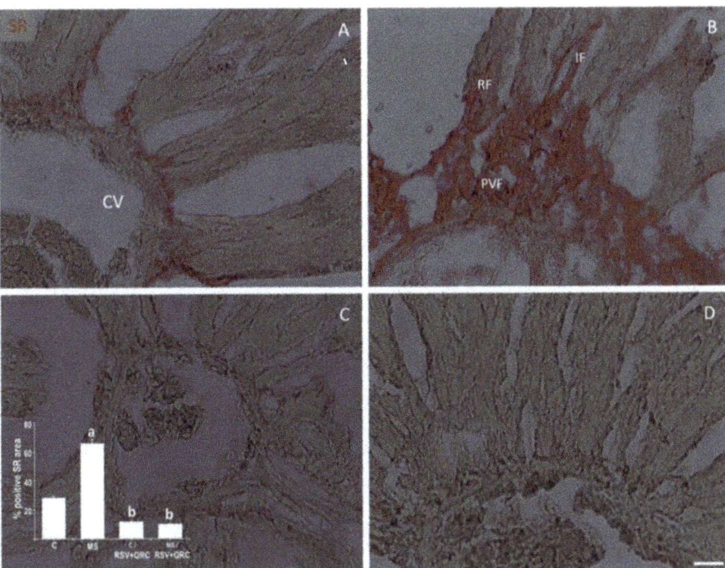

Figure 5. Effect of the administration of Resveratrol and Quercetin on fibrosis by Picrus-Sirius Red staining (SR) in livers from Control and MS rats. Representative images of the central vein (CV) of each condition are presented. Hepatocytes (RF) surrounding the vessels. The graph with the % SR positive (fibrosis) area in each group is shown in the lower-left corner. [a] $p < 0.05$ vs. Control: [b] $p < 0.01$ vs. MS group. Panel (**A**) = Control group, Panel (**B**) = metabolic syndrome (MS) group, Panel (**C**) = Control/RSV + QRC group, Panel (**D**) = MS/RSV + QRC group. Abbreviations: PVF = perivascular fibrosis; IF = interstitial fibrosis, RF = replacement fibrosis (hepatocytes surrounding the vessels). Bar = 50 μm.

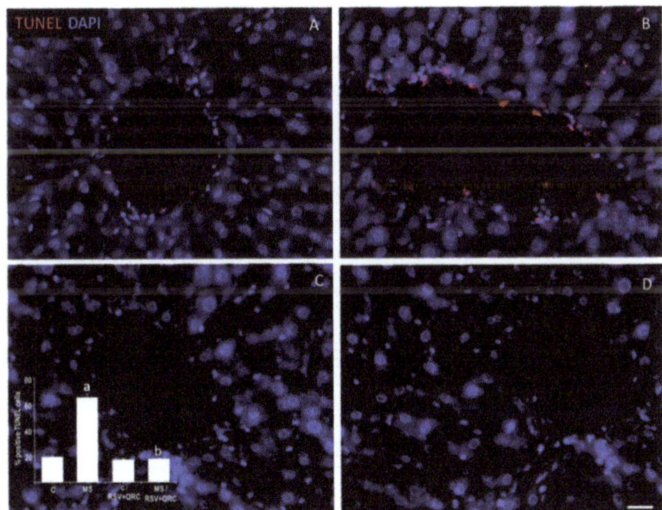

Figure 6. Resveratrol and quercetin treatment decreased apoptosis in livers from MS rats. The cells in apoptosis are located towards the lumen of the vessels as well as in the hepatocytes around the vessels, mainly in the central vein. The nuclei were marked with DAPI. The graph with the percentage of positive TUNEL positive cells is in the lower-left corner. [a] $p < 0.05$ vs. Control; [b] $p < 0.01$ vs. MS group. Panel (**A**) = Control group, Panel (**B**) = metabolic syndrome (MS) group, Panel (**C**) = Control/RSV + QRC group, Panel (**D**) = MS/RSV + QRC group. Bar = 50 μm.

The results shown in Figure 7 show the proliferative activity in liver tissue sections. Proliferating cell nuclear antigen (PCNA)-positive cells (brown color with a fine granular appearance) in the MS group were higher when compared with the Control group, and were localized as a part of infiltration. Our observations suggest that these cells could be Kupffer cells due to their localization in sinusoids (Figure 7A,B). The oral treatment with natural compounds significantly decreased the staining levels in the MS group; however, there were no significant differences in PCNA staining levels in the Control group.

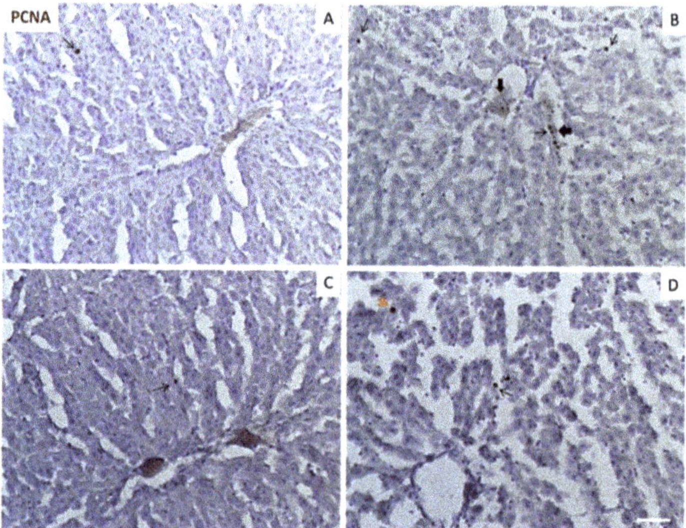

Figure 7. Expression of proliferating cell nuclear antigen (PCNA) in liver tissue from Control and MS rats. Thin arrows indicate the locations of PCNA positive cells. The thick arrows indicate cellular infiltrates and the orange arrow indicates a PCNA positive hepatocyte. Panel (**A**) = Control group, Panel (**B**) = metabolic syndrome (MS) group, Panel (**C**) = control/RSV + QRC group, Panel (**D**) = MS/RSV + QRC group. Bar = 100 μm.

2.7. Expression of PPARs

We also evaluated the expression of PPARs isotypes in the experimental groups because they play a major role in metabolism and the inflammation process in the liver. Western blot analyses revealed differences in the expression of PPAR-α and PPAR-γ in liver homogenates from all groups (Figure 8). As expected, PPAR-α expression was higher in Control rats when compared to MS rats and PPAR-γ was increased in liver form MS rats. Nevertheless, protein levels of both PPARs isotypes were not modified by the treatment with natural compounds in both, Control and MS groups.

2.8. Activity of Transaminases

Finally, serum transaminases activity was determined in all groups because liver disease is often reflected by biochemical abnormalities. ALT and ALP activity was significantly higher in the MS than in Control rats (Table 2). This indicated liver damage in MS rats; however, no significant difference was seen between Control and MS groups in AST and GGT activities. Treatment with natural compounds significantly reduced ALT and ALP activities in MS animals (66% and 32%, respectively).

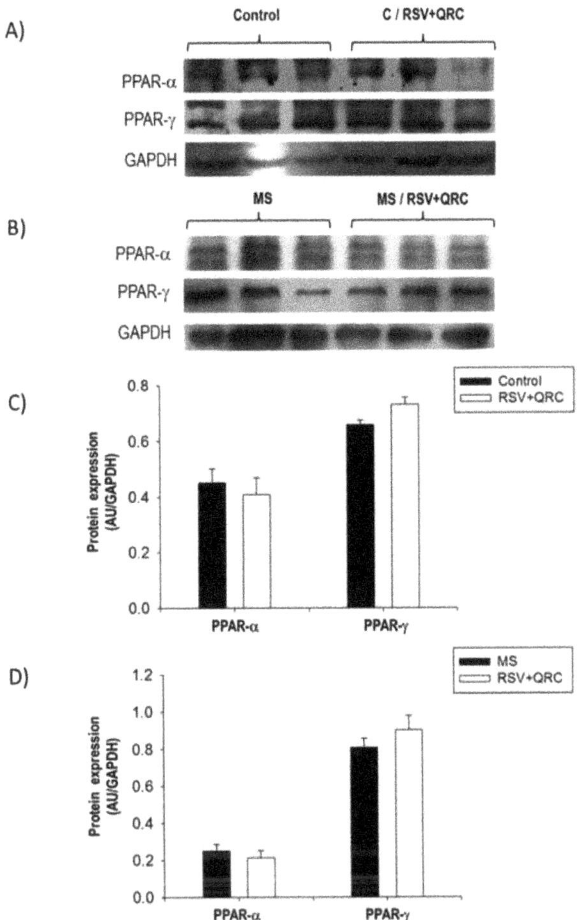

Figure 8. Effect of resveratrol and quercetin treatment on the expression of PPAR isotypes in the liver. Representative western blot from Control (**A**) and metabolic syndrome (MS) rats (**B**). (**C**) Expression of PPAR-α and PPAR-γ in Control group; (**D**) Expression of PPAR-αand PPAR-γ in MS group. Data represent mean ± SEM (n = 6 per group).

Table 2. Effects of polyphenols on liver function in experimental groups.

	Control	Control/RSV + QRC	MS	MS/RSV + QRC
ALT (U/L)	17.6 ± 3.2	21.0 ± 1.7	42.2 ± 9.1 ^{a}	14.2 ± 2.5 ^{b}
AST (U/L)	114.4 ± 8.4	87.2 ± 16.5	112.3 ± 15.6	103.7 ± 13.3
GGT (U/L)	4.8 ± 1.2	2.4 ± 1.5	4.8 ± 1.2	2.4 ± 1.5
ALP (U/L)	55.5 ± 3.7	60.0 ± 7.8	73.0 ± 5.4 ^{a}	50 ± 4.0 ^{b}

Values represented as means and standard errors, the different superscript letters mean a significant difference; ^{a} $p < 0.05$ vs. Control: ^{b} $p < 0.01$ vs. MS group. Abbreviations: MS: metabolic syndrome; ALT, alanine transaminase; AST, aspartate transaminase; GGT, gamma-glutamyltransferase: ALP, alkaline phosphatase.

3. Discussion

Extensive evidence has demonstrated the beneficial effect of polyphenols on the treatment of cardiometabolic disorders. The mechanisms underlying the beneficial effects of

RSV and QRC have not yet been fully elucidated and have mainly been related to epigenetic processes, intrinsic antioxidant activity, and anti-inflammatory mechanisms [23,24,28]. However, to the best of our knowledge, there are not reports on the effect of RSV and QRC on the pathways analyzed in this paper. In this work, we show that the decrease in TLR4 and NE expression are new mechanisms through which these natural compounds may have a protective role on liver damage associated with MS. These decreases are then followed by the diminution in liver fibrosis and apoptosis. Also, a novel finding presented in this study is that the treatment with RSV and QRC mixture is associated with the decrease in the expression of the P2Y2 receptor. Moreover, we evaluated if these anti-inflammatory effects are associated with differences in PPAR α and γ expression.

Table 1 shows that the administration of RSV + QRC reversed some signs of MS such as body weight, central adiposity, hypertension, IR, and dyslipidemia without affecting the concentration of glucose, total cholesterol, adiponectin, and leptin; these results are in accordance with our previous reports [22,24,29].

Numerous studies have shown that different mechanisms exert synergic action in the development or progression of liver disease linked to MS such as accumulation of fatty acids, oxidative stress, and inflammation [24,28]. On this aspect, some authors have reported that TLR4 is involved in obesity and liver damage [7,30]. We analyzed the expression of the TLR4 receptor in livers from all experimental groups (Figure 1). Our results on the anti-inflammatory effects of polyphenols by decreasing TLR4 expression and its signaling pathway are in line with those reported by other authors [31,32]. Zhang [33] reported that non-esterified fatty acids (NEFAs), a crucial source of energy in the liver, may activate TLR4. On this aspect, in a previous study from our group, we found that circulating levels of NEFAs were higher in MS rats and that the RSV + QRC treatment reduces these levels [22]. This effect could be added to the decrease of TLR4 caused by FFA as a result of the presence of polyphenols since RSV plus QRC reduce the amount of FFA by increasing their oxidation in the liver [34]. Therefore, further studies are needed to support our hypothesis.

Another mechanism that mediates liver damage controlled by TLR4 is the activation of neutrophils and the release of NE. Although this enzyme might be a potential target to treat liver disorders linked to obesity and MS, there are still few studies reporting the effect of polyphenols on NE expression [11,13,35]. We observed that the livers from MS rats had an increase in this enzyme compared to livers from Control rats and that the RSV plus QRC mixture was able to abolish the expression of NE; however, it would be interesting to evaluate the effect of polyphenols treatment on NE activity.

Some authors have proposed that ATP serves as a messenger that links inflammation and metabolic derangements through its binding to the P2Y2 receptor [13,14,17]. Moreover, some reports have shown the therapeutic role of polyphenols by acting as antagonists of the P2Y2 receptors [25–27]. A novel finding of the present study is that the treatment with RSV plus QRC can decrease the expression of P2Y2 in livers from MS rats; hence, we suggest that this is a new mechanism for the therapeutic role of polyphenols in liver disease without ruling out their other pleiotropic effects.

Liver inflammatory and purinergic signaling modulate several physio-pathological processes such as proliferation, differentiation, migration, and death in response to injury [36,37]. Therefore, we analyzed the effect of the administration of polyphenols on the levels of fibrosis, apoptosis, and proliferation in livers from Control and MS animals (Figures 4–7). Figures 4 and 5 show that MS is associated with an increase in liver fibrosis compared to the Control group, and the administration of RSV + QRC mixture reverted this effect. Our results are in line with experimental and clinical evidence which suggests that RSV and QRC attenuate liver inflammation and fibrosis [28,38,39].

Results in Figure 6 demonstrate that livers from MS animals showed significantly higher levels of hepatocytes in apoptosis compared to the Control group. This effect was abolished by the treatment with polyphenols (Figure 6C,D). Our data are in accordance with previous studies that showed the association of apoptosis and liver disease as well as the anti-apoptotic role of polyphenols [38,40]. Subsequently, we studied if there were differences in the proliferative levels in the tissue from all experimental groups, due to the regenerative capacity of the liver in response to injury. However, we found that the positive nuclear immunoreactivity was limited to Kupffer cells in the liver sections from MS animals and that the administration of RSV plus QRC decreased the levels of proliferation (Figure 7). These results suggest that the pathway through which the RSV + QRC treatment reverses apoptosis and fibrosis generated by MS is related to the decreased expression of P2Y2 and TLR4 receptors thus diminishing inflammation. This is linked to the fact that there are cellular infiltrates and higher levels of NE. The differences that we found in the analysis of cellular proliferation with those of other studies could be due to the experimental model of MS and the tested doses of RSV + QRC used here as the regenerative response of the liver has been previously reported by other authors in experimental models of acute hepatic damage [41].

PPAR α and PPAR γ play a pivotal role in the control of several cardiometabolic diseases including liver diseases and these nuclear receptors bind FFA as their physiological ligands. Indeed, activated PPARs exert anti-inflammatory activities in several models through their ability to antagonize other signaling pathways [18]. They interact with other proteins, including Nuclear factor kappa B, Activated protein-1, and AMP-activated protein kinase, and they also downregulate TLR4 [20,42]. We found that livers from MS rats show a decreased expression of PPAR α and that PPAR γ is upregulated (Figure 8). These results suggest a relationship between the increase in inflammatory components, such as TLR4 and NE, and were consistent with other studies which indicated that PPARs receptors play a protective role in attenuating liver fibrosis [43–45]. Furthermore, there are conflicting reports on the contribution of PPAR α in various liver cell types to regulate cell proliferation [46,47]. Although the treatment with these concentrations of polyphenols did not affect the PPARs expression in MS animals, RSV + QRC could be regulating PPARs activity. To further clarify this point, it would be important to evaluate the effect of RSV + QRC administration on gene expression of transcriptional targets of PPARs in livers.

Finally, we analyzed the levels of transaminases to evaluate if the damage observed in the liver was reflected in their leak into the bloodstream. In this study we observed a significant increase in serum ALT and ALP activities in the MS rats when compared to Control rats and that the polyphenol mixture improved levels of these hepatic markers. These results are consistent with observations by other authors showing the hepatoprotective effects of natural compounds [4,39,48].

4. Materials and Methods

4.1. Animals and Surgical Procedures

All of the experiments were conducted in accordance with the ethical guidelines of the Instituto Nacional de Cardiología Ignacio Chávez (protocol #14-860). Male Wistar rats, 25 days old and weighing 45 $\pm$ 9 g, were randomly separated into two groups of 12 animals: group 1, Control rats that were given tap water for drinking, and group 2, MS rats that received 30% sugar in their drinking water during 20 weeks. Half of each group of rats (Control or MS) received their sucrose solution or drinking water with a mixture of RSV and QRC every day for four weeks in a dose 50–0.95 mg/kg/day, respectively (provided by ResVitalé^{TM}, which contains 20 mg of QRC per 1050 mg of RSV). Groups without RSV + QRC treatment only received the vehicle (n = 6 per group). The mixture of RSV and QRC had been previously dissolved in 1 mL ethanolic solution (20%). The animals were maintained under standard conditions of light and temperature with water and food (LabDiet 5001; Richmond, IN, USA) ad libitum. At the end of the treatment, the animals were weighed and systolic arterial blood pressure was determined in conscious animals by a plethysmographic method previously described [22]. After overnight fasting, rats were euthanized by decapitation by guillotine. The intra-abdominal white adipose tissue (retroperitoneal fat pad) was carefully dissected with scissors, wet weight was determined, and then the tissue was discarded. The livers were excised and divided for histological analyses while fresh.

4.2. Measurement of Serum Biochemical Parameters

The fasting measurements of glucose, total cholesterol, and triglycerides were performed with commercial enzymatic kits (RANDOX Laboratories Ltd., Crumlin, Country Antrim, UK). Serum insulin levels were measured using a rat-specific insulin radioimmunoassay (Linco Research, Inc., Saint Charles, MO, USA). IR was estimated from the homeostasis model (HOMA-IR), as previously described [24].

Serum glutamic-oxaloacetic transaminase (SGOT/AST), glutamic pyruvic transaminase (SGPT/ALT), alkaline phosphatase (ALP), and γ-glutamyl transferase (GGT) activities were determinate spectrophotometrically using UV-test, International Federation of Clinical Chemistry [IFCC] (Roche Cobas C-501, Roche Diagnostics, IN, USA) [49].

4.3. Liver Tissue Preparation and Histological Examinations

The liver tissue of each group was processed to make frozen sections (10 µm). Sections for colorimetric staining (Picro-Sirius Red (SR)) and Masson's trichrome (MT) were placed on gelatinized slides. The sections for purinergic receptor P2Y2, TLR4, and NE, and PCNA were placed on electro-charged slides. The photomicrographs for MT and SR were taken with a QIMAGING Micropuplisher 5.0 camera with Real-Time Viewing (RTV) coupled to an Olympus BX5 microscope. The images for PCNA were acquired with a Carl Zeiss microscope (Carl Zeiss Microscopy GmbH, Jena, Germany). Analysis and quantification of the area with collagen deposits (MT and SR) and with a signal for P2Y2, TLR4, NE, and the percentage of PCNA and TUNEL positive cells was performed with Image-Pro Premier Version 9.0 software (Media Cybernetics, Inc., Rockville, MD, USA). Four fields of each animal ($n = 6$) were analyzed, for a total of 16 fields of each condition in 20× photomicrographs.

4.3.1. Fibrosis Detection

For the detection of fibrosis, the staining was performed with Accustain Trichrome Stain (MT) Kit (Sigma-Aldrich, HT15) and Picro-SR solution (ab246832; Abcam PLC, Cambridge, UK) following the manufacturer's instructions.

4.3.2. Immunofluorescence

The sections were incubated in blocking solution for 1 h at room temperature. The incubation with the primary antibodies at a dilution of 1: 500 [(anti-P2Y2 (sc-518121], anti-TLR4 (sc-518121) and anti-NE (sc-55549) was carried out overnight at 4 °C. A 1:200 dilution of the secondary antibody m-IgGκBP (sc-516141) was used for NE and P2Y2; while for TLR4 a 1:400 dilution of mouse anti-rabbit IgG (sc-3753) with overnight incubation at 4 °C was done (all from Santa Cruz Biotechnology, CA, USA). Nuclei were labeled with DAPI. Observation and photographs for fluorescence images were obtained with a Cell Imaging Station (Life Technologies, Carlsbad, CA, USA).

4.3.3. Apoptosis and Proliferation Analysis

The apoptosis of liver cells was detected using the In Situ Cell Death Detection Kit, TMR (tetramethylrhodamine-5-dUTP) red, version 12 (12156792910; Roche Applied Science, Mannheim, Germany) according to the manufacturer's instructions. Sections were mounted with DAPI and observed in fluorescence microscopy (FLoid™ Cell Imaging Station). The percentage of TUNEL positive cells was calculated.

For PCNA, the incubation for 48 h at 4 °C with mouse monoclonal antibody (13-3900-Invitrogen Biotechnology, Waltham, MA, USA) (1:50) and the incubation for 1 h at 37 °C with m-IgGκ BP-HRP:sc-516102 (Santa Cruz Biotechnology, CA, USA) (1:500) as secondary antibody was carried out. The signal was revealed with the 3,3′-Diaminobenzidine

(DAB)/chromogen substrate and hematoxylin. The images were captured in a Carl Zeiss microscope (Carl Zeiss Microscopy GmbH, Jena, Germany).

4.4. Western Blotting Analysis

The livers were homogenized in a lysis buffer pH = 8 (25 mM Hepes, 100 mM NaCl, 15 mM Imidazole, 10% glycerol, 1% Triton X-100) and protease inhibitor cocktail. The homogenate was centrifuged at 19,954× g for 10 min at 4 °C; the supernatant was separated and stored at −70 °C. The Bradford method was used to determine the total proteins [50].

A total of 50 µg protein was separated on an SDS-PAGE (12% bis-acrylamide-laemmli gel) and transferred to a polyvinylidene difluoride (PVDF) membrane. Blots were blocked for 1 h at room temperature using Tris-buffered saline (TBS)-0.01% Tween (TBS-T 0.01%) plus 5% non-fat milk. The membranes were incubated overnight at 4 °C with rabbit primary polyclonal antibodies PPAR-α, and PPAR-γ from Santa Cruz Biotechnology (Santa Cruz, CA, USA) as previously described [26]. All blots were incubated with Glyceraldehyde 3-phosphate dehydrogenase (GAPDH) antibody as a loading control. Images from films were digitally obtained by GS-800 densitometer with the Quantity One software (Bio-Rad Laboratories, Inc. Hercules, CA, USA) and they are reported as arbitrary units (AU).

4.5. Statistical Analysis

Results are expressed as mean ± standard error of the mean (SEM). Differences were considered statistically significant when $p < 0.05$. The different letters (a and b) in tables and figures indicate significant differences. We applied a one-way analysis of variance (ANOVA) followed by a Bonferroni post hoc test using the SigmaPlot program version 11 (Jandel Scientific, San Jose, CA, USA).

5. Conclusions

The most important outcome of the present study was that there is a downregulation of the expression of TLR4, NE, and P2Y2. This is a new mechanism underlying the beneficial effects of RSV and QRC against inflammation in liver damage associated with MS. This effect leads to a decrease in apoptosis and fibrosis with no changes in hepatocytes proliferation. In addition, PPAR alpha and gamma expressions were altered in MS but their expression was not affected by the treatment with the natural compounds.

Author Contributions: A.C.-M. was responsible for planning and performing the experiments, capture, quantification of images, data analysis, and writing the paper; R.B.-P. was responsible for performing immunofluorescence assays; I.F.-F. was responsible for performing proliferation assay; V.C.-T. was responsible for western blot analysis, I.P.-T. was responsible for performing some physiological experiments; E.C.-T. and E.D.-D. were responsible for serum biochemical analysis; V.G.-L. revised the paper; M.E.R.-R. was responsible for study conception and design, data analysis, and writing the paper. All authors have read and agreed to the published version of the manuscript.

Funding: This work was partially supported by a research grant of Consejo Nacional de Ciencia y Tecnología (CONACyT-169736) to A.C.-M.

Institutional Review Board Statement: All the experiments were conducted in accordance with our Institutional Ethical Guidelines (Ministry of Agriculture, SAGARPA, NOM-062-ZOO-1999, Mexico) (protocol #14-860).

Informed Consent Statement: Not applicable.

Data Availability Statement: The data in our study are available from the corresponding author upon reasonable request.

Acknowledgments: The authors would like to thank Héctor Vázquez Meza for providing a factual review and Jhony Pérez for the excellent technical assistance. This study was supported by Fondos del Gasto Directo Autorizado a la Subdirección de Investigación Básica, INC "Ignacio Chávez".

Conflicts of Interest: The authors declare no conflict of interest.

References

1. Castrejón-Téllez, V.; Villegas-Romero, M.; Rubio-Ruiz, M.E.; Pérez-Torres, I.; Carreón-Torres, E.; Díaz-Díaz, E.; Guarner-Lans, V. Effect of a Resveratrol/Quercetin mixture on the reversion of hypertension induced by a short-term exposure to high sucrose levels near weaning and a long-term exposure that leads to metabolic syndrome in rats. *Int. J. Mol Sci.* **2020**, *21*, 2231. [CrossRef] [PubMed]
2. Rubio-Ruiz, M.E.; Guarner-Lans, V. Inflammation and the use of anti-inflammatory agents in signs and cardiovascular consequences of metabolic syndrome. In *Handbook on Metabolic Syndrome. Classification, Risk Factors and Health Impact*; Lopez-García, C., Pérez-González, P., Eds.; Nova Biomedical, Nova Science Publishers: New York, NY, USA, 2012; pp. 169–188.
3. Porras, D.; Nistal, E.; Martínez-Flórez, S.; Pisonero-Vaquero, S.; Olcoz, J.L.; Jover, R.; González-Gallego, J.; García-Mediavilla, M.V.; Sánchez-Campos, S. Protective effect of quercetin on high-fat diet-induced non-alcoholic fatty liver disease in mice is mediated by modulating intestinal microbiota imbalance and related gut-liver axis activation. *Free Radic. Biol. Med.* **2017**, *102*, 188–202. [CrossRef] [PubMed]
4. Lasker, S.; Rahman, M.M.; Parvez, F.; Zamila, M.; Miah, P.; Nahar, K.; Kabir, F.; Sharmin, S.B.; Subhan, N.; Ahsan, G.U.; et al. High-fat diet-induced metabolic syndrome and oxidative stress in obese rats are ameliorated by yogurt supplementation. *Sci. Rep.* **2019**, *9*, 20026. [CrossRef] [PubMed]
5. Jia, L.; Vianna, C.R.; Fukuda, M.; Berglund, E.D.; Liu, C.; Tao, C.; Sun, K.; Liu, T.; Harper, M.J.; Lee, C.E.; et al. Hepatocyte Toll-like receptor 4 regulates obesity-induced inflammation and insulin resistance. *Nat. Commun.* **2014**, *5*, 3878. [CrossRef]
6. Davis, J.E.; Gabler, N.K.; Walker-Daniels, J.; Spurlock, M.E. Tlr-4 deficiency selectively protects against obesity induced by diets high in saturated fat. *Obesity* **2008**, *16*, 1248–1255. [CrossRef]
7. Chaurasia, B.; Talbot, C.L.; Summers, S.A. Adipocyte ceramides-the nexus of inflammation and metabolic disease. *Front. Immunol.* **2020**, *11*, 576347. [CrossRef]
8. Saghazadeh, A.; Rezaei, N. *Introductory Chapter: Toll-like Receptors*; IntechOpen: London, UK, 2020. [CrossRef]
9. Kim, J.J.; Sears, D.D. TLR4 and insulin resistance. *Gastroenterol. Res. Pract.* **2010**, 212563. [CrossRef]
10. Melzig, M.F.; Löser, B.; Ciesielski, S. Inhibition of neutrophil elastase activity by phenolic compounds from plants. *Pharmazie* **2001**, *56*, 967–970.
11. Mansuy-Aubert, V.; Zhou, Q.L.; Xie, X.; Gong, Z.; Huang, J.Y.; Khan, A.R.; Aubert, G.; Candelaria, K.; Thomas, S.; Shin, D.J.; et al. Imbalance between neutrophil elastase and its inhibitor α1-antitrypsin in obesity alters insulin sensitivity, inflammation, and energy expenditure. *Cell Metab.* **2013**, *17*, 534–548. [CrossRef]
12. Talukdar, S.; Oh, D.Y.; Bandyopadhyay, G.; Li, D.; Xu, J.; McNelis, J.; Lu, M.; Li, P.; Yan, Q.; Zhu, Y.; et al. Neutrophils mediate insulin resistance in mice fed a high-fat diet through secreted elastase. *Nat. Med.* **2012**, *18*, 1407–1412. [CrossRef]
13. Huang, Z.; Xie, N.; Illes, P.; Di Virgilio, F.; Ulrich, H.; Semyanov, A.; Verkhratsky, A.; Sperlagh, B.; Yu, S.G.; Huang, C.; et al. From purines to purinergic signalling: Molecular functions and human diseases. *Signal Transduct. Target. Ther.* **2021**, *6*, 162. [CrossRef]
14. Linden, J.; Koch-Nolte, F.; Dahl, G. Purine release, metabolism, and signaling in the inflammatory response. *Annu. Rev. Immunol.* **2019**, *37*, 325–347. [CrossRef] [PubMed]
15. Säve, S.; Persson, K. Extracellular ATP and P2Y receptor activation induce a proinflammatory host response in the human urinary tract. *Infect. Immun.* **2010**, *78*, 3609–3615. [CrossRef]
16. Merz, J.; Albrecht, P.; von Garlen, S.; Ahmed, I.; Dimanski, D.; Wolf, D.; Hilgendorf, I.; Härdtner, C.; Grotius, K.; Willecke, F.; et al. Purinergic receptor Y2 (P2Y2)- dependent VCAM-1 expression promotes immune cell infiltration in metabolic syndrome. *Basic Res. Cardiol.* **2018**, *113*, 45. [CrossRef] [PubMed]
17. Zhang, Y.; Ecelbarger, C.M.; Lesniewski, L.A.; Müller, C.E.; Kishore, B.K. P2Y2 receptor promotes high-fat diet-induced obesity. *Front. Endocrinol.* **2020**, *11*, 341. [CrossRef] [PubMed]
18. Wang, Y.; Nakajima, T.; Gonzalez, F.J.; Tanaka, N. PPARs as metabolic regulators in the liver: Lessons from liver-specific PPAR-null mice. *Int. J. Mol Sci.* **2020**, *21*, 2061. [CrossRef] [PubMed]
19. Mirza, A.Z.; Althagafi, I.I.; Shamshad, H. Role of PPAR receptor in different diseases and their ligands: Physiological importance and clinical implications. *Eur. J. Med. Chem.* **2019**, *166*, 502–513. [CrossRef]
20. Shen, W.; Gao, Y.; Lu, B.; Zhang, Q.; Hu, Y.; Chen, Y. Negatively regulating TLR4/NF-κB signaling via PPARα in endotoxin-induced uveitis. *Biochim. Biophys. Acta.* **2014**, *1842*, 1109–1120. [CrossRef]
21. Necela, B.M.; Su, W.; Thompson, E.A. Toll-like receptor 4 mediates cross-talk between peroxisome proliferator-activated receptor gamma and nuclear factor-kappaB in macrophages. *Immunology* **2008**, *125*, 344–358. [CrossRef]
22. Peredo-Escárcega, A.E.; Guarner-Lans, V.; Pérez-Torres, I.; Ortega-Ocampo, S.; Carreón-Torres, E.; Castrejón-Tellez, V.; Díaz-Díaz, E.; Rubio-Ruiz, M.E. The combination of resveratrol and quercetin attenuates Metabolic Syndrome in rats by modifying the serum fatty acid composition and by upregulating SIRT 1 and SIRT 2 expression in white adipose tissue. *Evid. Based Complement. Alternat. Med.* **2015**, *2015*, 474032. [CrossRef]
23. Faghihzadeh, F.; Hekmatdoost, A.; Adibi, P. Resveratrol and liver: A systematic review. *J. Res. Med. Sci.* **2015**, *20*, 797–810. [CrossRef] [PubMed]
24. Rubio-Ruiz, M.E.; Guarner-Lans, V.; Cano-Martínez, A.; Díaz-Díaz, E.; Manzano-Pech, L.; Gamas-Magaña, A.; Castrejón-Tellez, V.; Tapia-Cortina, C.; Pérez-Torres, I. Resveratrol and Quercetin administration improves antioxidant DEFENSES and reduces fatty liver in metabolic syndrome rats. *Molecules* **2019**, *24*, 1297. [CrossRef] [PubMed]
25. Burnstock, G. Purinergic signalling: Therapeutic developments. *Front. Pharmacol.* **2017**, *8*, 661. [CrossRef] [PubMed]

26. Kaulich, M.; Streicher, F.; Mayer, R.; Müller, I.; Müller, E. Flavonoids F novel lead compounds for the development of P2Y2 receptor antagonists. *Drug Dev. Res.* **2003**, *59*, 72–81. [CrossRef]
27. Faria, R.; Ferreira, L.; Bezerra, R.; Frutuoso, V.; Alves, L. Action of natural products on p2 receptors: A reinvented era for drug discovery. *Molecules* **2012**, *17*, 13009–13025. [CrossRef]
28. Izzo, C.; Annunziata, M.; Melara, G.; Sciorio, R.; Dallio, M.; Masarone, M.; Federico, A.; Persico, M. The role of resveratrol in liver disease: A comprehensive review from in vitro to clinical trials. *Nutrients* **2021**, *13*, 933. [CrossRef] [PubMed]
29. Castrejón-Tellez, V.; Rodríguez-Pérez, J.M.; Pérez-Torres, I.; Pérez-Hernández, N.; Cruz-Lagunas, A.; Guarner-Lans, V.; Vargas-Alarcón, G.; Rubio-Ruiz, M.E. The effect of resveratrol and quercetin treatment on PPAR mediated uncoupling protein (UCP-) 1, 2, and 3 expression in visceral white adipose tissue from metabolic syndrome rats. *Int. J. Mol. Sci.* **2016**, *17*, 1069. [CrossRef]
30. Benomar, Y.; Taouis, M. Molecular mechanisms underlying obesity-induced hypothalamic inflammation and insulin resistance: Pivotal role of resistin/TLR4 pathways. *Front. Endocrinol.* **2019**, *10*, 140. [CrossRef]
31. Ma, J.Q.; Li, Z.; Xie, W.R.; Liu, C.M.; Liu, S.S. Quercetin protects mouse liver against CCl_4-induced inflammation by the TLR2/4 and MAPK/NF-κB pathway. *Int. Immunopharmacol.* **2015**, *28*, 531–539. [CrossRef]
32. Xiong, G.; Ji, W.; Wang, F.; Zhang, F.; Xue, P.; Cheng, M.; Sun, Y.; Wang, X.; Zhang, T. Quercetin inhibits inflammatory response induced by LPS from Porphyromonas gingivalis in human gingival fibroblasts via suppressing NF-κB signaling pathway. *BioMed Res. Int.* **2019**, *2019*, 6282635. [CrossRef] [PubMed]
33. Zhang, Y.; Li, X.; Zhang, H.; Zhao, Z.; Peng, Z.; Wang, Z.; Liu, G.; Li, X. Non-Esterified Fatty Acids over-activate the TLR2/4-NF-Kb signaling pathway to increase inflammatory cytokine synthesis in neutrophils from ketotic cows. *Cell Physiol. Biochem.* **2018**, *48*, 827–837. [CrossRef] [PubMed]
34. Gimeno-Mallench, L.; Mas-Bargues, C.; Inglés, M.; Olaso, G.; Borras, C.; Gambini, J.; Vina, J. Resveratrol shifts energy metabolism to increase lipid oxidation in healthy old mice. *Biomed. Pharmacother.* **2019**, *118*, 109130. [CrossRef] [PubMed]
35. Uchida, Y.; Freitas, M.C.; Zhao, D.; Busuttil, R.W.; Kupiec-Weglinski, J.W. The inhibition of neutrophil elastase ameliorates mouse liver damage due to ischemia and reperfusion. *Liver Transpl.* **2009**, *15*, 939–947. [CrossRef] [PubMed]
36. Coutinho-Silva, R.; Stahl, L.; Cheung, K.K.; de Campos, N.E.; de Oliveira Souza, C.; Ojcius, D.M.; Burnstock, G. P2X and P2Y purinergic receptors on human intestinal epithelial carcinoma cells: Effects of extracellular nucleotides on apoptosis and cell proliferation. *Am. J. Physiol. Gastrointest. Liver Physiol.* **2005**, *288*, G1024–G1035. [CrossRef]
37. Wang, P.; Jia, J.; Zhang, D. Purinergic signalling in liver diseases: Pathological functions and therapeutic opportunities. *JHEP Rep.* **2020**, *2*, 100165. [CrossRef]
38. Li, X.; Jin, Q.; Yao, Q.; Xu, B.; Li, L.; Zhang, S.; Tu, C. The flavonoid quercetin ameliorates liver inflammation and fibrosis by regulating hepatic macrophages activation and polarization in mice. *Front. Pharmacol.* **2018**, *9*, 72. [CrossRef]
39. Araújo Miguel, N.; Franco Andrade, S.; Nai, G.; Braga Laposy, C.; Franco Nascimento, F.; Rangel Dinallo, H.; Melchert, A. Effects of resveratrol on liver function of obese female wistar rats. *Cienc. Anim. Bras.* **2016**, *17*, 402–410. [CrossRef]
40. Zhu, M.; Zhou, X.; Zhao, J. Quercetin prevents alcohol-induced liver injury through targeting of PI3K/Akt/nuclear factor-κB and STAT3 signaling pathway. *Exp. Ther. Med.* **2017**, *14*, 6169–6175. [CrossRef]
41. Okay, E.; Simsek, T.; Subasi, C.; Gunes, A.; Duruksu, G.; Gurbuz, Y.; Gacar, G.; Karaoz, E. Cross effects of resveratrol and mesenchymal stem cells on liver regeneration and homing in partially hepatectomized rats. *Stem. Cell Rev. Rep.* **2015**, *11*, 322–331. [CrossRef]
42. Sun, H.; Zhu, X.; Lin, W.; Zhou, Y.; Cai, W.; Qiu, L. Interactions of TLR4 and PPARγ, dependent on AMPK signalling pathway contribute to anti-inflammatory effects of Vaccariae hypaphorine in endothelial cells. *Cell Physiol. Biochem.* **2017**, *42*, 1227–1239. [CrossRef]
43. Hassan, N.F.; Nada, S.A.; Hassan, A.; El-Ansary, M.R.; Al-Shorbagy, M.Y.; Abdelsalam, R.M. Saroglitazar deactivates the hepatic LPS/TLR4 signaling pathway and ameliorates adipocyte dysfunction in rats with high-fat emulsion/LPS model-induced Non-alcoholic steatohepatitis. *Inflammation* **2019**, *42*, 1056–1070. [CrossRef] [PubMed]
44. Montagner, A.; Polizzi, A.; Fouché, E.; Ducheix, S.; Lippi, Y.; Lasserre, F.; Barquissau, V.; Régnier, M.; Lukowicz, C.; Benhamed, F.; et al. Liver PPARα is crucial for whole-body fatty acid homeostasis and is protective against NAFLD. *Gut* **2016**, *65*, 1202–1214. [CrossRef]
45. Li, J.; Guo, C.; Wu, J. The agonists of Peroxisome Proliferator-Activated Receptor-γ for liver fibrosis. *Drug Des. Devel. Ther.* **2021**, *15*, 2619–2628. [CrossRef]
46. Stienstra, R.; Saudale, F.; Duval, C.; Keshtkar, S.; Groener, J.E.; van Rooijen, N.; Staels, B.; Kersten, S.; Müller, M. Kupffer cells promote hepatic steatosis via interleukin-1beta-dependent suppression of peroxisome proliferator-activated receptor alpha activity. *Hepatology* **2010**, *51*, 511–522. [CrossRef]
47. Brocker, C.N.; Yue, J.; Kim, D.; Qu, A.; Bonzo, J.A.; Gonzalez, F.J. Hepatocyte-specific PPARA expression exclusively promotes agonist-induced cell proliferation without influence from nonparenchymal cells. *Am. J. Physiol. Gastrointest. Liver Physiol.* **2017**, *312*, G283–G299. [CrossRef] [PubMed]
48. Xu, Y.; Han, J.; Dong, J.; Fan, X.; Cai, Y.; Li, J.; Wang, T.; Zhou, J.; Shang, J. Metabolomics characterizes the effects and mechanisms of Quercetin in nonalcoholic fatty liver disease development. *Int. J. Mol. Sci.* **2019**, *20*, 1220. [CrossRef]

49. Bergmeyer, H.U.; Hørder, M.; Rej, R. International Federation of Clinical Chemistry (IFCC) Scientific Committee, Analytical Section: Approved recommendation (1985) on IFCC methods for the measurement of catalytic concentration of enzymes. Part 2. IFCC method for aspartate aminotransferase (L-aspartate: 2-oxoglutarate aminotransferase, EC 2.6.1.1). *J. Clin. Chem. Clin. Biochem.* **1986**, *24*, 497–510.
50. Bradford, M.M. A rapid and sensitive method for the quantitation of microgram quantities of protein utilizing the principle of protein-dye binding. *Anal. Biochem.* **1976**, *72*, 248–254. [CrossRef]

International Journal of
Molecular Sciences

MDPI

Review

The PPAR β/δ-AMPK Connection in the Treatment of Insulin Resistance

David Aguilar-Recarte [1,2,3], Xavier Palomer [1,2,3], Walter Wahli [4,5,6] and Manuel Vázquez-Carrera [1,2,3,*]

1 Department of Pharmacology, Toxicology and Therapeutic Chemistry, Institute of Biomedicine of the University of Barcelona (IBUB), Faculty of Pharmacy and Food Sciences, University of Barcelona, Avinguda Joan XXIII 27-31, 08028 Barcelona, Spain; d.aguilarrecarte@gmail.com (D.A.-R.); xpalomer@ub.edu (X.P.)
2 Pediatric Research Institute-Hospital Sant Joan de Déu, 08950 Esplugues de Llobregat, Spain
3 Spanish Biomedical Research Centre in Diabetes and Associated Metabolic Diseases (CIBERDEM)-Instituto de Salud Carlos III, 28029 Madrid, Spain
4 Center for Integrative Genomics, University of Lausanne, CH-1015 Lausanne, Switzerland; walter.wahli@unil.ch
5 Lee Kong Chian School of Medicine, Nanyang Technological University Singapore, Singapore 308232, Singapore
6 ToxAlim (Research Center in Food Toxicology), INRAE, UMR1331, CEDEX, 31300 Toulouse, France
* Correspondence: mvazquezcarrera@ub.edu

Abstract: The current treatment options for type 2 diabetes mellitus do not adequately control the disease in many patients. Consequently, there is a need for new drugs to prevent and treat type 2 diabetes mellitus. Among the new potential pharmacological strategies, activators of peroxisome proliferator-activated receptor (PPAR)β/δ show promise. Remarkably, most of the antidiabetic effects of PPARβ/δ agonists involve AMP-activated protein kinase (AMPK) activation. This review summarizes the recent mechanistic insights into the antidiabetic effects of the PPARβ/δ-AMPK pathway, including the upregulation of glucose uptake, muscle remodeling, enhanced fatty acid oxidation, and autophagy, as well as the inhibition of endoplasmic reticulum stress and inflammation. A better understanding of the mechanisms underlying the effects resulting from the PPARβ/δ-AMPK pathway may provide the basis for the development of new therapies in the prevention and treatment of insulin resistance and type 2 diabetes mellitus.

Keywords: PPARβ/δ; AMPK; GDF15; insulin resistance; type 2 diabetes mellitus

Citation: Aguilar-Recarte, D.; Palomer, X.; Wahli, W.; Vázquez-Carrera, M. The PPAR β/δ-AMPK Connection in the Treatment of Insulin Resistance. *Int. J. Mol. Sci.* **2021**, *22*, 8555. https://doi.org/10.3390/ijms22168555

Academic Editor: Wolfgang Graier

Received: 22 July 2021
Accepted: 5 August 2021
Published: 9 August 2021

1. Insulin Resistance: A Major Determinant of Type 2 Diabetes Mellitus

The prevalence of type 2 diabetes mellitus has reached global epidemic proportions and is one of the medical challenges of the 21st century [1]. Type 2 diabetes mellitus is defined by the presence of fasting hyperglycemia, which is responsible for the development of long-term complications, a decreased quality of life, and premature death [1]. It should be noted that abnormal glucose regulation may begin more than 10 years before the diagnosis of type 2 diabetes mellitus with the development of obesity-associated insulin resistance, which is defined as an impairment in the ability of insulin to maintain glucose homeostasis. However, at this early stage, subjects are asymptomatic, with glycemic values near normal levels because pancreatic islets usually respond by increasing insulin secretion to maintain normoglycemia in a process known as β cell compensation. Over time, β cell compensation for insulin resistance fails, resulting in fasting hyperglycemia and the establishment of type 2 diabetes mellitus [2]. As insulin resistance precedes and predicts type 2 diabetes mellitus [3], the development of new effective pharmacological approaches that prevent or delay its progression to type 2 diabetes mellitus relies on targeting the underlying pathological mechanisms. This is of paramount importance as current treatment options do not adequately control hyperglycemia or prevent the negative impact of type 2 diabetes mellitus in all patients. Among the new pharmacological strategies for treating obesity-induced insulin resistance and type 2 diabetes mellitus, Peroxisome Proliferator-Activated Receptor (PPAR)β/δ agonists show promise [4–6]. Ligands of this nuclear receptor have been reported to ameliorate insulin resistance and type 2 diabetes mellitus mainly through the activation of AMP-activated protein kinase (AMPK), a central regulator of multiple metabolic pathways. This review summarizes the recent mechanistic insights into how PPARβ/δ activates AMPK to ameliorate insulin resistance and type 2 diabetes mellitus.

2. Basic PPARβ/δ and AMPK Features

PPARs are members of the nuclear receptor superfamily of ligand-inducible transcription factors. The PPAR subfamily comprises three isotypes: PPARα (NR1C1: nuclear receptor subfamily 1, group C, member 1, according to the nomenclature agreed by the NC-IUPHAR Subcommittee on Nuclear Hormone Receptors), PPARβ/δ (NR1C2), and PPARγ (NR1C3) [4–6]. The PPARβ/δ isotype is ubiquitously expressed, but is most abundant in metabolically active tissues/cells, mainly those associated with fatty acid (FA) metabolism such as skeletal and cardiac muscle, hepatocytes, and adipocytes, and in macrophages. Ligand binding and activation of PPARβ/δ lead to its heterodimerization with its obligate dimerization partner retinoic acid receptor (RXR or NR2B). These heterodimers then bind to peroxisome proliferator response elements (PPREs) located in the promoters of their target genes to regulate their transcription. PPARβ/δ also regulates gene expression through DNA-independent mechanisms via crosstalk with other transcription factors [4–6]. Furthermore, it has been proposed that PKCα is a binding partner of PPARβ/δ, suggesting it as a mechanism through which the receptor may impact platelet reactivity [7]. Another example for a non-genomic effect of PPARβ/δ is the ligand-dependent interaction of the receptor with T-cell protein tyrosine phosphatase 45 (TCPTP45), which enhances insulin signaling [8]. In addition, the physiological activation status of PPARβ/δ depends on the presence of tissue-enriched specific ligands and the recruitment of coactivators or corepressors. Many of the target genes regulated by PPARβ/δ are involved in lipid and glucose metabolism, tissue repair, and inflammation [4–6]. The natural ligands of all PPAR isotypes are polyunsaturated and saturated FAs and their derivatives, but most of them show little receptor isotype selectivity. The development of several synthetic ligands with a high affinity and specificity for PPARβ/δ (GW501516, GW0742, and L-165041) has helped the understanding of the functions and pharmacology of this nuclear receptor [6] (Figure 1). Although no selective PPARβ/δ agonists have yet been approved for human use, several ongoing clinical trials are studying the efficacy and safety of several compounds selectively targeting this nuclear receptor: ASP0367 and ASP1128 (Mitobridge/Astellas Pharma, Cambridge, USA), MBX-8025 or Seladelpar (CymaBay Therapeutics, Neward, NJ, USA), and REN-001 (Reneo Pharmaceuticals, San Diego, CA, USA).

Over the last twenty years, many studies have robustly demonstrated that PPARβ/δ is crucial in regulating lipid metabolism and glucose homeostasis. Consequently, its activation is especially helpful in experimental models to prevent insulin resistance, type 2 diabetes mellitus, and associated metabolic disorders. Interestingly, many of the antidiabetic effects of the PPARβ/δ activators involve the activation of AMPK [6].

AMPK is a protein kinase that protects against insulin resistance and is activated by a low cellular energy status and glucose starvation [9]. These conditions, which activate AMPK, are signaled by the rise of the cellular AMP/ATP and ADP/ATP ratios. Once activated, AMPK triggers catabolic pathways that generate ATP and inhibits anabolic pathways that consume ATP. The heterotrimeric structure of AMPK comprises the α catalytic subunit and the regulatory β and γ subunits [9–11]. The binding of AMP to the γ subunit promotes AMPK activation through the phosphorylation of a conserved threonine (Thr172) residue within the α subunit via three complementary mechanisms: (1) phosphorylation by the upstream kinases liver kinase B1 (LKB1), Ca^{2+}/calmodulin-dependent protein kinase kinase β (CaMKKβ), and transforming growth factor β-activated kinase 1 (TAK1); (2) inhibition of Thr172 dephosphorylation by protein phosphatases;

and (3) allosteric activation. In addition to AMP, ADP also activates AMPK through mechanisms 1 and 2, while ATP inhibits these three mechanisms [9–11].

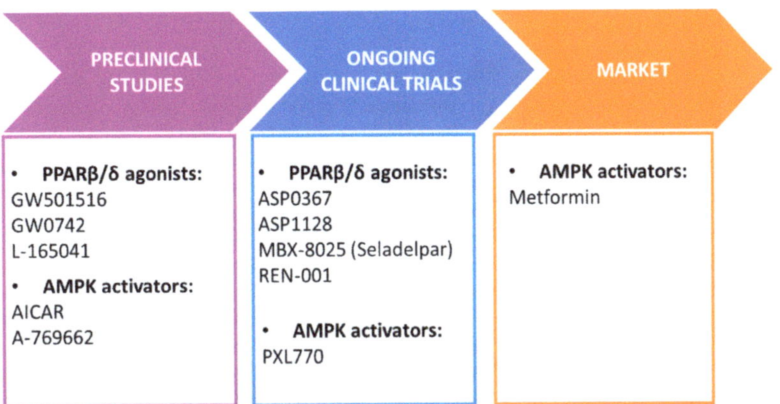

Figure 1. PPARβ/δ agonists and AMPK activators and current status in clinical pipeline. AICAR, 5-aminoimidazole-4-carboxamide ribonucleoside.

Given the importance of AMPK in lowering insulin resistance and associated metabolic disorders, many AMPK activators with different mechanisms of action have been developed. The most important AMPK activator is metformin, which is the most prescribed drug for type 2 diabetes mellitus treatment (Figure 1). However, its mechanism of action remains to be fully elucidated [12]. It has been reported that pharmacological metformin concentrations directly activate AMPK. By contrast, suprapharmacological metformin concentrations inhibit mitochondrial complex I, thereby reducing mitochondrial ATP production and increasing cellular AMP levels that subsequently activate AMPK [10,12]. A novel direct AMPK activator, PXL770 (Poxel), is being evaluated in an ongoing clinical trial (ClinicalTrials.gov 3 August 2021). In addition, many natural products, including resveratrol [13] and berberine [14], also indirectly activate AMPK by increasing cellular AMP levels. Another group of AMPK activators are AMP analogs, such as 5-aminoimidazole-4-carboxamide ribonucleoside (AICAR), which activate the γ subunit of AMPK [15]. A different group of ligands, exemplified by A-769662, includes synthetic direct activators that promote the allosteric activation of AMPK and the protection against Thr172 dephosphorylation [16,17]. Tetrahydrofolate analogs such as pemetrexed and methotrexate constitute another group of AMPK activators. These molecules inhibit the metabolism of ZMP, the phosphorylated form of AICAR, and promote its accumulation and subsequent activation of AMPK [18,19]. Finally, AMPK inhibitors are also useful in elucidating the effects mediated by this kinase. Compound C/dorsomorphin is an ATP-competitive AMPK inhibitor. However, this inhibitor is not specific for AMPK and shows AMPK-independent cellular effects [20]. More recently, a new direct inhibitor of AMPK has been characterized, SBI-0206965, with a 40-fold greater potency than compound C [21].

Once AMPK is activated, it phosphorylates key metabolic substrates and transcriptional regulators that affect many aspects of cellular metabolism, increasing glucose uptake, FA oxidation, mitochondrial oxidative capacity, and insulin sensitivity [22,23]. Interestingly, a high-fat diet (HFD) reduces AMPK phosphorylation levels in the skeletal muscle, liver, and other tissues, thereby indicating that restoration of the activity of this kinase can overcome metabolic alterations associated with the overconsumption of fat in animal models.

3. PPARβ/δ as a Major Regulator of Insulin Resistance through AMPK Activation

In the following sections of the review, we discuss studies that implicate AMPK activation in the antidiabetic effects of PPARβ/δ ligands in the main organs involved in insulin resistance.

3.1. Skeletal Muscle

The primary site of insulin resistance in obesity and type 2 diabetes mellitus is the skeletal muscle, as it accounts for around 80% of insulin-stimulated glucose disposal [24–26]. Activation of AMPK in skeletal muscle by contraction (a process that results in a significant decrease in cellular ATP levels) or by activators of this kinase is associated with an insulin-independent mechanism that stimulates glucose transporter 4 (GLUT4) vesicle trafficking to the plasma membrane, resulting in elevated glucose transport into muscle, which lowers plasma glucose levels. This mechanism involves the phosphorylation by AMPK of tre-2/USP6, BUB2, cdc16 domain family member 1 (TBC1D1) and TBC1D4 (also known as Akt substrate of 160 kDa, AS160) [27], and phosphatidylinositol 3-phosphate 5-kinase [28]. Contrary to what was initially believed, a recent study suggested a role for AMPK in the regulation of insulin-stimulated glucose uptake [29]. The PPARβ/δ agonist GW501516 was reported to upregulate basal and insulin-stimulated glucose uptake in cultured primary human skeletal myotubes through AMPK activation [30], providing a role for AMPK in the antidiabetic effects of PPARβ/δ agonists (Figure 2). The authors of the study later reported that the activation of AMPK by GW501516 could be due to a reduction of the cellular energy status, as they observed an increase in the AMP/ATP ratio [31] (Figure 3). Moreover, transgenic mice with muscle-specific overexpression of PPARβ/δ show increased levels of mitochondrial enzymes and oxidative muscle fibers, which are more resistant to fatigue than glycolytic fibers, resulting in enhanced running endurance [32]. Notably, this overexpression of PPARβ/δ is accompanied by AMPK activation, with GW501516 and exercise training synergistically increasing oxidative myofibers and running endurance [33] (Figure 2). In the skeletal muscle of these mice, there is an interaction between PPARβ/δ and AMPK that is accompanied by more glycogen stores, increased levels of GLUT4, and an augmented capacity for mitochondrial pyruvate oxidation [34]. Thus, PPARβ/δ mimics the effects of endurance exercise training and GW501516 could be used as an exercise mimetic. In fact, this compound, sold under the name of Cardarine, has been misused for performance enhancement [35] and was entered into the list of prohibited substances in 2009 by the World Anti-Doping Agency [36]. This effect of PPARβ/δ was initially reported not to be associated with an increase in the mRNA levels of PPARγ co-activator 1α (PGC-1α) [32]. PGC-1α mediates mitochondrial biogenesis and its upregulation is associated with adaptation to endurance exercise through increased muscle mitochondrial numbers. However, later studies confirmed that PPARβ/δ does increase the protein levels of this transcriptional co-activator [37,38]. More recently, an elegant study revealed the mechanisms by which PPARβ/δ increased PGC-1α levels and activated AMPK in skeletal muscle during exercise [39]. PPARβ/δ increased PGC-1α protein levels via a post-transcriptional mechanism by protecting it from degradation through binding to PGC-1α and limiting its ubiquitination. PPARβ/δ also promoted the transcription of nuclear respiratory factor 1 (NRF-1), resulting in increases in the mitochondrial respiratory chain and in the transcription of CaMKKβ, ultimately leading to AMPK activation [38] (Figure 3). Overall, these findings showed that PPARβ/δ is essential for the maintenance and increase in mitochondrial enzymes, unveiling a new mechanism through which this nuclear receptor activates AMPK. This conclusion is supported by the phenotype of mice in which PPARβ/δ is selectively ablated in skeletal muscle myocytes. This somatic mutation causes a muscle fiber-type switching toward lower oxidative capacity that results in markedly reduced capacity to sustain running exercise, obesity, and type 2 diabetes mellitus [37].

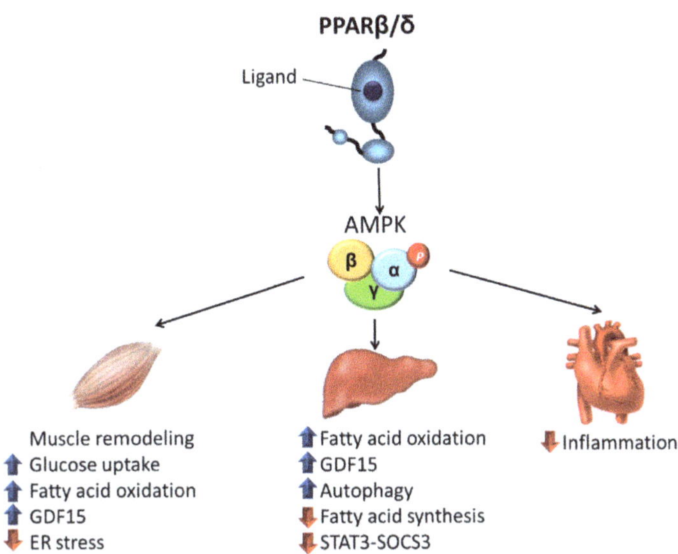

Figure 2. Antidiabetic effects of the PPARβ/δ-AMPK pathway in different organs. AMPK, AMP-activated protein kinase; ER, endoplasmic reticulum; GDF15, growth differentiation factor 15; PPARβ/δ: peroxisome proliferator-activated receptor β/δ; SOCS3: suppressor of cytokine signaling 3; STAT3: signal transducer and activator of transcription 3.

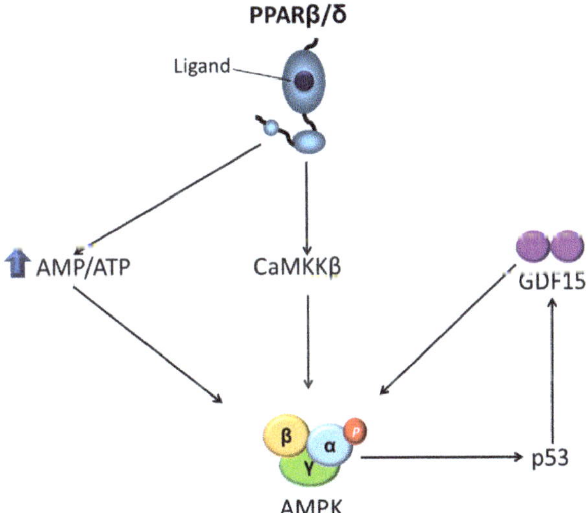

Figure 3. Mechanisms involved in the activation of AMPK by PPARβ/δ. AMPK is activated by PPARβ/δ through three mechanisms: (1) an increased AMP/ATP ratio; (2) an increased transcription of CaMKKβ; and (3) increased levels of GDF15 that sustain AMPK activation. AMPK, AMP-activated protein kinase; CaMKKβ, Ca^{2+}/calmodulin-dependent protein kinase kinase-β; GDF15, growth differentiation factor 15; PPARβ/δ: peroxisome proliferator-activated receptor β/δ.

In obesity, as the amount of visceral adipose tissue increases, so does the rate of lipolysis. This increases FA mobilization and raises the levels of circulating non-esterified FAs, which induce insulin resistance in skeletal muscle through activation of toll-like receptor (TLR)-dependent mechanisms or by promoting the accumulation of deleterious complex FA derivatives such as diacylglycerol (DAG) and ceramides. These pathways ultimately activate kinases (IκB kinase β, c-Jun N-terminal kinase 1, and protein kinase Cθ) that phosphorylate insulin receptor substrate 1 (IRS-1) on serine residues, attenuating the insulin signaling pathway [40]. The activation of PPARβ/δ in myotubes has been reported to transcriptionally upregulate the expression of target genes involved in FA β-oxidation such as pyruvate dehydrogenase kinase 4 (PDK4) and carnitine palmitoyltransferase-1β (CPT-1β). The increase in the expression of these genes promotes FA β-oxidation and reduces their availability to form complex lipids that induce insulin resistance [41] (Figure 2). CPT-1β, which catalyzes the rate-limiting step of mitochondrial FA oxidation, is inhibited by malonyl-CoA, a product of acetyl-CoA carboxylase (ACC) [22]. AMPK phosphorylates and inhibits ACC, thereby causing a decrease in intracellular malonyl-CoA levels, relieving CPT-1β inhibition and increasing FA oxidation. Therefore, PPARβ/δ activation in skeletal muscle increases mitochondrial FA oxidation by upregulating the expression of the target genes involved in this process as well as through increasing CPT-1β activity by phosphorylating AMPK.

In obese patients, the release of free FAs from visceral adipose tissue is also an important factor that triggers endoplasmic reticulum (ER) stress. This process induces insulin resistance by several mechanisms including the activation of inflammatory pathways, which activate the serine/threonine kinases that phosphorylate IRS-1 on serine residues [42]. PPARβ/δ ligands inhibit ER stress in skeletal muscle through a mechanism that seems to involve AMPK activation and the subsequent inhibition of extracellular signal-regulated kinase (ERK1/2) (Figure 2). In fact, AMPK activation protects against several deleterious processes by reducing ER stress [43–46]. Notably, there is inhibitory crosstalk between AMPK and ERK1/2 [47], with the inhibition of ERK1/2 promoting AMPK and Akt signaling and reversing ER stress-induced insulin resistance in skeletal muscle cells [48]. Therefore, PPARβ/δ ligands seem to require the activation of AMPK to inhibit ER stress, which strongly contributes to the antidiabetic effects of these compounds.

Recently, we reported that the metabolic effects caused by the pharmacological activation of PPARβ/δ may involve the stress-activated cytokine growth differentiation factor 15 (GDF15) [49]. This divergent member of the transforming growth factor β (TGFβ) superfamily [50] plays an important role in several biological processes, including the regulation of energy homeostasis [51]. In fact, overexpression of *Gdf15* in mice ameliorates glucose tolerance and insulin sensitivity and lowers body weight, although no difference in food intake was observed [52]. By contrast, administration of GDF15 to rodents reduces food intake and ameliorates glucose tolerance. Interestingly, a recent study reports that high pharmacological doses of GDF15 used in most studies reduce food intake, while physiological induction of endogenous circulating GDF15 levels does not affect it [53]. Although TGFβ receptors were initially reported to mediate the effects of GDF15, the presence of TGFβ contamination in recombinant GDF15 and the lack of a direct binding of GDF15 to known TGFβ receptors led to the search for the bona fide receptor of GDF15. Four independent groups reported in 2017 that GDF15 signals through the glial cell line-derived neurotrophic factor (GDNF)-like alpha-1 (glial cell-derived neurotrophic factor receptor alpha-like (GFRAL))/rearranged during transfection (RET) co-receptor complex [54–57]. The expression of GFRAL is limited to the central nervous system, specifically in the area postrema of the brainstem and parts of the nucleus of the solitary tract. Its activation by GDF15 in obesity improves glucose tolerance by reducing food intake. However, it has been reported that GDF15 also regulates metabolic parameters independently of changes in food intake [58], suggesting that GDF15 might also exert its effects via other receptors and peripheral mechanisms. We have reported recently that PPARβ/δ ligands increase GDF15 levels through an AMPK-p53-dependent mechanism [49]. Interestingly, the beneficial effects of the PPARβ/δ agonist GW501516 on glucose intolerance, FA oxidation, ER stress, inflammation, and AMPK activation in HFD-fed mice were abrogated by the injection of a GDF15-neutralizing antibody as well as in *Gdf15*$^{-/-}$ mice. More importantly, these findings demonstrated that the increase in GDF15 caused by PPARβ/δ activation resulted in AMPK activation that did not require central effects, as these effects were observed in cultured myotubes and isolated muscle, suggesting the presence of autocrine/paracrine effects for GDF15 in skeletal muscle for which the mediating receptor remains to be identified (Figure 3). Although additional studies are needed to reject the potential involvement of GFRAL on the GDF15-mediated antidiabetic effects of PPARβ/δ agonists, as *Gfral* mRNA is absent in C2C12 cells [49,55] and skeletal muscle [49,59], the GDF15-mediated activation of AMPK in isolated skeletal muscle and cultured myotubes seems to exclude this receptor. The question that remains unanswered is the identity of the new potential receptor responsible for the autocrine/paracrine effects of GDF15 in skeletal muscle. Future studies should shed light on this issue.

3.2. Liver

Alterations in liver function are frequently observed in insulin resistance and type 2 diabetes mellitus. In fact, many patients suffering these metabolic alterations present nonalcoholic fatty liver disease (NAFLD), defined by a hepatic lipid accumulation >5% of the liver weight [60]. Hepatic lipid accumulation can also trigger inflammation, resulting in more severe liver disorders such as nonalcoholic steatohepatitis (NASH), cirrhosis, and hepatocellular carcinoma (HCC). Intriguingly, although hepatic lipid accumulation results from insulin resistance, it also contributes to hepatic insulin resistance [61], thereby suggesting that reversing hepatic steatosis can delay the progression from prediabetes to overt type 2 diabetes mellitus. Unregulated lipogenesis and reduced FA oxidation contribute to lipid accumulation in the liver, with AMPK regulating both processes in hepatocytes. Thus, as mentioned above, AMPK-mediated ACC inhibition leads to a decrease in intracellular levels of malonyl-CoA, which is both a precursor for FA biosynthesis and a potent allosteric inhibitor of FA oxidation. Moreover, AMPK reduces the expression of lipogenic genes by phosphorylating transcription factors such as sterol regulatory element binding protein-1c (SREBP-1c) [62] and carbohydrate-responsive element-binding protein (ChREBP) [63]. It has been reported that HFDs reduce hepatic phospho-AMPK levels and increase phospho-ERK levels, with GW501516 treatment preventing these changes by a mechanism that may involve an increased AMP/ATP ratio and elevated plasma β-hydroxybutyrate levels, indicating enhanced hepatic FA oxidation [64] (Figure 2). Interestingly, a different study reported that GW501516 treatment stimulated AMPK and ACC phosphorylation and attenuated FA synthesis in wild-type hepatocytes, but not in AMPKβ$^{-/-}$ hepatocytes [65], thereby confirming the involvement of AMPK in these effects.

Autophagy is a catabolic process that delivers intracellular proteins and organelles to the lysosome during starvation for degradation and recycling, thereby promoting the redistribution of nutrients to maintain cellular energetic balance [66]. Notably, the inhibition of autophagy results in triglyceride accumulation and reduced FA oxidation in the liver, while drugs increasing autophagy alleviate liver steatosis in mice fed an HFD [67]. AMPK activation promotes autophagy through two different mechanisms: inhibition of the mammalian target of rapamycin (mTOR) protein kinase complex and direct phosphorylation of Unc-51-like kinase 1 (ULK1) [68]. Recently, it has been demonstrated that PPARβ/δ reduces hepatic steatosis and stimulates FA oxidation in the liver and hepatic cells by an autophagy-lysosomal pathway involving the AMPK-mTOR pathway [69] (Figure 2). More generally, the roles of PPARs and their novel ligands as potential drugs for the treatment of NAFLD have been reviewed recently [70].

Insulin resistance and type 2 diabetes mellitus are closely associated with a chronic low-grade inflammation characterized by an abnormal production of cytokines. Of these cytokines, interleukin 6 (IL-6) has been reported to induce hepatic insulin resistance [71]. IL-6 induces insulin resistance in the liver through the activation of signal transducer and activator of transcription 3 (STAT3) and the subsequent induction of suppressor of cytokine signaling 3 (SOCS3), which inhibits insulin signaling by interfering with insulin receptor activation, blocking IRS activation, and inducing IRS degradation [72]. In liver cells, PPARβ/δ activation was demonstrated to prevent IL-6-induced STAT3 activation and SOCS3 upregulation by counteracting the reduction in phospho-AMPK levels, which inhibits STAT3 phosphorylation [73] (Figure 2). Consistent with this, the livers of *Ppard*$^{-/-}$ mice show increased phospho-STAT3 levels. This action of PPARβ/δ prevents the reduction in IRS-1 and IRS-2 levels caused by exposure of hepatic cells to IL-6 [73].

3.3. Heart

The risk of developing heart failure is higher in patients with insulin resistance and type 2 diabetes mellitus, with inflammation being a key systemic factor contributing to this relationship [74]. Indeed, the progression of cardiac hypertrophy and heart failure usually entails a local rise in proinflammatory factors, which are under the transcriptional control of nuclear factor-κB (NF-κB). Notably, AMPK activation may block NF-κB signaling through suppressing IκB kinase activity [75]. It has been reported that PPARβ/δ activation reduces the lipid-induced expression of NF-κB-target genes in the hearts of mice and in human cardiac cells, with these effects involving an AMPK-dependent mechanism [76] (Figure 2). In addition, NF-κB activity has been reported to be increased in the hearts of PPARβ/δ-knockout mice compared with wild-type mice, which is consistent with the anti-inflammatory effects of PPARβ/δ activity.

ER stress contributes to the pathogenesis of diabetic cardiomyopathy by promoting apoptotic cell death in the myocardium [77]. PPARβ/δ activation prevents lipid-induced ER stress in the heart by inducing autophagy [78]. In addition, PPARβ/δ-knockout mice display a reduction in autophagic markers. However, in contrast to what has been reported for the liver [69], these effects of PPARβ/δ occur in an AMPK-independent manner.

4. Going the Other Way: The AMPK-PPARβ/δ Pathway

While previous sections of this review clearly demonstrate that many of the antidiabetic effects of PPARβ/δ agonists are mediated via AMPK activation, a few studies have reported the opposite, i.e., the regulation of PPARβ/δ by AMPK. In fact, a recent study proposed the existence of a positive loop between activated AMPK, PPARβ/δ, and myocyte enhancer factor 2A (MEF2A), the latter being a transcription factor that upregulates the expression of *Ppard* and *Glut4* [79]. The authors of this study demonstrated that AMPK activation increases PPARβ/δ levels via MEF2A [79]. As mentioned above, increased levels of PPARβ/δ would activate the NRF-1/CaMKKβ pathway, thereby leading to AMPK activation, ultimately closing the loop (Figure 4). Thus, PPARβ/δ activates AMPK and AMPK activity influences PPARβ/δ levels, establishing a mutual cooperation that regulates MEF2A promoter activity and *Glut4* expression.

More recently, it has been reported that AMPK regulates PPARβ/δ phosphorylation, modulating its activity [80]. The authors of this study observed that the AMPK agonist metformin induced the phosphorylation of PPARβ/δ at Ser^{50} through the common LXRXXSXXXL phosphorylation motif recognized by this kinase, which localizes in the short N-terminal A/B activation domain of this nuclear receptor. Of note, AMPK-mediated phosphorylation of PPARβ/δ at Ser^{50} resulted in an accumulation of the protein levels of this PPAR isotype, suggesting that its phosphorylation attenuated PPARβ/δ degradation. In fact, PPARβ/δ phosphorylation at Ser^{50} inhibits the p62-mediated misfolded PPARβ/δ autophagic degradation. Despite the increase in PPARβ/δ levels caused by AMPK activation, the findings of this study suggest that PPARβ/δ phosphorylation inhibits transcriptional activity as a PPARβ/δ-Ser^{50} mutant showed increased activity compared with wild-type PPARβ/δ. Although this study was conducted in cancer cell lines, the AMPK-mediated phosphorylation of PPARβ/δ attenuated glucose uptake by reducing the expression of *Glut1*, thereby suggesting that this pathway can have metabolic implications. Further studies are needed to confirm whether this pathway operates in metabolic tissues such as the liver and skeletal muscle and how it regulates metabolism.

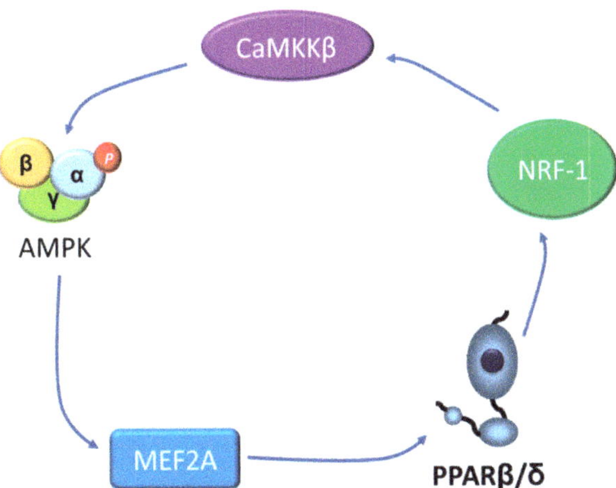

Figure 4. Potential positive loop between activated AMPK, PPARβ/δ, and MEF2A. AMPK, AMP-activated protein kinase; CaMKKβ, Ca^{2+}/calmodulin-dependent protein kinase kinase-β; MEF2A, myocyte enhancer factor 2A; NRF-1: nuclear respiratory factor 1; PPARβ/δ, peroxisome proliferator-activated receptor β/δ.

5. Conclusions and Perspectives

The development of novel drugs to treat type 2 diabetes mellitus continues to attract attention in the metabolism field. The PPARβ/δ-AMPK pathway is in the spotlight as it pharmacologically promotes the effects of exercise in skeletal muscle, such as increased glucose uptake and FA oxidation. This pathway also prevents lipid-induced ER stress and inflammation, thereby ameliorating insulin resistance. New specific molecular mechanisms indicating how this pathway ameliorates insulin resistance are beginning to emerge, such as the recently reported upregulation of GDF15 by PPARβ/δ agonists via AMPK. GDF15 upregulation activates AMPK, thereby implying that this mechanism contributes to the effects of PPARβ/δ agonists by sustaining AMPK activation. In addition, *Gdf15*$^{-/-}$ mice show reduced AMPK activation in skeletal muscle, whereas GDF15 administration results in AMPK activation in this organ. Interestingly, this effect of GDF15 in AMPK activation seems to be independent of the central receptor GFRAL, thereby suggesting that this cytokine exerts autocrine/paracrine effects through yet to be determined receptors. Future studies aimed at expanding the mechanisms of action of the PPARβ/δ-AMPK pathway may facilitate the development of new antidiabetic compounds with improved efficacy and minimal side effects for the treatment of insulin resistance and the prevention of its progression to type 2 diabetes mellitus. In fact, type 2 diabetic patients might benefit from the development of new antidiabetic drugs targeting both PPARβ/δ and AMPK given the positive feedback loop that potentiates them each other. This might result in a new generation of molecules for the prevention and treatment of obesity-induced insulin resistance and type 2 diabetes mellitus. It is noteworthy that PPARβ/δ, similar to PPARα and PPARγ, has been ascribed pro- and anti-tumor activities that have to be considered in the development of new candidate drugs [5,81,82]. Several factors can contribute to the highly debated functional role of PPARβ/δ in tumorigenesis or carcinogenesis. For instance, the tumor promoter effects of PPARβ/δ agonists have been mostly observed in animal models. Although these animal models are a valuable tool for basic tumor research, they show some limitations and the conclusions obtained from these studies are not always confirmed in human beings. Thus, the expression of the different PPAR isotypes is higher in rodent than in human cells and the regulation of these nuclear receptors is also different depending on the cell type studied [5]. These differences may explain why, after decades of treating patients with the PPARα activators fibrates, no incidence of carcinogenesis has been reported, whereas it is well-known that administration of these drugs to rodents leads to carcinogenesis. Either way, as controversy about the role of PPARβ/δ agonists in cancer still remains, to minimize side effects, the success of PPARβ/δ-based treatment of insulin resistance would benefit from the development of innovative strategies for organ- or cell-type-specific drug delivery or release systems.

Funding: This work was funded by the Ministerio de Economía y Competitividad of the Spanish Government (RTI2018-093999-B-100) and CIBER de Diabetes y Enfermedades Metabólicas Asociadas (CIBERDEM). CIBERDEM is an initiative of the Instituto de Salud Carlos III (IS-CIII)—Ministerio de Economía y Competitividad.

Institutional Review Board Statement: Not applicable.

Informed Consent Statement: Not applicable.

Data Availability Statement: Not applicable.

Acknowledgments: We thank Language Services of the University of Barcelona for revising the manuscript.

Conflicts of Interest: The authors declare no conflict of interest.

References

1. Zimmet, P.; Alberti, K.G.M.M.; Shaw, J. Global and societal implications of the diabetes epidemic. *Nature* **2001**, *414*, 782–787. [CrossRef]
2. Alejandro, E.U.; Gregg, B.; Blandino-Rosano, M.; Cras-Méneur, C.; Bernal-Mizrachi, E. Natural history of β-cell adaptation and failure in type 2 diabetes. *Mol. Asp. Med.* **2015**, *42*, 19–41. [CrossRef]
3. Tripathy, D.; Chavez, A.O. Defects in insulin secretion and action in the pathogenesis of type 2 diabetes mellitus. *Curr. Diab. Rep.* **2010**, *10*, 184–191. [CrossRef] [PubMed]
4. Giordiano Attianese, G.M.P.; Desvergne, B. Integrative and systemic approaches for evaluating PPARβ/δ (PPARD) function. *Nucl. Recept. Signal* **2015**, *13*, 13001. [CrossRef] [PubMed]
5. Tan, N.S.; Vázquez-Carrera, M.; Montagner, A.; Sng, M.K.; Guillou, H.; Wahli, W. Transcriptional control of physiological and pathological processes by the nuclear receptor PPARβ/δ. *Prog. Lipid Res.* **2016**, *64*, 98–122. [CrossRef]
6. Vázquez-Carrera, M. Unraveling the Effects of PPARβ/δ on Insulin Resistance and Cardiovascular Disease. *Trends Endocrinol. Metab.* **2016**, *27*, 319–334. [CrossRef] [PubMed]
7. Unsworth, A.J.; Flora, G.D.; Gibbins, J.M. Non-genomic effects of nuclear receptors: Insights from the anucleate platelet. *Cardiovasc. Res.* **2018**, *114*, 645–655. [CrossRef]
8. Yoo, T.; Ham, S.A.; Lee, W.J.; Hwang, S.I.; Park, J.A.; Hwang, J.S.; Hur, J.; Shin, H.C.; Han, S.G.; Lee, C.H.; et al. Ligand-Dependent Interaction of PPARδ With T-Cell Protein Tyrosine Phosphatase 45 Enhances Insulin Signaling. *Diabetes* **2018**, *67*, 360–371. [CrossRef] [PubMed]
9. Lin, S.-C.; Hardie, D.G. AMPK: Sensing Glucose as well as Cellular Energy Status. *Cell Metab.* **2018**, *27*, 299–313. [CrossRef] [PubMed]
10. Hardie, D.G.; Schaffer, B.E.; Brunet, A. AMPK: An Energy-Sensing Pathway with Multiple Inputs and Outputs. *Trends Cell Biol.* **2016**, *26*, 190–201. [CrossRef]
11. Day, E.A.; Ford, R.J.; Steinberg, G.R. AMPK as a Therapeutic Target for Treating Metabolic Diseases. *Trends Endocrinol. Metab.* **2017**, *28*, 545–560. [CrossRef]
12. He, L.; Wondisford, F.E. Metformin action: Concentrations matter. *Cell Metab.* **2015**, *21*, 159–162. [CrossRef]
13. Baur, J.A.; Pearson, K.J.; Price, N.L.; Jamieson, H.A.; Lerin, C.; Kalra, A.; Prabhu, V.V.; Allard, J.S.; Lopez-Lluch, G.; Lewis, K.; et al. Resveratrol improves health and survival of mice on a high-calorie diet. *Nature* **2006**, *444*, 337–342. [CrossRef] [PubMed]
14. Lee, Y.S.; Kim, W.S.; Kim, K.H.; Yoon, M.J.; Cho, H.J.; Shen, Y.; Ye, J.-M.; Lee, C.H.; Oh, W.K.; Kim, C.T.; et al. Berberine, a natural plant product, activates AMP-activated protein kinase with beneficial metabolic effects in diabetic and insulin-resistant states. *Diabetes* **2006**, *55*, 2256–2264. [CrossRef]
15. Corton, J.M.; Gillespie, J.G.; Hawley, S.A.; Hardie, D.G. 5-Aminoimidazole-4-carboxamide ribonucleoside: A specific method for activating AMP-activated protein kinase in intact cells? *Eur. J. Biochem.* **1995**, *229*, 558–565. [CrossRef] [PubMed]
16. Goransson, O.; McBride, A.; Hawley, S.A.; Ross, F.A.; Shpiro, N.; Foretz, M.; Viollet, B.; Hardie, D.G.; Sakamoto, K. Mechanism of action of A-769662, a valuable tool for activation of AMP-activated protein kinase. *J. Biol. Chem.* **2007**, *282*, 32549–32560. [CrossRef] [PubMed]
17. Sanders, M.J.; Ali, Z.S.; Hegarty, B.D.; Heath, R.; Snowden, M.A.; Carling, D. Defining the mechanism of activation of AMP-activated protein kinase by the small molecule A-769662, a member of the thienopyridone family. *J. Biol. Chem.* **2007**, *282*, 32539–32548. [CrossRef]

18. Racanelli, A.C.; Rothbart, S.B.; Heyer, C.L.; Moran, R.G. Therapeutics by cytotoxic metabolite accumulation: Pemetrexed causes ZMP accumulation, AMPK activation, and mammalian target of rapamycin inhibition. *Cancer Res.* **2009**, *69*, 5467–5474. [CrossRef]
19. Pirkmajer, S.; Kulkarni, S.S.; Tom, R.Z.; Ross, F.A.; Hawley, S.A.; Hardie, D.G.; Zierath, J.R.; Chibalin, A.V. Methotrexate promotes glucose uptake and lipid oxidation in skeletal muscle via AMPK activation. *Diabetes* **2015**, *64*, 360–369. [CrossRef]
20. Dasgupta, B.; Seibel, W. Compound C/Dorsomorphin: Its Use and Misuse as an AMPK Inhibitor. *Methods Mol. Biol.* **2018**, *1732*, 195–202.
21. Dite, T.A.; Langendorf, C.G.; Hoque, A.; Galic, S.; Rebello, R.J.; Ovens, A.J.; Lindqvist, L.M.; Ngoei, K.R.W.; Ling, N.X.Y.; Furicet, L.; et al. AMP-activated protein kinase selectively inhibited by the type II inhibitor SBI-0206965. *J. Biol. Chem.* **2018**, *293*, 8874–8885. [CrossRef] [PubMed]
22. Hardie, D.G.; Ross, F.A.; Hawley, S.A. AMPK: A nutrient and energy sensor that maintains energy homeostasis. *Nat. Rev. Mol. Cell Biol.* **2012**, *13*, 251–262. [CrossRef] [PubMed]
23. Ruderman, N.B.; Carling, D.; Prentki, M.; Cacicedo, J.M. AMPK, insulin resistance, and the metabolic syndrome. *J. Clin. Investig.* **2013**, *123*, 2764–2772. [CrossRef]
24. DeFronzo, R.A.; Tripathy, D. Skeletal muscle insulin resistance is the primary defect in type 2 diabetes. *Diabetes Care* **2009**, *32* (Suppl. 2), S157–S163. [CrossRef]
25. DeFronzo, R.A.; Ferrannini, E.; Sato, Y.; Felig, P.; Wahren, J. Synergistic interaction between exercise and insulin on peripheral glucose uptake. *J. Clin. Investig.* **1981**, *68*, 1468–1474. [CrossRef] [PubMed]
26. Gustafson, B.; Hedjazifar, S.; Gogg, S.; Hammarstedt, A.; Smith, U. Insulin resistance and impaired adipogenesis. *Trends Endocrinol. Metab.* **2015**, *26*, 193–200. [CrossRef]
27. Chen, Q.; Xie, B.; Zhu, S.; Rong, P.; Sheng, Y.; Ducommun, S.; Chen, L.; Quan, C.; Li, M.; Sakamoto, K.; et al. A Tbc1d1 Ser231Ala-knockin mutation partially impairs AICAR- but not exercise-induced muscle glucose uptake in mice. *Diabetologia* **2017**, *60*, 336–345. [CrossRef]
28. Liu, Y.; Lai, Y.C.; Hill, E.V.; Tyteca, T.; Carpentier, S.; Ingvaldsen, A.; Vertommen, D.; Lantier, L.; Foretz, M.; Dequiedt, F.; et al. Phosphatidylinositol 3-phosphate 5-kinase (PIKfyve) is an AMPK target participating in contraction-stimulated glucose uptake in skeletal muscle. *Biochem. J.* **2013**, *455*, 195–206. [CrossRef]
29. Jaiswal, N.; Gavin, M.G.; Quinn, W.J., 3rd; Luongo, T.S.; Gelfer, R.G.; Baur, J.A.; Titchenell, P.M. The role of skeletal muscle Akt in the regulation of muscle mass and glucose homeostasis. *Mol. Metab.* **2019**, *28*, 1–13. [CrossRef] [PubMed]
30. Krämer, D.K.; Al-Khalili, L.; Perrini, S.; Skogsberg, J.; Wretenberg, P.; Kannisto, K.; Wallberg-Henriksson, H.; Ehrenborg, E.; Zierath, J.R.; Krook, A. Direct activation of glucose transport in primary human myotubes after activation of peroxisome proliferator-activated receptor delta. *Diabetes* **2005**, *54*, 1157–1163. [CrossRef]
31. Krämer, D.K.; Al-Khalili, L.; Guigas, B.; Leng, Y.; Garcia-Roves, P.M.; Krook, A. Role of AMP kinase and PPARdelta in the regulation of lipid and glucose metabolism in human skeletal muscle. *J. Biol. Chem.* **2007**, *282*, 19313–19320. [CrossRef]
32. Wang, Y.X.; Zhang, C.L.; Yu, R.T.; Cho, H.K.; Nelson, M.C.; Bayuga-Ocampo, C.R.; Ham, J.; Kang, H.; Evans, R.M. Regulation of muscle fiber type and running endurance by PPARdelta. *PLoS Biol.* **2004**, *2*, e294. [CrossRef]
33. Narkar, V.A.; Downes, M.; Yu, R.T.; Embler, E.; Wang, Y.X.; Banayo, E.; Mihaylova, M.M.; Nelson, M.C.; Zou, Y.; Juguilon, H.; et al. AMPK and PPARdelta agonists are exercise mimetics. *Cell* **2008**, *134*, 405–415. [CrossRef]
34. Gan, Z.; Burkart-Hartman, E.M.; Han, D.-H.; Finck, B.; Leone, T.C.; Smith, E.Y.; Ayala, J.E.; Holloszy, J.; Kelly, D.P. The nuclear receptor PPARβ/δ programs muscle glucose metabolism in cooperation with AMPK and MEF2. *Genes Dev.* **2011**, *25*, 2619–2630. [CrossRef] [PubMed]
35. Australian Government Department of Health, Therapeutic Goods Administration. Available online: https://www.tga.gov.au/book-page/12-cardarine#fn8 (accessed on 2 July 2021).
36. The World Anti-Doping Agency. The World Anti-Doping Code: The 2009 Prohibited List International Standard. 2009, p. 6. Available online: https://www.wada-ama.org/sites/default/files/resources/files/WADA_Prohibited_List_2009_EN.pdf (accessed on 5 July 2021).
37. Schuler, M.; Ali, F.; Chambon, C.; Duteil, D.; Bornert, J.-M.; Tardivel, A.; Desvergne, B.; Wahli, W.; Chambon, P.; Metzger, D. PGC1alpha expression is controlled in skeletal muscles by PPARbeta, whose ablation results in fiber-type switching, obesity, and type 2 diabetes. *Cell Metab.* **2006**, *4*, 407–414. [CrossRef] [PubMed]
38. Hancock, C.R.; Han, D.H.; Chen, M.; Terada, S.; Yasuda, T.; Wright, D.C.; Holloszy, J.O. High-fat diets cause insulin resistance despite an increase in muscle mitochondria. *Proc. Natl. Acad. Sci. USA* **2008**, *105*, 7815–7820. [CrossRef]
39. Koh, J.H.; Hancock, C.R.; Terada, S.; Higashida, K.; Holloszy, J.O.; Han, D.H. PPARβ Is Essential for Maintaining Normal Levels of PGC-1α and Mitochondria and for the Increase in Muscle Mitochondria Induced by Exercise. *Cell Metab.* **2017**, *25*, 1176–1185. [CrossRef]
40. Coll, T.; Alvarez-Guardia, D.; Barroso, E.; Gómez-Foix, A.M.; Palomer, X.; Laguna, J.C.; Vázquez-Carrera, M. Activation of peroxisome proliferator-activated receptor-δ by GW501516 prevents fatty acid-induced nuclear factor-kB activation and insulin resistance in skeletal muscle cells. *Endocrinology* **2010**, *151*, 1560–1569. [CrossRef]
41. Petersen, M.C.; Shulman, G.I. Mechanisms of Insulin Action and Insulin Resistance. *Physiol Rev.* **2018**, *98*, 2133–2223. [CrossRef] [PubMed]
42. Salvadó, L.; Palomer, X.; Barroso, E.; Vázquez-Carrera, M. Targeting endoplasmic reticulum stress in insulin resistance. *Trends Endocrinol. Metab.* **2015**, *26*, 438–448. [CrossRef] [PubMed]

43. Terai, K.; Hiramoto, Y.; Masaki, M.; Sugiyama, S.; Kuroda, T.; Hori, M.; Kawase, I.; Hirota, H. AMP-activated protein kinase protects cardiomyocytes against hypoxic injury through attenuation of endoplasmic reticulum stress. *Mol. Cell. Biol.* **2005**, *25*, 9554–9575. [CrossRef] [PubMed]
44. Dong, Y.; Zhang, M.; Wang, S.; Liang, B.; Zhao, Z.; Liu, C.; Wu, M.; Choi, H.C.; Lyons, T.J.; Zou, M.H. Activation of AMP-activated protein kinase inhibits oxidized LDL-triggered endoplasmic reticulum stress in vivo. *Diabetes* **2010**, *59*, 1386–1396. [CrossRef] [PubMed]
45. Dong, Y.; Zhang, M.; Liang, B.; Xie, Z.; Zhao, Z.; Asfa, S.; Choi, H.C.; Zou, M.-H. Reduction of AMP-activated protein kinase alpha2 increases endoplasmic reticulum stress and atherosclerosis in vivo. *Circulation* **2010**, *121*, 792–803. [CrossRef]
46. Wang, Y.; Wu, Z.; Li, D.; Wang, D.; Wang, X.; Feng, X.; Xia, M. Involvement of oxygen-regulated protein 150 in AMP-activated protein kinase-mediated alleviation of lipid-induced endoplasmic reticulum stress. *J. Biol. Chem.* **2011**, *286*, 11119–11131. [CrossRef]
47. Du, J.; Guan, T.; Zhang, H.; Xia, Y.; Liu, F.; Zhang, Y. Inhibitory crosstalk between ERK and AMPK in the growth and proliferation of cardiac fibroblasts. *Biochem. Biophys. Res. Commun.* **2008**, *368*, 402–407. [CrossRef]
48. Hwang, S.L.; Jeong, Y.T.; Li, X.; Kim, Y.D.; Lu, Y.; Chang, Y.-C.; Lee, I.-K.; Chang, H.W. Inhibitory cross-talk between the AMPK and ERK pathways mediates endoplasmic reticulum stress induced insulin resistance in skeletal muscle. *Br. J. Pharmacol.* **2013**, *169*, 69–81. [CrossRef] [PubMed]
49. Aguilar-Recarte, D.; Barroso, E.; Gumà, A.; Pizarro-Delgado, J.; Peña, L.; Ruart, M.; Palomer, X.; Wahli, W.; Vázquez-Carrera, M. GDF15 mediates the metabolic effects of PPARbeta/delta by activating AMPK. *Cell Rep.* **2021**, in press.
50. Hsiao, E.C.; Koniaris, L.G.; Zimmers-Koniaris, T.; Sebald, S.M.; Huynh, T.V.; Lee, S.J. Characterization of growth-differentiation factor 15, a transforming growth factor beta superfamily member induced following liver injury. *Mol. Cell. Biol.* **2000**, *20*, 3742–3751. [CrossRef]
51. Tsai, V.W.W.; Husaini, Y.; Sainsbury, A.; Brown, D.A.; Breit, S.N. The MIC-1/GDF15-GFRAL Pathway in Energy Homeostasis: Implications for Obesity, Cachexia, and Other Associated Diseases. *Cell Metab.* **2018**, *28*, 353–368. [CrossRef]
52. Baek, S.J.; Eling, T. Growth differentiation factor 15 (GDF15): A survival protein with therapeutic potential in metabolic diseases. *Pharmacol. Ther.* **2019**, *198*, 46–58. [CrossRef]
53. Klein, A.B.; Nicolaisen, T.S.; Ørtenblad, N.; Gejl, K.D.; Jensen, R.; Fritzen, A.M.; Larsen, E.L.; Karstoft, K.; Poulsen, H.E.; Morville, T.; et al. Pharmacological but not physiological GDF15 suppresses feeding and the motivation to exercise. *Nat. Commun.* **2021**, *12*, 1041. [CrossRef]
54. Emmerson, P.J.; Wang, F.; Du, Y.; Liu, Q.; Pickard, R.T.; Gonciarz, M.D.; Coskun, T.; Hamang, M.J.; Sindelar, D.K.; Ballmanet, K.K.; et al. The metabolic effects of GDF15 are mediated by the orphan receptor GFRAL. *Nat. Med.* **2017**, *23*, 1215–1219. [CrossRef] [PubMed]
55. Yang, L.; Chang, C.-C.; Sun, Z.; Madsen, D.; Zhu, H.; Padkjær, S.B.; Wu, X.; Huang, T.; Hultman, K.; Paulsenet, S.J.; et al. GFRAL is the receptor for GDF15 and is required for the anti-obesity effects of the ligand. *Nat. Med.* **2017**, *23*, 1158–1166. [CrossRef]
56. Mullican, S.E.; Lin-Schmidt, X.; Chin, C.-N.; Chavez, J.A.; Furman, J.L.; Armstrong, A.A.; Beck, S.C.; South, V.J.; Dinh, T.Q.; Cash-Mason, T.D.; et al. GFRAL is the receptor for GDF15 and the ligand promotes weight loss in mice and nonhuman primates. *Nat. Med.* **2017**, *23*, 1150–1157. [CrossRef] [PubMed]
57. Hsu, J.Y.; Crawley, S.; Chen, M.; Ayupova, D.A.; Lindhout, D.A.; Higbee, J.; Kutach, A.; Joo, W.; Gao, Z.; Fu, D.; et al. Non-homeostatic body weight regulation through a brainstem-restricted receptor for GDF15. *Nature* **2017**, *550*, 255–259. [CrossRef] [PubMed]
58. Chung, H.K.; Ryu, D.; Kim, K.S.; Chang, J.Y.; Kim, Y.K.; Yi, H.-S.; Kang, S.G.; Choi, M.J.; Lee, S.E.; Jung, S.-B.; et al. Growth differentiation factor 15 is a myomitokine governing systemic energy homeostasis. *J. Cell Biol.* **2017**, *216*, 149–165. [CrossRef]
59. Laurens, C.; Parmar, A.; Murphy, E.; Carper, D.; Lair, B.; Maes, P.; Vion, J.; Boulet, N.; Fontaine, C.; Marquès, M.; et al. Growth and differentiation factor 15 is secreted by skeletal muscle during exercise and promotes lipolysis in humans. *JCI Insight* **2020**, *5*, e131870. [CrossRef]
60. Rinella, M.E. Nonalcoholic fatty liver disease: A systematic review. *JAMA* **2015**, *313*, 2263–2273. [CrossRef]
61. Smith, B.K.; Marcinko, K.; Desjardins, E.M.; Lally, J.S.; Ford, R.J.; Steinberg, G.R. Treatment of nonalcoholic fatty liver disease: Role of AMPK. *Am. J. Physiol. Endocrinol. Metab.* **2016**, *311*, E730–E740. [CrossRef]
62. Li, Y.; Xu, S.; Mihaylova, M.M.; Zheng, B.; Hou, X.; Jiang, B.; Park, O.; Luo, Z.; Lefai, E.; Shyy, J.Y.; et al. AMPK phosphorylates and inhibits SREBP activity to attenuate hepatic steatosis and atherosclerosis in diet-induced insulin-resistant mice. *Cell Metab.* **2011**, *13*, 376–388. [CrossRef] [PubMed]
63. Kawaguchi, T.; Osatomi, K.; Yamashita, H.; Kabashima, T.; Uyeda, K. Mechanism for fatty acid "sparing" effect on glucose-induced transcription: Regulation of carbohydrate-responsive element-binding protein by AMP-activated protein kinase. *J. Biol. Chem.* **2002**, *277*, 3829–3835. [CrossRef] [PubMed]
64. Barroso, E.; Rodríguez-Calvo, R.; Serrano-Marco, L.; Astudillo, A.M.; Balsinde, J.; Palomer, X.; Vázquez-Carrera, M. The PPARbeta/delta activator GW501516 prevents the down-regulation of AMPK caused by a high-fat diet in liver and amplifies the PGC-1alpha-Lipin 1-PPARalpha pathway leading to increased fatty acid oxidation. *Endocrinology* **2011**, *152*, 1848–1859. [CrossRef] [PubMed]

65. Bojic, L.A.; Telford, D.E.; Fullerton, M.D.; Ford, R.J.; Sutherland, B.G.; Edwards, J.Y.; Sawyez, C.G.; Gros, R.; Kemp, B.E.; Steinberg, G.R.; et al. PPARδ activation attenuates hepatic steatosis in Ldlr-/- mice by enhanced fat oxidation, reduced lipogenesis, and improved insulin sensitivity. *J. Lipid Res.* **2014**, *55*, 1254–1266. [CrossRef] [PubMed]
66. Yorimitsu, T.; Klionsky, D.J. Autophagy: Molecular machinery for self-eating. *Cell Death Differ.* **2005**, *12* (Suppl. 2), 1542–1552. [CrossRef]
67. Singh, R.; Kaushik, S.; Wang, Y.; Xiang, Y.; Novak, I.; Komatsu, M.; Tanaka, K.; Cuervo, A.M.; Czaja, M.J. Autophagy regulates lipid metabolism. *Nature* **2009**, *458*, 1131–1135. [CrossRef]
68. Kim, J.; Kundu, M.; Viollet, B.; Guan, K.L. AMPK and mTOR regulate autophagy through direct phosphorylation of Ulk1. *Nat. Cell Biol.* **2011**, *13*, 132–141. [CrossRef]
69. Tong, L.; Wang, L.; Yao, S.; Jin, L.; Yang, J.; Zhang, Y.; Ning, G.; Zhang, Z. PPARδ attenuates hepatic steatosis through autophagy-mediated fatty acid oxidation. *Cell Death Dis.* **2019**, *10*, 197. [CrossRef]
70. Fougerat, A.; Montagner, A.; Loiseau, N.; Guillou, H.; Wahli, W. Peroxisome Proliferator-Activated Receptors and Their Novel Ligands as Candidates for the Treatment of Non-Alcoholic Fatty Liver Disease. *Cells* **2020**, *9*, 1638. [CrossRef]
71. Yamaguchi, K.; Nishimura, T.; Ishiba, H.; Seko, Y.; Okajima, A.; Fujii, H.; Tochiki, N.; Umemura, A.; Moriguchi, M.; Sumida, Y.; et al. Blockade of interleukin 6 signalling ameliorates systemic insulin resistance through upregulation of glucose uptake in skeletal muscle and improves hepatic steatosis in high-fat diet fed mice. *Liver Int.* **2015**, *35*, 550–561. [CrossRef] [PubMed]
72. Galic, S.; Sachithanandan, N.; Kay, T.W.; Steinberg, G.R. Suppressor of cytokine signalling (SOCS) proteins as guardians of inflammatory responses critical for regulating insulin sensitivity. *Biochem. J.* **2014**, *461*, 177–188. [CrossRef]
73. Serrano-Marco, L.; Barroso, E.; Kochairi, I.E.; Palomer, X.; Michalik, L.; Wahli, W.; Vázquez-Carrera, M. The peroxisome proliferator-activated receptor (PPAR) β/δ agonist GW501516 inhibits IL-6-induced signal transducer and activator of transcription 3 (STAT3) activation and insulin resistance in human liver cells. *Diabetologia* **2012**, *55*, 743–751. [CrossRef] [PubMed]
74. Maack, C.; Lehrke, M.; Backs, J.; Heinzel, F.R.; Hulot, J.-S.; Marx, N.; Paulus, W.J.; Rossignol, P.; Taegtmeyer, H.; Bauersachs, J.; et al. Heart failure and diabetes: Metabolic alterations and therapeutic interventions: A state-of-the-art review from the Translational Research Committee of the Heart Failure Association-European Society of Cardiology. *Eur. Heart J.* **2018**, *39*, 4243–4254. [CrossRef] [PubMed]
75. Li, H.L.; Yin, R.; Chen, D.; Liu, D.; Wang, D.; Yang, Q.; Dong, Y.G. Long-term activation of adenosine monophosphate-activated protein kinase attenuates pressure-overload-induced cardiac hypertrophy. *J. Cell. Biochem.* **2007**, *100*, 1086–1099. [CrossRef] [PubMed]
76. Alvarez-Guardia, D.; Palomer, X.; Coll, T.; Serrano, L.; Rodríguez-Calvo, R.; Davidson, M.M.; Merlos, M.; Kochairi, I.E.; Michalik, L.; Wahli, W.; et al. PPARβ/δ activation blocks lipid-induced inflammatory pathways in mouse heart and human cardiac cells. *Biochim. Biophys. Acta* **2011**, *1811*, 59–67. [CrossRef]
77. Palomer, X.; Pizarro-Delgado, J.; Vázquez-Carrera, M. Emerging Actors in Diabetic Cardiomyopathy: Heartbreaker Biomarkers or Therapeutic Targets? *Trends Pharmacol. Sci.* **2018**, *39*, 452–467. [CrossRef] [PubMed]
78. Palomer, X.; Capdevila-Busquets, E.; Botteri, G.; Salvadó, L.; Barroso, E.; Davidson, M.M.; Michalik, L.; Wahli, W.; Vázquez-Carrera, M. PPARβ/δ attenuates palmitate-induced endoplasmic reticulum stress and induces autophagic markers in human cardiac cells. *Int. J. Cardiol.* **2014**, *174*, 110–118. [CrossRef]
79. Koh, J.H.; Hancock, C.R.; Han, D.H.; Holloszy, J.O.; Nair, K.S.; Dasari, S. AMPK and PPARbeta positive feedback loop regulates endurance exercise training-mediated GLUT4 expression in skeletal muscle. *Am. J. Physiol. Endocrinol. Metab.* **2019**, *316*, E931–E939. [CrossRef]
80. Ding, J.; Gou, Q.; Jia, X.; Liu, Q.; Jin, J.; Shi, J.; Hou, Y. AMPK phosphorylates PPARdelta to mediate its stabilization, promote glucose and glutamine uptake, and inhibit colon tumor growth. *J. Biol. Chem.* **2021**, 100954. [CrossRef]
81. Wagner, N.; Wagner, K.-D. PPARs and Angiogenesis-Implications in Pathology. *Int. J. Mol. Sci.* **2020**, *21*, 5723. [CrossRef]
82. Cheng, H.S.; Yip, Y.S.; Lim, E.K.Y.; Wahli, W.; Tan, N.S. PPARs and Tumor Microenvironment: The Emerging Roles of the Metabolic Master Regulators in Tumor Stromal-Epithelial Crosstalk and Carcinogenesis. *Cancers* **2021**, *13*, 2153. [CrossRef]

International Journal of
Molecular Sciences

MDPI

Review

PPAR-Targeted Therapies in the Treatment of Non-Alcoholic Fatty Liver Disease in Diabetic Patients

Naomi F. Lange [1,2,*], Vanessa Graf [3], Cyrielle Caussy [4,5] and Jean-François Dufour [6,7,*]

1 Department of Visceral Surgery and Medicine, Inselspital, Bern University Hospital, University of Bern, 3010 Bern, Switzerland
2 Graduate School for Health Sciences, University of Bern, 3012 Bern, Switzerland
3 Department of Diabetes, Endocrinology, Clinical Nutrition, and Metabolism, Inselspital, Bern University Hospital, University of Bern, 3010 Bern, Switzerland; vanessa.graf@insel.ch
4 Univ Lyon, CarMen Laboratory, INSERM, INRA, INSA Lyon, Université Claude Bernard Lyon 1, 69495 Pierre-Bénite, France; cyrielle.caussy@chu-lyon.fr
5 Département Endocrinologie, Diabète et Nutrition, Hôpital Lyon Sud, Hospices Civils de Lyon, 69495 Pierre-Bénite, France
6 Centre des Maladies Digestives, 1003 Lausanne, Switzerland
7 Swiss NASH Foundation, 3011 Bern, Switzerland
* Correspondence: naomi.lange@insel.ch (N.F.L.); jf.dufour@svmed.ch (J.-F.D.)

Abstract: Peroxisome proliferator-activated receptors (PPAR), ligand-activated transcription factors of the nuclear hormone receptor superfamily, have been identified as key metabolic regulators in the liver, skeletal muscle, and adipose tissue, among others. As a leading cause of liver disease worldwide, non-alcoholic fatty liver disease (NAFLD) and non-alcoholic steatohepatitis (NASH) cause a significant burden worldwide and therapeutic strategies are needed. This review provides an overview of the evidence on PPAR-targeted treatment of NAFLD and NASH in individuals with type 2 diabetes mellitus. We considered current evidence from clinical trials and observational studies as well as the impact of treatment on comorbid metabolic conditions such as obesity, dyslipidemia, and cardiovascular disease. Future areas of research, such as possible sexually dimorphic effects of PPAR-targeted therapies, are briefly reviewed.

Keywords: non-alcoholic fatty liver disease (NAFLD); non-alcoholic steatohepatitis (NASH); type 2 diabetes mellitus; peroxisome proliferator-activated receptors (PPAR)

Citation: Lange, N.F.; Graf, V.; Caussy, C.; Dufour, J.-F. PPAR-Targeted Therapies in the Treatment of Non-Alcoholic Fatty Liver Disease in Diabetic Patients. *Int. J. Mol. Sci.* **2022**, *23*, 4305. https://doi.org/10.3390/ijms23084305

Academic Editors: Manuel Vázquez-Carrera and Walter Wahli

Received: 21 February 2022
Accepted: 8 April 2022
Published: 13 April 2022

1. Introduction

As a leading cause of liver disease worldwide, non-alcoholic fatty liver disease (NAFLD) and non-alcoholic steatohepatitis (NASH) cause a significant burden [1]. NAFLD is a common comorbidity especially among individuals living with type 2 diabetes mellitus (T2DM) [2]. The complex bidirectional pathophysiological relationships between NAFLD and other metabolic diseases, particularly T2DM [3,4], demand a holistic and interdisciplinary approach to the treatment of NAFLD [5].

T2DM represents a major risk factor for NAFLD with over 55% of persons living with T2DM being affected by NAFLD [2]. T2DM furthermore predisposes individuals to advanced NAFLD, including development of NASH and liver fibrosis, and increases the risk of hepatocellular carcinoma [6,7]. NAFLD, in turn, increases the risk of incident T2DM [8]. Among NAFLD patients with advanced fibrosis, the majority have T2DM [9]. This complex population with multiple metabolic alterations such as NAFLD and T2DM should specifically be considered in the evaluation of potential pharmacological treatment strategies for NAFLD.

Peroxisome proliferator-activated receptors (PPAR), ligand-activated transcription factors of the nuclear hormone receptor superfamily, have been identified as key metabolic regulators in the liver, skeletal muscle, and adipose tissue, among others [10,11]. PPAR bolic regulators in the liver, skeletal muscle, and adipose tissue, among others [10,11]. PPAR modulation has long been employed in the pharmacological treatment of multiple conditions, predominantly metabolic diseases such as T2DM and dyslipidemia, but has also been examined in the context of liver disease [12–14]. Considering the pathophysiological and epidemiological links between these conditions and NAFLD, PPAR modulators are being examined regarding their effects on NAFLD [15,16].

In the following, we will review the clinical evidence on PPAR-directed therapy for NAFLD, focusing on results and considerations in patients with type 2 diabetes mellitus.

2. PPAR Agonists in the Treatment of NAFLD with Concomitant T2DM

The three PPAR isotypes, PPARα, PPARβ/δ, and PPARγ in humans compose the 1C subfamily of the nuclear hormone receptor superfamily, which encompasses a large group of ligand-regulated transcription factors that share a common modular structure [17]. Tissue expression and effects of activation vary by isotype and tissue, demonstrating both redundant as well as distinct effects (reviewed in detail by [11,18]). Figure 1 provides an overview of tissue-specific and systemic PPAR main functions. Overall, PPAR isotypes exert pleiotropic functions in multiple tissues and pathways relating mainly to metabolism and immunity, which are involved in the regulation of hepatic steatosis and inflammation, and improvement of liver histology in patients with NASH [18]. Selected PPAR agonists have demonstrated anti-fibrotic properties in the context of NAFLD [19,20]. As such, PPAR represent important targets in the treatment of NAFLD [15].

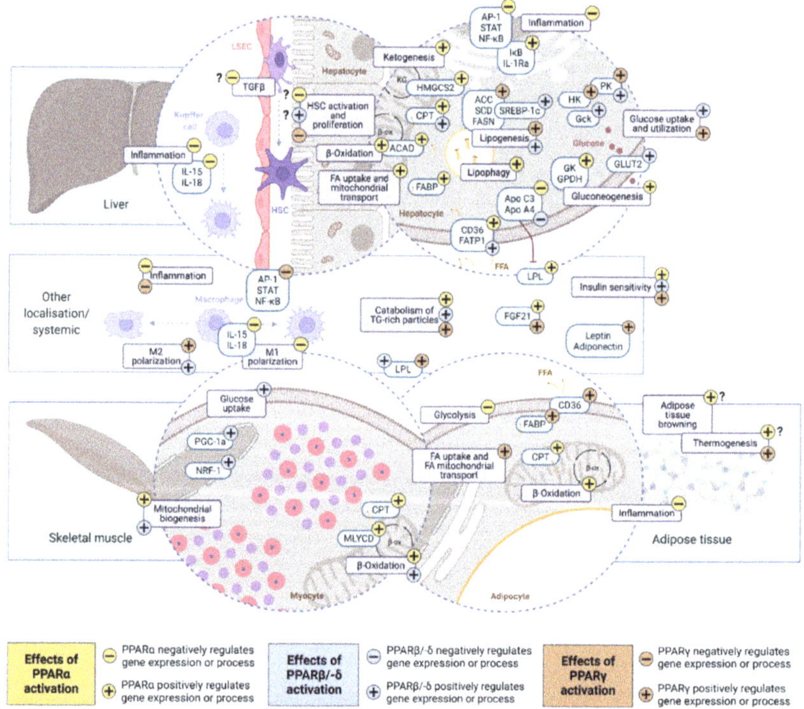

Figure 1. Overview of main tissue-specific and systemic effects of PPAR activation. Blue fields indicate proteins, purple fields indicate biochemical processes. Question marks indicate effects that are suspected but not confirmed. Red bars indicate inhibition. Created with BioRender.com. Abbreviations: β-ox, beta oxidation; ACAD, acyl-CoA dehydrogenases; ACC, acetyl-CoA carboxylase; AP-1, activator protein-1; Apo A4, apolipoprotein A4; Apo C3, apolipoprotein C3; CPT, carnitine palmitoyltransferases; FA, fatty acid; FABP, fatty acid binding protein; FASN, fatty acid synthase; FATP1, fatty acid transport protein-1; FFA, free fatty acid; FGF21, fibroblast growth factor 21; Gck, glucokinase;

GK, glycerol kinase; GLUT2, glucose transporter 2; GPDH, glycerol 3-phosphate dehydrogenase; HK, hexokinase; HMGCS2, 3-hydroxy-3-methylglutaryl-CoA synthase 2; HSC, hepatic stellate cell; IκB, inhibitor of nuclear factor kappa B; KG, ketogenesis; IL-15, interleukin 15; IL-18, interleukin 18; IL-1Ra, interleukin-1 receptor antagonist; LPL, lipoprotein lipase; LSEC, liver sinusoidal endothelial cell; MLYCD, malonyl-CoA decarboxylase; NF-κB, nuclear factor kappa B; NRF-1, nuclear respiratory factor 1; PK, pyruvate kinase; PGC-1a, PPARG coactivator 1 alpha; SCD, stearoyl-CoA desaturase; SREBP-1c, sterol regulatory element binding protein 1; STAT, signal transducer and activator of transcription family; TG, triglyceride; TGFβ, transforming growth factor β.

2.1. Molecular Basics of PPAR-Dependent Regulation

PPAR-dependent metabolic regulation of transcriptional activity occurs via several mechanisms. Firstly, ligand-dependent PPAR activation (ligand-dependent transactivation) prompts corepressor dissociation followed by heterodimerization with retinoid X receptors (RXR) and recruitment of a co-activator. The activated heterodimer proceeds to bind specific DNA sequences in the promotor regions of target genes, i.e., PPAR-responsive elements (PPREs) [10,21]. This PPRE-dependent mechanism leads to increased transcription of target genes. A multitude of both specific and shared ligands of PPARs has been identified, including natural as well as synthetic ligands [11].

PPAR may also regulate gene transcription negatively. Ligand-dependent transrepression describes a protein-protein interaction that leads to decreased transcription of predominantly inflammatory genes by interacting with transcription factors, such as members of the nuclear factor κB (NF-κB) family, and is independent of binding to a receptor-specific response element. Conversely, ligand-independent repression requires binding to PPRE, followed by recruitment of co-repressors. These mechanisms are comprehensively reviewed elsewhere [21]. Anti-inflammatory mechanisms of PPAR are mostly regulated through transrepression [18].

2.2. PPARα (NR1C1)

In 1990, the first isoform of PPAR was identified in humans and later classified as PPARα, which is encoded on the *PPARA* (*NR1C1*) gene [22]. This discovery was fueled by exploration of the pharmacological mechanisms of fibrates, which had been produced since the 1950s [12]. Multiple other synthetic and endogenous ligands for PPARα have since been characterized, including phospholipids and fatty acids and their derivatives, such as eicosanoids [11,23]

PPARα is a major regulator of cellular energy homeostasis and as such is expressed predominantly in oxidative tissues, such as the liver, adipose tissue, skeletal muscle, heart, and kidneys [11,24]. In the liver, the nuclear receptor is expressed mainly in hepatocytes but also non-parenchymal cells, namely stellate cells and liver sinusoidal endothelial cells [25].

While PPARα is active in both the fed and fasting state, it has a central role predominantly in the adaptive response to the latter [26,27]. Main functions include the transcriptional regulation of lipid catabolism by modulating expression of genes that mediate triglyceride hydrolysis, fatty acid transport, and β-oxidation in liver, skeletal muscle, and adipose tissue [23,28,29]. Additionally, PPARα regulates ketogenesis, which has been found to be severely impaired in the absence of PPARα [26,27].

Other functions of PPARα that are related to NAFLD include direct anti-inflammatory effects, which have been found to be independent of its metabolic functions in the liver [30]. Anti-fibrogenic effects of PPARα may be mediated through these anti-inflammatory effects as well as other mechanisms. Findings from pre-clinical mouse models of diet- and thioacetamide-induced fibrosing NASH suggest that PPARα agonism indirectly ameliorates liver fibrosis through modulation of hepatic stellate cell activation and related pro-fibrogenic pathways [31,32]. Interestingly, several findings indicate sexually dimorphic responses to PPARα activation, which warrants further exploration in the clinical context of NAFLD [33,34]. Diurnal cycling of nuclear receptor expression has been identified in several instances, notably including variable expression of PPARα [35]. A detailed review of PPARα functions can be found here [36].

2.3. *PPARδ (PPARβ; NR1C2)*

The isoform PPARδ (also: PPARβ), encoded on the *NR1C2* gene on chromosome 6, has previously been identified as a target for several metabolic conditions, including NAFLD [37,38], as receptor modulation was found to increase insulin sensitivity and improve lipid profile, while reducing obesity [39–41].

The receptor is expressed most abundantly in skeletal and cardiac muscle tissue, as well as in brown and white adipose tissue, macrophages, and the liver [11,17,24]. In the liver, the receptor further demonstrates a ubiquitous expression pattern, being present in hepatocytes, hepatic stellate cells, liver sinusoidal endothelial cells, and Kupffer cells [25]. Endogenous ligands of PPARδ include fatty acids and eicosanoids [42].

PPARδ exerts beneficial metabolic functions through maintaining oxidative capacity of skeletal muscle and mediating the adaptive response to exercise, enhancing mitochondrial biogenesis, fatty acid oxidation, and glucose utilization [41,43,44]. An increase in mitochondria and mitochondrial proteins in skeletal muscle is facilitated by PPARδ-mediated increase in peroxisome proliferator-activated receptor gamma coactivator 1-alpha (PGC-1a) concentrations and nuclear respiratory factor (NRF-1) expression [45]. PPARδ further has a critical role in the regulation of hepatic metabolism. In the pre-clinical setting, hepatic PPARδ overexpression led to increased liver glucose utilization and de novo lipogenesis while changing lipid profiles towards an increased ratio of monounsaturated to saturated fatty acids [46]. Despite lipid accumulation, PPARδ-overexpressing cells displayed less damage [46]. Overall, these findings indicate that PPARδ regulates hepatic glucose and fatty acid metabolism, thus playing a pivotal role in hepatic energy substrate homeostasis [46]. Further evidence suggests that liver-specific PPARδ activation also modulates energy substrate homeostasis in skeletal muscle towards fatty acid oxidation [47]. Hepatic PPARδ further regulates genes involved in lipoprotein metabolism, thus accounting for its beneficial effects on lipid profiles, as well as pathways related to inflammation and immunity, including promotion of anti-inflammatory macrophage polarization [48,49]. Recently, two detailed reviews have summarized the regulation of metabolism via PPARδ with a focus on NAFLD etiopathogenesis [37,38].

2.4. *PPARγ (NR1C3)*

PPARγ, which is encoded by *NR1C3* on chromosome 3, exerts its main metabolic effects in adipose tissue, being expressed in white and brown adipose tissue, as well as in macrophages [24]. Two isotypes of the PPARγ receptor, PPARγ 1 and PPARγ 2, have been identified [50]. Among these isoforms, PPARγ 1 demonstrates a broader expression pattern, while PPARγ 2 is predominantly expressed in adipose tissue [51].

Similar to other PPAR isoforms, endogenous ligands of PPARγ are fatty acids and eicosanoids [11]. Synthetic agonists of PPARγ include the anti-diabetic treatments rosiglitazone and pioglitazone, but also arachidonic acid metabolite anti-inflammatory drugs such as ibuprofen and indomethacin as well as the dual agonist saroglitazar [42,52].

PPARγ beneficially affects metabolism mainly by improving adipose tissue adipogenesis and adipose tissue fatty acid uptake and expenditure [52–54]. Adipose-tissue-specific PPARγ deletion in a pre-clinical model leads to severe lipoatrophy, highlighting the role of PPARγ in adipocyte development [55]. PPARγ deletion in adipose tissue and liver has furthermore been linked to insulin resistance [54–56]. Accordingly, PPARγ activation, for example with thiazolidinediones, displays insulin-sensitizing properties [52]. PPARγ activation has also been demonstrated to increase levels of adiponectin, an anti-atherogenic adipokine [48]. The receptor further possesses anti-inflammatory properties, acting via modulation of macrophage polarization and attenuation of the NF-κB pathway [49,57,58]. Regarding direct anti-fibrotic properties, PPARγ activity is linked to hepatic stellate cells displaying a quiescent phenotype and reduced hepatic stellate cell proliferation [59,60].

levels of adiponectin, an anti-atherogenic adipokine [48]. The receptor further possesses anti-inflammatory properties, acting via modulation of macrophage polarization and attenuation of the NF-κB pathway [49,57,58]. Regarding direct anti-fibrotic properties, PPARγ activity is linked to hepatic stellate cells displaying a quiescent phenotype and reduced hepatic stellate cell proliferation [59,60].

3. Pharmacologic PPAR-Targeted Therapies

Several PPAR-modulating agents, with varying degrees of affinity for the different PPAR isotypes, have been investigated for the therapy of NAFLD and NASH (Figure 2). Table 1 provides an overview of recent controlled clinical trials reporting liver-related outcomes in patients with NAFLD.

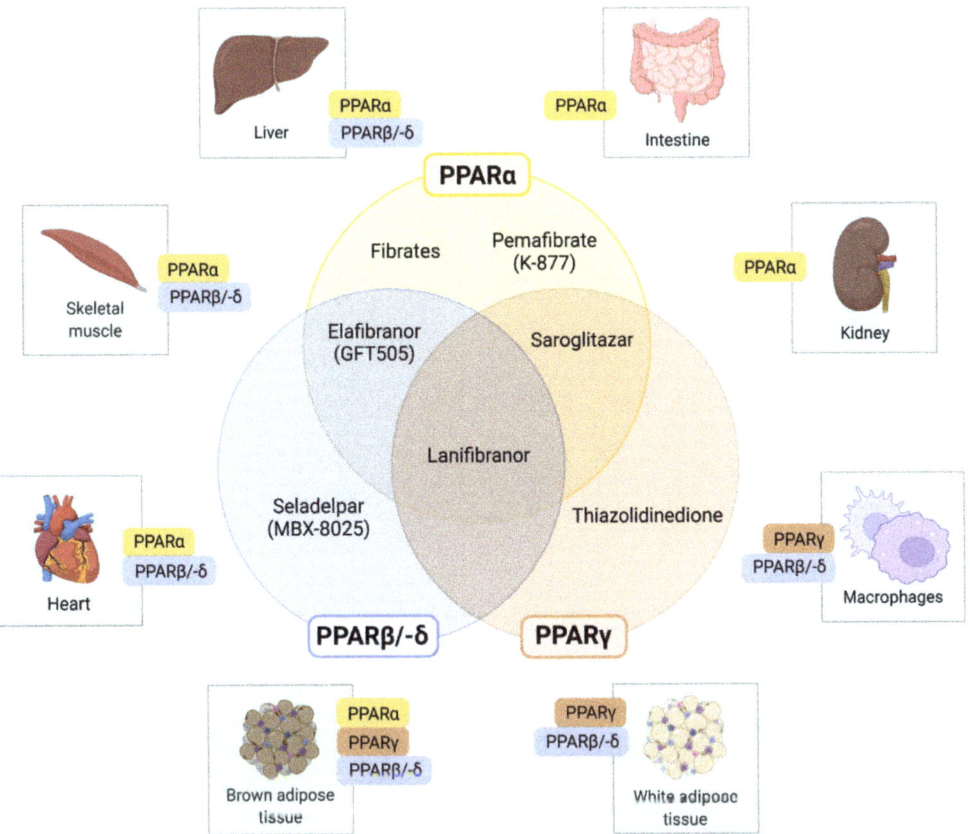

Figure 2. Overview of respective receptor profile of PPAR-modulating agents used in clinical trials and main tissue distribution of PPAR isotypes. Created with BioRender.com.

Table 1. Evidence from randomized, controlled trials reporting liver-specific outcomes of PPAR-modulating therapy in NAFLD patients, 2016–2021.

Drug Name; Target Pharmaceutical Company	Reference, Country	Study Type; Treatment Duration	Participants	Intervention	Results
Pemafibrate; PPARα Kowa Pharmaceutical	Nakajima et al., 2021 [61], Japan	RCT; 72 weeks	Adults w/NAFLD defined by liver fat content $\geq$ 10% (MRI-PDFF), liver stiffness $\geq$ 2.5 kPa (MRE), ALT > 40 U/L for men, >30 U/L for women Exclusion: poorly controlled T2DM (HbA1c $\geq$ 8%) N = 118 Female: 50 (42%) T2DM: 43 (36%)	Arm (1): Pemafibrate 0.2 mg/day Arm (2): Placebo	Liver-related outcomes ΔIHTG by MRI-PDFF (%): (1): −5.3 (2): −4.2 ΔLSM by MRE (%): (1): -7.3^{ab} (2): −1.1 Significant reductions of ALT, γ-GT, and ALP in (1). No significant changes in AST. Metabolic outcomes Significant reduction of TC, LDL-C, non-HDL-C, and TG but also of HDL-C (1).
	Yokote et al., 2021 [62], Japan	Pooled analysis of 6 RCTs; 12 weeks	Adult patients w/hypertriglyceridemia N = 1253 Female: 184 (15%) NAFLD: 534 (43%) T2DM: 449 (36%)	Arm (1): Pemafibrate 0.1 mg/day Arm (2): Pemafibrate 0.2 mg/day Arm (3): Pemafibrate 0.4 mg/day Arm (4): Placebo	Liver-related outcomes Non-significant reductions of AST (−45.5 to −58.1 U/L (1–3) vs. −33.3 U/L (4)). Dose-dependent reductions of ALT, significant for (2) and (3) (−58.5 and −67.2 U/L vs. −18.4 U/L (4)). Significant reductions in γ-GT (−61.1 to −80.6 U/L (1–3) vs. −10.9 U/L (4)). Significant reductions in ALP (1–3); non-significant reductions in bilirubin. Metabolic outcomes Significant reductions of fasting plasma glucose, fasting serum insulin, and HOMA-IR (1–3); no change in HbA1c. Significant reductions of TG and increases of HDL-C (1–3).
Pioglitazone and rosiglitaone; PPARγ	Mantovani et al., 2020 [63], international (USA, Europe, and Asia)	Systematic review of 8 RCTs (pioglitazone 6 and rosiglitazone (2)); 4 to 36 months	Adults w/NAFLD and thiazolidinedione treatment for NAFLD/NASH N = 828 Female: 43% T2DM: 15%	(1): Rosiglitazone 8 mg/day (2): Pioglitazone 30 to 45 mg/day	Liver-related outcomes Significant improvements of liver fat content, NASH, and serum ALT/AST levels (1–2). No significant change in fibrosis stage compared to control in all RCTs, except one (1).
	Musso et al., 2017 [19], international (USA, Europe, and Asia)	Meta-analysis of 8 RCTs (pioglitazone 5 and rosiglitazone (3)); 6 to 24 months	Adults w/NAFLD defined by radiological or histological evidence of steatosis N = 516 (in main analysis of primary outcome; N = 698 participants overall) Female: 333 (48%) T2DM: 142 (20%)	(1): Rosiglitazone 4 to 8 mg/day (2): Pioglitazone 30 to 45 mg/day	Liver-related outcomes Improvement of advanced fibrosis (F3-4 to F0-2; $\geq$2 stages improvement) in all participants (OR, 95% CI): (1): 1.30, 0.23–7.20 (2): 4.53, 1.52–13.52 Overall: 3.15, 1.25–7.93 Improvement of advanced fibrosis (F3-4 to F0-2; $\geq$2 stages improvement) in participants with advanced fibrosis (OR, 95% CI): (1): 1.84, 0.29–11.66 (2): 10.17, 2.83–36.54 Overall: 5.84, 2.04–16.71 Improvement of $\geq$1 fibrosis stage in all participants (OR, 95% CI): (1): 1.18, 0.43–3.25 (2): 4.53, 1.77, 1.15–2.72 Overall: 1.66, 1.12–2.47 NASH resolution in participants with NASH (OR, 95% CI): (1): 2.14, 0.94–4.86 (2): 3.65, 2.32–5.74 Overall: 3.22, 2.17–4.79

Table 1. *Cont.*

Drug Name; Target Pharmaceutical Company	Reference, Country	Study Type; Treatment Duration	Participants	Intervention	Results
Pioglitazone; PPARγ	Bril et al., 2019 [64], USA	RCT; 18 months	Adults w/NASH in liver histology and T2DM Exclusion: T1DM N = 105 Female: 12 (11%)	Arm (1): Pioglitazone 45 mg/day plus vitamin E 400 IU b.i.d. Arm (2) Vitamin E 400 IU b.i.d. Arm (3): Placebo	Liver-related outcomes Reduction in NAS ≥2 points w/o worsening of fibrosis (n%): (1): 54 [b] (Δ 35 [14 to 56]) (2): 31 (Δ 12 [−1 to 32]) (3): 19 Reduction of NASH w/o worsening of fibrosis (% of n): (1): 43 [b] (Δ 31 [11 to 50]) (2): 33 [b] (Δ 21 [2 to 40]) (3): 12 Significant improvements in steatosis, inflammation, and ballooning (1). No significant improvements in fibrosis. Significant reduction in IHTG by 1H-MRS in (1 and 2). Metabolic outcomes Significant improvement in HbA1c (1). Modest increase in HDL-C (1).
	Bril et al., 2019, and Cusi et al., 2018 [65,66], USA	RCT; 18 months	Adults w/NASH on liver histology, and T2DM/prediabetes Exclusion: T1DM, clinically significant renal, pulmonary, or cardiac disease N = 101 Female: 30 (30%) T2DM: 52 (51%)	Arm (1a): Pioglitazone 30 to 45 mg/day in T2DM patients Arm (1b): Pioglitazone 30 to 45 mg/day in prediabetic patients Arms (2a and b): Placebo in T2DM and prediabetic patients	Liver-related outcomes Reduction in NAS ≥2 points (≥2 different categories) w/o worsening of fibrosis (n%): (1 overall, 1a and b): 58 [b], 60 [ab], and 55 [a] (2 overall, 2a and b): 17, 16, and 29 ΔFibrosis stage (SD): (1a and b): −0.5 [0.9] [b] and −0.4 [0.9] (2a and b): 0.2 [1.2] and −0.2 [0.7] Significant reduction in IHTG by ^{1}H-MRS (1a and b). Metabolic outcomes Increases in hepatic and skeletal muscle insulin sensitivity (1 and b), and in adipose tissue insulin sensitivity (1a). Significant reduction in fasting plasma insulin, and significant increases in adiponectin (1a and b), significant reduction in HbA1c (1a). Improvements in HDL-C and TG (1 and b).
Saroglitazar (EVIDENCES IV); PPARα/γ Zydus Discovery	Gawrieh et al., 2021 [67], USA	RCT; 16 weeks	Adults w/NAFLD based on histology or imaging, ALT ≥ 50 U/L and BMI ≥ 25 kg/m^2 Exclusion: T1DM, poorly controlled T2DM (HbA1c ≥ 9.0%) N = 106 Female: 49 (46%) T2DM: 56 (53%)	Arm (1): Saroglitazar 1 mg Arm (2): Saroglitazar 2 mg Arm (3): Saroglitazar 4 mg Arm (4): Placebo	Liver-related outcomes Significant dose-dependent ALT reductions (−25.5 to −45.8% (1–3) vs. +3.4% (4)). ALT reduction ≥25% in 64–70% (1–3) of patients compared to 18% (4). ALT reduction ≥50% in 15–52% (1–3) of patients compared to 4% (4). ALT < ULN in 6 patients (3) compared to 0 (4). Significant reductions in AST (−25.4 to 34.9% (1–3) vs. +9.8% (4)), ALP (−17.0 to 35.7% (1–3) vs. +3.3 (4)), and γ-GT (−29.4 to −45.7 (1–3) vs. +10.9 (4)). Significant reduction of steatosis by MRI-PDFF (difference −23.8% (3)) with >30% reduction in 11 patients (3) compared to 2 (4). No significant changes in CK18, LSM, or CAP. Metabolic outcomes Significant improvements in TG (1 and 3), non-significant improvements in HDL-C and LDL-C (3). Significant improvement in HOMA-IR (3), and non-significant improvements in blood glucose, HbA1c, and insulin levels (1 and 3).

Table 1. *Cont.*

Drug Name; Target Pharmaceutical Company	Reference, Country	Study Type; Treatment Duration	Participants	Intervention	Results
Elafibranor (GOLDEN-505); PPARα/δ Genfit	Ratziu et al., 2016 [68], Europe and USA	RCT; 52 weeks	Adults w/NASH on liver histology N = 274 Female: 81 (33%) T2DM: 107 (39%)	Arm (1): Elafibranor 80 mg Arm (2): Elafibranor 120 mg Arm (3): Placebo	Liver-related outcomes Resolution of NASH w/o worsening of fibrosis according to current definition (OR, 95% CI): (1) 2.31, 1.02–5.24 [b] (2) 1.48, 0.7–3.14 Overall more pronounced effects in those with severe inflammation (NAS $\geq$ 4, (1): 3.52, 1.32–9.40) or presence of fibrosis (fibrosis any stage, (1): 3.75, 1.39–10.12). Improvements in γ-GT and ALP (1–2), no changes in ALT. Metabolic outcomes Significant improvements of HbA1, HOMA-IR, and fasting glucose (1) in diabetic patients, as well as improvement in plasma insulin (1–2). Dose-dependent increases in HDL-C (1–2). Decreases in LDL-C and TG (1–2).
Lanifibranor (NATIVE); Pan-PPAR Inventiva Pharma	Francque et al., 2021 [20], international (Europe, Canada, USA, and Australia)	RCT; 24 weeks	Adults w/NASH on liver histology Exclusion: cirrhosis (F4) N = 247 Female: 144 (58%) T2DM: 103 (2%)	Arm (1): Lanifibranor 800 mg Arm (2): Lanifibranor 1200 mg Arm (3): Placebo	Liver-related outcomes Decrease of $\geq$2 points from SAF activity score w/o worsening of fibrosis (RR, 95% CI): (1) 1.45, 1.00–2.10 [b] (2) 1.69, 1.22–2.34 [b] Improvement of NASH w/o worsening of fibrosis (RR, 95% CI): (1) 1.70, 1.07–2.71 [b] (2) 2.20, 1.49–3.26 [b] Improvement of fibrosis $\geq$1 stage w/o worsening of NASH (RR, 95% CI): (1) 1.15, 0.72–1.85 (2) 1.68, 1.15–2.46 [b] Significant reductions in AST, ALT, and γ-GT (1–2). Metabolic outcomes Significant improvements in HOMA-IR (−5.5 to −5.8 (1–2) vs. −1.47 (3)), insulin (−115 to −119 pmol/L (1–2) vs. −36 pmol/L (3)), HbA1c, and fasting glucose (1–2). Significant reduction of TG (1–2), no significant changes in TC, HDL-C, and LDL-C.

[a] Significant compared to baseline; [b] significant compared to placebo; non-significant if not mentioned separately. Abbreviations: Δ, difference; 1H-MRS, proton magnetic resonance spectroscopy; 95% CI, 95% confidence interval; ALP, alkaline phosphatase; ALT, alanine aminotransferase; AST, aspartate aminotransferase; BID, twice daily; BMI, body mass index; CAP, controlled attenuation parameter; CI, confidence interval; EMA, European Medicines Agency; γ-GT, gamma-glutamyltransferase; HbA1c, glycated hemoglobin A1c; HDL-C, high-density lipoprotein cholesterol; HOMA-IR, homeostatic model assessment of insulin resistance; IHTG, intrahepatic triglyceride content; LDL-C, low-density lipoprotein cholesterol; LSM, liver stiffness measurement; MRE, magnetic resonance elastography; MRI-PDFF, magnetic resonance imaging proton density fat fraction; na, not available; NAFLD, non-alcoholic fatty liver disease; NAS, NAFLD Activity Score; NASH, non-alcoholic steatohepatitis; OR, odds ratio; RCT, randomized controlled trial; RR, risk ratio; SAF, steatosis activity fibrosis scoring system; T1DM, type 1 diabetes mellitus; T2DM, type 2 diabetes mellitus; TC, total cholesterol; TG, triglycerides; ULN, upper limit of normal; USA, United States of America; w/, with; w/o, without.

3.1. Selective PPARα Modulator: Pemafibrate (K-877)

Pemafibrate is a selective PPARα modulator (SPPARMα), as among PPAR isotypes it is highly selective for PPARα [69]. Structural differences of the pemafibrate molecule compared to other PPARα agonists such as fenofibrate allow for this higher selectivity and agonistic activity at the receptor ligand-binding site [70]. The drug is currently approved and marketed in Japan for the treatment of dyslipidemia with high triglyceride (TG) and low high-density lipoprotein cholesterol (HDL-C) levels under the name Parmodia® [71,72]. Thus, most evidence on pemafibrate in NAFLD is derived from trials conducted in this target population.

Pre-clinical data indicate a beneficial effect of pemafibrate on some aspects of liver histology in NAFLD/NASH. Steatosis, measured by area of oil red O staining but not hepatic TG content, inflammatory activity, and fibrosis improved under pemafibrate in a mouse model of diet-induced NASH [73]. In a STAM mouse model, mimicking NASH with underlying diabetes, pemafibrate ameliorated inflammatory activity while again no effect on hepatic TG content was observed [74].

A double-blind, randomized controlled phase 2 trial including 224 Japanese patients with dyslipidemia treated with pemafibrate twice daily (0.025 mg, n = 34; 0.05 mg, n = 37; 0.1 mg, n = 36; 0.2 mg, n = 36) versus fenofibrate once daily (100 mg, n = 36) or placebo (n = 34) over the course of 12 weeks has assessed the efficacy and safety of pemafibrate for the treatment of dyslipidemia [75]. The study population included 12% patients with T2DM and 20% with fatty liver. However, participants with poorly controlled T2DM (glycated hemoglobin A1c (HbA1c) $\geq$ 8.4%), history of hepatic impairment, and aspartate aminotransferase (AST) or alanine aminotransferase (ALT) levels more than 2-fold above the upper limit of normal (ULN) were excluded [75]. Pemafibrate showed a dose-dependent, significant reduction of plasma TG and increase in HDL-C levels compared to baseline and placebo, while elevations of AST and ALT occurred less frequently compared to fenofibrate [75].

Only patients with T2DM, 54% of whom had concomitant NAFLD (N = 166), were included in the randomized, placebo-controlled phase 3 PROVIDE trial (pemafibrate 0.2 mg/day, 0.4 mg/day, or placebo over 24 weeks), to assess the effect on fasting serum TG and further lipid-related as well as glycemic parameters [76]. The study demonstrated a significant decrease in TG levels by around 45% in both treatment groups. The treatment groups experienced fewer liver-related adverse events [76].

These findings were supported by a randomized controlled phase 3 trial, which included 223 Japanese patients with dyslipidemia, who received either pemafibrate 0.2 mg, 0.4 mg, or fenofibrate 106.6 mg daily [77]. Besides improvements in TG and HDL-C levels, the pemafibrate groups furthermore showed significant decreases in AST, ALT, gamma-glutamyltransferase (γ-GT), and alkaline phosphatase (ALP) [77]. However, only 5% of the participants had T2DM, no data on NAFLD is reported, and patients with poorly controlled T2DM or liver impairment were excluded [77].

Recently, a large post hoc analysis has summarized evidence from six, randomized controlled clinical trials of pemafibrate for treatment of dyslipidemia regarding outcomes related to glycemic control and liver values [62]. This pooled analysis included 1253 patients, 36% and 43% of whom had T2DM and fatty liver disease, respectively [62]. Among individuals with liver function tests above ULN at baseline, the proportion of patients with normalization of ALT, γ-GT, and ALP was significantly higher in the group treated with high-dose pemafibrate compared to placebo [62]. A significant decrease of liver tests was observed in all groups, predominantly in the 0.4 mg per day group [62]. Markers of glucose homeostasis, namely fasting plasma glucose and insulin, and homeostatic model assessment of insulin resistance (HOMA-IR) decreased significantly among the pemafibrate groups compared to placebo [62]. These data suggest a potential beneficial effect on both liver outcomes and glycemic control.

Few studies have assessed the effect of pemafibrate on liver-related outcomes other than blood-based markers. A single-arm, prospective trial investigated the efficacy and safety of pemafibrate 0.2 mg daily in patients with sonographically assessed NAFLD and dyslipidemia (N = 20, 40% T2DM) over the course of 12 weeks [78]. Elevated ALT levels decreased significantly in all participants ($p = 0.001$) and normalized in around half of the patients [78]. Furthermore, liver stiffness measurement (LSM) by vibration-controlled transient elastography (VCTE) and controlled attenuation parameter (CAP) decreased, but this did not reach statistical significance [78]. Similar findings were reported by a retrospective study that evaluated the effect of pemafibrate in 31 patients with NAFLD/NASH (16% T2DM), assessed non-invasively by FibroScan-aspartate aminotransferase (FAST) score, who also received pemafibrate 0.2 mg daily [79]. A significant decrease in FAST score over 48 weeks was observed, with a moderate, non-significant decrease in LSM [79].

Several other studies have retrospectively investigated pemafibrate in NAFLD and NASH [80–82]. Pemafibrate 0.2 mg was associated with improvements in several hepatic markers after a follow-up of one year in non-diabetic NAFLD patients [82] and in patients with biopsy-proven NASH [80].

To date, one randomized, placebo-controlled trial of pemafibrate 0.2 mg daily in patients with NAFLD (N = 118), defined as a liver fat content of ≥10% measured by magnetic-resonance-imaging-estimated proton density fat fraction (MRI-PDFF), has been reported [61]. In total, 36% of participants had T2DM and 67% had metabolic syndrome [61]. While no significant change in liver fat content was observed over 72 weeks, LSM by magnetic resonance elastography (MRE) significantly decreased by −6.2% (95% confidence interval [CI] −11.5 to −0.8, $p = 0.024$) in the pemafibrate group compared to placebo [61]. Among the six serious adverse events reported, none were related to pemafibrate [61].

Overall, the evidence regarding pemafibrate's potential in the treatment of NAFLD patients with T2DM should be further explored. Currently only one randomized controlled trial has been reported and the number of patients with both NAFLD and T2DM in present studies is limited, hampering the extrapolation of findings to this population. Clinical trials evaluating the effect of pemafibrate on histological liver outcomes are currently missing.

3.2. PPARα Agonists: Fibrates

Fibrates were the first drug class utilizing PPARα agonism, albeit unknowingly, until discovery of the nuclear receptor [22]. Compared to pemafibrate, fibrates such as gemfibrozil, clofibrate, fenofibrate, and bezafibrate show relatively weak PPARα agonism [15,72]. While these substances are generally considered to be PPARα agonists, the individual receptor profile may differ between drugs. Bezafibrate, for instance, shows activity at all PPAR isotypes and may thus be classified as a pan-PPAR agonist [83]. Fibrates are currently used in the treatment of dyslipidemia and their potential to improve NAFLD has been explored for several decades [84]. In the field of liver disease, fibrates are further studied regarding use in primary biliary cholangitis (PBC) [85].

Among individuals with T2DM, serum levels of CC chemokine ligand 5 (CCL5) [86], a pro-inflammatory chemokine that has been implicated in advancing the development of fibrosis in NAFLD and NASH, decreased during fenofibrate treatment [87]. Recently, use of fibrates was identified as a protective factor against progression of NAFLD to advanced fibrosis (odds ratio [OR] 0.90, $p < 0.05$), defined as an increase in non-invasive markers, in a large cohort of American individuals with diabetes (N = 50,695) [7].

While promising results regarding hepatocellular damage and fibrosis have been reported in pre-clinical models [88,89], these findings have not translated into a clear benefit in clinical trials of NAFLD and NASH. In a small controlled clinical trial of gemfibrozil (N = 46), a decrease in transaminases, predominantly ALT, in NASH patients was noted [90]. In a prospective single-arm, dual biopsy trial of NAFLD patients (N = 16), only hepatocyte ballooning, but not other histological parameters, including steatosis, lobular inflammation, and fibrosis, changed significantly under fenofibrate (200 mg/day) over 48 weeks [91]. Treatment decreased TG, γ-GT, and ALP, but not transaminases [91]. Only one participant in this trial had T2DM [91].

Randomized, placebo-controlled trials have assessed the effect of fenofibrate in individuals without T2DM. Among 25 participants with insulin resistance and metabolic syndrome, fenofibrate reduced plasma TG as well as inflammatory markers interleukin-6 (IL-6) and high-sensitivity C-reactive protein (hsCRP), but did not improve insulin sensitivity [92]. Another trial assigned 27 patients with NAFLD (intrahepatic triglyceride (IHTG) content by magnetic resonance imaging ((MRI) $\geq$ 5.6%) and obesity to receive fenofibrate, niacin or placebo [93]. Fenofibrate decreased plasma TG and very low-density lipoprotein (VLDL) composition regarding TG and apolipoprotein B (apoB) content, while having no effect on insulin sensitivity or IHTG content [93]. Another clinical trial of non-diabetic NAFLD patients reported an increase in total liver volume and total liver fat volume under fenofibrate [94].

Thus, benefits of fibrates on liver-related outcomes in NAFLD currently seem limited, although data from clinical trials are largely based on individuals without T2DM. Previous trials failed to demonstrate insulin-sensitizing effects [92,93,95], but treatment might be warranted in certain dyslipidemic conditions (see Section 4.2.). Treatment seems safe regarding liver outcomes, as most commonly liver enzyme elevations are transient, although instances of acute liver injury during treatment with fibrates, mostly under fenofibrate, have been reported [96,97].

3.3. *Selective PPARδ/β Agonist: Seladelpar (MBX-8025)*

Seladelpar (MBX-8025) was developed as a selective PPARδ (PPARβ) agonist [98]. The molecule improved several parameters of glucose homeostasis and liver histology, including inflammation and steatosis, in a mouse model of NASH with T2DM and obesity [99]. In humans, seladelpar has previously demonstrated beneficial metabolic effects on atherogenic dyslipidemia (see Section 4.2), while no significant insulin-sensitizing effect was observed [98].

A randomized, placebo-controlled phase 2 trial (NCT03551522) of seladelpar in patients with histologically confirmed NASH provided preliminary results, demonstrating no effect on hepatic steatosis compared to placebo [18]. Final results, however, have not yet been published and development of the molecule has been halted for this indication after histological evaluation revealed findings of interface hepatitis [15].

3.4. *PPARγ Agonists: Thiazolidinediones*

Thiazolidinediones are a drug class of PPARγ agonists currently approved for the treatment of diabetes due to their insulin-sensitizing effects [52], with similar effects on glycemic control between different substances of the class [100]. Positive effects in NAFLD patients have been confirmed by several systematic reviews and meta-analyses [19,63,101,102]. A systematic review of eight randomized controlled trials of thiazolidinediones (pioglitazone and rosiglitazone) for treatment of NAFLD or NASH (15% T2DM), which was diagnosed based on histological criteria in the majority of cases, reported improvements in liver fat content and in serum levels of transaminases [63]. A meta-analysis of five clinical trials further indicated an improvement in lobular inflammation (risk ratio [RR] 1.72, 95% CI 1.33–2.22, $p < 0.0001$) and fibrosis (RR 1.39, 95% CI 1.01–1.90, $p = 0.04$) with thiazolidinedione (pioglitazone and rosiglitazone) treatment, although these latter findings did not consistently hold up in subgroup analyses [102]. Data obtained from another meta-analysis, which included eight randomized controlled clinical trials of patients with biopsy-confirmed NASH (overall N = 516 in analysis of primary outcome), indicated that only pioglitazone, but not rosiglitazone, leads to improvement of fibrosis $\geq$1 stage (OR 1.77, 95% CI 1.15–2.72, $p = 0.009$, and OR 1.18, 95% CI 0.43–3.25, $p = 0.74$, respectively) and NASH resolution (OR 2.14, 95% CI 0.94–4.86, $p < 0.001$, and OR 3.65, 95% CI 2.32–5.74, $p = 0.07$, respectively) [19].

3.4.1. Pioglitazone

Pioglitazone also exhibits weak PPARα agonism, which may explain the beneficial effect on NAFLD compared with the other PPARγ agonists mentioned above [103]. Pioglitazone may currently be considered for treatment of NASH according to several international clinical guidelines [104–106]. Specifically, the clinical practice guidelines by the European Association for the Study of the Liver (EASL) and the American Association for the Study of the Liver (AASLD) state that treatment with pioglitazone may be discussed in patients with confirmed NASH with and without concomitant diabetes [107,108]. Recently, a network meta-analysis of 30 studies (N overall = 2356) found pioglitazone to be the most effective therapy for (NAS) reduction along with rosiglitazone and gastric bypass [109].

In a murine model of NAFLD, high-fat-diet-induced steatosis was ameliorated through pioglitazone administration by PPARγ- and PPARα-dependent increases of lipolysis, β-oxidation, and autophagy [110]. Improvement of hepatic steatosis by pioglitazone was found to be impaired in adiponectin knockout mice, indicating adiponectin involvement in the mechanisms exerted by pioglitazone [111]. Among human patients with T2DM, pioglitazone increased adiponectin levels, which correlated with improvements in parameters of glucose homeostasis [112].

Data on the anti-fibrotic effect of pioglitazone are not conclusive [19,102]. In different rat models of fibrosis (carbon tetrachloride, bile duct ligation, choline-deficient diet), the effect of pioglitazone treatment on fibrosis varied by the type of injury as well as stage of fibrosis at administration [113]. In humans, genetic factors have been implicated in the variation of response to pioglitazone [114] and in fibrosis regression among the Pioglitazone versus Vitamin E versus Placebo for the Treatment of Non-Diabetic Patients with Nonalcoholic Steatohepatitis (PIVENS) trial participants [115].

In the randomized, placebo-controlled PIVENS trial (N = 247), the effects of vitamin E and pioglitazone in NASH without T2DM were evaluated [116]. While significant reductions of transaminases compared to placebo were observed, the primary endpoint of a composite improvement in NASH histological features was not met in the pioglitazone 30 mg group after 96 weeks (34% vs. 19%, $p = 0.04$, pre-specified significance level of 0.025) [117]. While NAS improved significantly, fibrosis stage did not [117]. Histological resolution of steatohepatitis was found to be associated with fibrosis regression among participants of the PIVENS trial (OR 3.9, 95% CI 2.0 to 7.6, $p < 0.001$) [115].

The data suggest that a beneficial effect of pioglitazone is also present in T2DM patients and may even be more pronounced in this group compared to patients without T2DM. A placebo-controlled proof-of-concept study of 55 NASH patients with T2DM or prediabetes confirmed a beneficial effect on several metabolic and histological features, including steatosis and inflammatory activity [118]. Similarly, significantly more patients in the pioglitazone group reached the primary endpoint of a ≥2-point reduction in NAS compared to placebo (difference 41%, 95% CI 23% to 59%, $p < 0.001$) in a randomized controlled trial of patients with T2DM or prediabetes (N = 101) after 18 months [65]. In this trial, the mean change in fibrosis score was greater in the treatment group (difference −0.5, 95% CI −0.9 to 0) [65].

The data from this randomized placebo-controlled trial were evaluated specifically with regard to the effect of pioglitazone in T2DM compared to prediabetes [66]. NASH resolution under pioglitazone compared to placebo occurred significantly more often only in the T2DM patients (60% vs. 1%, $p = 0.002$), but not in patients with prediabetes (55% vs. 29%, $p = 0.12$), owing also to a high resolution rate under placebo in the prediabetes group [66]. Similarly, the improvement in fibrosis stage was significant only in the T2DM treatment group (-0.5 ± 0.9 vs. 0.2 ± 1.2, $p = 0.042$), but, as would be expected, this group also presented significantly higher baseline fibrosis scores [66]. An 18-month proof-of-concept study of the combination therapy of vitamin E 400 international units (IU) and pioglitazone 30–45 mg in patients with T2DM and bioptically confirmed NASH (N = 105) showed that a ≥2-point improvement in NAS (difference 35%, 95% CI 14–56%, $p = 0.003$) and NASH resolution (difference 31%, 95% CI 11–50%, $p = 0.005$) occurred more often in the combination than the placebo group [64]. The proportion of patients achieving an improvement in fibrosis stage did not differ significantly between groups (52% vs. 30%, $p = 0.07$) [64].

The effect of pioglitazone on fibrosis improvement in T2DM and prediabetes thus remains inconclusive. A meta-analysis of three of the above-mentioned trials along with one Chinese trial of pioglitazone versus berberine indicated significant improvements in steatosis, lobular inflammation, and ballooning, but not fibrosis [119].

Besides effects on NAFLD, pioglitazone is furthermore a strong insulin sensitizer and has a protective effect on beta-cell function, delaying the onset of T2DM in individuals with impaired glucose tolerance and impaired fasting glucose [120,121]. This was demonstrated in several clinical trials, for example the Actos Now for Prevention of Diabetes (ACT NOW) [122] and the Insulin Resistance Intervention After Stroke (IRIS) trials [123], which showed a 72% ($p < 0.001$) and 52% ($p < 0.0001$) risk reduction of development of overt T2DM under pioglitazone, respectively.

Despite strong indications that pioglitazone exerts multiple metabolic benefits in patients with T2DM, widespread use of pioglitazone is hampered by adverse effects. Besides weight gain (see Section 4.1), pioglitazone has been implicated in increasing the risk of bladder cancer [124–126] and fractures by decreasing bone mineral density [127]. However, evidence on the association of bladder cancer with pioglitazone remains inconclusive. One meta-analysis concluded that the risk of bladder cancer was not increased significantly with pioglitazone use (hazard ratio [HR] 1.07, 95% CI 0.96 to 1.18) [126]. A more recent systematic review and meta-analysis came to a similar conclusion when assessing data from randomized controlled clinical trials, but did find a significantly increased risk among subjects in observational studies (OR 1.13, 95% CI 1.03 to 1.25) [124]. Another systematic review of observational trials stated that existing data were too heterogeneous to derive a reliable conclusion [125]. An increased risk of bladder cancer was not described for rosiglitazone, indicating a possible adverse effect specific to pioglitazone rather than the drug class of thiazolidinediones [128].

Among individuals with NASH and T2DM/prediabetes, randomized to receive either pioglitazone 45 mg or placebo, use of pioglitazone was associated with a decrease of bone mineral density at the level of the lumbar spine at 18 months (−3.5%, $p = 0.002$) [127]. Bone mineral density did not change during the extension phase until 36 months and no low-energy fractures were reported [127]. Overall, pioglitazone might be a viable option for treatment of NASH in patients with T2DM, but individual risks and benefits should be carefully weighed. Several trials of pioglitazone in NAFLD treatment are currently ongoing (Table 2).

Table 2. Ongoing interventional trials of PPAR-targeted therapy in NAFLD/NASH.

Drug Name; Target Pharmaceutical Company	Country	Trial Registration ID Trial Name	Participants (Randomization)	Intervention	Treatment Duration (Weeks)	Primary Outcome(s) Secondary Outcome(s)
Pioglitazone; PPARγ	South Korea	NCT03646292	Adults w/NAFLD (diagnosed on ultrasound and other modalities) and T2DM N = 60 (1:1:1)	Arm (1): Empagliflozin 10 mg Arm (2): Pioglitazone 15 mg Arm (3): Empagliflozin 10 mg + pioglitazone 15 mg	24	Change in hepatic fat content by MRI-PDFF. Change in liver fibrosis by MRE, changes in lipid profiles, liver enzymes, glucose metabolism, and inflammatory biomarkers.
	Pakistan	NCT04976283	Adults w/NAFLD (diagnosed by FibroScan) and T2DM N = 123 (1:1:1)	Arm (1): Pioglitazone 15 mg Arm (2): Empagliflozin 5–12.5 mg Arm (3): Empagliflozin 10 mg + Pioglitazone 15 mg	52	Change in radiologic liver parameters. Changes in liver enzymes, liver fibrosis scores, body weight, body composition, glucose metabolism, and lipid profiles.
	USA	NCT04501406	Adults w/NASH (histologically confirmed) and T2DM N = 138 (1:1)	Arm (1): Pioglitazone 15 mg Arm (2): Placebo	72	Proportion of patients achieving ≥2 points improvement in NAS w/o worsening of fibrosis. Resolution of NASH w/o worsening of fibrosis, improvements in SAF score and NAS, and change in fibrosis.

Table 2. *Cont.*

Drug Name; Target Pharmaceutical Company	Country	Trial Registration ID Trial Name	Participants (Randomization)	Intervention	Treatment Duration (Weeks)	Primary Outcome(s) Secondary Outcome(s)
Saroglitazar; PPARα/γ Zydus Discovery	USA	NCT03639623 EVIDENCES VIII	Adults 6 months post-transplantation for NASH N = 15	Arm (1): Saroglitazar 4 mg	24	Safety (adverse events). Changes in hepatic fat, metabolic flexibility, lipid profiles, liver enzymes, glucose metabolism, pharmacokinetics, and quality of life.
	USA	NCT05011305	Adults w/NASH (histologically confirmed) N = 240 (1:1:1)	Arm (1): Saroglitazar 2 mg Arm (2): Saroglitazar 4 mg Arm (3): Placebo	76	Resolution of NASH w/o worsening of fibrosis. Improvements in fibrosis, NAS and SAF score, changes in lipid profiles, liver enzymes, glucose metabolism, and body weight.
	USA	NCT03617263	Adult females w/NAFLD and PCOSN = 90 (1:1)	Arm (1): Saroglitazar 4 mg Arm (2): Placebo	34	Hepatic fat content by MRI-PDFF. Changes in liver enzymes, liver steatosis, liver fibrosis, BMI, body composition, glucose metabolism, pharmacokinetics, ovarian function, and free androgen index.
	India	NCT04193982	Adults w/NAFLD (histologically confirmed) N = 250 (1:1:1:1)	Arm (1): Saroglitazar 4 mg Arm (2): Vitamin E 400 mg Arm (3): Saroglitazar 4 mg + vitamin E 400 mg Arm (4): Lifestyle intervention	24	Change in NFS. Changes in liver enzymes, lipid profiles, liver fibrosis in histology, NAS, and HbA1c.
Lanifibranor; Pan-PPAR Inventiva Pharma	USA	NCT03459079	Adults w/NAFLD and T2DM N = 44 (1:1)	Arm (1): Lanifibranor 800 mg Arm (2): Placebo	24	Change in IHTG by 1H-MRS. Proportion patients w/ ≥ 30% decrease in IHTG and NAFLD resolution, changes in insulin sensitivity, lipid profiles, glucose metabolism, and biomarkers of fibrosis.
	USA	NCT04849728 NATIV3	Adults w/NASH (histologically confirmed) and fibrosis stages 2–3 N = 2000 (1:1:1)	Arm (1): Lanifibranor 800 mg Arm (2): Lanifibranor 1200 mg Arm (3): Placebo	72	Resolution of NASH and improvement of fibrosis, and time to first clinical outcome event. Resolution of NASH w/o worsening of fibrosis, improvement of fibrosis w/o worsening of NASH, changes in liver enzymes, lipid profiles, glucose metabolism, and quality of life.

Abbreviations: 1H-MRS, proton magnetic resonance spectroscopy; BMI, body mass index; HbA1c, glycated hemoglobin A1c; IHTG, intrahepatic triglyceride content; MRE, magnetic resonance elastography; MRI-PDFF, magnetic resonance imaging proton density fat fraction; NAFLD, non-alcoholic fatty liver disease; NAS, NAFLD Activity Score; NASH, non-alcoholic steatohepatitis; NFS, non-alcoholic fatty liver fibrosis score; PCOS, polycystic ovary syndrome; SAF, steatosis activity fibrosis scoring system; T2DM, type 2 diabetes mellitus; USA, United States of America; w/, with; w/o, without.

3.4.2. Rosiglitazone

The thiazolidinedione rosiglitazone has been withdrawn from the European market and its use is restricted in the United States of America (USA) due to concerns regarding cardiovascular safety (see Section 4.3). With regard to NAFLD, positive results have previously been reported.

In a small, single-arm study of NASH patients (N = 30), 50% of whom had impaired glucose tolerance or T2DM, resolution of NASH was observed in 10 out of 22 participants (45%) with consecutive biopsies under rosiglitazone 8 mg daily after 48 weeks [129]. Serum ALT levels improved significantly [129]. Similarly, in a single-arm trial that included only T2DM-NAFLD patients (N = 68), rosiglitazone treatment over 24 weeks led to a reduction of liver enzymes and improvement in glycemic control [130].

The Fatty Liver Improvement with Rosiglitazone (FLIRT) randomized placebo-controlled trial was conducted in 63 patients with histologically confirmed NASH [131]. This trial showed significant steatosis improvement (≥30%) in 47% of participants under rosiglitazone versus 16% under placebo ($p = 0.014$), but failed to demonstrate a benefit regarding other histologic outcomes after one year [131]. Interestingly, absence of diabetes was identified as a predictor of treatment response in this trial [131]. Among 44 patients who completed an open-label, one-year extension phase of this trial, prolonged treatment with rosiglitazone did not improve fibrosis stage or inflammatory activity [132].

The effect of rosiglitazone was further explored in combination therapies. Rosiglitazone 8 mg daily alone or with either metformin or the angiotensin receptor blocker losartan was compared in a randomized, open-label trial of paired biopsies in patients with confirmed NASH (N = 135) [133]. This trial showed no difference in NASH histology, including steatosis, hepatocellular inflammation, and fibrosis, among the three treatment groups [133]. Contrary to rosiglitazone alone and in combination with losartan, no weight gain was observed in the group taking rosiglitazone in combination with metformin, but this difference failed to reach statistical significance, indicating that metformin did not sufficiently ameliorate weight gain under rosiglitazone [133].

More recently, an analysis of hepatic gene expression patterns in the treatment group of the FLIRT trial revealed increased expression of hepatic PPARγ and pro-inflammatory genes, indicating a potentially detrimental long-term effect of rosiglitazone treatment [134]. Given these findings and the concerns regarding cardiovascular adverse effects, rosiglitazone's role in the treatment of NASH is currently limited.

3.4.3. Lobeglitazone

Lobeglitazone is a more recently developed thiazolidinedione that along with PPARγ agonism also exerts partial PPARα-agonism, similarly to pioglitazone. Lobeglitazone is currently approved and marketed in Korea as an anti-diabetes drug under the name Duvie® [42].

In a murine model of diet-induced NAFLD with obesity (high-fat diet), lobeglitazone administration for 4 weeks improved glucose homeostasis, hepatic steatosis, and serum lipid profile, accompanied by upregulation of hepatic gene expression related to fatty acid β-oxidation and decrease of genes involved in lipid synthesis and hepatic gluconeogenesis [135].

Data in human NAFLD is sparse. In a single-arm trial, 50 participants with T2DM and NAFLD, defined as controlled-attenuation parameter (CAP) over 250 dB/m, received lobeglitazone 0.5 mg for 24 weeks [136]. A modest but significant decline in hepatic steatosis, assessed non-invasively by CAP, compared to baseline was observed (313.4 dB/m vs. 297.8 dB/m, $p = 0.016$) [136]. Patients furthermore showed an improvement in glycemic control and atherogenic dyslipidemia [136].

3.5. *Dual PPARα and -γ Agonist: Saroglitazar*

Given the positive findings regarding dual agonism at PPARα and PPARγ with drugs such as pioglitazone, therapies specifically targeting both receptors for treatment of metabolic conditions were developed [137]. Recently one dual agonist, saroglitazar, was approved for NASH treatment in India (Lipaglyn®) [42]. The drug has previously been approved and marketed in India for treatment of diabetic dyslipidemia [138,139]. Compared to pioglitazone, saroglitazar exerts potent PPARα and only modest PPARγ agonism [140].

Data from pre-clinical, in vivo studies in rodent models indicate an improvement of insulin sensitivity, lipid profile, and other metabolic parameters with saroglitazar administration, while exhibiting a good safety profile [140]. In in vitro models of NASH (palmitic-acid-treated HepG2 and HepG2-LX2 co-cultures), saroglitazar showed a beneficial effect on several mechanisms involved in NASH pathogenesis [141]. In a mouse model of diet-induced NASH (high-fat, choline-deficient diet), saroglitazar demonstrated a more pronounced improvement in NAS compared to pioglitazone or fenofibrate [141]. Observed beneficial effects regarding fibrosis and fibrotic biomarkers were further confirmed in a mouse model of carbon tetrachloride (CCl_4)-induced fibrosis [141]. These observations are in line with findings reported from a diet-induced mouse model of NASH induced by Western high-fat diet and sugar water [142]. In this model, saroglitazar improved all histologic features of NASH, including fibrosis stage, and led to resolution of NASH in the treatment group [142].

In clinical trials in participants with diabetes, saroglitazar has demonstrated beneficial effects on atherogenic dyslipidemia and insulin sensitivity. Decreases of TG, low-density lipoprotein cholesterol (LDL-C), and fasting plasma glucose were observed in a placebo-controlled trial of saroglitazar 2 mg or 4 mg in T2DM patients (N = 302) [143]. In a three-arm trial of individuals with T2DM (N = 122), higher-dose saroglitazar (4 mg) showed significant improvements in TG and LDL-C compared to pioglitazone [144]. No serious adverse events were observed under saroglitazar [144]. Insulin-sensitizing effects in T2DM were demonstrated more recently in a small randomized, placebo-controlled trial (N = 30) [145].

A systematic review evaluated the effect of saroglitazar in three clinical trials, currently published as abstracts, demonstrating liver-related outcomes in NAFLD patients with dyslipidemia [139]. Saroglitazar was shown to improve hepatic steatosis, assessed non-invasively by CAP, and plasma ALT levels [139]. Further evidence of saroglitazar in NAFLD exists from observational studies. In two prospective observational studies, patients with T2DM and NAFLD on ultrasound, who received saroglitazar 4 mg for 24 weeks, showed a significant improvement in transaminases, LSM, and steatosis, measured by CAP [146,147].

Promising findings from two phase 2 clinical trials of saroglitazar in NAFLD/NASH have recently been published [67,148]. A paired biopsy, controlled trial randomized 16 patients with histologically confirmed NASH to receive either saroglitazar 2 mg or 4 mg, or placebo over 24 weeks [148]. NAS decreased in both treatment groups (-1.5 ± 0.84, $p = 0.77$ in 2 mg; -1.9 ± 1.57, $p = 0.60$ in 4 mg), but differences were not statistically significant compared to placebo (-1.33 ± 0.58) [148]. NASH resolution without worsening of fibrosis occurred in three (4 mg) and four (2 mg) patients of the treatment groups compared to none under placebo [148]. In the four-arm, double-blind, randomized, controlled EVIDENCES IV trial (NCT0306172), 106 patients (52% T2DM) with obesity and NAFLD according to imaging or biopsy were randomized to receive saroglitazar (1 mg, 2 mg, or 4 mg) or placebo for 16 weeks [67]. Patients in all treatment arms achieved significant reductions in ALT, AST, ALP, and γ-GT [67]. Liver fat content on MRI-PDFF decreased significantly in the high saroglitazar dose compared to placebo (difference −23.8%, 95% CI −39.9 to −7.7, $p = 0.004$) [67]. Fibrosis markers decreased but did not differ significantly from placebo [67,148].

Saroglitazar has exhibited a favorable safety profile [139]. For other molecules of this drug class, adverse events were similar to pioglitazone, including edema and weight gain [149]. Data on the use of saroglitazar currently seem promising although evidence from larger trials with histological endpoints are lacking. Several phase 2 and 3 clinical trials are currently ongoing to evaluate the use of saroglitazar in NAFLD (Table 2).

3.6. Dual PPARα and -δ Agonist: Elafibranor (GFT505)

Elafibranor (GFT505) is a dual agonist of PPARα and PPARδ, with predominant activity on the former [150]. In several rodent models of NAFLD/NASH, elafibranor administration decreased expression of pro-fibrotic and pro-inflammatory genes and improved various histological outcomes, including steatosis, inflammation, and fibrosis [151]. In an in vitro model of NASH, elafibranor was found to exert the strongest anti-NASH effects compared to seven other PPAR-modulating agents [152].

In humans, elafibranor enhanced insulin sensitivity in liver and muscle tissue, and reduced plasma TG, LDL-C, and ALT levels [153]. In the phase 2, randomized controlled GOLDEN-505 study of 274 patients with histologically confirmed NASH (39% T2DM), individuals treated with elafibranor 120 mg over 52 weeks had higher rates of NASH resolution without worsening of fibrosis compared to placebo (19% vs. 12%, OR 2.31, 95% CI 1.02 to 5.24, $p = 0.045$) [68]. The effect was more pronounced in individuals with NAS of $\geq$4 at baseline (OR 3.52, 95% CI 1.32 to 9.40, $p = 0.013$) [68]. Importantly, these analyses were performed according to the revised definition of treatment response while the protocol-defined primary endpoint was not met [68].

Elafibranor subsequently went on to the phase 3 RESOLVE-IT trial (NCT02704403), but the development has been halted after an interim analysis failed to achieve the primary endpoint of NASH resolution without worsening of fibrosis [154,155].

3.7. Pan-PPAR-Agonist: Lanifibranor (IVA337)

Lanifibranor is an indole sulfonamide derivative and a balanced pan-PPAR agonist that has demonstrated strong therapeutic potential in pre-clinical models of NAFLD/NASH [156]. Specifically, lanifibranor ameliorated insulin resistance and improved histological features of NASH, including steatosis, ballooning, and inflammation, in diet-induced and genetic animal models [157]. Lanifibranor showed both therapeutic as well as preventive anti-fibrotic properties in a CCl_4-induced model of fibrosis, inhibiting the expression of pro-fibrotic and inflammatory genes [157]. In a mouse model of diet-induced NASH, the ameliorative effects of lanifibranor on certain aspects of NASH histology were greater than those observed with agonists of individual PPARs [158]. While macrophage infiltration due to acute CCl_4-induced injury remained unchanged under lanifibranor, macrophages displayed a metabolically activated phenotype, decreasing inflammation [158]. In vivo and in vitro models further indicate a beneficial effect on portal hypertension. In rat models of cirrhotic liver disease (bile duct ligation, thioacetamide exposure), lanifibranor lowered portal pressure, improved microvascular function, and attenuated fibrosis [159]. These findings indicate potential in the treatment of advanced chronic liver disease.

The impact of lanifibranor in human NASH has been evaluated in the randomized, placebo-controlled phase 2 NATIVE trial (NCT03008070) [160]. In total, 247 patients with non-cirrhotic (fibrosis stages F2–3), histologically active NASH (42% T2DM) were randomized to receive either lanifibranor (1200 mg or 800 mg) or placebo over 24 weeks [20]. The primary endpoint of decrease in histological activity ($\geq$2 points in the activity score SAF-A) without worsening of fibrosis was significantly more likely in the higher dosage treatment group compared to placebo (55% vs. 33%, RR 1.69, 95% CI 1.22 to 2.34, $p = 0.007$), while no significant improvement was observed with the lower dose (48% vs. 33%, RR 1.45, 95% CI 1.00 to 2.10 $p = 0.07$) [20]. An improvement of $\geq$1 fibrosis stage without worsening of NASH also occurred more often in the high-dose lanifibranor group compared to placebo (RR 1.68, 95% CI 1.15 to 2.46) [20]. A network meta-analysis of pharmacologic therapies for NAFLD ranked the probability of achieving an improvement of $\geq$1 fibrosis stage as being highest with lanifibranor (OR 2.38, 95% CI 1.21 to 4.67) [161]. Lanifibranor is one of the two pharmacological therapies that have demonstrated an improvement in fibrosis stage in clinical trials. The efficacy of lanifibranor is currently investigated in a phase 3 clinical trial in patients with NASH (NATiV3; NCT0484972).

4. Comorbidities of the Metabolic Syndrome in the PPAR-Targeted Treatment of Diabetic NAFLD Patients

Care of individuals with T2DM must consider co-existing conditions [162], and the presence of NAFLD or NASH adds further complexity to this population with multiple metabolic comorbidities [5]. The following chapter provides an overview of common comorbid conditions of the metabolic syndrome in T2DM patients and the possible impact of PPAR-directed therapies on these conditions. Findings are summarized in Table 3. Given the large volume of evidence published on these topics, we focused on pivotal trials and recent works summarizing previous findings.

Table 3. Effects of PPAR-directed therapy on other comorbidities of the metabolic syndrome.

PPAR Target	Drug Name	Overweight and Obesity	Effects on Lipid Levels	Cardiovascular Comorbidities
PPARα	Pemafibrate	No effect	↑ HDL-C ↓(↓) Triglycerides	Unknown; phase 3 trial (PROMINENT) ongoing
	Fibrates	No effect	↑ HDL-C ↓ Triglycerides	Modest decrease of cardiovascular risk in primary and secondary prevention
PPARδ	Seladelpar	No effect	↑ HDL-C ↓ LDL-C ↓ Triglycerides	No data on cardiovascular outcomes; good cardiovascular safety profile
PPARγ	Pioglitazone and other thiazolidine-dione	Weight gain (3–7% of body weight)	↑ HDL-C (↑) LDL-C ↓ Triglycerides	Risk reduction for several cardiovascular outcomes; causing fluid retention and edema; possibly increased cardiovascular risk with rosiglitazone
PPARα/γ	Saroglitazar	No effect	(↑) HDL-C (↓) LDL-C ↓ Triglycerides	No data on cardiovascular outcomes; good cardiovascular safety profile
PPARα/δ	Elafibranor	No effect	(↑) HDL-C ↓ Triglycerides	No data on cardiovascular outcomes; good cardiovascular safety profile
Pan-PPAR	Lanifibranor	Mild weight gain (3% of body weight)	↑ HDL-C ↓ Triglycerides	No data on cardiovascular outcomes; good cardiovascular safety profile

Abbreviations: HDL-C, high-density lipoprotein cholesterol; LDL-C, low-density lipoprotein cholesterol; ↑, increases; ↓, lowers; brackets denote conflicting or unclear results.

4.1. Overweight and Obesity

Both NAFLD and T2DM are closely linked with obesity. The prevalence of obesity has been estimated at 51% among NAFLD patients, rising to 82% in patients with NASH [1]. Thus, pharmacological treatments for NAFLD should be evaluated with regard to their effects on weight, especially in diabetic patients.

Among the discussed PPAR-directed therapies, weight gain has consistently been reported for thiazolidinediones [100]. However, conflicting data exist as to whether this weight gain is predominantly associated with fluid retention or an increase in adipose tissue mass [163–165]. Possible cardiac implications of fluid retention are discussed in chapter 4.3. Recent data from obese women treated with pioglitazone 30 mg over 16 weeks compared to placebo indicate an increase in adipogenesis in the subcutaneous femoral adipose tissue depot, which is considered beneficial for metabolic health compared to other depots [166], while reducing visceral adipose tissue [167]. These findings are in line with other evidence demonstrating improved adipose tissue metabolism [165] and an overall beneficial cardiovascular effect of pioglitazone (see Section 4.3).

In the three-arm PIVENS trial of pioglitazone or vitamin E versus placebo, only the pioglitazone group demonstrated a significant weight gain of 4.7 kg ($p < 0.001$) [117]. Overall, trials have consistently reported a considerable increase of around 3–7% of body weight during thiazolidinedione treatment [117,118,129,168,169], which was also confirmed in participants with T2DM and prediabetes [65]. In a study by Bril et al. (2019), individuals with T2DM, who received combination therapy with vitamin E and pioglitazone, demonstrated a significant weight gain (5.7 ± 5.4 kg, $p < 0.001$) after 18 months compared to no significant changes in the vitamin E and placebo groups [64]. Weight gain was not ameliorated by combining pioglitazone with instructions regarding a hypocaloric diet [65,118]. Weight gain among NASH patients in a 48-week trial of pioglitazone partially remained at the 6-month post-treatment follow-up [129].

Inconsistent findings regarding weight gain have been reported from trials of dual or pan-PPAR agonists, which exert PPARγ agonism. Both bezafibrate and saroglitazar have demonstrated no effect on body weight [139,170]. In the context of bezafibrate, it has been discussed that this might be due to concomitant PPARδ activation, ameliorating PPARγ-mediated weight gain [170]. In contrast, weight gain has been observed for lanifibranor, the pan-PPAR agonist currently under investigation in NAFLD [20]. Francque et al. (2021) reported a 3% increase in body weight in both the low- and high-dose treatment group of the phase 2 NATIVE trial [20].

No clinically relevant changes in body weight have been reported for seladelpar [99], elafibranor [68], and fibrates, including fenofibrate [91] and pemafibrate [61].

4.2. Dyslipidemia

PPAR agonists have demonstrated effects mostly in the treatment of atherogenic dyslipidemia, which is a common comorbidity in T2DM patients [16,171]. Atherogenic dyslipidemia is defined by low plasma levels of HDL-C with elevated levels of TG and small and dense LDL-C [172]. Atherogenic dyslipidemia represents a major risk factor for cardiovascular disease. Effects of PPAR-agonists on cardiovascular outcomes are discussed in Section 4.3.

As fibrates reduce TG and, to a lesser extent, improve levels of HDL-C [36], the use of fibrates to reduce residual cardiovascular risk in persistent atherogenic dyslipidemia despite lifestyle or statin treatment in patients with T2DM has been evaluated [173]. The Fenofibrate Intervention and Event Lowering in Diabetes (FIELD) trial included 9795 T2DM patients without lipid-lowering treatment at baseline and without clear indication for the former [174]. At 2 years, TG levels in the fenofibrate treatment group compared to placebo were 21% and 29% lower in the subgroups of patients with and without other lipid-lowering treatment during the study period, respectively [174]. In the large randomized, controlled Action to Control Cardiovascular Risk in Diabetes (ACCORD) trial, T2DM patients with dyslipidemia (N = 5518) were treated with fenofibrate or placebo along with open-label simvastatin [175]. A mild increase of HDL-C, paralleling that in the placebo group, and regression in TG were observed in the fenofibrate group [175].

The selective PPARα agonist pemafibrate is currently approved in Japan for the treatment of hyperlipidemia [72]. A thorough, detailed review of the role of pemafibrate in the treatment of atherogenic dyslipidemia can be found here [176]. In the placebo-controlled, phase 3 PROVIDE trial, the use of pemafibrate led to a significant decrease of fasting TG compared to placebo ($p < 0.001$) [76], an effect that was stable during the open-label extension period [177]. A pooled analysis of six placebo-controlled phase 2 and 3 trials in a large cohort of 1253 patients further confirmed these findings in combination therapy [178]. After 12 weeks, TG levels in both statin users and non-users significantly declined by 45–50%, in a dose-dependent manner with pemafibrate doses ranging from 0.1 mg/day to 0.4 mg/day, while no significant changes were observed in the placebo groups ($p < 0.001$ vs. placebo) [178]. Currently available data indicate that the lipid-lowering effects of pemafibrate are comparable or superior to those of fibrates [77,179].

Lipid-modulating effects of pioglitazone were observed in the Pioglitazone Effect on Regression of Intravascular Sonographic Coronary Obstruction Prospective Evaluation (PERISCOPE) [180] and Carotid Intima-Media Thickness in Atherosclerosis Using Pioglitazone (CHICAGO) [181] trials. In the PERISCOPE trial, HDL-C levels significantly increased and TG levels decreased in 543 T2DM patients who received pioglitazone 15–45 mg/day versus glimepiride 1–4 mg/day [180]. This was accompanied by a reduction of coronary atherosclerosis progression with pioglitazone as measured by a decrease in percent atheroma volume in intravascular ultrasound [180]. A post hoc analysis revealed that atheroma regression was associated with changes in lipid levels [182]. In the CHICAGO trial, pioglitazone compared to glimepiride reduced carotid intima artery intima-media thickness in 462 patients with T2DM, which was found to be associated with improvements in HDL-C in a post hoc analysis [181,183].

Beneficial effects of saroglitazar regarding TG levels in T2DM patients have been reported in both randomized controlled trials as well as observational cohorts [184,185]. Saroglitazar is currently approved in India for treatment of atherogenic dyslipidemia in

T2DM [185]. Recently, the randomized, controlled phase 3 PRESS XII trial evaluated the effect of saroglitazar 2 mg or 4 mg compared to pioglitazone on glycemic control and lipid profiles in 1155 patients with T2DM over 56 weeks [186]. Similarly, to participants in the pioglitazone arm, participants in the saroglitazar groups experienced a significant reduction of TG and LDL-C while HDL-C increased [186]. Information on use of other lipid-lowering agents, however, is not reported [186]. A recent meta-analysis of five randomized controlled trials confirms a benefit regarding TG reduction with saroglitazar compared to placebo or pioglitazone, but not compared to active control with other lipid-lowering agents (atorvastatin or fenofibrate) after 12 weeks [185]. Changes in HbA1c, LDL-C, or HDL-C levels were not significant [185].

Improvements in lipid profiles have also been demonstrated in the now discontinued agents seladelpar and elafibranor. In a randomized, placebo-controlled trial, seladelpar with or without atorvastatin significantly lowered LDL-C and TG, and increased HDL-C in individuals (N = 183) with dyslipidemia and abdominal obesity [98]. An improvement of lipoprotein subfractions was observed in another randomized, placebo-controlled trial with seladelpar alone or in combination with atorvastatin [187]. Compared to placebo, elafibranor significantly reduced fasting TG and increased HDL-C in a randomized, placebo-controlled trial of 141 patients with prediabetes or dyslipidemia, while LDL-lowering effects were only observed in the prediabetes group [188]. Similar findings were reported in the GOLDEN-505 trial in NASH patients [68].

Dyslipidemia is highly prevalent in individuals with NAFLD and T2DM, most of whom benefit from statin therapy, given the pleiotropic beneficial cardiovascular [189] as well as liver-related [190,191] effects of statins. Both FIELD and ACCORD trials showed overall low rates of myopathy under fenofibrate alone and under combination of fenofibrate with statins [174,175]. Overall, incidence rates of rhabdomyolysis in combination therapy were found to be lowest for fenofibrate combinations, although risk was higher in older and T2DM patients [192]. Among fibrates, gemfibrozil is associated with a higher risk of muscle-related adverse events in combination therapy with statins due to different pharmacokinetics, resulting in impaired statin metabolism [193]. Currently, statin-fibrate combination therapy may be considered in select patients with severe or refractory mixed dyslipidemia, intact renal function, and careful clinical follow-up [42].

In this context, the development of newer PPARα agonists for treatment of NASH further leads one to question the safety of these treatments, especially in combination with statins. Saroglitazar is specifically marketed for treatment of residual atherogenic dyslipidemia under statin treatment and has demonstrated a favorable safety profile regarding myopathy in the phase 3 PRESS VI trial, where it was combined with atorvastatin 10 mg [144]. Likewise, the SPPARMα pemafibrate demonstrated a good safety profile, regardless of statin use and mild renal dysfunction, in a pooled analysis of several randomized trials [178]. Effects of the Pan-PPAR agonist lanifibranor on lipid profiles were overall modest with no muscle-related adverse events in the phase 2 NATIVE trial [20].

4.3. Cardiovascular Comorbidities

T2DM, along with comorbid obesity and dyslipidemia, constitutes a major risk factor for cardiovascular disease (CVD) [194]. Moreover, several studies have provided evidence that NAFLD could be an independent CVD risk factor with a potential synergistic increased risk in patients with NAFLD and T2DM [195]. NAFLD patients are at risk of excess mortality from CVD, with the risk increasing with more advanced disease [196]. Treatment strategies in NAFLD should thus be considered with regard to their effect on cardiovascular conditions [197]. As PPAR modulation improves metabolism as well as endothelial dysfunction and inflammation [198], several PPAR-targeted therapies have been assessed regarding their potential to ameliorate cardiovascular disease and prevent cardiovascular events (reviewed in [23,42]).

Given their role in atherogenic dyslipidemia and their long-standing market approval, a lot of evidence exists regarding the effects of fibrates on cardiovascular outcomes [23].

The previously mentioned FIELD and ACCORD trials are two landmark studies of fibrates in the prevention of cardiovascular events in T2DM patients [174,175]. The randomized, controlled FIELD trial (N = 9795) included individuals with T2DM both with and without previous cardiovascular disease (approximately 1:4) and without specific indication for dyslipidemia treatment or presence of NAFLD [174]. While fenofibrate did not reduce the risk of major coronary events, it did reduce the incidence of non-fatal myocardial infarction and microvascular-associated complications [174]. Furthermore, events were significantly reduced in a subgroup analysis of those with dyslipidemia [174]. However, the ACCORD trial failed to demonstrate a reduction of the CVD risk compared to statin therapy alone in patients with T2DM at high risk for CVD [175]. In a meta-analysis of six primary prevention trials, including ACCORD and FIELD, it was determined that fibrates lower the risk of cardiovascular events (coronary heart disease death or non-fatal myocardial infarction) in primary prevention, although the absolute effect was rather modest with an absolute risk reduction of merely <1% [97]. The majority of patients included in the overall cohort had T2DM [97].

Regarding secondary prevention, a systematic review and meta-analysis concluded that fibrates were effective in the prevention of the composite outcome of non-fatal stroke, non-fatal myocardial infarction, and vascular death [199]. This analysis, however, included data on the drug clofibrate, which has been withdrawn from the market [199]. Whether these findings can be extrapolated to currently available fibrates is unclear [199].

As described above, the SPPARMα pemafibrate has demonstrated beneficial effects on atherogenic dyslipidemia. The effects of pemafibrate on reduction of cardiovascular events in diabetic patients are currently being investigated in the clinical Pemafibrate to Reduce Cardiovascular Outcomes by Reducing Triglycerides (PROMINENT) trial, which plans to enroll 10,000 subjects in 24 countries [200].

Another class of drugs that has been extensively studied for potential cardiovascular outcomes is thiazolidinediones, especially pioglitazone [120]. Pioglitazone has been demonstrated to improve certain parameters of cardiac metabolism and function in T2DM subjects, including myocardial insulin sensitivity, left ventricular diastolic function, and systolic function [201,202]. In the PERISCOPE trial of patients with coronary artery disease and T2DM, pioglitazone furthermore slowed the progression of coronary atherosclerotic lesions, assessed by intravascular ultrasound [180].

In the phase 3 PROspective pioglitAzone Clinical Trial In macroVascular Events (PROactive) trial, the use of pioglitazone in the high-risk group of diabetic patients with prior evidence of macrovascular disease was assessed [203]. After a mean follow-up of almost 3 years, pioglitazone failed to significantly improve the composite primary outcome, which included lower extremity revascularization among other cardiovascular endpoints such as mortality and myocardial infarction (HR 0.90, 95% CI 0.80 to 1.02, $p = 0.095$) [203]. Regarding the narrower secondary composite outcome of all-cause mortality, non-fatal myocardial infarction, and stroke, however, pioglitazone was superior to placebo (HR 0.84, 95% CI 0.72 to 0.98, $p = 0.027$) [203]. An individual patient data meta-analysis of 16,390 T2DM patients from 19 trials, including the PROactive trial, further confirmed this observation of risk reduction in the composite endpoint of mortality, myocardial infarction, and stroke (HR 0.82; 95% CI 0.72 to 0.94; $p = 0.005$) [204]. In the randomized, placebo-controlled Insulin Resistance Intervention After Stroke (IRIS) trial, pioglitazone has been shown to reduce the risk of stroke and myocardial infarction after a previous recent cerebrovascular event in patients with insulin resistance but without diabetes (HR 0.76; 95% CI 0.62 to 0.93; $p = 0.007$) [123,169]. In patients with prediabetes and good adherence to pioglitazone treatment (≥80%), the risk for acute coronary syndrome was reduced by 53% (95% CI 74% to 15%; $p = 0.01$) [205].

However, subjects receiving pioglitazone have also been found to be more likely to develop edema and difficulty breathing in the IRIS study [169]. PPARγ prompts fluid retention by increasing sodium avidity in the renal collecting ducts [206]. Data further indicate that fluid retention is a class effect of thiazolidinediones rather than an effect of individual drugs [207]. Among NASH patients, however, data suggest that weight gain may be attributable to an increase in adipose tissue rather than fluid retention, possibly indicating that this adverse effect might be less pronounced in this patient group [163].

As sodium and fluid retention exert deleterious effects on the cardiovascular system, the relationship between thiazolidinediones and heart failure has long been a matter of debate. In a large individual patient data meta-analysis of Lincoff et al. (2007), subjects in the pioglitazone group experienced serious heart failure significantly more often (HR 1.41; 95% CI 1.14 to 1.76; $p = 0.002$) [204]. However, this did not translate into an increased risk of overall mortality (HR 0.92; 95% CI 0.76 to 1.11; $p = 0.38$) [204]. A secondary analysis of the IRIS trial concluded that the risk of heart failure was not increased in individuals with non-diabetic insulin resistance after cerebrovascular events under pioglitazone compared to placebo (4.1% vs. 4.2%) [208]. Among patients with prediabetes and good adherence ($\geq$ 80%), the risk of the composite endpoint stroke, myocardial infarction, and hospitalization for heart failure was reduced (HR 0.61; 95% CI 0.42 to 0.88; $p = 0.008$) despite a significantly higher rate of edema (37% vs. 25%; $p < 0.001$) [205]. In the IRIS trial, patients with pre-existing heart failure were excluded, participants were closely monitored by their providers, and dosage adjustments were performed where necessary, indicating that this complication of pioglitazone treatment may be managed clinically without increased risk of a negative outcome [208]. It has been hypothesized that weight gain may lead to overt heart failure only in patients with underlying, sub-clinical cardiac dysfunction rather than development of heart failure [197]. This seems plausible, given the high baseline prevalence of cardiac dysfunction in the group of patients with T2DM [209].

A controversy regarding increased risk of cardiovascular mortality with rosiglitazone has long been ongoing. A meta-analysis of 42 trials revealed a significantly increased odds ratio of 1.43 (95% CI 1.03 to 1.98; $p = 0.03$) for myocardial infarction as well as an increased, albeit not significant, odds ratio for death from cardiovascular causes 1.64 (95% CI 0.98 to 2.74, $p = 0.06$) [210]. Updated meta-analyses have supported these findings [211,212]. In contrast, a large open-label randomized controlled trial (N = 4447) of patients with type 2 diabetes, who received either rosiglitazone or a combination therapy with metformin and sulphonylureas, showed non-inferiority of rosiglitazone compared to the active control regarding the composite primary endpoint of cardiovascular hospitalization or cardiovascular death (HR 0.99, 95% CI 0.85 to 1.16) [213]. While the American FDA has lifted restrictions on the use of rosiglitazone, the approval of rosiglitazone by the EMA ended in 2010 [18].

As detailed below, other PPAR-modulating agents have demonstrated favorable effects on lipid profiles in diabetic patients, thus indicating possible beneficial effects in cardiovascular disease (see Section 3.3), although long-term cardiovascular safety has not been established. Notably, the partial PPARγ agonist saroglitazar has demonstrated a satisfactory cardiovascular safety profile in the short term [186]. Likewise, no cardiovascular safety concerns were raised for elafibranor [68].

In the context of cardiovascular conditions, it is worth noting that the antihypertensive agent telmisartan, an angiotensin receptor blocker, further exerts agonistic effects at PPARγ and -α. The role of telmisartan in the treatment of NAFLD has thus been evaluated for both its renin-angiotensin-system (RAS)- and PPAR-modulating properties [214]. In T2DM human subjects with arterial hypertension, telmisartan attenuated liver-spleen ratio, indicating an improvement in hepatic steatosis [215]. In transcriptome analyses, telmisartan was shown to ameliorate development of NASH in a mouse model of diabetic NASH (STAM) [216]. Further studies are needed to determine the effect of telmisartan in patients with NASH and liver fibrosis.

5. Outlook and Further Areas of Research

As outlined in the previous chapters, PPAR modulates a wide range of metabolic functions and elicits pleiotropic effects in multiple tissues. Adding further complexity, specific effects can be elicited and combined by the use of molecules with distinct activity profiles on multiple PPAR isotypes [11]. This presents major challenges for research into PPAR therapies for NAFLD, but also offers considerable opportunities. One aspect of NAFLD therapy that has elicited attention is the prospect of possible combination therapies, simultaneously acting on several targets and thus offering synergistic treatment effects [217]. Combination of PPAR-targeted therapies with other pharmacological agents in the treatment of NAFLD will warrant careful exploration, given the multi-systemic effects of PPAR modulation [217]. This holds true also for concomitant treatments targeted towards other components of the metabolic syndrome such as the combination of statin and fibrate therapy for dyslipidemia.

Similarly, an aspect of PPAR-targeted therapy needing further investigation is the interplay of pharmacologic agents with PPAR modulation derived from the individuals' environment. Among the identified ligands of PPARs are so-called endocrine-disrupting chemicals (EDCs), which are defined as exogenous chemicals or mixtures of chemicals that interfere with any aspect of hormone action [218]. EDCs have been demonstrated to deregulate the activity of nuclear hormone receptors, such as PPAR isotypes and their heterodimerization partner RXR [219]. Subsequently, modulation of PPARs by EDCs has been discussed in the etiopathogenesis of NAFLD [220], obesity [221], and T2DM [222]. Although this may be far-reaching from a clinical point of view, the interaction between PPAR-modulating pharmacological agents and EDCs in NAFLD patients with metabolic conditions presents an interesting aspect for future research, especially given the high worldwide prevalence of NAFLD.

Other PPAR-directed environmental factors may be immediately influenced by lifestyle adjustments. As mentioned previously, nutrition-derived fatty acids and their metabolites have been identified as ligands for all PPAR isotypes [11]. As a key regulator of energy source homeostasis, PPARα, for example, mediates the response to acute fasting, while its involvement in the adaptive response to intermittent fasting is not fully elucidated [26,223,224]. PPAR activity may thus be directly or indirectly influenced by adjustments in nutrition and dietary patterns, especially fasting, as well as modulation of the gut microbiome [225,226]. Combination treatments of lifestyle interventions with pharmacologic PPAR-targeted therapy thus present an interesting area for future research. Ideally, clinical trials should consider these lifestyle-related factors to further elucidate possible synergistic mechanisms with PPAR-targeted therapies.

Closely related to their function as key regulators in metabolism and energy homeostasis is the diurnal cycling of several PPAR isotypes. Specifically, PPARα and PPARδ demonstrate diurnal expression and activity patterns, related to feeding status [35,47,227]. Circadian rhythm, encompassing the diurnal activity of several nuclear receptors, plays a pivotal role in metabolic homeostasis and disturbances of the former have been linked to NAFLD development [228]. Differences in the activation patterns of these receptors, as would be prompted by pharmacologic therapies, might elicit metabolic responses different to those observed with the natural fluctuation of PPAR activity. The extent to which this affects overall metabolism, circadian rhythm, and treatment effects will warrant further exploration [228].

Another aspect of research that has recently gathered interest is the sexual dimorphism of several metabolic conditions, including amongst others NAFLD and T2DM [229–233], although previous research in the field of NAFLD has often neglected to take these sex differences into account [234,235]. While the biological and social factors contributing to sex differences in metabolic and cardiovascular conditions as well as liver metabolism are complex and manifold (as reviewed here [236] and here [237]), one particular target of NAFLD treatment that has been identified as eliciting sexually dimorphic responses is PPARα [11]. In previous research, PPARα SUMOylation in females has been described as protecting the liver from estrogen-mediated intrahepatic cholestasis of pregnancy [34]. Recently, sexually dimorphic responses to PPARα activation by pemafibrate have been described in a rodent model [33]. Four models of diet-induced NAFLD elicited distinctly different responses in male compared to female mice, with transcriptome analysis indicating marked differences in genes regulated by PPARα [33]. Sexually dimorphic gene expression related to PPARα was subsequently demonstrated in human liver tissue samples from patients with NAFLD [33].

Because PPARα signaling is involved in the response to fasting, it could be hypothesized that differential responses to fasting in male and female rodents [238] as well as to dietary interventions in humans [239] might be mediated to some extent by PPARα. However, the degree to which these differences are conferred by sexual dimorphism in PPARα activation remains to be elucidated further, as numerous other factors and mechanisms, including estrogen signaling, strongly influence sex differences in the response to feeding and fasting [240,241]. Interestingly, several clinical trials of PPARα agonists in humans support a possible sexually dimorphic effect, although the results are inconclusive and the magnitude as well as the direction of the effect remains unclear. A subgroup analysis of the previously described ACCORD trial revealed a possible differential treatment effect, with sex showing a significant interaction with treatment, resulting in more favorable effects in men [175]. While improvement in lipid profiles was more pronounced in females in the FIELD trial, this did not translate into a significant difference regarding cardiovascular outcomes or significant interaction with treatment effect [174,242].

Data regarding possible sex differences in the other PPAR isoforms PPARδ and PPARγ are scarce. While glucose homeostasis and insulin sensitivity have been described as sexually dimorphic factors [243], the role of PPARγ as a main regulator of these processes in the context of these sex-specific findings is not well-described. Previous research clearly indicates sex hormone signaling as a central mediator of these processes [243], with PPARγ interacting with these pathways through estrogen receptor β (ERβ) [244]. In vitro findings indicate an inhibition of PPARγ transcriptional activity through ERβ, which was in accordance with increased PPARγ activity displayed by ERβ-deficient mice [244]. A rodent model of PPARγ deficiency confirmed a sexually dimorphic response to PPARγ activation by rosiglitazone [245]. The implications of these findings for PPAR-targeted therapy in human diabetic NAFLD require further research.

6. Summary

Due to the central role of PPARs in metabolism, the use of PPAR-agonists in T2DM patients offers unique challenges along with opportunities. PPAR-targeted therapies in the field of NAFLD and NASH have demonstrated pleiotropic beneficial effects, both on NAFLD-specific outcomes as well as on a multitude of metabolic functions.

The findings of the NATIVE trial of lanifibranor in particular represent a noteworthy exception in the field of NASH pharmacotherapy, as regression of fibrosis has been demonstrated. These findings, however, need to be further confirmed in the phase 3 NATiV3 trial. While lanifibranor was safe with regard to muscle-related adverse events, reported weight gain of around 3% may hamper use in NAFLD patients with metabolic comorbidities.

While data indicate a possible positive effect on steatosis, no anti-fibrotic effects have been demonstrated for saroglitazar and effects regarding inflammatory activity remain inconclusive. Several clinical trials in NAFLD and NASH are currently ongoing. Overall, saroglitazar has demonstrated a favorable profile regarding metabolic effects and adverse events, although a benefit regarding cardiovascular outcomes remains to be established.

Among the currently available treatment options, pioglitazone is recommended by several NAFLD guidelines. While data on the anti-fibrotic effect in T2DM patients are not fully conclusive, pioglitazone has shown positive effects on NASH inflammatory activity and glucose homeostasis. There has been considerable debate regarding the cardiovascular risk profile of pioglitazone, mainly revolving around the risk of heart failure due to weight gain and fluid retention. The extent to which a positive effect on dyslipidemia translates into overall cardiovascular risk reduction with pioglitazone is therefore unclear.

The role of fibrates both in the treatment of NAFLD and dyslipidemia seems limited. While combination with statins may be safe, fibrates do not offer a relevant benefit regarding cardiovascular outcomes. Results from the PROMINENT trial will offer insights into the cardiovascular benefit of the selective PPARα agonist pemafibrate, where data on histological outcomes in NAFLD are currently lacking. However, since PPARα agonism has been shown to elicit sexually dimorphic effects, both fibrates and pemafibrate for NAFLD should be reviewed with regard to this aspect.

Overall, sexually dimorphic effects of PPARα agonism—and possibly other PPAR isotypes—clearly warrant further exploration. Reporting of trial results stratified by sex might provide further cues and insights into the complex mechanisms of PPAR agonists. Precise phenotyping of trial participants with regard to not only sex but also comorbid conditions, concomitant medications, and lifestyle is needed to adequately capture the multi-systemic effects of PPAR-targeted therapies.

Author Contributions: All authors contributed substantially to the conception and design of the work. N.F.L. and V.G. performed the literature review and drafted the manuscript. C.C. and J.-F.D. reviewed the manuscript critically for important intellectual content. All authors have read and agreed to the published version of the manuscript.

Funding: N.F.L. receives financial support through a scholarship from the Swiss Liver Foundation, and a grant from the Gottfried and Julia Bangerter-Rhyner Foundation and Swiss Academy of Medical Sciences (SAMS). The authors have received no other financial support pertaining to this project.

Acknowledgments: N.F.L. would like to thank the Swiss Liver Foundation, and the Gottfried and Julia Bangerter-Rhyner Foundation and Swiss Academy of Medical Sciences (SAMS) for supporting her work.

Conflicts of Interest: The authors declare no conflict of interest.

References

1. Younossi, Z.M.; Koenig, A.B.; Abdelatif, D.; Fazel, Y.; Henry, L.; Wymer, M. Global epidemiology of nonalcoholic fatty liver disease-Meta-analytic assessment of prevalence, incidence, and outcomes. *Hepatology* **2016**, *64*, 73–84. [CrossRef] [PubMed]
2. Younossi, Z.M.; Golabi, P.; de Avila, L.; Paik, J.M.; Srishord, M.; Fukui, N. The global epidemiology of NAFLD and NASH in patients with type 2 diabetes: A systematic review and meta-analysis. *J. Hepatol.* **2019**, *71*, 793–801. [CrossRef] [PubMed]
3. Gastaldelli, A.; Cusi, K. From NASH to diabetes and from diabetes to NASH: Mechanisms and treatment options. *JHEP Rep.* **2019**, *1*, 312–328. [CrossRef] [PubMed]
4. Tanase, D.M.; Gosav, E.M.; Costea, C.F.; Ciocoiu, M.; Lacatusu, C.M.; Maranduca, M.A. The Intricate Relationship between Type 2 Diabetes Mellitus (T2DM), Insulin Resistance (IR), and Nonalcoholic Fatty Liver Disease (NAFLD). *J. Diabetes Res.* **2020**, *2020*, 3920196. [CrossRef]
5. Kanwal, F.; Shubrook, J.H.; Younossi, Z.; Natarajan, Y.; Bugianesi, E.; Rinella, M.E. Preparing for the NASH Epidemic: A Call to Action. *Gastroenterology* **2021**, *161*, 1030–1042.e8. [CrossRef]
6. El Serag, H.B.; Tran, T.; Everhart, J.E. Diabetes increases the risk of chronic liver disease and hepatocellular carcinoma. *Gastroenterology* **2004**, *126*, 460–468. [CrossRef]
7. Noureddin, N.; Noureddin, M.; Singh, A.; Alkhouri, N. Progression of Nonalcoholic Fatty Liver Disease-Associated Fibrosis in a Large Cohort of Patients with Type 2 Diabetes. *Dig. Dis. Sci.* **2022**, *67*, 1379–1388. [CrossRef]
8. Mantovani, A.; Byrne, C.D.; Bonora, E.; Targher, G. Nonalcoholic Fatty Liver Disease and Risk of Incident Type 2 Diabetes: A Meta-analysis. *Diabetes Care* **2018**, *41*, 372–382. [CrossRef]
9. Bril, F. Nonalcoholic fatty liver disease in type 2 diabetes: Awareness is the first step toward change. *Hepatobiliary Surg. Nutr.* **2020**, *9*, 493–496. [CrossRef]
10. Feige, J.N.; Gelman, L.; Michalik, L.; Desvergne, B.; Wahli, W. From molecular action to physiological outputs: Peroxisome proliferator-activated receptors are nuclear receptors at the crossroads of key cellular functions. *Prog. Lipid Res.* **2006**, *45*, 120–159. [CrossRef]
11. Fougerat, A.; Montagner, A.; Loiseau, N.; Guillou, H.; Wahli, W. Peroxisome Proliferator-Activated Receptors and Their Novel Ligands as Candidates for the Treatment of Non-Alcoholic Fatty Liver Disease. *Cells* **2020**, *9*, 1638. [CrossRef] [PubMed]
12. Lalloyer, F.; Staels, B. Fibrates, glitazones, and peroxisome proliferator-activated receptors. *Arter. Thromb. Vasc. Biol.* **2010**, *30*, 894–899. [CrossRef] [PubMed]
13. Hong, F.; Xu, P.; Zhai, Y. The Opportunities and Challenges of Peroxisome Proliferator-Activated Receptors Ligands in Clinical Drug Discovery and Development. *Int. J. Mol. Sci.* **2018**, *19*, 2189. [CrossRef] [PubMed]
14. Monroy-Ramirez, H.C.; Galicia-Moreno, M.; Sandoval-Rodriguez, A.; Meza-Rios, A.; Santos, A.; Armendariz-Borunda, J. PPARs as Metabolic Sensors and Therapeutic Targets in Liver Diseases. *Int. J. Mol. Sci.* **2021**, *22*, 8298. [CrossRef] [PubMed]
15. Boeckmans, J.; Natale, A.; Rombaut, M.; Buyl, K.; Rogiers, V.; De Kock, J. Anti-NASH Drug Development Hitches a Lift on PPAR Agonism. *Cells* **2019**, *9*, 37. [CrossRef] [PubMed]
16. Gross, B.; Pawlak, M.; Lefebvre, P.; Staels, B. PPARs in obesity-induced T2DM, dyslipidaemia and NAFLD. *Nat. Rev. Endocrinol.* **2017**, *13*, 36–49. [CrossRef] [PubMed]

17. Weikum, E.R.; Liu, X.; Ortlund, E.A. The nuclear receptor superfamily: A structural perspective. *Protein Sci.* **2018**, *27*, 1876–1892. [CrossRef]
18. Francque, S.; Szabo, G.; Abdelmalek, M.F.; Byrne, C.D.; Cusi, K.; Dufour, J.-F. Nonalcoholic steatohepatitis: The role of peroxisome proliferator-activated receptors. *Nat. Rev. Gastroenterol. Hepatol.* **2021**, *18*, 24–39. [CrossRef]
19. Musso, G.; Cassader, M.; Paschetta, E.; Gambino, R. Thiazolidinediones and Advanced Liver Fibrosis in Nonalcoholic Steatohepatitis: A Meta-analysis. *JAMA Intern. Med.* **2017**, *177*, 633–640. [CrossRef]
20. Francque, S.M.; Bedossa, P.; Ratziu, V.; Anstee, Q.M.; Bugianesi, E.; Sanyal, A.J. A Randomized, Controlled Trial of the Pan-PPAR Agonist Lanifibranor in NASH. *N. Engl. J. Med.* **2021**, *385*, 1547–1558. [CrossRef]
21. Ricote, M.; Glass, C.K. PPARs and molecular mechanisms of transrepression. *Biochim. Biophys. Acta (BBA)—Mol. Cell Biol. Lipids* **2007**, *1771*, 926–935. [CrossRef] [PubMed]
22. Issemann, I.; Green, S. Activation of a member of the steroid hormone receptor superfamily by peroxisome proliferators. *Nature* **1990**, *347*, 645–650. [CrossRef]
23. Han, L.; Shen, W.-J.; Bittner, S.; Kraemer, F.B.; Azhar, S. PPARs: Regulators of metabolism and as therapeutic targets in cardiovascular disease. Part I: PPAR-α. *Future Cardiol.* **2017**, *13*, 259–278. [CrossRef]
24. Braissant, O.; Foufelle, F.; Scotto, C.; Dauça, M.; Wahli, W. Differential expression of peroxisome proliferator-activated receptors (PPARs): Tissue distribution of PPAR-alpha, -beta, and -gamma in the adult rat. *Endocrinology* **1996**, *137*, 354–366. [CrossRef] [PubMed]
25. Tailleux, A.; Wouters, K.; Staels, B. Roles of PPARs in NAFLD: Potential therapeutic targets. *Biochim. Biophys. Acta (BBA)—Mol. Cell Biol. Lipids* **2011**, *1821*, 809–818. [CrossRef] [PubMed]
26. Kersten, S.; Seydoux, J.; Peters, J.M.; Gonzalez, F.J.; Desvergne, B.; Wahli, W. Peroxisome proliferator-activated receptor alpha mediates the adaptive response to fasting. *J. Clin. Investig.* **1999**, *103*, 1489–1498. [CrossRef]
27. Régnier, M.; Polizzi, A.; Lippi, Y.; Fouché, E.; Michel, G.; Lukowicz, C. Insights into the role of hepatocyte PPARα activity in response to fasting. *Mol. Cell. Endocrinol.* **2018**, *471*, 75–88. [CrossRef]
28. Ribet, C.; Montastier, E.; Valle, C.; Bezaire, V.; Mazzucotelli, A.; Mairal, A. Peroxisome proliferator-activated receptor-alpha control of lipid and glucose metabolism in human white adipocytes. *Endocrinology* **2010**, *151*, 123–133. [CrossRef]
29. Muoio, D.M.; Way, J.M.; Tanner, C.J.; Winegar, D.A.; Kliewer, S.A.; Houmard, J.A. Peroxisome proliferator-activated receptor-alpha regulates fatty acid utilization in primary human skeletal muscle cells. *Diabetes* **2002**, *51*, 901–909. [CrossRef]
30. Pawlak, M.; Baugé, E.; Bourguet, W.; De Bosscher, K.; Lalloyer, F.; Tailleux, A. The transrepressive activity of peroxisome proliferator-activated receptor alpha is necessary and sufficient to prevent liver fibrosis in mice. *Hepatology* **2014**, *60*, 1593–1606. [CrossRef]
31. Ip, E.; Farrell, G.; Hall, P.; Robertson, G.; Leclercq, I. Administration of the potent PPARalpha agonist, Wy-14,643, reverses nutritional fibrosis and steatohepatitis in mice. *Hepatology* **2004**, *39*, 1286–1296. [CrossRef]
32. Chen, L.; Li, L.; Chen, J.; Li, L.; Zheng, Z.; Ren, J. Oleoylethanolamide, an endogenous PPAR-α ligand, attenuates liver fibrosis targeting hepatic stellate cells. *Oncotarget* **2015**, *6*, 42530–42540. [CrossRef] [PubMed]
33. Smati, S.; Polizzi, A.; Fougerat, A.; Ellero-Simatos, S.; Blum, Y.; Lippi, Y. Integrative study of diet-induced mouse models of NAFLD identifies PPARα as a sexually dimorphic drug target. *Gut* **2021**, *71*, 807–821. [CrossRef] [PubMed]
34. Leuenberger, N.; Pradervand, S.; Wahli, W. Sumoylated PPARalpha mediates sex-specific gene repression and protects the liver from estrogen-induced toxicity in mice. *J. Clin. Investig.* **2009**, *119*, 3138–3148. [CrossRef] [PubMed]
35. Yang, X.; Downes, M.; Yu, R.T.; Bookout, A.L.; He, W.; Straume, M. Nuclear receptor expression links the circadian clock to metabolism. *Cell* **2006**, *126*, 801–810. [CrossRef] [PubMed]
36. Bougarne, N.; Weyers, B.; Desmet, S.J.; Deckers, J.; Ray, D.W.; Staels, B. Molecular Actions of PPARα in Lipid Metabolism and Inflammation. *Endocr. Rev.* **2018**, *39*, 760–802. [CrossRef]
37. Chen, J.; Montagner, A.; Tan, N.S.; Wahli, W. Insights into the Role of PPARβ/δ in NAFLD. *Int. J. Mol. Sci.* **2018**, *19*, 1893. [CrossRef]
38. Zarei, M.; Aguilar-Recarte, D.; Palomer, X.; Vázquez-Carrera, M. Revealing the role of peroxisome proliferator-activated receptor β/δ in nonalcoholic fatty liver disease. *Metabolism* **2021**, *114*, 154342. [CrossRef]
39. Takahashi, S.; Tanaka, T.; Kodama, T.; Sakai, J. Peroxisome proliferator-activated receptor delta (PPARdelta), a novel target site for drug discovery in metabolic syndrome. *Pharmacol. Res.* **2006**, *53*, 501–507. [CrossRef]
40. Seedorf, U.; Aberle, J. Emerging roles of PPARdelta in metabolism. *Biochim. Biophys. Acta* **2007**, *1771*, 1125–1131. [CrossRef]
41. Palomer, X.; Barroso, E.; Pizarro-Delgado, J.; Peña, L.; Botteri, G.; Zarei, M. PPARβ/δ: A Key Therapeutic Target in Metabolic Disorders. *Int. J. Mol. Sci.* **2018**, *19*, 913. [CrossRef] [PubMed]
42. Han, L.; Shen, W.-J.; Bittner, S.; Kraemer, F.B.; Azhar, S. PPARs: Regulators of metabolism and as therapeutic targets in cardiovascular disease. Part II: PPAR-β/δ and PPAR-γ. *Future Cardiol.* **2017**, *13*, 279–296. [CrossRef] [PubMed]
43. Manickam, R.; Wahli, W. Roles of Peroxisome Proliferator-Activated Receptor β/δ in skeletal muscle physiology. *Biochimie* **2017**, *136*, 42–48. [CrossRef]
44. Schuler, M.; Ali, F.; Chambon, C.; Duteil, D.; Bornert, J.-M.; Tardivel, A. PGC1alpha expression is controlled in skeletal muscles by PPARbeta, whose ablation results in fiber-type switching, obesity, and type 2 diabetes. *Cell Metab.* **2006**, *4*, 407–414. [CrossRef]

45. Koh, J.-H.; Hancock, C.R.; Terada, S.; Higashida, K.; Holloszy, J.O.; Han, D.-H. PPARβ Is Essential for Maintaining Normal Levels of PGC-1α and Mitochondria and for the Increase in Muscle Mitochondria Induced by Exercise. *Cell Metab.* **2017**, *25*, 1176–1185.e5. [CrossRef]
46. Liu, S.; Hatano, B.; Zhao, M.; Yen, C.-C.; Kang, K.; Reilly, S.M. Role of peroxisome proliferator-activated receptor {delta}/{beta} in hepatic metabolic regulation. *J. Biol. Chem.* **2011**, *286*, 1237–1247. [CrossRef]
47. Liu, S.; Brown, J.D.; Stanya, K.J.; Homan, E.; Leidl, M.; Inouye, K. A diurnal serum lipid integrates hepatic lipogenesis and peripheral fatty acid use. *Nature* **2013**, *502*, 550–554. [CrossRef]
48. Sanderson, L.M.; Boekschoten, M.V.; Desvergne, B.; Müller, M.; Kersten, S. Transcriptional profiling reveals divergent roles of PPARalpha and PPARbeta/delta in regulation of gene expression in mouse liver. *Physiol. Genom.* **2010**, *41*, 42–52. [CrossRef]
49. Kazankov, K.; Jørgensen, S.M.D.; Thomsen, K.L.; Møller, H.J.; Vilstrup, H.; George, J. The role of macrophages in nonalcoholic fatty liver disease and nonalcoholic steatohepatitis. *Nat. Rev. Gastroenterol. Hepatol.* **2019**, *16*, 145–159. [CrossRef]
50. Fajas, L.; Auboeuf, D.; Raspé, E.; Schoonjans, K.; Lefebvre, A.M.; Saladin, R. The organization, promoter analysis, and expression of the human PPARgamma gene. *J. Biol. Chem.* **1997**, *272*, 18779–18789. [CrossRef]
51. Michalik, L.; Auwerx, J.; Berger, J.P.; Chatterjee, V.K.; Glass, C.K.; Gonzalez, F.J. International Union of Pharmacology. LXI. Peroxisome proliferator-activated receptors. *Pharmacol. Rev.* **2006**, *58*, 726–741. [CrossRef] [PubMed]
52. Soccio, R.E.; Chen, E.R.; Lazar, M.A. Thiazolidinediones and the promise of insulin sensitization in type 2 diabetes. *Cell Metab.* **2014**, *20*, 573–591. [CrossRef] [PubMed]
53. Cusi, K. Role of obesity and lipotoxicity in the development of nonalcoholic steatohepatitis: Pathophysiology and clinical implications. *Gastroenterology* **2012**, *142*, 711–725.e6. [CrossRef] [PubMed]
54. Ma, X.; Wang, D.; Zhao, W.; Xu, L. Deciphering the Roles of PPARγ in Adipocytes via Dynamic Change of Transcription Complex. *Front. Endocrinol.* **2018**, *9*, 473. [CrossRef]
55. Wang, F.; Mullican, S.E.; DiSpirito, J.R.; Peed, L.C.; Lazar, M.A. Lipoatrophy and severe metabolic disturbance in mice with fat-specific deletion of PPARγ. *Proc. Natl. Acad. Sci. USA* **2013**, *110*, 18656–18661. [CrossRef]
56. Gavrilova, O.; Haluzik, M.; Matsusue, K.; Cutson, J.J.; Johnson, L.; Dietz, K.R. Liver peroxisome proliferator-activated receptor gamma contributes to hepatic steatosis, triglyceride clearance, and regulation of body fat mass. *J. Biol. Chem.* **2003**, *278*, 34268–34276. [CrossRef]
57. Maeda, N.; Takahashi, M.; Funahashi, T.; Kihara, S.; Nishizawa, H.; Kishida, K. PPARgamma ligands increase expression and plasma concentrations of adiponectin, an adipose-derived protein. *Diabetes* **2001**, *50*, 2094–2099. [CrossRef]
58. Odegaard, J.I.; Ricardo-Gonzalez, R.R.; Goforth, M.H.; Morel, C.R.; Subramanian, V.; Mukundan, L. Macrophage-specific PPARgamma controls alternative activation and improves insulin resistance. *Nature* **2007**, *447*, 1116–1120. [CrossRef]
59. Hazra, S.; Xiong, S.; Wang, J.; Rippe, R.A.; Krishna, V.; Chatterjee, K. Peroxisome proliferator-activated receptor gamma induces a phenotypic switch from activated to quiescent hepatic stellate cells. *J. Biol. Chem.* **2004**, *279*, 11392–11401. [CrossRef]
60. Galli, A.; Crabb, D.W.; Ceni, E.; Salzano, R.; Mello, T.; Svegliati–Baroni, G. Antidiabetic thiazolidinediones inhibit collagen synthesis and hepatic stellate cell activation in vivo and in vitro. *Gastroenterology* **2002**, *122*, 1924–1940. [CrossRef]
61. Nakajima, A.; Eguchi, Y.; Yoneda, M.; Imajo, K.; Tamaki, N.; Suganami, H. Randomised clinical trial: Pemafibrate, a novel selective peroxisome proliferator-activated receptor α modulator (SPPARMα), versus placebo in patients with non-alcoholic fatty liver disease. *Aliment. Pharm. Ther.* **2021**, *54*, 1263–1277. [CrossRef]
62. Yokote, K.; Yamashita, S.; Arai, H.; Araki, E.; Matsushita, M.; Nojima, T. Effects of pemafibrate on glucose metabolism markers and liver function tests in patients with hypertriglyceridemia: A pooled analysis of six phase 2 and phase 3 randomized double-blind placebo-controlled clinical trials. *Cardiovasc. Diabetol.* **2021**, *20*, 96. [CrossRef] [PubMed]
63. Mantovani, A.; Byrne, C.D.; Scorletti, E.; Mantzoros, C.; Targher, G. Efficacy and safety of anti-hyperglycaemic drugs in patients with non-alcoholic fatty liver disease with or without diabetes: An updated systematic review of randomized controlled trials. *Diabetes Metab.* **2020**, *46*, 427–441. [CrossRef] [PubMed]
64. Bril, F.; Biernacki, D.M.; Kalavalapalli, S.; Lomonaco, R.; Subbarayan, S.K.; Lai, J. Role of Vitamin E for Nonalcoholic Steatohepatitis in Patients with Type 2 Diabetes: A Randomized Controlled Trial. *Diabetes Care* **2019**, *42*, 1481–1488. [CrossRef] [PubMed]
65. Cusi, K.; Orsak, B.; Bril, F.; Lomonaco, R.; Hecht, J.; Ortiz-Lopez, C. Long-Term Pioglitazone Treatment for Patients with Nonalcoholic Steatohepatitis and Prediabetes or Type 2 Diabetes Mellitus. *Ann. Intern. Med.* **2016**, *165*, 305–315. [CrossRef] [PubMed]
66. Bril, F.; Kalavalapalli, S.; Clark, V.C.; Lomonaco, R.; Soldevila-Pico, C.; Liu, I.-C.; Orsak, B.; Tio, F.; Cusi, K. Response to Pioglitazone in Patients with Nonalcoholic Steatohepatitis With vs Without Type 2 Diabetes. *Clin. Gastroenterol. Hepatol.* **2018**, *16*, 558–566.e2. [CrossRef]
67. Gawrieh, S.; Noureddin, M.; Loo, N.; Mohseni, R.; Awasty, V.; Cusi, K. Saroglitazar, a PPAR-α/γ Agonist, for Treatment of Nonalcoholic Fatty Liver Disease: A Randomized Controlled Double-Blind Phase 2 Trial. *Hepatology* **2021**, *74*, 1809–1824. [CrossRef]
68. Ratziu, V.; Harrison, S.A.; Francque, S.; Bedossa, P.; Lehert, P.; Serfaty, L. Elafibranor, an Agonist of the Peroxisome Proliferator-Activated Receptor-α and-δ, Induces Resolution of Nonalcoholic Steatohepatitis without Fibrosis Worsening. *Gastroenterology* **2016**, *150*, 1147–1159.e5. [CrossRef]

69. Raza-Iqbal, S.; Tanaka, T.; Anai, M.; Inagaki, T.; Matsumura, Y.; Ikeda, K. Transcriptome Analysis of K-877 (a Novel Selective PPARα Modulator (SPPARMα))-Regulated Genes in Primary Human Hepatocytes and the Mouse Liver. *J. Atheroscler. Thromb.* **2015**, *22*, 754–772. [CrossRef]
70. Fruchart, J.-C.; Santos, R.D.; Aguilar-Salinas, C.; Aikawa, M.; Al Rasadi, K.; Amarenco, P. The selective peroxisome proliferator-activated receptor alpha modulator (SPPARMα) paradigm: Conceptual framework and therapeutic potential: A consensus statement from the International Atherosclerosis Society (IAS) and the Residual Risk Reduction Initiative (R3i) Foundation. *Cardiovasc. Diabetol.* **2019**, *18*, 71.
71. Ida, S.; Kaneko, R.; Murata, K. Efficacy and safety of pemafibrate administration in patients with dyslipidemia: A systematic review and meta-analysis. *Cardiovasc. Diabetol.* **2019**, *18*, 38. [CrossRef] [PubMed]
72. Blair, H.A. Pemafibrate: First Global Approval. *Drugs* **2017**, *77*, 1805–1810. [CrossRef] [PubMed]
73. Honda, Y.; Kessoku, T.; Ogawa, Y.; Tomeno, W.; Imajo, K.; Fujita, K. Pemafibrate, a novel selective peroxisome proliferator-activated receptor alpha modulator, improves the pathogenesis in a rodent model of nonalcoholic steatohepatitis. *Sci. Rep.* **2017**, *7*, 42477. [CrossRef] [PubMed]
74. Sasaki, Y.; Asahiyama, M.; Tanaka, T.; Yamamoto, S.; Murakami, K.; Kamiya, W. Pemafibrate, a selective PPARα modulator, prevents non-alcoholic steatohepatitis development without reducing the hepatic triglyceride content. *Sci. Rep.* **2020**, *10*, 7818. [CrossRef]
75. Ishibashi, S.; Yamashita, S.; Arai, H.; Araki, E.; Yokote, K.; Suganami, H. Effects of K-877, a novel selective PPARα modulator (SPPARMα), in dyslipidaemic patients: A randomized, double blind, active- and placebo-controlled, phase 2 trial. *Atherosclerosis* **2016**, *249*, 36–43. [CrossRef]
76. Araki, E.; Yamashita, S.; Arai, H.; Yokote, K.; Satoh, J.; Inoguchi, T. Effects of Pemafibrate, a Novel Selective PPARα Modulator, on Lipid and Glucose Metabolism in Patients with Type 2 Diabetes and Hypertriglyceridemia: A Randomized, Double-Blind, Placebo-Controlled, Phase 3 Trial. *Diabetes Care* **2018**, *41*, 538–546. [CrossRef]
77. Ishibashi, S.; Arai, H.; Yokote, K.; Araki, E.; Suganami, H.; Yamashita, S. Efficacy and safety of pemafibrate (K-877), a selective peroxisome proliferator-activated receptor α modulator, in patients with dyslipidemia: Results from a 24-week, randomized, double blind, active-controlled, phase 3 trial. *J. Clin. Lipidol.* **2018**, *12*, 173–184. [CrossRef]
78. Seko, Y.; Yamaguchi, K.; Umemura, A.; Yano, K.; Takahashi, A.; Okishio, S. Effect of pemafibrate on fatty acid levels and liver enzymes in non-alcoholic fatty liver disease patients with dyslipidemia: A single-arm, pilot study. *Hepatol. Res. Off. J. Jpn. Soc. Hepatol.* **2020**, *50*, 1328–1336. [CrossRef]
79. Hatanaka, T.; Kosone, T.; Saito, N.; Takakusagi, S.; Tojima, H.; Naganuma, A. Effect of 48-week pemafibrate on non-alcoholic fatty liver disease with hypertriglyceridemia, as evaluated by the FibroScan-aspartate aminotransferase score. *JGH Open* **2021**, *5*, 1183–1189. [CrossRef]
80. Hatanaka, T.; Kakizaki, S.; Saito, N.; Nakano, Y.; Nakano, S.; Hazama, Y. Impact of Pemafibrate in Patients with Hypertriglyceridemia and Metabolic Dysfunction-associated Fatty Liver Disease Pathologically Diagnosed with Non-alcoholic Steatohepatitis: A Retrospective, Single-arm Study. *Intern. Med. Tokyo Jpn.* **2021**, *2021*, 6574-20. [CrossRef]
81. Shinozaki, S.; Tahara, T.; Lefor, A.K.; Ogura, M. Pemafibrate decreases markers of hepatic inflammation in patients with non-alcoholic fatty liver disease. *Clin. Exp. Hepatol.* **2020**, *6*, 270–274. [CrossRef] [PubMed]
82. Shinozaki, S.; Tahara, T.; Lefor, A.K.; Ogura, M. Pemafibrate improves hepatic inflammation, function and fibrosis in patients with non-alcoholic fatty liver disease: A one-year observational study. *Clin. Exp. Hepatol.* **2021**, *7*, 172–177. [CrossRef] [PubMed]
83. Tenenbaum, A.; Motro, M.; Fisman, E.Z. Dual and pan-peroxisome proliferator-activated receptors (PPAR) co-agonism: The bezafibrate lessons. *Cardiovasc. Diabetol.* **2005**, *4*, 14. [CrossRef] [PubMed]
84. Laurin, J.; Lindor, K.D.; Crippin, J.S.; Gossard, A.; Gores, G.J.; Ludwig, J. Ursodeoxycholic acid or clofibrate in the treatment of non-alcohol-induced steatohepatitis: A pilot study. *Hepatology* **1996**, *23*, 1464–1467. [CrossRef]
85. Honda, A.; Tanaka, A.; Kaneko, T.; Komori, A.; Abe, M.; Inao, M. Bezafibrate Improves GLOBE and UK-PBC Scores and Long-Term Outcomes in Patients With Primary Biliary Cholangitis. *Hepatology* **2019**, *70*, 2035–2046. [CrossRef]
86. Feng, X.; Gao, X.; Jia, Y.; Zhang, H.; Xu, Y.; Wang, G. PPAR-α Agonist Fenofibrate Decreased RANTES Levels in Type 2 Diabetes Patients with Hypertriglyceridemia. *Med. Sci. Monit.* **2016**, *22*, 743–751. [CrossRef]
87. Li, B.-H.; He, F.-P.; Yang, X.; Chen, Y.-W.; Fan, J.-G. Steatosis induced CCL5 contributes to early-stage liver fibrosis in nonalcoholic fatty liver disease progress. *Transl. Res.* **2017**, *180*, 103–117.e4. [CrossRef]
88. Nikam, A.; Patankar, J.V.; Somlapura, M.; Lahiri, P.; Sachdev, V.; Kratky, D. The PPARα Agonist Fenofibrate Prevents Formation of Protein Aggregates (Mallory-Denk bodies) in a Murine Model of Steatohepatitis-like Hepatotoxicity. *Sci. Rep.* **2018**, *8*, 12964. [CrossRef]
89. Rodríguez-Vilarrupla, A.; Laviña, B.; García-Calderó, H.; Russo, L.; Rosado, E.; Roglans, N. PPARα activation improves endothelial dysfunction and reduces fibrosis and portal pressure in cirrhotic rats. *J. Hepatol.* **2012**, *56*, 1033–10369. [CrossRef]
90. Basaranoglu, M.; Acbay, O.; Sonsuz, A. A controlled trial of gemfibrozil in the treatment of patients with nonalcoholic steatohepatitis. *J. Hepatol.* **1999**, *31*, 384. [CrossRef]
91. Fernández-Miranda, C.; Pérez-Carreras, M.; Colina, F.; López-Alonso, G.; Vargas, C.; Solís-Herruzo, J.A. A pilot trial of fenofibrate for the treatment of non-alcoholic fatty liver disease. *Dig. Liver Dis.* **2008**, *40*, 200–205. [CrossRef] [PubMed]

92. Belfort, R.; Berria, R.; Cornell, J.; Cusi, K. Fenofibrate reduces systemic inflammation markers independent of its effects on lipid and glucose metabolism in patients with the metabolic syndrome. *J. Clin. Endocrinol. Metab.* **2010**, *95*, 829–836. [CrossRef] [PubMed]
93. Fabbrini, E.; Mohammed, B.S.; Korenblat, K.M.; Magkos, F.; McCrea, J.; Patterson, B.W. Effect of fenofibrate and niacin on intrahepatic triglyceride content, very low-density lipoprotein kinetics, and insulin action in obese subjects with nonalcoholic fatty liver disease. *J. Clin. Endocrinol. Metab.* **2010**, *95*, 2727–2735. [CrossRef] [PubMed]
94. Oscarsson, J.; Önnerhag, K.; Risérus, U.; Sundén, M.; Johansson, L.; Jansson, P.-A. Effects of free omega-3 carboxylic acids and fenofibrate on liver fat content in patients with hypertriglyceridemia and non-alcoholic fatty liver disease: A double-blind, randomized, placebo-controlled study. *J. Clin. Lipidol.* **2018**, *12*, 1390–1403.e4. [CrossRef] [PubMed]
95. Karhapää, P.; Uusitupa, M.; Voutilainen, E.; Laakso, M. Effects of bezafibrate on insulin sensitivity and glucose tolerance in subjects with combined hyperlipidemia. *Clin. Pharmacol. Ther.* **1992**, *52*, 620–626. [CrossRef]
96. LiverTox. Fibrates. In *LiverTox: Clinical and Research Information on Drug-Induced Liver Injury*; National Institute of Diabetes and Digestive and Kidney Diseases: Bethesda, MD, USA. Available online: http://www.ncbi.nlm.nih.gov/books/NBK547893/ (accessed on 10 October 2021).
97. Jakob, T.; Nordmann, A.J.; Schandelmaier, S.; Ferreira-González, I.; Briel, M. Fibrates for primary prevention of cardiovascular disease events. *Cochrane Database Syst. Rev.* **2016**, *11*, CD009753. [CrossRef]
98. Bays, H.E.; Schwartz, S.; Littlejohn, T.; Kerzner, B.; Krauss, R.M.; Karpf, D.B. MBX-8025, a novel peroxisome proliferator receptor-delta agonist: Lipid and other metabolic effects in dyslipidemic overweight patients treated with and without atorvastatin. *J. Clin. Endocrinol. Metab.* **2011**, *96*, 2889–2897. [CrossRef]
99. Haczeyni, F.; Wang, H.; Barn, V.; Mridha, A.R.; Yeh, M.M.; Haigh, W.G. The selective peroxisome proliferator-activated receptor-delta agonist seladelpar reverses nonalcoholic steatohepatitis pathology by abrogating lipotoxicity in diabetic obese mice. *Hepatol. Commun.* **2017**, *1*, 663–674. [CrossRef]
100. Chiquette, E.; Ramirez, G.; Defronzo, R. A meta-analysis comparing the effect of thiazolidinediones on cardiovascular risk factors. *Arch. Intern. Med.* **2004**, *164*, 2097–2104. [CrossRef]
101. Boettcher, E.; Csako, G.; Pucino, F.; Wesley, R.; Loomba, R. Meta-analysis: Pioglitazone improves liver histology and fibrosis in patients with non-alcoholic steatohepatitis. *Aliment. Pharmacol. Ther.* **2012**, *35*, 66–75. [CrossRef]
102. He, L.; Liu, X.; Wang, L.; Yang, Z. Thiazolidinediones for nonalcoholic steatohepatitis: A meta-analysis of randomized clinical trials. *Medicine* **2016**, *95*, e4947. [CrossRef] [PubMed]
103. Sakamoto, J.; Kimura, H.; Moriyama, S.; Odaka, H.; Momose, Y.; Sugiyama, Y. Activation of human peroxisome proliferator-activated receptor (PPAR) subtypes by pioglitazone. *Biochem. Biophys. Res. Commun.* **2000**, *278*, 704–711. [CrossRef] [PubMed]
104. Duseja, A.; Singh, S.P.; Saraswat, V.A.; Acharya, S.K.; Chawla, Y.K.; Chowdhury, S. Non-alcoholic Fatty Liver Disease and Metabolic Syndrome-Position Paper of the Indian National Association for the Study of the Liver, Endocrine Society of India, Indian College of Cardiology and Indian Society of Gastroenterology. *J. Clin. Exp. Hepatol.* **2015**, *5*, 51–68. [CrossRef] [PubMed]
105. Kang, S.H.; Lee, H.W.; Yoo, J.-J.; Cho, Y.; Kim, S.U.; Lee, T.H. KASL clinical practice guidelines: Management of nonalcoholic fatty liver disease. *Clin. Mol. Hepatol.* **2021**, *27*, 363–401. [CrossRef] [PubMed]
106. Arab, J.P.; Dirchwolf, M.; Álvares-da-Silva, M.R.; Barrera, F.; Benítez, C.; Castellanos-Fernandez, M. Latin American Association for the study of the liver (ALEH) practice guidance for the diagnosis and treatment of non-alcoholic fatty liver disease. *Ann. Hepatol.* **2020**, *19*, 674–690. [CrossRef]
107. EASL–EASD–EASO. EASL–EASD–EASO Clinical Practice Guidelines for the management of non-alcoholic fatty liver disease. *J. Hepatol.* **2016**, *64*, 1388–1402. [CrossRef]
108. Chalasani, N.; Younossi, Z.; Lavine, J.E.; Charlton, M.; Cusi, K.; Rinella, M. The diagnosis and management of nonalcoholic fatty liver disease: Practice guidance from the American Association for the Study of Liver Diseases. *Hepatology* **2018**, *67*, 328–357. [CrossRef]
109. Panunzi, S.; Maltese, S.; Verrastro, O.; Labbate, L.; De Gaetano, A.; Pompili, M. Pioglitazone and bariatric surgery are the most effective treatments for non-alcoholic steatohepatitis: A hierarchical network meta-analysis. *Diabetes Obes. Metab.* **2021**, *23*, 980–990. [CrossRef]
110. Hsiao, P.-J.; Chiou, H.-Y.C.; Jiang, H.-J.; Lee, M.-Y.; Hsieh, T.-J.; Kuo, K.-K. Pioglitazone Enhances Cytosolic Lipolysis, β-oxidation and Autophagy to Ameliorate Hepatic Steatosis. *Sci. Rep.* **2017**, *7*, 9030. [CrossRef]
111. de Mendonça, M.; Dos Santos, B.D.A.C.; de Sousa, É.; Rodrigues, A.C. Adiponectin is required for pioglitazone-induced improvements in hepatic steatosis in mice fed a high-fat diet. *Mol. Cell. Endocrinol.* **2019**, *493*, 110480. [CrossRef]
112. Gastaldelli, A.; Miyazaki, Y.; Mahankali, A.; Berria, R.; Pettiti, M.; Buzzigoli, E. The effect of pioglitazone on the liver: Role of adiponectin. *Diabetes Care* **2006**, *29*, 2275–2281. [CrossRef] [PubMed]
113. Leclercq, I.A.; Sempoux, C.; Stärkel, P.; Horsmans, Y. Limited therapeutic efficacy of pioglitazone on progression of hepatic fibrosis in rats. *Gut* **2006**, *55*, 1020–1029. [CrossRef] [PubMed]
114. Kawaguchi-Suzuki, M.; Cusi, K.; Bril, F.; Gong, Y.; Langaee, T.; Frye, R.F. A Genetic Score Associates with Pioglitazone Response in Patients With Non-alcoholic Steatohepatitis. *Front. Pharmacol.* **2018**, *9*, 752. [CrossRef] [PubMed]
115. Brunt, E.M.; Kleiner, D.E.; Wilson, L.A.; Sanyal, A.J.; Neuschwander-Tetri, B.A. Improvements in Histologic Features and Diagnosis Associated with Improvement in Fibrosis in Nonalcoholic Steatohepatitis: Results from the Nonalcoholic Steatohepatitis Clinical Research Network Treatment Trials. *Hepatology* **2019**, *70*, 522–531. [PubMed]

116. Chalasani, N.P.; Sanyal, A.J.; Kowdley, K.V.; Robuck, P.R.; Hoofnagle, J.; Kleiner, D.E. Pioglitazone versus vitamin E versus placebo for the treatment of non-diabetic patients with non-alcoholic steatohepatitis: PIVENS trial design. *Contemp. Clin. Trials* **2009**, *30*, 88–96. [CrossRef] [PubMed]
117. Sanyal, A.J.; Chalasani, N.; Kowdley, K.V.; McCullough, A.; Diehl, A.M.; Bass, N.M. Pioglitazone, Vitamin E, or Placebo for Nonalcoholic Steatohepatitis. *N. Engl. J. Med.* **2010**, *362*, 1675–1685. [CrossRef]
118. Belfort, R.; Harrison, S.A.; Brown, K.; Darland, C.; Finch, J.; Hardies, J. A Placebo-Controlled Trial of Pioglitazone in Subjects with Nonalcoholic Steatohepatitis. *N. Engl. J. Med.* **2006**, *355*, 2297–2307. [CrossRef]
119. Lian, J.; Fu, J. Pioglitazone for NAFLD Patients with Prediabetes or Type 2 Diabetes Mellitus: A Meta-Analysis. *Front. Endocrinol.* **2021**, *12*, 615409. [CrossRef]
120. DeFronzo, R.A.; Inzucchi, S.; Abdul-Ghani, M.; Nissen, S.E. Pioglitazone: The forgotten, cost-effective cardioprotective drug for type 2 diabetes. *Diab. Vasc. Dis. Res.* **2019**, *16*, 133–143. [CrossRef]
121. Defronzo, R.A.; Tripathy, D.; Schwenke, D.C.; Banerji, M.; Bray, G.A.; Buchanan, T.A. Prevention of diabetes with pioglitazone in ACT NOW: Physiologic correlates. *Diabetes* **2013**, *62*, 3920–3926. [CrossRef]
122. DeFronzo, R.A.; Tripathy, D.; Schwenke, D.C.; Banerji, M.; Bray, G.A.; Buchanan, T.A. Pioglitazone for diabetes prevention in impaired glucose tolerance. *N. Engl. J. Med.* **2011**, *364*, 1104–1115. [CrossRef] [PubMed]
123. Inzucchi, S.E.; Viscoli, C.M.; Young, L.H.; Furie, K.L.; Gorman, M.; Lovejoy, A.M. Pioglitazone Prevents Diabetes in Patients with Insulin Resistance and Cerebrovascular Disease. *Diabetes Care* **2016**, *39*, 1684–1692. [CrossRef] [PubMed]
124. Tang, H.; Shi, W.; Fu, S.; Wang, T.; Zhai, S.; Song, Y. Pioglitazone and bladder cancer risk: A systematic review and meta-analysis. *Cancer Med.* **2018**, *7*, 1070–1080. [CrossRef] [PubMed]
125. Ripamonti, E.; Azoulay, L.; Abrahamowicz, M.; Platt, R.; Suissa, S. A systematic review of observational studies of the association between pioglitazone use and bladder cancer. *Diabet. Med.* **2018**, *36*, 22–35. [CrossRef]
126. Filipova, E.; Uzunova, K.; Kalinov, K.; Vekov, T. Pioglitazone and the Risk of Bladder Cancer: A Meta-Analysis. *Diabetes Ther.* **2017**, *8*, 705–726. [CrossRef]
127. Portillo-Sanchez, P.; Bril, F.; Lomonaco, R.; Barb, D.; Orsak, B.; Bruder, J.M. Effect of pioglitazone on bone mineral density in patients with nonalcoholic steatohepatitis: A 36-month clinical trial. *J. Diabetes* **2019**, *11*, 223–231. [CrossRef]
128. Tuccori, M.; Filion, K.B.; Yin, H.; Yu, O.H.; Platt, R.W.; Azoulay, L. Pioglitazone use and risk of bladder cancer: Population based cohort study. *BMJ* **2016**, *352*, i1541. [CrossRef]
129. Neuschwander-Tetri, B.A.; Brunt, E.M.; Wehmeier, K.R.; Oliver, D.; Bacon, B.R. Improved nonalcoholic steatohepatitis after 48 weeks of treatment with the PPAR-gamma ligand rosiglitazone. *Hepatology* **2003**, *38*, 1008–1017. [CrossRef]
130. Wang, C.-H.; Leung, C.-H.; Liu, S.-C.; Chung, C.-H. Safety and effectiveness of rosiglitazone in type 2 diabetes patients with nonalcoholic Fatty liver disease. *J. Formos. Med. Assoc.* **2006**, *105*, 743–752. [CrossRef]
131. Ratziu, V.; Giral, P.; Jacqueminet, S.; Charlotte, F.; Hartemann-Heurtier, A.; Serfaty, L. Rosiglitazone for nonalcoholic steatohepatitis: One-year results of the randomized placebo-controlled Fatty Liver Improvement with Rosiglitazone Therapy (FLIRT) Trial. *Gastroenterology* **2008**, *135*, 100–110. [CrossRef]
132. Ratziu, V.; Charlotte, F.; Bernhardt, C.; Giral, P.; Halbron, M.; Lenaour, G. Long-term efficacy of rosiglitazone in nonalcoholic steatohepatitis: Results of the fatty liver improvement by rosiglitazone therapy (FLIRT 2) extension trial. *Hepatology* **2010**, *51*, 445–453. [CrossRef] [PubMed]
133. Torres, D.M.; Jones, F.J.; Shaw, J.C.; Williams, C.D.; Ward, J.A.; Harrison, S.A. Rosiglitazone versus rosiglitazone and metformin versus rosiglitazone and losartan in the treatment of nonalcoholic steatohepatitis in humans: A 12-month randomized, prospective, open- label trial. *Hepatology* **2011**, *54*, 1631–1639. [CrossRef] [PubMed]
134. Lemoine, M.; Serfaty, L.; Cervera, P.; Capeau, J.; Ratziu, V. Hepatic molecular effects of rosiglitazone in human non-alcoholic steatohepatitis suggest long-term pro-inflammatory damage. *Hepatol. Res.* **2014**, *44*, 1241–1247. [CrossRef]
135. Choung, S.; Joung, K.H.; You, B.R.; Park, S.K.; Kim, H.J.; Ku, B.J. Treatment with Lobeglitazone Attenuates Hepatic Steatosis in Diet-Induced Obese Mice. *PPAR Res.* **2018**, *2018*, 4292509. [CrossRef] [PubMed]
136. Lee, Y.H.; Kim, J.H.; Kim, S.R.; Jin, H.Y.; Rhee, E.J.; Cho, Y.M. Lobeglitazone, a Novel Thiazolidinedione, Improves Non-Alcoholic Fatty Liver Disease in Type 2 Diabetes: Its Efficacy and Predictive Factors Related to Responsiveness. *J. Korean Med. Sci.* **2017**, *32*, 60–69. [CrossRef] [PubMed]
137. Fiévet, C.; Fruchart, J.-C.; Staels, B. PPARalpha and PPARgamma dual agonists for the treatment of type 2 diabetes and the metabolic syndrome. *Curr. Opin. Pharmacol.* **2006**, *6*, 606–614. [CrossRef] [PubMed]
138. Agrawal, R. The first approved agent in the Glitazar's Class: Saroglitazar. *Curr. Drug Targets* **2014**, *15*, 151–155. [CrossRef]
139. Kaul, U.; Parmar, D.; Manjunath, K.; Shah, M.; Parmar, K.; Patil, K.P. New dual peroxisome proliferator activated receptor agonist-Saroglitazar in diabetic dyslipidemia and non-alcoholic fatty liver disease: Integrated analysis of the real world evidence. *Cardiovasc. Diabetol.* **2019**, *18*, 80. [CrossRef]
140. Jain, M.R.; Giri, S.R.; Trivedi, C.; Bhoi, B.; Rath, A.; Vanage, G. Saroglitazar, a novel PPARα/γ agonist with predominant PPARα activity, shows lipid-lowering and insulin-sensitizing effects in preclinical models. *Pharmacol. Res. Perspect.* **2015**, *3*, e00136. [CrossRef]
141. Jain, M.R.; Giri, S.R.; Bhoi, B.; Trivedi, C.; Rath, A.; Rathod, R. Dual PPARα/γ agonist saroglitazar improves liver histopathology and biochemistry in experimental NASH models. *Liver Int.* **2018**, *38*, 1084–1094. [CrossRef]

142. Kumar, D.P.; Caffrey, R.; Marioneaux, J.; Santhekadur, P.K.; Bhat, M.; Alonso, C. The PPAR α/γ Agonist Saroglitazar Improves Insulin Resistance and Steatohepatitis in a Diet Induced Animal Model of Nonalcoholic Fatty Liver Disease. *Sci. Rep.* **2020**, *10*, 9330. [CrossRef] [PubMed]
143. Jani, R.H.; Pai, V.; Jha, P.; Jariwala, G.; Mukhopadhyay, S.; Bhansali, A. A Multicenter, Prospective, Randomized, Double-Blind Study to Evaluate the Safety and Efficacy of Saroglitazar 2 and 4 mg Compared with Placebo in Type 2 Diabetes Mellitus Patients Having Hypertriglyceridemia Not Controlled with Atorvastatin Therapy (PRESS VI). *Diabetes Technol. Ther.* **2014**, *16*, 63–71. [PubMed]
144. Pai, V.; Paneerselvam, A.; Mukhopadhyay, S.; Bhansali, A.; Kamath, D.; Shankar, V. A Multicenter, Prospective, Randomized, Double-blind Study to Evaluate the Safety and Efficacy of Saroglitazar 2 and 4 mg Compared to Pioglitazone 45 mg in Diabetic Dyslipidemia (PRESS V). *J. Diabetes Sci. Technol.* **2014**, *8*, 132–141. [CrossRef] [PubMed]
145. Jain, N.; Bhansali, S.; Kurpad, A.V.; Hawkins, M.; Sharma, A.; Kaur, S. Effect of a Dual PPAR α/γ agonist on Insulin Sensitivity in Patients of Type 2 Diabetes with Hypertriglyceridemia- Randomized double-blind placebo-controlled trial. *Sci. Rep.* **2019**, *9*, 19017. [CrossRef] [PubMed]
146. Mitra, A. An Observational Study of Reduction in Glycemic Parameters and Liver Stiffness by Saroglitazar 4 mg in Patients with Type 2 Diabetes Mellitus and Nonalcoholic Fatty Liver Disease. *Cureus* **2020**, *12*, e9065. [CrossRef] [PubMed]
147. Goyal, O.; Nohria, S.; Goyal, P.; Kaur, J.; Sharma, S.; Sood, A. Saroglitazar in patients with non-alcoholic fatty liver disease and diabetic dyslipidemia: A prospective, observational, real world study. *Sci. Rep.* **2020**, *10*, 21117. [CrossRef] [PubMed]
148. Siddiqui, M.S.; Idowu, M.O.; Parmar, D.; Borg, B.B.; Denham, D.; Loo, N.M. A Phase 2 Double Blinded, Randomized Controlled Trial of Saroglitazar in Patients with Nonalcoholic Steatohepatitis. *Clin. Gastroenterol. Hepatol.* **2021**, *19*, 2670–2672. [CrossRef]
149. Rubin, C.J.; Viraswami-Appanna, K.; Fiedorek, F.T. Efficacy and safety of muraglitazar: A double-blind, 24-week, dose-ranging study in patients with type 2 diabetes. *Diab. Vasc. Dis. Res.* **2009**, *6*, 205–215. [CrossRef]
150. Westerouen Van Meeteren, M.J.; Drenth, J.P.H.; Tjwa, E.T.T.L. Elafibranor: A potential drug for the treatment of nonalcoholic steatohepatitis (NASH). *Exp. Opin. Investig. Drugs* **2020**, *29*, 117–123. [CrossRef]
151. Staels, B.; Rubenstrunk, A.; Noel, B.; Rigou, G.; Delataille, P.; Millatt, L.J. Hepatoprotective effects of the dual peroxisome proliferator-activated receptor alpha/delta agonist, GFT505, in rodent models of nonalcoholic fatty liver disease/nonalcoholic steatohepatitis. *Hepatology* **2013**, *58*, 1941–1952. [CrossRef]
152. Boeckmans, J.; Natale, A.; Rombaut, M.; Buyl, K.; Cami, B.; De Boe, V. Human hepatic in vitro models reveal distinct anti-NASH potencies of PPAR agonists. *Cell Biol. Toxicol.* **2021**, *37*, 293–311. [CrossRef] [PubMed]
153. Cariou, B.; Hanf, R.; Lambert-Porcheron, S.; Zaïr, Y.; Sauvinet, V.; Noël, B. Dual peroxisome proliferator-activated receptor α/δ agonist GFT505 improves hepatic and peripheral insulin sensitivity in abdominally obese subjects. *Diabetes Care* **2013**, *36*, 2923–2930. [CrossRef] [PubMed]
154. Drenth, J.P.H.; Schattenberg, J.M. The nonalcoholic steatohepatitis (NASH) drug development graveyard: Established hurdles and planning for future success. *Exp. Opin. Investig. Drugs* **2020**, *29*, 1365–1375. [CrossRef] [PubMed]
155. Vuppalanchi, R.; Noureddin, M.; Alkhouri, N.; Sanyal, A.J. Therapeutic pipeline in nonalcoholic steatohepatitis. *Nat. Rev. Gastroenterol. Hepatol.* **2021**, *18*, 373–392. [CrossRef] [PubMed]
156. Boubia, B.; Poupardin, O.; Barth, M.; Binet, J.; Peralba, P.; Mounier, L. Design, Synthesis, and Evaluation of a Novel Series of Indole Sulfonamide Peroxisome Proliferator Activated Receptor (PPAR) $\alpha/\gamma/\delta$ Triple Activators: Discovery of Lanifibranor, a New Antifibrotic Clinical Candidate. *J. Med. Chem.* **2018**, *61*, 2246–2265. [CrossRef] [PubMed]
157. Wettstein, G.; Luccarini, J.; Poekes, L.; Faye, P.; Kupkowski, F.; Adarbes, V. The new generation pan-peroxisome proliferator-activated receptor agonist IVA337 protects the liver from metabolic disorders and fibrosis. *Hepatol. Commun.* **2017**, *1*, 524–537. [CrossRef]
158. Lefere, S.; Puengel, T.; Hundertmark, J.; Penners, C.; Frank, A.K.; Guillot, A. Differential effects of selective- and pan-PPAR agonists on experimental steatohepatitis and hepatic macrophages. *J. Hepatol.* **2020**, *73*, 757–770. [CrossRef] [PubMed]
159. Boyer-Diaz, Z.; Aristu-Zabalza, P.; Andrés-Rozas, M.; Robert, C.; Ortega-Ribera, M.; Fernández-Iglesias, A. Pan-PPAR agonist lanifibranor improves portal hypertension and hepatic fibrosis in experimental advanced chronic liver disease. *J. Hepatol.* **2021**, *74*, 1188–1199. [CrossRef]
160. Francque, S.; Bedossa, P.; Abdelmalek, M.F.; Anstee, Q.; Bugianesi, E.; Ratziu, V. A randomised, double-blind, placebo-controlled, multi-centre, dose-range, proof-of-concept, 24-week treatment study of lanifibranor in adult subjects with non-alcoholic steatohepatitis: Design of the NATIVE study. *Contemp. Clin. Trials* **2020**, *98*, 106170.
161. Majzoub, A.M.; Nayfeh, T.; Barnard, A.; Munaganuru, N.; Dave, S.; Singh, S. Systematic review with network meta-analysis: Comparative efficacy of pharmacologic therapies for fibrosis improvement and resolution of NASH. *Aliment. Pharm. Ther.* **2021**, *54*, 880–889. [CrossRef]
162. American Diabetes Association. Standards of Medical Care in Diabetes—2021. Diabetes Care. Available online: https://care.diabetesjournals.org/content/44/Supplement_1,/content/44/Supplement_1 (accessed on 25 October 2021).
163. Balas, B.; Belfort, R.; Harrison, S.A.; Darland, C.; Finch, J.; Schenker, S. Pioglitazone treatment increases whole body fat but not total body water in patients with non-alcoholic steatohepatitis. *J. Hepatol.* **2007**, *47*, 565–570. [CrossRef] [PubMed]
164. Basu, A.; Jensen, M.D.; McCann, F.; Mukhopadhyay, D.; Joyner, M.J.; Rizza, R.A. Effects of pioglitazone versus glipizide on body fat distribution, body water content, and hemodynamics in type 2 diabetes. *Diabetes Care* **2006**, *29*, 510–514. [CrossRef] [PubMed]

165. Miyazaki, Y.; Mahankali, A.; Wajcberg, E.; Bajaj, M.; Mandarino, L.J.; DeFronzo, R.A. Effect of pioglitazone on circulating adipocytokine levels and insulin sensitivity in type 2 diabetic patients. *J. Clin. Endocrinol. Metab.* **2004**, *89*, 4312–4319. [CrossRef] [PubMed]
166. Harvey, I.; Boudreau, A.; Stephens, J.M. Adipose tissue in health and disease. *Open Biol.* **2020**, *10*, 200291. [CrossRef]
167. White, U.; Fitch, M.D.; Beyl, R.A.; Hellerstein, M.K.; Ravussin, E. Adipose depot-specific effects of 16 weeks of pioglitazone on in vivo adipogenesis in women with obesity: A randomised controlled trial. *Diabetologia* **2021**, *64*, 159–167. [CrossRef]
168. Aithal, G.P.; Thomas, J.A.; Kaye, P.V.; Lawson, A.; Ryder, S.D.; Spendlove, I. Randomized, placebo-controlled trial of pioglitazone in nondiabetic subjects with nonalcoholic steatohepatitis. *Gastroenterology* **2008**, *135*, 1176–1184. [CrossRef]
169. Kernan, W.N.; Viscoli, C.M.; Furie, K.L.; Young, L.H.; Inzucchi, S.E.; Gorman, M. Pioglitazone after Ischemic Stroke or Transient Ischemic Attack. *N. Engl. J. Med.* **2016**, *374*, 1321–1331. [CrossRef]
170. Tenenbaum, A.; Fisman, E.Z. Balanced pan-PPAR activator bezafibrate in combination with statin: Comprehensive lipids control and diabetes prevention? *Cardiovasc. Diabetol.* **2012**, *11*, 140. [CrossRef]
171. Botta, M.; Audano, M.; Sahebkar, A.; Sirtori, C.R.; Mitro, N.; Ruscica, M. PPAR Agonists and Metabolic Syndrome: An Established Role? *Int. J. Mol. Sci.* **2018**, *19*, 1197. [CrossRef]
172. Grundy, S.M. Small LDL, atherogenic dyslipidemia, and the metabolic syndrome. *Circulation* **1997**, *95*, 1–4. [CrossRef]
173. Staels, B.; Maes, M.; Zambon, A. Fibrates and future PPARalpha agonists in the treatment of cardiovascular disease. *Nat. Clin. Pract. Cardiovasc. Med.* **2008**, *5*, 542–553. [CrossRef]
174. Keech, A.; Simes, R.J.; Barter, P.; Best, J.; Scott, R.; Taskinen, M.R. Effects of long-term fenofibrate therapy on cardiovascular events in 9795 people with type 2 diabetes mellitus (the FIELD study): Randomised controlled trial. *Lancet Lond. Engl.* **2005**, *366*, 1849–1861. [CrossRef]
175. ACCORD Study Group; Ginsberg, H.N.; Elam, M.B.; Lovato, L.C.; Crouse, J.R.; Leiter, L.A. Effects of combination lipid therapy in type 2 diabetes mellitus. *N. Engl. J. Med.* **2010**, *362*, 1563–1574. [PubMed]
176. Fruchart, J.-C.; Hermans, M.P.; Fruchart-Najib, J.; Kodama, T. Selective Peroxisome Proliferator-Activated Receptor Alpha Modulators (SPPARMα) in the Metabolic Syndrome: Is Pemafibrate Light at the End of the Tunnel? *Curr. Atheroscler. Rep.* **2021**, *23*, 3. [CrossRef] [PubMed]
177. Araki, E.; Yamashita, S.; Arai, H.; Yokote, K.; Satoh, J.; Inoguchi, T. Efficacy and safety of pemafibrate in people with type 2 diabetes and elevated triglyceride levels: 52-week data from the PROVIDE study. *Diabetes Obes. Metab.* **2019**, *21*, 1737–1744. [CrossRef] [PubMed]
178. Yamashita, S.; Arai, H.; Yokote, K.; Araki, E.; Matsushita, M.; Nojima, T. Efficacy and Safety of Pemafibrate, a Novel Selective Peroxisome Proliferator-Activated Receptor α Modulator (SPPARMα): Pooled Analysis of Phase 2 and 3 Studies in Dyslipidemic Patients with or without Statin Combination. *Int. J. Mol. Sci.* **2019**, *20*, 5537. [CrossRef]
179. Arai, H.; Yamashita, S.; Yokote, K.; Araki, E.; Suganami, H.; Ishibashi, S. Efficacy and Safety of Pemafibrate Versus Fenofibrate in Patients with High Triglyceride and Low HDL Cholesterol Levels: A Multicenter, Placebo-Controlled, Double-Blind, Randomized Trial. *J. Atheroscler. Thromb.* **2018**, *25*, 521–538. [CrossRef]
180. Nissen, S.E.; Nicholls, S.J.; Wolski, K.; Nesto, R.; Kupfer, S.; Perez, A. Comparison of pioglitazone vs glimepiride on progression of coronary atherosclerosis in patients with type 2 diabetes: The PERISCOPE randomized controlled trial. *JAMA* **2008**, *299*, 1561–1573. [CrossRef]
181. Mazzone, T.; Meyer, P.M.; Feinstein, S.B.; Davidson, M.H.; Kondos, G.T.; D'Agostino, R.B. Effect of pioglitazone compared with glimepiride on carotid intima-media thickness in type 2 diabetes: A randomized trial. *JAMA* **2006**, *296*, 2572–2581. [CrossRef]
182. Nicholls, S.J.; Tuzcu, E.M.; Wolski, K.; Bayturan, O.; Lavoie, A.; Uno, K. Lowering the triglyceride/high-density lipoprotein cholesterol ratio is associated with the beneficial impact of pioglitazone on progression of coronary atherosclerosis in diabetic patients: Insights from the PERISCOPE (Pioglitazone Effect on Regression of Intravascular Sonographic Coronary Obstruction Prospective Evaluation) study. *J. Am. Coll. Cardiol.* **2011**, *57*, 153–159.
183. Davidson, M.; Meyer, P.M.; Haffner, S.; Feinstein, S.; D'Agostino, R.; Kondos, G.T. Increased high-density lipoprotein cholesterol predicts the pioglitazone-mediated reduction of carotid intima-media thickness progression in patients with type 2 diabetes mellitus. *Circulation* **2008**, *117*, 2123–2130. [CrossRef] [PubMed]
184. Chatterjee, S.; Majumder, A.; Ray, S. Observational study of effects of Saroglitazar on glycaemic and lipid parameters on Indian patients with type 2 diabetes. *Sci. Rep.* **2015**, *5*, 7706. [CrossRef]
185. Dutta, D.; Bhattacharya, S.; Surana, V.; Aggarwal, S.; Singla, R.; Khandelwal, D. Efficacy and safety of saroglitazar in managing hypertriglyceridemia in type-2 diabetes: A meta-analysis. *Diabetes Metab. Syndr.* **2020**, *14*, 1759–1768. [CrossRef] [PubMed]
186. Krishnappa, M.; Patil, K.; Parmar, K.; Trivedi, P.; Mody, N.; Shah, C. Effect of saroglitazar 2 mg and 4 mg on glycemic control, lipid profile and cardiovascular disease risk in patients with type 2 diabetes mellitus: A 56-week, randomized, double blind, phase 3 study (PRESS XII study). *Cardiovasc. Diabetol.* **2020**, *19*, 93. [CrossRef]
187. Choi, Y.-J.; Roberts, B.K.; Wang, X.; Geaney, J.C.; Naim, S.; Wojnoonski, K. Effects of the PPAR-δ agonist MBX-8025 on atherogenic dyslipidemia. *Atherosclerosis* **2012**, *220*, 470–476. [CrossRef] [PubMed]
188. Cariou, B.; Zaïr, Y.; Staels, B.; Bruckert, E. Effects of the New Dual PPARα/δ Agonist GFT505 on Lipid and Glucose Homeostasis in Abdominally Obese Patients with Combined Dyslipidemia or Impaired Glucose Metabolism. *Diabetes Care* **2011**, *34*, 2008–2014. [CrossRef]
189. Adhyaru, B.B.; Jacobson, T.A. Safety and efficacy of statin therapy. *Nat. Rev. Cardiol.* **2018**, *15*, 757–769. [CrossRef]

190. Bernardi, M.; Caraceni, P. Novel perspectives in the management of decompensated cirrhosis. *Nat. Rev. Gastroenterol. Hepatol.* **2018**, *15*, 753–764. [CrossRef]
191. Torres-Peña, J.D.; Martín-Piedra, L.; Fuentes-Jiménez, F. Statins in Non-alcoholic Steatohepatitis. *Front. Cardiovasc. Med.* **2021**, *8*, 777131. [CrossRef] [PubMed]
192. Graham, D.J.; Staffa, J.A.; Shatin, D.; Andrade, S.E.; Schech, S.D.; La Grenade, L. Incidence of hospitalized rhabdomyolysis in patients treated with lipid-lowering drugs. *JAMA* **2004**, *292*, 2585–2590. [CrossRef]
193. Davidson, M.H.; Armani, A.; McKenney, J.M.; Jacobson, T.A. Safety considerations with fibrate therapy. *Am. J. Cardiol.* **2007**, *99*, 3C–18C. [CrossRef] [PubMed]
194. Ng, A.C.T.; Delgado, V.; Borlaug, B.A.; Bax, J.J. Diabesity: The combined burden of obesity and diabetes on heart disease and the role of imaging. *Nat. Rev. Cardiol.* **2021**, *18*, 291–304. [CrossRef] [PubMed]
195. Caussy, C.; Aubin, A.; Loomba, R. The Relationship Between Type 2 Diabetes, NAFLD, and Cardiovascular Risk. *Curr. Diab. Rep.* **2021**, *21*, 15. [CrossRef] [PubMed]
196. Simon, T.G.; Roelstraete, B.; Khalili, H.; Hagström, H.; Ludvigsson, J.F. Mortality in biopsy-confirmed nonalcoholic fatty liver disease: Results from a nationwide cohort. *Gut* **2021**, *70*, 1375–1382. [CrossRef]
197. Dennis, B.B.; Sallam, S.; Perumpail, B.J.; Shah, N.D.; Kim, D.; Cholankeril, G. Management of Cardiometabolic Complications in Patients with Nonalcoholic Fatty Liver Disease: A Review of the Literature With Recommendations. *J. Clin. Gastroenterol.* **2021**, *55*, 747–756. [CrossRef]
198. Montaigne, D.; Butruille, L.; Staels, B. PPAR control of metabolism and cardiovascular functions. *Nat. Rev. Cardiol.* **2021**, *18*, 809–823. [CrossRef]
199. Wang, D.; Liu, B.; Tao, W.; Hao, Z.; Liu, M. Fibrates for secondary prevention of cardiovascular disease and stroke. *Cochrane Database Syst Rev.* **2015**, *2015*, CD009580. [CrossRef]
200. Pradhan, A.D.; Paynter, N.P.; Everett, B.M.; Glynn, R.J.; Amarenco, P.; Elam, M. Rationale and design of the Pemafibrate to Reduce Cardiovascular Outcomes by Reducing Triglycerides in Patients with Diabetes (PROMINENT) study. *Am. Heart J.* **2018**, *206*, 80–93. [CrossRef]
201. van der Meer, R.W.; Rijzewijk, L.J.; de Jong, H.W.A.M.; Lamb, H.J.; Lubberink, M.; Romijn, J.A. Pioglitazone improves cardiac function and alters myocardial substrate metabolism without affecting cardiac triglyceride accumulation and high-energy phosphate metabolism in patients with well-controlled type 2 diabetes mellitus. *Circulation* **2009**, *119*, 2069–2077. [CrossRef] [PubMed]
202. Clarke, G.D.; Solis-Herrera, C.; Molina-Wilkins, M.; Martinez, S.; Merovci, A.; Cersosimo, E. Pioglitazone Improves Left Ventricular Diastolic Function in Subjects with Diabetes. *Diabetes Care* **2017**, *40*, 1530–1536. [CrossRef]
203. Dormandy, J.A.; Charbonnel, B.; Eckland, D.J.A.; Erdmann, E.; Massi-Benedetti, M.; Moules, I.K. Secondary prevention of macrovascular events in patients with type 2 diabetes in the PROactive Study (PROspective pioglitAzone Clinical Trial in macroVascular Events): A randomised controlled trial. *Lancet Lond. Engl.* **2005**, *366*, 1279–1289. [CrossRef]
204. Lincoff, A.M.; Wolski, K.; Nicholls, S.J.; Nissen, S.E. Pioglitazone and Risk of Cardiovascular Events in Patients with Type 2 Diabetes Mellitus: A Meta-analysis of Randomized Trials. *JAMA* **2007**, *298*, 1180. [CrossRef] [PubMed]
205. Spence, J.D.; Viscoli, C.M.; Inzucchi, S.E.; Dearborn-Tomazos, J.; Ford, G.A.; Gorman, M. Pioglitazone Therapy in Patients with Stroke and Prediabetes: A Post Hoc Analysis of the IRIS Randomized Clinical Trial. *JAMA Neurol.* **2019**, *76*, 526–535. [CrossRef] [PubMed]
206. Guan, Y.; Hao, C.; Cha, D.R.; Rao, R.; Lu, W.; Kohan, D.E. Thiazolidinediones expand body fluid volume through PPARgamma stimulation of ENaC-mediated renal salt absorption. *Nat. Med.* **2005**, *11*, 861–866. [CrossRef] [PubMed]
207. Lago, R.M.; Singh, P.P.; Nesto, R.W. Congestive heart failure and cardiovascular death in patients with prediabetes and type 2 diabetes given thiazolidinediones: A meta-analysis of randomised clinical trials. *Lancet Lond. Engl.* **2007**, *370*, 1129–1136. [CrossRef]
208. Young, L.H.; Viscoli, C.M.; Schwartz, G.G.; Inzucchi, S.E.; Curtis, J.P.; Gorman, M.J. Heart Failure After Ischemic Stroke or Transient Ischemic Attack in Insulin-Resistant Patients Without Diabetes Mellitus Treated with Pioglitazone. *Circulation* **2018**, *138*, 1210–1220. [CrossRef]
209. Lehrke, M.; Marx, N. Diabetes Mellitus and Heart Failure. *Am. J. Cardiol.* **2017**, *120*, S37–S47. [CrossRef]
210. Nissen, S.E.; Wolski, K. Effect of rosiglitazone on the risk of myocardial infarction and death from cardiovascular causes. *N. Engl. J. Med.* **2007**, *356*, 2457–2471. [CrossRef]
211. Nissen, S.E.; Wolski, K. Rosiglitazone revisited: An updated meta-analysis of risk for myocardial infarction and cardiovascular mortality. *Arch. Intern. Med.* **2010**, *170*, 1191–1201. [CrossRef]
212. Graham, D.J.; Ouellet-Hellstrom, R.; MaCurdy, T.E.; Ali, F.; Sholley, C.; Worrall, C. Risk of acute myocardial infarction, stroke, heart failure, and death in elderly Medicare patients treated with rosiglitazone or pioglitazone. *JAMA* **2010**, *304*, 411–418. [CrossRef]
213. Home, P.D.; Pocock, S.J.; Beck-Nielsen, H.; Curtis, P.S.; Gomis, R.; Hanefeld, M. Rosiglitazone evaluated for cardiovascular outcomes in oral agent combination therapy for type 2 diabetes (RECORD): A multicentre, randomised, open-label trial. *Lancet Lond. Engl.* **2009**, *373*, 2125–2135. [CrossRef]
214. Borém, L.M.A.; Neto, J.F.R.; Brandi, I.V.; Lelis, D.F.; Santos, S.H.S. The role of the angiotensin II type I receptor blocker telmisartan in the treatment of non-alcoholic fatty liver disease: A brief review. *Hypertens. Res.* **2018**, *41*, 394–405. [CrossRef] [PubMed]

215. Hirata, T.; Tomita, K.; Kawai, T.; Yokoyama, H.; Shimada, A.; Kikuchi, M. Effect of Telmisartan or Losartan for Treatment of Nonalcoholic Fatty Liver Disease: Fatty Liver Protection Trial by Telmisartan or Losartan Study (FANTASY). *Int. J. Endocrinol.* **2013**, *2013*, 587140. [CrossRef] [PubMed]
216. Park, J.G.; Mok, J.S.; Han, Y.I.; Park, T.S.; Kang, K.W.; Choi, C.S. Connectivity mapping of angiotensin-PPAR interactions involved in the amelioration of non-alcoholic steatohepatitis by Telmisartan. *Sci. Rep.* **2019**, *9*, 4003. [CrossRef] [PubMed]
217. Dufour, J.-F.; Caussy, C.; Loomba, R. Combination therapy for non-alcoholic steatohepatitis: Rationale, opportunities and challenges. *Gut* **2020**, *69*, 1877–1884. [CrossRef]
218. Zoeller, R.T.; Brown, T.R.; Doan, L.L.; Gore, A.C.; Skakkebaek, N.E.; Soto, A.M. Endocrine-disrupting chemicals and public health protection: A statement of principles from The Endocrine Society. *Endocrinology* **2012**, *153*, 4097–4110. [CrossRef] [PubMed]
219. le Maire, A.; Bourguet, W.; Balaguer, P. A structural view of nuclear hormone receptor: Endocrine disruptor interactions. *Cell Mol. Life Sci.* **2010**, *67*, 1219–1237. [CrossRef]
220. Cano, R.; Pérez, J.L.; Dávila, L.A.; Ortega, Á.; Gómez, Y.; Valero-Cedeño, N.J. Role of Endocrine-Disrupting Chemicals in the Pathogenesis of Non-Alcoholic Fatty Liver Disease: A Comprehensive Review. *Int. J. Mol. Sci.* **2021**, *22*, 4807. [CrossRef]
221. Amato, A.A.; Wheeler, H.B.; Blumberg, B. Obesity and endocrine-disrupting chemicals. *Endocr. Connect.* **2021**, *10*, R87–R105. [CrossRef]
222. Lind, P.M.; Lind, L. Endocrine-disrupting chemicals and risk of diabetes: An evidence-based review. *Diabetologia* **2018**, *61*, 1495–1502. [CrossRef]
223. Li, G.; Brocker, C.N.; Yan, T.; Xie, C.; Krausz, K.W.; Xiang, R. Metabolic adaptation to intermittent fasting is independent of peroxisome proliferator-activated receptor alpha. *Mol. Metab.* **2018**, *7*, 80–89. [CrossRef] [PubMed]
224. Pawlak, M.; Lefebvre, P.; Staels, B. Molecular mechanism of PPARα action and its impact on lipid metabolism, inflammation and fibrosis in non-alcoholic fatty liver disease. *J. Hepatol.* **2015**, *62*, 720–733. [CrossRef] [PubMed]
225. Lombardi, R.; Iuculano, F.; Pallini, G.; Fargion, S.; Fracanzani, A.L. Nutrients, Genetic Factors, and Their Interaction in Non-Alcoholic Fatty Liver Disease and Cardiovascular Disease. *Int. J. Mol. Sci.* **2020**, *21*, 8761. [CrossRef]
226. Wu, L.; Li, J.; Feng, J.; Ji, J.; Yu, Q.; Li, Y. Crosstalk between PPARs and gut microbiota in NAFLD. *Biomed. Pharm. Biomed. Pharm.* **2021**, *136*, 111255. [CrossRef] [PubMed]
227. Lemberger, T.; Saladin, R.; Vázquez, M.; Assimacopoulos, F.; Staels, B.; Desvergne, B. Expression of the peroxisome proliferator-activated receptor alpha gene is stimulated by stress and follows a diurnal rhythm. *J. Biol. Chem.* **1996**, *271*, 1764–1769. [CrossRef] [PubMed]
228. Saran, A.R.; Dave, S.; Zarrinpar, A. Circadian Rhythms in the Pathogenesis and Treatment of Fatty Liver Disease. *Gastroenterology* **2020**, *158*, 1948–1966.e1. [CrossRef] [PubMed]
229. Ballestri, S.; Nascimbeni, F.; Baldelli, E.; Marrazzo, A.; Romagnoli, D.; Lonardo, A. NAFLD as a Sexual Dimorphic Disease: Role of Gender and Reproductive Status in the Development and Progression of Nonalcoholic Fatty Liver Disease and Inherent Cardiovascular Risk. *Adv. Ther.* **2017**, *34*, 1291–1326. [CrossRef]
230. Mauvais-Jarvis, F. Epidemiology of Gender Differences in Diabetes and Obesity. *Adv. Exp. Med. Biol.* **2017**, *1043*, 3–8.
231. Lonardo, A.; Nascimbeni, F.; Ballestri, S.; Fairweather, D.; Win, S.; Than, T.A. Sex Differences in Nonalcoholic Fatty Liver Disease: State of the Art and Identification of Research Gaps. *Hepatology* **2019**, *70*, 1457–1469. [CrossRef]
232. Tramunt, B.; Smati, S.; Grandgeorge, N.; Lenfant, F.; Arnal, J.-F.; Montagner, A. Sex differences in metabolic regulation and diabetes susceptibility. *Diabetologia* **2020**, *63*, 453–461. [CrossRef]
233. Burra, P.; Bizzaro, D.; Gonta, A.; Shalaby, S.; Gambato, M.; Morelli, M.C. Clinical impact of sexual dimorphism in non-alcoholic fatty liver disease (NAFLD) and non-alcoholic steatohepatitis (NASH). *Liver Int.* **2021**, *41*, 1713–1733. [CrossRef] [PubMed]
234. Lonardo, A.; Suzuki, A. Sexual Dimorphism of NAFLD in Adults. Focus on Clinical Aspects and Implications for Practice and Translational Research. *J. Clin. Med.* **2020**, *9*, 1278. [CrossRef] [PubMed]
235. Lonardo, A.; Suzuki, A. Nonalcoholic fatty liver disease: Does sex matter? *Hepatobiliary Surg. Nutr.* **2019**, *8*, 164–166. [CrossRef] [PubMed]
236. Mauvais-Jarvis, F.; Bairey Merz, N.; Barnes, P.J.; Brinton, R.D.; Carrero, J.-J.; DeMeo, D.L. Sex and gender: Modifiers of health, disease, and medicine. *Lancet Lond. Engl.* **2020**, *396*, 565–582. [CrossRef]
237. Lefebvre, P.; Staels, B. Hepatic sexual dimorphism—Implications for non-alcoholic fatty liver disease. *Nat. Rev. Endocrinol.* **2021**, *17*, 662–670. [CrossRef]
238. Kane, A.E.; Sinclair, D.A.; Mitchell, J.R.; Mitchell, S.J. Sex differences in the response to dietary restriction in rodents. *Curr. Opin. Physiol.* **2018**, *6*, 28–34. [CrossRef]
239. Williams, R.L.; Wood, L.G.; Collins, C.E.; Callister, R. Effectiveness of weight loss interventions—Is there a difference between men and women: A systematic review. *Obes. Rev.* **2015**, *16*, 171–186. [CrossRef]
240. Wang, C.; Xu, Y. Mechanisms for Sex Differences in Energy Homeostasis. *J. Mol. Endocrinol.* **2019**, *62*, R129–R143. [CrossRef]
241. Della Torre, S. Non-alcoholic Fatty Liver Disease as a Canonical Example of Metabolic Inflammatory-Based Liver Disease Showing a Sex-Specific Prevalence: Relevance of Estrogen Signaling. *Front. Endocrinol.* **2020**, *11*, 572490. [CrossRef]
242. d'Emden, M.C.; Jenkins, A.J.; Li, L.; Zannino, D.; Mann, K.P.; Best, J.D. Favourable effects of fenofibrate on lipids and cardiovascular disease in women with type 2 diabetes: Results from the Fenofibrate Intervention and Event Lowering in Diabetes (FIELD) study. *Diabetologia* **2014**, *57*, 2296–2303. [CrossRef] [PubMed]
243. Mauvais-Jarvis, F. Gender differences in glucose homeostasis and diabetes. *Physiol. Behav.* **2018**, *187*, 20–23. [CrossRef] [PubMed]

244. Foryst-Ludwig, A.; Clemenz, M.; Hohmann, S.; Hartge, M.; Sprang, C.; Frost, N. Metabolic actions of estrogen receptor beta (ERbeta) are mediated by a negative cross-talk with PPARgamma. *PLoS Genet.* **2008**, *4*, e1000108. [CrossRef] [PubMed]
245. Duan, S.Z.; Usher, M.G.; Foley, E.L., IV; Milstone, D.S.; Brosius, F.C., III; Mortensen, R.M. Sex dimorphic actions of rosiglitazone in generalised peroxisome proliferator-activated receptor-gamma (PPAR-gamma)-deficient mice. *Diabetologia* **2010**, *53*, 1493–1505. [CrossRef] [PubMed]

International Journal of
Molecular Sciences

MDPI

Article

Sex Dimorphism of Nonalcoholic Fatty Liver Disease (NAFLD) in *Pparg*-Null Mice

Mariano Schiffrin [1], CarineWinkler [1], Laure Quignodon [1], Aurélien Naldi [1], Martin Trötzmüller [2], Harald Köfeler [2], Hugues Henry [3], Paolo Parini [4], Béatrice Desvergne [1] and Federica Gilardi [1,5,*]

1 Center of Integrative Genomics, Genopode, Lausanne Faculty of Biology and Medicine, CH-1015 Lausanne, Switzerland; mariano.schiffrin@gmail.com (M.S.); carine.winkler@unil.ch (C.W.); laurelausanne@gmail.com (L.Q.); aurelien.naldi@gmail.com (A.N.); Beatrice.Desvergne@unil.ch (B.D.)
2 Core Facility Mass Spectrometry, Medical University of Graz, 8036 Graz, Austria; martin.troetzmueller@medunigraz.at (M.T.); harald.koefeler@klinikum-graz.at (H.K.)
3 Centre Hospitalier Universitaire Vaudois (CHUV), Lausanne Faculty of Biology and Medicine, CH-1011 Lausanne, Switzerland; Hugues.Henry@chuv.ch
4 CardioMetabolic Unit, Department of Medicine and Department of Laboratory Medicine, Karolinska Insititutet and Theme Inflammation and Ageing Karolinska University Hospital Huddinge, 14186 Stockholm, Sweden; paolo.parini@ki.se
5 Faculty Unit of Toxicology, University Center of Legal Medicine, Faculty of Biology and Medicine, Lausanne University Hospital, CH-1000 Lausanne, Switzerland
* Correspondence: Federica.gilardi@chuv.ch; Tel.: +41-22-379-55-78

Abstract: Men with nonalcoholic fatty liver disease (NAFLD) are more exposed to nonalcoholic steatohepatitis (NASH) and liver fibrosis than women. However, the underlying molecular mechanisms of NALFD sex dimorphism are unclear. We combined gene expression, histological and lipidomic analyses to systematically compare male and female liver steatosis. We characterized hepatosteatosis in three independent mouse models of NAFLD, *ob/ob* and lipodystrophic fat-specific (*PpargF*$^{\Delta/\Delta}$) and whole-body PPARγ-null (*Pparg*$^{\Delta/\Delta}$) mice. We identified a clear sex dimorphism occurring only in *Pparg*$^{\Delta/\Delta}$ mice, with females showing macro- and microvesicular hepatosteatosis throughout their entire life, while males had fewer lipid droplets starting from 20 weeks. This sex dimorphism in hepatosteatosis was lost in gonadectomized *Pparg*$^{\Delta/\Delta}$ mice. Lipidomics revealed hepatic accumulation of short and highly saturated TGs in females, while TGs were enriched in long and unsaturated hydrocarbon chains in males. Strikingly, sex-biased genes were particularly perturbed in both sexes, affecting lipid metabolism, drug metabolism, inflammatory and cellular stress response pathways. Most importantly, we found that the expression of key sex-biased genes was severely affected in all the NAFLD models we tested. Thus, hepatosteatosis strongly affects hepatic sex-biased gene expression. With NAFLD increasing in prevalence, this emphasizes the urgent need to specifically address the consequences of this deregulation in humans.

Keywords: nonalcoholic fatty liver disease (NAFLD); sex dimorphism; lipidomics; hepatic sex-biased gene expression

Citation: Schiffrin, M.;Winkler, C.; Quignodon, L.; Naldi, A.; Trötzmüller, M.; Köfeler, H.; Henry, H.; Parini, P.; Desvergne, B.; Gilardi, F. Sex Dimorphism of Nonalcoholic Fatty Liver Disease (NAFLD) in *Pparg*-Null Mice. *Int. J. Mol. Sci.* **2021**, *22*, 9969. https://doi.org/10.3390/ijms22189969

Academic Editors: Manuel Vázquez-Carrera and Walter Wahli

Received: 19 July 2021
Accepted: 11 September 2021
Published: 15 September 2021

1. Introduction

Nonalcoholic fatty liver disease (NAFLD) is considered as the hepatic manifestation of the metabolic syndrome and is associated with obesity, insulin resistance and diabetes. Therefore, its clinical prevalence has grown in recent years, due to the obesity epidemic. NAFLD is characterized by an excessive accumulation of triglycerides (TGs) and cholesterol esters in hepatocytes, also referred to as hepatosteatosis. The simple accumulation of fat is *per se* harmless, and the incidence of NAFLD in the adult population in Western countries is estimated to be around 25%. However, approximately 20% of patients with NAFLD develop liver inflammation, which is the hallmark of nonalcoholic steatohepatitis (NASH) [1,2].

Once developed, NASH can progress towards fibrosis and ultimately cirrhosis, which is an important risk factor for hepatocellular carcinoma [3].

One factor that seems to play a role in NASH incidence is gender, although epidemiological studies on this topic are still scarce and limited, and the available information contradictory. Whereas earlier observations suggested that NAFLD/NASH was a female-predominant condition, recent data suggest a higher prevalence in men [3]. In addition, hepatocellular carcinoma, which can be triggered by advanced liver fibrosis, is clearly sexually dimorphic in both rodents and humans, with a significantly higher incidence in males [4,5]. Finally, several reports suggest that the sex-specific fat distribution, which favors subcutaneous versus visceral depots in women, could be one of the factors contributing to the lower global metabolic risk observed in women [6,7].

However, it must be considered that so far male subjects have been favored in human and animal biomedical research, whereas women or nonhuman females have been underrepresented [8]. Thus, the current observations may mostly reflect a lack of knowledge of the sex dimorphism of NAFLD. The liver is a highly sexually dimorphic organ in the situation of normal health, with hundreds of genes being differentially expressed between the two sexes [9,10]. It is thus expected that not only the development but also the consequences of NAFLD might be sex-dimorphic.

In this study, peroxisome proliferator-activated receptor gamma null mice ($Pparg^{\Delta/\Delta}$) were used as a new model of NAFLD. PPARγ is a nuclear receptor required for adipocyte differentiation and maturation [11]. $Pparg^{\Delta/\Delta}$ mice were obtained as described by Nadra et al. (2010) [12]. As expected from the critical role of PPARγ in adipogenesis, $Pparg^{\Delta/\Delta}$ mice are totally deprived of adipose tissue [13] and spontaneously develop hepatosteatosis. Herein we show that, in this mouse model, hepatosteatosis evolves differently in males and females. Using a combination of transcriptomics, lipidomics and further in vivo experiments, we systematically characterize the liver phenotype of $Pparg^{\Delta/\Delta}$ mice in both sexes in order to gain insight into the molecular mechanisms underlying sex dimorphism in NAFLD. We also pay particular attention to the expression of sex-biased genes in $Pparg^{\Delta/\Delta}$ mice as well as distinct mouse models of hepatosteatosis, revealing an important perturbation of the sex dimorphism pattern of the liver upon steatosis.

2. Results

2.1. PPARγ-Null Mice Represent a New Model of NAFLD Exhibiting Sex Dimorphism

PPARγ-null mice, hereafter called $Pparg^{\Delta/\Delta}$ mice, are totally deprived of adipose tissue [14]. Due to the impossibility of storing lipids in adipose tissue, lipodystrophy typically triggers fat accumulation in the liver [15]. Accordingly, both male and female $Pparg^{\Delta/\Delta}$ mice showed a massive enlargement of the liver and developed hepatic steatosis, as demonstrated by the presence of numerous lipid droplets at 7 weeks. In contrast, at 20 weeks, the liver of $Pparg^{\Delta/\Delta}$ males exhibited fewer lipid droplets and lower triglyceride (TG) accumulation compared to $Pparg^{\Delta/\Delta}$ females (Figure 1A,B). FFAs increased similarly in both $Pparg^{\Delta/\Delta}$ males and females, while neither total hepatic cholesterol nor cholesterol esters were increased in $Pparg^{\Delta/\Delta}$ mice (Supplementary Figure S1). We compared this model with two other models of NAFLD: the obese and diabetic *ob/ob* mice and the lipodystrophic *Adipoq-Cre*$^{tg/+}$;*Pparg*$^{fl/fl}$ mice (hereafter called $PpargF^{\Delta/\Delta}$), in which *Pparg* is deleted in preadipocytes but is present in the rest of the body [16]. As expected, hematoxylin and eosin staining showed that both sexes had a high number of hepatic lipid droplets in these two other mouse models (Figure 1C), but with no apparent dimorphism. Consistently, neither *ob/ob* mice nor $PpargF^{\Delta/\Delta}$ mice showed sex dimorphism in hepatic TG levels (Figure 1D). Thus, the sex dimorphism of hepatosteatosis with higher TG storage in female vs. male is specific to the $Pparg^{\Delta/\Delta}$ mice.

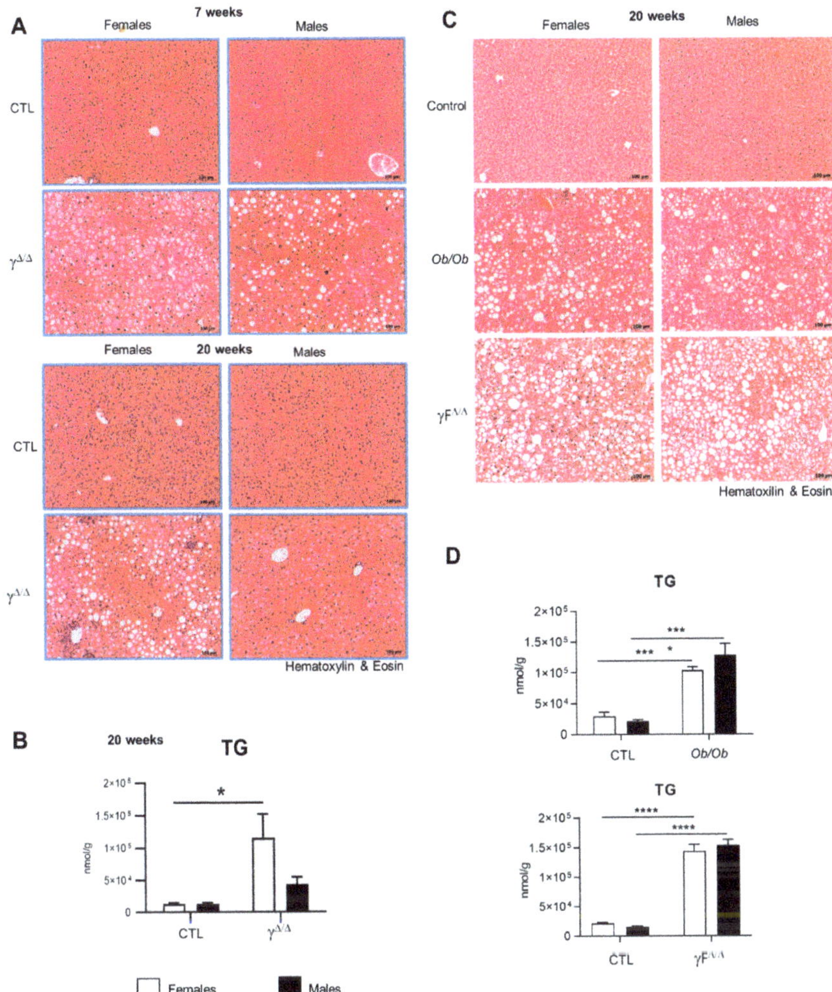

Figure 1. *Pparg*$^{\Delta/\Delta}$ mice are a new model of NAFLD exhibiting sex dimorphism. (**A**) Hematoxylin and eosin staining of liver sections of *Pparg*$^{\Delta/\Delta}$ mice and their control littermates at 7 weeks (up panels) and at 20 weeks (bottom panels). (**B**) Total hepatic TG measured in *Pparg*$^{\Delta/\Delta}$ mice and their control littermates at 20 weeks. n = 3–6. (**C**) Hematoxylin and eosin staining of liver sections of *ob/ob* mice, *Adipoq-Cre*$^{tg/+}$;*Pparg*$^{fl/fl}$ (*PpargF*$^{\Delta/\Delta}$) mice and control mice at 20 weeks. (**D**) Total hepatic TG measured in *ob/ob* mice, *PpargF*$^{\Delta/\Delta}$) at 20 weeks. n = 3–5. For (**A**,**C**), black bar corresponds to 100 μm. In (**B**,**D**), white bars are female and black bars are male data. All data were statistically treated by two-way ANOVA and Bonferroni multiple comparisons. *p* values: * <0.05, *** <0.001 and **** <0.0001.

To gain insights into the development of this sex-related phenotype, we fully characterized the lipid species accumulating in the livers of male and female *Pparg*$^{\Delta/\Delta}$ mice at 7 weeks, when hepatic lipid content is similar in males and females, and at 20 weeks, when the hepatic lipid content shows sexual dimorphism.

In control mice at 7 weeks, female livers had an overall slightly higher content of each TG species compared to male livers. At 20 weeks, most of these differences disappeared (Figure 2A and Supplementary Figure S2A,B). The same analyses in *Pparg*$^{\Delta/\Delta}$ mice at 7 weeks revealed a remarkable pattern with a higher amount of polyunsaturated long-chain TGs in *Pparg*$^{\Delta/\Delta}$ males but a higher amount of short-chain and more saturated TGs in

The profile of hepatic FFAs, from which TGs are synthesized, showed only few differences at 7 weeks between $Pparg^{\Delta/\Delta}$ males and females. In contrast, the FFA profile at 20 weeks reproduced the same pattern found in TGs (Figure 2C and Supplementary Figure S2C).

Similar analyses were performed in 20 weeks *ob/ob* mice. Unlike in $Pparg^{\Delta/\Delta}$ mice, short-hydrocarbon-chain TGs were more concentrated in *ob/ob* males compared to females, while FFA species did not show sex dimorphism (data not shown).

In summary, $Pparg^{\Delta/\Delta}$ males and females showed sexual dimorphism in the hepatic content of TG and FFA species. In addition, females exhibit more short and/or saturated hydrocarbon chain TGs and FFAs whereas males have more long and/or polyunsaturated TGs and FFAs.

$Pparg^{\Delta/\Delta}$ females, whereas no sex dimorphism was observed in that respect in control mice. The same phenotype was accentuated at 20 weeks, while $Pparg^{\Delta/\Delta}$ females had higher total TG content compared to males, suggesting that this pattern is independent of total hepatic TG content (Figure 2A,B).

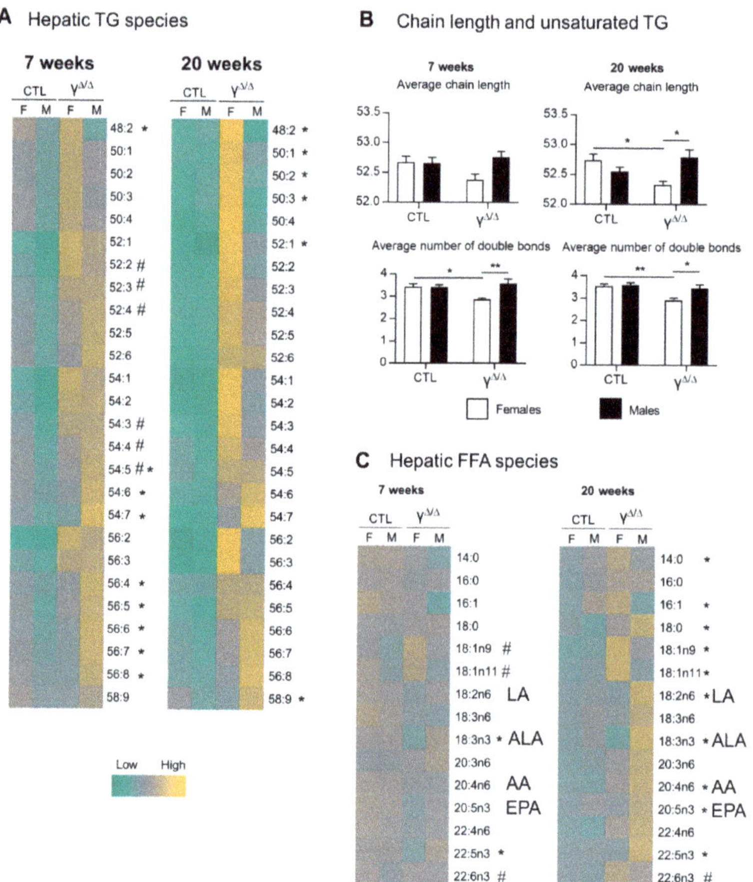

Figure 2. Hepatic triglyceride and FFA species in $Pparg^{\Delta/\Delta}$ mice. (**A**) Heat map showing the different TG species. For each line, corresponding to one lipid species, the absolute values are centered to 1, and relative changes between each group are expressed in log2 (log(V/mean;2)). # and * indicate statistically significant difference ($p < 0.05$) between females and males in control and $Pparg^{\Delta/\Delta}$ mice, respectively (Student's t-test). (**B**) TG chain length and TG unsaturated average at 20 weeks, n = 5. p values (* <0.05, ** <0.01) were calculated by two-way ANOVA and Bonferroni multiple comparisons. (**C**) Heat map showing the different FFA species. LA, linoleic acid (18:2n6); ALA, α-linolenic acid (18:3n3); AA, arachidonic acid (20:4n6); EPA, eicosapentaenoic acid (20:5n3); n = 5. # and * indicate statistically significant difference ($p < 0.05$) between females and males in control and $Pparg^{\Delta/\Delta}$ mice, respectively (Student's t-test).

The profile of hepatic FFAs, from which TGs are synthesized, showed only few differences at 7 weeks between $Pparg^{\Delta/\Delta}$ males and females. In contrast, the FFA profile at 20 weeks reproduced the same pattern found in TGs (Figure 2C and Supplementary Figure S2C).

Similar analyses were performed in 20 weeks *ob/ob* mice. Unlike in $Pparg^{\Delta/\Delta}$ mice, short-hydrocarbon-chain TGs were more concentrated in *ob/ob* males compared to females, while FFA species did not show sex dimorphism (data not shown).

In summary, $Pparg^{\Delta/\Delta}$ males and females showed sexual dimorphism in the hepatic content of TG and FFA species. In addition, females exhibit more short and/or saturated hydrocarbon chain TGs and FFAs whereas males have more long and/or polyunsaturated TGs and FFAs.

2.2. Distribution Pattern of Sex-Biased Genes in the Liver of CTL and $Pparg^{\Delta/\Delta}$ Mice

To define the signature of the steatotic liver in male and female $Pparg^{\Delta/\Delta}$ mice and the genes/mechanisms underlying the observed sex dimorphism of NAFLD, microarray analyses were performed at 20 weeks.

The hepatic expression of many genes is physiologically different between males and females. The genes more expressed in females compared to males and the opposite are referred to as "female-biased" or "male-biased" genes, respectively. Disruption of this natural dimorphism may lead to physiopathological disorders. We thus compared the sets of female-biased and male-biased genes in CTL and $Pparg^{\Delta/\Delta}$ mice. An intriguing pattern emerged, as most of the hepatic sex-biased genes in CTL mice lose their sex dimorphism in $Pparg^{\Delta/\Delta}$ mice (e.g., less than one-third of female-biased genes in CTL mice remain female-biased in $Pparg^{\Delta/\Delta}$ mice). Reciprocally, an important number of non-sex-dimorphic genes in CTL become sex-biased in $Pparg^{\Delta/\Delta}$ mice (Figure 3A).

In order to find the main biological pathways impacted by this particular pattern of sex-biased genes in $Pparg^{\Delta/\Delta}$ mice at 20 weeks, we performed gene ontology (GO) analysis, using the bioinformatics tool DAVID GO (https://david.ncifcrf.gov/summary.jsp, analyses performed from September 2015 to September 2016). We divided the overall set of sex-biased genes into four subsets, as shown in Figure 3A. Table 1 lists the main GO terms represented in each subset. Further analyses also take into account the specific genes that GO groups in these categories, as listed in Supplementary Table S2.

Table 1. Gene ontology analysis of hepatic sex-dimorphic genes in CTL and $Pparg^{\Delta/\Delta}$ mice at 20 weeks.

Group	Gene GO Term	No. of Genes	*p* Value
I (208 genes) Female-biased genes in CTL mice	Microsome	16	2.24×10^{-9}
	Drug Metabolism	11	1.91×10^{-8}
	Cytochrome P450	9	2.58×10^{-6}
	Arachidonic acid metabolism	7	5.16×10^{-4}
	Endoplasmic reticulum	19	6.23×10^{-4}
	Linoleic acid metabolism	4	0.01
II (145 genes) Male-biased genes in CTL mice	ncRNA metabolic process	7	3.22×10^{-3}
	Tubulin	4	3.1×10^{-4}
	Steroid dehydrogenase activity	4	6.18×10^{-4}
	Metal ion binding	38	0.04
	Nucleolus	8	5.78×10^{-3}
III (292 genes) Female-biased genes in $\gamma^{\Delta/\Delta}$ mice	Cell surface	24	1.02×10^{-9}
	Immune response	11	4.06×10^{-9}
	Immunoglobulin domain	28	4.59×10^{-9}
	Defense response	26	1.37×10^{-8}
	Neutrophil-mediated immunity	3	4.43×10^{-3}
	Fatty acid biosynthesis	5	5.2×10^{-3}
IV (115 genes) Male-biased genes in in $\gamma^{\Delta/\Delta}$ mice	Cell fraction	10	9.87×10^{-3}
	Cell cycle process	8	5.46×10^{-3}
	Cellular response to stress	9	1.54×10^{-3}
	Oxidation reduction	10	0.01
	Mitochondrion	10	0.02

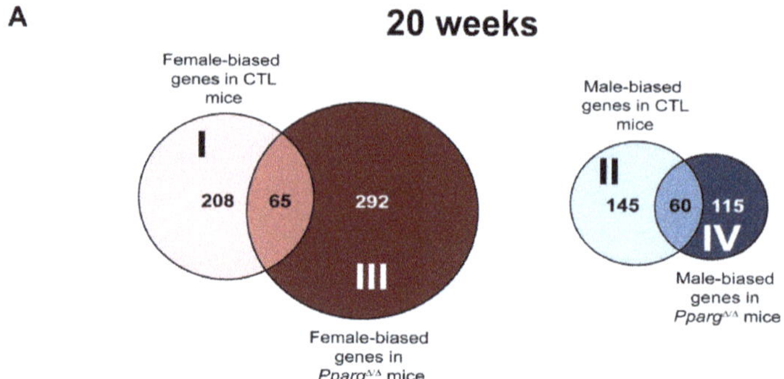

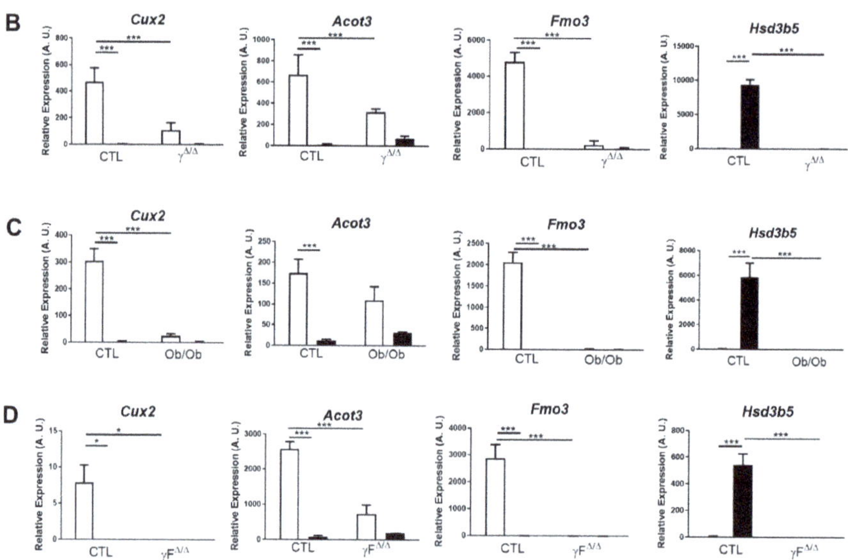

Figure 3. Perturbation of hepatic sex-biased gene expression by NAFLD. (**A**) Distribution of hepatic sex-biased genes measured by microarray analysis in $Pparg^{\Delta/\Delta}$ mice at 20 weeks. Global computation of *p*-value adjustment was performed for the two comparisons, with adjusted *p* value < 0.05 and no cutoff with fold change. "I" represents the group of genes that are female-biased only in CTL mice. "II" represents the group of genes that are male-biased only in CTL mice. "III" represents the group of genes that are not sex-biased in CTL but become female-biased in $Pparg^{\Delta/\Delta}$ mice. "IV" represents the group of genes that are not sex-biased in CTL but become male-biased in $Pparg^{\Delta/\Delta}$ mice. (**B**) Hepatic expression profile of *Cux2, Acot3, Fmo3* and *Hsd3b5* was confirmed by RT-qPCR in $Pparg^{\Delta/\Delta}$ mice at 20 weeks (n = 4–7) and was measured in (**C**) *Ob/Ob* mice (n = 5) and in (**D**) $Adipoq\text{-}Cre^{tg/+};Pparg^{fl/fl}$ ($PpargF^{\Delta/\Delta}$) (n = 3). White bars are female and black bars are male data. *p* values (* <0.05, *** <0.001) were calculated by two-way ANOVA and Bonferroni multiple comparisons.

2.3. Perturbation of the Physiological Sex-Biased Gene Expression by NAFLD

Subsets I and II include genes that are physiologically gender-biased in WT but lose their sex dimorphism in $Pparg^{\Delta/\Delta}$ mice. In particular, *subset I corresponds* to female-biased genes in WT but not in $Pparg^{\Delta/\Delta}$ mice. In this group, we found *Cux2*, which is a highly female-specific liver transcription factor, involved in male-biased gene repression and female-biased gene induction [17]. *Cux2* expression was reduced by more than 65% in

Pparg$^{\Delta/\Delta}$ females compared to control females at 20 weeks (Figure 3B). In addition, there is a major representation of genes involved in drug metabolism such as the cytochrome P450 family including *Cyp4a10*, the flavin-containing monooxygenases (*Fmo1*, *2*, *3* and *4*) and the glutathione S transferase. The same genes are also found under the GO terms arachidonic acid and linoleic acid metabolism. This strong sex dimorphism in drug metabolism-related genes in the healthy liver is known (reviewed by DJ Waxman and MG Holloway [9]), while its disruption in *Pparg*$^{\Delta/\Delta}$ mice raises questions about possible alteration of the normal drug metabolism.

Subset II corresponds to male-biased genes in CTL but not in *Pparg*$^{\Delta/\Delta}$ mice. One major observation within this subset concerns the steroid dehydrogenase activity that includes genes of the *Hsd3b* family (*Hsd3b2, Hsd3b4, Hsd3b5*). These genes are important for the biosynthesis of active steroid hormones. In particular, *Hsd3b5* is highly expressed in the liver in a male-specific manner [18] and is dramatically downregulated in *Pparg*$^{\Delta/\Delta}$ males, reaching the very low levels observed in females (Figure 3B).

Interestingly, the sex-biased expression profile of a panel of these sex-biased genes, including *Cux2*, *Acot3*, *Fmo3* and *Hsd3b5*, was similarly dampened in the two other models of NAFLD previously used in this study, namely the *ob/ob* and *PpargF*$^{\Delta/\Delta}$ mice (Figure 3C,D). These results seem against a possible involvement of these gene sets in the development of the sex-dimorphic lipid accumulation observed in *Pparg*$^{\Delta/\Delta}$ mice, while they suggest that NAFLD, rather than PPARγ, has an impact on the physiological hepatic gender dimorphism of gene expression. Given the involvement of these genes in drug metabolism, our observations raise questions about the possible consequences of NAFLD on pharmacological responses, which would deserve further studies.

2.4. Modulation of Pathways Involved in Lipid Droplet Formation, Storage and Secretion in *Pparg*$^{\Delta/\Delta}$ Mice

The two remaining subsets highlighted by microarray analysis (III and IV) include genes that are not physiologically sex-biased but acquired a sex-dimorphic gene expression in *Pparg*$^{\Delta/\Delta}$ mice.

Subset III corresponds to female-biased genes in *Pparg*$^{\Delta/\Delta}$ but not in CTL mice. Three main domains, immune response, cell activation and lipid metabolism, are associated with this subset, which is principally composed of genes involved in immune responses and cell activation and genes involved in mono- and polyunsaturated fatty acids (*Scd2, Fads1* and *Fads2*), as well as including genes involved in arachidonic acid metabolism (*Tbxas1* and *Hgpds*). This subset is likely to play an important role in the sex dimorphism of the fatty liver in *Pparg*$^{\Delta/\Delta}$ mice.

Subset IV corresponds to male-biased genes in *Pparg*$^{\Delta/\Delta}$ but not in CTL mice. The GO categories in this subset regroup cellular damages at the level of membranes, but also at the DNA level, converging towards the P53 pathway, as represented by *Aen* (apoptosis-enhancing nuclease) and *Jmy* (junction-mediating and regulatory protein) genes. This subset also includes genes involved in oxidation and mitochondrial functions (Supplementary Table S2).

Given the higher steatosis development in *Pparg*$^{\Delta/\Delta}$ female mice, we further explored the genes and/or pathways particularly highlighted in subset III. More particularly, the expression of genes involved in de novo lipogenesis and in lipid droplet formation, such as *G0s2, Plin2, Gpam, Scd1, Crat, G6pdx, Acacb* and *Elovl5*, were all upregulated only in *Pparg*$^{\Delta/\Delta}$ females and/or became female-biased in *Pparg*$^{\Delta/\Delta}$ mice at 20 weeks. This increased expression was validated by qRT-PCR, as shown in Figure 4A. In contrast, aldolase C fructose-bisphosphate (*Aldoc*) was only downregulated in *Pparg*$^{\Delta/\Delta}$ females. This could favor the use of glycogen and/or glucose to feed the pentose phosphate pathway. The adipose triglyceride lipase ATGL (*pnpla2*), which is involved in intracellular degradation of TGs, showed female-biased expression in control mice, but not in *Pparg*$^{\Delta/\Delta}$ mice. Finally, the hypoxia-inducible lipid droplet-associated gene (*Hilpda*), which inhibits hepatic triglyceride secretion [19], was upregulated only in *Pparg*$^{\Delta/\Delta}$ females (Figure 4B). The profile of the two latter genes is contributing, at least in part, to the higher hepatic TG content in females compared to males.

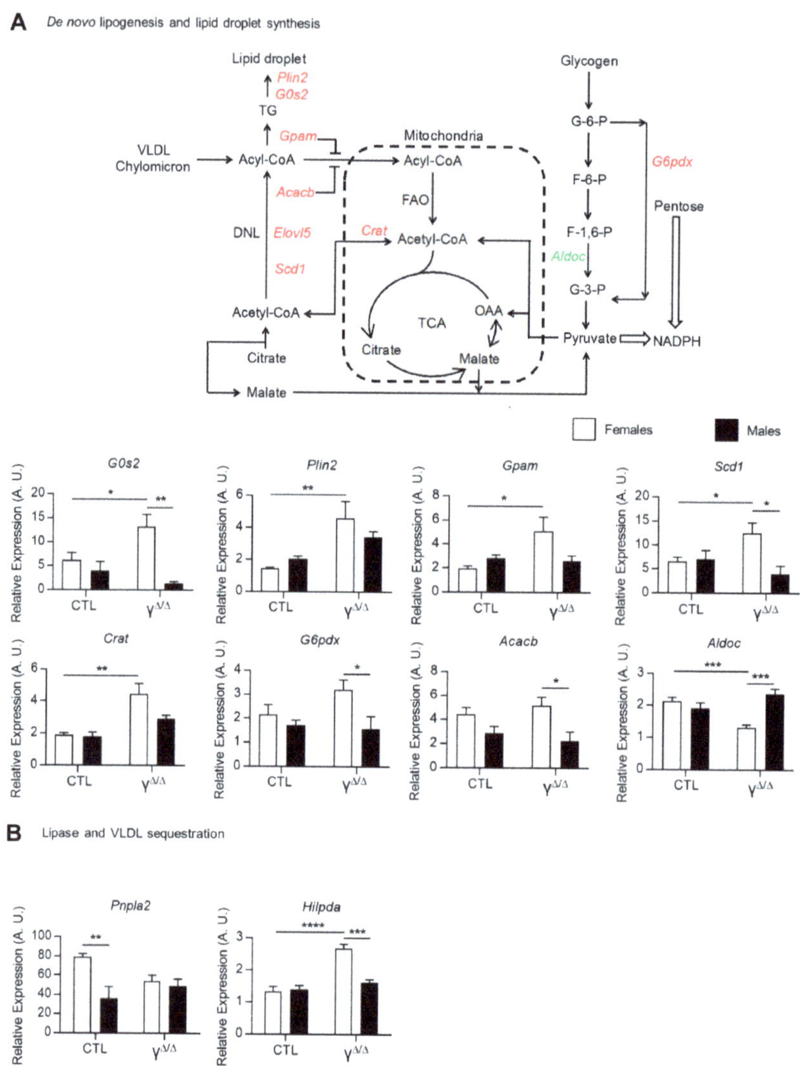

Figure 4. Pathways involved in lipid droplet formation, storage and secretion. (**A**) Representative scheme of the biochemical pathways of de novo lipogenesis and lipid droplet synthesis; upregulated and downregulated genes in *Pparg*$^{\Delta/\Delta}$ females, but not in *Pparg*$^{\Delta/\Delta}$ males, are in red and green, respectively. DNL, de novo lipogenesis; TCA, citric acid cycle; OAA, oxaloacetate; VLDL, very-low-density lipoprotein; G-6-P, glucose 6-phosphate; F-6-P, fructose 6-phosphate; F-1,6-P, fructose 1,6-bisphosphate; G-3-P, glyceraldehyde 3-phosphate. Gene expression of perilipin 2 (*Plin2*); G0/G1 switch gene 2 (*G0s2*); mitochondrial glycerol-3-phosphate acyltransferase (*Gpam*); acetyl-CoA carboxylase beta (*Acacb*); stearoyl-Coenzyme A desaturase 1 (*Scd1*); ELOVL family member 5, elongation of long-chain fatty acids (*Elovl5*); carnitine acetyltransferase (*Crat*); the glucose-6-phosphate dehydrogenase X-linked (*G6pdx*); and aldolase C, fructose-bisphosphate (*Aldoc*). (**B**) Adipose triglyceride lipase (ATGL or *Pnpla2*) and hypoxia-inducible lipid droplet-associated (*Hilpda*) gene expression measured by RT-qPCR at 20 weeks. n = 3–9. White bars are female and black bars are male data. *p* values (* <0.05, ** <0.01, *** <0.001 and **** <0.0001) were calculated by two-way ANOVA and Bonferroni multiple comparisons.

Altogether, gene expression analysis showed that dysregulations of all aspects of lipid synthesis, storage and secretion concur in determining higher hepatic TG levels in *Pparg*$^{\Delta/\Delta}$ females vs. males.

2.5. Analyses of Lipids and Lipid Pathways Involved in Cell Signaling and Inflammation

Subset III of genes (sex-biased in female *Pparg*$^{\Delta/\Delta}$ mice but not in female CTL mice) also highlighted genes involved in immune response and cell activation. Using the combination of lipidomics and transcriptomics, we thus further analyzed the lipids involved in cell signaling and inflammation. Ceramides, which are found in high concentrations in cell membranes, participate in a variety of cellular signaling pathways in differentiation, proliferation and cell death [20]. At 7 weeks, total ceramide content in the liver of control mice was significantly higher in males compared to females, a sex dimorphism that was attenuated in *Pparg*$^{\Delta/\Delta}$ mice. The same pattern was seen at 20 weeks, in both control and *Pparg*$^{\Delta/\Delta}$ mice, although the high variability precludes statistical differences for most species (Supplementary Figure S3A).

Among the FFA species, we more specifically analyzed omega-3 and omega-6 fatty acids because of their correlation with obesity and the progression to steatohepatitis [21]. At 20 weeks, the concentrations of omega-3 and omega-6 and the omega-6/omega-3 ratio were higher in *Pparg*$^{\Delta/\Delta}$ males compared to *Pparg*$^{\Delta/\Delta}$ females, suggesting that males could be more prone to progress to steatohepatitis (Supplementary Figure S3B).

Eicosanoids represent a third class of lipids with important roles in inflammation. As we indeed see in Figure 2C, the hepatic levels of two essential FFAs, linoleic acid (LA, 18:2n-6) and α-linolenic acid (ALA, 18:3n3), are lower in *Pparg*$^{\Delta/\Delta}$ females compared to *Pparg*$^{\Delta/\Delta}$ males. LA, the major vegetal dietary n-6 PUFA and precursor of arachidonic acid (AA), is considered as a proinflammatory compound, whereas ALA can be metabolized into anti-inflammatory molecules such as eicosapentaenoic acid (EPA, 20:5n3). We thus explored the expression of genes involved in LA and ALA metabolism. mRNA levels of delta-5-desaturase (*Fads1*), delta-6-desaturase (*Fads2*) and elongase-5 (*Elovl5*), which are involved in AA and EPA formation from LA and ALA, respectively, were upregulated only in *Pparg*$^{\Delta/\Delta}$ females (Figure 5A). Accordingly, the ratios AA/LA and EPA/ALA, which reflect the total activity of these enzymes [22], were higher in *Pparg*$^{\Delta/\Delta}$ females compared to *Pparg*$^{\Delta/\Delta}$ males (Figure 5B). Nonetheless, EPA and AA, which are the final products of these enzymes, showed a lower hepatic content in *Pparg*$^{\Delta/\Delta}$ females compared to *Pparg*$^{\Delta/\Delta}$ males (Figure 5B). In parallel, we found that 5-lipoxygenase (*Alox5*), 5-lipoxygenase activating protein (*Alox5ap*) and leukotriene-A4-hydrolase (*Lta4h*), are upregulated in *Pparg*$^{\Delta/\Delta}$ females (Figure 5C), suggesting a possible increased conversion of AA into eicosanoids, known to play an important role in the onset and progression of inflammation in the liver [23,24]. We thus measured the full set of eicosanoids and other FFA derivatives in the livers of control and *Pparg*$^{\Delta/\Delta}$ mice at 20 weeks. There were no significant differences in the levels of AA derivatives taken individually, as observed for leukotriene B4 (LTB4) (Figure 5D). However, the sum of AA derivatives synthesized through the Cox pathway, including prostaglandins, was higher in *Pparg*$^{\Delta/\Delta}$ females compared to control females. This was mainly due to the increased levels of the most abundant hepatic prostaglandin, PGF2a. In contrast, no differences in the Lox pathway were found comparing *Pparg*$^{\Delta/\Delta}$ to control males (Supplementary Figure S3C).

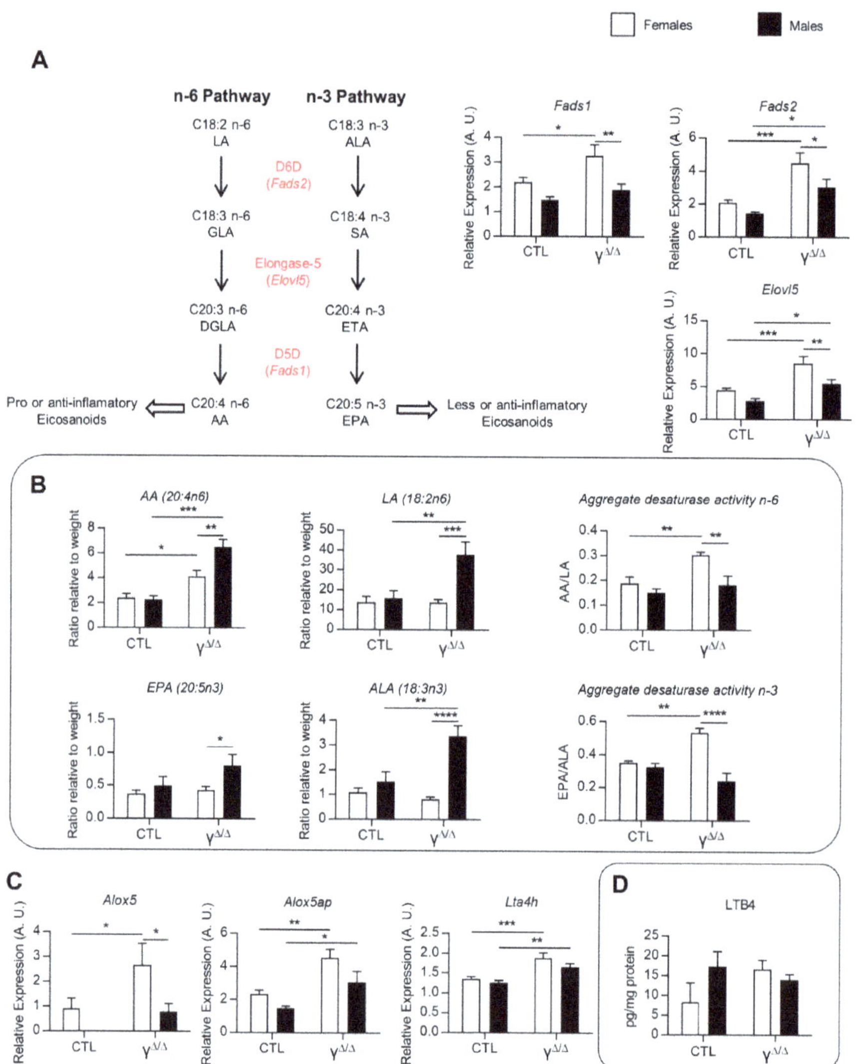

Figure 5. Linoleic acid (LA) and arachidonic acid (AA) metabolism and eicosanoids. (**A**) Left graphs: Pathways with main genes involved in essential FFA transformation and their hepatic gene expression. ALA, α-linolenic acid (18:3n3); SA, stearidonic acid; ETA, eicosatetraenoic acid; EPA, eicosapentaenoic acid (20:5n3); LA, linoleic acid (18:2n6); GLA, γ-linolenic acid; DGLA, dihomo-γ-linolenic acid; AA, arachidonic acid (20:4n6). D5D (Δ5-desaturase), D6D (Δ6-desaturase) and elongase-5 enzymes are encoded by *Fads1, Fads2* and *Elovl5*, respectively. Right graphs: Relative expression of the three genes corresponding to these enzymes, n = 3–9. (**B**) Aggregate desaturase activity, calculated as the AA/LA ratio and the EPA/ALA ratio, reflects the FADS1 and FADS2 activity [22], n = 3–5. (**C**) Hepatic expression of genes involved in leukotriene B4 (LTB4) synthesis from AA, n = 9. ND: not detectable. *Alox5* was almost not detectable by RT-qPCR. 5-Lipoxygenase-associated protein (*Alox5ap*) and leukotriene A4 hydrolase (*Lta4h*). (**D**) Hepatic LTB4, n = 5–9. White bars are female and black bars are male data. *p* values (* <0.05, ** <0.01, *** <0.001 and **** <0.0001) were calculated by two-way ANOVA and Bonferroni multiple comparisons.

Altogether, ceramides and the omega-6/omega-3 fatty acid ratio suggested that males may be more prone to steatohepatitis. However, females exhibited a higher activity of the Cox pathway. This might explain at least in part the lower AA level in the liver of *Pparg*$^{\Delta/\Delta}$ females compared to *Pparg*$^{\Delta/\Delta}$ males, via a higher transformation of AA into eicosanoid derivatives.

Given the importance of lipids and some derivatives in the modulation of the inflammatory response, we thus explored whether the observed changes influence the progression of *Pparg*$^{\Delta/\Delta}$ hepatosteatosis to the more severe states. As shown in Supplementary Figure S4, plasmatic levels of aspartate (ASAT) and alanine aminotransferases (ALAT), both markers of liver damage, as well as the expression levels of proinflammatory genes, were increased, although modestly. *Elane* and osteopontin (*Spp1*), which are linked to neutrophil infiltration, were more particularly increased.

In humans, a major complication of nonalcoholic steatohepatitis (NASH), following NAFLD, is liver fibrosis. We thus challenged the mice with a profibrotic diet for 6 weeks. In control mice, the diet induced hepatosteatosis and collagen deposition in both sexes and an upregulation of fibrotic markers such as *Acta2, Col1a1, Mmp13* and *Timp1*, with no statistical differences between males and females (Supplementary Figure S5). In *Pparg*$^{\Delta/\Delta}$ mice, collagen deposition and fibrotic markers were already increased under chow diet. The profibrotic diet triggered a further modest upregulation of fibrotic markers, but without sex dimorphism in collagen deposition. Thus, a profibrotic diet provoked further signs of moderate hepatic inflammation and fibrosis in both male and female *Pparg*$^{\Delta/\Delta}$ mice, with no detectable sex dimorphism. The fibrotic phenotype induced by the profibrotic diet in *Pparg*$^{\Delta/\Delta}$ mice was comparable to that obtained with the same diet in CTL mice.

2.6. Relationship between Hormonal Status and Sex Dimorphism of Hepatic Lipid Accumulation

Sex hormones are important regulators of hepatic lipid metabolism [25]. We thus explored the hormonal status of *Pparg*$^{\Delta/\Delta}$ mice by looking at the plasmatic levels of the various steroid hormones. Intriguingly, testosterone and androstenedione are significantly reduced in *Pparg*$^{\Delta/\Delta}$ male mice compared to control mice, while progesterone, deoxycorticosterone and corticosterone are increased (Figure 6A), indicating some perturbation of the sex-hormone homeostasis.

To determine the role of sex steroid hormones in *Pparg*$^{\Delta/\Delta}$ sex dimorphism, gonadectomy was performed between the ages of 4 and 6 weeks, prior to sexual maturation, and the resulting phenotype was analyzed at 20 weeks. The effectiveness of both castration and ovariectomy was demonstrated by the profound decrease in testosterone, progesterone and androstenedione in males and by the lack of estrogen cycle in female mice (Supplementary Figure S6). Importantly, hepatic TG content at 20 weeks was no longer dimorphic in gonadectomized mice, due to an increase in the lipid load in males and a decrease in the lipid load in females (Figure 6B,C). This confirmed the role of sex hormones in liver lipid accumulation and suggests a potential cross-talk between sex hormones and PPARγ in the onset of the sex-dependent hepatic lipid accumulation observed in *Pparg*$^{\Delta/\Delta}$.

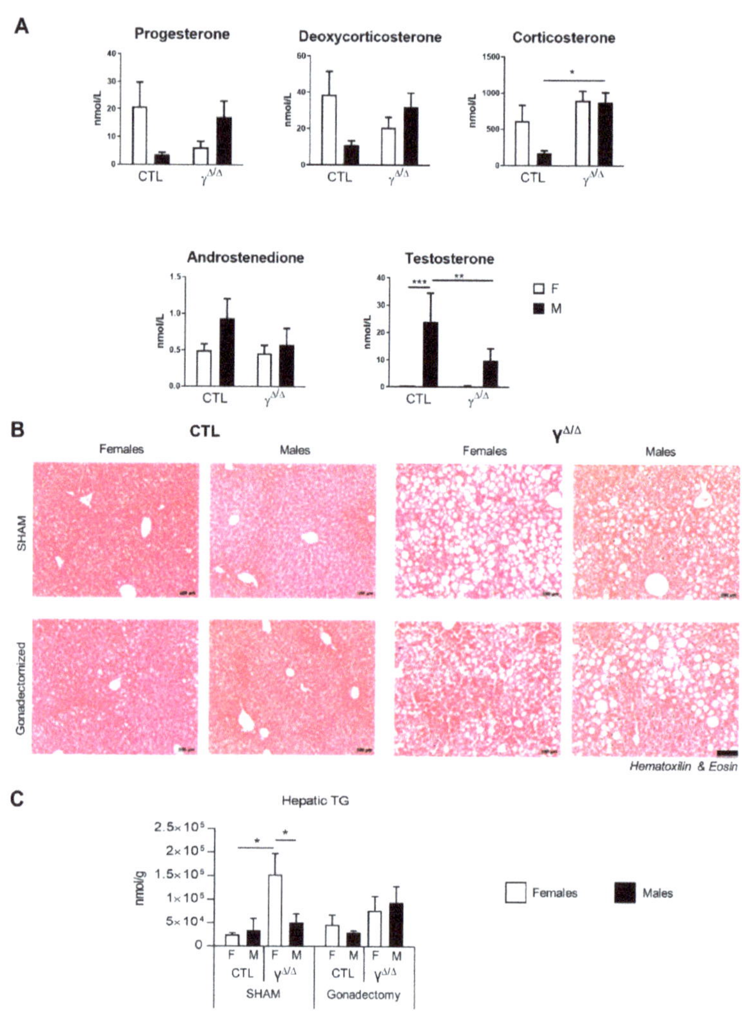

Figure 6. Role of sex hormones in the hepatosteatotic phenotype in *Pparg*$^{\Delta/\Delta}$ mice. (**A**) Plasmatic steroid hormones measured by mass spectrometry. n = 6–9. Progesterone and deoxycorticosterone show a *p* value lower than 0.05 for the 2-way ANOVA interaction. (**B**) H&E staining of liver sections and (**C**) hepatic TG content measured in ovariectomized and castrated mice. n = 4–9. In (**B**), black bar corresponds to 100 µm. In (**A**,**C**), white bars are female and black bars are male data. *p* values (* <0.05, ** <0.01 and *** <0.001) were calculated by two-way ANOVA and Bonferroni multiple comparisons.

3. Discussion

Our study highlights a progressive sex dimorphism of NAFLD in a new lipodystrophic mouse model. NAFLD development becomes sex-dimorphic at 20 weeks, with *Pparg*$^{\Delta/\Delta}$ females showing high levels of hepatic lipid droplets and triglycerides while *Pparg*$^{\Delta/\Delta}$ males present limited hepatosteatosis. Such sex-dimorphic phenotype is observed only in the presence of sex hormones in *Pparg*$^{\Delta/\Delta}$ mice, suggesting a cross-regulation between PPARγ and sex hormones in liver lipid metabolism. Whereas the female-biased liver steatosis was not reproduced in other mouse models of NAFLD, sex-dependent deregulation in the liver of sex-biased genes occurs in all the hepatosteatosis models that were tested.

Limitations of this study mainly stand along the fact that the liver sex dimorphism we herein characterized seems specific to one mouse model. However, it still provides some means to address the complexity of this quite uncharted biological phenomenon, and it gives even more importance to the features shared by all three NAFLD mouse models we have tested.

Lipidomics data in mouse liver addressing the differences between males and females are still quite limited. A recent report showed a transient difference in the saturation index of fatty acids in the livers of wild-type C57BL/6 males and females [26]. In our study, control females show higher abundance of several medium-chain TG species compared to males, and these differences disappeared at 20 weeks. Most interestingly, this pattern is different in *Pparg*$^{\Delta/\Delta}$ mice, with higher levels of long-chain TG species in males at 7 weeks, while, at 20 weeks, the higher hepatic TG content in *Pparg*$^{\Delta/\Delta}$ females mainly relies on TGs with short-and/or highly saturated hydrocarbon-chains. Variations in chain length and saturation of hepatic TGs were also observed when comparing fasting and high-fat diet (HFD) conditions in C57BL/6 male mice [27], and such differences were associated with the different energy status of these conditions. Indeed, the high rate of mitochondrial β-oxidation upon fasting could explain the reduction in fasted livers of short fatty acids, which can be directly oxidized. Along this line, the profile of hepatic TG species in *Pparg*$^{\Delta/\Delta}$ males is similar to that observed in fasted mice, whereas the TG pattern of *Pparg*$^{\Delta/\Delta}$ females is closer to that of HFD-fed mice. However, hepatic gene expression of markers of β-oxidation upon fasting does not reveal sex dimorphism in *Pparg*$^{\Delta/\Delta}$ (data not shown). Nevertheless, *Pparg*$^{\Delta/\Delta}$ female mice seem more prone to store efficiently the hepatic lipid overload compared to males, as shown by the female-specific overexpression of genes involved in lipid storage.

Sex effects were reported also for the production of fatty acid derivatives, which are of particular interest because of their ability to modulate inflammation, but studies mainly focused on the kidney and on one or two eicosanoids [28,29]. However, a recent report comprehensively characterized oxylipins in male and female rat livers and found sex effects in the abundance of 40% of them, with most of them higher in males [30]. At 20 weeks, we found a similar sex-dependent trend in control mice and an increase in the sum of AA derivatives synthesized through the Cox pathway in *Pparg*$^{\Delta/\Delta}$ females, whereas no alterations were found in *Pparg*$^{\Delta/\Delta}$ males.

Sex steroid hormones influence hepatic lipid metabolism through the activation of sex hormone receptors [25]. The differences between males and females in the circulating levels of sex hormones were slightly dampened in *Pparg*$^{\Delta/\Delta}$ mice. Sex-hormone activity has a direct effect on hepatic lipid accumulation. Estrogens decrease liver cholesterol and triglyceride concentrations only in females [31], while tamoxifen, a potent estrogen receptor antagonist, causes severe steatosis progressing towards NASH [32]. Along this line, male but not female mice with aromatase gene deletion develop hepatic steatosis that can be rescued by estrogen treatment [33]. On the one hand, these reports suggest that estrogen receptor signaling is negatively correlated with level of hepatosteatosis in both sexes. On the other hand, androgen receptor (AR) signaling seems protective against hepatosteatosis in a sex-dimorphic manner. Male but not female mice lacking AR in the liver develop hepatosteatosis and insulin resistance upon HFD [34]. Interestingly, in *Pparg*$^{\Delta/\Delta}$ mice, the impairment of sex-hormone activity through gonadectomy does not worsen hepatosteatosis in females, while it increases hepatic lipid droplet accumulation in males and suppresses sex-related differences in hepatosteatosis. These observations suggest that the hepatic phenotype of *Pparg*$^{\Delta/\Delta}$ mice depends on the sex-hormone activity and highlight a cross-regulation between PPARγ and sex hormones.

An important feature of the liver is the strong sex dimorphism affecting gene expression [10]. The growth hormone (GH) secretory patterns, highly pulsatile in males and more continuous in females, determine the hepatic sex-biased expression of a high number of genes [35]. Interestingly, patients with NAFLD have low GH production and/or hepatic GH resistance [36,37]. In mice, GH inhibits de novo lipogenesis through inhibition

4.8. Statistical Analysis

Values, expressed as mean ± SEM, were analyzed using Prism 5.0 (GraphPad Software, San Diego, CA, USA). Unless mentioned, two-way ANOVA and Bonferroni post-test for multiple group comparisons were used to assess statistical significance. p values: * <0.05, ** <0.01, *** <0.001 and **** <0.0001.

Supplementary Materials: Supplementary methods, figures and tables are available online at https://www.mdpi.com/article/10.3390/ijms22189969/s1.

Author Contributions: F.G. and B.D. conceived and supervised the study; M.S., F.G., C.W. and L.Q. performed the experiments; C.W. bred and genotyped the mice; M.S., F.G. and B.D. designed the experiments; A.N. performed bioinformatic analysis; H.K. and M.T. performed lipidomics; H.H. performed steroid hormone measurement; P.P. performed plasmatic measurements; F.G., M.S. and B.D. wrote the manuscript. All authors have read and agreed to the published version of the manuscript.

Funding: This work was supported by the Etat de Vaud, the SNSF (31003A_135583 and 310030_156771) and a grant from the Faculty of Biology and Medicine of University of Lausanne (8466) to B.D.

Institutional Review Board Statement: All animal experiments and procedures were approved by the Swiss Veterinary Office (VD-1453.4, VD-2560 and VD-2887).

Informed Consent Statement: Not applicable.

Data Availability Statement: Microarray data were deposited in GEO and are accessible at the following link: https://www.ncbi.nlm.nih.gov/geo/query/acc.cgi?acc=GSE176226 (from 13 September 2021).

Acknowledgments: We are indebted to Darius Moradpour, Markus Heim, Luigi Terracciano and Frederic Preitner for valuable discussion, and we thank Catherine Moret for the help in histological analyses. The authors also thank the Genome Technologies Facility (Center for Integrative Genomics, Lausanne), where the microarrays were performed.

Conflicts of Interest: The authors declare no conflict of interest.

References

1. Harrison, S.A.; Torgerson, S.; Hayashi, P.H. The natural history of nonalcoholic fatty liver disease: A clinical histopathological study. *Am. J. Gastroenterol.* **2003**, *98*, 2042–2047. [CrossRef] [PubMed]
2. Day, C.P. Natural History of NAFLD: Remarkably Benign in the Absence of Cirrhosis. *Gastroenterology* **2005**, *129*, 375–378. [CrossRef]
3. Farrell, G.C.; Larter, C.Z. Nonalcoholic fatty liver disease: From steatosis to cirrhosis. *Hepatology* **2006**, *43*, S99–S112. [CrossRef]
4. Parkin, D.M.; Bray, F.; Ferlay, J.; Pisani, P. Global cancer statistics, 2002. *CA Cancer J. Clin.* **2005**, *55*, 74–108. [CrossRef]
5. Li, Z.; Tuteja, G.; Schug, J.; Kaestner, K.H. Foxa1 and Foxa2 Are Essential for Sexual Dimorphism in Liver Cancer. *Cell* **2012**, *148*, 72–83. [CrossRef]
6. Matsuzawa, Y.; Shimomura, I.; Nakamura, T.; Keno, Y.; Kotani, K.; Tokunaga, K. Pathophysiology and pathogenesis of visceral fat obesity. *Obes Res.* **1995**, *3* (Suppl. 2), 187S–194S. [CrossRef] [PubMed]
7. White, U.A.; Tchoukalova, Y.D. Sex dimorphism and depot differences in adipose tissue function. *Biochim. Biophys. Acta Mol. Basis Dis.* **2014**, *1842*, 377–392. [CrossRef] [PubMed]
8. Beery, A.; Zucker, I. Sex bias in neuroscience and biomedical research. *Neurosci. Biobehav. Rev.* **2011**, *35*, 565–572. [CrossRef] [PubMed]
9. Waxman, D.J.; Holloway, M.G. Sex Differences in the Expression of Hepatic Drug Metabolizing Enzymes. *Mol. Pharmacol.* **2009**, *76*, 215–228. [CrossRef] [PubMed]
10. Mode, A.; Gustafsson, J.A. Sex and the liver—A journey through five decades. *Drug Metab. Rev.* **2006**, *38*, 197–207. [CrossRef]
11. Rosen, E.D.; Sarraf, P.; Troy, A.E.; Bradwin, G.; Moore, K.; Milstone, D.S.; Spiegelman, B.M.; Mortensen, R.M. PPAR gamma is required for the differentiation of adipose tissue in vivo and in vitro. *Mol. Cell* **1999**, *4*, 611–617. [CrossRef]
12. Nadra, K.; Quignodon, L.; Sardella, C.; Joye, E.; Mucciolo, A.; Chrast, R.; Desvergne, B. PPARgamma in placental angiogenesis. *Endocrinology* **2010**, *151*, 4969–4981. [CrossRef] [PubMed]
13. Gilardi, F.; Winkler, C.; Quignodon, L.; Diserens, J.G.; Toffoli, B.; Schiffrin, M.; Sardella, C.; Preitner, F.; Desvergne, B. Systemic PPARgamma deletion in mice provokes lipoatrophy, organomegaly, severe type 2 diabetes and metabolic inflexibility. *Metabolism* **2019**, *95*, 8–20. [CrossRef]

14. Sardella, C.; Winkler, C.; Quignodon, L.; Hardman, J.A.; Toffoli, B.; Attianese, G.M.P.G.; Hundt, J.E.; Michalik, L.; Vinson, C.R.; Paus, R.; et al. Delayed Hair Follicle Morphogenesis and Hair Follicle Dystrophy in a Lipoatrophy Mouse Model of Pparg Total Deletion. *J. Investig. Dermatol.* **2018**, *138*, 500–510. [CrossRef] [PubMed]
15. Moitra, J.; Mason, M.M.; Olive, M.; Krylov, D.; Gavrilova, O.; Marcus-Samuels, B.; Feigenbaum, L.; Lee, E.; Aoyama, T.; Eckhaus, M.; et al. Life without white fat: A transgenic mouse. *Genes Dev.* **1998**, *12*, 3168–3181. [CrossRef] [PubMed]
16. Wang, F.; Mullican, S.E.; DiSpirito, J.R.; Peed, L.C.; Lazar, M.A. Lipoatrophy and severe metabolic disturbance in mice with fat-specific deletion of PPARgamma. *Proc. Natl. Acad. Sci. USA* **2013**, *110*, 18656–18661. [CrossRef]
17. Conforto, T.L.; Zhang, Y.; Sherman, J.; Waxman, D.J. Impact of CUX2 on the Female Mouse Liver Transcriptome: Activation of Female-Biased Genes and Repression of Male-Biased Genes. *Mol. Cell. Biol.* **2012**, *32*, 4611–4627. [CrossRef]
18. Abbaszade, I.G.; Clarke, T.R.; Park, C.H.; Payne, A.H. The mouse 3 beta-hydroxysteroid dehydrogenase multigene family includes two functionally distinct groups of proteins. *Mol. Endocrinol.* **1995**, *9*, 1214–1222. [CrossRef] [PubMed]
19. Mattijssen, F.; Georgiadi, A.; Andasarie, T.; Szalowska, E.; Zota, A.; Krones-Herzig, A.; Heier, C.; Ratman, D.; De Bosscher, K.; Qi, L.; et al. Hypoxia-inducible Lipid Droplet-associated (HILPDA) Is a Novel Peroxisome Proliferator-activated Receptor (PPAR) Target Involved in Hepatic Triglyceride Secretion. *J. Biol. Chem.* **2014**, *289*, 19279–19293. [CrossRef]
20. Pagadala, M.; Kasumov, T.; McCullough, A.J.; Zein, N.N.; Kirwan, J.P. Role of ceramides in nonalcoholic fatty liver disease. *Trends Endocrinol. Metab.* **2012**, *23*, 365–371. [CrossRef]
21. Araya, J.; Rodrigo, R.; Videla, L.A.; Thielemann, L.; Orellana, M.; Pettinelli, P.; Poniachik, J. Increase in long-chain polyunsaturated fatty acid n−6/n−3 ratio in relation to hepatic steatosis in patients with non-alcoholic fatty liver disease. *Clin. Sci.* **2004**, *106*, 635–643. [CrossRef]
22. Merino, D.M.; Johnston, H.; Clarke, S.; Roke, K.; Nielsen, D.; Badawi, A.; El-Sohemy, A.; Ma, D.; Mutch, D.M. Polymorphisms in FADS1 and FADS2 alter desaturase activity in young Caucasian and Asian adults. *Mol. Genet. Metab.* **2011**, *103*, 171–178. [CrossRef]
23. Treffkorn, L.; Scheibe, R.; Maruyama, T.; Dieter, P. PGE2 exerts its effect on the LPS-induced release of TNF-alpha, ET-1, IL-1alpha, IL-6 and IL-10 via the EP2 and EP4 receptor in rat liver macrophages. *Prostagland. Other Lipid. Mediat.* **2004**, *74*, 113–123. [CrossRef] [PubMed]
24. Takeda, T.; Komiya, Y.; Koga, T.; Ishida, T.; Ishii, Y.; Kikuta, Y.; Nakaya, M.; Kurose, H.; Yokomizo, T.; Shimizu, T.; et al. Dioxin-induced increase in leukotriene B4 biosynthesis through the aryl hydrocarbon receptor and its relevance to hepatotoxicity owing to neutrophil infiltration. *J. Biol. Chem.* **2017**, *292*, 10586–10599. [CrossRef]
25. Palmisano, B.T.; Zhu, L.; Eckel, R.H.; Stafford, J.M. Sex differences in lipid and lipoprotein metabolism. *Mol. Metab.* **2018**, *15*, 45–55. [CrossRef]
26. Soares, A.F.; Paz-Montoya, J.; Lei, H.; Moniatte, M.; Gruetter, R. Sexual dimorphism in hepatic lipids is associated with the evolution of metabolic status in mice. *NMR Biomed.* **2017**, *30*, e3761. [CrossRef]
27. Chitraju, C.; Trötzmüller, M.; Hartler, J.; Wolinski, H.; Thallinger, G.G.; Lass, A.; Zechner, R.; Zimmermann, R.; Köfeler, H.; Spener, F. Lipidomic analysis of lipid droplets from murine hepatocytes reveals distinct signatures for nutritional stress. *J. Lipid Res.* **2012**, *53*, 2141–2152. [CrossRef] [PubMed]
28. Sullivan, J.C.; Sasser, J.M.; Pollock, D.M.; Pollock, J.S. Sexual Dimorphism in Renal Production of Prostanoids in Spontaneously Hypertensive Rats. *Hypertension* **2005**, *45*, 406–411. [CrossRef]
29. Cagen, L.M.; Baer, P.G. Effects of gonadectomy and steroid treatment on renal prostaglandin 9-ketoreductase activity in the rat. *Life Sci.* **1987**, *40*, 95–100. [CrossRef]
30. Leng, S.; Winter, T.; Aukema, H.M. Dietary LA and sex effects on oxylipin profiles in rat kidney, liver, and serum differ from their effects on PUFAs. *J. Lipid. Res.* **2017**, *58*, 1702–1712. [CrossRef]
31. Lemieux, C.; Phaneuf, D.; Labrie, F.; Giguère, V.; Richard, D.; Deshaies, Y. Estrogen receptor α-mediated adiposity-lowering and hypocholesterolemic actions of the selective estrogen receptor modulator acolbifene. *Int. J. Obes.* **2005**, *29*, 1236–1244. [CrossRef] [PubMed]
32. Cole, L.K.; Jacobs, R.; Vance, D.E. Tamoxifen induces triacylglycerol accumulation in the mouse liver by activation of fatty acid synthesis. *Hepatology* **2010**, *52*, 1258–1265. [CrossRef] [PubMed]
33. Hewitt, K.N.; Pratis, K.; Jones, M.; Simpson, E.R. Estrogen Replacement Reverses the Hepatic Steatosis Phenotype in the Male Aromatase Knockout Mouse. *Endocrinology* **2004**, *145*, 1842–1848. [CrossRef] [PubMed]
34. Lin, H.-Y.; Yu, I.-C.; Wang, R.-S.; Chen, Y.-T.; Liu, N.-C.; Altuwaijri, S.; Hsu, C.-L.; Ma, W.-L.; Jokinen, J.; Sparks, J.D.; et al. Increased hepatic steatosis and insulin resistance in mice lacking hepatic androgen receptor. *Hepatology* **2008**, *47*, 1924–1935. [CrossRef] [PubMed]
35. Brie, B.; Ramirez, M.C.; De Winne, C.; Vicchi, F.L.; Villarruel, L.; Sorianello, E.; Catalano, P.; Ornstein, A.M.; Becu-Villalobos, D. Brain Control of Sexually Dimorphic Liver Function and Disease: The Endocrine Connection. *Cell. Mol. Neurobiol.* **2019**, *39*, 169–180. [CrossRef]
36. Gardner, C.J.; Irwin, A.J.; Daousi, C.; McFarlane, I.A.; Joseph, F.; Bell, J.D.; Thomas, E.L.; Adams, V.L.; Kemp, G.J.; Cuthbertson, D.J. Hepatic steatosis, GH deficiency and the effects of GH replacement: A Liverpool magnetic resonance spectroscopy study. *Eur. J. Endocrinol.* **2012**, *166*, 993–1002. [CrossRef]
37. Laron, Z.; Ginsberg, S.; Webb, M. Nonalcoholic fatty liver in patients with Laron syndrome and GH gene deletion—Preliminary report. *Growth Horm. IGF Res.* **2008**, *18*, 434–438. [CrossRef] [PubMed]

38. Cordoba-Chacon, J.; Majumdar, N.; List, E.O.; Diaz-Ruiz, A.; Frank, S.J.; Manzano, A.; Bartrons, R.; Puchowicz, M.; Kopchick, J.J.; Kineman, R.D. Growth Hormone Inhibits Hepatic De Novo Lipogenesis in Adult Mice. *Diabetes* **2015**, *64*, 3093–3103. [CrossRef]
39. Sarmento-Cabral, A.; Del Rio-Moreno, M.; Vazquez-Borrego, M.C.; Mahmood, M.; Gutierrez-Casado, E.; Pelke, N.; Guzman, G.; Subbaiah, P.V.; Cordoba-Chacon, J.; Yakar, S.; et al. GH directly inhibits steatosis and liver injury in a sex-dependent and IGF1-independent manner. *J. Endocrinol.* **2021**, *248*, 31–44. [CrossRef]
40. Choi, H.K.; Waxman, D.J. Plasma Growth Hormone Pulse Activation of Hepatic JAK-STAT5 Signaling: Developmental Regulation and Role in Male-Specific Liver Gene Expression. *Endocrinology* **2000**, *141*, 3245–3255. [CrossRef] [PubMed]
41. Zhang, Y.; Laz, E.V.; Waxman, D.J. Dynamic, Sex-Differential STAT5 and BCL6 Binding to Sex-Biased, Growth Hormone-Regulated Genes in Adult Mouse Liver. *Mol. Cell. Biol.* **2012**, *32*, 880–896. [CrossRef]
42. Oshida, K.; Waxman, D.J.; Corton, J.C. Chemical and Hormonal Effects on STAT5b-Dependent Sexual Dimorphism of the Liver Transcriptome. *PLoS ONE* **2016**, *11*, e0150284.
43. Smati, S.; Polizzi, A.; Fougerat, A.; Ellero-Simatos, S.; Blum, Y.; Lippi, Y.; Régnier, M.; Laroyenne, A.; Huillet, M.; Arif, M.; et al. Integrative study of diet-induced mouse models of NAFLD identifies PPARα as a sexually dimorphic drug target. *Gut* **2021**. [CrossRef]
44. Bruce, S.J.; Rey, F.; Béguin, A.; Berthod, C.; Werner, D.; Henry, H. Discrepancy between radioimmunoassay and high performance liquid chromatography tandem-mass spectrometry for the analysis of androstenedione. *Anal. Biochem.* **2014**, *455*, 20–25. [CrossRef]
45. Parini, P.; Johansson, L.; Broijersen, A.; Angelin, B.; Rudling, M. Lipoprotein profiles in plasma and interstitial fluid analyzed with an automated gel-filtration system. *Eur. J. Clin. Investig.* **2006**, *36*, 98–104. [CrossRef]
46. Matyash, V.; Liebisch, G.; Kurzchalia, T.V.; Shevchenko, A.; Schwudke, D. Lipid extraction by methyl-tert-butyl ether for high-throughput lipidomics. *J. Lipid Res.* **2008**, *49*, 1137–1146. [CrossRef] [PubMed]
47. Fauland, A.; Köfeler, H.; Trötzmüller, M.; Knopf, A.; Hartler, J.; Eberl, A.; Chitraju, C.; Lankmayr, E.; Spener, F. A comprehensive method for lipid profiling by liquid chromatography-ion cyclotron resonance mass spectrometry. *J. Lipid Res.* **2011**, *52*, 2314–2322. [CrossRef]
48. Hartler, J.; Trötzmüller, M.; Chitraju, C.; Spener, F.; Köfeler, H.; Thallinger, G.G. Lipid Data Analyzer: Unattended identification and quantitation of lipids in LC-MS data. *Bioinformatics* **2011**, *27*, 572–577. [CrossRef] [PubMed]
49. Liebisch, G.; Vizcaino, J.A.; Köfeler, H.; Trötzmüller, M.; Griffiths, W.; Schmitz, G.; Spener, F.; Wakelam, M. Shorthand notation for lipid structures derived from mass spectrometry. *J. Lipid Res.* **2013**, *54*, 1523–1530. [CrossRef] [PubMed]
50. Le Faouder, P.; Baillif, V.; Spreadbury, I.; Motta, J.-P.; Rousset, P.; Chêne, G.; Guigné, C.; Tercé, F.; Vanner, S.; Vergnolle, N.; et al. LC–MS/MS method for rapid and concomitant quantification of pro-inflammatory and pro-resolving polyunsaturated fatty acid metabolites. *J. Chromatogr. B* **2013**, *932*, 123–133. [CrossRef] [PubMed]

International Journal of
Molecular Sciences

MDPI

Review

The PPARα and PPARγ Epigenetic Landscape in Cancer and Immune and Metabolic Disorders

Jesús Porcuna †, Jorge Mínguez-Martínez † and Mercedes Ricote *

Myocardial Pathophisiology Area, Centro Nacional de Investigaciones Cardiovasculares (CNIC), 28029 Madrid, Spain; jesus.porcuna@cnic.es (J.P.); jorge.minguez@cnic.es (J.M.-M.)
* Correspondence: mricote@cnic.es
† These authors contributed equally to this work.

Abstract: Peroxisome proliferator-activated receptors (PPARs) are ligand-modulated nuclear receptors that play pivotal roles in nutrient sensing, metabolism, and lipid-related processes. Correct control of their target genes requires tight regulation of the expression of different PPAR isoforms in each tissue, and the dysregulation of PPAR-dependent transcriptional programs is linked to disorders, such as metabolic and immune diseases or cancer. Several PPAR regulators and PPAR-regulated factors are epigenetic effectors, including non-coding RNAs, epigenetic enzymes, histone modifiers, and DNA methyltransferases. In this review, we examine advances in PPARα and PPARγ-related epigenetic regulation in metabolic disorders, including obesity and diabetes, immune disorders, such as sclerosis and lupus, and a variety of cancers, providing new insights into the possible therapeutic exploitation of PPAR epigenetic modulation.

Keywords: PPARs; cancer; immunity; obesity; diabetes; miRNA; DNA methylation; histone modification

Citation: Porcuna, J.; Mínguez-Martínez, J.; Ricote, M. The PPARα and PPARγ Epigenetic Landscape in Cancer and Immune and Metabolic Disorders. *Int. J. Mol. Sci.* **2021**, *22*, 10573. https://doi.org/10.3390/ijms221910573

Academic Editor: Manuel Vázquez-Carrera

Received: 8 September 2021
Accepted: 28 September 2021
Published: 30 September 2021

1. Introduction

1.1. Peroxisome Proliferator Activated Receptors

Peroxisome proliferator-activated receptors (PPARs) are a group of nuclear receptors (NRs) that act as ligand-activated transcription factors (TFs) [1]. Upon ligand binding, PPARs assemble with retinoid-X-receptors (RXRs), generating dimeric complexes that bind response elements in target genes to exert important regulatory functions [2]. PPARs are well known for their important functions in lipid and glucose homeostasis, nutrient sensing, inflammation, cellular differentiation, and development [3]. There are three PPAR isoforms: PPARα (NR1C1), PPARβ/δ (NR1C2), and PPARγ (NR1C3). The three PPAR isoforms are differentially expressed in distinct tissues and, more importantly, play different and contrasting roles upon ligand activation [4,5]. PPARα is expressed in tissues with high rates of fatty-acid catabolism, such as the liver, where it is mainly expressed. PPARα decreases glycolysis and lipogenesis, while enhancing glucose uptake, glycogen synthesis, and fatty acid oxidation. Although the PPARβ/δ isoform is expressed ubiquitously, its expression is prominent in the gastrointestinal tract and muscle, where it controls metabolism, glucose utilization, and lipid transport. PPARγ is mostly expressed in adipose tissue, where it promotes lipogenesis and adipocyte differentiation. It also improves insulin secretion by pancreatic β-cells, skeletal muscle sensitization to insulin, and gluconeogenesis in the liver.

Like other NRs, PPARs have a well-conserved structure. Between the N-terminal and C-terminal ends are a DNA binding domain (DBD), a flexible hinge, and a ligand-binding domain (LBD) [2]. The DBD includes a structure containing two zinc-fingers that recognize specific DNA sequences in the peroxisome proliferator response elements. These sequences consist of direct repeats of the hexanucleotide AGGTCA separated by a single nucleotide spacer [6]. The LBD contains 13 alpha helices and one four-stranded beta sheet and can interact with several ligands that activate or repress PPAR action [5,7]. Many natural and synthetic lipophilic acids are PPAR ligands, prominent among which are a wide variety of unsaturated fatty acids (docosahexaenoic and eicosapentanoic acids) and eicosanoids. Natural ligands include leukotriene B4 for PPARα and prostaglandin PGJ2 for PPARγ [5]. PPARα is also stimulated by a family of chemicals known as fibrates, such as fenofibrate and clofibrate. Similarly, PPARγ binds a group of synthetic molecules called thiazolidinediones (TZDs), including rosiglitazone and pioglitazone.

PPARs regulate energy metabolism and inflammation, exerting anti-fibrotic and anti-inflammatory effects in diverse conditions, including cancer, autoimmune diseases, liver steatosis, and type 2 diabetes (T2D) [8–10]. PPARs stimulate the expression of anti-inflammatory molecules and inhibit the production of extracellular matrix proteins and pro-inflammatory cytokines, as well as modulating the response and phenotype of immune cells such as macrophages and lymphocytes [10]. The activation of all three isoforms has been demonstrated to polarize macrophages to an anti-inflammatory M2 phenotype and to regulate CD4+ T cell survival and differentiation towards different Th and Treg lineages [11,12]. PPARγ has been demonstrated to act as a key transcription factor in alveolar macrophage and osteoclast identity and ontogeny [13]. While the PPAR pathways implicated in the control of these processes are well characterized [10], little is known about the epigenetic modulation of or by PPARα and PPARγ. Nevertheless, recent research has begun to identify features of the PPAR epigenome in different diseases (Figure 1). These advances, together with the amenability of PPARs to ligand-modulation and the increasing availability of synthetic ligands [5], are driving the study of the complex transcriptional and epigenetic regulation of PPARs in specific diseases. Here, we discuss recent advances focused on the PPARα- and PPARγ-related regulation of non-coding RNAs, histone modification, and DNA methylation in the context of cancer and metabolic and immune-related disorders, as well as the emerging therapeutic potential of these processes in these diseases.

1.2. Epigenetics

The term epigenetics was coined by Conrad Hal Waddington in 1942 to explain the link between genes and the environment. Epigenetics is the study of mechanisms of stable and heritable gene regulation that require no changes to DNA sequence and can be defined as the set of environmental influences that determine a phenotype [14]. The three main epigenetic mechanisms are DNA methylation, histone modification, and the binding of non-coding RNAs to regulatory elements. These mechanisms perpetually modulate gene expression states in order to ensure the correct cellular fate and state without altering the DNA sequence.

1.2.1. Major Epigenetic Modifications

DNA Methylation

DNA methylation, one of the most studied epigenetic modifications, is the addition of a methyl group (-CH3) to the fifth carbon atom of the cytosine ring, generating 5-methylcytosine (5meC). DNA methylation inhibits gene transcription [15] and is catalysed by the m5C DNA methyltransferase (DNMT) family of enzymes. These are classified into three groups, DNMT1, DNMT3a, and DNMT3b, which together establish and sustain the correct DNA methylation patterns. DNMTs, together with partners such as UHRF1 (ubiquitin like with PHD and ring finger domains 1), must be tightly regulated to avoid pathological outcomes, for instance the expression of oncogenes [16,17]. DNA methylation is a reversible epigenetic mark, and the removal of methyl groups is catalysed by the ten-eleven translocation (TET) methylcytosine dioxygenases [18].

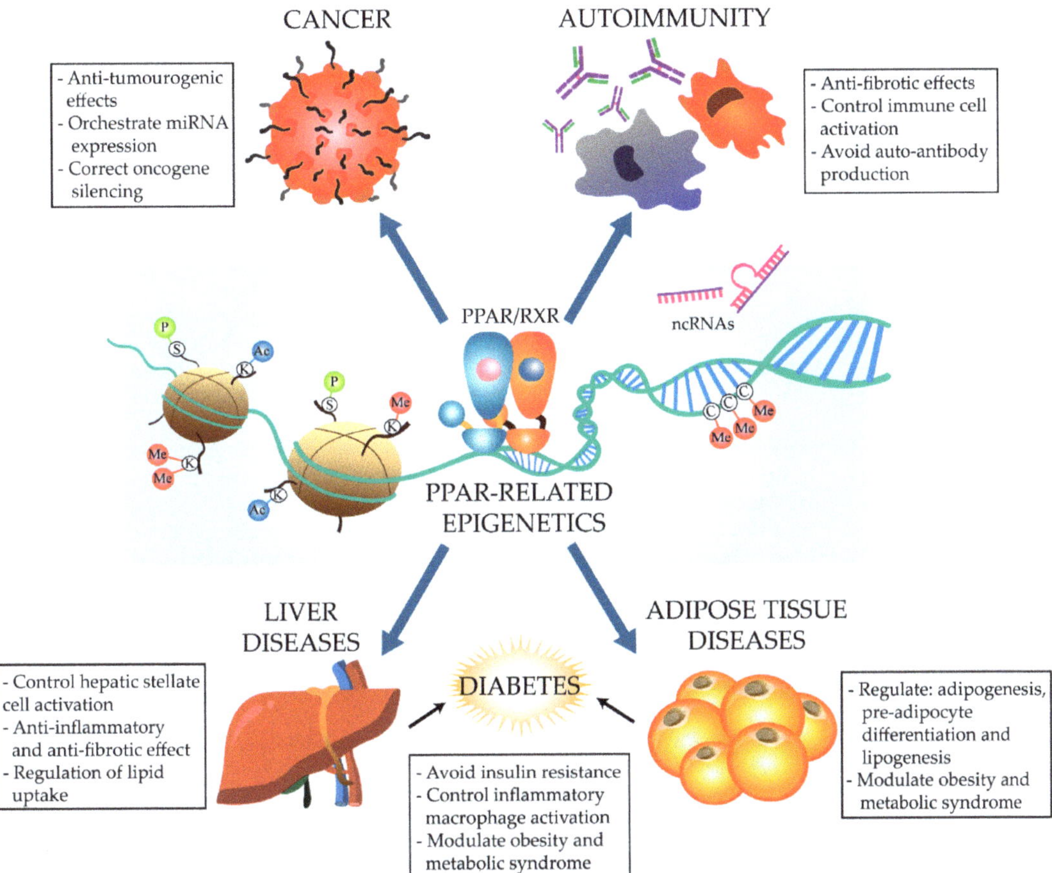

Figure 1. Scheme: PPAR epigenome implications in diseases. C, cytosine; Me, methyl group, ncRNAs, non-coding RNAs; K, histone lysine; S, histone serine; Ac, acetylation mark; P, phosphorylation mark.

Histone Modification

Histones are composed of the protein subunits H2A, H3, H3B, and H4 and act as cores around which DNA winds to form nucleosomes, the building blocks of chromatin. Histones can be modified by acetylation, methylation, ubiquitination, or phosphorylation. These specific modifications, or a combination of them, change the nucleosome conformation, thereby regulating access by the transcriptional machinery to the coding DNA [19–21]. The most well-studied histone modifications are acetylation and methylation [22]. Histone acetylation is the most frequent histone modification and is regulated by histone acetyltransferases (HATs) and histone deacetylases (HDACs). When histones are acetylated, the chromatin adopts a more relaxed and open conformation, allowing access to the gene transcription machinery. Histones can be methylated on lysines and arginines by histone methyltransferases (HMTs), with the reverse reaction catalysed by histone demethylases (HDMs) [23,24]. Histone methylation most often induces gene silencing by promoting the recruitment of DNMTs, followed by methyl-binding proteins and finally HDACs [25,26]. Nevertheless, histone methylation can promote the activity of positive transcriptional regulatory elements, such as de novo and pre-disposed enhancers or promoters [27–29].

Non-Coding RNAs

Non-coding RNAs (ncRNAs) are RNA molecules that do not translate into proteins but instead play important roles in gene expression regulation both transcriptionally, at the DNA level, and post-transcriptionally, at the mRNA level [30]. ncRNAs are a diverse group of molecules, and it is difficult to make general statements about their function and regulation [31]. The most widely studied ncRNAs in relation to epigenetics are micro RNAs (miRNAs) and long ncRNAs (lncRNAs). miRNAs are typically 18–24 nucleotides long and bind to complementary sequences in target mRNAs, marking them for degradation and thus preventing their translation into protein. lncRNAs, which can exceed 200 nucleotides, have a diverse interactome that includes DNA, proteins, peptides, mRNAs, and miRNAs, through which they regulate both transcription and translation [32].

The epigenome—the complete set of epigenetic marks—must be tightly regulated not only to sustain development and cell fate, but also to prevent pathogenic conditions that could otherwise arise at any moment during life [33]. The dynamic gene regulation afforded by epigenetics ensures a locally appropriate accessibility of chromatin to TFs and, therefore, the execution of a precise transcriptional program [34]. Given the importance of epigenetics for sustained gene transcription, all epigenetics programs within a cell need to work correctly in order to maintain cell function and phenotype, and to prevent possible inflammatory conditions derived from an altered epigenetic landscape [35]. Much recent research interest therefore focuses on the roles of specific epigenetic proteins, such as enzymes and TFs, in a range of biological processes, such as immune metabolism, inflammation, disease, and differentiation. Aberrant DNA and histone methylation, abnormal histone acetylation patterns, and altered ncRNA regulation have been linked to conditions, such as aging, neurological and metabolic disorders, allergies and other autoimmune diseases, and cancer [36–47]. Finding molecules able to modulate the epigenome would open up opportunities to specifically treat these conditions.

2. The PPARα and PPARγ Epigenetic Landscape in Disease

2.1. Cancer

PPARs have well known anti-tumourogenic effects [8]. PPARα activation can induce apoptosis and tumour cell death, preventing tumour expansion and inflammation. PPAR-related effects on tumour development have historically been linked to cell-cycle blockade genes such as p18, p21, and p27, leading to apoptosis through the inhibition of B-cell lymphoma 2 (Bcl-2) and reduced angiogenesis through the inhibition of vascular endothelial growth factor (VEGF) [48,49]. The implication of PPAR-related miRNAs and DNA modifications in tumour development has spurred interest in their potential as biomarkers and therapeutic targets. However, the evidence is disputed for some cancers and PPAR isoforms (Table 1).

2.1.1. Colorectal Cancer

Colorectal cancer (CRC) accounts for 7–10% of incident cancers and 3.2% of all cancer-related deaths worldwide, and the incidence is increasing in developed countries [50]. Several studies have shown that PPARγ plays a protective role in CRC and have described the pathways involved downstream of PPAR, opening up the possibility of using PPAR agonists to treat CRC [51]. However, less is known about the upstream pathways and epigenetic mechanisms involved in the action of PPARs in CRC, and research in this area is ongoing.

CRC is often associated with obesity, and the tissue hypoxia characteristic of obesity has been linked to altered expression of typical CRC miRNAs [52]. In 2017, Motawi et al. reported that PPARγ epigenetic regulation contributes to the CRC risk of obese patients [53], showing that obese CRC patients have upregulated expression of the miRNAs miR-27b, miR-130b, and miR-138. In line with the anti-tumourogenic role of PPARγ in CRC, the expression level of these miRNAs correlated negatively with PPARγ mRNA and protein expression, possibly as a result of direct targeting of PPARγ mRNA [53]. Motawi and coworker's findings are strongly supported by several previous studies [54–58]. However, others reported downregulated expression of miR-27b and miR-138 in colonic cancer cells and tissues [59–61], although none of these studies discussed PPARγ. Interestingly, miR-506, which is frequently dysregulated in cancer, has been shown to inhibit PPARα expression in the hydroxicamptothecin-resistant colon cancer cell line SW1116 [62]. Moreover, targeted downregulation of PPAR signalling pathway by a set of miRNAs has been reported in CRC-derived liver metastasis [63]. Together, these findings suggest the therapeutic potential of targeting PPAR-interacting miRNAs in CRC.

There is also evidence for a role in CRC of PPAR-related DNA methylation. UHRF1 was demonstrated to foster *Pparg* promoter methylation and repressive histone modifications that suppress PPARγ expression in human-derived CRC cell lines [64]. These in vitro results are in step with studies in CRC patients reporting an association between increased methylation of Pparg [65] and PPARγ target genes [66] and decreased PPARγ expression [67]. Furthermore, hypermethylation of the Pparg promoter suppressed PPARγ expression and was associated with CRC regardless of patient body weight [53]. Interestingly, PPARα acts as a suppressor of colon carcinogenesis in mice and is downregulated in mouse colonic tumours. Mice lacking PPARα had increased expression of DNMT1 and protein arginine methyltransferase 6 (PRMT6), resulting in methylation of the tumour suppressor genes P21 and p27, respectively [68]. However, recent evidence indicates that PPARα, along with PPARδ, is overexpressed in human CRC [69]. The inconsistency between these studies could be explained by the significant differences in PPARα expression and function between mice and humans [70].

2.1.2. Liver Cancer

The most common type of primary liver cancer in humans is hepatocellular carcinoma (HCC), which is the third deadliest cancer in the world. A recent analysis of mouse and human single and bulk RNA-seq data revealed that PPARγ controls the expression of a set of antifibrotic miRNAs, including miR-30, miR-29c, and miR-338, that are important for the maintenance of low profibrotic protein levels during HCC-related liver fibrosis [71]. Conversely, other studies have reported that miRNA regulation of the PPAR pathway may contribute to HCC progression. For example, miR-27a inhibits the expression of PPARγ in hepatocarcinoma cells [72]. Interestingly, miR-27a also inhibits RXRα, possibly contributing to cell proliferation in rhabdomyosarcoma [73]. Given that PPAR forms obligate heterodimers with RXRs to regulate transcription, RXR-targeting miRNAs, like miR-27a and miR-34a [74], might also modify the binding capacity and activity of PPAR indirectly. One of the most differentially expressed miRNAs in human HCC samples is miR-9 [72,75], which has been shown to favour tumour growth and aggressiveness. Moreover, bioinformatic analysis identified putative miR-9 binding sites in the PPARα 3′UTR. However, it remains uncertain whether miR-9 contributes to the regulation of PPARα expression in HCC [75].

2.1.3. Other Cancers

PPARγ has been proposed as a therapeutic target in thyroid cancer [76], but although attempts have been made to correlate PPARγ expression with miR-27a, as yet there is no firm evidence linking miRNAs and PPARs in this type of cancer [77]. PPARs are also plausible therapeutic targets in lung cancer. In canine primary lung cancer cells, the Pparg promoter shows a significant loss of 5′-methylation. However, although PPARγ is highly expressed in canine non-small lung cancer cells, this change in the methylation pattern was unrelated to the observed changes in PPARγ protein expression [78]. PPARγ is also dysregulated in gingivo-buccal oral squamous cell carcinoma (OSCC-GB), with OSCC-GB patients showing significant differential methylation of the PPAR pathway genes Cd36, Cyp27a1, Olr1, and Pparg itself. The anti-cancer potential of targeting PPARs is highlighted by the finding that synthetic PPARγ ligands can reduce the incidence of carcinogen-induced tongue tumours [79]. However, current PPARγ ligands are cytotoxic.

As an alternative, interest has emerged in the epigenetic action of DNA methyltransferase inhibitors (DNMTI), which is able to renew the transcription of key silenced genes in this cancer, including Pparg [79]. However, as yet, there have been no reports on the molecular mechanism underlying DNA methylation and PPARγ regulation in these tumours. In 1.25-dihydroxyvitamin D3-treated human prostate adenocarcinoma cells, expression of miR-17/92 correlated with PPARα downregulation. However, a direct effect of miR-17/92 on PPARα expression has not been demonstrated experimentally [80].

Table 1. PPAR epigenetics in different cancers.

Condition	PPAR Isoform	Epigenetic Player	Effect	References
Colorectal cancer	PPARα	miR-506	PPARα expression inhibition in a hydroxicamptothecin resistant colon cancer cell line.	[62]
		DNMT1	Absence of PPARα caused P21 and P27 methylation by DNMT1.	[68]
	PPARγ	miR-27b, miR-130b and miR-138	Potential downregulation of PPARγ.	[53]
		UHRF1	Epigenetic PPARγ inactivation in human-derived CRC cell lines.	[64]
		Promoter hypermethylation	Hypermethylation of *Pparg* promoter caused PPARγ suppression.	[53]
Hepatocellular carcinoma	PPARα	miR-9	Putative biding sites to PPARα 3′ UTR.	[75]
	PPARγ	miR-30, miR-29c and miR-338	Antifibrotic miRNAs regulated by PPARγ during HCC-related liver fibrosis.	[71]
		miR-27a	PPARγ inhibition in hepatocarcinoma cells.	[72]
Thyroid cancer	PPARγ	miR-27a	no relation obsrved yet.	[77]
Lung cancer	PPARγ	Promoter methylation	Significantly loss of 5′-methylation.	[78]
Gingivo-buccal oral squamous cell carcinoma	PPARγ	DNMTs	DNA methyltransferase inhibitors could renew PPARγ transcription.	[79]
Prostate cancer	PPARα	miR-17/92	Possible direct PPARα targetting and dowregulation.	[80]

Thus, although research is uncovering new PPAR epigenetics-related factors with potential for the treatment of different types of cancer, much of the evidence has been obtained in vitro or consists of observational data obtained from patient samples. Much further research is therefore needed before the field can contemplate moving to cancer clinical trials of therapies based on the modulation of PPAR epigenetics.

2.2. Immune Disorders

PPARs, especially PPARγ, contribute to the suppression of key pro-inflammatory genes such as NF-kB, INFγ, TNFα, TGFβ, and the interleukins IL-1a and IL-6 [1,10]. These actions are related to the key roles played by PPARs in autoimmune diseases, such as celiac disease [81] and lupus [82]. In sepsis patients and in LPS-treated THP-1 cells PPARγ has been shown to upregulate miR-142-3p. This miRNA targets the 3′-UTR of high mobility group box-1 (HMGB1), a protein with increased expression in many autoimmune diseases, and through miR-142-3p, PPARγ thus contributes to maintaining reduced HMGB1 expression [10,83]. Moreover, several studies have demonstrated PPAR-related regulation of histone and DNA modifications in asthma [84] and lupus [85]. PPARs thus regulate immune-related diseases and have the potential to serve as therapeutic targets in these diseases (Table 2).

2.2.1. Asthma

Asthma is an immune disorder characterized by hyper-responsiveness and inflammation of the airways and involving various immune cell types, such as Th2 lymphocytes or eosinophils and inflammatory cytokines. Asthma affects approximately 300 million people worldwide. Although several treatments are available, including corticoids, not all of them are effective and some can have adverse effects in some individuals. Luckily, accumulating evidence is starting to show that PPARs are not only involved in asthma pathogenesis, but could also serve as targets to reduce asthma symptoms [86].

A well-known cause of asthma is exposure to nicotine. Human primary lung fibroblasts from smokers and mouse primary lung fibroblasts from mice exposed to nicotine both show reduced PPARγ protein levels [87]. In the nicotine-exposed mice, treatment with the PPARγ pathway activator rosiglitazone restores the expression level of miR-98, a miRNA that negatively regulates the expression of airway remodelling proteins associated with collagen deposition and fibrosis [87]. Similarly, pioglitazone-mediated PPARγ activation in rats inhibits airway smooth muscle cell proliferation and remodelling by supressing the Smad-TGFβ1-miR-21 signalling pathway [88]. Human miR-21 is known to target phosphatase and tensin homolog deleted on chromosome ten (PTEN), thereby promoting airway smooth muscle cell proliferation [88]. However, the proposed beneficial role of PPARs in asthma was brought into question by the recent finding that IgE promotes airway inflammatory remodelling in asthma patients by activating the PPARγ pathway [89]. Moreover, there is currently a lack of specific mouse models for studying the implication of immune cells in asthma, thus impeding the identification of immune regulators linked to PPARs and associated miRNAs such as miR-98.

Research into the PPAR epigenetic regulatory network in asthma has also identified a group of lncRNAs in sputa from patients with eosinophilic asthma (the most common type of asthma) that appear to target and modulate PPAR target-gene mRNAs [90]. However, this report did not specify whether the effect was to increase or decrease PPAR pathway activity, and the samples came from a small pool of just six patients [90]. A study of the leukocyte methylome in asthma patients detected PPARα pathway enriched in differentially methylated regions [84], but the study design did not permit identification of the specific cell types affected.

The proposed anti-inflammatory actions of PPARs in asthma thus point pointing to the therapeutic potential of PPAR agonists in asthma-related disorders [86]. However, although evidence of PPAR-related epigenetic mechanisms in asthma is beginning to emerge, the roles of miRNAs, lncRNAs, and DNA methylation in these processes remains largely unknown.

2.2.2. Systemic Lupus Erythematosus

Systemic lupus erythematosus (SLE) is an autoimmune disease in which dysfunctional immune cells, such as antigen presenting cells, T cells, and B cells, lead to a multiple organ malfunction characteristic of each patient [91]. Among several advances in SLE research, PPARγ has emerged as a promising target, and the PPARγ agonists pioglitazone and rosiglitazone have yielded hopeful results in mouse models of the disease [82,92].

Monomethylation of the 20th lysine of histone 4 (H4K20) at the Pparg promoter has been demonstrated to increase the expression of the histone deacetylase HDAC9 [93]. Subsequent analysis of SLE patient samples and mouse models showed that histone modifications at the Pparg promoter influence cytokine and autoantibody production [94]. The authors showed that HDAC9 deletion in mouse CD4+ T cells increased H3K9ac and H3K18ac in the Pparg promoter, prompting a shift in T cell cytokine production towards a more anti-inflammatory class, accompanied by reduced anti-dsDNA autoantibody production by B cells, and therefore protection against proteinuria and renal disease [94].

In a very recent study of CD14+ monocytes from SLE patients, Liu Yu et al. reported the emergence of an immunosuppressive M2-phenotype upon TLR-induced epigenetic activation of PPARγ expression [85]. In these experiments, TLR2 activation with the synthetic ligand Pam3CSK4 triggered decreased expression and binding of the deacetylase Sirt1 to the Pparg promoter. ChIP-qPCR revealed that reduced Sirt1 binding leads to increased histone 3 acetylation in the Pparg promoter, with no changes in histone 4 acetylation, resulting in increased PPARγ protein expression and thus allowing the monocytic transition towards a M2 phenotype [85]. These findings are in line with increased Sirt1 expression in the CD4+ T cells of active SLE patients [95].

Taken together, these results highlight the importance of the epigenetic modulation of PPARγ in autoimmune diseases such as lupus, the protective role of TLR-Sirt1-PPARγ signalling in SLE, and the therapeutic potential of targeting this pathway and histone deactelyases in SLE.

2.2.3. Systemic Sclerosis (Scleroderma)

Systemic sclerosis (SSc), or scleroderma, is a rare and severe autoimmune disease featuring diffuse fibrosis and vascular abnormalities in organs, joints, and skin. Of SSc patients, 30% die within 10 years of diagnosis. One of the main challenges of SSc is the rapid worsening of the disease due to uncontrolled inflammation, collagen deposition, and dysregulation of fibroblast growth [96].

PPARγ expression is low in SSc lesions [97], and in SSc animal models, ligand activation of PPARγ reduces both TGFβ-dependent fibrogenesis and fibroblast hyperactivation [98]. In line with these findings, PPARγ has been shown to reduce Smad-dependent fibroblast activation and differentiation [99], and PPARγ activation blocks recruitment to DNA of the histone acetyl transferase p300 [100]. p300 is required for interaction with Smad3, activation of the pro-fibrogenic Smad3 pathway [101], and histone 4 hyperacetylation at the Col1a2 locus [100]. PPARγ activation thus leads to Smad3 pathway blockade and reduced collagen production, resulting in diminished inflammation and fibrosis [100,102]. Although no effective therapies have yet been devised for SSc [103], epigenetic-based strategies are being postulated as promising future SSc treatments [104–106]. The pharmacological modulation of PPARγ is one of the strategies being considered as a means of epigenetically reducing the fibrotic response in SSc patients.

Table 2. PPAR epigenetics in autoimmune diseases.

Condition	PPAR Isoform	Epigenetic Player	Effect	References
Asthma	PPARα	DNA methylation	Human white blood cells showed DNA methylation in several PPAR pathway.	[84]
	PPARγ	miR-21	The profibroti Smad-TGFβ1-miR-21c axis was supress upon PPARγ pioglitazone activation.	[87]
		miR-98	This profibrotic miRNA was downregulated upon PPARγ rosiglitazone activation.	[80]
	Not specified	set of lncRNAs	Modulation of PPAR signalling pathway in sputa from eosinophilic asthma patients.	[90]
Systemic Lupus Erythematosus	PPARγ	H4K20me1 and HDAC9	Decreased H3K9ac and H3K18ac in the Pparg promoter leading to pro-inflammatory T cell cytokines and B cell auto-antibodies.	[93,94]
	PPARγ	Sirt1	Reduced PPARγ expression due to H3 deacetylation, avoiding M2 monocytic transition.	[85]
Systemic sclerosis	PPARγ	p300	Ligand-activated PPARγ blocks histone acetylatransferase p300 avoiding Smad3 pathway activation and Col1a2 locus histone 4 hyperacetylation.	[99–101]

2.3. Metabolism-Related Diseases

Metabolism-related diseases are a broad class of medical conditions, caused by both genetic and non-genetic defects, which lead to altered metabolic processes. These dysfunctions form a group of diseases that frequently derive from widespread nutritionally poor and unhealthy lifestyles [107]. Overnutrition or low-quality nutrition can lead to a wide range of symptoms converging in the pathologic condition called metabolic syndrome [108]. Some of these symptoms are high blood pressure, high levels of triglycerides, low high-density lipoprotein (HDL) concentrations, increased liver fat, non-alcoholic fatty liver disease (NAFLD), elevated amounts of visceral adipose tissue, insulin resistance and diabetes, high inflammatory state, and even cancer [107,109].

Much PPAR research in this area has focused on direct or indirect activation with natural or synthetic ligands [110]. For example, the important role of PPARs in glucose metabolism and effective insulin signaling prompted research into the use of PPARγ-activating TZDs as insulin-sensitizing drugs in T2D [111,112]. More recent approaches have sought to unravel the regulatory networks controlling PPAR expression and function. PPARs clearly play roles spanning many interconnected metabolic disorders. Given the profound effects of transcriptional and epigenetic modulation of PPARs in diverse diseases, new epigenetic targets may have promising therapeutic potential. Here, we focus on the underlying epigenetic mechanisms involving PPARs in three distinct but intimately related metabolic disorders: liver diseases, adipose tissue diseases, and T2D.

2.3.1. Liver Diseases

NAFLD includes a group of liver diseases unrelated to significant alcohol intake. Although the global prevalence and the development of these liver disorders are influenced by ethnicity and geographic origin, there is significant evidence linking NAFLD to poor dietary habits, obesity, adipose tissue dysregulation, and insulin resistance [113]. NAFLD progresses from diet-induced steatosis to a severe inflammatory state, resulting in hepatocyte damage and death that triggers the transdifferentiation of hepatic stellate cells (HSCs) into extracellular matrix-producing myofibroblast-like cells [114]. HSC activation is generally followed by a shift from adipogenesis to a fibrogenic state. This shift is accompanied by a downregulation in the expression of PPARs, which have an anti-inflammatory and protective action in the liver. The shift to fibrogenesis can lead to non-alcoholic steatohepatitis (NASH) and potentially to end-stage liver diseases such as hepatocellular carcinoma.

Several studies have explored epigenetic changes taking place during hepatic metabolic diseases and how they might regulate the expression of PPARs or modulate their binding to promoter and regulatory regions [115–118] (Table 3). Many epigenetic modifications take place during the progression of steatosis and inflammation and when HSC transdifferentiation begins. For example, many metabolic, proinflammatory, and fibrogenic pathways are regulated by miR-21. This miRNA, which is strongly overexpressed in NASH, represses PPARα expression by direct mRNA targeting and induces HSC activation [119]. Much research into the role PPARs in the hepatic response to dietary fat has focused on the balance between DNA methylation and demethylation and how this determines chromatin accessibility and subsequent changes in gene expression patterns. High dietary fat decreases the methylation of the *Ppara* promoter, resulting in increased PPARα protein expression and the consequent upregulation of carnitine palmitoyl transferase-1 and downregulation of fatty acid synthase, two important lipid metabolism-related enzymes [120,121]. These changes ensure adequate lipid metabolism in response to high dietary fat intake and reveal the important anti-inflammatory role of PPARα in liver diseases and the complex downstream network it controls.

Table 3. PPAR epigenetics in non-alcoholic steatohepatitis.

Condition	PPAR Isoform	Epigenetic Player	Effect	References
NASH	PPARα	miR-21	Diminished PPARα expression and activation of HSCs in obesogenic models	[119]
		TET1 and TET2	Downregulated enzymes under high fat diet conditions, promoting *Ppara* hypermethylation	[121]
		Ascorbic acid	Cofactor of TET enzymes. Its lack promotes PPARα target genes hypermethylation	[122]
		JMJD3	Phosphorylated upon fasting-induced FGF21 signaling. Direct interaction with PPARα for the upregulation of autophagy-related genes	[123]
		PRMT5	Downregulation of *Ppara* expression	[124]
	PPARγ	miR-132	miR-132 downregulation induces the expression of MeCP2 in HSCs	[125]
		miR-29a	Expressed upon Rosiglitazone-mediated PPARγ activation. Repression of profibrotic genes	[126]
		MeCP2	H3K9 and H3K27 methylation and HP1α repressor recruitment in *Pparg* locus of HSCs. MeCP2 also induces the expression of EZH2 and ASH1 in HSCs.	[125]
		Pparg promoter CpG methylation	Downregulation of PPARγ. Potential non-invasive fibrosis marker in cell-free DNA in plasma.	[127]
		PRMT6	Repression of PPARγ activity	[128]
		JMJD1A and JMJD2B	Upregulation of *Pparg* and increased lipid uptake	[129,130]
		LncRNA-H19	Control of hepatic lipogenesis through mi-130A/PPARγ axis	[126]

In newborn and suckling mice, PPARα regulates increased liver DNA demethylation and an accompanying increase in the mRNA expression of β-oxidation-related genes [122]. The molecular mechanism underlying this process has not been thoroughly described. Nonetheless, this metabolic transition makes sense given the high dietary fat intake during suckling. A recent study of the livers of fetal and adult offspring of mice fed a high-fat diet during gestation revealed downregulation of the ten-eleven translocation (TET) enzymes TET1 and TET2, together with hypermethylation of *Ppara* and correspondingly lower levels of PPARα protein expression [131]. These findings suggest that dietary alterations during gestation and lactation could downregulate TET enzyme expression in offspring, favouring the hypermethylation of *Ppara* and decreased expression of its lipid metabolism-related target genes. However, further studies are needed to confirm this. TET enzymes require ascorbic acid as a cofactor, and ascorbic acid deficiency during the suckling period increases the hypermethylation of PPARα-dependent lipid metabolism genes such as fibroblast growth factor 21 (*Fgf21*) [132]. FGF21 is a mainly liver-secreted peptide hormone that stimulates adipocytes to take up glucose from the blood [133,134]. In adult mice, fasting-induced FGF21 signalling triggers further epigenetic modifications, such as phosphorylation of the histone demethylase Jumonji-D3 (JMJD3). Phosphorylated JMJD3 interacts directly with PPARα to upregulate the expression of autophagy-related genes [123]. Since this induced process is closely related to triglyceride hydrolysis and ketone body production, PPARα-dependent FGF21–JMJD3 autophagy signalling emerges as an important endocrine regulator and a potential therapeutic target in metabolic disorders [135–137].

Other histone modifying enzymes include protein arginine methyltransferase 5 (PRMT5), which regulates gene expression via the dimethylation of histone residues H4R3, H3R8, and H2R3. These methylation marks induce gene silencing through the recruitment of DNA methyltransferase 3a (DNMT3a). PRMT5 is abundant in the liver of fat-fed mice and is implicated in the development of hepatic steatosis [124]. Reduced or annulled PRMT5 expression triggers the overexpression of PPARα and an increased mitochondrial biogenesis [124]. Similarly, the methyltransferase PRMT6 has shown to be a repressor of

PPARγ activity [128]. The repression of PPARs by PRMT activity thus presents a further possible target for the treatment of fatty liver.

Although PPARγ is more weakly expressed in the liver than PPARα, it is essential for liver function, and the DNA methylation status of the *Pparg* gene has been identified as a marker of liver disease progression. Analysis of the *Pparg* promoter in plasma cell-free DNA has identified differential DNA methylation patterns in specific CpGs that distinguish between mild and severe fibrosis in NAFLD patients [127]. This cell-free DNA is believed to originate in dying hepatocytes that release their genomic content to the systemic circulation, and thus could provide a noninvasive means of measuring liver status [116]. Taken together, these findings open up new prospective research directions and possibilities for the early diagnosis, screening, and treatment of NAFLD.

PPARγ modulates the expression of lipid uptake and metabolism genes and is a well characterized and important negative regulator of HSC transdifferentiation [125,138]. During this process, downregulation of miR-132 enhances the expression of methyl-CpG binding domain protein 2 (MeCP2), which binds to the 5′ region of *Pparg*, promoting H3K9 methylation and recruitment of the transcriptional repressor heterochromatin protein 1 (HP1α). MeCP2 additionally promotes expression of the H3K27 methyltransferase EZH2 (enhancer zeste homolog 2), generating a repression complex at the 3′ region of *Pparg*. Furthermore, MeCP2 induces the expression of the H3K4 methyltransferase ASH1 (absent small and homeotic disks protein 1), which opposes the action of PPARγ by positively regulating the expression of profibrogenic genes [139,140]. In line with these results, miR-132 was recently linked to human NAFLD [141], and strategies targeting MeCP2 and EZH2 have succeeded in decreasing fibrogenic markers characteristics [142,143]. Additionally, a novel mechanism was shown to promote hepatic lipogenesis through the lncRNA-H19/mi-130a/PPARγ axis [126], becoming a potential target to treat NAFLD.

Other miRNAs involved in PPARγ regulation include miR-29a, which is expressed upon rosiglitazone-induced PPARγ activation in a human HSC cell line and results in the inhibition of fibrosis-related genes [144]. Both miR-29a and miR-652 have been shown to contribute to the resolution of liver fibrosis by modulating the activity of CD4+ T cells and HSCs [145,146]. However, as yet, no relationship has been established between the prevention of HSC activation by miR-652 and PPARγ activity.

PPARγ is also involved in the regulation of adipogenic metabolism by certain demethylases that act as essential modulators of hepatic lipid homeostasis. For example, the H3K9-specific Jumonji demethylases JMJD1A and JMJD2B have been reported to bind to the *Pparg* promoter, and the loss of these enzymes resulted in an increase in the number of H3K9me2 marks in this region, leading to *Pparg* repression and higher levels of fibrosis markers [129]. Conversely, overexpression of these demethylases upregulated *Pparg* expression and increased lipid uptake and intracellular triglyceride accumulation, thus favouring adipogenesis and steatosis [130].

2.3.2. Adipose Tissue Diseases

Evidence accumulated over the past 20 years has established that adipose tissue is an endocrine organ involved in a wide array of metabolic and immune processes [147]. Defects in adipose tissue are typically related to obesity, diabetes and insulin resistance, cardiovascular diseases, cancer, longevity, and even fertility [148,149]. The main transcriptional modulators in adipose tissue are CCAAT/enhancer binding proteins (C/EBP) and PPARγ (specifically PPARγ2), which cooperate in fatty acid uptake and in preadipocyte differentiation to the mature adipocyte phenotype [150,151]. Given the important role of PPARγ in lipid homeostasis, there is intense interest in not only the transcriptional, but also the epigenetic regulation of PPARγ in the development and function of adipose tissue (Tables 4 and 5).

The methylation status of the *Pparg* promoter undergoes characteristic changes during adipogenesis and obesity. *Pparg* promoter methylation correlates with low expression of PPARγ in preadipocytes of the mouse cell line 3T3-L1 [152], and preadipocyte differentiation to mature adipocytes is accompanied by progressive *Pparg* promoter demethylation as the expression of PPARγ protein increases, whereas obesity is associated with the reverse effect, with *Pparg* methylation increasing as PPARγ expression decreases [152].

Table 4. PPARα epigenetics in adipose tissue diseases.

Condition	PPAR Isoform	Epigenetic Player	Effect	References
Adipose tissue diseases	PPARα	Lsd1	Targets PPARα to control beige adipocyte numbers	[153]
		Bta-miR-199a-3p, -154c, -320a and -432	Control lipid metabolism through PPARα	[154]
		miR-519d	Suppresses PPARα protein translation in obese patients	[155]

The expression and function of PPARγ in adipose tissue is determined by insertions of histone variants and histone modifications. A crucial protein in adipocyte differentiation is the complex formed by E1A-binding protein p400 and bromo-containing protein 8 (p400/Brd8). The p400/Brd8 complex can incorporate the histone variant H2A.Z, which preferentially locates within transcriptional regulatory sequences, into the promoter regions of PPARγ target genes [156]. In line with this finding, knockdown of Brd8 or H2A.Z results in cell arrest at the immature preadipocyte stage [156] because the PPARγ target genes involved in differentiation are incorrectly expressed. Histone modifications have been investigated in a genome-wide analysis in mouse and human adipocytes during adipogenesis, demonstrating enrichment of the H3K4me2/me3 and H3K27ac active histone marks in the promoters of *Pparg*1 and 2 [157]. Interestingly, a recent study showed that *Pparg* is repressed by the action of piperine, a major component of black pepper, resulting in the inhibition of various adipogenic genes [158]. In contrast, *Pparg* expression and lipogenesis are enhanced upon H3K4 methylation by the methyltransferases mixed-lineage leukemia proteins 3 and 4 (MLL3 and MLL4), which form a complex with ASC-2 and are recruited by C/EBPβ to the *Pparg* locus [159]. Another study reported that MLL4 induces H3K4me3 marks in the promoters of both C/EBPα and PPARγ through a process requiring the histone methylation regulator PTIP [160]. Moreover, MLL4 itself interacts with some adipogenic TFs, such as tonicity-responsive enhancer binding protein (TonEBP), enabling it to bind the *Pparg* promoter region, increase H3K9me2 marks, and thereby decrease PPARγ expression [161]. Another important methyltransferase in adipocyte differentiation is EZH2, which adds H3K27me marks to the promoter region of the histone deacetylase HDAC9c in adipose tissue, downregulating its expression [162]. Proposals to target EZH2–HDAC9c interaction for the treatment of age-associated osteoporosis and obesity are supported by the report that HDAC9c attenuates adipogenesis by interfering with PPARγ transcriptional activity [163]. Two other methyltransferases of interest are the H3K36 methyltransferase Nsd2 and the lysine methyltransferase 5 (KMT5A, also known as SETD8). Deletion of Nsd2 alters PPARγ target gene expression, adipogenesis, and adipose tissue function [164], whereas KMT5A, a PPARγ target gene expressed during adipocyte differentiation, boosts the expression of PPARγ and the levels of H4K20me marks in other PPARγ target genes in a positive feedback loop [93]. Research has also addressed the role of demethylases in PPARγ regulation in adipose tissue [165], with the histone demethylase JMJD2C reported to downregulate PPARγ transcriptional activation and decrease preadipocyte differentiation, and the H3K9-specific demethylase JHDM2A shown to facilitate the recruitment of PPARγ and RXRα while promoting brown adipogenesis [166–168].

Epigenetic analysis of the the *Pparg* gene has revealed increases in H3K9 and H3K27 acetylation marks, paralleling increased PPARγ expression during the differentiation from preadipocytes to mature adipocytes [169]. PPARγ expression is also increased upon the recruitment of C/EBP and the glucocorticoid receptor (GR) to the *Pparg* enhancer by a complex formed between RNA polymerase II transcription subunit 1 (MED1) and the histone acetyltransferase p300 [170]. Another study reported that the *Pparg* promoter and PPARγ target genes are bound by poly(ADP-Ribose)-Polymerase-1 (PARP1), which enhances their expression and thus acts as an adipogenic modulator [171]. However, in contrast with these results, p300 is known to interact with cyclin D1, which inhibits its acetyltransferase activity and thereby reduces *Pparg* expression [172]. These results provide evidence for a central role of PPARγ in the fine epigenetic regulation of adipocyte differentiation, development, and proliferation

Histone deacetylases regulated during adipogenesis include the fasting-induced NAD-dependent histone deacetylase sirtuin-1 (SIRT1). SIRT1 blocks PPARγ activity by docking with the NR co-repressor (NCoR) and the silencing mediator of retinoid and thyroid hormone receptors (SMRT). The resulting complex occupies PPAR binding sites, inhibiting the expression lipogenesis-related genes [173,174]. This finding has prompted interest in SIRT1 as a potential pharmacological target for obesity and obesity-related diseases [173,175]. Recent studies in mouse models of obesity have already demonstrated that HDAC inhibitors stimulate adipose tissue function and oxidative potential, improving the metabolic profile [176–178]. Additionally, epigenetic changes upon PPARγ-ligand binding have been studied in relation to their effects on adipogenesis. Rosiglitazone-induced PPARγ activation was found to require the methylcytosine dioxygenase TET2, which is important for demethylation. TET2 enhances the expression of PPARγ target genes and thus participates as an epigenetic regulator and a transcriptional modulator in adipocytes [179]. In 2017, Duteil and colleagues revealed that the lysine-specific demethylase 1 (Lsd1) targeted *Ppara*, maintaining the transcriptional program that sustains beige adipocyte homeostasis. PPARα pharmacological intervention could be used to fight obesity by preventing beige-to-white transition [153].

Research in the past few years has uncovered essential roles of ncRNAs in PPARγ regulation in adipose tissue. The levels of specific ncRNAs have been found to oscillate during adipogenesis and obesity, cell commitment, and adipocyte differentiation. For instance, in vitro studies showed that lncRNA U90926 inhibits *Pparg* promoter activity and therefore decreases its expression [180], whereas nuclear enriched abundant transcript 1 (NEAT1) regulates *Pparg* splicing [181], and the HOX antisense intergenic RNA (HOTAIR) enhances *Pparg* expression and adipocyte differentiation [182]. Another study in obese mice showed that lncRNA taurine upregulated gene1 (TUG1) diminished fatty acid accumulation, insulin intolerance, and inflammation by attenuating miR-204 and promoting GLUT4/PPARγ/AKT pathway [183].

MiRNAs described to have an epigenetic effect on *Pparg* include miR-155, miR-221, and miR-122. These miRNAs are downregulated during adipogenesis in human bone-marrow-derived stromal cells, and their overexpression results in lower levels of PPARγ [184]. Moreover, bovine fat-enriched miRNAs, Bta-miR-199a-3p, -154c, -320a, and -432, targeted both *Ppara* and *Pparg* in order to control lipid metabolism [154]. Similarly, miR-540 acts as a negative regulator of adipogenesis in adipose tissue-derived stromal cells through binding to the 3′-UTR region of *Pparg* transcripts, blocking their expression [185]. Studies in the 3T3-L1 preadipocyte mouse cell line identified miR-27a/b, miR-31, miR-130/b, miR-301a, miR-302a, and miR-548d5p as negative regulators of *Pparg* expression and thereby inhibitors of adipogenesis [186,187]. In contrast, the expression of miR-103, miR-143, miR-200a, miR-335, and miR-375 accounts for the upregulation of *Pparg* under high-fat diet conditions [187–189]. Interestingly, miR-519d has been shown to be upregulated in obese patients and to suppress PPARα protein translation, resulting in an increased lipid accumulation during pre-adipocyte differentiation [155].

Together, these results demonstrate the importance and complexity of the epigenetic regulation of PPARs in the control of adipogenesis and adipocyte differentiation in homeostatic and pathological conditions. Since the mechanisms by which adipocytes acquire their specific identity are well known, the quest for new therapeutic applications appears to be very promising. Although some of studies cited here were carried out in human preadipocytes and human multipotent adipose-derived stem cells, most research has been performed in adipocytes from mouse models of obesity. Further research into the epigenetic control of PPARs in human studies is thus needed to move the field towards therapeutic applications in obesity and adipose tissue disorders.

Table 5. PPARγ epigenetics in adipose tissue diseases.

Condition	PPAR Isoform	Epigenetic Player	Effect	References
Adipose tissue diseases	PPARγ	U90926	Inhibition of *Pparg* transcription activity	[180]
		NEAT1	Regulation of *Pparg* splicing	[178]
		HOTAIR	Increased expression of PPARγ	[182]
		miR-155, miR-221 and miR-122	Decreased expression of PPARγ in human bone-marrow-derived stromal cells	[184]
		miR-540	Decreased expression of PPARγ in adipose tissue-derived stromal cells	[185]
		miR-27a/b, miR-31, miR-130/b, miR301a, miR-302a and miR-548d5p	Negative regulation of PPARγ and adipogenesis	[186,187]
		miR-103, miR-143, miR-200a, miR-335 and miR-375	Upregulation of *Pparg*	[187,188]
		p400/Brd8 complex	Incorporation of the histone variant H2A.Z, which facilitates the expression of PPARγ target genes	[156]
		MLL3 and MLL4	Complex with ASC-2. Migration to the *Pparg* locus and methylation of H3K4, promoting enhanced *Pparg* expression	[159]
		EZH2	H3K27 methylation in the *Hdac9c* promoter. Enhanced adipogenesis	[162]
		SETD8 (KMT5A)	Enhanced H4K20me marks in PPARγ target genes.	[93]
		JMJD2C	Downregulation of PPARγ transcriptional activation	[166]
		JHDM2A (JMJD1A)	Decreased H3K9me2 marks and facilitated recruitment of PPARγ, RXRα and PGC1α	[167,168]
		Cyclin D1	Interaction with p300 and HDACs to inhibit *Pparg* expression	[172]
		SIRT1	Blocked PPARγ mechanism of action	[173,174]
		LncRNA TUG1 and miR-294	Control fatty acid accumulation through GLUT4/PPARγ/AKT axis	[183]

2.3.3. Insulin Sensitivity and Resistance: Type 2 Diabetes

Diabetes is a metabolic disorder characterized by an inability to properly clear glucose from the blood. The most common form is T2D, in which two related features converge: insufficient insulin production by pancreatic β-cells and progressive insulin resistance [190]. T2D is intimately associated with obesity, inflammation, ageing, and steroid use, and over the past decades its incidence has worryingly increased in children [191–193]. Although research has traditionally focused on insulin signaling defects, some studies have emphasized the transcriptional and epigenetic basis of chronic inflammation in insulin resistance and T2D [194] (Table 6), and others have identified NRs, such as the glucocorticoid and vitamin D receptors, as common mediators of insulin resistance [195].

Although NRs require activating ligands, some researchers have concluded that post-translational modifications such as acetylation increase NR activity in the absence of external ligand [196]. Some histone deacetylases have been implicated in post-translational modifications of PPARs and their activity. High expression of the deacetylase HDAC3 correlated with high levels of proinflammatory markers and insulin resistance in peripheral blood mononuclear cells from T2D patients and hepatocytes from fat-fed E3 rats, which develop metabolic syndrome [197,198]. Inhibition of HDAC3 in adipocytes increased PPARγ acetylation and the expression of PPARγ target genes, including adipokines and adipocyte protein 2, resulting in decreased insulin resistance. These adipokines include adiponectin, which facilitates hepatic glucose output, and leptins, which are important regulators of feeding behaviour [196,199]. Adipose tissue-specific knockout of SIRT 1 triggers a hyperacetylated PPARγ state and enhanced PPARγ activity, leading to increased insulin sensitivity [175]. These results suggest that HDAC inhibitors have the potential to improve insulin sensitivity through a variety of actions. For example, HDAC inhibitors might release PPAR binding sites, as described for SIRT1, and promote maintenance of the acetylated state of PPARs and PPAR target genes. These inhibitors could also stimulate significant PPARγ activation. Recent studies have begun to explore the therapeutic potential of HDAC inhibitors in insulin resistance and obesity [200–202]. However, their application to human disease requires further research.

T2D is also closely related to immunity. During diabetes, adipose tissue macrophages (ATMs) are activated and shift to a pro-inflammatory phenotype, contributing to the propagation of the altered metabolic state by expressing the pro-inflammatory cytokines TNFα, IL-6, and MCP-1 [203,204]. Macrophage activation during T2D is in part mediated by epigenetic mechanisms [205]. The regulation of ATM alternative activation and insulin sensitivity correlate with PPARγ activation [206–208], and ATM alternative activation is held in check by DNA methylation at the *Pparg* promoter. DNA methylation blockade at the *Pparg* promoter boosts macrophage alternative activation, whereas DNA hypermethylation promotes inflammatory responses and insulin resistance [209]. In another study, DNMT3b downregulation in ATMs was found to promote an anti-inflammatory state and enhanced insulin sensitivity, revealing the contribution of DNMT3b-mediated methylation at the *Pparg* promoter to increased inflammatory conditions and insulin resistance [210]. Studies have also reported the contribution of other DNMTs to the epigenetic control of PPARγ target genes. For instance, hypermethylation of *FGF21* by DNMT3a in human adipocytes decreased its expression and correlated with insulin resistance in patients [211]. In another study, methylation of the adiponectin promoter by DNMT1 reduced adiponectin expression in obese mice, and DNMT1 inhibition increased insulin sensitivity and ameliorated glucose intolerance [212]. DNMT inhibitors are thus able to lower DNA methylation that directly affects *Pparg* and PPARγ target genes, identifying these inhibitors as a promising potential treatment for T2D.

The adipogenesis inhibiting miRNA miR-27a has also been reported to promote insulin resistance [213], acting as a glucose metabolism mediator that regulates the PI3K–Akt–GLUT4 signalling pathway by targeting the 3′UTR region of *Pparg* transcripts, promoting insulin resistance [213]. MiR27a is also upregulated during obesity and induces ATM proinflammatory activation by targeting *Pparg* [214].

Table 6. PPAR epigenetics in type 2 diabetes.

Condition	PPAR Isoform	Epigenetic Player	Effect	References
Insulin sensitivity and resistance: Type 2 Diabetes	PPARγ	miR27-a	Target of *Pparg* transcripts, promoting insulin resistance. Induction of inflammatory ATM activation in obesity	[213,214]
		HDAC3	Decreased expression of PPARγ in E3 rat livers. Correlated with inflammation and insulin resistance	[196–198]
		SIRT1	Control of the PPARγ acetylation status and its activity	[175]
		DNMT3b	*Pparg* promoter methylation. Increased inflammatory macrophage activation and insulin resistance	[209,210]
		DNMT3a	Fgf21 hypermethylation in human adipocytes, insulin resistance	[211]
		DNMT1	Adiponectin promoter methylation in obese mice. Glucose intolerance	[212]

Further epigenetic studies have focused on the PPAR coactivator 1α (PGC1α). This protein binds and modulates the activity of PPARγ and PPARα, thereby indirectly regulating the expression of PPAR target genes and functions [215,216]. Like PPARγ, PGC1α can be regulated by reversible acetylation. Its protein sequence contains 13 lysine acetylation sites, and acetylation/deacetylation of these sites depends on the cell energy state [217]. PGC1α can be activated by deacetylation mediated by SIRT1 [218,219]. This activation promotes the expression of PPAR target genes and increased expression of gluconeogenic genes [220]. In contrast, PGC1α is inactivated by acetylation by p300, SRC1/3, GCN5, or hepatic PCAF, producing the opposite effect [220]. Epigenetic changes thus not only control *Pparg* expression directly, but also regulate the availability and activity of obligate PPARγ coactivators. These studies increase the relevance of *Pparg* epigenetic modulation and underline the importance of continuing to develop new therapeutic approaches to apply these observations to the treatment of T2D.

3. Conclusions

Despite the importance of PPARs in the control of inflammation and lipid homeostasis in different disease contexts, efforts to decipher the diversity of PPAR-related epigenetic modulation are still at an early stage. This review provides a broad overview of PPAR biology and epigenetics in different diseases. PPARs have a complex and tightly regulated transcriptional network that when dysregulated can lead to disease conditions such as metabolic disorders, autoimmune diseases, or cancer. Although research in this area has characterized several factors of the PPAR regulatory network, the epigenetic effectors and regulators remain largely unknown. For instance, many studies discussed have established a correlation between PPAR and epigenetics in different diseases but have failed to establish a clear causal relationship. Nonetheless, the current evidence establishes that cancer-related, immune, and metabolic disorders have an epigenetic regulatory basis, in which PPARs act as central regulators of inflammation, fibrosis, immune responses, as well as lipid and glucose homeostasis. Some lines of research suggest a potential for therapeutic strategies based on PPAR epigenetics. For instance, HDAC and DNMT inhibitors could serve as therapies in PPAR-dependent inflammatory diseases such as obesity or cancer. Moreover, some PPAR network epigenetic effectors such as miRNAs could be used as early biomarkers of specific disorders. The PPAR epigenetic network is a fascinating emerging field of study that is beginning to identify promising targets for the treatment of cancer, immune, and metabolic disorders.

Funding: This research was funded by the Ministerio de Ciencia, Innovación y Universidades (MCNU) (SAF2017-90604-REDT-NurCaMeIn, RTI2018-095928-BI00) and the Comunidad de Madrid (MOIR-B2017/BMD-3684) to MR; the MCNU fellowships to JP (FPU17/01731) and to JM-M (PRE2019-087964). The CNIC is supported by the MCNU and the Pro CNIC Foundation.

Acknowledgments: We thank Simon Bartlett for English editing, and María Piedad Menéndez Gutierrez for advice on the manuscript.

Conflicts of Interest: The authors declare no conflict of interest.

References

1. Mirza, A.Z.; AlThagafi, I.I.; Shamshad, H. Role of PPAR receptor in different diseases and their ligands: Physiological importance and clinical implications. *Eur. J. Med. Chem.* **2019**, *166*, 502–513. [CrossRef] [PubMed]
2. Amber-Vitos, O.; Chaturvedi, N.; Nachliel, E.; Gutman, M.; Tsfadia, Y. The effect of regulating molecules on the structure of the PPAR-RXR complex. *Biochim. Biophys. Acta (BBA)-Mol. Cell Biol. Lipids* **2016**, *1861*, 1852–1863. [CrossRef] [PubMed]
3. Feige, J.N.; Gelman, L.; Michalik, L.; Desvergne, B.; Wahli, W. From molecular action to physiological outputs: Peroxisome proliferator-activated receptors are nuclear receptors at the crossroads of key cellular functions. *Prog. Lipid Res.* **2006**, *45*, 120–159. [CrossRef] [PubMed]
4. Hong, F.; Pan, S.; Guo, Y.; Xu, P.; Zhai, Y. PPARs as Nuclear Receptors for Nutrient and Energy Metabolism. *Molecules* **2019**, *24*, 2545. [CrossRef] [PubMed]
5. Grygiel-Gorniak, B. Peroxisome proliferator-activated receptors and their ligands: Nutritional and clinical implications—A review. *Nutr. J.* **2014**, *13*, 17. [CrossRef]
6. Hall, J.A.; Rusten, M.; AbuGhazaleh, R.D.; Wuertz, B.; Souksavong, V.; Escher, P.; Ondrey, F. Effects of PPAR-gamma agonists on oral cancer cell lines: Potential horizons for chemopreventives and adjunctive therapies. *Head Neck* **2020**, *42*, 2542–2554. [CrossRef] [PubMed]

7. Zoete, V.; Grosdidier, A.; Michielin, O. Peroxisome proliferator-activated receptor structures: Ligand specificity, molecular switch and interactions with regulators. *Biochim. Biophys. Acta* **2007**, *1771*, 915–925. [CrossRef] [PubMed]
8. Font-Diaz, J.; Jimenez-Panizo, A.; Caelles, C.; Vivanco, M.D.; Perez, P.; Aranda, A.; Estébanez-Perpina, E.; Castrillo, A.; Ricote, M.; Valledor, A.F. Nuclear receptors: Lipid and hormone sensors with essential roles in the control of cancer development. *Semin. Cancer Biol.* **2021**, *73*, 58–75. [CrossRef]
9. Fougerat, A.; Montagner, A.; Loiseau, N.; Guillou, H.; Wahli, W. Peroxisome Proliferator-Activated Receptors and Their Novel Ligands as Candidates for the Treatment of Non-Alcoholic Fatty Liver Disease. *Cells* **2020**, *9*, 1638. [CrossRef]
10. Liu, Y.; Wang, J.; Luo, S.; Zhan, Y.; Lu, Q. The roles of PPARgamma and its agonists in autoimmune diseases: A comprehensive review. *J. Autoimmun.* **2020**, *113*, 102510. [CrossRef]
11. Nobs, S.P.; Natali, S.; Pohlmeier, L.; Okreglicka, K.; Schneider, C.; Kurrer, M.; Sallusto, F.; Kopf, M. PPARgamma in dendritic cells and T cells drives pathogenic type-2 effector responses in lung inflammation. *J. Exp. Med.* **2017**, *214*, 3015–3035. [CrossRef] [PubMed]
12. Housley, W.J.; Adams, C.O.; Vang, A.G.; Brocke, S.; Nichols, F.C.; Lacombe, M.; Rajan, T.V.; Clark, R.B. Peroxisome Proliferator-Activated Receptor gamma Is Required for CD4+ T Cell-Mediated Lymphopenia-Associated Autoimmunity. *J. Immunol.* **2011**, *187*, 4161–4169. [CrossRef]
13. Porcuna, J.; Menendez-Gutierrez, M.P.; Ricote, M. Molecular control of tissue-resident macrophage identity by nuclear receptors. *Curr. Opin. Pharmacol.* **2020**, *53*, 27–34. [CrossRef] [PubMed]
14. Tronick, E.; Hunter, R.G. Waddington, Dynamic Systems, and Epigenetics. *Front. Behav. Neurosci.* **2016**, *10*, 107. [CrossRef] [PubMed]
15. Moore, L.D.; Le, T.; Fan, G. DNA Methylation and Its Basic Function. *Neuropsychopharmacology* **2013**, *38*, 23–38. [CrossRef]
16. Rajabi, H.; Tagde, A.; Alam, M.; Bouillez, A.; Pitroda, S.; Suzuki, Y.; Kufe, D. DNA methylation by DNMT1 and DNMT3b methyltransferases is driven by the MUC1-C oncoprotein in human carcinoma cells. *Oncogene* **2016**, *35*, 6439–6445. [CrossRef] [PubMed]
17. Oh, Y.M.; Mahar, M.; Ewan, E.E.; Leahy, K.M.; Zhao, G.; Cavalli, V. Epigenetic regulator UHRF1 inactivates REST and growth suppressor gene expression via DNA methylation to promote axon regeneration. *Proc. Natl. Acad. Sci. USA* **2018**, *115*, E12417–E12426. [CrossRef]
18. Ginno, P.A.; Gaidatzis, D.; Feldmann, A.; Hoerner, L.; Imanci, D.; Burger, L.; Zilbermann, F.; Peters, A.; Edenhofer, F.; Smallwood, S.A.; et al. A genome-scale map of DNA methylation turnover identifies site-specific dependencies of DNMT and TET activity. *Nat. Commun.* **2020**, *11*, 2680. [CrossRef]
19. Giaimo, B.D.; Ferrante, F.; Herchenrother, A.; Hake, S.B.; Borggrefe, T. The histone variant H2A.Z in gene regulation. *Epigenetics Chromatin* **2019**, *12*, 37. [CrossRef]
20. Miller, J.L.; Grant, P.A. The Role of DNA Methylation and Histone Modifications in Transcriptional Regulation in Humans. *Subcell Biochem.* **2013**, *61*, 289–317. [CrossRef] [PubMed]
21. Rossetto, D.; Avvakumov, N.; Cote, J. Histone phosphorylation: A chromatin modification involved in diverse nuclear events. *Epigenetics* **2012**, *7*, 1098–1108. [CrossRef] [PubMed]
22. Bannister, A.J.; Kouzarides, T. Regulation of chromatin by histone modifications. *Cell Res.* **2011**, *21*, 381–395. [CrossRef]
23. Blythe, S.A.; Cha, S.W.; Tadjuidje, E.; Heasman, J.; Klein, P.S. beta-Catenin Primes Organizer Gene Expression by Recruiting a Histone H3 Arginine 8 Methyltransferase, Prmt2. *Dev. Cell* **2010**, *19*, 220–231. [CrossRef] [PubMed]
24. Svobodova Kovarikova, A.; Legartova, S.; Krejci, J.; Bartova, E. H3K9me3 and H4K20me3 represent the epigenetic landscape for 53BP1 binding to DNA lesions. *Aging (Albany NY)* **2018**, *10*, 2585–2605. [CrossRef]
25. Hervouet, E.; Peixoto, P.; Delage-Mourroux, R.; Boyer-Guittaut, M.; Cartron, P.-F. Specific or not specific recruitment of DNMTs for DNA methylation, an epigenetic dilemma. *Clin. Epigenet.* **2018**, *10*, 17. [CrossRef]
26. Di Croce, L.; Raker, V.; Corsaro, M.; Fazi, F.; Fanelli, M.; Faretta, M.; Fuks, F.; Lo Coco, F.; Kouzarides, T.; Nervi, C.; et al. Methyltransferase Recruitment and DNA Hyermethylation of Target Promoters by an Oncogenic Transcription Factor. *Science* **2002**, *295*, 1079–1082. [CrossRef]
27. Ferrari, K.J.; Scelfo, A.; Jammula, S.; Cuomo, A.; Barozzi, I.; Stutzer, A.; Fischle, W.; Bonaldi, T.; Pasini, D. Polycomb-Dependent H3K27me1 and H3K27me2 Regulate Active Transcription and Enhancer Fidelity. *Mol. Cell* **2014**, *53*, 49–62. [CrossRef]
28. Kaikkonen, M.U.; Spann, N.J.; Heinz, S.; Romanoski, C.E.; Allison, K.A.; Stender, J.D.; Chun, H.B.; Tough, D.F.; Prinjha, R.K.; Benner, C.; et al. Remodeling of the Enhancer Landscape during Macrophage Activation Is Coupled to Enhancer Transcription. *Mol. Cell* **2013**, *51*, 310–325. [CrossRef]
29. Lara-Astiaso, D.; Weiner, A.; Lorenzo-Vivas, E.; Zaretsky, I.; Jaitin, D.A.; David, E.; Keren-Shaul, H.; Mildner, A.; Winter, D.; Jung, S.; et al. Immunogenetics. Chromatin state dynamics during blood formation. *Science* **2014**, *345*, 943–949. [CrossRef]
30. Zhang, P.; Wu, W.; Chen, Q.; Chen, M. Non-Coding RNAs and their Integrated Networks. *J. Integr. Bioinform.* **2019**, *16*. [CrossRef]
31. Frias-Lasserre, D.; Villagra, C.A. The Importance of ncRNAs as Epigenetic Mechanisms in Phenotypic Variation and Organic Evolution. *Front. Microbiol.* **2017**, *8*, 2483. [CrossRef]
32. Ratti, M.; Lampis, A.; Ghidini, M.; Salati, M.; Mirchev, M.B.; Valeri, N.; Hahne, J.C. MicroRNAs (miRNAs) and Long Non-Coding RNAs (lncRNAs) as New Tools for Cancer Therapy: First Steps from Bench to Bedside. *Target. Oncol.* **2020**, *15*, 261–278. [CrossRef]
33. Zhang, W.; Qu, J.; Liu, G.-H.; Belmonte, J.C.I. The ageing epigenome and its rejuvenation. *Nat. Rev. Mol. Cell Biol.* **2020**, *21*, 137–150. [CrossRef]

34. Zhao, Y.; Zheng, D.; Cvekl, A. Profiling of chromatin accessibility and identification of general cis-regulatory mechanisms that control two ocular lens differentiation pathways. *Epigenet. Chromatin* **2019**, *12*, 27. [CrossRef]
35. Chen, S.; Yang, J.; Wei, Y.; Wei, X. Epigenetic regulation of macrophages: From homeostasis maintenance to host defense. *Cell. Mol. Immunol.* **2020**, *17*, 36–49. [CrossRef]
36. Elnady, H.G.; Sherif, L.S.; Kholoussi, N.M.; Ali Azzam, M.; Foda, A.R.; Helwa, I.; Sabry, R.N.; Eissa, E.; Fahmy, R.F. Aberrant Expression of Immune-related MicroRNAs in Pediatric Patients with Asthma. *Int. J. Mol. Cell Med.* **2020**, *9*, 246–255.
37. Neganova, M.E.; Klochkov, S.G.; Aleksandrova, Y.R.; Aliev, G. Histone modifications in epigenetic regulation of cancer: Perspectives and achieved progress. In *Seminars in Cancer Biology*; Academic Press: Cambridge, MA, USA, 2020. [CrossRef]
38. Islam, A.; Mohammad, E.; Khan, M.A. Aberration of the modulatory functions of intronic microRNA hsa-miR-933 on its host gene ATF2 results in type II diabetes mellitus and neurodegenerative disease development. *Hum. Genom.* **2020**, *14*, 34. [CrossRef]
39. Yang, Y.; Wicks, J.; Haitchi, H.M.; Powell, R.M.; Manuyakorn, W.; Howarth, P.H.; Holgate, S.T.; Davies, D.E. Regulation of A Disintegrin and Metalloprotease-33Expression by Transforming Growth Factor-β. *Am. J. Respir. Cell Mol. Biol.* **2012**, *46*, 633–640. [CrossRef]
40. Harb, H.; Raedler, D.; Ballenberger, N.; Bock, A.; Kesper, D.A.; Renz, H.; Schaub, B. Childhood allergic asthma is associated with increased IL-13 and FOXP3 histone acetylation. *J. Allergy Clin. Immunol.* **2015**, *136*, 200–202. [CrossRef]
41. Le, T.N.; Williams, S.R.; Alaimo, J.T.; Elsea, S.H. Genotype and phenotype correlation in 103 individuals with 2q37 deletion syndrome reveals incomplete penetrance and supports HDAC4 as the primary genetic contributor. *Am. J. Med Genet. Part A* **2019**, *179*, 782–791. [CrossRef]
42. Wu, Y.; Cui, W.; Zhang, D.; Wu, W.; Yang, Z. The shortening of leukocyte telomere length relates to DNA hypermethylation of LINE-1 in type 2 diabetes mellitus. *Oncotarget* **2017**, *8*, 73964–73973. [CrossRef]
43. Sun, Z.H.; Liu, Y.H.; Liu, J.D.; Xu, D.D.; Li, X.F.; Meng, X.M.; Ma, T.T.; Huang, C.; Li, J. MeCP2 Regulates PTCH1 Expression Through DNA Methylation in Rheumatoid Arthritis. *Inflammation* **2017**, *40*, 1497–1508. [CrossRef]
44. Raveche, E.S.; Salerno, E.; Scaglione, B.J.; Manohar, V.; Abbasi, F.; Lin, Y.C.; Fredrickson, T.; Landgraf, P.; Ramachandra, S.; Huppi, K.; et al. Abnormal microRNA-16 locus with synteny to human 13q14 linked to CLL in NZB mice. *Blood* **2007**, *109*, 5079–5086. [CrossRef]
45. Zernecke, A.; Bidzhekov, K.; Noels, H.; Shagdarsuren, E.; Gan, L.; Denecke, B.; Hristov, M.; Koppel, T.; Jahantigh, M.N.; Lutgens, E.; et al. Delivery of MicroRNA-126 by Apoptotic Bodies Induces CXCL12-Dependent Vascular Protection. *Sci. Signal.* **2009**, *2*, ra81. [CrossRef]
46. Rana, Z.; Diermeier, S.; Hanif, M.; Rosengren, R.J. Understanding Failure and Improving Treatment Using HDAC Inhibitors for Prostate Cancer. *Biomedicines* **2020**, *8*, 22. [CrossRef]
47. Dovey, O.M.; Foster, C.T.; Conte, N.; Edwards, S.A.; Edwards, J.M.; Singh, R.; Vassiliou, G.; Bradley, A.; Cowley, S.M. Histone deacetylase 1 and 2 are essential for normal T-cell development and genomic stability in mice. *Blood* **2013**, *121*, 1335–1344. [CrossRef]
48. Tian, J.; Hu, L.; Li, X.; Geng, J.; Dai, M.; Bai, X. MicroRNA-130b promotes lung cancer progression via PPARγ/VEGF-A/BCL-2-mediated suppression of apoptosis. *J. Exp. Clin. Cancer Res.* **2016**, *35*, 105. [CrossRef]
49. Koga, H.; Sakisaka, S.; Harada, M.; Takagi, T.; Hanada, S.; Taniguchi, E.; Kawaguchi, T.; Sasatomi, K.; Kimura, R.; Hashimoto, O.; et al. Involvement of p21(WAF1/Cip1), p27(Kip1), and p18(INK4c) in troglitazone-induced cell-cycle arrest in human hepatoma cell lines. *Hepatology* **2001**, *33*, 1087–1097. [CrossRef]
50. Rawla, P.; Sunkara, T.; Barsouk, A. Epidemiology of colorectal cancer: Incidence, mortality, survival, and risk factors. *Prz. Gastroenterol.* **2019**, *14*, 89–103. [CrossRef]
51. Park, J.-I.; Kwak, J.-Y. The Role of Peroxisome Proliferator-Activated Receptors in Colorectal Cancer. *PPAR Res.* **2012**, *2012*, 876418. [CrossRef] [PubMed]
52. Tait, S.; Baldassarre, A.; Masotti, A.; Calura, E.; Martini, P.; Vari, R.; Scazzocchio, B.; Gessani, S.; Del Corno, M. Integrated Transcriptome Analysis of Human Visceral Adipocytes Unravels Dysregulated microRNA-Long Non-coding RNA-mRNA Networks in Obesity and Colorectal Cancer. *Front. Oncol.* **2020**, *10*, 1089. [CrossRef]
53. Motawi, T.K.; Shaker, O.G.; Ismail, M.F.; Sayed, N.H. Peroxisome Proliferator-Activated Receptor Gamma in Obesity and Colorectal Cancer: The Role of Epigenetics. *Sci. Rep.* **2017**, *7*, 10714. [CrossRef]
54. Baffa, R.; Fassan, M.; Volinia, S.; O'Hara, B.; Liu, C.G.; Palazzo, J.P.; Gardiman, M.; Rugge, M.; Gomella, L.G.; Croce, C.M.; et al. MicroRNA expression profiling of human metastatic cancers identifies cancer gene targets. *J. Pathol.* **2009**, *219*, 214–221. [CrossRef]
55. Colangelo, T.; Fucci, A.; Votino, C.; Sabatino, L.; Pancione, M.; Laudanna, C.; Binaschi, M.; Bigioni, M.; Maggi, C.A.; Parente, D.; et al. MicroRNA-130b Promotes Tumor Development and Is Associated with Poor Prognosis in Colorectal Cancer. *Neoplasia* **2013**, *15*, 1086–1099. [CrossRef]
56. Karbiener, M.; Fischer, C.; Nowitsch, S.; Opriessnig, P.; Papak, C.; Ailhaud, G.; Dani, C.; Amri, E.Z.; Scheideler, M. microRNA miR-27b impairs human adipocyte differentiation and targets PPARγ. *Biochem. Biophys. Res. Commun.* **2009**, *390*, 247–251. [CrossRef]
57. Lin, Q.; Gao, Z.; Alarcon, R.M.; Ye, J.; Yun, Z. A role of miR-27 in the regulation of adipogenesis. *FEBS J.* **2009**, *276*, 2348–2358. [CrossRef]

58. Yi, R.; Li, Y.; Wang, F.; Gu, J.; Isaji, T.; Li, J.; Qi, R.; Zhu, X.; Zhao, Y. Transforming growth factor (TGF) β1 acted through miR-130b to increase integrin α5 to promote migration of colorectal cancer cells. *Tumor Biol.* **2016**, *37*, 10763–10773. [CrossRef]
59. Matsuyama, R.; Okuzaki, D.; Okada, M.; Oneyama, C. Micro RNA -27b suppresses tumor progression by regulating ARFGEF 1 and focal adhesion signaling. *Cancer Sci.* **2016**, *107*, 28–35. [CrossRef]
60. Qin, Y.Z.; Xie, X.C.; Liu, H.Z.; Lai, H.; Qiu, H.; Ge, L.Y. Screening and preliminary validation of miRNAs with the regulation of hTERT in colorectal cancer. *Oncol. Rep.* **2015**, *33*, 2728–2736. [CrossRef]
61. Zhao, L.; Yu, H.; Yi, S.; Peng, X.; Su, P.; Xiao, Z.; Liu, R.; Tang, A.; Li, X.; Liu, F.; et al. The tumor suppressor miR-138-5p targets PD-L1 in colorectal cancer. *Oncotarget* **2016**, *7*, 45370–45384. [CrossRef]
62. Tong, J.L.; Zhang, C.P.; Nie, F.; Xu, X.T.; Zhu, M.M.; Xiao, S.D.; Ran, Z.H. MicroRNA 506 regulates expression of PPAR alpha in hydroxycamptothecin-resistant human colon cancer cells. *FEBS Lett.* **2011**, *585*, 3560–3568. [CrossRef]
63. Liu, J.; Li, H.; Sun, L.; Shen, S.; Zhou, Q.; Yuan, Y.; Xing, C. Epigenetic Alternations of MicroRNAs and DNA Methylation Contribute to Liver Metastasis of Colorectal Cancer. *Dig. Dis. Sci.* **2019**, *64*, 1523–1534. [CrossRef] [PubMed]
64. Sabatino, L.; Fucci, A.; Pancione, M.; Carafa, V.; Nebbioso, A.; Pistore, C.; Babbio, F.; Votino, C.; Laudanna, C.; Ceccarelli, M.; et al. UHRF1 coordinates peroxisome proliferator activated receptor gamma (PPARG) epigenetic silencing and mediates colorectal cancer progression. *Oncogene* **2012**, *31*, 5061–5072. [CrossRef] [PubMed]
65. Lee, B.B.; Lee, E.J.; Jung, E.H.; Chun, H.K.; Chang, D.K.; Song, S.Y.; Park, J.; Kim, D.H. Aberrant Methylation of APC, MGMT, RASSF2A, and Wif-1 Genes in Plasma as a Biomarker for Early Detection of Colorectal Cancer. *Clin. Cancer Res.* **2009**, *15*, 6185–6191. [CrossRef]
66. Yang, Y.; Chu, F.H.; Xu, W.R.; Sun, J.Q.; Sun, X.; Ma, X.M.; Yu, M.W.; Yang, G.W.; Wang, X.M. Identification of regulatory role of DNA methylation in colon cancer gene expression via systematic bioinformatics analysis. *Medicine* **2017**, *96*, e8487. [CrossRef] [PubMed]
67. Pancione, M.; Sabatino, L.; Fucci, A.; Carafa, V.; Nebbioso, A.; Forte, N.; Febbraro, A.; Parente, D.; Ambrosino, C.; Normanno, N.; et al. Epigenetic Silencing of Peroxisome Proliferator-Activated Receptor γ Is a Biomarker for Colorectal Cancer Progression and Adverse Patients' Outcome. *PLoS ONE* **2010**, *5*, e14229. [CrossRef]
68. Luo, Y.; Xie, C.; Brocker, C.N.; Fan, J.; Wu, X.; Feng, L.; Wang, Q.; Zhao, J.; Lu, D.; Tandon, M.; et al. Intestinal PPARα Protects Against Colon Carcinogenesis via Regulation of Methyltransferases DNMT1 and PRMT6. *Gastroenterology* **2019**, *157*, 744–759.e4. [CrossRef] [PubMed]
69. Yaghoubizadeh, M.; Pishkar, L.; Basati, G. Aberrant Expression of Peroxisome Proliferator-Activated Receptors in Colorectal Cancer and Their Association with Cancer Progression and Prognosis. *Gastrointest. Tumors* **2020**, *7*, 11–20. [CrossRef] [PubMed]
70. Yang, Q.; Nagano, T.; Shah, Y.; Cheung, C.; Ito, S.; Gonzalez, F.J. The PPARα-Humanized Mouse: A Model to Investigate Species Differences in Liver Toxicity Mediated by PPARα. *Toxicol. Sci.* **2008**, *101*, 132–139. [CrossRef]
71. Winkler, I.; Bitter, C.; Winkler, S.; Weichenhan, D.; Thavamani, A.; Hengstler, J.G.; Borkham-Kamphorst, E.; Kohlbacher, O.; Plass, C.; Geffers, R.; et al. Identification of Pparγ-modulated miRNA hubs that target the fibrotic tumor microenvironment. *Proc. Natl. Acad. Sci. USA* **2020**, *117*, 454–463. [CrossRef] [PubMed]
72. Li, S.; Li, J.; Fei, B.-Y.; Shao, D.; Pan, Y.; Mo, Z.-H.; Sun, B.-Z.; Zhang, D.; Zheng, X.; Zhang, M.; et al. MiR-27a Promotes Hepatocellular Carcinoma Cell Proliferation Through Suppression of its Target Gene Peroxisome Proliferator-activated Receptor γ. *Chin. Med. J.* **2015**, *128*, 941–947. [CrossRef] [PubMed]
73. Tombolan, L.; Zampini, M.; Casara, S.; Boldrin, E.; Zin, A.; Bisogno, G.; Rosolen, A.; De Pitta, C.; Lanfranchi, G. MicroRNA-27a Contributes to Rhabdomyosarcoma Cell Proliferation by Suppressing RARA and RXRA. *PLoS ONE* **2015**, *10*, e0125171. [CrossRef]
74. Oda, Y.; Nakajima, M.; Tsuneyama, K.; Takamiya, M.; Aoki, Y.; Fukami, T.; Yokoi, T. Retinoid X receptor α in human liver is regulated by miR-34a. *Biochem. Pharmacol.* **2014**, *90*, 179–187. [CrossRef]
75. Cui, M.; Xiao, Z.; Wang, Y.; Zheng, M.; Song, T.; Cai, X.; Sun, B.; Ye, L.; Zhang, X. Long Noncoding RNA HULC Modulates Abnormal Lipid Metabolism in Hepatoma Cells through an miR-9–Mediated RXRA Signaling Pathway. *Cancer Res.* **2015**, *75*, 846–857. [CrossRef]
76. Catalano, M.G.; Poli, R.; Pugliese, M.; Fortunati, N.; Boccuzzi, G. Emerging molecular therapies of advanced thyroid cancer. *Mol. Asp. Med.* **2010**, *31*, 215–226. [CrossRef] [PubMed]
77. Toraih, E.A.; Fawzy, M.S.; Abushouk, A.I.; Shaheen, S.; Hobani, Y.H.; Alruwetei, A.M.; Mansouri, O.A.; Kandil, E.; Badran, D.I. Prognostic value of the miRNA-27a and PPAR/RXRα signaling axis in patients with thyroid carcinoma. *Epigenomics* **2020**, *12*, 1825–1843. [CrossRef]
78. Herrera, C.L.; Kim, D.Y.; Kumar, S.R.; Bryan, J.N. Peroxisome proliferator activated receptor γ protein expression is asymmetrically distributed in primary lung tumor and metastatic to lung osteosarcoma samples and does not correlate with gene methylation. *BMC Veter-Res.* **2015**, *11*, 230. [CrossRef]
79. Das, D.; Ghosh, S.; Maitra, A.; Biswas, N.K.; Panda, C.K.; Roy, B.; Sarin, R.; Majumder, P.P. Epigenomic dysregulation-mediated alterations of key biological pathways and tumor immune evasion are hallmarks of gingivo-buccal oral cancer. *Clin. Epigenet.* **2019**, *11*, 178. [CrossRef]
80. Wang, W.-L.W.; Welsh, J.; Tenniswood, M. 1,25-Dihydroxyvitamin D3 modulates lipid metabolism in prostate cancer cells through miRNA mediated regulation of PPARA. *J. Steroid Biochem. Mol. Biol.* **2013**, *136*, 247–251. [CrossRef]
81. Mohan, M.; Okeoma, C.M.; Sestak, K. Dietary Gluten and Neurodegeneration: A Case for Preclinical Studies. *Int. J. Mol. Sci.* **2020**, *21*, 5407. [CrossRef] [PubMed]

82. Aprahamian, T.; Bonegio, R.G.; Weitzner, Z.; Gharakhanian, R.; Rifkin, I.R. Peroxisome proliferator-activated receptor gamma agonists in the prevention and treatment of murine systemic lupus erythematosus. *Immunology* **2014**, *142*, 363–373. [CrossRef] [PubMed]
83. Yuan, Z.; Luo, G.; Li, X.; Chen, J.; Wu, J.; Peng, Y. PPARγ inhibits HMGB1 expression through upregulation of miR-142-3p in vitro and in vivo. *Cell. Signal.* **2016**, *28*, 158–164. [CrossRef]
84. Jeong, A.; Imboden, M.; Ghantous, A.; Novoloaca, A.; Carsin, A.-E.; Kogevinas, M.; Schindler, C.; Lovison, G.; Herceg, Z.; Cuenin, C.; et al. DNA Methylation in Inflammatory Pathways Modifies the Association between BMI and Adult-Onset Non-Atopic Asthma. *Int. J. Environ. Res. Public Health* **2019**, *16*, 600. [CrossRef] [PubMed]
85. Liu, Y.; Luo, S.; Zhan, Y.; Wang, J.; Zhao, R.; Li, Y.; Zeng, J.; Lu, Q. Increased Expression of PPAR-γ Modulates Monocytes Into a M2-Like Phenotype in SLE Patients: An Implicative Protective Mechanism and Potential Therapeutic Strategy of Systemic Lupus Erythematosus. *Front. Immunol.* **2021**, *11*, 579372. [CrossRef]
86. Banno, A.; Reddy, A.; Lakshmi, S.; Reddy, A.R.C. PPARs: Key Regulators of Airway Inflammation and Potential Therapeutic Targets in Asthma. *Nucl. Recept. Res.* **2018**, *5*, 101306. [CrossRef]
87. Wongtrakool, C.; Ko, J.; Jang, A.J.; Grooms, K.; Chang, S.; Sylber, C.; Kosmider, B.; Bahmed, K.; Blackburn, M.R.; Sutliff, R.L.; et al. MicroRNA-98 reduces nerve growth factor expression in nicotine-induced airway remodeling. *J. Biol. Chem.* **2020**, *295*, 18051–18064. [CrossRef]
88. Liu, L.; Pan, Y.; Zhai, C.; Zhu, Y.; Ke, R.; Shi, W.; Wang, J.; Yan, X.; Su, X.; Song, Y.; et al. Activation of peroxisome proliferation–activated receptor-γ inhibits transforming growth factor-β1-induced airway smooth muscle cell proliferation by suppressing Smad–miR-21 signaling. *J. Cell. Physiol.* **2018**, *234*, 669–681. [CrossRef] [PubMed]
89. Fang, L.; Wang, X.; Sun, Q.; Papakonstantinou, E.; S'Ng, C.; Tamm, M.; Stolz, D.; Roth, M. IgE Downregulates PTEN through MicroRNA-21-5p and Stimulates Airway Smooth Muscle Cell Remodeling. *Int. J. Mol. Sci.* **2019**, *20*, 875. [CrossRef] [PubMed]
90. Zhu, Y.-J.; Mao, D.; Gao, W.; Hu, H. Peripheral whole blood lncRNA expression analysis in patients with eosinophilic asthma. *Medicine* **2018**, *97*, e9817. [CrossRef]
91. Moulton, V.R.; Suárez-Fueyo, A.; Meidan, E.; Li, H.; Mizui, M.; Tsokos, G.C. Pathogenesis of Human Systemic Lupus Erythematosus: A Cellular Perspective. *Trends Mol. Med.* **2017**, *23*, 615–635. [CrossRef]
92. Rőszer, T.; Menendez-Gutierrez, M.P.; Lefterova, M.I.; Alameda, D.; Nunez, V.; Lazar, M.A.; Fischer, T.; Ricote, M. Autoimmune kidney disease and impaired engulfment of apoptotic cells in mice with macrophage peroxisome proliferator-activated receptor gamma or retinoid X receptor alpha deficiency. *J. Immunol.* **2011**, *186*, 621–631. [CrossRef]
93. Wakabayashi, K.-I.; Okamura, M.; Tsutsumi, S.; Nishikawa, N.S.; Tanaka, T.; Sakakibara, I.; Kitakami, J.-I.; Ihara, S.; Hashimoto, Y.; Hamakubo, T.; et al. The Peroxisome Proliferator-Activated Receptor γ/Retinoid X Receptor α Heterodimer Targets the Histone Modification Enzyme PR-Set7/Setd8 Gene and Regulates Adipogenesis through a Positive Feedback Loop. *Mol. Cell. Biol.* **2009**, *29*, 3544–3555. [CrossRef]
94. Yan, K.; Cao, Q.; Reilly, C.; Young, N.; Garcia, B.A.; Mishra, N. Histone Deacetylase 9 Deficiency Protects against Effector T Cell-mediated Systemic Autoimmunity. *J. Biol. Chem.* **2011**, *286*, 28833–28843. [CrossRef]
95. Hu, N.; Qiu, X.; Luo, Y.; Yuan, J.; Li, Y.; Lei, W.; Zhang, G.; Zhou, Y.; Su, Y.; Lu, Q. Abnormal histone modification patterns in lupus CD4+ T cells. *J. Rheumatol.* **2008**, *35*, 804–810.
96. Xu, X.; Ramanujam, M.; Visvanathan, S.; Assassi, S.; Liu, Z.; Li, L. Transcriptional insights into pathogenesis of cutaneous systemic sclerosis using pathway driven meta-analysis assisted by machine learning methods. *PLoS ONE* **2020**, *15*, e0242863. [CrossRef]
97. Wei, J.; Ghosh, A.K.; Sargent, J.L.; Komura, K.; Wu, M.; Huang, Q.-Q.; Jain, M.; Whitfield, M.L.; Feghali-Bostwick, C.; Varga, J. PPARγ Downregulation by TGFß in Fibroblast and Impaired Expression and Function in Systemic Sclerosis: A Novel Mechanism for Progressive Fibrogenesis. *PLoS ONE* **2010**, *5*, e13778. [CrossRef]
98. Kohno, S.; Endo, H.; Hashimoto, A.; Hayashi, I.; Murakami, Y.; Kitasato, H.; Kojima, F.; Kawai, S.; Kondo, H. Inhibition of skin sclerosis by 15deoxy Δ12,14-prostaglandin J2 and retrovirally transfected prostaglandin D synthase in a mouse model of bleomycin-induced scleroderma. *Biomed. Pharmacother.* **2006**, *60*, 18–25. [CrossRef]
99. Nuwormegbe, S.A.; Sohn, J.H.; Kim, S.W. A PPAR-Gamma Agonist Rosiglitazone Suppresses Fibrotic Response in Human Pterygium Fibroblasts by Modulating the p38 MAPK Pathway. *Investig. Opthalmol. Vis. Sci.* **2017**, *58*, 5217–5226. [CrossRef]
100. Ghosh, A.K.; Bhattacharyya, S.; Wei, J.; Kim, S.; Barak, Y.; Mori, Y.; Varga, J. Peroxisome proliferator-activated receptor-γ abrogates Smad-dependent collagen stimulation by targeting the p300 transcriptional coactivator. *FASEB J.* **2009**, *23*, 2968–2977. [CrossRef]
101. Chen, J.; Xia, Y.; Lin, X.; Feng, X.-H.; Wang, Y. Smad3 signaling activates bone marrow-derived fibroblasts in renal fibrosis. *Lab. Investig.* **2014**, *94*, 545–556. [CrossRef]
102. Ghosh, A.K.; Bhattacharyya, S.; Lakos, G.; Chen, S.-J.; Mori, Y.; Varga, J. Disruption of transforming growth factor ? Signaling and profibrotic responses in normal skin fibroblasts by peroxisome proliferator-activated receptor? *Arthritis Rheum.* **2004**, *50*, 1305–1318. [CrossRef]
103. Connolly, M.K. Systemic sclerosis (scleroderma): Remaining challenges. *Ann. Transl. Med.* **2021**, *9*, 438. [CrossRef]
104. Ding, W.; Pu, W.; Wang, L.; Jiang, S.; Zhou, X.; Tu, W.; Yu, L.; Zhang, J.; Guo, S.; Liu, Q.; et al. Genome-Wide DNA Methylation Analysis in Systemic Sclerosis Reveals Hypomethylation of IFN-Associated Genes in CD4+ and CD8+ T Cells. *J. Investig. Dermatol.* **2018**, *138*, 1069–1077. [CrossRef]

105. Hemmatazad, H.; Rodrigues, H.M.; Maurer, B.; Brentano, F.; Pileckyte, M.; Distler, J.H.W.; Gay, R.E.; Michel, B.A.; Gay, S.; Huber, L.C.; et al. Histone deacetylase 7, a potential target for the antifibrotic treatment of systemic sclerosis. *Arthritis Rheum.* **2009**, *60*, 1519–1529. [CrossRef]
106. Huber, L.C.; Distler, J.H.W.; Moritz, F.; Hemmatazad, H.; Hauser, T.; Michel, B.A.; Gay, R.E.; Matucci-Cerinic, M.; Gay, S.; Distler, O.; et al. Trichostatin A prevents the accumulation of extracellular matrix in a mouse model of bleomycin-induced skin fibrosis. *Arthritis Rheum.* **2007**, *56*, 2755–2764. [CrossRef]
107. Saklayen, M.G. The Global Epidemic of the Metabolic Syndrome. *Curr. Hypertens. Rep.* **2018**, *20*, 12. [CrossRef]
108. Sherling, D.H.; Perumareddi, P.; Hennekens, C.H. Metabolic Syndrome. *J. Cardiovasc. Pharmacol. Ther.* **2017**, *22*, 365–367. [CrossRef]
109. McCracken, E.; Monaghan, M.; Sreenivasan, S. Pathophysiology of the metabolic syndrome. *Clin. Dermatol.* **2018**, *36*, 14–20. [CrossRef]
110. Hong, F.; Xu, P.; Zhai, Y. The Opportunities and Challenges of Peroxisome Proliferator-Activated Receptors Ligands in Clinical Drug Discovery and Development. *Int. J. Mol. Sci.* **2018**, *19*, 2189. [CrossRef]
111. Bansal, G.; Thanikachalam, P.V.; Maurya, R.K.; Chawla, P.; Ramamurthy, S. An overview on medicinal perspective of thiazolidine-2,4-dione: A remarkable scaffold in the treatment of type 2 diabetes. *J. Adv. Res.* **2020**, *23*, 163–205. [CrossRef] [PubMed]
112. Hevener, A.L.; Olefsky, J.M.; Reichart, D.; Nguyen, M.T.; Bandyopadyhay, G.; Leung, H.Y.; Watt, M.J.; Benner, C.; Febbraio, M.A.; Nguyen, A.K.; et al. Macrophage PPAR gamma is required for normal skeletal muscle and hepatic insulin sensitivity and full antidiabetic effects of thiazolidinediones. *J. Clin. Investig.* **2007**, *117*, 1658–1669. [CrossRef]
113. Brunt, E.M.; Wong, V.W.-S.; Nobili, V.; Day, C.P.; Sookoian, S.; Maher, J.J.; Bugianesi, E.; Sirlin, C.B.; Neuschwander-Tetri, B.A.; Rinella, M.E. Nonalcoholic fatty liver disease. *Nat. Rev. Dis. Prim.* **2015**, *1*, 15080. [CrossRef]
114. Tsuchida, T.; Friedman, S.L. Mechanisms of hepatic stellate cell activation. *Nat. Rev. Gastroenterol. Hepatol.* **2017**, *14*, 397–411. [CrossRef]
115. Hajri, T.; Zaiou, M.; Fungwe, T.; Ouguerram, K.; Besong, S. Epigenetic Regulation of Peroxisome Proliferator-Activated Receptor Gamma Mediates High-Fat Diet-Induced Non-Alcoholic Fatty Liver Disease. *Cells* **2021**, *10*, 1355. [CrossRef]
116. Moran-Salvador, E.; Mann, J. Epigenetics and Liver Fibrosis. *Cell. Mol. Gastroenterol. Hepatol.* **2017**, *4*, 125–134. [CrossRef]
117. Claveria-Cabello, A.; Colyn, L.; Arechederra, M.; Urman, J.M.; Berasain, C.; Avila, M.A.; Fernandez-Barrena, M.G. Epigenetics in Liver Fibrosis: Could HDACs Be a Therapeutic Target? *Cells* **2020**, *9*, 2321. [CrossRef]
118. Sodum, N.; Kumar, G.; Bojja, S.L.; Kumar, N.; Rao, C.M. Epigenetics in NAFLD/NASH: Targets and therapy. *Pharmacol. Res.* **2021**, *167*, 105484. [CrossRef]
119. Rodrigues, P.M.; Rodrigues, C.; Castro, R.E. Modulation of liver steatosis by miR-21/PPARα. *Cell Death Discov.* **2018**, *4*, 9. [CrossRef]
120. Moody, L.; Xu, G.B.; Chen, H.; Pan, Y.-X. Epigenetic regulation of carnitine palmitoyltransferase 1 (Cpt1a) by high fat diet. *Biochim. Biophys. Acta (BBA)-Bioenerg.* **2018**, *1862*, 141–152. [CrossRef]
121. Wang, L.; Chen, L.; Tan, Y.; Wei, J.; Chang, Y.; Jin, T.; Zhu, H. Betaine supplement alleviates hepatic triglyceride accumulation of apolipoprotein E deficient mice via reducing methylation of peroxisomal proliferator-activated receptor alpha promoter. *Lipids Health Dis.* **2013**, *12*, 34. [CrossRef]
122. Ehara, T.; Kamei, Y.; Yuan, X.; Takahashi, M.; Kanai, S.; Tamura, E.; Tsujimoto, K.; Tamiya, T.; Nakagawa, Y.; Shimano, H.; et al. Ligand-Activated PPARα-Dependent DNA Demethylation Regulates the Fatty Acid β-Oxidation Genes in the Postnatal Liver. *Diabetes* **2014**, *64*, 775–784. [CrossRef]
123. Byun, S.; Seok, S.; Kim, Y.-C.; Zhang, Y.; Yau, P.; Iwamori, N.; Xu, H.E.; Ma, J.; Kemper, B.; Kemper, J.K. Fasting-induced FGF21 signaling activates hepatic autophagy and lipid degradation via JMJD3 histone demethylase. *Nat. Commun.* **2020**, *11*, 807. [CrossRef]
124. Huang, L.; Liu, J.; Zhang, X.-O.; Sibley, K.; Najjar, S.M.; Lee, M.M.; Wu, Q. Inhibition of protein arginine methyltransferase 5 enhances hepatic mitochondrial biogenesis. *J. Biol. Chem.* **2018**, *293*, 10884–10894. [CrossRef]
125. Mann, J.; Chu, D.C.; Maxwell, A.; Oakley, F.; Zhu, N.; Tsukamoto, H.; Mann, D.A. MeCP2 Controls an Epigenetic Pathway That Promotes Myofibroblast Transdifferentiation and Fibrosis. *Gastroenterology* **2010**, *138*, 705–714.e4. [CrossRef]
126. Liu, J.; Tang, T.; Wang, G.-D.; Liu, B. LncRNA-H19 promotes hepatic lipogenesis by directly regulating miR-130a/PPARγ axis in non-alcoholic fatty liver disease. *Biosci. Rep.* **2019**, *39*, BSR20181722. [CrossRef]
127. Hardy, T.; Zeybel, M.; Day, C.P.; Dipper, C.; Masson, S.; McPherson, S.; Henderson, E.; Tiniakos, D.; White, S.; French, J.; et al. Plasma DNA methylation: A potential biomarker for stratification of liver fibrosis in non-alcoholic fatty liver disease. *Gut* **2016**, *66*, 1321–1328. [CrossRef]
128. Hwang, J.W.; So, Y.; Bae, G.; Kim, S.; Kim, Y.K. Protein arginine methyltransferase 6 suppresses adipogenic differentiation by repressing peroxisome proliferator-activated receptor γ activity. *Int. J. Mol. Med.* **2019**, *43*, 2462–2470. [CrossRef]
129. Jiang, Y.; Wang, S.; Zhao, Y.; Lin, C.; Zhong, F.; Jin, L.; He, F.; Wang, H. Histone H3K9 demethylase JMJD1A modulates hepatic stellate cells activation and liver fibrosis by epigenetically regulating peroxisome proliferator-activated receptor γ. *FASEB J.* **2015**, *29*, 1830–1841. [CrossRef]
130. Kim, J.-H.; Jung, D.Y.; Nagappan, A.; Jung, M.H. Histone H3K9 demethylase JMJD2B induces hepatic steatosis through upregulation of PPARγ2. *Sci. Rep.* **2018**, *8*, 13734. [CrossRef]

131. Pang, H.; Ling, D.; Cheng, Y.; Akbar, R.; Jin, L.; Ren, J.; Wu, H.; Chen, B.; Zhou, Y.; Zhu, H.; et al. Gestational high-fat diet impaired demethylation of Ppar α and induced obesity of offspring. *J. Cell. Mol. Med.* **2021**, *25*, 5404–5416. [CrossRef] [PubMed]
132. Kawahori, K.; Kondo, Y.; Yuan, X.; Kawasaki, Y.; Hanzawa, N.; Tsujimoto, K.; Wada, F.; Kohda, T.; Ishigami, A.; Yamada, T.; et al. Ascorbic acid during the suckling period is required for proper DNA demethylation in the liver. *Sci. Rep.* **2020**, *10*, 21228. [CrossRef]
133. Kharitonenkov, A.; Shiyanova, T.L.; Koester, A.; Ford, A.M.; Micanovic, R.; Galbreath, E.; Sandusky, G.E.; Hammond, L.J.; Moyers, J.S.; Owens, R.A.; et al. FGF-21 as a novel metabolic regulator. *J. Clin. Investig.* **2005**, *115*, 1627–1635. [CrossRef]
134. Jimenez, V.; Jambrina, C.; Casana, E.; Sacristan, V.; Muñoz, S.; Darriba, S.; Rodó, J.; Mallol, C.; Garcia, M.; León, X.; et al. FGF21 gene therapy as treatment for obesity and insulin resistance. *EMBO Mol. Med.* **2018**, *10*, e8791. [CrossRef]
135. Longo, R.; Peri, C.; Cricrì, D.; Coppi, L.; Caruso, D.; Mitro, N.; De Fabiani, E.; Crestani, M. Ketogenic Diet: A New Light Shining on Old but Gold Biochemistry. *Nutrients* **2019**, *11*, 2497. [CrossRef]
136. Seok, S.; Kim, Y.-C.; Byun, S.; Choi, S.; Xiao, Z.; Iwamori, N.; Zhang, Y.; Wang, C.; Ma, J.; Ge, K.; et al. Fasting-induced JMJD3 histone demethylase epigenetically activates mitochondrial fatty acid β-oxidation. *J. Clin. Investig.* **2018**, *128*, 3144–3159. [CrossRef]
137. Yuan, X.; Tsujimoto, K.; Hashimoto, K.; Kawahori, K.; Hanzawa, N.; Hamaguchi, M.; Seki, T.; Nawa, M.; Ehara, T.; Kitamura, Y.; et al. Epigenetic modulation of Fgf21 in the perinatal mouse liver ameliorates diet-induced obesity in adulthood. *Nat. Commun.* **2018**, *9*, 636. [CrossRef]
138. Hazra, S.; Miyahara, T.; Rippe, R.A.; Tsukamoto, H. PPAR Gamma and Hepatic Stellate Cells. *Comp. Hepatol.* **2004**, *3* (Suppl. 1), S7. [CrossRef]
139. Perugorria, M.J.; Wilson, C.L.; Zeybel, M.; Walsh, M.; Amin, S.; Robinson, S.; White, S.A.; Burt, A.D.; Oakley, F.; Tsukamoto, H.; et al. Histone methyltransferase ASH1 orchestrates fibrogenic gene transcription during myofibroblast transdifferentiation. *Hepatology* **2012**, *56*, 1129–1139. [CrossRef]
140. Byrd, K.N.; Shearn, A. ASH1, a Drosophila trithorax group protein, is required for methylation of lysine 4 residues on histone H3. *Proc. Natl. Acad. Sci. USA* **2003**, *100*, 11535–11540. [CrossRef]
141. Zong, Y.; Yan, J.; Jin, L.; Xu, B.; He, Z.; Zhang, R.; Hu, C.; Jia, W. Relationship between circulating miR-132 and non-alcoholic fatty liver disease in a Chinese population. *Hereditas* **2020**, *157*, 22. [CrossRef] [PubMed]
142. Lau-Corona, D.; Bae, W.K.; Hennighausen, L.; Waxman, D.J. Sex-biased genetic programs in liver metabolism and liver fibrosis are controlled by EZH1 and EZH2. *PLoS Genet.* **2020**, *16*, e1008796. [CrossRef]
143. Moran-Salvador, E.; Garcia-Macia, M.; Sivaharan, A.; Sabater, L.; Zaki, M.Y.; Oakley, F.; Knox, A.; Page, A.; Luli, S.; Mann, J.; et al. Fibrogenic Activity of MECP2 Is Regulated by Phosphorylation in Hepatic Stellate Cells. *Gastroenterology* **2019**, *157*, 1398–1412.e9. [CrossRef]
144. Jing, F.; Geng, Y.; Xu, X.-Y.; Xu, H.-Y.; Shi, J.-S.; Xu, Z.-H. MicroRNA29a Reverts the Activated Hepatic Stellate Cells in the Regression of Hepatic Fibrosis through Regulation of ATPase H+ Transporting V1 Subunit C1. *Int. J. Mol. Sci.* **2019**, *20*, 796. [CrossRef] [PubMed]
145. Xuan, J.; Guo, S.-L.; Huang, A.; Xu, H.-B.; Shao, M.; Yang, Y.; Wen, W. MiR-29a and miR-652 Attenuate Liver Fibrosis by Inhibiting the Differentiation of CD4+ T Cells. *Cell Struct. Funct.* **2017**, *42*, 95–103. [CrossRef]
146. Matsumoto, Y.; Itami, S.; Kuroda, M.; Yoshizato, K.; Kawada, N.; Murakami, Y. MiR-29a Assists in Preventing the Activation of Human Stellate Cells and Promotes Recovery From Liver Fibrosis in Mice. *Mol. Ther.* **2016**, *24*, 1848–1859. [CrossRef]
147. Galic, S.; Oakhill, J.; Steinberg, G.R. Adipose tissue as an endocrine organ. *Mol. Cell. Endocrinol.* **2010**, *316*, 129–139. [CrossRef] [PubMed]
148. Khandekar, M.J.; Cohen, P.; Spiegelman, B.M. Molecular mechanisms of cancer development in obesity. *Nat. Rev. Cancer* **2011**, *11*, 886–895. [CrossRef] [PubMed]
149. Heydari, H.; Ghiasi, R.; Ghaderpour, S.; Keyhanmanesh, R. The Mechanisms Involved in Obesity-Induced Male Infertility. *Curr. Diabetes Rev.* **2021**, *17*, 259–267. [CrossRef]
150. Siersbaek, R.; Nielsen, R.; Mandrup, S. PPARγ in adipocyte differentiation and metabolism—Novel insights from genome-wide studies. *FEBS Lett.* **2010**, *584*, 3242–3249. [CrossRef]
151. Lefterova, M.I.; Zhang, Y.; Steger, D.J.; Schupp, M.; Schug, J.; Cristancho, A.; Feng, D.; Zhuo, D.; Stoeckert, J.C.J.; Liu, X.S.; et al. PPAR and C/EBP factors orchestrate adipocyte biology via adjacent binding on a genome-wide scale. *Genes Dev.* **2008**, *22*, 2941–2952. [CrossRef] [PubMed]
152. Fujiki, K.; Kano, F.; Shiota, K.; Murata, M. Expression of the peroxisome proliferator activated receptor γ gene is repressed by DNA methylation in visceral adipose tissue of mouse models of diabetes. *BMC Biol.* **2009**, *7*, 38. [CrossRef]
153. Duteil, D.; Tosic, M.; Willmann, D.; Georgiadi, A.; Kanouni, T.; Schüle, R. Lsd1 prevents age-programed loss of beige adipocytes. *Proc. Natl. Acad. Sci. USA* **2017**, *114*, 5265–5270. [CrossRef] [PubMed]
154. Sun, J.; Zhang, B.; Lan, X.; Zhang, C.; Lei, C.; Chen, H. Comparative Transcriptome Analysis Reveals Significant Differences in MicroRNA Expression and Their Target Genes between Adipose and Muscular Tissues in Cattle. *PLoS ONE* **2014**, *9*, e102142. [CrossRef] [PubMed]
155. Martinelli, R.; Nardelli, C.; Pilone, V.; Buonomo, T.; Liguori, R.; Castanò, I.; Buono, P.; Masone, S.; Persico, G.; Forestieri, P.; et al. miR-519d Overexpression Is Associated With Human Obesity. *Obesity* **2010**, *18*, 2170–2176. [CrossRef] [PubMed]

156. Couture, J.-P.; Nolet, G.; Beaulieu, E.; Blouin, R.; Gévry, N. The p400/Brd8 Chromatin Remodeling Complex Promotes Adipogenesis by Incorporating Histone Variant H2A.Z at PPARγ Target Genes. *Endocrinology* **2012**, *153*, 5796–5808. [CrossRef]
157. Mikkelsen, T.S.; Xu, Z.; Zhang, X.; Wang, L.; Gimble, J.M.; Lander, E.S.; Rosen, E.D. Comparative Epigenomic Analysis of Murine and Human Adipogenesis. *Cell* **2010**, *143*, 156–169. [CrossRef]
158. Park, U.; Hwang, J.; Youn, H.; Kim, E.; Um, S. Piperine inhibits adipocyte differentiation via dynamic regulation of histone modifications. *Phytother. Res.* **2019**, *33*, 2429–2439. [CrossRef]
159. Lee, J.-E.; Wang, C.; Xu, S.; Cho, Y.-W.; Wang, L.; Feng, X.; Baldridge, A.; Sartorelli, V.; Zhuang, L.; Peng, W.; et al. H3K4 mono- and di-methyltransferase MLL4 is required for enhancer activation during cell differentiation. *Elife* **2013**, *2*, e01503. [CrossRef]
160. Cho, Y.-W.; Hong, S.; Jin, Q.; Wang, L.; Lee, J.-E.; Gavrilova, O.; Ge, K. Histone Methylation Regulator PTIP Is Required for PPARγ and C/EBPα Expression and Adipogenesis. *Cell Metab.* **2009**, *10*, 27–39. [CrossRef]
161. Lee, J.H.; Lee, H.H.; Ye, B.J.; Lee-Kwon, W.; Choi, S.Y.; Kwon, H.M. TonEBP suppresses adipogenesis and insulin sensitivity by blocking epigenetic transition of PPARγ2. *Sci. Rep.* **2015**, *5*, 10937. [CrossRef] [PubMed]
162. Chen, Y.-H.; Chung, C.-C.; Liu, Y.-C.; Yeh, S.-P.; Hsu, J.L.; Hung, M.-C.; Su, H.-L.; Li, L.-Y. Enhancer of Zeste Homolog 2 and Histone Deacetylase 9c Regulate Age-Dependent Mesenchymal Stem Cell Differentiation into Osteoblasts and Adipocytes. *Stem Cells* **2016**, *34*, 2183–2193. [CrossRef] [PubMed]
163. Chen, Y.-H.; Yeh, F.-L.; Yeh, S.-P.; Ma, H.-T.; Hung, S.-C.; Hung, M.-C.; Li, L.-Y. Myocyte Enhancer Factor-2 Interacting Transcriptional Repressor (MITR) Is a Switch That Promotes Osteogenesis and Inhibits Adipogenesis of Mesenchymal Stem Cells by Inactivating Peroxisome Proliferator-activated Receptor γ-2. *J. Biol. Chem.* **2011**, *286*, 10671–10680. [CrossRef] [PubMed]
164. Zhuang, L.; Jang, Y.; Park, Y.-K.; Lee, J.-E.; Jain, S.; Froimchuk, E.; Broun, A.; Liu, C.; Gavrilova, O.; Ge, K. Depletion of Nsd2-mediated histone H3K36 methylation impairs adipose tissue development and function. *Nat. Commun.* **2018**, *9*, 1796. [CrossRef] [PubMed]
165. Nic-Can, G.I.; Rodas-Junco, B.A.; Carrillo-Cocom, L.M.; Zepeda-Pedreguera, A.; Peñaloza-Cuevas, R.; Aguilar-Ayala, F.J.; Rojas-Herrera, R.A. Epigenetic Regulation of Adipogenic Differentiation by Histone Lysine Demethylation. *Int. J. Mol. Sci.* **2019**, *20*, 3918. [CrossRef] [PubMed]
166. Lizcano, F.; Romero, C.; Vargas, D. Regulation of adipogenesis by nucelar receptor PPARγ is modulated by the histone demethylase JMJD2C. *Genet. Mol. Biol.* **2010**, *34*, 19–24. [CrossRef]
167. Tateishi, K.; Okada, Y.; Kallin, E.M.; Zhang, Y. Role of Jhdm2a in regulating metabolic gene expression and obesity resistance. *Nature* **2009**, *458*, 757–761. [CrossRef]
168. Inagaki, T.; Tachibana, M.; Magoori, K.; Kudo, H.; Tanaka, T.; Okamura, M.; Naito, M.; Kodama, T.; Shinkai, Y.; Sakai, J. Obesity and metabolic syndrome in histone demethylase JHDM2a-deficient mice. *Genes Cells* **2009**, *14*, 991–1001. [CrossRef]
169. Lee, J.-E.; Ge, K. Transcriptional and epigenetic regulation of PPARγ expression during adipogenesis. *Cell Biosci.* **2014**, *4*, 29. [CrossRef]
170. Steger, D.J.; Grant, G.; Schupp, M.; Tomaru, T.; Lefterova, M.I.; Schug, J.; Manduchi, E.; Stoeckert, C.J.; Lazar, M.A. Propagation of adipogenic signals through an epigenomic transition state. *Genes Dev.* **2010**, *24*, 1035–1044. [CrossRef]
171. Erener, S.; Hesse, M.; Kostadinova, R.; Hottiger, M.O. Poly(ADP-Ribose)Polymerase-1 (PARP1) Controls Adipogenic Gene Expression and Adipocyte Function. *Mol. Endocrinol.* **2012**, *26*, 79–86. [CrossRef] [PubMed]
172. Fu, M.; Rao, M.; Bouras, T.; Wang, C.; Wu, K.; Zhang, X.; Li, Z.; Yao, T.-P.; Pestell, R.G. Cyclin D1 Inhibits Peroxisome Proliferator-activated Receptor γ-mediated Adipogenesis through Histone Deacetylase Recruitment. *J. Biol. Chem.* **2005**, *280*, 16934–16941. [CrossRef] [PubMed]
173. Jin, Q.; Zhang, F.; Yan, T.; Liu, Z.; Wang, C.; Ge, X.; Zhai, Q. C/EBPα regulates SIRT1 expression during adipogenesis. *Cell Res.* **2010**, *20*, 470–479. [CrossRef] [PubMed]
174. Picard, F.; Kurtev, M.; Chung, N.; Topark-Ngarm, A.; Senawong, T.; Machado De Oliveira, R.; Leid, M.; McBurney, M.W.; Guarente, L. Sirt1 promotes fat mobilization in white adipocytes by repressing PPAR-gamma. *Nature* **2004**, *429*, 771–776. [CrossRef]
175. Mayoral, R.; Osborn, O.; McNelis, J.; Johnson, A.M.; Oh, D.Y.; Izquierdo, C.L.; Chung, H.; Li, P.; Traves, P.G.; Bandyopadhyay, G.; et al. Adipocyte SIRT1 knockout promotes PPARγ activity, adipogenesis and insulin sensitivity in chronic-HFD and obesity. *Mol. Metab.* **2015**, *4*, 378–391. [CrossRef]
176. Ferrari, A.; Fiorino, E.; Longo, R.; Barilla, S.; Mitro, N.; Cermenati, G.; Giudici, M.; Caruso, D.; Mai, A.; Guerrini, U.; et al. Attenuation of diet-induced obesity and induction of white fat browning with a chemical inhibitor of histone deacetylases. *Int. J. Obes.* **2016**, *41*, 289–298. [CrossRef]
177. Aguilar, E.C.; Silva, J.; Navia-Pelaez, J.M.; Leonel, A.J.; Lopes, L.G.; Menezes-Garcia, Z.; Ferreira, A.; Capettini, L.; Teixeira, L.G.; Lemos, V.S.; et al. Sodium butyrate modulates adipocyte expansion, adipogenesis, and insulin receptor signaling by upregulation of PPAR-γ in obese Apo E knockout mice. *Nutrition* **2017**, *47*, 75–82. [CrossRef]
178. Yoon, G.-E.; Jung, J.K.; Lee, Y.-H.; Jang, B.-C.; Kim, J.I. Histone deacetylase inhibitor CG200745 ameliorates high-fat diet-induced hypertension via inhibition of angiotensin II production. *Naunyn-Schmiedeberg's Arch. Pharmacol.* **2019**, *393*, 491–500. [CrossRef]
179. Bian, F.; Ma, X.; Villivalam, S.D.; You, D.; Choy, L.R.; Paladugu, A.; Fung, S.; Kang, S. TET2 facilitates PPARγ agonist–mediated gene regulation and insulin sensitization in adipocytes. *Metabolism* **2018**, *89*, 39–47. [CrossRef]
180. Chen, J.; Liu, Y.; Lu, S.; Yin, L.; Zong, C.; Cui, S.; Qin, D.; Yang, Y.; Guan, Q.; Li, X.; et al. The role and possible mechanism of lncRNA U90926 in modulating 3T3-L1 preadipocyte differentiation. *Int. J. Obes.* **2016**, *41*, 299–308. [CrossRef]

181. Cooper, D.R.; Carter, G.; Li, P.; Patel, R.; Watson, J.E.; Patel, N.A. Long Non-Coding RNA NEAT1 Associates with SRp40 to Temporally Regulate PPARγ2 Splicing during Adipogenesis in 3T3-L1 Cells. *Genes* **2014**, *5*, 1050–1063. [CrossRef] [PubMed]
182. Divoux, A.; Karastergiou, K.; Xie, H.; Guo, W.; Perera, R.J.; Fried, S.K.; Smith, S.R. Identification of a novel lncRNA in gluteal adipose tissue and evidence for its positive effect on preadipocyte differentiation. *Obesity* **2014**, *22*, 1781–1785. [CrossRef] [PubMed]
183. Zhang, Y.; Gu, M.; Ma, Y.; Peng, Y. LncRNA TUG1 reduces inflammation and enhances insulin sensitivity in white adipose tissue by regulating miR-204/SIRT1 axis in obesity mice. *Mol. Cell. Biochem.* **2020**, *475*, 171–183. [CrossRef] [PubMed]
184. Skårn, M.; Namløs, H.M.; Noordhuis, P.; Wang, M.-Y.; Meza-Zepeda, L.A.; Myklebost, O. Adipocyte Differentiation of Human Bone Marrow-Derived Stromal Cells Is Modulated by MicroRNA-155, MicroRNA-221, and MicroRNA-222. *Stem Cells Dev.* **2012**, *21*, 873–883. [CrossRef] [PubMed]
185. Chen, L.; Chen, Y.; Zhang, S.; Ye, L.; Cui, J.; Sun, Q.; Li, K.; Wu, H.; Liu, L. MiR-540 as a Novel Adipogenic Inhibitor Impairs Adipogenesis Via Suppression of PPARγ. *J. Cell. Biochem.* **2015**, *116*, 969–976. [CrossRef]
186. Li, H.; Xue, M.; Xu, J.; Qin, X. MiR-301a is involved in adipocyte dysfunction during obesity-related inflammation via suppression of PPARgamma. *Pharmazie* **2016**, *71*, 84–88.
187. Huang, Q.; Ma, C.; Chen, L.; Luo, D.; Chen, R.; Liang, F. Mechanistic Insights into the Interaction between Transcription Factors and Epigenetic Modifications and the Contribution to the Development of Obesity. *Front. Endocrinol.* **2018**, *9*, 370. [CrossRef]
188. McGregor, R.A.; Choi, M.S. microRNAs in the Regulation of Adipogenesis and Obesity. *Curr. Mol. Med.* **2011**, *11*, 304–316. [CrossRef]
189. Xie, H.; Lim, B.; Lodish, H.F. MicroRNAs Induced During Adipogenesis that Accelerate Fat Cell Development Are Downregulated in Obesity. *Diabetes* **2009**, *58*, 1050–1057. [CrossRef]
190. Kolb, H.; Martin, S. Environmental/lifestyle factors in the pathogenesis and prevention of type 2 diabetes. *BMC Med.* **2017**, *15*, 131. [CrossRef]
191. Ojo, O. Dietary Intake and Type 2 Diabetes. *Nutrients* **2019**, *11*, 2177. [CrossRef] [PubMed]
192. Henning, R.J. Type-2 diabetes mellitus and cardiovascular disease. *Future Cardiol.* **2018**, *14*, 491–509. [CrossRef]
193. Ogurtsova, K.; da Rocha Fernandes, J.D.; Huang, Y.; Linnenkamp, U.; Guariguata, L.; Cho, N.; Cavan, D.; Shaw, J.; Makaroff, L. IDF Diabetes Atlas: Global estimates for the prevalence of diabetes for 2015 and 2040. *Diabetes Res. Clin. Pract.* **2017**, *128*, 40–50. [CrossRef] [PubMed]
194. Naidoo, V.; Naidoo, M.; Ghai, M. Cell- and tissue-specific epigenetic changes associated with chronic inflammation in insulin resistance and type 2 diabetes mellitus. *Scand. J. Immunol.* **2018**, *88*, e12723. [CrossRef] [PubMed]
195. Kang, S.; Tsai, L.; Zhou, Y.; Evertts, A.G.; Xu, S.; Griffin, M.; Issner, R.; Whitton, H.J.; Garcia, B.A.; Epstein, C.B.; et al. Identification of nuclear hormone receptor pathways causing insulin resistance by transcriptional and epigenomic analysis. *Nat. Cell. Biol.* **2015**, *17*, 44–56. [CrossRef] [PubMed]
196. Jiang, X.; Ye, X.; Guo, W.; Lu, H.; Gao, Z. Inhibition of HDAC3 promotes ligand-independent PPARγ activation by protein acetylation. *J. Mol. Endocrinol.* **2014**, *53*, 191–200. [CrossRef]
197. Li, D.; Wang, X.; Ren, W.; Ren, J.; Lan, X.; Wang, F.; Zhang, F.; Han, Y.; Song, T.; Holmdahl, R.; et al. High expression of liver histone deacetylase 3 contributes to high-fat-diet-induced metabolic syndrome by suppressing the PPAR-γ and LXR-α-pathways in E3 rats. *Mol. Cell. Endocrinol.* **2011**, *344*, 69–80. [CrossRef]
198. Sathishkumar, C.; Prabu, P.; Balakumar, M.; Lenin, R.; Prabhu, D.; Anjana, R.M.; Mohan, V.; Balasubramanyam, M. Augmentation of histone deacetylase 3 (HDAC3) epigenetic signature at the interface of proinflammation and insulin resistance in patients with type 2 diabetes. *Clin. Epigenet.* **2016**, *8*, 125. [CrossRef]
199. Ahmadian, M.; Suh, J.M.; Hah, N.; Liddle, C.; Atkins, A.R.; Downes, M.; Evans, R.M. PPARγ signaling and metabolism: The good, the bad and the future. *Nat. Med.* **2013**, *19*, 557–566. [CrossRef]
200. Makkar, R.; Behl, T.; Arora, S. Role of HDAC inhibitors in diabetes mellitus. *Curr. Res. Transl. Med.* **2020**, *68*, 45–50. [CrossRef]
201. Hu, Y.; Liu, J.; Yuan, Y.; Chen, J.; Cheng, S.; Wang, H.; Xu, Y. Sodium butyrate mitigates type 2 diabetes by inhibiting PERK-CHOP pathway of endoplasmic reticulum stress. *Environ. Toxicol. Pharmacol.* **2018**, *64*, 112–121. [CrossRef] [PubMed]
202. Rafehi, H.; Kaspi, A.; Ziemann, M.; Okabe, J.; Karagiannis, T.C.; El-Osta, A. Systems approach to the pharmacological actions of HDAC inhibitors reveals EP300 activities and convergent mechanisms of regulation in diabetes. *Epigenetics* **2017**, *12*, 991–1003. [CrossRef] [PubMed]
203. Fujisaka, S. The role of adipose tissue M1/M2 macrophages in type 2 diabetes mellitus. *Diabetol. Int.* **2020**, *12*, 74–79. [CrossRef] [PubMed]
204. Lu, J.; Zhao, J.; Meng, H.; Zhang, X. Adipose Tissue-Resident Immune Cells in Obesity and Type 2 Diabetes. *Front. Immunol.* **2019**, *10*, 1173. [CrossRef] [PubMed]
205. Ahmed, M.; de Winther, M.P.; Bossche, J. Epigenetic mechanisms of macrophage activation in type 2 diabetes. *Immunobiology* **2017**, *222*, 937–943. [CrossRef]
206. Bouhlel, M.A.; Derudas, B.; Rigamonti, E.; Dievart, R.; Brozek, J.; Haulon, S.; Zawadzki, C.; Jude, B.; Torpier, G.; Marx, N.; et al. PPARgamma activation primes human monocytes into alternative M2 macrophages with anti-inflammatory properties. *Cell Metab.* **2007**, *6*, 137–143. [CrossRef]

207. Bouhlel, M.A.; Brozek, J.; Derudas, B.; Zawadzki, C.; Jude, B.; Staels, B.; Chinetti-Gbaguidi, G. Unlike PPARgamma, PPARalpha or PPARbeta/delta activation does not promote human monocyte differentiation toward alternative macrophages. *Biochem. Biophys. Res. Commun.* **2009**, *386*, 459–462. [CrossRef]
208. Charo, I.F. Macrophage Polarization and Insulin Resistance: PPARγ in Control. *Cell Metab.* **2007**, *6*, 96–98. [CrossRef]
209. Wang, X.; Cao, Q.; Yu, L.; Shi, H.; Xue, B.; Shi, H. Epigenetic regulation of macrophage polarization and inflammation by DNA methylation in obesity. *JCI Insight* **2016**, *1*, e87748. [CrossRef]
210. Yang, X.; Wang, X.; Liu, D.; Yu, L.; Xue, B.; Shi, H. Epigenetic Regulation of Macrophage Polarization by DNA Methyltransferase 3b. *Mol. Endocrinol.* **2014**, *28*, 565–574. [CrossRef]
211. You, D.; Nilsson, E.; Tenen, D.E.; Lyubetskaya, A.; Lo, J.C.; Jiang, R.; Deng, J.; Dawes, B.A.; Vaag, A.; Ling, C.; et al. Dnmt3a is an epigenetic mediator of adipose insulin resistance. *Elife* **2017**, *6*, e30766. [CrossRef]
212. Kim, A.Y.; Park, Y.J.; Pan, X.; Shin, K.C.; Kwak, S.-H.; Bassas, A.F.; Sallam, R.M.; Park, K.S.; Alfadda, A.; Xu, A.; et al. Obesity-induced DNA hypermethylation of the adiponectin gene mediates insulin resistance. *Nat. Commun.* **2015**, *6*, 7585. [CrossRef]
213. Chen, T.; Zhang, Y.; Liu, Y.; Zhu, D.; Yu, J.; Li, G.; Sun, Z.; Wang, W.; Jiang, H.; Hong, Z. MiR-27a promotes insulin resistance and mediates glucose metabolism by targeting PPAR-γ-mediated PI3K/AKT signaling. *Aging (Albany NY)* **2019**, *11*, 7510–7524. [CrossRef]
214. Yao, F.; Yu, Y.; Feng, L.; Li, J.; Zhang, M.; Lan, X.; Yan, X.; Liu, Y.; Guan, F.; Zhang, M.; et al. Adipogenic miR-27a in adipose tissue upregulates macrophage activation via inhibiting PPARγ of insulin resistance induced by high-fat diet-associated obesity. *Exp. Cell Res.* **2017**, *355*, 105–112. [CrossRef]
215. Liang, H.; Ward, W.F. PGC-1α: A key regulator of energy metabolism. *Adv. Physiol. Educ.* **2006**, *30*, 145–151. [CrossRef] [PubMed]
216. Fernandez-Marcos, P.J.; Auwerx, J. Regulation of PGC-1α, a nodal regulator of mitochondrial biogenesis. *Am. J. Clin. Nutr.* **2011**, *93*, 884S–890S. [CrossRef]
217. Jeninga, E.H.; Schoonjans, K.; Auwerx, J. Reversible acetylation of PGC-1: Connecting energy sensors and effectors to guarantee metabolic flexibility. *Oncogene* **2010**, *29*, 4617–4624. [CrossRef] [PubMed]
218. Rodgers, J.; Lerin, C.; Gerhart-Hines, Z.; Puigserver, P. Metabolic adaptations through the PGC-1α and SIRT1 pathways. *FEBS Lett.* **2007**, *582*, 46–53. [CrossRef] [PubMed]
219. Gerhart-Hines, Z.; Rodgers, J.; Bare, O.; Lerin, C.; Kim, S.-H.; Mostoslavsky, R.; Alt, F.W.; Wu, Z.; Puigserver, P. Metabolic control of muscle mitochondrial function and fatty acid oxidation through SIRT1/PGC-1α. *EMBO J.* **2007**, *26*, 1913–1923. [CrossRef] [PubMed]
220. Sharabi, K.; Lin, H.; Tavares, C.D.; Dominy, J.E.; Camporez, J.P.; Perry, R.J.; Schilling, R.; Rines, A.K.; Lee, J.; Hickey, M.; et al. Selective Chemical Inhibition of PGC-1α Gluconeogenic Activity Ameliorates Type 2 Diabetes. *Cell* **2017**, *169*, 148–160.e15. [CrossRef] [PubMed]

International Journal of
Molecular Sciences

MDPI

Review

The PPARg System in Major Depression: Pathophysiologic and Therapeutic Implications

Philip W. Gold

National Institutes of Health, Bethesda, MD 20892, USA; philipgold@mail.nih.gov; Tel.: +1-301-605-5902

Abstract: To an exceptional degree, and through multiple mechanisms, the PPARg system rapidly senses cellular stress, and functions in the CNS in glial cells, neurons, and cerebrovascular endothelial cell in multiple anti-inflammatory and neuroprotective ways. We now know that depression is associated with neurodegeneration in the subgenual prefrontal cortex and hippocampus, decreased neuroplasticity, and defective neurogenesis. Brain-derived neurotrophic factor (BDNF) is markedly depleted in these areas, and is thought to contribute to the neurodegeneration of the subgenual prefrontal cortex and the hippocampus. The PPARg system strongly increases BDNF levels and activity in these brain areas. The PPARg system promotes both neuroplasticity and neurogenesis, both via effects on BDNF, and through other mechanisms. Ample evidence exists that these brain areas transduce many of the cardinal features of depression, directly or through their projections to sites such as the amygdala and nucleus accumbens. Behaviorally, these include feelings of worthlessness, anxiety, dread of the future, and significant reductions in the capacity to anticipate and experience pleasure. Physiologically, these include activation of the CRH and noradrenergic system in brain and the sympathetic nervous system and hypothalamic–pituitary–adrenal axis in the periphery. Patients with depression are also insulin-resistant. The PPARg system influences each of these behavioral and physiological in ways that would ameliorate the manifestations of depressive illness. In addition to the cognitive and behavioral manifestations of depression, depressive illness is associated with the premature onsets of coronary artery disease, stroke, diabetes, and osteoporosis. As a consequence, patients with depressive illness lose approximately seven years of life. Inflammation and insulin resistance are two of the predominant processes that set into motion these somatic manifestations. PPARg agonists significantly ameliorate both pathological processes. In summary, PPARg augmentation can impact positively on multiple significant pathological processes in depression. These include loss of brain tissue, defective neuroplasticity and neurogenesis, widespread inflammation in the central nervous system and periphery, and insulin resistance. Thus, PPARg agonists could potentially have significant antidepressant effects.

Keywords: depression; PPARg; inflammation; neuropathology; corticotropin releasing hormone; norepinephrine; subgenual prefrontal cortex; amygdala; nucleus accumbens

Citation: Gold, P.W. The PPARg System in Major Depression: Pathophysiologic and Therapeutic Implications. *Int. J. Mol. Sci.* **2021**, *22*, 9248. https://doi.org/10.3390/ijms22179248

Academic Editor: Manuel Vázquez-Carrera

Received: 1 July 2021
Accepted: 13 August 2021
Published: 26 August 2021

1. Introduction

The World Health Organization ranks depression as the second greatest cause of disability worldwide. In addition to its potentially crippling affective and cognitive elements, depression is associated with the premature onset of multiple systemic diseases, including coronary artery disease [1], stroke [2], diabetes [3,4], and osteoporosis [5]. These disorders are primarily mediated by the same changes in the central nervous system that transduce the cognitive and affective components of depressive illness. Overall, patients with depressive illness lose approximately seven years of their lives because of these somatic manifestations of the disorder [6].

Multiple lines of evidence indicate that depression represents a stress system that has become highly dysregulated [7–11]. Among the key regulators of the stress response in the brain and periphery is the peroxisome proliferator activated receptor gamma (PPARg)

system, which I postulate plays a major role in depression, especially in one of the most pronounced concomitants of depression, namely central and peripheral inflammation [12]. Inflammation in these loci contributes to the cognitive and affective components, as well as to their somatic manifestations [9]. This paper will first document that the stress system runs awry in depression. I will also describe the organization of the stress system and the structural changes of the stress system during stress, provide a brief overview of the behavioral manifestations of depression, and detail the changes in structure and function that the stress system undergoes in encoding the clinical and biochemical manifestations of depressive illness. Finally, I will provide an overview of the biological changes associated with depression, with particular emphasis on the extensive roles of the PPARg system in depressive illness. This roles of stress system dysregulation in depression and of the PPARg's role in regulating the stress system response and its dysregulation in depression are extremely important, and hence will be the highlights of this publication [13].

There are two principal subtypes of depression that I will show have different manifestation of stress system dysfunctional activity. These are melancholic and atypical depression [9]. Melancholic depression is inconsistent with the word depression in that it is often a state of hyperarousal and anxiety, often attached to the self and experienced as an anguished sense of worthlessness compounded by insomnia, anorexia, and a host of metabolic and physiologic perturbations that interfere with the quality of life while significantly reducing the lifespan. Patients with melancholia have increased secretion of the stress hormones cortisol and norepinephrine. Their symptoms are worse in the morning when the stress system is at its peak. As an example of an activated stress system in depression, Figure 1 shows striking elevations in CSF and plasma norepinephrine and epinephrine levels in hourly samples drawn for thirty consecutive hours. Figure 2 shows increased norepinephrine spillover in mild–moderately depressed patients with melancholia.

Atypical depression seems to be the antithesis of melancholic depression and is associated with lethargy, fatigue, hypersomnia, and hyperphagia. The symptoms of atypical depression peak in the evening, when the stress system is most quiescent.

Rene Spitz made important observations regarding developmental abnormalities that afflicted infants placed soon after birth in understaffed orphanages, where they had little or no sustained human contact. Initially, the infants cried bitterly for hours until attended to. Subsequently, they withdrew and ceased crying at all, even if they were left alone or had gone without eating for hours. In addition, they lost interest in their environment. Thus, it were as if their early deprivation had led to a virtual shutdown of their stress response and affective existence to protect them from enormous distress [14]. Subsequent studies in non-human primates who were removed from their mothers and raised by peers reveal a similar behavioral withdrawal in association with significant inhibition of the HPA axis [15]. These may represent very severe forms of atypical depression.

Much less is known about atypical depression than about melancholia, thus much of this paper is focused on this depressive subtype. Overall, the weight of available data indicates that melancholic depression reflects a stress system that is pathologically activated, while atypical depression reflects a stress system that has been pathologically suppressed.

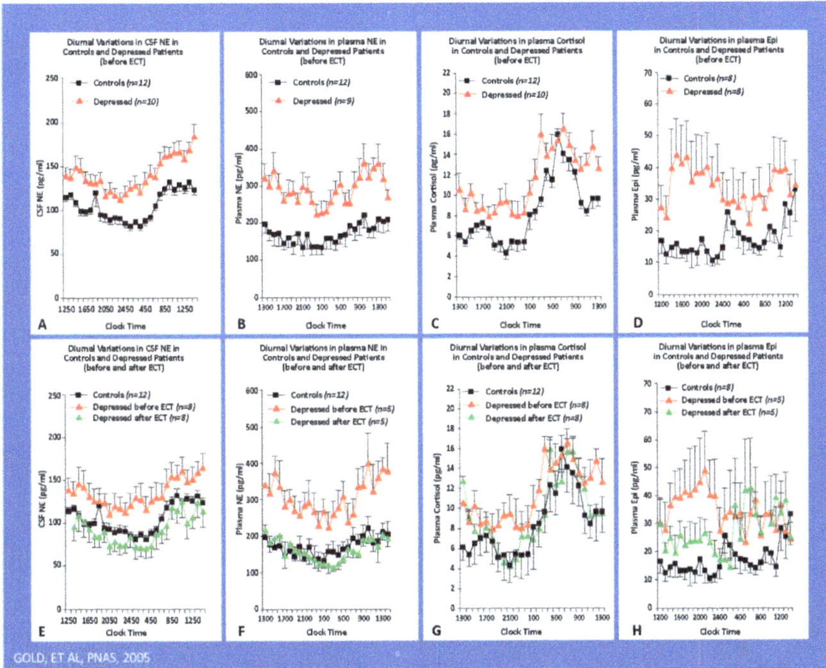

Figure 1. The hypernoradrenergic state of depression as assessed by around-the-clock hourly CSF and plasma sampling. The hypernoradrenergic state of melancholic depression. Hourly around-the-clock sampling of CSF NE, plasma NE, plasma cortisol, and plasma epinephrine in severely depressed, medication-free patients with melancholic depression. We studied patients during depression and after ECT-induced remission. Severely depressed patients had concomitant elevations of the hourly 24 h levels of CSF NE, plasma NE, plasma and CSF epinephrine, and plasma cortisol. These levels all fell to normal levels after ECT. The diurnal variations of CSF NE, plasma NE, and plasma cortisol were virtually superimposable and highly correlated with one another. Their arithmetic means also were highly correlated. These arousal-producing compound levels all peaked at 08:00 to 09:00, a time when melancholic symptoms are at their worst. Their peaks also coincide with the time for maximal susceptibility to myocardial infarction and sudden death. CSF NE and plasma norepinephrine correlate with one another, yet they derive from different sets of neurons. Patients with the Shy Drager syndrome have very low plasma norepinephrine levels in association with robust CSF norepinephrine concentrations. Excessive central norepinephrine secretion in melancholia exerts several adverse effects. Norepinephrine inhibits critical structures in the prefrontal cortex such as the subgenual and dorsolateral prefrontal cortices. NE stimulates the amygdala and the CRH/HPA axis. Central noradrenergic excess also contributes to hypertension, activation of the HPA axis, and the sympathetic nervous system. Plasma and CSF epinephrine were also elevated around the clock.

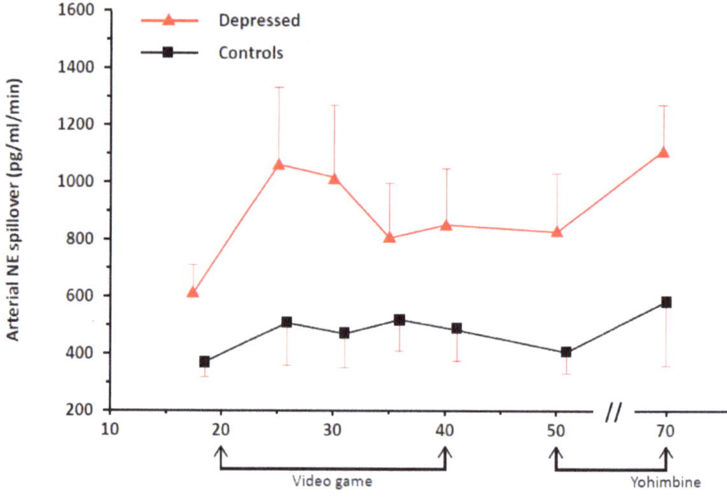

Figure 2. Norepinephrine spillover into arterial plasma corrected for norepinephrine clearance at baseline, during the stress of a video game, and after the infusion of yohimbine, an a-2 noradrenergic antagonist that increases the secretion of norepinephrine by blocking the inhibitory norepinephrine a-2 receptor ($p < 0.01$) for all findings.

2. Organization of the Normal Stress System: Template for Depressive Illness

Anxiety is a cardinal manifestation of the stress response and is essential for survival. The PPARg system is highly activated during stress and plays multiple roles [16], including an anxiolytic one [17]. PPARg receptors are widely distributed in the amygdala and in the medial prefrontal cortex [18].

The CRH system is activated during stress. CRH is primarily in the hypothalamus and amygdala, and transduces activation of the hypothalamic–pituitary–adrenal axis, while amygdala CRH transduces anxiety and conditioned fear. Accordingly, CRH sets into motion multiple behavioral and physiological phenomena during stress. These include anxiety, hyperarousal, fear-related behaviors, activation of the sympathomedullary system, and hypothalamic–pituitary–adrenal activation. CRH is a potent stimulus to the activation of inflammation, which I shall show is highly increased during physical and psychological stress and in depression [9]. Inflammation is activated during stress as a premonitory response to the likely contingency of injury during stressful confrontations [9]. CRH diminishes sleep, food intake, sexual activity, and the capacity to anticipate or experience pleasure [9]. These actions serve many functions. One of their most important functions is to prevent distraction during threatening situations. The PPARg system restrains the CRH system and the activation of the sympathetic nervous system [19]. Thus, attention is directed primarily to the danger at hand.

As noted earlier, as examples of the activation of the stress response, we first noted that both CSF and plasma norepinephrine and epinephrine are elevated around the clock in drug-free patients with melancholia in hourly samples taken for 30 consecutive hours (Figure 1). Both compounds are arousal-producing and anxiogenic in brain and transduce multiple autonomic and metabolic aspects of the stress response [8,10,20]. We also found norepinephrine spillover was significantly increased in mild–moderately melancholic depressed patients [20] (Figure 2).

The dorsolateral prefrontal cortex is modestly inhibited, leading to a decrease in the cognitive control of anxiety [21]. Emotional memories of past confrontations with stress or danger are readily retrieved to support survival in the present threatening situation.

Stress is accompanied by a small, but significant down-regulation of the subgenual prefrontal cortex [9,22]. This structure restrains the amygdala fear system; estimates the likelihood of punishment and reward; helps prime the nucleus accumbens or reward center of the brain, thereby increasing the capacity to anticipate and experience pleasure. The subgenual prefrontal cortex also restrains the CRH and locus ceruleus-norepinephrine system and the sympathetic nervous system [9]. Taken together, this modest reduction in the size and functional activity of the subgenual prefrontal cortex leads to increased anxiety, increased expectation of harm, decreased capacity to anticipate or experience pleasure, activation of the locus ceruleus, and disinhibition of the CRH system and hypothalamic–pituitary–adrenal axis. As I will note, the PPARg system intersects with each of these important processes.

The amygdala, necessary for the cognitive experience of fear, is modestly disinhibited, partially through the decrease in the activity of the subgenual prefrontal cortex. Anxiety is kept manageable to prevent its interference with a successful stress response. The amygdala down-regulates the nucleus accumbens reward system to prevent distraction. It is difficult to experience pleasure when you are afraid. An activated amygdala also stimulates the CRH and sympathetic nervous systems via a CRH projection from the amygdala. As noted, the amygdala is amply supplied by PPARg receptors [17,18].

The hippocampus responds to cortisol and norepinephrine, which promote the encoding and retrieval of negatively charged emotional memories [23,24]. Mineralocorticoid receptors in the hippocampus exert negative feedback effects on the CRH system [25]. A direct projection from the anterior hippocampus to the subgenual prefrontal cortex modifies the subgenual prefrontal cortex during a stress response. The hippocampus also contains ample numbers of PPARg receptors [18].

The insula helps to control the shifting between the default mode network and the salient mode network. When the default mode system is activated in the context of an activated amygdala, as in depression, attention turns inward, leading to adverse self-assessments and contributing to the feelings of worthlessness, which are such critical elements in depression [17,26]. The default mode network is significantly hypoactive in depression and returns to its normal level of activation with successful antidepressant treatment. The insula has ample PPARg receptors that activate its capacity to promote a successful stress response [27].

As noted, the nucleus accumbens is modestly downregulated during stress, but not sufficiently to impede an effective stress response [28,29]. PPARg receptors are plentiful in the nucleus accumbens [18].

The locus ceruleus is the principal site of norepinephrine synthesis in the brain. It produces a state of general alarm, promotes anxiety, and plays an important role in activating the physiological aspects of the stress response.

During normal stress, there is a highly significant increase in neuroplasticity in the subgenual prefrontal cortex, amygdala, and hippocampus, and an increase in neurogenesis, all of which assist in effective response to stressful and rapidly changing circumstances [12]. Stress can also be associated with damage to brain cells that may be alleviated by PPARg agonists [12]. PPARg agonists are potent stimuli to both neuroplasticity and neurogenesis, which, I will show, are both markedly decreased in depression [12].

3. Behavioral and Cognitive Manifestations of Melancholic Depression

Melancholic depression is often associated with significant anxiety, hyperarousal, decreased appetite, and decreased sleep. As noted, the focus on the melancholic patient is on the inner self with active self-assessment, which, because of amygdala activation, is highly negative. Attention is focused on sad stimuli and there is significant difficulty in effectively disengaging from them. There is preferential access to negatively charged emotional memories that are highly resistant to extinction. Prior stressful events that are encoded in emotional memory and the affect associated with them are prominent in melancholia and are reinforced by cortisol and norepinephrine [9]. PPARg agonists down-regulate cortisol levels [12,30].

Recent data reveal that deletion of neuronal PPARg enhances the emotional response to acute stress and exacerbates anxiety. Importantly, these effects are reversed by rescue of amygdala PPARg function [17].

4. Dysregulation of the Stress System in Melancholic Depression: Evidence for a Pathological Activation of the Stress System

The size of the subgenual prefrontal cortex is reduced in melancholia by as much as 40% [31], owing to a loss of glial cells, with a marked decrease in its neuroplasticity, a decrease in the size of its neurons, and substantial loss of key synaptic proteins [32] (Figure 3). Stress and hypercortisolism are two factors that decrease the size and activity of this crucial structure [32]. As noted, one of the functions of the subgenual prefrontal cortex, an important component of the default mode network, is to participate in the process of self-assessment, and its impairment leads to a loss of self-esteem. Earlier, I pointed out that the subgenual prefrontal cortex is an important component of the medial prefrontal cortex, and although PPARg receptors have not been assessed in the subgenual prefrontal cortex, they are abundantly present in the medial prefrontal cortex [18].

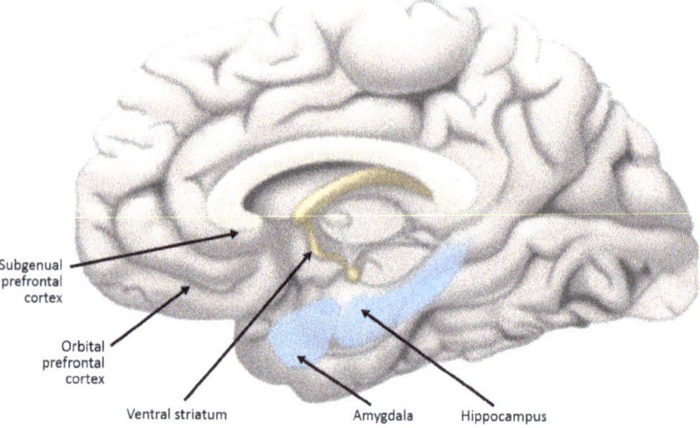

Figure 3. Sagittal section of the human brain. Structures playing a particularly important role in the pathophysiology of depression that are the targets of PPARg-amelioration of multiple core components of depressive illness. Please see text for descriptions, roles, and connections.

As noted earlier, we first demonstrated that the CRH system was activated in melancholic patients [11], as were indices of increased norepinephrine secretion, as assessed hourly through indwelling canullae in the antecubital space and spinal canal (Figure 1). The highly elevated levels in melancholia fell to normal after electroconvulsive shock induced remission [8,10,20]. We also showed that melancholic depressed patients have increased norepinephrine spillover into arterial plasma at rest, in response to a video game, and in response to yohimbine, an alpha-2 noradrenergic antagonist (Figure 2) [20]. Along with CRH hypersecretion in melancholia, these lead to further anxiety and arousal, decreased sleep, and decreased food intake and sexual activity. Once again, chronic stress in experimental animals leads to these changes.

The nucleus accumbens (Figure 3), ordinarily primed by the subgenual prefrontal cortex, becomes enlarged and much less responsive to pleasurable stimuli. This leads to anhedonia, one of the cardinal manifestations of depression. Hypercortisolism contributes to this phenomenon [9]. The nucleus accumbens is amply supplied with PPARg receptors.

The hippocampus is decreased in size in depression (Figure 3). Neurogenesis is markedly reduced. Neuroplasticity is also significantly diminished. There is a decrement in BDNF that is a principal cause of these phenomena [33]. Ample data report that PPARg agonists significantly increase BDNF levels. The BDNF deficiency in depression is an important component of its pathophysiology. Some feel that it is among the most critical pathophysiological mediators in depressive illness [33,34]. Thus, PPARg agonists' capacity to increase BDNF could be an important component of its possible therapeutic effects in depressive illness [35].

In addition, the dorsolateral prefrontal cortex is reduced in size in depression. Rich in PPARg receptors, the dorsolateral prefrontal cortex loses capacity to exert emotional control over cognition.

Depressive illness is not only associated with changes in structure and the function of sites altered in depression, but also with synaptic loss and deficits in functional connectivity [36–39]. Post-mortem research has demonstrated lower numbers of synapses and, correspondingly, lower expression of synaptic function-related genes in the dorsolateral prefrontal cortex in patients with depressive illness, consistent with the loss of lower levels of synaptic signaling proteins in these patients [40]. Lower synaptic density is associated with higher severity of depressive symptoms [40].

Almost all the manifestations of melancholia are corrected, either partially or fully, by effective antidepressant treatment [34,36,37,41–45]. These include partial restoration of the volume of the subgenual PFC and its functions [46], restoration of the hypoactivity of the dorsolateral prefrontal cortex [21,40], reduction in the size and activity of the amygdala [47], restoration of the volume of the hippocampus, normalization of its neuroplasticity, and restoration of neurogenesis [48]. There is normalization of the activations of the CRH and catecholaminergic systems, as well as of the growth hormone and reproductive axes [49,50]. BDNF levels rise significantly in the PFC and the hippocampus [33]. In addition, there is a decrease in the size of the nucleus accumbens and restoration of the normal capacity to experience pleasure. Cognitive function improves significantly [35,51]. Inflammation, which I will discuss extensively in the next section as a key component of depressive illness, also falls to normal levels [52].

5. Inflammation in Depression

Inflammation is an important component of depressive pathophysiology. Over 10,000 papers have been written on inflammation in depression. The PPARg system has highly significant effects in the restraint of multiple form of inflammation. For this reason, I will cover inflammation in depression in more detail.

Inflammation in Depression: A Key Factor that Contributes to Making PPARg Agonists of Relevance to Depression

Inflammation is an inherent component of depressive illness. This inflammation both influences the brain and is widespread in the rest of the body [53,54]. Thousands of papers have been written concerning the connection between inflammation and depression. Recent studies of animal models of stress such as social defeat, predatory stress, and resident intruders show that such stressors induce central nervous system inflammation, characterized by the secretion of cytokines, and evidence of neuronal inflammation [54]. This inflammation occurred in many areas of the brain thought to be involved in depression such as the subgenual prefrontal cortex and the amygdala [12,26]. The administration of blockers to inflammatory compounds blocked the impact of the stressors on behavior, including a depression-like picture. The administration of antidepressants prior to severe stress prevented any signs of neuroinflammation. Similarly, antidepressants in humans correct the evidence of peripheral and central inflammation [44,45,53].

Before discussing the many manifestations of inflammation in depressive illness, I would first like to discuss the evolutionary and biological roots in environments long ago that have fostered the close connection between depression and inflammation [53,54]. Almost all stressful situations that mammals encountered included risks implicit in being hunted, hunting, or competing for reproductive status or access. Thus, there was a premonitory activation of the inflammatory system anticipating possible injury and infection, because the risk of pathogen exposure and a consequent infection was highly increased [53,54]. Thus, in ancient environments, the connection between the perception of stress or danger and the risk of subsequent tissue injury was so expectable that evolution favored organisms that activated responses of inflammatory systems to a wide variety of environmental stressors, including psychosocial stressors [53,54]. Inflammatory mediators are among the most potent stimuli leading to the activation of prominent stress mediators such as activations of the sympathetic nervous system and hypothalamic–pituitary–adrenal axis. Data show that concentrations of IL-6 as low as 10^{-18} molar activated hypothalamic CRH neurons and the hypothalamic–pituitary–adrenal axis [9].

CRH plays a major role in neuron–microglial interactions in the CNS [47]. Among the most powerful microglia-activating factors is CRH, which plays many other roles in inflammation. CRH converts resting microglia into activated microglia, an effect suppressed by antidepressants [55], which we have shown to consistently down-regulate the CRH system [25,56]. Activated microglia lead to inflammation mediated by multiple cytokines and other proinflammatory compounds such as those associated with oxidative stress [55]. As noted, CRH is also among the most potent stimuli to mast cell degranulation in the CNS, the blood brain barrier, and the periphery [57–59]. Thus, antalarmin, our CRH antagonist, seems like an ideal candidate to address a primary mechanism in this CNS inflammation. As noted, PPARg agonists have consistent and multiple anti-inflammatory effects in the CNS, to be covered in more detail below.

Consistent, with the evolutionary advantages of the partnership between the brain and the immune system, inflammatory mediators in the brain, including CRH and cytokines, influence brain areas that regulate motivation, motor activity, areas promoting social avoidance and energy conservation, as well as arousal and anxiety, and fear, providing warning against attack [9]. Inflammatory mediators have also been associated with reduction in reward responsiveness, thus decreasing the adverse consequences of distraction by pleasurable stimuli such as sexual activity and food consumption [9].

Stress-mediated activation of the CRH system also leads to the release of CRH from sympathetic terminals in the periphery [50]. CRH in a potent inflammatory mediator, in the periphery. This provides an explanation for the mechanism of stress-induced skin disorders that occur because of stress, including urticaria.

Women have a greater behavioral response to endotoxin and a much higher depression rate in the context of gamma-interferon administration [53]. Thus, by being more responsive to inflammation-induced depressive symptoms, women may have benefitted more from the protection conferred by these symptoms in terms of coping with infection, healing wounds, and subsequent exposure to pathogens [54]. Inflammation also inhibits fertility. This may have protected women from unwanted pregnancy during times of adversity. It is well-known that depression occurs almost twice as frequently in women than in men. It should be noted that inflammation in the context of stress in both women and men is a sterile information [12], unrelated to direct exposures to pathogens or the induction of physical injury. The sterile nature of inflammatory stress may work through substances that, for instance, induce oxidative stress. Sterile inflammation is also very sensitive to catecholamines.

Another term for sterile inflammation is parainflammation. Among the manifestations of parainflammation is a low level, persistent activation of inflammation characterized by modest elevation in the levels of the inflammatory marker CRP, well-known to be a predictor of heart disease [12,26].

Parainflammation occurs in response to stressors such as overfeeding or aging that were not present during our early evolutionary history, and for which we are not adequately prepared. These also include alterations in the light/dark cycle and exposure to novel foodstuffs or chemicals [12]. I have suggested that the kind of frequent daily psychological stressors we encounter in our lives now was not also present in our early evolutionary history, and that stress-responsive illnesses like depression may be a parainflammatory disorder as well [12]. Parainflammation is likely to contribute to the chronic inflammatory conditions associated with modern human diseases, resulting from stimuli to which we were not exposed early in our evolution.

The difference between classical inflammation and parainflammation is that the latter does not occur in response to pathogens or tissue damage, but rather from alterations from the normal set-point in tissues in response to stressors such as those involving nutrient sensing, energy metabolism, oxidant burden, endocrine regulation, and autonomic stability [12]. Parainflammatory mediators have effects beyond those on inflammatory phenomena, and help coordinate endocrine, metabolic, and autonomic activity as well [50]. Chronic stress is likely to set into motion alterations in the normal set-point for endocrine, metabolic, and inflammatory processes that build to such a degree that, in individuals who are genetically susceptible, depression develops.

6. Manifestation of Inflammation in Patients with Depressive Illness

Inflammation in the body and the brain is a prominent component of depressive illness, to the point that multiple studies conclude that anti-inflammatory agents are useful in the treatment of depression [53,54]. Patients with major depression have been found to have increased peripheral blood inflammatory biomarkers, including inflammatory cytokines, compounds produced by immune cells that activate other immune cells to encode a significant inflammatory response. These peripheral cytokines have been shown to access the brain and interact with virtually every pathophysiologic domain known to be involved in depression. These include neurotransmitter metabolism, neuroendocrine function, neurogenesis, and neural plasticity, as well as inflammation [53,54]. Indeed, activation of inflammatory pathways within the brain is believed to contribute to a confluence of decreased birth of new neurons, decreased neuroplasticity, increased glutamate release, as well as oxidative stress, leading to destruction of neurons and loss of glial elements, cells that provide nutritional and other support to nerve cells.

Depressed patients with increased inflammatory biomarkers have been found to be more likely to exhibit treatment resistance [54]. Moreover, multiple studies in depressed patients have demonstrated that antidepressant therapy leads to decreased inflammatory responses. Finally, preliminary data from patients with inflammatory disorders, as well as medically healthy depressed patients, suggest that inhibiting proinflammatory cytokines or their signaling pathways may improve depressed mood and increase treatment response to conventional antidepressant medication [45]. These findings include the possibility of identifying relevant patient populations, applying immune-targeted therapies, and monitoring therapeutic efficacy at the level of the immune system in addition to behavior.

Stressed experimental animals have central nervous system inflammation, characterized by the secretion of cytokines and evidence of neuronal inflammation. This may occur, in part, because of increased glutamate activity in depression [39]. Glutamate is a potent stimulus to central nervous system inflammation. This inflammation occurs in many areas of the brain thought to be involved in depression such as the subgenual prefrontal cortex and the amygdala.

7. More on the PPARg System

The PPARg receptor is a nuclear receptor found in neurons, glia, and cerebrovascular vessels in the frontal cortex, nucleus accumbens, striatum, hippocampus, hypothalamus, and substantia nigra. To an exceptional degree, and through multiple mechanisms, the PPARg system in the brain is widespread and rapidly senses CNS cellular stress [57] and functions in the CNS as a potent anti-inflammatory agent that protects neurons, glial cells, and cerebrovascular endothelial cells [58–65] from many inflammatory mediators of the innate immune system such as NF-kB, TNF-a, IFNg, IL-1b, and IL-6, as well as COX2, MCP1, and VCAM1.

In addition to these pleiotropic anti-inflammatory properties, the PPARg system is neuroprotective and affords protection from oxidative stress [60] and inappropriate apoptosis owing to excessive endoplasmic reticulum stress [61,62] and other stimuli. Thus, interference with the central PPARg system is associated with a pronounced down regulation in the availability of SOD-1 and glutathione S transferase [63], klotho, which functions as an anti-oxidant [64–67], and a marked increase in susceptibility to the adverse effects of excessive NMDA neurotransmission [19,68,69]. Interference with chaperones of the endoplasmic reticulum stress response, which promote a successful response, impairs the capacity for the endoplasmic reticulum stress response to handle oxidative stress and other neuronal stressors [70]. The PPARg system also promotes neuroplasticity and neurogenesis during periods of neuronal stress [71].

In addition to directly mediating anti-inflammatory effects and neuroprotection directly, the PPARg system is essential to the anti-inflammatory effects of compounds like angiotensin receptor-1 antagonists [72], whose anti-inflammatory responses are completed abolished by antagonism of neuronal PPARg receptors [72].

PPARg agonists also inhibit the CRH system [19], as well as the neurotoxic effects of norepinephrine [73] and glucocorticoid excess [19]. The striking anti-neuroinflammatory and neuroprotective effects of PPARg agonists have led to the recent initiation of pioglitazone, a PPARg agonist that crosses the blood brain barrier, to either treat or delay the progression of a variety of neurodegenerative diseases including Alzheimer's disease [74,75], Parkinson's disease [75], Huntington's disease [75], Friedrichs' ataxia [75], and the demyelination of multiple sclerosis [75]. In the periphery, an intact PPARg system is essential for ketamine-induced suppression of the innate immune system.

Prior to the elucidation of PPARg-mediated anti-inflammatory and cellular protective mechanisms in both the periphery and the CNS, the therapeutic efficacy of PPARg agonists was thought to reside solely in its capacity to significantly increase peripheral insulin sensitivity, protect pancreatic beta cells, and indirectly provide cardioprotection by ameliorating insulin resistance [76]. Thus, initially, PPARg agonists have been used primarily in the treatment type II diabetes, which is almost always associated with atherosclerosis and widespread inflammation [77]. Thus, because insulin is a potent stimulus to inflammation, pioglitazone also exerts a marked anti-inflammatory response in the periphery and brain.

It is now firmly established that insulin in the CNS plays a pronounced role in promoting adaptive neuroplasticity and in protecting from oxidative and glutaminergic stress via the widely distributed insulin receptor substrate p53, which plays a key role in modulating the actin cytoskeleton and the remodeling of dendritic extensions [78–80]. Insulin in the brain derives solely from the periphery via active transport across the blood brain barrier. In states of insulin resistance and peripheral hyperinsulinemia, insulin transport is reduced because of the saturation of insulin receptors in the blood brain barrier [81]. Thus, peripheral insulin resistance, which we see in our depressed patients, is likely to be associated with a CNS insulin deficiency and associated with disturbed neuroplasticity, increased susceptibility to oxidative and glutaminergic stress, and inflammation.

We now know that overfeeding in mice results in a primary neuroinflammation, associated with activation of the intraneuronal NF-kB system [82]. This form of autonomous intraneuronally-mediated inflammation is unique, in contrast to canonical neuroinflammation that is activated via proinflammatory mediators deriving from glial immunocompetent cells (the canonical form of neuroinflammation) [82]. Blockade of this response in hypothalamic neurons by NF-kB antagonists eliminates the effect of overfeeding on peripheral insulin resistance, thus establishing the brain as the primary initiation site in the kind of insulin resistance we see in patients with major depression [82].

This autonomous intraneuronal inflammation has been designated as parainflammation [12]. Parainflammatory responses are unique in that they occur in the context of stressors for which we were evolutionarily unprepared, including not only overfeeding, but also marked decreases in physical activity, aging, disturbances due to loss of exposure to the naturalistic light–dark cycle and sleep deprivation, as well as novel foods and drugs [12]. We postulate that repeated acute social stressors that may not have been present during our early evolution may also set into motion parainflammatory responses related to increased NF-kB activation. Insulin and PPARg receptors in the CNS are among the most inhibitory modulators of neuronal NF-kb activity [12].

A key marker for parainflammation in the periphery is a smoldering, subtle 50–75% elevation in the level of acute phase proteins such as CRP, in contrast to CRP responses to infection that rise quickly 100-fold or more. Smoldering CRP elevations are seen in coronary artery disease, which was likely to be rare in our early history, as well as in states of major depression.

We have found that, compared with unmedicated, remitted patients with major depression, remitted patients receiving specific serotonin uptake inhibitors (SSRI) treatment are insulin-resistant, hyperinsulinemic, and have significantly higher levels of plasma glucose (preliminary observations). Two well-controlled epidemiological studies have shown that patients on SSRI treatment have a 2–3-fold increase in the incidence of type II diabetes. These data indicate that there might be a dissociation between the positive impact of SSRIs on the affective and cognitive components of the depressive syndrome from the systemic stigmata that were assumed to occur only during the depressed state.

PPARg agonists have activity on a striking multiplicity of interrelated pathophysiological CNS processes we now know occur in patients with major depression. Therefore, we propose a placebo-controlled, double-blind trial of pioglitazone, a safe and potent PPARg antagonist that crosses the blood brain barrier. Pioglitazone's CNS anti-inflammatory, neuroprotective, and neurotropic effects intersect with virtually every known pathophysiologic parameter identified in patients with major depression. Moreover, its insulin-sensitizing effects in the brain would complement these actions, and the amelioration of insulin resistance in the periphery would not only correct plasma hyperinsulinemia and decreased availability of insulin in the CSF, but also the highly pathogenic sequalae of insulin resistance on the quality of health and the lifespan. Data in experimental animals reveal that PPARg agonists exert behavioral effects interpreted as antidepressive. One open trial without placebo control [83] reported that pioglitazone has antidepressant properties in depressed patients.

In summary, PPARg augmentation can impact multiple significant pathophysiological inflammatory, neurotransmitter, and neuroendocrine processes involving peripheral and central insulin regulation, as well as intracellular processes critical to transducing the clinical and biological manifestations of depressive illness. These include processes such as parainflammation, multiple inflammatory pathways in the brain and periphery, endoplasmic reticulum stress, neuroplasticity, neurogenesis, BDNF mediated processes, neutralization of oxidative stress, the sequela of glutamate toxicity, and the consequences of hypercortisolism (Figure 4).

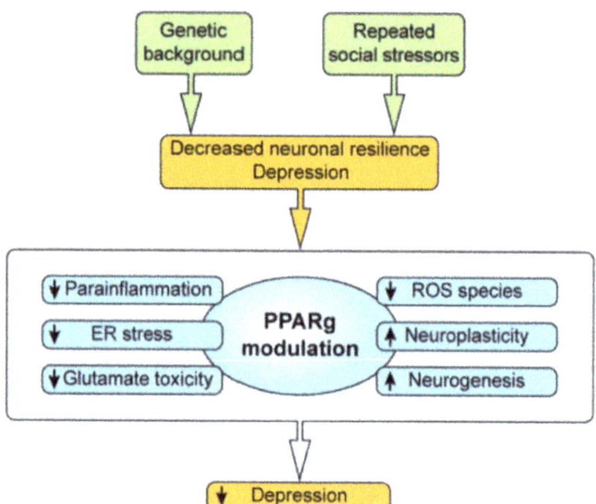

Figure 4. Repeated social and other stressors plus genetic predisposition lead to decreased neuronal resilience that can, in turn, lead to depression. Processes set into motion include parainflammation, extreme endoplasmic reticulum stress responses, glutamate toxicity, BDNF function, and the regulation of central and peripheral insulin dynamics. The PPARg system can modulate and diminish each of these pathologic drivers, and others as well, as noted in the text.

Funding: This research received no external funding.

Conflicts of Interest: The author declare no conflict of interests.

References

1. Barefoot, J.C.; Schroll, M. Symptoms of depression, acute myocardial infarction, and total mortality in a community sample. *Circulation* **1996**, *93*, 1976–1980. [CrossRef]
2. Ford, C.D.; Gray, M.S.; Crowther, M.R.; Wadley, V.G.; Austin, A.L.; Crowe, M.G.; Pulley, L.; Unverzagt, F.; Kleindorfer, D.O.; Kissela, B.M.; et al. Depressive symptoms and risk of stroke in a national cohort of blacks and whites from RE-GARDS. *Neurol. Clin. Pract.* **2020**, *6*, 108–112.
3. Silva, N.D.M.L.E.; Lam, M.P.; Soares, C.N.; Munoz, D.P.; Milev, R.; De Felice, F.G. Insulin Resistance as a Shared Pathogenic Mechanism Between Depression and Type 2 Diabetes. *Front. Psychiatry* **2019**, *10*, 57. [CrossRef]
4. Knol, M.J.; Twisk, J.W.R.; Beekman, A.T.F.; Heine, R.J.; Snoek, F.J.; Pouwer, F. Depression as a risk factor for the onset of type II diabetes. *Diabetologia* **2005**, *49*, 837–845. [CrossRef]
5. Michelson, D.; Stratakis, C.; Hill, L.; Reynolds, J.; Galliven, E.; Chrousos, G.; Gold, P. Bone mineral density in women with depression. *N. Engl. J. Med.* **1996**, *335*, 1176–1181. [CrossRef] [PubMed]
6. Whooley, M.A.; de Jonge, P.; Vittinghoff, E.; Otte, C.; Moos, R. Depressive symptoms, health behaviors, and risk of cardiovascular events in patients with coronary heart disease. *JAMA* **2008**, *3*, 2379–2388. [CrossRef] [PubMed]
7. Gold, P.W.; Goodwin, F.K.; Chrousos, G.P. Clinical and biochemical manifestations of depression. Relation to the neurobiology of stress (2). *N. Engl. J. Med.* **1988**, *319*, 413–420. [CrossRef] [PubMed]
8. Gold, P.W.; Wong, M.-L. Re-assessing the catecholamine hypothesis of depression: The case of melancholic depression. *Mol. Psychiatry* **2021**, 1–4. [CrossRef]
9. Gold, P.W. The organization of the stress system and its dysregulation in depressive illness. *Mol. Psychiatry* **2015**, *20*, 32–47. [CrossRef]
10. Wong, M.L.; Kling, M.A.; Munson, P.J.; Listwak, S.; Licinio, J.; Karp, B.; McCutcheon, E.; Geracioti, T.; DeBellis, M.; Oldfield, E.; et al. Pronounced and sustained central hypernoradrenergic function in major depression with melancholic features: Relation to hypercortisolism and corticotropin-releasing hormone. *Proc. Natl. Acad. Sci. USA* **2000**, *97*, 325–330. [CrossRef]
11. Gold, P.W.; Loriaux, D.L.; Roy, A.; Kling, M.A.; Calabrese, J.; Kellner, C.; Post, R.M.; Gold, P.W. Responses to corticotropin-releasing hormone in the hypercortisolism of depression and Cushing's disease. Pathophysiologic and diagnostic implications. *N. Engl. J. Med.* **1986**, *314*, 1329–1335. [CrossRef]
12. Gold, P.W.; Licinio, J.; Pavlatou, M.G. Pathological parainflammation and endoplasmic reticulum stress in depression: Potential translational targets through the insulin, klotho and PPAR-gamma systems. *Mol. Psychiatry* **2013**, *18*, 154–165. [CrossRef] [PubMed]

13. Sternberg, E.M.; Chrousos, G.P.; Wilder, R.L.; Gold, P.W. The Stress Response and the Regulation of Inflammatory Disease. *Ann. Intern. Med.* **1992**, *117*, 854–866. [CrossRef] [PubMed]
14. Spitz, R.A. Hospitalism: An inquiry into the genesis of psychiatric illness in early childhood. *Psychoanal. Study Child* **1945**, *1*, 53–74. [CrossRef] [PubMed]
15. Feng, X.; Wang, L.; Yang, S.; Qin, D.; Wang, J.; Li, C.; Lv, L.; Ma, Y.; Hu, X. Maternal separation produces long-lasting changes in cortisol and behavior in rhesus monkeys. *Proc. Natl. Acad. Med. USA* **2010**, *108*, 14312–14317. [CrossRef] [PubMed]
16. Villapol, S. Roles of Peroxisome Proliferator-Activated Receptor Gamma on Brain and Peripheral Inflammation. *Cell Mol. Neurobiol.* **2018**, *38*, 121–132. [CrossRef] [PubMed]
17. Domi, E.; Uhrig, S.; Soverchia, L.; Hansen, A.; Barbier, E.; Heilig, M.; Ubaldi, M. Genetic Deletion of Neuronal PPARgamma Enhances the Emotional Response to Acute Stress Exacerbates Anxiety: An Effect Reversed by Rescue of Amygdala PPARgamma Function. *J. Neurosci. Off. J. Soc. Neurosci.* **2016**, *36*, 12611–12623. [CrossRef]
18. Warden, A.T.J.; Merriman, M.; Truitt, J.; Pomomareva, O.; Jameson, K.; Ferguson, L.B.; Mayfild, R.D.; Harris, R.A. Localization of PPARg isotypes in the adult mouse and human brain. *Sci. Rep.* **2016**, *6*, 1–15. [CrossRef] [PubMed]
19. García-Bueno, B.; Madrigal, J.L.; Pérez-Nievas, B.G.; Leza, J.C. Stress mediators regulate brain prostaglandin synthesis and peroxisome proliferator-activated receptor-γ activation after stress in rats. *Endocrinology* **2008**, *149*, 1969–1978. [CrossRef]
20. Gold, P.W.; Wong, M.L.; Goldstein, D.S.; Gold, H.K.; Ronsaville, D.S.; Esler, M.; Alesci, S.; Masood, A.; Licinio, J.; Geracioti, T.D.; et al. Cardiac implications of increased arterial entry and reversible 24-h central and peripheral norepinephrine levels in melancholia. *Proc. Natl. Acad. Sci. USA* **2005**, *102*, 8303–8308. [CrossRef]
21. Akiyama, T.; Koeda, M.; Okubo, Y.; Kimura, M. Hypofunction of left dorsolateral prefrontal cortex in depression during verbal fluency task: A multi-channel near-infrared spectroscopy study. *J. Affect. Disord.* **2018**, *231*, 83–90. [CrossRef]
22. Simpson, D.W.C., Jr.; Snyder, A.Z.; Gusnard, D.A.; Raichle, M.E. Emotion-induced changes in human medial prefrontal cortex: II. During anticipatory anxiety. *Proc. Nat. Acad. Sci. USA* **2001**, *98*, 688–693. [CrossRef]
23. Roozendaal, B.; McEwen, B.S.; Chattarji, S. Stress, memory and the amygdala. *Nat. Rev. Neurosci.* **2009**, *10*, 423–433. [CrossRef] [PubMed]
24. Roozendaal, B.; Barsegyan, A.; Lee, S. Adrenal stress hormones, amygdala activation, and memory for emotionally arousing experiences. *Prog. Brain Res.* **2008**, *167*, 79–97.
25. Brady, L.S.; Whitfield, H.J., Jr.; Fox, R.J.; Gold, P.W.; Herkenham, M. Long-term antidepressant administration alters corticotropin-releasing hormone, tyrosine hydroxylase, and mineralocorticoid receptor gene expression in rat brain. Therapeutic implications. *J. Clin. Investig.* **1991**, *87*, 831–837. [CrossRef] [PubMed]
26. Gold, P.W. Endocrine Factors in Key Structural and Intracellular Changes in Depression. *Trends Endocrinol. Metab.* **2021**, *32*, 212–223. [CrossRef]
27. Berger, J.P.H.; Woods, J.; Hayes, N.S.; Parent, S.A.; Clemas, J.; Leibowitz, M.D.; Ebrecht, A.; Rachubinski, R.A.; Capone, J.P.; Moller, D.E. A PPARg mutant serves as a dominant negative inhibitor of PPARg signaling and is localized in the nucleus. *Mol. Cell Endocrinol.* **2000**, *162*, 57–67. [CrossRef]
28. Pizzagalli, D.A.; Holmes, A.J.; Dillon, D.G.; Goetzz, E.L.; Birk, J.; Bogdan, R.; Dugherty, D.D.; Iosefescu, D.V.; Rouch, S.L.; Fava, M. Reduced caudate and nucleus accumbens response to rewards in unmedicated individuals with major depressive disorder. *Am. J. Psychiatry* **2009**, *166*, 702–710. [CrossRef] [PubMed]
29. Nauczyciel, C.; Robic, S.; Dondaine, T. The nucleus accumbens: A target for deep brain stimulation in resistant major depressive disorder. *J. Mol. Psychiatry* **2013**, *1*, 17. [CrossRef]
30. Ulrich-Lai, Y.M.; Ryan, K.K. PPARgamma and stress: Implications for aging. *Exp. Gerontol.* **2013**, *48*, 671–676. [CrossRef] [PubMed]
31. Drevets, W.C.; Price, J.L.; Simpson, J.R., Jr.; Todd, R.D.; Reich, T.; Vannier, M.; Raichle, M.E. Subgenual prefrontal cortex abnormalities in mood disorders. *Nature* **1997**, *386*, 824–827. [CrossRef]
32. Drevets, W.C.; Savitz, J.; Trimble, M.R.D.; Reich, T.; Vannier, M.; Raichle, M.E. The subgenual anterior cingulate cortex in mood disorders. *CNS Spectr.* **2008**, *13*, 663–681. [CrossRef] [PubMed]
33. Bjorkholm, C.; Monteggia, L.M. BDNF–a key transducer of antidepressant effects. *Neuropharmacology* **2016**, *102*, 72–79. [CrossRef] [PubMed]
34. Lee, B.-H.; Kim, Y.-K. The Roles of BDNF in the Pathophysiology of Major Depression and in Antidepressant Treatment. *Psychiatry Investig.* **2010**, *7*, 231–235. [CrossRef] [PubMed]
35. Kariharan, T.; Nanayakkara, G.; Parameshwaran, K.; Bagasrawala, I.; Ahuja, M.; Abdel-Rahman, E.; Amin, A.T.; Dhanasekaran, M.; Suppiramaniam, V.; Amin, R.H. Central activation of PPAR-gamma ameliorates diabetes induced cognitive dysfunction and improves BDNF expression. *Neurobiol. Aging* **2015**, *36*, 1451–1461. [CrossRef] [PubMed]
36. Duman, R.S.; Aghajanian, G.K. Synaptic Dysfunction in Depression: Potential Therapeutic Targets. *Science* **2012**, *338*, 68–72. [CrossRef]
37. Duman, R.S.; Aghajanian, G.K.; Sanacora, G.; Krystal, J.H. Synaptic plasticity and depression: New insights from stress and rapid-acting antidepressants. *Nat. Med.* **2016**, *22*, 238–249. [CrossRef]
38. Holmes, S.E.; Scheinost, D.; Finnema, S.J.; Naganawa, M.; Davis, M.T.; Della Gioia, N.; Nabulsi, N.; Matuskey, D.; Angaria, G.A.; Pietrzak, R.H.; et al. Lower synaptic density is associated with depression severity and network alterations. *Nat. Commun.* **2019**, *10*, 1529. [CrossRef]

39. Popoli, M.; Yan, Z.; McEwen, B.S.; Sanacora, G. The stressed synapse: The impact of stress and glucocorticoids on glutamate transmission. *Nat. Rev. Neurosci.* **2011**, *13*, 22–37. [CrossRef]
40. Fales, C.L.; Barch, D.M.; Rundle, M.M.; Mintun, M.A.; Mathews, J.; Snyder, A.Z.; Sheline, Y.I. Antidepressant treatment normalizes hypoactivity in dorsolateral prefrontal cortex during emotional interference processing in major depression. *J. Affect. Disord.* **2009**, *112*, 206–211. [CrossRef]
41. Duman, R.S. Neuronal damage and protection in the pathophysiology and treatment of psychiatric illness: Stress and depression. *Dialogues Clin. Neurosci.* **2009**, *11*, 239–255.
42. Pittenger, C.D.R. Stress, deprression, and neuroplasticity: A comvergence of mechanisms. *Neuropsychopharmacology* **2008**, *33*, 88–109. [CrossRef] [PubMed]
43. Laje, G.; Lally, N.; Mathews, D.; Brutche, N.; Chemerinski, A.; Akula, N.; Kelmendi, B.; Simen, A.; McMahon, F.J.; Sanacora, G.S.; et al. Brain-derived neurotrophic factor Val66Met polymorphism and antidepressant efficacy of ketamine in depressed patients. *Biol. Psychiatry* **2012**, *72*, e27–e28. [CrossRef] [PubMed]
44. Banasr, M.; Dwyer, J.M.; Duman, R.S. Cell atrophy and loss in depression: Reversal by antidepressant treatment. *Curr. Opin. Cell Biol.* **2011**, *23*, 730–737. [CrossRef] [PubMed]
45. D'Sa, C.; Duman, R.S. Antidepressants and neuroplasticity. *Bipolar Disord.* **2002**, *4*, 183–194. [CrossRef]
46. Chang, C.H.; Chen, M.C.; Lu, J. Effect of antidepressant drugs on the vmPFC-limbic circuitry. *Neuropharmacology* **2015**, *92*, 116–124. [CrossRef]
47. McEwen, B.S.; Nasca, C.; Gray, J.D. Stress Effects on Neuronal Structure: Hippocampus, Amygdala, and Prefrontal Cortex. *Neuropsychopharmacology* **2016**, *41*, 3–23. [CrossRef]
48. Warner-Schmidt, J.L.; Duman, R.S. Hippocampal neurogenesis: Opposing effects of stress and antidepressant treatment. *Hippocampus* **2006**, *16*, 239–249. [CrossRef]
49. Gold, P.W.; Chrousos, G.P. The endocrinology of melancholic and atypical depression: Relation to neurocircuitry and somatic consequences. *Proc. Assoc. Am. Physicians* **1999**, *111*, 22–34. [CrossRef]
50. Gold, P.W.; Wong, M.L.; Chrousos, G.P.; Licinio, J. Stress system abnormalities in melancholic and atypical depression: Molecular, pathophysiological, and therapeutic implications. *Mol. Psychiatry* **1996**, *1*, 257–264.
51. Disner, S.G.; Beevers, C.G.; Haigh, E.A.; Beck, A.T. Neural mechanisms of the cognitive model of depression. *Nat. Rev. Rev. Neurosci.* **2011**, *12*, 467–477. [CrossRef]
52. Wohleb, E.S.; Franklin, T.; Iwata, M.; Duman, R.S. Integrating neuroimmune systems in the neurobiology of depression. *Nat. Rev. Neurosci.* **2016**, *17*, 497–511. [CrossRef]
53. Miller, A.H.; Maletic, V.; Raison, C.L. Inflammation and its discontents: The role of cytokines in the pathophysiology of major depression. *Biol. Psychiatry* **2009**, *65*, 732–741. [CrossRef]
54. Miller, A.H.; Raison, C.L. The role of inflammation in depression: From evolutionary imperative to modern treatment target. *Nat. Rev. Immunol.* **2016**, *16*, 22–34. [CrossRef]
55. Kritas, S.K.; Saggini, A.; Cerulli, G. Corticotropin-releasing hormone, microglia and mental disorders. *Int. J. Immunopathol. Pharmacol.* **2014**, *27*, 163–167. [CrossRef]
56. Brady, L.S.; Gold, P.W.; Herkenham, M.; Lynn, A.B.; Whitfield, H.J., Jr. The antidepressants fluoxetine, idazoxan and phenelzine alter corticotropin-releasing hormone and tyrosine hydroxylase mRNA levels in rat brain: Therapeutic implications. *Brain Res.* **1992**, *572*, 117–125. [CrossRef]
57. Ayyadurai, S.; Gibson, A.J.; D'Costa, S.; Overman, E.; Somerville, L.J.; Poopal, A.; MacKey, E.; Yihang, L.; Moeser, A.J. Frontline Science: Corticotropin-releasing factor receptor subtype 1 is a critical modulator of mast cell degranulation and stress-induced pathophysiology. *J. Leukoc. Biol.* **2017**, *102*, 1299–1312. [CrossRef] [PubMed]
58. Esposito, P.; Chandler, N.; Kandere, K.; Basli, S.; Jacobson, S.; Connolly, R.; Tutor, D.; Theoharides, T.C. Corticotropin-releasing hormone and brain mast cells regulate blood-brain-barrier permeability induced by acute stress. *J. Pharmacol. Exp. Ther.* **2002**, *303*, 1061–1066. [CrossRef]
59. Theoharides, T.C.; Singh, L.K.; Boucher, W.; Pang, X.; Letourneau, R.; Webster, E.; Chrousos, G.P. Corticotropin-releasing hormone induces skin mast cell degranulation and increased vascular permeability, a possible explanation for its proinflammatory effects. *Endocrinology* **1998**, *139*, 403–413. [CrossRef]
60. Aoun, P.; Watson, D.G.; Simpkins, J.W. Neuroprotective effects of PPARgamma agonists against oxidative insults in HT-22 cells. *Eur. J. Pharmacol.* **2003**, *472*, 65–71. [CrossRef]
61. Boden, G. Endoplasmic reticulum stress: Another link between obesity and insulin resistance/inflammation? *Diabetes* **2009**, *58*, 518–519. [CrossRef]
62. Kim, I.; Xu, W.; Reed, J.C. Cell death and endoplasmic reticulum stress: Disease relevance and therapeutic opportunities. *Nat. Rev. Drug Discov.* **2008**, *7*, 1013–1030. [CrossRef]
63. Zhao, X.; Strong, R.; Zhang, J.; Sun, G.; Tsien, J.Z.; Cui, Z.; Grotta, J.G.; Aronowski, J. Neuronal PPARgamma deficiency increases susceptibility to brain damage after cerebral ischemia. *J. Neurosci.* **2009**, *29*, 6186–6195. [CrossRef]
64. Zhang, R.; Zheng, F. PPAR-gamma and aging: One link through klotho? *Kidney Int.* **2008**, *74*, 702–704. [CrossRef]
65. Yamamoto, M.; Clark, J.D.; Pastor, J.V. Regulation of oxidative stress by the anti-aging hormone klotho. *J. Biol. Chem.* **2005**, *280*, 38029–38034. [CrossRef] [PubMed]
66. Kuro-O, M. Klotho as a regulator of oxidative stress and senescence. *Biol. Chem.* **2008**, *389*, 233–241. [CrossRef]

67. Nagai, T.; Yamada, K.; Kim, H.C.; Kim, Y.S.; Yukihiro, N.; Imura, A.; Nabachima, N.; Nebaschima, T. Cognition impairment in the genetic model of aging klotho gene mutant mice: A role of oxidative stress. *FASEB J.* **2003**, *17*, 50–52. [CrossRef]
68. Luna-Medina, R.; Cortes-Canteli, M.; Sanchez-Galiano, S. NP031112, a thiadiazolidinone compound, prevents inflammation and neurodegeneration under excitotoxic conditions: Potential therapeutic role in brain disorders. *J. Neurosci. Off. J. Soc. Neurosci.* **2007**, *27*, 5766–5776. [CrossRef] [PubMed]
69. Aoun, P.; Simpkins, J.W.; Agarwal, N. Role of PPAR-gamma ligands in neuroprotection against glutamate-induced cytotoxicity in retinal ganglion cells. *Investig. Ophthalmol. Vis. Sci.* **2003**, *44*, 2999–3004. [CrossRef] [PubMed]
70. Kitao, Y.; Ozawa, K.; Miyazaki, M. Expression of the endoplasmic reticulum molecular chaperone (ORP150) rescues hippocampal neurons from glutamate toxicity. *J. Clin. Investig.* **2001**, *108*, 1439–1450. [CrossRef] [PubMed]
71. Cimini, A.; Ceru, M.P. Emerging roles of peroxisome proliferator-activated receptors (PPARs) in the regulation of neural stem cells proliferation and differentiation. *Stem Cell Rev.* **2008**, *4*, 293–303. [CrossRef]
72. Imayama, I.; Ichiki, T.; Inanaga, K.; Ohtsubo, H.; Fukayama, K.; Ono, H.; Hashaguchi, Y.; Sunagawa, K. Telmisartan downregulates angiotensin II type 1 receptor through activation of peroxisome proliferator-activated receptor gamma. *Cardiovasc. Res.* **2006**, *72*, 184–190. [CrossRef]
73. Festuccia, W.T.; Oztezcan, S.; Laplante, M.; Berthiaume Chantal, M.; Doghu, S.; Denis, R.G.; Brito, M.N.; Brito, N.A.; Miller, D.S. Peroxisome proliferator-activated receptor-gamma-mediated positive energy balance in the rat is associated with reduced sympathetic drive to adipose tissues and thyroid status. *Endocrinology* **2008**, *149*, 2121–2130. [CrossRef] [PubMed]
74. Craft, S. Insulin resistance syndrome and Alzheimer disease: Pathophysiologic mechanisms and therapeutic implications. *Alzheimer Dis. Assoc. Disord.* **2006**, *20*, 298–301. [CrossRef] [PubMed]
75. Wójtowicz, S.; Strosznajder, A.K.; Jeżyna, M.; Strosznajder, J.B. The Role of PPAR in the brain: Promising target in therapy of Alzheimer's Disease and other neurodegenerative disorders. *Neurochem. Res.* **2020**, *45*, 972–988. [CrossRef] [PubMed]
76. Fahmida, A.A.A.; Mafauzy, M. Efficacy and safety of pioglitizone monotherapy an type II diabetes:A systematic review and meta-analysis of randomized controlled trials. *Sci. Rep.* **2019**, *9*, 5839.
77. Klutzky, J.K. PPARs as Therapeutic Agents in Reverse Cardiology and Endocrinology. *Science* **2010**, *2010*.
78. Chiu, S.L.; Chen, C.M.; Cline, H.T. Insulin receptor signaling regulates synapse number, dendritic plasticity, and circuit function in vivo. *Neuron* **2008**, *58*, 708–719. [CrossRef]
79. Choi, J.; Ko, J.; Racz, B.; Burette, A.; Lee, J.; Kim, S.; Na, M.; Lee, H.W.; Kim, S.; Weinberg, R.J.; et al. Regulation of dendritic spine morphogenesis by insulin receptor substrate 53, a downstream effector of Rac1 and Cdc42 small GTPases. *J. Neurosci. Off. J. Soc. Neurosci.* **2005**, *25*, 869–879. [CrossRef]
80. Nelson, T.J.; Sun, M.K.; Hongpaisan, J.; Alkon, D.L. Insulin, PKC signaling pathways and synaptic remodeling during memory storage and neuronal repair. *Eur. J. Pharmacol.* **2008**, *585*, 76–87. [CrossRef]
81. Schwartz, M.W.; Porte, D., Jr. Diabetes, obesity, and the brain. *Science* **2005**, *307*, 375–379. [CrossRef] [PubMed]
82. Zhang, X.; Zhang, G.; Zhang, H.; Karin, M.; Bai, H.; Cai, D. Hypothalamic IKKbeta/NF-kappaB and ER stress link overnutrition to energy imbalance and obesity. *Cell* **2008**, *135*, 61–73. [CrossRef] [PubMed]
83. Kemp, D.E.; Schinagle, M.; Gao, K.; Calabrese, J. PPAR-gamma agonism as a modulator of mood: Proof-of-concept for pioglitazone in bipolar depression. *CNS Drugs* **2014**, *28*, 571–581. [CrossRef] [PubMed]

International Journal of
Molecular Sciences

MDPI

Review

Essential Roles of PPARs in Lipid Metabolism during Mycobacterial Infection

Kazunari Tanigawa [1], Yuqian Luo [2,3], Akira Kawashima [2], Mitsuo Kiriya [2], Yasuhiro Nakamura [1], Ken Karasawa [1] and Koichi Suzuki [2,*]

[1] Department of Molecular Pharmaceutics, Faculty of Pharma-Science, Teikyo University, Itabashi-ku, Tokyo 173-8605, Japan; tanigawa@pharm.teikyo-u.ac.jp (K.T.); nakayasu@pharm.teikyo-u.ac.jp (Y.N.); karasawa@pharm.teikyo-u.ac.jp (K.K.)
[2] Department of Clinical Laboratory Science, Faculty of Medical Technology, Teikyo University, Itabashi-ku, Tokyo 173-8605, Japan; yuqianluo31@foxmail.com (Y.L.); akirak5243@gmail.com (A.K.); mkiriya0226@med.teikyo-u.ac.jp (M.K.)
[3] Department of Laboratory Medicine, Nanjing Drum Tower Hospital, Nanjing University Medical School, Nanjing 210008, China
* Correspondence: koichis0923@med.teikyo-u.ac.jp; Tel.: +81-3-3964-1211

Abstract: The mycobacterial cell wall is composed of large amounts of lipids with varying moieties. Some mycobacteria species hijack host cells and promote lipid droplet accumulation to build the cellular environment essential for their intracellular survival. Thus, lipids are thought to be important for mycobacteria survival as well as for the invasion, parasitization, and proliferation within host cells. However, their physiological roles have not been fully elucidated. Recent studies have revealed that mycobacteria modulate the peroxisome proliferator-activated receptor (PPAR) signaling and utilize host-derived triacylglycerol (TAG) and cholesterol as both nutrient sources and evasion from the host immune system. In this review, we discuss recent findings that describe the activation of PPARs by mycobacterial infections and their role in determining the fate of bacilli by inducing lipid metabolism, anti-inflammatory function, and autophagy.

Keywords: mycobacteria; *M. tuberculosis*; *M. leprae*; PPARs; lipid droplets

Citation: Tanigawa, K.; Luo, Y.; Kawashima, A.; Kiriya, M.; Nakamura, Y.; Karasawa, K.; Suzuki, K. Essential Roles of PPARs in Lipid Metabolism during Mycobacterial Infection. *Int. J. Mol. Sci.* **2021**, *22*, 7597. https://doi.org/10.3390/ijms22147597

Academic Editors: Manuel Vázquez-Carrera and Walter Wahli

Received: 25 June 2021
Accepted: 13 July 2021
Published: 15 July 2021

1. Introduction

The *Mycobacterium* genus was one of the first bacterial genera described. The most characteristic feature of mycobacteria is resistance to acid alcohol, which is utilized for Ziehl–Neelsen staining. Pathogenic mycobacteria can be categorized into three groups: *Mycobacterium tuberculosis* (*M. tuberculosis*) complex, which causes tuberculosis; *M. leprae* and *M. lepromatosis*, which both cause leprosy; and atypical mycobacteria or nontuberculous mycobacteria (NTM), which are mycobacteria responsible for a wide range of diseases. Mycobacterial cell walls consist of large amounts of lipids (30% to 40% of the total weight) that form a complex tripartite structure. The lipids are major effector molecules that affect the physiology of both the host cells and the bacilli by modulating their metabolism and stimulating immune responses to the bacilli. Most pathogenic mycobacteria, including *M. leprae*, utilize lipids from the host as a source of nutrients and to evade the immunity from the host, enabling the bacteria to both hide and replicate within host cells.

The transcription factors known as peroxisome proliferator-activated receptors (PPARs) were discovered in 1990 as enhancers of peroxisome proliferation in rodents [1] and belong to the ligand-activating nuclear hormone receptor (NR) superfamily. PPARs form heterodimers with retinoid X receptors (RXRs), enabling them to bind PPAR-responsive regulatory elements (PPRE) located in the promoter regions of their target genes. Three types of PPARs have been identified in mammals: PPAR-α (NR1C1), PPAR-β/δ (NR1C2), and PPAR-γ (NR1C3) [1,2]. Each PPAR is encoded by a separate gene and is expressed in amphibians [3], rodents [4,5], and humans [6,7]. PPAR-α and PPAR-γ are conserved proteins expressed in wide varieties of species, whereas PPAR-β/δ has diverged considerably [5]. PPARs respond to ligands and regulate the transcription of target genes. The role of PPARs is to modulate the expression of genes central to regulating glucose, lipid, and cholesterol metabolism.

It has been reported that the induction of PPAR-γ by the Middle East respiratory syndrome coronavirus (MERS-CoV) is necessary for infection. The PPAR-γ activation is mediated by the MERS-CoV-derived S glycoprotein along with concurrent inhibition of macrophage responses and the suppression of proinflammatory cytokines [8]. It has also been reported that PPAR-γ activation maintains the viral infection by inducing lipid metabolism. Monocytes isolated from coronavirus disease 2019 (COVID-19) patients show an accumulation of lipid droplets compared with other donors [9]. Infection with SARS-CoV-2 modulates lipid uptake and synthesis pathways by inducing PPAR-γ expression in monocytes; lipid droplet formation is also triggered in multiple human cell lines [9].

Recently, the lipid metabolism pathway used by mycobacteria in host cells has been revealed, and the involvement of PPARs clarified. In this review, we focus on the involvement of PPARs in host–mycobacteria crosstalk.

2. Activation of PPARs by Mycobacteria

PPARs are activated by endogenous and exogenous compounds. For instance, eicosanoids and long-chain fatty acids (LCFAs) are the endogenous ligands for PPAR-α and PPAR-β/δ [10,11]. PPAR-γ is activated by metabolites of arachidonic acid, such as 5-oxo-eicosatetraenoic acid (5-oxo-ETE) and 5-oxo-15(S)-hydroxyeicosatetraenoic acids (5-oxo-15(S)-HETE) [12,13], in addition to oxidized low-density lipoprotein (oxLDL) derivatives [14]. Several exogenous compounds are highly specific activators and modulators for mammalian PPAR subtypes: PPAR-α by the hypolipidemic drugs clofibrate and fenofibrate and the synthetic ligand Wy-14643 and PPAR-γ by the thiazolidinedione (TZD) group of antidiabetic drugs (including rosiglitazone, ciglitazone, troglitazone, and pioglitazone) [15]. GW501516, GW0742, and bezafibrate are highly selective PPAR-β/δ agonists, while GW1929 and GW2090 are specific PPAR-γ activators [16].

PPARs are also activated by mycobacterial infection; however, the bacterial component(s) responsible are not well understood. Organisms that naturally produce unsaturated fatty acids at the C10 position are relatively rare in nature, while several mycobacteria species, including *M. vaccae*, are able to accomplish this desaturation [17–20]. The mycobacteria-derived 10 (Z)-hexadecenoic acid upregulates genes in the PPAR signaling pathway and represses the proinflammatory cytokines in macrophages [21]. Furthermore, 10 (Z)-hexadecenoic acid and monoacylglycerol (MAG), which contains 10 (Z)-hexadecenoic acid, both activate PPAR-α but have no effect on PPAR-γ or PPAR-δ. The observed effects are blocked by PPAR-α antagonists and absent in PPAR-α-deficient mice. Recently, we found that PPAR-γ and PPAR-δ are activated in *M. leprae*-infected macrophages [22]. Infection with a recombinant strain of *M. bovis* BCG that produces phenolic glycolipid-1 (PGL-1) of *M. leprae* activates PPAR-γ in primary cultures of human Schwann cells [23].

Mannose-capped lipoarabinomannan (ManLAM) is present in the members of the *M. tuberculosis* complex, which interact with the mannose receptor (MR) in alveolar macrophages (AMs). High levels of PPAR-γ are expressed in activated AMs and macrophage-derived foam cells [24,25]. ManLAM upregulates PPAR-γ expression in human macrophages, consistent with *M. tuberculosis* infection. Furthermore, activation by ManLAM is suppressed by MR siRNA. These results indicate that the activation of PPAR-γ by *M. tuberculosis* is due to the interaction between its cell wall component ManLAM and host MRs.

Several molecules are known to bind to PPARs, including polyunsaturated fatty acids (PUFAs), such as certain ω3-PUFAs (e.g., docosahexaenoic acid with C22:6 and α-linolenic acid with C18:3) and certain ω6-PUFAs (e.g., arachidonic acid with C20:4 and linoleic acid with C18:2) [26,27]. Saturated fatty acids, such as stearic acid with C18:0 and myristic acid with C14:0, also bind to PPAR-α. *M. leprae* cell wall lipids also contain mycolic acids, other types of LCFAs typical for mycobacteria, such as alpha-mycolic acids and ketomycolic acids [28]. However, whether or not this lipid could be a ligand for PPARs is not known.

3. The Roles of PPARs in Lipid Metabolism

The function of PPARs, including PPAR-α, PPAR-β/δ, and PPAR-γ, is closely involved in lipogenesis and lipid metabolism in triacylglycerol (TAG) and cholesterol synthesis. PPAR-α plays an important role in the regulation of cholesterol and metabolism of bile acid. In the fasting state, PPAR-α accelerates fatty acid formation in the liver by regulating apolipoprotein expression. This increases high-density lipoprotein cholesterol (HDL-C) in the plasma and reduces low-density lipoprotein cholesterol (LDL-C) levels [29,30]. In addition, PPAR-α mediates cholesterol transport by enhancing the expression of the apolipoprotein AI (Apo-AI). Thus, PPAR-α stimulates the expression of the liver X receptor (LXR), which regulates ATP-binding cassette transporter A1 (ABCA1) expression, increases the production of HDL (rich in Apo-AI), and induces outflow of cholesterol from macrophages [31]. Therefore, fibrates that activate PPAR-α have a beneficial effect on reducing TAG and LDL-C (arteriosclerotic lipids), as well as increasing HDL-C levels in plasma [29,32].

Apolipoprotein E (ApoE)$^{-/-}$ mice fed a high cholesterol diet have high plasma concentrations of LDL-C and develop atherosclerosis. The potential PPAR-γ agonist Danshensu Bingpian Zhi (DBZ) prevents atherosclerosis by modulating the expression of LXR to inhibit foam cell formation and inflammatory response [33,34]. PPAR-β/δ agonists seem to have similar effects as PPAR-α and PPAR-γ agonists by increasing plasma HDL levels while lowering LDL. These effects have been tested in both primate and rodent models [35,36]. In addition, PPAR-β/δ reduces the expression of Niemann-Pick C1-like 1 (NPC1L1), a cholesterol importer in the intestinal cells; reduces cholesterol absorption; and improves intestinal cholesterol outflow [37].

Intracellular TAG synthesis requires fatty acid metabolism and glucose homeostasis regulation. PPAR-α promotes glycolysis and de novo synthesis of fatty acid, while it decreases gluconeogenesis. Thus PPAR-α has an antagonistic function in glucose homeostasis to reduce lipid accumulation by suppressing glycolysis and enhance glycogen synthesis and fatty acid oxidation (FAO) [38]. These effects of PPAR-α were observed following PPAR-α overexpression in mouse skeletal muscle, which resulted in increased glucose and insulin levels in the plasma [39].

PPAR-β/δ has an important role in improving glycolysis and glucose uptake as well as glycogen storage, while suppressing gluconeogenesis [40,41]. PPAR-β/δ synergistically improves the catabolism of fatty acid and suppresses lipogenesis [42]. It has also been reported in the liver that PPAR-β/δ reduces the stability of sterol regulatory element-binding protein C (SREBP1C), which enhances lipogenesis by activating insulin-induced gene 1 (Insig-1) and preventing lipid accumulation [43]. Furthermore, PPAR-β/δ enhances the thermogenesis of brown adipose tissue (BAT) by regulating the transcription of FAO enzymes and uncoupling protein 1 (UCP-1) [44].

PPAR-γ directly binds the promoter region of various adipogenic genes, suggesting that it is an essential factor for adipogenesis [45]. Activated PPAR-γ reduces the amount of free fatty acid to increase the storage of TAG in adipose tissue [46]. PPAR-γ stimulates the differentiation of preadipocytes into adipocytes [47] and regulates the sensitivity of insulin in tissues and fatty acid storage by modulating genes involved in the release, transport, and storage of fatty acids in mature adipocytes. Furthermore, PPAR-γ transcriptionally activates the genes encoding c-Cbl-associated protein (CAP) and glucose transporter type 4 (Glut4), and contributes to glucose metabolism [48]. This evidence is consistent with our previous reports that *M. leprae* promotes the activation of PPAR-γ and increases intracellular TAG levels in THP-1 cells. Therefore, it is suggested that the activation of PPAR-γ is important for the increase of TAG and cholesterol in the formation of lipid droplets following mycobacterial infection.

4. Emerging Roles of PPARs in Lipid Metabolism during Mycobacteria Infection

Mycobacterial infection induces lipid droplet formation in macrophages. These lipids are essential for mycobacterial survival and are presumed to be a carbon source. In several different models, *M. tuberculosis* has been shown to use accumulated lipids as a carbon source at various stages of the infectious process [49–52]. *M. tuberculosis*-induced lipid droplets in macrophages primarily contain cholesterol esters and TAG. The cholesterol is transported through the bacterial cell membrane by Mce4, a bacterial lipid transporter required for cholesterol import and its utilization [53,54]. Many of the active compounds that limit *M. tuberculosis* growth in macrophages have been found to inhibit cholesterol-related processes, indicating that cholesterol is central to *M. tuberculosis* infection [55]. Fatty acids are also an abundant lipid in human granulomas [56]. Although it has been thought that *M. tuberculosis* assimilates and metabolizes fatty acids, recent genome sequencing has identified many putative fatty acid β-oxidation genes [57].

Since *M. leprae* has lost the *mce4* operon, *M. leprae* seems to use cholesterol oxidase (ML1492) in order to convert cholesterol to cholestenone for survival [58]. In leprosy skin tissue sections, *M. leprae*-containing histiocytes and Schwann cells are filled with cholesterol [59,60]. This has been confirmed with the observation of cholesterol accumulation in *M. leprae*-infected primary macrophage [60,61]. Furthermore, the expression of cholesterol synthase, HMG-CoA reductase, was increased following infection, and when de novo cholesterol synthesis was inhibited by lovastatin, viability of *M. leprae* was reduced [61].

Conversely, high-performance thin-layer chromatography (HPTLC) analysis demonstrates that TAG is the main component of the lipid in *M. leprae*-infected human monocytic THP-1 cells [62]. It has been reported in Schwann cells that *M. leprae* infection enhances glucose uptake and stimulates the pentose phosphate pathway, which is required for TAG synthesis [63]. The accumulated TAGs are maintained by the enhanced expression of adipose differentiation-related protein (ADRP) and perilipin and by the reduced expression of hormone-sensitive lipase (HSL), which contributes to lipid degradation [64,65]. Glycerol-3-phosphate acyltransferase 3 (GPAT3) is an important rate-limiting enzyme for TAG synthesis [66]; accordingly, the internalization and viability of bacilli are lower in *GPAT3* knockout cells [62]. Furthermore, clofazimine, a therapeutic agent for leprosy, reduces the accumulation of lipid in *M. leprae*-infected THP-1 cells and promotes the production of interferon (IFN)-β and IFN-γ [67]. Therefore, mycobacterial viability is hypothesized to be closely related to lipid metabolism in host cells, especially the accumulation of TAG and cholesterol.

A recent study demonstrated that PPAR-mediated lipid metabolism is a key process in foamy cell formation following *M. leprae* infection. Among PPARs, the involvement of PPAR-γ in mycobacterial infections has been studied. Infection with *M. tuberculosis* modulates homeostasis of host lipid and induces foamy macrophages, which is necessary for intracellular parasitization and growth [68,69]. The virulent H37Rv strain of *M. tuberculosis* induces PPAR-γ expression [25], while attenuated *M. bovis* BCG slightly upregulates PPAR-γ [25,70]. In vitro interference with PPAR-γ signaling in *M. tuberculosis*-infected macrophages decreases intracellular lipid accumulation and increases mycobacterium killing [71]. Pretreatment with a PPAR-γ antagonist significantly suppressed mycobacterial (*M. bovis* BCG and *M. tuberculosis*) induction of intracellular lipid droplet accumulation [70–72]. In addition, *M. tuberculosis* growth was attenuated in human lung macrophages after PPAR-γ deletion or isolation from PPAR-γ-deficient mice. Taken together, these data suggest that PPAR-γ is required for foam cell formation in tuberculous granulomas, which is related to bacilli survival.

Recently, in *M. leprae*-infected THP-1 cells, we reported that the increased expression of PPAR-γ and PPAR-δ coincided with the induction of intracellular lipid droplet formation [22]. Further, the expression of the PPAR-γ target genes *ADRP*, scavenger receptor *CD36*, fatty acid-binding protein 4 (*FABP4*), and apolipoprotein C-1 (*APOC1*) were significantly increased. Activation of the PPAR-γ signaling pathway is responsible for the upregulation of *Gpat3* expression during adipocyte differentiation [73–75]. We also found that GPAT3 expression is induced in THP-1 cells infected with *M. leprae*, suggesting that the mechanism of intracellular TAG accumulation is triggered by PPAR-γ activation [62].

The expression of CD36, an essential receptor for LDL-C incorporation, is also induced by *M. tuberculosis* through PPAR-γ in THP-1 macrophages [71]. CD36 can interact with surfactant in the lungs and promote the proliferation of *M. tuberculosis* in human macrophages in vitro [76]. CD36 directly interacts with TLR2 in macrophages infected with *M. bovis* BCG, as demonstrated by co-immunoprecipitation [77]. The neutralization of CD36 subsequently decreases PPAR-γ expression and lipid droplet formation and prostaglandin E2 (PGE2) secretion. In addition, *M. tuberculosis* upregulates the expression of GLUT1 and GLUT3 on the cell membrane by PPAR-γ activation of glucose metabolism. Its activation is suppressed by the PPAR-γ inhibitor T0070907 but enhanced by the agonist pioglitazone [78]. These data suggest that the activation of PPAR-γ promotes cholesterol and TAG uptake, both of which are components of the lipid droplets in mycobacteria-infected macrophages. Cholesterol accumulation in infected macrophages reduces cell wall permeability to rifampin, one of the first-line antituberculosis drugs, and masks surface antigens of mycobacteria [79]. Thus, lipids also play a role in drug resistance.

On the other hand, PPAR-α is known to promote the metabolism of lipids accumulated in *M. tuberculosis*-infected macrophages and suppress lipid droplet formation. Following infection with *M. tuberculosis*, PPAR-$\alpha^{-/-}$ bone marrow-derived macrophages decrease the activation of the transcription factor EB (TFEB), a responsible factor for the regulation of autophagy, and increase lipid droplet formation. Conversely, PPAR-α activation significantly reduces the amount of lipid droplets in mycobacteria-infected macrophages, suggesting that PPAR-α promotes lipid catabolism in mycobacterial infection [80]. Thus, PPAR-α and PPAR-γ may have opposed roles in the host defense during mycobacterial infection.

5. Anti-Inflammatory Effects Are Mediated by PPARs during Mycobacterial Infection

PPAR-α and PPAR-γ may also have opposing functions in the immune response to mycobacterial infections. PPAR-α deficiency leads to excessive proinflammatory cytokine and chemokine production after lung and macrophage infection. The deletion of PPAR-α in mice increases the expression of interleukin (IL)-6 and tumor necrosis factor (TNF)-α as well as neutrophil recruitment following *M. tuberculosis* infection [80]. Infection of PPAR-α-deficient mice with *M. abscessus* also increased the intracellular bacterial load and histopathological damage [81].

However, PPAR-γ suppresses IL-1β, IL-6, and TNF-α in phorbol-12-myristate-13-acetate (PMA)-stimulated human monocytes [82]. Phagocytosis of *M. tuberculosis* by human macrophages activates PPAR-γ via the mannose receptor (CD206) [25], which reduces the proinflammatory response. Deletion of PPAR-γ in pulmonary macrophages enhanced proinflammatory cytokines and reduced *M. tuberculosis* growth in murine models [83]. Similarly, *M. bovis* BCG infection enhances PPAR-γ expression through TLR2 in mouse macrophages, which regulates lipid droplet formation and PGE2 production [70]. Synthetic aptamers (ZXL1) against ManLAM inhibits immunosuppression of CD11c and enhances presentation of *M. tuberculosis* antigens in dendritic cells (DCs) [84]. In this process, PPAR-γ expression is downregulated, thereby enhancing mRNA expression and cytokine production of IL-1β and IL-12 and decreasing anti-inflammatory cytokine IL-10 production in ManLAM-treated macrophages [84,85]. Therefore, the PPAR-γ expression induced by mycobacterium infection is considered important for the suppression of the immune response in the host cells.

6. Mycobacteria-Induced PPAR-Mediated Autophagy

Autophagy, a cytoplasmic degradation system, plays an important role in the host defense against intracellular bacteria. PPAR-α induces autophagy and ameliorates inflammatory and injurious conditions in many cell types [86,87]. PPAR-α modulates antimicrobial responses to *M. tuberculosis*, *M. bovis* BCG, and *M. abscessus* by TFEB [80,81,88]. TFEB is an important regulator of autophagy, lipid catabolism, and lysosomal function [89–91]. PPAR-α transactivates autophagy-related genes (ATGs) to promote autophagy [80,92]. Importantly, there is considerable evidence for crosstalk between PPAR-α and TFEB [80,89,90].

A recent study showed that sirtuin 3 (SIRT3) induced antibacterial autophagy during *M. tuberculosis* infection through PPAR-α [88], indicating that PPAR-α may function in the host defense against intracellular *M. tuberculosis* by mediating autophagy.

PPAR-γ is less studied for its involvement in autophagy during mycobacterial infection. A DNA microarray analysis of mouse macrophage J774 cells treated with ManLAM alone or with the PPAR-γ inhibitor GW9662 showed that the inhibitor downregulates the expression of the AMP-activated protein kinase (AMPK) regulatory 2 subunit *(Prkag2)*, which activates AMPK [93]. AMPK, an essential regulator of autophagy [94], is required for the phagosome–lysosome fusion [95]. These data suggest a regulatory role for PPAR-γ signaling in autophagy. In addition, recent studies have revealed that either an AMPK activator (5-aminoimidazole-4-carboxamide 1-β-D-ribofuranoside, AICAR), a SIRT1 activator (e.g., resveratrol), or a SIRT3 activator (konokiol) can promote anti-mycobacterial activity through autophagy induction or AMPK activation [88,95–97]. It was shown in a previous study that lipid droplets are transported to lysosomes through the autophagy pathway, thus presenting the possibility that lysosomal acid lipase hydrolyzes lipid droplets [98]. Therefore, autophagy may be an essential mechanism in the regulation of lipid metabolism in macrophages during *M. tuberculosis* infection.

7. Conclusions

PPARs are important to the host-dependent mechanism of lipid metabolism and accumulation during mycobacterial infection. After the infection of macrophages by *M. leprae* or *M. tuberculosis*, PPAR-γ is activated and translocated into the nucleus to regulate genes that contribute to lipid metabolism, accumulation, and uptake (Figure 1). During cholesterol accumulation, APOC1 promotes the formation of LDL-C through the lipidation of very-low-density lipoprotein (VLDL) and chylomicron (CM), which is then imported by CD36. CD36 also imports fatty acids, which are transported to several intracellular organelles by FABP4. Glycerol-3-phosphate (G3P) is synthesized from glucose taken up by GLUT1/3, and GPAT3 esterifies fatty acids to promote TAG synthesis.

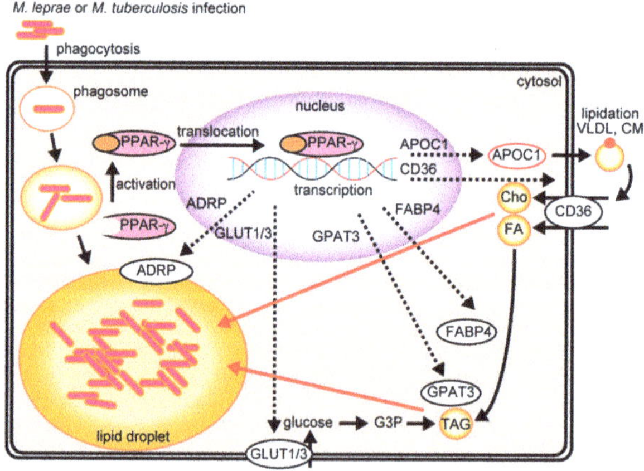

Figure 1. *M. leprae* and *M. tuberculosis* infections activate PPAR-γ to induce lipid droplet formation in the host cell. APOC1 binds to extracellular VLDL cholesterol or CM and is followed by LDL cholesterol uptake via CD36, thus accumulating intracellular cholesterol. Intracellular TAG accumulation is induced by two PPAR-γ-mediated pathways. FABP4 acylates extracellular fatty acids taken up by CD36 and is utilized by GPAT3 for TAG synthesis. GLUT1/3 induces intracellular glucose uptake and is subsequently utilized by GPAT3 for TAG synthesis. APOC1, apolipoprotein C-1; VLDL, very-low-density lipoprotein; CM, chylomicron; LDL, low-density lipoprotein; FABP4, fatty acid-binding protein 4; FA, fatty acid; TAG, triacylglycerol; GPAT3, glycerol-3-phosphate acyltransferase 3; GLUT1/3, glucose transport protein type 1/3.

Since the cell wall lipids of mycobacteria are complex constructions of several lipid components, it is possible that the lipids that accumulate in the host cells are used for both immune evasion and the construction of the mycobacterial cell wall. Although many endogenous and exogenous PPAR ligands have been identified, the ligands essential for mycobacteria infection are still poorly understood. Since the host transcriptional profiles conducted after inoculation of live or dead bacteria are significantly different (unpublished observation), such an approach may be appropriate for identifying the unknown factor(s) specific in live mycobacteria. As described, lipid accumulation in the host is associated with mycobacterial survival; therefore, existing lipid metabolism inhibitors may be potential antimycobacterial agents. The PPARs that play important roles in lipid metabolism, anti-inflammatory action, and autophagy may be novel therapeutic targets in mycobacterial infections.

Author Contributions: Conceptualization, K.T.; writing—original draft preparation, K.T.; writing—review and editing, K.T., Y.L., A.K., M.K., Y.N., K.K., and K.S.; supervision, K.S. All authors have read and agreed to the published version of the manuscript.

Funding: This work was supported by Grant-in-Aid for Scientific Research (C) Grant Number 21K07012 (to K.T.) and Grant-in-Aid for Early-Career Scientists Grant Number 18K15150 (to K.T.).

Institutional Review Board Statement: Not applicable.

Informed Consent Statement: Not applicable.

Conflicts of Interest: The authors declare no conflict of interest.

References

1. Issemann, I.; Green, S. Activation of a member of the steroid hormone receptor superfamily by peroxisome proliferators. *Nature* **1990**, *347*, 645–650. [CrossRef]
2. Dreyer, C.; Krey, G.; Keller, H.; Givel, F.; Helftenbein, G.; Wahli, W. Control of the peroxisomal beta-oxidation pathway by a novel family of nuclear hormone receptors. *Cell* **1992**, *68*, 879–887. [CrossRef]
3. Kawada, T. Lipid metabolism related nuclear receptor—The structure, function, expression and classification of peroxisome proliferation-activated receptor (PPAR). *Nihon Rinsho* **1998**, *56*, 1722–1728.
4. Gottlicher, M.; Widmark, E.; Li, Q.; Gustafsson, J.A. Fatty acids activate a chimera of the clofibric acid-activated receptor and the glucocorticoid receptor. *Proc. Natl. Acad. Sci. USA* **1992**, *89*, 4653–4657. [CrossRef]
5. Kliewer, S.A.; Forman, B.M.; Blumberg, B.; Ong, E.S.; Borgmeyer, U.; Mangelsdorf, D.J.; Umesono, K.; Evans, R.M. Differential expression and activation of a family of murine peroxisome proliferator-activated receptors. *Proc. Natl. Acad. Sci. USA* **1994**, *91*, 7355–7359. [CrossRef] [PubMed]
6. Greene, M.E.; Blumberg, B.; McBride, O.W.; Yi, H.F.; Kronquist, K.; Kwan, K.; Hsieh, L.; Greene, G.; Nimer, S.D. Isolation of the human peroxisome proliferator activated receptor gamma cDNA: Expression in hematopoietic cells and chromosomal mapping. *Gene Expr.* **1995**, *4*, 281–299.
7. Sher, T.; Yi, H.F.; McBride, O.W.; Gonzalez, F.J. cDNA cloning, chromosomal mapping, and functional characterization of the human peroxisome proliferator activated receptor. *Biochemistry* **1993**, *32*, 5598–5604. [CrossRef]
8. Al-Qahtani, A.A.; Lyroni, K.; Aznaourova, M.; Tseliou, M.; Al-Anazi, M.R.; Al-Ahdal, M.N.; Alkahtani, S.; Sourvinos, G.; Tsatsanis, C. Middle east respiratory syndrome corona virus spike glycoprotein suppresses macrophage responses *via* DPP4-mediated induction of IRAK-M and PPARgamma. *Oncotarget* **2017**, *8*, 9053–9066. [CrossRef] [PubMed]
9. Dias, S.S.G.; Soares, V.C.; Ferreira, A.C.; Sacramento, C.Q.; Fintelman-Rodrigues, N.; Temerozo, J.R.; Teixeira, L.; Nunes da Silva, M.A.; Barreto, E.; Mattos, M.; et al. Lipid droplets fuel SARS-CoV-2 replication and production of inflammatory mediators. *PLoS Pathog.* **2020**, *16*, e1009127. [CrossRef] [PubMed]
10. Yu, K.; Bayona, W.; Kallen, C.B.; Harding, H.P.; Ravera, C.P.; McMahon, G.; Brown, M.; Lazar, M.A. Differential activation of peroxisome proliferator-activated receptors by eicosanoids. *J. Biol. Chem.* **1995**, *270*, 23975–23983. [CrossRef] [PubMed]
11. Narala, V.R.; Adapala, R.K.; Suresh, M.V.; Brock, T.G.; Peters-Golden, M.; Reddy, R.C. Leukotriene B4 is a physiologically relevant endogenous peroxisome proliferator-activated receptor-alpha agonist. *J. Biol. Chem.* **2010**, *285*, 22067–22074. [CrossRef]
12. O'Flaherty, J.T.; Rogers, L.C.; Paumi, C.M.; Hantgan, R.R.; Thomas, L.R.; Clay, C.E.; High, K.; Chen, Y.Q.; Willingham, M.C.; Smitherman, P.K.; et al. 5-Oxo-ETE analogs and the proliferation of cancer cells. *Biochim. Biophys. Acta* **2005**, *1736*, 228–236. [CrossRef]
13. Altmann, R.; Hausmann, M.; Spottl, T.; Gruber, M.; Bull, A.W.; Menzel, K.; Vogl, D.; Herfarth, H.; Scholmerich, J.; Falk, W.; et al. 13-Oxo-ODE is an endogenous ligand for PPARgamma in human colonic epithelial cells. *Biochem. Pharmacol.* **2007**, *74*, 612–622. [CrossRef]
14. Tontonoz, P.; Nagy, L.; Alvarez, J.G.; Thomazy, V.A.; Evans, R.M. PPARgamma promotes monocyte/macrophage differentiation and uptake of oxidized LDL. *Cell* **1998**, *93*, 241–252. [CrossRef]

15. Yki-Jarvinen, H. Thiazolidinediones. *N. Engl. J. Med.* **2004**, *351*, 1106–1118. [CrossRef] [PubMed]
16. Marion-Letellier, R.; Savoye, G.; Ghosh, S. Fatty acids, eicosanoids and PPAR gamma. *Eur. J. Pharmacol.* **2016**, *785*, 44–49. [CrossRef]
17. Suutari, M.; Laakso, S. Effect of growth temperature on the fatty acid composition of *Mycobacterium phlei*. *Arch. Microbiol.* **1993**, *159*, 119–123. [CrossRef] [PubMed]
18. Springer, B.; Kirschner, P.; Rost-Meyer, G.; Schroder, K.H.; Kroppenstedt, R.M.; Bottger, E.C. Mycobacterium interjectum, a new species isolated from a patient with chronic lymphadenitis. *J. Clin. Microbiol.* **1993**, *31*, 3083–3089. [CrossRef]
19. Pacifico, C.; Fernandes, P.; de Carvalho, C. Mycobacterial response to organic solvents and possible implications on cross-resistance with antimicrobial agents. *Front. Microbiol.* **2018**, *9*, 1–12. [CrossRef]
20. Chou, S.; Chedore, P.; Kasatiya, S. Use of gas chromatographic fatty acid and mycolic acid cleavage product determination to differentiate among *Mycobacterium genavense*, *Mycobacterium fortuitum*, *Mycobacterium simiae*, and *Mycobacterium tuberculosis*. *J. Clin. Microbiol.* **1998**, *36*, 577–579. [CrossRef] [PubMed]
21. Smith, D.G.; Martinelli, R.; Besra, G.S.; Illarionov, P.A.; Szatmari, I.; Brazda, P.; Allen, M.A.; Xu, W.; Wang, X.; Nagy, L.; et al. Identification and characterization of a novel anti-inflammatory lipid isolated from *Mycobacterium vaccae*, a soil-derived bacterium with immunoregulatory and stress resilience properties. *Psychopharmacology* **2019**, *236*, 1653–1670. [CrossRef]
22. Luo, Y.; Tanigawa, K.; Kawashima, A.; Ishido, Y.; Ishii, N.; Suzuki, K. The function of peroxisome proliferator-activated receptors PPAR-gamma and PPAR-delta in *Mycobacterium leprae*-induced foam cell formation in host macrophages. *PLoS Negl. Trop. Dis.* **2020**, *14*, e0008850. [CrossRef]
23. Diaz Acosta, C.C.; Dias, A.A.; Rosa, T.; Batista-Silva, L.R.; Rosa, P.S.; Toledo-Pinto, T.G.; Costa, F.; Lara, F.A.; Rodrigues, L.S.; Mattos, K.A.; et al. PGL I expression in live bacteria allows activation of a CD206/PPARgamma cross-talk that may contribute to successful *Mycobacterium leprae* colonization of peripheral nerves. *PLoS Pathog.* **2018**, *14*, e1007151. [CrossRef]
24. Arnett, E.; Weaver, A.M.; Woodyard, K.C.; Montoya, M.J.; Li, M.; Hoang, K.V.; Hayhurst, A.; Azad, A.K.; Schlesinger, L.S. PPARgamma is critical for *Mycobacterium tuberculosis* induction of Mcl-1 and limitation of human macrophage apoptosis. *PLoS Pathog.* **2018**, *14*, e1007100. [CrossRef]
25. Rajaram, M.V.; Brooks, M.N.; Morris, J.D.; Torrelles, J.B.; Azad, A.K.; Schlesinger, L.S. *Mycobacterium tuberculosis* activates human macrophage peroxisome proliferator-activated receptor gamma linking mannose receptor recognition to regulation of immune responses. *J. Immunol.* **2010**, *185*, 929–942. [CrossRef]
26. Luo, W.; Xu, Q.; Wang, Q.; Wu, H.; Hua, J. Effect of modulation of PPAR-gamma activity on Kupffer cells M1/M2 polarization in the development of non-alcoholic fatty liver disease. *Sci. Rep.* **2017**, *7*, 44612. [CrossRef] [PubMed]
27. Lian, M.; Luo, W.; Sui, Y.; Li, Z.; Hua, J. Dietary n-3 PUFA protects mice from con A induced liver injury by modulating regulatory T cells and PPAR-gamma expression. *PLoS ONE* **2015**, *10*, e0132741. [CrossRef] [PubMed]
28. Minnikin, D.E.; Dobson, G.; Goodfellow, M.; Draper, P.; Magnusson, M. Quantitative comparison of the mycolic and fatty acid compositions of *Mycobacterium leprae* and *Mycobacterium gordonae*. *J. Gen. Microbiol.* **1985**, *131*, 2013–2021. [CrossRef] [PubMed]
29. Rubins, H.B.; Robins, S.J.; Collins, D.; Fye, C.L.; Anderson, J.W.; Elam, M.B.; Faas, F.H.; Linares, E.; Schaefer, E.J.; Schectman, G.; et al. Gemfibrozil for the secondary prevention of coronary heart disease in men with low levels of high-density lipoprotein cholesterol. Veterans affairs high-density lipoprotein cholesterol intervention trial study group. *N. Engl. J. Med.* **1999**, *341*, 410–418. [CrossRef]
30. Group, A.S.; Ginsberg, H.N.; Elam, M.B.; Lovato, L.C.; Crouse, J.R., 3rd; Leiter, L.A.; Linz, P.; Friedewald, W.T.; Buse, J.B.; Gerstein, H.C.; et al. Effects of combination lipid therapy in type 2 diabetes mellitus. *N. Engl. J. Med.* **2010**, *362*, 1563–1574.
31. Hossain, M.A.; Tsujita, M.; Gonzalez, F.J.; Yokoyama, S. Effects of fibrate drugs on expression of ABCA1 and HDL biogenesis in hepatocytes. *J. Cardiovasc. Pharmacol.* **2008**, *51*, 258–266. [CrossRef]
32. Manninen, V.; Tenkanen, L.; Koskinen, P.; Huttunen, J.K.; Manttari, M.; Heinonen, O.P.; Frick, M.H. Joint effects of serum triglyceride and LDL cholesterol and HDL cholesterol concentrations on coronary heart disease risk in the Helsinki Heart Study. Implications for treatment. *Circulation* **1992**, *85*, 37–45. [CrossRef] [PubMed]
33. Xu, P.; Wang, J.; Hong, F.; Wang, S.; Jin, X.; Xue, T.; Jia, L.; Zhai, Y. Melatonin prevents obesity through modulation of gut microbiota in mice. *J. Pineal Res.* **2017**, *62*, e12399. [CrossRef]
34. Wang, J.; Xu, P.; Xie, X.; Li, J.; Zhang, J.; Wang, J.; Hong, F.; Li, J.; Zhang, Y.; Song, Y.; et al. DBZ (Danshensu Bingpian Zhi), a novel natural compound derivative, attenuates atherosclerosis in apolipoprotein E-deficient mice. *J. Am. Heart Assoc.* **2017**, *6*, 1–15. [CrossRef]
35. Bays, H.E.; Schwartz, S.; Littlejohn, T., 3rd; Kerzner, B.; Krauss, R.M.; Karpf, D.B.; Choi, Y.J.; Wang, X.; Naim, S.; Roberts, B.K. MBX-8025, a novel peroxisome proliferator receptor-delta agonist: Lipid and other metabolic effects in dyslipidemic overweight patients treated with and without atorvastatin. *J. Clin. Endocrinol. Metab.* **2011**, *96*, 2889–2897. [CrossRef] [PubMed]
36. Oliver, W.R., Jr.; Shenk, J.L.; Snaith, M.R.; Russell, C.S.; Plunket, K.D.; Bodkin, N.L.; Lewis, M.C.; Winegar, D.A.; Sznaidman, M.L.; Lambert, M.H.; et al. A selective peroxisome proliferator-activated receptor delta agonist promotes reverse cholesterol transport. *Proc. Natl. Acad. Sci. USA* **2001**, *98*, 5306–5311. [CrossRef] [PubMed]
37. Vrins, C.L.; van der Velde, A.E.; van den Oever, K.; Levels, J.H.; Huet, S.; Oude Elferink, R.P.; Kuipers, F.; Groen, A.K. Peroxisome proliferator-activated receptor delta activation leads to increased transintestinal cholesterol efflux. *J. Lipid. Res.* **2009**, *50*, 2046–2054. [CrossRef]
38. Peeters, A.; Baes, M. Role of PPARalpha in hepatic carbohydrate metabolism. *PPAR Res.* **2010**, *2010*, 1–12. [CrossRef] [PubMed]

39. Finck, B.N.; Bernal-Mizrachi, C.; Han, D.H.; Coleman, T.; Sambandam, N.; LaRiviere, L.L.; Holloszy, J.O.; Semenkovich, C.F.; Kelly, D.P. A potential link between muscle peroxisome proliferator- activated receptor-alpha signaling and obesity-related diabetes. *Cell Metab.* **2005**, *1*, 133–144. [CrossRef]
40. Liu, S.; Hatano, B.; Zhao, M.; Yen, C.C.; Kang, K.; Reilly, S.M.; Gangl, M.R.; Gorgun, C.; Balschi, J.A.; Ntambi, J.M.; et al. Role of peroxisome proliferator-activated receptor delta/beta in hepatic metabolic regulation. *J. Biol. Chem.* **2011**, *286*, 1237–1247. [CrossRef]
41. Lee, C.H.; Olson, P.; Hevener, A.; Mehl, I.; Chong, L.W.; Olefsky, J.M.; Gonzalez, F.J.; Ham, J.; Kang, H.; Peters, J.M.; et al. PPARdelta regulates glucose metabolism and insulin sensitivity. *Proc. Natl. Acad. Sci. USA* **2006**, *103*, 3444–3449. [CrossRef] [PubMed]
42. Derosa, G.; Sahebkar, A.; Maffioli, P. The role of various peroxisome proliferator-activated receptors and their ligands in clinical practice. *J. Cell Physiol.* **2018**, *233*, 153–161. [CrossRef] [PubMed]
43. Bojic, L.A.; Telford, D.E.; Fullerton, M.D.; Ford, R.J.; Sutherland, B.G.; Edwards, J.Y.; Sawyez, C.G.; Gros, R.; Kemp, B.E.; Steinberg, G.R.; et al. PPARdelta activation attenuates hepatic steatosis in Ldlr-/- mice by enhanced fat oxidation, reduced lipogenesis, and improved insulin sensitivity. *J. Lipid Res.* **2014**, *55*, 1254–1266. [CrossRef]
44. Corrales, P.; Vidal-Puig, A.; Medina-Gomez, G. PPARs and metabolic disorders associated with challenged adipose tissue plasticity. *Int. J. Mol. Sci.* **2018**, *19*, 2124. [CrossRef] [PubMed]
45. Lefterova, M.I.; Zhang, Y.; Steger, D.J.; Schupp, M.; Schug, J.; Cristancho, A.; Feng, D.; Zhuo, D.; Stoeckert, C.J., Jr.; Liu, X.S.; et al. PPARgamma and C/EBP factors orchestrate adipocyte biology via adjacent binding on a genome-wide scale. *Genes Dev.* **2008**, *22*, 2941–2952. [CrossRef]
46. Gross, B.; Pawlak, M.; Lefebvre, P.; Staels, B. PPARs in obesity-induced T2DM, dyslipidaemia and NAFLD. *Nat. Rev. Endocrinol.* **2017**, *13*, 36–49. [CrossRef] [PubMed]
47. Rosen, E.D.; Spiegelman, B.M. PPARgamma: A nuclear regulator of metabolism, differentiation, and cell growth. *J. Biol. Chem.* **2001**, *276*, 37731–37734. [CrossRef]
48. Wu, Z.; Xie, Y.; Morrison, R.F.; Bucher, N.L.; Farmer, S.R. PPARgamma induces the insulin-dependent glucose transporter GLUT4 in the absence of C/EBPalpha during the conversion of 3T3 fibroblasts into adipocytes. *J. Clin. Investig.* **1998**, *101*, 22–32. [CrossRef]
49. Gouzy, A.; Larrouy-Maumus, G.; Wu, T.D.; Peixoto, A.; Levillain, F.; Lugo-Villarino, G.; Guerquin-Kern, J.L.; de Carvalho, L.P.; Poquet, Y.; Neyrolles, O. *Mycobacterium tuberculosis* nitrogen assimilation and host colonization require aspartate. *Nat. Chem. Biol.* **2013**, *9*, 674–676. [CrossRef]
50. Trujillo, C.; Blumenthal, A.; Marrero, J.; Rhee, K.Y.; Schnappinger, D.; Ehrt, S. Triosephosphate isomerase is dispensable in vitro yet essential for *Mycobacterium tuberculosis* to establish infection. *mBio* **2014**, *5*, e00085. [CrossRef]
51. Daniel, J.; Maamar, H.; Deb, C.; Sirakova, T.D.; Kolattukudy, P.E. *Mycobacterium tuberculosis* uses host triacylglycerol to accumulate lipid droplets and acquires a dormancy-like phenotype in lipid-loaded macrophages. *PLoS Pathog.* **2011**, *7*, e1002093. [CrossRef]
52. Griffin, J.E.; Gawronski, J.D.; Dejesus, M.A.; Ioerger, T.R.; Akerley, B.J.; Sassetti, C.M. High-resolution phenotypic profiling defines genes essential for mycobacterial growth and cholesterol catabolism. *PLoS Pathog.* **2011**, *7*, e1002251. [CrossRef]
53. Pandey, A.K.; Sassetti, C.M. Mycobacterial persistence requires the utilization of host cholesterol. *Proc. Natl. Acad. Sci. USA* **2008**, *105*, 4376–4380. [CrossRef]
54. Griffin, J.E.; Pandey, A.K.; Gilmore, S.A.; Mizrahi, V.; McKinney, J.D.; Bertozzi, C.R.; Sassetti, C.M. Cholesterol catabolism by *Mycobacterium tuberculosis* requires transcriptional and metabolic adaptations. *Chem. Biol.* **2012**, *19*, 218–227. [CrossRef]
55. VanderVen, B.C.; Fahey, R.J.; Lee, W.; Liu, Y.; Abramovitch, R.B.; Memmott, C.; Crowe, A.M.; Eltis, L.D.; Perola, E.; Deininger, D.D.; et al. Novel inhibitors of cholesterol degradation in *Mycobacterium tuberculosis* reveal how the bacterium's metabolism is constrained by the intracellular environment. *PLoS Pathog.* **2015**, *11*, e1004679. [CrossRef] [PubMed]
56. Kim, M.J.; Wainwright, H.C.; Locketz, M.; Bekker, L.G.; Walther, G.B.; Dittrich, C.; Visser, A.; Wang, W.; Hsu, F.F.; Wiehart, U.; et al. Caseation of human tuberculosis granulomas correlates with elevated host lipid metabolism. *EMBO Mol. Med.* **2010**, *2*, 258–274. [CrossRef] [PubMed]
57. Cole, S.T.; Brosch, R.; Parkhill, J.; Garnier, T.; Churcher, C.; Harris, D.; Gordon, S.V.; Eiglmeier, K.; Gas, S.; Barry, C.E., 3rd; et al. Deciphering the biology of *Mycobacterium tuberculosis* from the complete genome sequence. *Nature* **1998**, *393*, 537–544. [CrossRef] [PubMed]
58. Marques, M.A.; Berredo-Pinho, M.; Rosa, T.L.; Pujari, V.; Lemes, R.M.; Lery, L.M.; Silva, C.A.; Guimaraes, A.C.; Atella, G.C.; Wheat, W.H.; et al. The essential role of cholesterol metabolism in the intracellular survival of *Mycobacterium leprae* is not coupled to central carbon metabolism and energy production. *J. Bacteriol.* **2015**, *197*, 3698–3707. [CrossRef]
59. Chatterjee, K.R.; Das Gupta, N.N.; De, M.L. Electron microscopic observations on the morphology of *Mycobacterium leprae*. *Exp. Cell Res.* **1959**, *18*, 521–527. [CrossRef]
60. Mattos, K.A.; Lara, F.A.; Oliveira, V.G.; Rodrigues, L.S.; D'Avila, H.; Melo, R.C.; Manso, P.P.; Sarno, E.N.; Bozza, P.T.; Pessolani, M.C. Modulation of lipid droplets by *Mycobacterium leprae* in Schwann cells: A putative mechanism for host lipid acquisition and bacterial survival in phagosomes. *Cell Microbiol.* **2011**, *13*, 259–273. [CrossRef]
61. Mattos, K.A.; Oliveira, V.C.; Berredo-Pinho, M.; Amaral, J.J.; Antunes, L.C.; Melo, R.C.; Acosta, C.C.; Moura, D.F.; Olmo, R.; Han, J.; et al. *Mycobacterium leprae* intracellular survival relies on cholesterol accumulation in infected macrophages: A potential target for new drugs for leprosy treatment. *Cell Microbiol.* **2014**, *16*, 797–815. [CrossRef] [PubMed]

62. Tanigawa, K.; Hayashi, Y.; Hama, K.; Yamashita, A.; Yokoyama, K.; Luo, Y.; Kawashima, A.; Maeda, Y.; Nakamura, Y.; Harada, A.; et al. *Mycobacterium leprae* promotes triacylglycerol de novo synthesis through induction of GPAT3 expression in human premonocytic THP-1 cells. *PLoS ONE* **2021**, *16*, e0249184. [CrossRef]
63. Medeiros, R.C.; Girardi, K.D.; Cardoso, F.K.; Mietto, B.S.; Pinto, T.G.; Gomez, L.S.; Rodrigues, L.S.; Gandini, M.; Amaral, J.J.; Antunes, S.L.; et al. Subversion of schwann cell glucose metabolism by *Mycobacterium leprae*. *J. Biol. Chem.* **2016**, *291*, 21375–21387. [CrossRef] [PubMed]
64. Tanigawa, K.; Degang, Y.; Kawashima, A.; Akama, T.; Yoshihara, A.; Ishido, Y.; Makino, M.; Ishii, N.; Suzuki, K. Essential role of hormone-sensitive lipase (HSL) in the maintenance of lipid storage in *Mycobacterium leprae*-infected macrophages. *Microb. Pathog.* **2012**, *52*, 285–291. [CrossRef] [PubMed]
65. Tanigawa, K.; Suzuki, K.; Nakamura, K.; Akama, T.; Kawashima, A.; Wu, H.; Hayashi, M.; Takahashi, S.; Ikuyama, S.; Ito, T.; et al. Expression of adipose differentiation-related protein (ADRP) and perilipin in macrophages infected with *Mycobacterium leprae*. *FEMS Microbiol. Lett.* **2008**, *289*, 72–79. [CrossRef]
66. Karasawa, K.; Tanigawa, K.; Harada, A.; Yamashita, A. Transcriptional Regulation of Acyl-CoA:Glycerol-sn-3-Phosphate Acyltransferases. *Int. J. Mol. Sci.* **2019**, *20*, 964. [CrossRef]
67. Degang, Y.; Akama, T.; Hara, T.; Tanigawa, K.; Ishido, Y.; Gidoh, M.; Makino, M.; Ishii, N.; Suzuki, K. Clofazimine modulates the expression of lipid metabolism proteins in *Mycobacterium leprae*-infected macrophages. *PLoS Negl. Trop. Dis.* **2012**, *6*, e1936. [CrossRef]
68. Peyron, P.; Vaubourgeix, J.; Poquet, Y.; Levillain, F.; Botanch, C.; Bardou, F.; Daffe, M.; Emile, J.F.; Marchou, B.; Cardona, P.J.; et al. Foamy macrophages from tuberculous patients' granulomas constitute a nutrient-rich reservoir for *M. tuberculosis* persistence. *PLoS Pathog.* **2008**, *4*, e1000204. [CrossRef]
69. Singh, V.; Jamwal, S.; Jain, R.; Verma, P.; Gokhale, R.; Rao, K.V. *Mycobacterium tuberculosis*-driven targeted recalibration of macrophage lipid homeostasis promotes the foamy phenotype. *Cell Host Microbe* **2012**, *12*, 669–681. [CrossRef]
70. Almeida, P.E.; Silva, A.R.; Maya-Monteiro, C.M.; Torocsik, D.; D'Avila, H.; Dezso, B.; Magalhaes, K.G.; Castro-Faria-Neto, H.C.; Nagy, L.; Bozza, P.T. *Mycobacterium bovis* bacillus Calmette-Guerin infection induces TLR2-dependent peroxisome proliferator-activated receptor gamma expression and activation: Functions in inflammation, lipid metabolism, and pathogenesis. *J. Immunol.* **2009**, *183*, 1337–1345. [CrossRef]
71. Mahajan, S.; Dkhar, H.K.; Chandra, V.; Dave, S.; Nanduri, R.; Janmeja, A.K.; Agrewala, J.N.; Gupta, P. *Mycobacterium tuberculosis* modulates macrophage lipid-sensing nuclear receptors PPARgamma and TR4 for survival. *J. Immunol.* **2012**, *188*, 5593–5603. [CrossRef] [PubMed]
72. D'Avila, H.; Melo, R.C.; Parreira, G.G.; Werneck-Barroso, E.; Castro-Faria-Neto, H.C.; Bozza, P.T. *Mycobacterium bovis* bacillus Calmette-Guerin induces TLR2-mediated formation of lipid bodies: Intracellular domains for eicosanoid synthesis in vivo. *J. Immunol.* **2006**, *176*, 3087–3097. [CrossRef]
73. Cao, J.; Li, J.L.; Li, D.; Tobin, J.F.; Gimeno, R.E. Molecular identification of microsomal acyl-CoA:glycerol-3-phosphate acyltransferase, a key enzyme in de novo triacylglycerol synthesis. *Proc. Natl. Acad. Sci. USA* **2006**, *103*, 19695–19700. [CrossRef]
74. Gorga, A.; Rindone, G.M.; Regueira, M.; Pellizzari, E.H.; Camberos, M.C.; Cigorraga, S.B.; Riera, M.F.; Galardo, M.N.; Meroni, S.B. PPARgamma activation regulates lipid droplet formation and lactate production in rat Sertoli cells. *Cell Tissue Res.* **2017**, *369*, 611–624. [CrossRef]
75. Lee, M.J.; Jash, S.; Jones, J.E.C.; Puri, V.; Fried, S.K. Rosiglitazone remodels the lipid droplet and britens human visceral and subcutaneous adipocytes ex vivo. *J. Lipid Res.* **2019**, *60*, 856–868. [CrossRef] [PubMed]
76. Dodd, C.E.; Pyle, C.J.; Glowinski, R.; Rajaram, M.V.; Schlesinger, L.S. CD36-Mediated Uptake of Surfactant Lipids by Human Macrophages Promotes Intracellular Growth of *Mycobacterium tuberculosis*. *J. Immunol.* **2016**, *197*, 4727–4735. [CrossRef] [PubMed]
77. Almeida, P.E.; Roque, N.R.; Magalhaes, K.G.; Mattos, K.A.; Teixeira, L.; Maya-Monteiro, C.; Almeida, C.J.; Castro-Faria-Neto, H.C.; Ryffel, B.; Quesniaux, V.F.; et al. Differential TLR2 downstream signaling regulates lipid metabolism and cytokine production triggered by *Mycobacterium bovis* BCG infection. *Biochim. Biophys. Acta* **2014**, *1841*, 97–107. [CrossRef]
78. Dasgupta, S.; Rai, R.C. PPAR-gamma and Akt regulate GLUT1 and GLUT3 surface localization during *Mycobacterium tuberculosis* infection. *Mol. Cell Biochem.* **2018**, *440*, 127–138. [CrossRef]
79. Brzostek, A.; Pawelczyk, J.; Rumijowska-Galewicz, A.; Dziadek, B.; Dziadek, J. *Mycobacterium tuberculosis* is able to accumulate and utilize cholesterol. *J. Bacteriol.* **2009**, *191*, 6584–6591. [CrossRef]
80. Kim, Y.S.; Lee, H.M.; Kim, J.K.; Yang, C.S.; Kim, T.S.; Jung, M.; Jin, H.S.; Kim, S.; Jang, J.; Oh, G.T.; et al. PPAR-alpha Activation Mediates Innate Host Defense through Induction of TFEB and Lipid Catabolism. *J. Immunol.* **2017**, *198*, 3283–3295. [CrossRef]
81. Kim, Y.S.; Kim, J.K.; Hanh, B.T.B.; Kim, S.Y.; Kim, H.J.; Kim, Y.J.; Jeon, S.M.; Park, C.R.; Oh, G.T.; Park, J.W.; et al. The peroxisome proliferator-activated receptor alpha- agonist gemfibrozil promotes defense against *Mycobacterium abscessus* infections. *Cells* **2020**, *9*, 648. [CrossRef]
82. Jiang, C.; Ting, A.T.; Seed, B. PPAR-gamma agonists inhibit production of monocyte inflammatory cytokines. *Nature* **1998**, *391*, 82–86. [CrossRef]
83. Guirado, E.; Rajaram, M.V.; Chawla, A.; Daigle, J.; La Perle, K.M.; Arnett, E.; Turner, J.; Schlesinger, L.S. Deletion of PPARgamma in lung macrophages provides an immunoprotective response against *M. tuberculosis* infection in mice. *Tuberculosis* **2018**, *111*, 170–177. [CrossRef]

84. Pan, Q.; Wang, Q.; Sun, X.; Xia, X.; Wu, S.; Luo, F.; Zhang, X.L. Aptamer against mannose-capped lipoarabinomannan inhibits virulent *Mycobacterium tuberculosis* infection in mice and rhesus monkeys. *Mol. Ther.* **2014**, *22*, 940–951. [CrossRef] [PubMed]
85. Pan, Q.; Yan, J.; Liu, Q.; Yuan, C.; Zhang, X.L. A single-stranded DNA aptamer against mannose-capped lipoarabinomannan enhances anti-tuberculosis activity of macrophages through downregulation of lipid-sensing nuclear receptor peroxisome proliferator-activated receptor gamma expression. *Microbiol. Immunol.* **2017**, *61*, 92–102. [CrossRef]
86. Jiao, M.; Ren, F.; Zhou, L.; Zhang, X.; Zhang, L.; Wen, T.; Wei, L.; Wang, X.; Shi, H.; Bai, L.; et al. Peroxisome proliferator-activated receptor alpha activation attenuates the inflammatory response to protect the liver from acute failure by promoting the autophagy pathway. *Cell Death Dis.* **2014**, *5*, e1397. [CrossRef] [PubMed]
87. Zhou, Y.; Chen, X.; Qu, N.; Zhang, B.; Xia, C. Chondroprotection of PPARalpha activation by WY14643 *via* autophagy involving Akt and ERK in LPS-treated mouse chondrocytes and osteoarthritis model. *J. Cell Mol. Med.* **2019**, *23*, 2782–2793. [CrossRef] [PubMed]
88. Kim, T.S.; Jin, Y.B.; Kim, Y.S.; Kim, S.; Kim, J.K.; Lee, H.M.; Suh, H.W.; Choe, J.H.; Kim, Y.J.; Koo, B.S.; et al. SIRT3 promotes antimycobacterial defenses by coordinating mitochondrial and autophagic functions. *Autophagy* **2019**, *15*, 1356–1375. [CrossRef] [PubMed]
89. Ghosh, A.; Jana, M.; Modi, K.; Gonzalez, F.J.; Sims, K.B.; Berry-Kravis, E.; Pahan, K. Activation of peroxisome proliferator-activated receptor alpha induces lysosomal biogenesis in brain cells: Implications for lysosomal storage disorders. *J. Biol. Chem* **2015**, *290*, 10309–10324. [CrossRef]
90. Settembre, C.; De Cegli, R.; Mansueto, G.; Saha, P.K.; Vetrini, F.; Visvikis, O.; Huynh, T.; Carissimo, A.; Palmer, D.; Klisch, T.J.; et al. TFEB controls cellular lipid metabolism through a starvation-induced autoregulatory loop. *Nat. Cell Biol.* **2013**, *15*, 647–658. [CrossRef] [PubMed]
91. Brady, O.A.; Martina, J.A.; Puertollano, R. Emerging roles for TFEB in the immune response and inflammation. *Autophagy* **2018**, *14*, 181–189. [CrossRef]
92. Lee, J.M.; Wagner, M.; Xiao, R.; Kim, K.H.; Feng, D.; Lazar, M.A.; Moore, D.D. Nutrient-sensing nuclear receptors coordinate autophagy. *Nature* **2014**, *516*, 112–115. [CrossRef] [PubMed]
93. Halder, P.; Kumar, R.; Jana, K.; Chakraborty, S.; Ghosh, Z.; Kundu, M.; Basu, J. Gene expression profiling of *Mycobacterium tuberculosis* lipoarabinomannan-treated macrophages: A role of the Bcl-2 family member A1 in inhibition of apoptosis in mycobacteria-infected macrophages. *IUBMB Life* **2015**, *67*, 726–736. [CrossRef]
94. Mihaylova, M.M.; Shaw, R.J. The AMPK signalling pathway coordinates cell growth, autophagy and metabolism. *Nat. Cell Biol.* **2011**, *13*, 1016–1023. [CrossRef] [PubMed]
95. Yang, C.S.; Kim, J.J.; Lee, H.M.; Jin, H.S.; Lee, S.H.; Park, J.H.; Kim, S.J.; Kim, J.M.; Han, Y.M.; Lee, M.S.; et al. The AMPK-PPARGC1A pathway is required for antimicrobial host defense through activation of autophagy. *Autophagy* **2014**, *10*, 785–802. [CrossRef]
96. Kim, S.Y.; Yang, C.S.; Lee, H.M.; Kim, J.K.; Kim, Y.S.; Kim, Y.R.; Kim, J.S.; Kim, T.S.; Yuk, J.M.; Dufour, C.R.; et al. ESRRA (estrogen-related receptor alpha) is a key coordinator of transcriptional and post-translational activation of autophagy to promote innate host defense. *Autophagy* **2018**, *14*, 152–168. [CrossRef]
97. Cheng, C.Y.; Gutierrez, N.M.; Marzuki, M.B.; Lu, X.; Foreman, T.W.; Paleja, B.; Lee, B.; Balachander, A.; Chen, J.; Tsenova, L.; et al. Host sirtuin 1 regulates mycobacterial immunopathogenesis and represents a therapeutic target against tuberculosis. *Sci. Immunol.* **2017**, *2*, 1–27. [CrossRef]
98. Ouimet, M.; Franklin, V.; Mak, E.; Liao, X.; Tabas, I.; Marcel, Y.L. Autophagy regulates cholesterol efflux from macrophage foam cells *via* lysosomal acid lipase. *Cell Metab.* **2011**, *13*, 655–667. [CrossRef] [PubMed]

International Journal of
Molecular Sciences

MDPI

Review

PPAR Gamma and Viral Infections of the Brain

Pierre Layrolle, Pierre Payoux and Stéphane Chavanas *

Toulouse NeuroImaging Center (ToNIC), INSERM/UPS UMR 1214, CHU Toulouse-Purpan, 31024 Toulouse, France; pierre.layrolle@inserm.fr (P.L.); pierre.payoux@inserm.fr (P.P.)
* Correspondence: stephane.chavanas@inserm.fr; Tel.: +33-562-74-6178

Abstract: Peroxisome Proliferator-Activated Receptor gamma (PPARγ) is a master regulator of metabolism, adipogenesis, inflammation and cell cycle, and it has been extensively studied in the brain in relation to inflammation or neurodegeneration. Little is known however about its role in viral infections of the brain parenchyma, although they represent the most frequent cause of encephalitis and are a major threat for the developing brain. Specific to viral infections is the ability to subvert signaling pathways of the host cell to ensure virus replication and spreading, as deleterious as the consequences may be for the host. In this respect, the pleiotropic role of PPARγ makes it a critical target of infection. This review aims to provide an update on the role of PPARγ in viral infections of the brain. Recent studies have highlighted the involvement of PPARγ in brain or neural cells infected by immunodeficiency virus 1, Zika virus, or human cytomegalovirus. They have provided a better understanding on PPARγ functions in the infected brain, and revealed that it can be a double-edged sword with respect to inflammation, viral replication, or neuronogenesis. They unraveled new roles of PPARγ in health and disease and could possibly help designing new therapeutic strategies.

Keywords: PPAR gamma; brain; neural stem cells; infection; neuroinflammation; HIV; Zika; cytomegalovirus; neurogenesis; microglia

Citation: Layrolle, P.; Payoux, P.; Chavanas, S. PPAR Gamma and Viral Infections of the Brain. *Int. J. Mol. Sci.* **2021**, *22*, 8876. https://doi.org/10.3390/ijms22168876

Academic Editors: Manuel Vázquez Carrera and Walter Wahli

Received: 22 July 2021
Accepted: 6 August 2021
Published: 18 August 2021

1. Introduction

Peroxisome Proliferator-Activated Receptor gamma (PPARγ) was discovered and cloned almost 30 years ago, as a new member of a family of receptors activated in response to treatment of liver cells by an heterogeneous group of chemicals, namely peroxysome proliferators [1]. Since then, an ever growing body of research has provided us with a better knowledge about PPARγ, which is now known as a master regulator of gene expression in lipid and glucose metabolism, adipogenesis, inflammation, cell proliferation and cancer [2].

It has been almost 25 years since PPARγ transcripts were detected in brain of rat embryos [3]. This early finding suggested that PPARγ might be of importance in brain development; an assumption that was strengthened thereafter by the observation of a «disorganized brain» in *Pparg* knock-out mouse embryos [4]. PPARγ in the brain has been extensively studied in relation to inflammation or neurodegeneration [5]. A wealth of in vitro, in vivo and clinical studies have shown that PPARγ plays a beneficial role on brain injury [6] and neurodegenerative disorders such as Multiple Sclerosis, Alzheimer's disease and Amyotrophic Lateral Sclerosis [7]. Also, on the bases of encouraging preclinical studies, PPARγ has been proposed as a possible therapeutic target for psychiatric disorders [8] or drug addiction and substance abuse [9]. Although the role of PPARγ in the regulation of the immune response and inflammation is well established, little is known however about its role in infections of the brain parenchyma, particularly viral infections.

A wide range of different neurotropic viruses cause infections of the adult or developing brain and underlie acute or chronic neuropathies worldwide [10]. Viral infections of the brain represent the most frequent cause of encephalitis, a neurological disorder characterized by acute fever, seizures, neurologic deficits and/or altered behaviour, which affects 7 people out of 100,000 in the U.S.A. each year [11]. Viral congenital infections may have devastating outcomes on the structure and function of the developing brain, or may result in mild to severe lifelong unabilities [12].

A better understanding on the role of PPARγ in the infected brain may help designing new therapeutic strategies. Furthermore, specific to viral infections is the ability to subvert signaling pathways of the host cell in order to ensure viral replication and spread, as deleterious as the consequences may be for the host. For example, many viruses have evolved mechanisms to regulate positively or negatively activity of the nuclear factor kB (NF-kB) to facilitate their replication, host cell survival, or immuno-evasion [13]. In this respect, the pleiotropic role of PPARγ makes it an expected critical target of infection. Thus, investigating PPARγ in neural cell infections can provide insight on the molecular and cellular outcomes of PPARγ activity in the healthy cell as well as the infected cell.

This review aims to provide for the first time an update on our knowledge of the role of PPARγ in viral infections of the brain parenchyma. It will update the current knowledge on PPARγ molecular aspects and brain expression, point out recent advances about PPARγ focusing on specific brain issues, and, finally, summarise and discuss knowledge on PPARγ and viral infections of the brain parenchyma.

2. PPARγ Molecular Levers

Peroxysome proliferator-activated receptors (PPARs) are members of the nuclear receptor superfamily [14]. As such, they are activated by lipophilic, membrane-permeant, ligands. Upon ligand binding, nuclear receptors form homo- or hetero-dimers and translocate to the nucleus to regulate gene transcription. PPAR family comprises three members, PPARα, PPARβ/δ, and PPARγ. They share a common structure containing six highly conserved functional domains: a first transcription activation function domain (AF-1), a two zinc-fingers DNA binding domain (DBD), a hinge domain, a ligand binding domain (LBD) and a second activation function domain (AF-2) that modulates binding to either co-activator or repressor factors in a ligand-dependent fashion [14,15]. The gene encoding PPARγ, namely *PPARG*, has a complex pattern of expression. Two alternative promoters and alternative splicing events can generate seven *PPARG* transcripts translated to two PPARγ isoforms: the widely expressed PPARγ1 and the adipocyte-restricted PPARγ2 [2].

Transactivation and transrepression refer to positive or negative gene transcriptional regulation by PPARγ, respectively. Transactivation requires both DNA-binding and agonist-binding whereas transrepression may require or not DNA-binding (Figure 1).

PPARγ forms dimers with another nuclear receptor, namely the retinoid X receptor alpha (RXRα) whose ligand is 9-cis retinoic acid [16]. The PPARγ-RXRα dimer translocates to the nucleus and binds cognate DNA sequences named PPAR Responsive Elements (PPRE) [17]. For transactivation, the dimer formed by agonist-bound PPARγ and RXRα recruits coactivators such as PPARγ coactivator 1-α (PGC-1α), E1A binding protein p300 (EP300), or steroid receptor coactivator (SRC1), and histone acetyl transferases (HAT) to assemble a permissive complex on target gene promoters or enhancers, what results in focal chromatin relaxation and enhanced transcription of the cognate gene [2] (Figure 1). This is how PPARγ transactivates expression of a wealth of neuroprotective genes critical for mitochondria, microglial regulation and oxidative stress management [2]. Transrepression occurs differently depending on whether PPARγ is bound to a ligand or not. When PPARγ is unbound or bound to an antagonist (or a so-called inverse agonist), the PPARγ-RXRα dimer recruits corepressors as nuclear receptor corepressor 1 alpha (NCoR1α) or silencing mediator of retinoid and thyroid receptors (SMRT), and histone deacetylase 3 (HDAC3) to assemble a repressive complex on the target gene promoters, what impairs chromatin relaxation and inhibits transcription of the cognate gene [2]. PPARγ also exerts DNA-binding independent transrepression. When activated by an agonist, PPARγ bound to corepressors can bind to other transcription factors such as nuclear factor kB (NF-kB) or activating protein 1 (AP-1) to prevent them from activating inflammatory gene transcription [2] (Figure 1). The PPARγ-corepressor complex can also promote NF-kB degradation or export out of the nucleus [6,18]. These transrepressive mecanisms underlie the anti-inflammatory action of PPARγ [2].

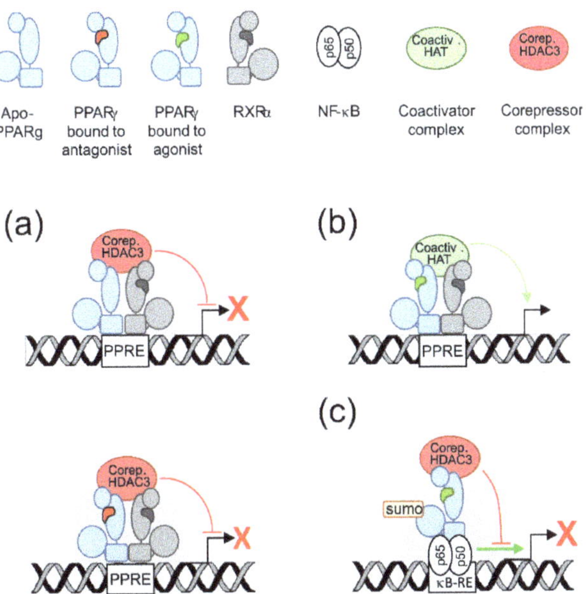

Figure 1. Graphical summary of PPARγ transactivating and transrepressing activities. (**a**) DNA binding-dependent transrepression: unbound PPARγ (top) or PPARγ bound to an antagonist (bottom) forms a dimer with RXRα, binds to a cognate response element (PPRE) and recruits corepressors (Corep.) and HDAC3 to assemble a repressive complex which blocks transcription of the cognate gene (broken arrow). (**b**) Transactivation: agonist-bound PPARγ and RXRα, bound to a PPRE, recruit coactivators (Coactiv.) and HAT to assemble a permissive complex which enhances transcription of the cognate gene. (**c**) DNA binding-independent transrepression: ligand-activated PPARγ and corepressor complex bind to a target transcription factor as NF-kB to prevent it from activating cognate gene transcription. Sumoylation (sumo) increases the stability of the complex PPARγ-corepressor. kB-RE: NF-kB response element.

Post-translational modifications regulate PPARγ activity. Ligand-bound PPARγ may undergo sumoylation which favours its stable binding to the corepressor [19]. PPARγ serine residues may be phosphorylated by the extracellular regulated kinases (ERK) or p38 MAP kinase pathways, what inhibits PPARγ activity by blocking ligand or cofactor binding [20]. In addition, PPARγ has been shown to undergo ubiquitination which increases its stability, or lysine acetylation which stabilizes its binding to co -activators or -repressors, or glycosylation with β-*O*-linked *N*-acetylglucosamine (*O*-GlcNAcylation) which decreases its transactivating ability [20].

PPARγ agonists and antagonists have been widely used in studies which shed light on PPARγ role in health and disease. The best known endogenous agonists of PPARγ are fatty acids such as 15-deoxy-$\Delta^{12,14}$ prostaglandin (PG) J2 (15d-PGJ_2), 15-hydroxyeicosatetraenoic acid (15-HETE), 9- or 13- hydroxyoctadecadienoic acid (9/13-HODE), all derived from oxidation cascades of poly-unsaturated fatty acids (PUFA) as linoleic acid or arachidonic acid [21,22]. Other natural PPARγ agonists are the phospholipids lysophosphatidic acid and hexadecylazelaoyl phosphatidylcholine, nitroalkenes and some dietary lipids such as isoflavones and flavonoids [23]. Recent studies have disclosed that astragaladoside IV from herbal extract [24], alliin from garlic [25] or cannabidiol from cannabis [26] were PPARγ agonists. Synthetic PPARγ agonists are thiazolidinediones (TZDs, e.g., rosiglitazone, pioglitazone, troglitazone) which share as common structural motifs a cyclic tail, an aromatic core, and an acidic head [23]. Noteworthy, the ability of TZDs to cross the brain blood barrier is controversial [6] and some receptor-independent effects of TZDs treatment have been reported [6,22]. Saroglitazar [27] and lanifibranor [28] were recently designed as efficient PPARγ agonists but they also activate PPARα and/or PPARβ/δ. In addition to activating the receptor, some ligands have been shown to upregulate PPARγ expression levels, like pioglitazone in embryonic rat brain cells [22], 15d-PGJ2 in rat primary microglia cells [29] and in a model of neonatal rat cerebral hemorrhage [30], and 9-HODE in human neural stem cells [31], in U937 monocytic cell line [32] and in kidney mesangial cells [33]. Structural studies have recently revealed that PPARγ so-called antagonists such as T0070907 [34] or the novel, rosiglitazone-derived, compound 31 [35] function as inverse agonists: their binding to PPARγ LBD results in conformational changes which increase the receptor affinity to corepressors and decrease its affinity to coactivators, what finally enhances PPARγ transrepressive activity.

3. PPARγ Expression in the Brain

In a founder study, in situ hybridization analyses of embryonic rat brains revealed transient PPARγ mRNA expression in forebrain, midbrain and, at higher levels, hindbrain, from E13.5, to before E18.5 [3]. Immunohistological exploration of PPARγ localization in the brain of adult rats revealed a heterogeneous pattern. PPARγ was detected in basal ganglia including substantia nigra, in hippocampus, in hypothalamus and in some parts of the cerebellum and of the cerebral cortex (cortex) [36]. Strikingly, in the latter, PPARγ expression appeared restricted to three out of the six cortical layers, and only in the frontal and parietal parts, suggesting a complex regulation of expression.

In the adult mouse brain, PPARγ immunoreactivity was observed specifically in prefrontal cortex, nucleus accumbens, amygdala and ventral tegmental area, four brain regions known to be involved in the pathophysiology of neurodegenerative diseases or of addiction [37]. In another study based on quantitative RTPCR and in situ hybridization on laser-microdissected mouse brain sections, PPARγ transcripts were detected in cortex, olfactory bulb and cerebellum, but not in caudate putamen or brain stem [38]. At the cell level, PPARγ was detected in both neurons and astrocytes of mouse or rat [36,37], and was only detectable in microglia after lipopolysaccharide (LPS) stimulation [37].

Few data are available on PPARγ expression in human brain due to its limited accessibility. We explored PPARγ expression by immunohistological analysis using fetal brain slices from elective abortion [31]. The cases were 23 to 28 gestational weeks and presented with conditions non related to brain such as (1) Digeorges syndrome (ie cardiopathy, endocrinopathy, facial dysplasia), (2) chorioamniotitis and anamnios (i.e., loss of amniotic liquid due to inflammation and premature rupture of membranes), (3) renal failure and (4) atrioventricular canal (heart dysplasia) and omphalocele (defective development of the abdominal wall). In any cases, no PPARγ was detected in any area of the brain parenchyma whereas it was detected in brain blood vessel cells. Soon after, immunofluorescence analysis of superior frontal gyrus (a part of the frontal cortex) from postmortem adult human brain has shown PPARγ expression in neurons and astrocytes but not in microglia [37]. Together those studies underscore that PPARγ is not evenly expressed in the brain, nor is it expressed in the same way in the fetal or adult brain, which raises the possibility that it exerts specific functions apart from its anti-inflammatory and metabolic functions.

4. PPARγ Responds to Specific Issues of the Brain Cell

4.1. Energy Supply, Oxidative Stress, and Mitochondria

The brain is particularly sensitive to changes in the energy supply: at baseline, the brain consumes over 20% of the oxygen and 25% of the glucose in the body, although it makes up only 2% of the body's weight. This energy is dedicated to housekeeping neural cell functions, synaptic plasticity, neurotransmitter release and recycling, management of action and resting potentials, and neuronal computation and information processing [39].

Such high activity of energy metabolism and corresponding redox reactions lead to a high production of harmful reactive oxygen species (ROS) such as hydroxyl ($HO^{\bullet}$) and superoxide ($^{\bullet}O_2^-$) radical anions, hydroperoxyl radical ($HO_2^{\bullet}$) and peroxyl radicals ($ROO^{\bullet}$) [40]. Neurons, as long lasting, postmitotic, cells, are more sensitive to the accumulation of oxidative damage in the long run as compared to dividing cells [41]. Thus, brain is highly sensitive to oxidative stress and this is exacerbated in neurodegenerative [40] or presumably psychiatric [39] disorders. PPARγ and/or PPARγ agonists were shown to exert antioxidant functions by upregulating the antioxidant enzymes haem oxygenase-1 (HO-1), catalase or copper/zinc superoxide dismutase (SOD) and downregulating the pro-oxydative enzymes inducible nitric oxide synthase (iNOS) or cyclooxygenase 2 (COX2) (reviewed in [5,7]). Rosiglitazone was also shown to prevent apoptosis related to amyloid [42] or tumor necrosis factor alpha (TNF-α) [43] in human neural stem cells by normalization of oxidative stress and mitochondrial function. Indeed, PPARγ protective role is further supported by its positive effect on mitochondria, that, beyond the cell powerhouse, are key regulators of redox balance [44]. A wealth of in vitro studies reviewed in [5,7] have shown that PPARγ and/or its agonists improved mitochondrial functions in human lymphocytes, adipocytes, astrocytes, neuroblastoma (SH-SY5Y) or neuronal (NT2) cell lines and hippocampal neurons, as shown by increased mitochondrial membrane potential ($\Delta\Psi_m$), increased mitochondrial DNA (mtDNA) copy number, modulation of mitochondrial fusion-fission events and/or expression of factors beneficial to mitochondrial biogenesis and homeostasis, namely the co-activator PGC1-α [45], the mitochondrial transcription factor A (TFAM) [46] or the nuclear factor erythroid-derived 2-like 2 Nrf2 [47].

Recent studies have provided further insight on the role of PPARγ in an oxidative context in brain cells. Pioglitazone has been shown to inhibit, significantly for all, albeit moderately for some, the decrease in total thiol, SOD and catalase levels and the increase in malondialdehyde (MDA, a marker of PUFA peroxidation) levels in hippocampal and cortical extracts, in a rat model of hypothyroidism, a phenotype known to cause neurological damage [48]. Pioglitazone has also been shown to induce expression of TFAM and PGC-1α along with increased mitochondrial biogenesis and to restore mitochondrial membrane potential after challenge with rotenone, an inhibitor of the mitochondrial transport chain complex 1, in rat oligodendrocyte cultures [49]. Recent reports also documented a similar role of PPARγ and agonists in non brain tissues or cells. A C-terminally truncated form of PPARγ2 has been recently shown to localize in the mitochondrial matrix and to bind the D-loop region of mtDNA in brown adipocytes, what strongly suggested that PPARγ transactivated mitochondrial electron transport chain genes [50]. Lentivirally- expressed PPARγ has been shown to restore expression of the antioxidant uncoupling protein 1 (UCP1) in mouse tubular epithelial cells treated with hypoxia, concommitantly to inhibition of ROS generation, whereas pioglitazone administrated to mouse with experimental kidney hypoxia caused reduction of MDA levels and increase of UCP1 mRNA levels in kidney [51]. Pioglitazone has been also shown to increase catalase activity and levels of reduced gluthatione in a PPARγ-dependent manner in a rat model of hypertension [52]. Rosiglitazone was also found to decrease oxidative stress in MDCK canine kidney cells challenged with oxalate in a PPARγ-dependent way [53] and mitochondrial ROS levels, mitochondrial dysfunction and expression of the NLR family pyrin domain containing 3 (NLRP3) inflammasome in C_2C_{12} myotubes and in a mouse model [54].

4.2. Neuroinflammation

Neuroinflammation represents the innate immune response specific to the nervous system. It is mediated by glial cells (i.e., astrocytes and the macrophage-like microglia cells), which activation underlies pathogenesis of neuroinflammation [55]. Noteworthingly, neuroinflammation is linked to oxidative stress since ROS are signaling messengers for inflammation [56]. Neuroinflammation has been widely documented and PPARγ and/or agonists have been shown to decrease neuroinflammation in a wealth of studies, as reviewed in [5].

Recent findings have provided better knowledge on the protective role of PPARγ in neuroinflammation. PPARγ has been shown to mediate suppression of inflammation by the anesthetic propofol in rat astrocytes [57]. To note, this effect is associated with PPARγ-dependent inhibition of the Wnt/β-catenin pathway, an important pathway which enhances neuroinflammation and has a mutual positive regulation with NF-kB [58]. It has been shown that translocator protein (TSPO) inhibited microglia activation by interleukin (IL-) 4 through PPARγ activity in a primary microglia polarization model [59]. Rice bran extract (which is rich in PUFA) as well as pioglitazone have been reported to protect against inflammation induced by lipopolysaccharides (LPS) in a mouse model, decreasing TNF-a and COX2 levels in brain, reducing striatal plaque formation and suppressing cortical and hippocampal tissue damage, all effects requiring PPARγ activity [60]. Other recent studies converged to support positive, PPARγ-dependent, role against neuroinflammation of PPARγ agonists as rosiglitazone which induced IL-10 in primary rat astrocytes exposed to LPS [61], or pioglitazone in a rat model of chronic intermittent hypoxia [62]. Other studies did not assess PPARγ involvement but still reported a protective role of its agonists against neuroinflammation, as rosiglitazone in a mouse model of epilepsy [63] and pioglitazone in rat models of autism [64], Parkinson's disease [65], or neuroinflammation due to intracerebroventricular administration of LPS [66].

4.3. Neurogenesis

Brain is the most complex organ of the body, with a sophisticated tissue architecture. Neurogenesis and brain development rely on finely spatially and temporally tuned cell processes as differentiation, maturation, migration and acquisition of regional identities, whether they involve neural stem cells (NSCs), neural intermediate progenitor cells (NPCs) and/or their neuronal or glial progeny [67]. In the embryo, PPARγ has been shown to support NPC proliferation, trigger astrogliogenesis, inhibit neuron production (neuronogenesis) and enhance neurite outgrowth of differentiating neurons, whereas in the adult brain, PPARγ has been reported to enhance NSC self-renewal and differentiation [68]. A wealth of studies recently reviewed in [69] showed that PPARγ supports NSC growth, survival and stemness maintenance and positively regulates neuronogenesis and neurite outgrowth in maturing neurons. More recently, pioglitazone was shown to promote differentiation of rat primary oligodendrocytes [49].

Besides, neural progenitor/stem cells have specific metabolic needs: they have been shown to require predominantly glycolytic activity to maintain stemness and fatty acids as their energy source, whereas inhibition of lipogenic pathway was reported to decrease their proliferative potential [70]. Indeed, mitochondria are especially important in the regulation of NSC fate decisions, in embryonic and adult brains, as reviewed in [71]. It has been demonstrated that enhanced mitochondrial fragmentation was associated with increased levels of ROS which, as signalling messengers, promote Nrf2-mediated transcriptional upregulation of genes that activate differentiation and prevent self-renewal of NSCs [72]. By the way, a number of mitochondrial diseases or conditions with mitochondrial dysfunction result in neurological outcomes from mild cognitive impairment to severe psychiatric conditions [71]. Although these studies do not investigate the possible link between PPARγ and these processes, it is highly likely that the latter is involved given its importance for mitochondria and metabolism.

5. PPARγ in the Infected Adult or Developing Brain

Brain parenchyma can be infected by a large and heterogeneous range of viruses such as human immunodeficiency virus 1 (HIV), herpesviruses as herpes simplex virus (HSV), varicella-zoster virus (VZV), human cytomegalovirus (HCMV) or herpes virus 6 (HHV6 [73]), Zika virus (ZIKV), japanese encephalitis virus (JEV), West Nile virus (WNV) [11], Borna-disease virus [74] or SARS-CoV-2 [75]. However, the impact of viral infection on PPARγ activity has been investigated for only a small minority of these pathogens. In this respect, most of our knowledge comes from studies on HIV, ZIKV and HCMV infections.

5.1. PPARγ, the Adult Brain and Human Immunodeficiency Virus 1

Human immunodeficiency virus 1 (HIV, genus: *Lentivirus*, family: *Retroviridae*) bears a positive-sense, single-stranded RNA genome spanning around 9700 nucleotides and consisting of 9 genes encoding 19 proteins. HIV is predominantly transmitted by sexual contact across mucosal surfaces, by maternal-infant exposure in the absence of prophylaxis, or by percutaneous inoculation [76]. HIV infection is the causative factor of Acquired Immuno-Deficiency Syndrome (AIDS) that remains a major health issue worldwide [77]. Highly active anti-retroviral therapy (HAART) dramatically decreased mortality and morbidity of infected people through efficient inhibition of both viral replication and opportunistic infections, without, however, eradicating the virus from its lifelong latent reservoirs. A major consequence of persistent HIV infection is the development of HIV-Associated Neurocognitive Disorders (HAND), which are estimated to impact 30–60% of infected people [78], including individuals on successful HAART with undetectable plasma viral load [79]. Subjects with HAND may present paucisymptomatic neurocognitive impairment, or neurocognitive disorder with deficits in concentration, attention and memory, or HIV-associated dementia in the severely affected [80].

Resident brain cells show discrepant sensitivity to infection. Glia cells (astrocytes and microglia), but not neurons, are sensitive to HIV infection. Notably, two recent studies showed that microglial cells are highly permissive to HIV, i.e., they strongly support productive infection and virus spread [81], and that they constitute a stable population of slowly dividing, long-living (up to two decades) cells [82]. Together with other works reviewed in [83], those studies strongly suggested that microglia are the main HIV cell reservoir in the brain. In contrast, astrocytes were shown recently to be non permissive to HIV [79]. Upon infection, glial cells have been shown to release inflammatory cytokines (e.g., TNFα, interleukin-1β or interferon-γ), neurotoxic mediators (e.g., ROS, nitric oxide or glutamate) and viral proteins (namely « virotoxins », as the HIV glycoprotein gp120), resulting in an inflammatory, neurotoxic, and oxidative context, harmful and possibly lethal for neurons and deleterious for synaptic plasticity and astrocyte neuroprotective functions [84]. Unsurprisingly in this context, the anti-inflammatory action of PPARγ is found at the forefront and PPARγ agonists have been shown in a bundle of studies (reviewed in [22]) to be efficient regulators of microglia activation by inhibiting the synthesis of nitric oxide, prostaglandins, inflammatory cytokines and chemokines by microglia and by inducing apoptosis of activated microglia.

More recent studies have converged to highlight the beneficial role of PPARγ activation in HIV-infected brain. It has been disclosed that insulin treatment upregulated PPARγ expression in HIV-infected primary cultures of human microglia as well as in the cortex, but not in the striatum, of cats infected with feline immunodefiency virus, along with antiviral, anti-inflammatory, and neuroprotective outcomes [85]. Rosiglitazone was found to inhibit NF-κB as well as the release of inflammatory mediators (TNFα, IL-1β) or of iNOS and to prevent downregulation of the mouse ortholog of the glutamate transporter EAAT2 (excitatory amino acid transporter 2) caused by recombinant gp120 in primary mixed cultures of rat astrocytes and microglia or in rat after intracranial injection [86]. Interestingly, the same study reported a decrease in PPARγ transcript levels associated with gp120 treatment. EcoHIV is a chimeric HIV harboring gp80 from murine leukemia virus in place of gp120, thereby allowing for the infection of mouse cells and the onset of some molecular change observed in HAND [87]. Rosiglitazone and pioglitazone were demonstrated to reverse the increase in inflammatory mediators (TNFα, IL-1β, the chemokines CCL2, CCL3, CXCL10) and iNOS levels induced by EcoHIV in primary cultures of mouse glial cells and in mouse brains after intracranial injection [88]. In the same study, the two thiazolidinediones were also found to reduce in vivo EcoHIV p24 protein levels in the brain, what strongly supported an antiviral activity of the two agonists. Since then, similar results were obtained by the same group with the novel, non-thiazolidinedione, PPARγ agonist, INT131 [89]. PPARγ activity was however not assessed in these three reports.

Another role of PPARγ apart from neuroinflammatory modulation, has been highlighted in the context of HIV infection. Blood-brain barrier (BBB) is critical for HIV entry into the brain, and tight junction proteins are key structural and functional elements of integrity and efficiency of the BBB. In an in vitro BBB model, loss of barrier efficiency caused by HIV-infected human monocytes was shown to be reduced by overexpression of PPARγ in monocytes, in particular through repression of HIV-induced matrix metalloproteases (MMP) -2 and -9 activities [90]. Further, rosiglitazone has been shown to reduce astrogliosis, neuronal loss and disruption of BBB permeability caused by exposure to the HIV protein Tat, in a PPARγ-dependent fashion, in a mouse model [91]. Similarly, more recent works demonstrated that metabolites of the flavonoid quercetin suppressed MMP-2 activity and invasion of a lung cancer cell line in a PPARγ-dependent manner [92], and that PPARγ blocked the increase in activities of MMP-2 and MMP-9 due to *Toxoplasma Gondii* infection in astrocytes [93]. PPARγ could possibly hinder MMP expression by NF-kB transrepression since NF-κB has been shown to upregulate MMP-2 in murine melanoma cells [94] and MMP-9 in a rat model of intracerebral hemorrhage [95]. Those studies underscored the role of PPARγ in the management of both extracellular matrix and cell to cell adhesion.

On the virus side, NF-kB activity is known to be subverted to stimulate viral replication in the host cell by using the two NF-kB responsive elements within the promoter enhancer region of the long terminal repeat sequence (LTR) of the HIV genome [13]. Hence, by counteracting NF-kB through transrepression, PPARγ hampers not only inflammatory mediators release but also viral replication. Indeed, PPARγ activity was shown to suppress HIV LTR promoter activity, to decrease NF-κB occupancy of the LTR in infected cell, and, finally, to impair HIV replication in brain macrophages of an humanized mouse model of HIV encephalitis [96].

Together those studies converged to show that PPARγ has a beneficial role in the brain of HIV carriers, by counteracting both neuroinflammation and virus replication and by managing proteolysis-mediated regulation of the BBB.

5.2. PPARγ, the Developing Brain and Zika Virus

Zika virus (ZIKV, genus: *Flavivirus*, family: *Flaviviridae*) has a single-stranded RNA genome spanning around 11,000 nucleotides and consisting of a single open reading frame (ORF) and 5′ and 3′ noncoding regions. ZIKV is an arthropod-borne virus (*arbovirus*), predominantly transmitted by mosquitoes, but it can also be transmitted sexually or from mother to fetus [97]. Infected adults may present with mild symptoms or more severe neurological manifestations (eg Guillain-Barré Syndrome or encephalitis) whereas congenital infections may result in severe neurodevelopmental sequelae as microcephaly [97]. Although Zika pandemics outbreak in Brazil in 2016 is relatively recent, key findings on ZIKV neuropathogenesis have been published since. A wealth of recent studies have highlighted various neuropathogenic mechanisms of ZIKV infection, including neural cell receptors, altered gene expression, host RNA modifications or autophagy (reviewed in [97]). Brain organoid studies showed that ZIKV infection caused depletion of NPCs, because of either proliferation arrest and cell death or of premature differentiation (reviewed in [98]).

Notably, PPARγ transcript levels were found to be increased in human NPCs derived from induced pluripotent stem cells (iPSC), as revealed by RNA-seq, along with productive infection, proliferation arrest and apoptosis ([99], and supplemental data therein). A more recent study used quantitative proteomics and transcriptomics in ZIKV-infected human NPCs and revealed, however, decreased levels of PPARγ mRNA [100]. The same study reported upregulation of RXRγ, of a positive regulator of PPARγ activity (Signal transducer and activator of transcription [STAT] 5 [101]) and of two negative regulators of PPARγ activity (FGR, a member of the Src family of tyrosine protein kinases [102], and the AP-1 transcription factor c-Jun [103]), whereas nuclear receptor coactivator 1 (NCOA1), a coactivator of both RXR and PPARγ [104], was found to be downregulated.

Together the diversity of these regulations and their apparently contradictory consequences on PPARγ activity underscore the wide spectrum of cell signaling alterations caused by the infection. Those studies have paved the way to further investigations about the role of PPARγ in ZIKV infection of NPCs.

5.3. PPARγ, the Developing Brain and Human Cytomegalovirus

Human cytomegalovirus (HCMV, genus: *Cytomegalovirus*, family: *Herpesviridae*) is a beta herpes virus bearing a large genome (235-kb double stranded DNA) and that has remarkably co-evolved with humans. As all herpes viruses, it is able to establish lifelong latency after primo infection. Prevalence of HCMV ranges from 50–90% worldwide. HCMV is transmitted by body fluids. Although infection of immunocompetent adult subjects by HCMV is usually benign, congenital infection by HCMV is a leading cause of permanent abnormalities of the central nervous system [105]. About 1% of newborns are congenitally infected by HCMV each year in the U.S.A., as a result of either primary infection of a seronegative pregnant mother, or reinfection or viral reactivation in a seropositive pregnant mother. Among congenitally infected newborns, 10% are symptomatic at birth and present with neurological sequelae; in addition, 10 to 15% of those asymptomatic at birth will display neurological sequelae with onset later in infancy [106]. The most severely affected cases present with brain developmental abnormalities such as microcephaly or brain gyration defects whereas the most frequent sequelae include mental disabilities, sensorineural hearing or vision loss, and/or spastic cerebral palsies [105,106].

Infection of neural progenitor cells in the developing brain is thought to be a primary cause of the neurological sequelae due to HCMV congenital infection ([31] and references therein). In vitro studies showed that HCMV infection of progenitors disrupted self-renewal and polarization [107], apoptosis [108], differentiation [107–112] or migratory abilities [113]. Because PPARγ had been shown previously to be upregulated in human placenta cells infected by HCMV [114], NSCs from human embryonic stem cells were as a model to investigate the outcomes on PPARγ activity of the infection of neural progenitors by HCMV (Figure 2) [31].

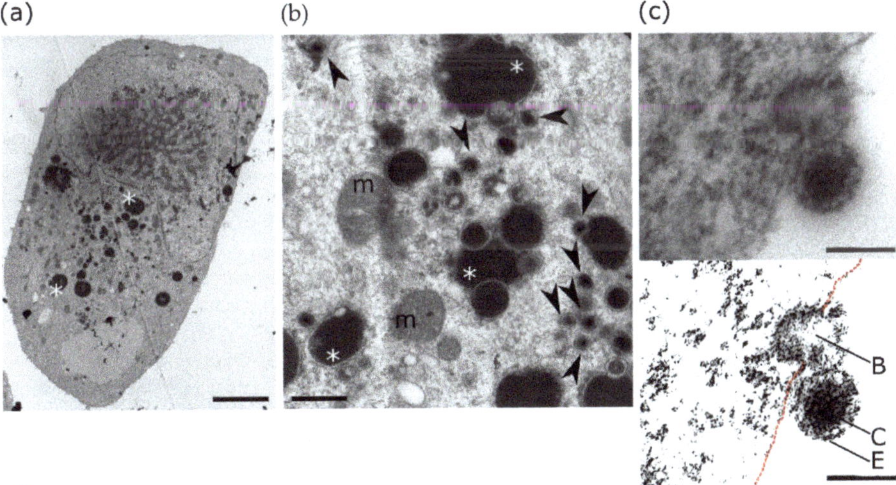

Figure 2. Transmission electron microscopy of human NSC cultures infected by HCMV. (a) Representative HCMV-infected NSC, displaying numerous electron-dense lipid droplets (asterisks) consistent with active PPARγ. Scale bar: 5 µm. (b) Representative view of the cytoplasm of an infected NSC, containing morphologically mature viral particles (arrowheads), mitochondria (m) and still lipid droplets. Scale bar: 0.5 µm. (c) Viral HCMV particle shedding from an infected cell (top) and the same view after image processing (bottom) to highlight plasma membrane (red dotted line), the exocytosis cavity (B), the viral capsid (C) containing the electron-dense viral chromatin and the viral envelope (E). Scale bar: 0.2 µm.

Infection by HCMV was found to dramatically impair neuronal differentiation of

Infection by HCMV was found to dramatically impair neuronal differentiation of NSCs [31]. PPARγ was barely detectable in uninfected NSCs whereas nuclei of infected NSCs showed strong immunoreactivity to PPARγ, indicating increased expression and activity of PPARγ [31]. This result was confirmed by chromatin immunoprecipitation, reporter gene assay or cellular lipid droplet staining. More importantly, this finding was strongly supported by the immunodetection of nuclear PPARγ specifically in the brain germinative zones of congenitally infected fetuses (N = 20) but not in control samples [31]. Lipidomic analysis revealed that levels of 9-HODE were significantly and specifically increased in infected NSCs, indicating that 9-HODE was the agonist associated with PPARγ activation. 9-HODE was also found to dramatically increase PPARγ levels and activity in uninfected NSCs, recapitulating the effect of infection [31]. Furthermore, 9-HODE treatment and/or single-out expression of PPARγ were sufficient to impair neuronogenesis of uninfected NSCs, whereas treatment of HCMV-infected NSCs with the PPARγ antagonist T0070907 restored a normal rate of differentiation [31]. Together these findings revealed that PPARγ exerts a negative role on NSC differentiation to neurons, should they be infected by HCMV or not. This has been supported soon after in another study which demonstrated that conditionally forced expression of Pparγ in mouse neural progenitors resulted in severe microcephaly and brain malformation [115].

The high level of 9-HODE biosynthesis could result from an interesting feature of HCMV particles. Indeed, the production of 9-HODE results from the oxidation of linoleic acid by cellular lipoxygenase 15-LOX, and linoleic acid is released from membrane phospholipids by viral, onboarded, phospholipase A2 (oPLA2) during infection. oPLA2 is a cell-derived phospholipase A2, packaged in the tegument of the virion during its release from the cell, and subsequently injected in the new host cell during viral particle entry [116]. In other words, HCMV particles carry oPLA2 as a ready to use tool for efficient 9-HODE biosynthesis in the host cell. HCMV infection has also been shown to inhibit Wnt/β-catenin signaling in dermal fibroblasts and placental extravillous trophoblasts [117], and this could also account for increased PPARγ activity in HCMV-infected NSCs since Wnt/β-catenin inhibits PPARγ [58].

Increased viral replication was observed in HCMV-infected NSCs exposed to 9-HODE [31]. Indeed, it had been demonstrated in human placenta cells that PPARγ exerted a positive role on HCMV replication by transactivating HCMV major immediate early promoter (MIEP) through the use of two PPREs [118]. Furthermore, neural progenitors require predominantly fatty acids as their energy source [119], and productive infection requires a large energetic supply and enhanced biosynthesis of fatty acids in the host cell to allow efficient viral replication and envelope assembly [120]. Increased PPARγ activity could thus be beneficial to both virus replication and host cell survival, given its role on fatty acid metabolism and mitochondria. This seems of particular importance in infection by HCMV since HCMV, as the other beta herpes viruses, undergoes in his host a long replicative cycle which numbers in days, and which, to be completed, requires prolonged survival of the host cell in spite of the metabolic storm caused by the infection.

6. Conclusions

Investigations about the outcomes of viral infection in the brain shed new light on PPARγ in the developing and adult brain. Recent studies underscored that expression and/or activity of such a master regulator as PPARγ must be finely tuned in time and space, especially during brain development.

Probably because of its multifaceted role at the crossroads of inflammation, metabolism and cell differentiation, PPARγ can be a double-edged sword in viral infections of neural cells: besides its role in both moderating inflammation and supporting host cell survival, it can be deleterious to neuronal differentiation of progenitors, and either inhibit or support viral replication (Figure 3).

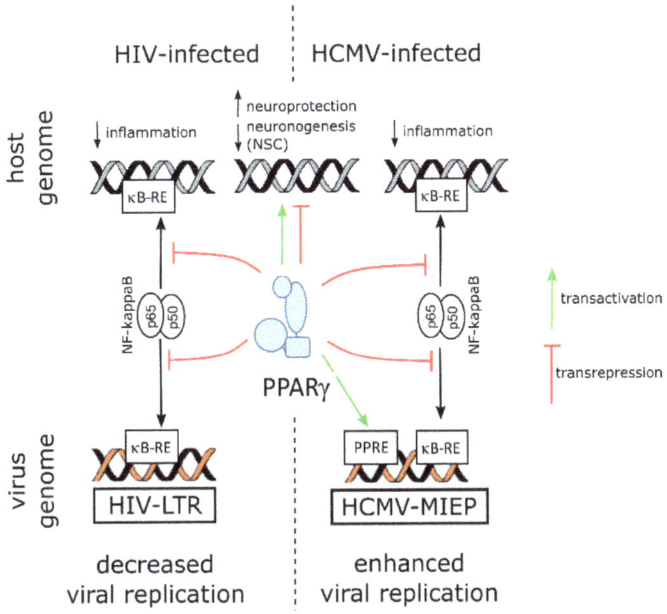

Figure 3. Graphical summary of PPARγ involvement during infection by HIV (**left**) or HCMV (**right**). In HIV-infected cells, PPARγ inhibits NF-kB by transrepression (red lines), thereby downregulating inflammatory genes and decreasing the efficiency of viral replication. In contrast, in HCMV-infected cells, PPARγ enhances viral replication by transactivation (green arrow) of the HCMV major immediate early promoter (MIEP) through two PPAR responsive elements (PPRE). In both cases, PPARγ regulates expression of the host cell genome, contributing to neuroprotection and, in neural stem cells (NSC), inhibition of neuronogenesis. kB-RE: NF-γB responsive element.

In the infected adult brain, the role of PPARγ in the host response to infection appeared beneficial against inflammation, oxidative stress and viral replication, as exemplified in HIV infection (Figure 3). PPARγ agonists have been proposed to be promising candidate drugs in the treatment of HIV-1 brain inflammation and neurocognitive outcomes [86], especially as they are already being used in treatment of HIV-associated lipodystrophy [121]. In contrast, in the developing brain, PPARγ activation has deleterious outcomes on neurogenesis, as shown in HCMV infection, and possibly in ZIKV infection. Notably, the activation of PPARγ in infection by HCMV is beneficial to viral replication (Figure 3).

Viruses undergo evolutionary pressure which optimizes both their spreading efficiency and the survival of their host. Whereas both the genomes of HIV and HCMV contain responsive elements to NF-kB, HCMV genome has evolved to gain two PPAR responsive elements within its major promoter. These responsive elements allow the subversion of PPARγ activity in the benefit of HCMV replication. Moreover, NF-kB transrepression by activated PPARγ accounts for immune evasion.

Yet, it is important to recall how variable the severity of neurological sequelae of HCMV infection may be. Host genetic factors still to be discovered may be important determinants of the severity of the sequelae, as, for example, cis-acting transcriptional regulators of PPARγ gene expression, or reciprocally, putative PPRE within PPARγ target genes.

Author Contributions: Conceptualization, investigation, writing—original draft preparation, writing—review and editing, S.C., P.P. and P.L. All authors have read and agreed to the published version of the manuscript.

Funding: This research received no external funding.

Institutional Review Board Statement: Not applicable.

Informed Consent Statement: Not applicable.

Data Availability Statement: Not applicable.

Conflicts of Interest: The authors declare no conflict of interest.

References

1. Zhu, Y.; Alvares, K.; Huang, Q.; Rao, M.S.; Reddy, J.K. Cloning of a New Member of the Peroxisome Proliferator-Activated Receptor Gene Family from Mouse Liver. *J. Biol. Chem.* **1993**, *268*, 26817–26820. [CrossRef]
2. Hernandez-Quiles, M.; Broekema, M.F.; Kalkhoven, E. PPARgamma in Metabolism, Immunity, and Cancer: Unified and Diverse Mechanisms of Action. *Front. Endocrinol.* **2021**, *12*, 624112. [CrossRef]
3. Braissant, O.; Wahli, W. Differential Expression of Peroxisome Proliferator-Activated Receptor-Î±, -Î², and -Î³ during Rat Embryonic Development. *Endocrinology* **1998**, *139*, 2748–2754. [CrossRef] [PubMed]
4. Wada, K.; Nakajima, A.; Katayama, K.; Kudo, C.; Shibuya, A.; Kubota, N.; Terauchi, Y.; Tachibana, M.; Miyoshi, H.; Kamisaki, Y.; et al. Peroxisome Proliferator-Activated Receptor Gamma-Mediated Regulation of Neural Stem Cell Proliferation and Differentiation. *J. Biol. Chem.* **2006**, *281*, 12673–12681. [CrossRef]
5. Villapol, S. Roles of Peroxisome Proliferator-Activated Receptor Gamma on Brain and Peripheral Inflammation. *Cell. Mol. Neurobiol.* **2018**, *38*, 121–132. [CrossRef]
6. Cai, W.; Yang, T.; Liu, H.; Han, L.; Zhang, K.; Hu, X.; Zhang, X.; Yin, K.J.; Gao, Y.; Bennett, M.V.L.; et al. Peroxisome Proliferator-Activated Receptor γ (PPARγ): A Master Gatekeeper in CNS Injury and Repair. *Prog. Neurobiol.* **2018**, *163-164*, 27–58. [CrossRef] [PubMed]
7. Corona, J.C.; Duchen, M.R. PPARγ as a Therapeutic Target to Rescue Mitochondrial Function in Neurological Disease. *Free Radic. Biol. Med.* **2016**, *100*, 153–163. [CrossRef]
8. Tufano, M.; Pinna, G. Is There a Future for PPARs in the Treatment of Neuropsychiatric Disorders? *Molecules* **2020**, *25*, 1062. [CrossRef] [PubMed]
9. Cheng, H.S.; Tan, W.R.; Low, Z.S.; Marvalim, C.; Lee, J.Y.H.; Tan, N.S. Exploration and Development of PPAR Modulators in Health and Disease: An Update of Clinical Evidence. *Int. J. Mol. Sci.* **2019**, *20*, 5055. [CrossRef]
10. Soung, A.; Klein, R.S. Viral Encephalitis and Neurologic Diseases: Focus on Astrocytes. *Trends Mol. Med.* **2018**, *24*, 950–962. [CrossRef]
11. Tyler, K.L. Acute Viral Encephalitis. *N. Engl. J. Med.* **2018**, *379*, 557–566. [CrossRef]
12. Cordeiro, C.N.; Tsimis, M.; Burd, I. Infections and Brain Development. *Obstet. Gynecol. Surv.* **2015**, *70*, 644–655. [CrossRef] [PubMed]
13. Hiscott, J.; Kwon, H.; Génin, P. Hostile Takeovers: Viral Appropriation of the NF-KB Pathway. *J. Clin. Investig.* **2001**, *107*, 143–151. [CrossRef] [PubMed]
14. Weikum, E.R.; Liu, X.; Ortlund, E.A. The Nuclear Receptor Superfamily: A Structural Perspective. *Protein Sci.* **2018**, *27*, 1876–1892. [CrossRef]
15. Shang, J.; Mosure, S.A.; Zheng, J.; Brust, R.; Bass, J.; Nichols, A.; Solt, L.A.; Griffin, P.R.; Kojetin, D.J. A Molecular Switch Regulating Transcriptional Repression and Activation of PPARγ. *Nat. Commun.* **2020**, *11*. [CrossRef] [PubMed]
16. Tontonoz, P.; Graves, R.A.; Budavari, A.I.; Erdjument-bromage, H.; Lui, M.; Hu, E.; Tempst, P.; Spiegelman, B.M. Adipocyte-Specific Transcription Factor ARF6 Is a Heterodimeric Complex of Two Nuclear Hormone Receptors, PPAR7 and RXRa. *Nucleic Acids Res.* **1994**, *22*, 5628–5634. [CrossRef]
17. Lemay, D.G.; Hwang, D.H. Genome-Wide Identification of Peroxisome Proliferator Response Elements Using Integrated Computational Genomics. *J. Lipid Res.* **2006**, *47*, 1583–1587. [CrossRef]
18. Wahli, W. A Gut Feeling of the PXR, PPAR and NF-kappaB Connection. *J. Int. Med.* **2008**, *263*, 613–619. [CrossRef]
19. Pascual, G.; Fong, A.L.; Ogawa, S.; Gamliel, A.; Li, A.C.; Perissi, V.; Rose, D.W.; Willson, T.M.; Rosenfeld, M.G.; Glass, C.K. A SUMOylation-Dependent Pathway Mediates Transrepression of Inflammatory Response Genes by PPAR-γ. *Nature* **2005**, *437*, 759–763. [CrossRef]
20. Brunmeir, R.; Xu, F. Functional Regulation of PPARs through Post-Translational Modifications. *Int. J. Mol. Sci.* **2018**, *19*, 1738. [CrossRef]
21. Marion-Letellier, R.; Savoye, G.; Ghosh, S. Fatty Acids, Eicosanoids and PPAR Gamma. *Eur. J. Pharmacol.* **2016**, *785*, 44–49. [CrossRef] [PubMed]
22. Bernardo, A.; Minghetti, L. PPAR-Gamma Agonists as Regulators of Microglial Activation and Brain Inflammation. *Curr. Pharm. Des.* **2007**, *12*, 93–109. [CrossRef] [PubMed]
23. Prashantha Kumar, B.R.; Kumar, A.P.; Jose, J.A.; Prabitha, P.; Yuvaraj, S.; Chipurupalli, S.; Jeyarani, V.; Manisha, C.; Banerjee, S.; Jeyabalan, J.B.; et al. Minutes of PPAR-γ Agonism and Neuroprotection. *Neurochem. Int.* **2020**, *140*, 104814. [CrossRef] [PubMed]
24. Wang, X.; Wang, Y.; Hu, J.-P.; Yu, S.; Li, B.-K.; Cui, Y.; Ren, L.; Zhang, L.-D. Astragaloside IV, a Natural PPARγ Agonist, Reduces Aβ Production in Alzheimer's Disease Through Inhibition of BACE1. *Mol. Neurobiol.* **2017**, *54*, 2939–2949. [CrossRef]

25. Shi, L.; Lin, Q.; Li, X.; Nie, Y.; Sun, S.; Deng, X.; Wang, L.; Lu, J.; Tang, Y.; Luo, F. Alliin, a Garlic Organosulfur Compound, Ameliorates Gut Inflammation through MAPK-NF-KB/AP-1/STAT-1 Inactivation and PPAR-γ Activation. *Mol. Nutr. Food Res.* **2017**, *61*, 1601013. [CrossRef]
26. Vallée, A.; Lecarpentier, Y.; Guillevin, R.; Vallée, J.-N. Effects of Cannabidiol Interactions with Wnt/β-Catenin Pathway and PPARγ on Oxidative Stress and Neuroinflammation in Alzheimer's Disease. *Acta Biochim. Biophys. Sin.* **2017**, *49*, 853–866. [CrossRef] [PubMed]
27. Makled, M.N.; Sharawy, M.H.; El-Awady, M.S. The Dual PPAR-α/γ Agonist Saroglitazar Ameliorates Thioacetamide-Induced Liver Fibrosis in Rats through Regulating Leptin. *Naunyn. Schmiedebergs Arch. Pharmacol.* **2019**, *392*, 1569–1576. [CrossRef]
28. Boubia, B.; Poupardin, O.; Barth, M.; Binet, J.; Peralba, P.; Mounier, L.; Jacquier, E.; Gauthier, E.; Lepais, V.; Chatar, M.; et al. Design, Synthesis, and Evaluation of a Novel Series of Indole Sulfonamide Peroxisome Proliferator Activated Receptor (PPAR) α/γ/δ Triple Activators: Discovery of Lanifibranor, a New Antifibrotic Clinical Candidate. *J. Med. Chem.* **2018**, *61*, 2246–2265. [CrossRef]
29. Bernardo, A.; Levi, G.; Minghetti, L. Role of the Peroxisome Proliferator-Activated Receptor-γ (PPAR-γ) and Its Natural Ligand 15-Deoxy-Δ(12,14)-Prostaglandin J2 in the Regulation of Microglial Functions. *Eur. J. Neurosci.* **2000**, *12*, 2215–2223. [CrossRef]
30. Flores, J.J.; Klebe, D.; Rolland, W.B.; Lekic, T.; Krafft, P.R.; Zhang, J.H. PPARγ-Induced Upregulation of CD36 Enhances Hematoma Resolution and Attenuates Long-Term Neurological Deficits after Germinal Matrix Hemorrhage in Neonatal Rats. *Neurobiol. Dis.* **2016**, *87*, 124–133. [CrossRef]
31. Rolland, M.; Li, X.; Sellier, Y.; Martin, H.; Perez-Berezo, T.; Rauwel, B.; Benchoua, A.; Bessières, B.; Aziza, J.; Cenac, N.; et al. PPARγ Is Activated during Congenital Cytomegalovirus Infection and Inhibits Neuronogenesis from Human Neural Stem Cells. *PLoS Pathog.* **2016**, *12*, e1005547. [CrossRef]
32. Hampel, J.K.; Brownrigg, L.M.; Vignarajah, D.; Croft, K.D.; Dharmarajan, A.M.; Bentel, J.M.; Puddey, I.B.; Yeap, B.B. Differential Modulation of Cell Cycle, Apoptosis and PPARgamma2 Gene Expression by PPARgamma Agonists Ciglitazone and 9-Hydroxyoctadecadienoic Acid in Monocytic Cells. *Prostaglandins Leukot Essent. Fat. Acids* **2006**, *74*, 283–293. [CrossRef]
33. Negishi, M.; Shimizu, H.; Okada, S.; Kuwabara, A.; Okajima, F.; Mori, M. 9HODE Stimulates Cell Proliferation and Extracellular Matrix Synthesis in Human Mesangial Cells via PPARgamma. *Exp. Biol. Med.* **2004**, *229*, 1053–1060. [CrossRef]
34. Brust, R.; Shang, J.; Fuhrmann, J.; Mosure, S.A.; Bass, J.; Cano, A.; Heidari, Z.; Chrisman, I.M.; Nemetchek, M.D.; Blayo, A.L.; et al. A Structural Mechanism for Directing Corepressor-Selective Inverse Agonism of PPARγ. *Nat. Commun.* **2018**, *9*. [CrossRef]
35. Toyota, Y.; Nomura, S.; Makishima, M.; Hashimoto, Y.; Ishikawa, M. Structure-Activity Relationships of Rosiglitazone for Peroxisome Proliferator-Activated Receptor Gamma Transrepression. *Bioorgan. Med. Chem. Lett.* **2017**, *27*, 2776–2780. [CrossRef]
36. Moreno, S.; Farioli-Vecchioli, S.; CerÃ1, M.P. Immunolocalization of Peroxisome Proliferator-Activated Receptors and Retinoid x Receptors in the Adult Rat CNS. *Neuroscience* **2004**, *123*, 131–145. [CrossRef] [PubMed]
37. Warden, A.; Truitt, J.; Merriman, M.; Ponomareva, O.; Jameson, K.; Ferguson, L.B.; Mayfield, R.D.; Harris, R.A. Localization of PPAR Isotypes in the Adult Mouse and Human Brain. *Sci. Rep.* **2016**, *6*, 27618. [CrossRef]
38. Gofflot, F.; Chartoire, N.; Vasseur, L.; Heikkinen, S.; Dembele, D.; Le Merrer, J.; Auwerx, J. Systematic Gene Expression Mapping Clusters Nuclear Receptors According to Their Function in the Brain. *Cell* **2007**, *131*, 405–418. [CrossRef]
39. Kim, Y.; Vadodaria, K.C.; Lenkei, Z.; Kato, T.; Gage, F.H.; Marchetto, M.C.; Santos, R. Mitochondria, Metabolism, and Redox Mechanisms in Psychiatric Disorders. *Antioxid. Redox Signal.* **2019**, *31*, 275–317. [CrossRef]
40. Singh, A.; Kukreti, R.; Saso, L.; Kukreti, S. Oxidative Stress: A Key Modulator in Neurodegenerative Diseases. *Molecules* **2019**, *24*, 1583. [CrossRef] [PubMed]
41. Terman, A.; Kurz, T.; Navratil, M.; Arriaga, E.A.; Brunk, U.T. Mitochondrial Turnover and Aging of Long-Lived Postmitotic Cells: The Mitochondrial-Lysosomal Axis Theory of Aging. *Antioxid. Redox Signal.* **2010**, *12*, 503–535. [CrossRef] [PubMed]
42. Chiang, M.C.; Nicol, C.J.; Cheng, Y.C.; Lin, K.H.; Yen, C.H.; Lin, C.H. Rosiglitazone Activation of PPARγ-Dependent Pathways Is Neuroprotective in Human Neural Stem Cells against Amyloid-Beta-Induced Mitochondrial Dysfunction and Oxidative Stress. *Neurobiol. Aging* **2016**, *40*, 181–190. [CrossRef] [PubMed]
43. Chiang, M.C.; Cheng, Y.C.; Lin, K.H.; Yen, C.H. PPARγ Regulates the Mitochondrial Dysfunction in Human Neural Stem Cells with Tumor Necrosis Factor Alpha. *Neuroscience* **2013**, *229*, 118–129. [CrossRef]
44. Cenini, G.; Lloret, A.; Cascella, R. Oxidative Stress in Neurodegenerative Diseases: From a Mitochondrial Point of View. *Oxid. Med. Cell. Longev.* **2019**, *2019*, 2105607. [CrossRef]
45. Scarpulla, R.C. Metabolic Control of Mitochondrial Biogenesis through the PGC-1 Family Regulatory Network. *Biochim. Biophys. Acta Mol. Cell Res.* **2011**, *1813*, 1269–1278. [CrossRef] [PubMed]
46. Kang, I.; Chu, C.T.; Kaufman, B.A. The Mitochondrial Transcription Factor TFAM in Neurodegeneration: Emerging Evidence and Mechanisms. *FEBS Lett.* **2018**, *592*, 793–811. [CrossRef] [PubMed]
47. Kang, T.C. Nuclear Factor-Erythroid 2-Related Factor 2 (Nrf2) and Mitochondrial Dynamics/Mitophagy in Neurological Diseases. *Antioxidants* **2020**, *9*, 617. [CrossRef]
48. Baghcheghi, Y.; Salmani, H.; Beheshti, F.; Shafei, M.N.; Sadeghnia, H.R.; Soukhtanloo, M.; Ebrahimzadeh Bideskan, A.; Hosseini, M. Effects of PPAR-γ Agonist, Pioglitazone on Brain Tissues Oxidative Damage and Learning and Memory Impairment in Juvenile Hypothyroid Rats. *Int. J. Neurosci.* **2019**, *129*, 1024–1038. [CrossRef] [PubMed]

49. De Nuccio, C.; Bernardo, A.; Troiano, C.; Brignone, M.S.; Falchi, M.; Greco, A.; Rosini, M.; Basagni, F.; Lanni, C.; Serafini, M.M.; et al. Nrf2 and Ppar-γ Pathways in Oligodendrocyte Progenitors: Focus on Ros Protection, Mitochondrial Biogenesis and Promotion of Cell Differentiation. *Int. J. Mol. Sci.* **2020**, *21*, 7216. [CrossRef]
50. Chang, J.S.; Ha, K. A Truncated PPAR Gamma 2 Localizes to Mitochondria and Regulates Mitochondrial Respiration in Brown Adipocytes. *PLoS ONE* **2018**, *13*, e0195007. [CrossRef]
51. Jia, P.; Wu, X.; Pan, T.; Xu, S.; Hu, J.; Ding, X. Uncoupling Protein 1 Inhibits Mitochondrial Reactive Oxygen Species Generation and Alleviates Acute Kidney Injury. *EBioMedicine* **2019**, *49*, 331–340. [CrossRef]
52. Soliman, E.; Behairy, S.F.; El-maraghy, N.N.; Elshazly, S.M. PPAR-γ Agonist, Pioglitazone, Reduced Oxidative and Endoplasmic Reticulum Stress Associated with L-NAME-Induced Hypertension in Rats. *Life Sci.* **2019**, *239*, 117047. [CrossRef]
53. Liu, Y.D.; Yu, S.L.; Wang, R.; Liu, J.N.; Jin, Y.S.; Li, Y.F.; An, R.H. Rosiglitazone Suppresses Calcium Oxalate Crystal Binding and Oxalate-Induced Oxidative Stress in Renal Epithelial Cells by Promoting PPAR- γ Activation and Subsequent Regulation of TGF-β 1 and HGF Expression. *Oxid. Med. Cell. Longev.* **2019**, *2019*, 4826525. [CrossRef]
54. Liu, Y.; Bi, X.; Zhang, Y.; Wang, Y.; Ding, W. Mitochondrial Dysfunction/NLRP3 Inflammasome Axis Contributes to Angiotensin II-Induced Skeletal Muscle Wasting via PPAR-γ. *Lab. Investig.* **2020**, *100*, 712–726. [CrossRef] [PubMed]
55. Yang, Q.-Q.; Zhou, J.-W. Neuroinflammation in the Central Nervous System: Symphony of Glial Cells. *Glia* **2019**, *67*, 1017–1035. [CrossRef] [PubMed]
56. Forrester, S.J.; Kikuchi, D.S.; Hernandes, M.S.; Xu, Q.; Griendling, K.K. Reactive Oxygen Species in Metabolic and Inflammatory Signaling. *Circ. Res.* **2018**, *122*, 877–902. [CrossRef]
57. Jiang, P.; Jiang, Q.; Yan, Y.; Hou, Z.; Luo, D. Propofol Ameliorates Neuropathic Pain and Neuroinflammation through PPAR γ Up-Regulation to Block Wnt/β-Catenin Pathway. *Neurol. Res.* **2021**, *43*, 71–77. [CrossRef] [PubMed]
58. Vallée, A.; Vallée, J.N.; Guillevin, R.; Lecarpentier, Y. Interactions Between the Canonical WNT/Beta-Catenin Pathway and PPAR Gamma on Neuroinflammation, Demyelination, and Remyelination in Multiple Sclerosis. *Cell. Mol. Neurobiol.* **2018**, *38*, 783–795. [CrossRef]
59. Zhou, D.; Ji, L.; Chen, Y. TSPO Modulates IL-4-Induced Microglia/Macrophage M2 Polarization via PPAR-γ Pathway. *J. Mol. Neurosci.* **2020**, *70*, 542–549. [CrossRef]
60. Abd El Fattah, M.A.; Abdelhamid, Y.A.; Elyamany, M.F.; Badary, O.A.; Heikal, O.A. Rice Bran Extract Protected against LPS-Induced Neuroinflammation in Mice through Targeting PPAR-γ Nuclear Receptor. *Mol. Neurobiol.* **2021**, *58*, 1504–1516. [CrossRef]
61. Chistyakov, D.V.; Astakhova, A.A.; Goriainov, S.V.; Sergeeva, M.G. Comparison of PPAR Ligands as Modulators of Resolution of Inflammation, via Their Influence on Cytokines and Oxylipins Release in Astrocytes. *Int. J. Mol. Sci.* **2020**, *21*, 9577. [CrossRef] [PubMed]
62. Zhang, X.; Li, N.; Lu, L.; Lin, Q.; Li, L.; Dong, P.; Yang, B.; Li, D.; Fei, J. Pioglitazone Prevents Sevoflurane-induced Neuroinflammation and Cognitive Decline in a Rat Model of Chronic Intermittent Hypoxia by Upregulating Hippocampal PPAR-γ. *Mol. Med. Rep.* **2019**, *49*, 3815–3822. [CrossRef] [PubMed]
63. Peng, J.; Wang, K.; Xiang, W.; Li, Y.; Hao, Y.; Guan, Y. Rosiglitazone Polarizes Microglia and Protects against Pilocarpine-Induced Status Epilepticus. *CNS Neurosci. Ther.* **2019**, *25*, 1363–1372. [CrossRef] [PubMed]
64. Mirza, R.; Sharma, B. A Selective Peroxisome Proliferator-Activated Receptor-γ Agonist Benefited Propionic Acid Induced Autism-like Behavioral Phenotypes in Rats by Attenuation of Neuroinflammation and Oxidative Stress. *Chem. Biol. Interact.* **2019**, *311*, 108758. [CrossRef] [PubMed]
65. Machado, M.M.F.; Bassani, T.B.; Cóppola-Segovia, V.; Moura, E.L.R.; Zanata, S.M.; Andreatini, R.; Vital, M.A.B.F. PPAR-γ Agonist Pioglitazone Reduces Microglial Proliferation and NF-KB Activation in the Substantia Nigra in the 6-Hydroxydopamine Model of Parkinson's Disease. *Pharmacol. Rep.* **2019**, *71*, 556–564. [CrossRef]
66. Justin, A.; Ashwini, P.; Jose, J.A.; Jeyarani, V.; Dhanabal, S.P.; Manisha, C.; Mandal, S.P.; Bhavimani, G.; Prabitha, P.; Yuvaraj, S.; et al. Two Rationally Identified Novel Glitazones Reversed the Behavioral Dysfunctions and Exhibited Neuroprotection Through Ameliorating Brain Cytokines and Oxy-Radicals in ICV-LPS Neuroinflammatory Rat Model. *Front. Neurosci.* **2020**, *14*, 530148. [CrossRef]
67. Xu, W.; Lakshman, N.; Morshead, C.M. Building a Central Nervous System: The Neural Stem Cell Lineage Revealed. *Neurogenesis* **2017**, *4*, e1300037. [CrossRef]
68. Stergiopoulos, A.; Politis, P.K. The Role of Nuclear Receptors in Controlling the Fine Balance between Proliferation and Differentiation of Neural Stem Cells. *Arch. Biochem. Biophys.* **2013**, *534*, 27–37. [CrossRef] [PubMed]
69. Gkikas, D.; Tsampoula, M.; Politis, P.K. Nuclear Receptors in Neural Stem/Progenitor Cell Homeostasis. *Cell. Mol. Life Sci.* **2017**, *74*, 4097–4120. [CrossRef]
70. Knobloch, M.; Braun, S.M.G.; Zurkirchen, L.; Von Schoultz, C.; Zamboni, N.; Araúzo-Bravo, M.J.; Kovacs, W.J.; Karalay, Ö.; Suter, U.; MacHado, R.A.C.; et al. Metabolic Control of Adult Neural Stem Cell Activity by Fasn-Dependent Lipogenesis. *Nature* **2013**, *493*, 226–230. [CrossRef]
71. Khacho, M.; Harris, R.; Slack, R.S. Mitochondria as Central Regulators of Neural Stem Cell Fate and Cognitive Function. *Nat. Rev. Neurosci.* **2019**, *20*, 34–48. [CrossRef]

72. Khacho, M.; Clark, A.; Svoboda, D.S.; Azzi, J.; MacLaurin, J.G.; Meghaizel, C.; Sesaki, H.; Lagace, D.C.; Germain, M.; Harper, M.E.; et al. Mitochondrial Dynamics Impacts Stem Cell Identity and Fate Decisions by Regulating a Nuclear Transcriptional Program. *Cell Stem Cell* **2016**, *19*, 232–247. [CrossRef] [PubMed]
73. Santpere, G.; Telford, M.; Andrés-Benito, P.; Navarro, A.; Ferrer, I. The Presence of Human Herpesvirus 6 in the Brain in Health and Disease. *Biomolecules* **2020**, *10*, 1520. [CrossRef] [PubMed]
74. Bétourné, A.; Szelechowski, M.; Thouard, A.; Abrial, E.; Jean, A.; Zaidi, F.; Foret, C.; Bonnaud, E.M.; Charlier, C.M.; Suberbielle, E.; et al. Hippocampal Expression of a Virus-Derived Protein Impairs Memory in Mice. *Proc. Natl. Acad. Sci. USA* **2018**, *115*, 1611. [CrossRef] [PubMed]
75. Jafari Khaljiri, H.; Jamalkhah, M.; Amini Harandi, A.; Pakdaman, H.; Moradi, M.; Mowla, A. Comprehensive Review on Neuro-COVID-19 Pathophysiology and Clinical Consequences. *Neurotox. Res.* **2021**, *1*, 3. [CrossRef]
76. Shaw, G.M.; Hunter, E. HIV Transmission. *Cold Spring Harb. Perspect. Med.* **2012**, *2*, a006965. [CrossRef]
77. Deeks, S.G.; Overbaugh, J.; Phillips, A.; Buchbinder, S. HIV Infection. *Nat. Rev. Dis. Primers* **2015**, *1*, 1–22. [CrossRef] [PubMed]
78. Cotto, B.; Natarajanseenivasan, K.; Langford, D. HIV-1 Infection Alters Energy Metabolism in the Brain: Contributions to HIV-Associated Neurocognitive Disorders. *Prog. Neurobiol.* **2019**, *181*, 101616. [CrossRef]
79. Ko, A.; Kang, G.; Hattler, J.B.; Galadima, H.I.; Zhang, J.; Li, Q.; Kim, W.-K. Macrophages but Not Astrocytes Harbor HIV DNA in the Brains of HIV-1-Infected Aviremic Individuals on Suppressive Antiretroviral Therapy. *J. Neuroimmune Pharmacol.* **2019**, *14*, 110–119. [CrossRef]
80. Eggers, C.; Arendt, G.; Hahn, K.; Husstedt, I.W.; Maschke, M.; Neuen-Jacob, E.; Obermann, M.; Rosenkranz, T.; Schielke, E.; Straube, E. HIV-1-Associated Neurocognitive Disorder: Epidemiology, Pathogenesis, Diagnosis, and Treatment. *J. Neurol.* **2017**, *264*, 1715–1727. [CrossRef]
81. Cenker, J.J.; Stultz, R.D.; McDonald, D. Brain Microglial Cells Are Highly Susceptible to HIV-1 Infection and Spread. *AIDS Res. Hum. Retroviruses* **2017**, *33*, 1155–1165. [CrossRef]
82. Réu, P.; Khosravi, A.; Bernard, S.; Mold, J.E.; Salehpour, M.; Alkass, K.; Perl, S.; Tisdale, J.; Possnert, G.; Druid, H.; et al. The Lifespan and Turnover of Microglia in the Human Brain. *Cell Rep.* **2017**, *20*, 779–784. [CrossRef] [PubMed]
83. Wallet, C.; De Rovere, M.; Van Assche, J.; Daouad, F.; De Wit, S.; Gautier, V.; Mallon, P.W.G.; Marcello, A.; Van Lint, C.; Rohr, O.; et al. Microglial Cells: The Main HIV-1 Reservoir in the Brain. *Front. Cell. Infect. Microbiol.* **2019**, *9*, 362. [CrossRef]
84. Lindl, K.A.; Marks, D.R.; Kolson, D.L.; Jordan-Sciutto, K.L. HIV-Associated Neurocognitive Disorder: Pathogenesis and Therapeutic Opportunities. *J. Neuroimmune Pharmacol.* **2010**, *5*, 294–309. [CrossRef] [PubMed]
85. Mamik, M.K.; Asahchop, E.L.; Chan, W.F.; Zhu, Y.; Branton, W.G.; McKenzie, B.A.; Cohen, E.A.; Power, C. Insulin Treatment Prevents Neuroinflammation and Neuronal Injury with Restored Neurobehavioral Function in Models of HIV/AIDS Neurodegeneration. *J. Neurosci.* **2016**, *36*, 1683–1695. [CrossRef] [PubMed]
86. Omeragic, A.; Hoque, M.T.; Choi, U.; Bendayan, R. Peroxisome Proliferator-Activated Receptor-Gamma: Potential Molecular Therapeutic Target for HIV-1-Associated Brain Inflammation. *J. Neuroinflamm.* **2017**, *14*. [CrossRef]
87. Potash, M.J.; Chao, W.; Bentsman, G.; Paris, N.; Saini, M.; Nitkiewicz, J.; Belem, P.; Sharer, L.; Brooks, A.I.; Volsky, D.J. A Mouse Model for Study of Systemic HIV-1 Infection, Antiviral Immune Responses, and Neuroinvasiveness. *Proc. Natl. Acad. Sci. USA* **2005**, *102*, 3760–3765. [CrossRef] [PubMed]
88. Omeragic, A.; Kara-Yacoubian, N.; Kelschenbach, J.; Sahin, C.; Cummins, C.L.; Volsky, D.J.; Bendayan, R. Peroxisome Proliferator-Activated Receptor-Gamma Agonists Exhibit Anti-Inflammatory and Antiviral Effects in an EcoHIV Mouse Model. *Sci. Rep.* **2019**, *9*. [CrossRef]
89. Omeragic, A.; Saikali, M.F.; Currier, S.; Volsky, D.J.; Cummins, C.L.; Bendayan, R. Selective Peroxisome Proliferator-Activated Receptor-Gamma Modulator, INT131 Exhibits Anti-Inflammatory Effects in an EcoHIV Mouse Model. *FASEB J.* **2020**, *34*, 1996–2010. [CrossRef] [PubMed]
90. Huang, W.; Eum, S.Y.; András, I.E.; Hennig, B.; Toborek, M. PPARα and PPARγ Attenuate HIV-induced Dysregulation of Tight Junction Proteins by Modulations of Matrix Metalloproteinase and Proteasome Activities. *FASEB J.* **2009**, *23*, 1596–1606. [CrossRef]
91. Huang, W.; Chen, L.; Zhang, B.; Park, M.; Toborek, M. PPAR Agonist-Mediated Protection against HIV Tat-Induced Cerebrovascular Toxicity Is Enhanced in MMP-9-Deficient Mice. *J. Cereb. Blood Flow Metab.* **2014**, *34*, 646–653. [CrossRef] [PubMed]
92. Chuang, C.H.; Yeh, C.L.; Yeh, S.L.; Lin, E.S.; Wang, L.Y.; Wang, Y.H. Quercetin Metabolites Inhibit MMP-2 Expression in A549 Lung Cancer Cells by PPAR-γ Associated Mechanisms. *J. Nutr. Biochem.* **2016**, *33*, 45–53. [CrossRef]
93. Shyu, L.-Y.; Chen, K.-M.; Lu, C.-Y.; Lai, S.-C. Regulation of Proinflammatory Enzymes by Peroxisome Proliferator–Activated Receptor Gamma in Astroglia Infected with *Toxoplasma Gondii*. *J. Parasitol.* **2020**, *106*, 564–571. [CrossRef]
94. Philip, S.; Kundu, G.C. Osteopontin Induces Nuclear Factor KB-Mediated Promatrix Metalloproteinase-2 Activation through IκBα/IKK Signaling Pathways, and Curcumin (Diferulolylmethane) down-Regulates These Pathways. *J. Biol. Chem.* **2003**, *278*, 14487–14497. [CrossRef]
95. Song, Y.; Yang, Y.; Cui, Y.; Gao, J.; Wang, K.; Cui, J. Lipoxin A4 Methyl Ester Reduces Early Brain Injury by Inhibition of the Nuclear Factor Kappa B (NF-KB)-Dependent Matrix Metallopeptidase 9 (MMP-9) Pathway in a Rat Model of Intracerebral Hemorrhage. *Med. Sci. Monit.* **2019**, *25*, 1838–1847. [CrossRef]

96. Potula, R.; Ramirez, S.H.; Knipe, B.; Leibhart, J.; Schall, K.; Heilman, D.; Morsey, B.; Mercer, A.; Papugani, A.; Dou, H.; et al. Peroxisome Proliferator-Activated Receptor-γ Activation Suppresses HIV-1 Replication in an Animal Model of Encephalitis. *AIDS* **2008**, *22*, 1539–1549. [CrossRef] [PubMed]
97. Christian, K.M.; Song, H.; Ming, G.L. Pathophysiology and Mechanisms of Zika Virus Infection in the Nervous System. *Annu. Rev. Neurosci.* **2019**, *42*, 249–269. [CrossRef]
98. Qian, X.; Nguyen, H.N.; Jacob, F.; Song, H.; Ming, G.L. Using Brain Organoids to Understand Zika Virus-Induced Microcephaly. *Development* **2017**, *144*, 952–957. [CrossRef] [PubMed]
99. Tang, H.; Hammack, C.; Ogden, S.C.; Wen, Z.; Qian, X.; Li, Y.; Yao, B.; Shin, J.; Zhang, F.; Lee, E.M.; et al. Zika Virus Infects Human Cortical Neural Progenitors and Attenuates Their Growth. *Cell Stem Cell* **2016**, *18*, 587–590. [CrossRef] [PubMed]
100. Thulasi Raman, S.N.; Latreille, E.; Gao, J.; Zhang, W.; Wu, J.; Russell, M.S.; Walrond, L.; Cyr, T.; Lavoie, J.R.; Safronetz, D.; et al. Dysregulation of Ephrin Receptor and PPAR Signaling Pathways in Neural Progenitor Cells Infected by Zika Virus. *Emerg. Microbes Infect.* **2020**, *9*, 2046–2060. [CrossRef] [PubMed]
101. Sharma, R.; Luong, Q.; Sharma, V.M.; Harberson, M.; Harper, B.; Colborn, A.; Berryman, D.E.; Jessen, N.; Jørgensen, J.O.L.; Kopchick, J.J.; et al. Growth Hormone Controls Lipolysis by Regulation of FSP27 Expression. *J. Endocrinol.* **2018**, *239*, 289–301. [CrossRef] [PubMed]
102. Hua, T.N.M.; Kim, M.K.; Vo, V.T.A.; Choi, J.W.; Choi, J.H.; Kim, H.W.; Cha, S.K.; Park, K.S.; Jeong, Y. Inhibition of Oncogenic Src Induces FABP4-Mediated Lipolysis via PPARγ Activation Exerting Cancer Growth Suppression. *EBioMedicine* **2019**, *41*, 134–145. [CrossRef] [PubMed]
103. Ban, K.; Peng, Z.; Lin, W.; Kozar, R.A. Arginine Decreases Peroxisome Proliferator-Activated Receptor-γ Activity via c-Jun. *Mol. Cell. Biochem.* **2012**, *362*, 7–13. [CrossRef]
104. Triki, M.; Lapierre, M.; Cavailles, V.; Mokdad-Gargouri, R. Expression and Role of Nuclear Receptor Coregulators in Colorectal Cancer. *World J. Gastroenterol.* **2017**, *23*, 4480. [CrossRef]
105. Cannon, M.J. Congenital Cytomegalovirus (CMV) Epidemiology and Awareness. *J Clin Virol* **2009**, *46* (Suppl. 4), S6–S10. [CrossRef] [PubMed]
106. Cheeran, M.C.J.; Lokensgard, J.R.; Schleiss, M.R. Neuropathogenesis of Congenital Cytomegalovirus Infection: Disease Mechanisms and Prospects for Intervention. *Clin Microbiol Rev* **2009**, *22*, 99–126. [CrossRef]
107. Han, D.; Byun, S.-H.; Kim, J.; Kwon, M.; Pleasure, S.J.; Ahn, J.-H.; Yoon, K. Human Cytomegalovirus IE2 Protein Disturbs Brain Development by the Dysregulation of Neural Stem Cell Maintenance and the Polarization of Migrating Neurons. *J. Virol.* **2017**, *91*. [CrossRef] [PubMed]
108. Odeberg, J.; Wolmer, N.; Falci, S.; Westgren, M.; Seiger, A.; Söderberg-Nauclér, C. Human Cytomegalovirus Inhibits Neuronal Differentiation and Induces Apoptosis in Human Neural Precursor Cells. *J. Virol.* **2006**, *80*, 8929–8939. [CrossRef] [PubMed]
109. Luo, M.H.; Hannemann, H.; Kulkarni, A.S.; Schwartz, P.H.; O'Dowd, J.M.; Fortunato, E.A. Human Cytomegalovirus Infection Causes Premature and Abnormal Differentiation of Human Neural Progenitor Cells. *J. Virol.* **2010**, *84*, 3528–3541. [CrossRef]
110. Belzile, J.P.; Stark, T.J.; Yeo, G.W.; Spector, D.H. Human Cytomegalovirus Infection of Human Embryonic Stem Cell-Derived Primitive Neural Stem Cells Is Restricted at Several Steps but Leads to the Persistence of Viral DNA. *J. Virol.* **2014**, *88*, 4021–4039. [CrossRef]
111. Odeberg, J.; Wolmer, N.; Falci, S.; Westgren, M.; Sundtröm, E.; Seiger, Å.; Söderberg-Nauclér, C. Late Human Cytomegalovirus (HCMV) Proteins Inhibit Differentiation of Human Neural Precursor Cells into Astrocytes. *J. Neurosci. Res.* **2007**, *85*, 583–593. [CrossRef]
112. D'Aiuto, L.; Di Maio, R.; Heath, B.; Raimondi, G.; Milosevic, J.; Watson, A.M.; Bamne, M.; Parks, W.T.; Yang, L.; Lin, B.; et al. Human Induced Pluripotent Stem Cell-Derived Models to Investigate Human Cytomegalovirus Infection in Neural Cells. *PLoS ONE* **2012**, *7*, e49700. [CrossRef]
113. Rolland, M.; Martin, H.; Bergamelli, M.; Sellier, Y.; Bessières, B.; Aziza, J.; Benchoua, A.; Leruez-Ville, M.; Gonzalez-Dunia, D.; Chavanas, S. Human Cytomegalovirus Infection Is Associated with Increased Expression of the Lissencephaly Gene PAFAH1B1 Encoding LIS1 in Neural Stem Cells and Congenitally Infected Brains. *J. Pathol.* **2021**, *254*, 92–102. [CrossRef]
114. Leghmar, K.; Cenac, N.; Rolland, M.; Martin, H.; Rauwel, B.; Bertrand-Michel, J.; Le Faouder, P.; Bénard, M.; Casper, C.; Davrinche, C.; et al. Cytomegalovirus Infection Triggers the Secretion of the PPARgamma Agonists 15-Hydroxyeicosatetraenoic Acid (15-HETE) and 13-Hydroxyoctadecadienoic Acid (13-HODE) in Human Cytotrophoblasts and Placental Cultures. *PLoS ONE* **2015**, *10*, e0132627. [CrossRef] [PubMed]
115. Stump, M.; Guo, D.F.; Lu, K.T.; Mukohda, M.; Cassell, M.D.; Norris, A.W.; Rahmouni, K.; Sigmund, C.D. Nervous System Expression of PPARγ and Mutant PPARγ Has Profound Effects on Metabolic Regulation and Brain Development. *Endocrinology* **2016**, *157*, 4266–4275. [CrossRef]
116. Allal, C.; Buisson-Brenac, C.; Marion, V.; Claudel-Renard, C.; Faraut, T.; Dal Monte, P.; Streblow, D.; Record, M.; Davignon, J.L. Human Cytomegalovirus Carries a Cell-Derived Phospholipase A2 Required for Infectivity. *J. Virol.* **2004**, *78*, 7717–7726. [CrossRef] [PubMed]
117. Angelova, M.; Zwezdaryk, K.; Ferris, M.B.; Shan, B.; Morris, C.A.; Sullivan, D.E. Human Cytomegalovirus Infection Dysregulates the Canonical Wnt/β-Catenin Signaling Pathway. *PLoS Pathog.* **2012**, *8*, e1002959. [CrossRef]

118. Rauwel, B.; Mariamé, B.; Martin, H.; Nielsen, R.; Allart, S.; Pipy, B.; Mandrup, S.; Devignes, M.D.; Evain-Brion, D.; Fournier, T.; et al. Activation of Peroxisome Proliferator-Activated Receptor Gamma by Human Cytomegalovirus for de Novo Replication Impairs Migration and Invasiveness of Cytotrophoblasts from Early Placentas. *J. Virol.* **2010**, *84*, 2946–2954. [CrossRef] [PubMed]
119. Maffezzini, C.; Calvo-Garrido, J.; Wredenberg, A.; Freyer, C. Metabolic Regulation of Neurodifferentiation in the Adult Brain. *Cell. Mol. Life Sci.* **2020**, *77*, 2483–2496. [CrossRef]
120. Munger, J.; Bajad, S.U.; Coller, H.A.; Shenk, T.; Rabinowitz, J.D. Dynamics of the Cellular Metabolome during Human Cytomegalovirus Infection. *PLoS Pathog.* **2006**, *2*, e132. [CrossRef] [PubMed]
121. Kamin, D.; Hadigan, C.; Lehrke, M.; Mazza, S.; Lazar, M.A.; Grinspoon, S. Resistin Levels in Human Immunodeficiency Virus-Infected Patients with Lipoatrophy Decrease in Response to Rosiglitazone. *J. Clin. Endocrinol. Metab.* **2005**, *90*, 3423–3426. [CrossRef] [PubMed]

MDPI
St. Alban-Anlage 66
4052 Basel
Switzerland
Tel. +41 61 683 77 34
Fax +41 61 302 89 18
www.mdpi.com

International Journal of Molecular Sciences Editorial Office
E-mail: ijms@mdpi.com
www.mdpi.com/journal/ijms

www.ingramcontent.com/pod-product-compliance
Lightning Source LLC
LaVergne TN
LVHW071621170726
843515LV00009B/2452

* 9 7 8 3 0 3 6 5 4 1 9 2 1 *